Advances in Experimental Medicine and Biology

Volume 1026

More information about this series at http://www.springer.com/series/5584

Erwei Song • Hai Hu
Editors

Translational Research in Breast Cancer

Biomarker Diagnosis, Targeted Therapies and Approaches to Precision Medicine

Springer

Editors
Erwei Song
Sun Yat-sen Memorial Hospital
Sun Yat-sen University
Guangzhou, Guangdong, China

Hai Hu
Sun Yat-sen Memorial Hospital
Sun Yat-sen University
Guangzhou, Guangdong, China

ISSN 0065-2598 ISSN 2214-8019 (electronic)
Advances in Experimental Medicine and Biology
ISBN 978-981-10-6019-9 ISBN 978-981-10-6020-5 (eBook)
https://doi.org/10.1007/978-981-10-6020-5

Library of Congress Control Number: 2017959644

Printed on acid-free paper

This Springer imprint is published by Springer Nature
The registered company is Springer Nature Singapore Pte Ltd.
The registered company address is: 152 Beach Road, #21-01/04 Gateway East, Singapore 189721, Singapore

Preface

Breast cancer is the most frequently diagnosed cancer and the leading cause of cancer death among females, accounting for 23% of the total cancer cases and 14% of the cancer deaths worldwide. Breast cancers are genetically heterogeneous, which can be generally classified into four categories with distinct patterns of molecular profile and different treatment strategies. Besides, about 5–10% of cases are inherited mainly due to the germline mutations of BRCA1 and BRCA2, which are key DNA repair regulators, and thus lead to the DNA repair-targeting therapy with PARP inhibitors. Unraveling the biological heterogeneity of breast cancer in its natural history and its responsiveness to therapy from one patient to another will help to translate new approaches for breast cancer prevention and treatment and improve the quality of care offered to breast cancer patients.

There is a broad consensus that cancer is a genetic disease and that accumulation of molecular alterations in the genome of somatic cells is the basis of cancer progression. In breast cancer, the accumulated mutations often result in the amplification of growth signal followed by the activation of PI3K/AKT/mTOR pathway and RAS/MEK/ERK pathway and thus cause the agitation of downstream transcription, metabolic reprogramming, etc., leading to the increase of breast cancer stem cell self-renewal and acceleration of cell cycle or less apoptosis. The understanding of genomic changes and oncogenic signaling cascade of breast cancer has inspired the development of targeting treatments, such as the clinical trials in PI3K, AKT, and mTORC1 inhibitors.

On the other hand, breast cancer, like other cancers, occurs because of an interaction between an environmental (external) factor and the genetically susceptible host. The immuno-environment has been demonstrated as a barrier of the clinical cancer. Overcoming the restriction of immune checkpoints is an essential step for cancer development. Therefore, recovering these checkpoints may prevent cancer progression. In addition, stromal cells such as tumor-associated macrophages (TAMs) promote tumorigenesis by limiting immuno-response or directly promoting cancer metastasis. The knowledge has been translated into successful approaches of immuno-therapies such as the application of anti-PD1 and anti-PDL1 strategies for malignancy treatment.

Together, elucidating the molecular landscape of breast cancers has facilitated the development of diagnostic, prognostic, and predictive biomarkers for clinical oncology. In addition, a burst of knowledge in cancer biology, immunology, genomics, metabolism, and so on has broken new grounds for designing innovative therapeutic approaches and selecting appropriate treatments according to the precise information of an individual cancer patient. The present book is an endeavor to convey a comprehensive knowledge of the translational efforts in breast cancer and address the latest approaches of precision medicine based on the understanding of breast cancer.

Contents

1 The Dawning of Translational Breast Cancer: From Bench to Bedside

Xueman Chen, Siting Fan, and Erwei Song

Abstract

Breast cancer is one of the world's leading causes of death in women. Although tumor initiation and progression are predominantly driven by somatic or acquired (epi) genetic alterations that govern signaling abnormalities, growing evidence suggests that the inflammatory microenvironments of cancer also play a role. Molecular characterization of breast cancer biology is essential for high-efficient management of this disease in clinical practice. Translating basic research into clinically valuable biomarkers for diagnosis, prognosis, and prediction of response to treatment and into precisely targeted therapies is crucial for the development of precision medicine in breast cancer. Such a process is known as "from bench to bedside." In this chapter, we will present an overview of breast cancer pathogenesis and selected translational advances in multistage clinical settings and aim to illustrate the dawning of precision medicine implementation in managing human breast malignancies.

Keywords

Breast cancer • Biological hallmark • Translational oncology • Precision medicine

X. Chen • S. Fan
Breast Tumor Center, Sun Yat-Sen Memorial Hospital, Sun Yat-Sen University, Guangzhou 510120, China

Guangdong Provincial Key Laboratory of Malignant Tumor Epigenetics and Gene Regulation, Sun Yat-Sen Memorial Hospital, Sun Yat-Sen University, Guangzhou 510120, China

E. Song (✉)
Sun Yat-sen Memorial Hospital, Sun Yat-sen University, Guangzhou, Guangdong 510120, China
e-mail: songew@mail.sysu.edu.cn

1.1 Introduction

Breast cancer (BrCa) is one of the most common types of invasive cancers in women, which accounts for a majority of female cancer-related deaths worldwide. It is essentially a deregulated organogenetic disorder of human breast with high heterogeneity that requires accurate molecular classification. There are thus far at least five definitive molecular subtypes delineating distinct

E. Song, H. Hu (eds.), *Translational Research in Breast Cancer*, Advances in Experimental Medicine and Biology 1026, https://doi.org/10.1007/978-981-10-6020-5_1

biological features of BrCa, which include luminal A, luminal B, HER2-positive, basal-like/triple-negative, and "claudin-low" or "normal-like" [1, 2]. Such molecularly-defined categories, largely determined by the status of ER or PgR and HER2 locus [1, 3], also offer a rationale for treatment orientation in clinics other than surgery. Especially for endocrine therapies, HER2-targeted therapies, cytotoxic chemotherapies or radiotherapy alone, or in combination, pre-/post-operative BrCa patients should be preferentially matched based on the molecular subtypes prior to the appropriate treatment.

While the multistep breast tumorigenesis is driven by progressively accumulated genetic (and epigenetic) abnormalities that might prime oncogenic transformation in one (or a few) stem-like cells, subsequent clonal expansion and selection of these transformed cells come to presage cancer progression. The last several decades have witnessed remarkable advances in the molecular pathology characterizing BrCa by virtue of high-throughput sequencing technologies, proteomic profiling, and development of "big data" sets [4, 5]. Accompanying translational genomics consist of a vast number of valuable predictive biomarkers for diagnosis, prognosis, and therapeutic response, among which the profile of 21 genes has been fully validated to predict the recurrence risk of patients with hormone receptor-positive, node-negative BrCa [6]. Additionally, overwhelming translational studies have identified numerous key governors in regulating tumor cell biology and programing malignant traits, which have provided opportunities for the development of precision medicine, particularly molecule-targeted therapies for cancer treatment.

Metastasis is generally known as the leading cause of death in BrCa patients. According to the "seed and soil" hypothesis, seeding of neoplastic cells is a prerequisite while the mechanistic determinant for metastatic colonization belongs to a tumor-promoting microenvironment, which is the rich soil not only fertilized by but also breeding the tumor seeds [7, 8]. Together with extracellular matrix (ECM), a diversity of stromal cells including fibroblasts, macrophages, neutrophils, and other immune cells constitute the tumor microenvironment, surrounding the tumor cells and emitting pro-tumor signals. Although most, if not all, stromal components have been found to be attractive prognostic indicators or accessory therapeutic/preventive targets for metastatic breast cancer (MBC), there is still a long way to go in making the leap from bench to bedside.

In this chapter, we will provide a summarized overview of BrCa biology and hallmark features based on genetic alterations and signaling networks, highlighting the role of microenvironmental constituents in assisting tumor progression and their potentials as candidate predictive biomarkers and/or therapeutic targets. Additionally, we will briefly introduce the critical translational oncology and future perspective of precision medicine in treatment of this malignant disease.

1.2 Oncogenic Changes in Genome: Inherent Power for BrCa Evolution

Breast cancer shares inherent similarity with other types of cancers, carrying somatic mutations in the genomes. Among these mutations, proto-oncogenes proceed into oncogenes that ultimately drive tumorigenesis via conferring clonal selective advantage on neoplastic cells. With the tumor suppressor genes dying away, an unsupervised state of cancerous breast is favored (Fig. 1.1).

1.2.1 Human Epidermal Growth Factor Receptor-2, HER2

HER2 is an essential oncogene in driving BrCa pathogenesis. Amplification of the HER2/ErbB2/neu gene and/or resultant HER2 protein overexpression occur in 18–25% of all primary invasive breast cancers, which are casually classified as HER2-positive BrCa [9, 10]. The encoded HER2 oncoprotein belongs to a family of four closely related transmembrane receptors (EGFR/ErbB1, HER2, HER3/ErbB3, and HER4/ErbB4), structured by two cysteine-rich extracellular ligand-binding domains and a transmembrane

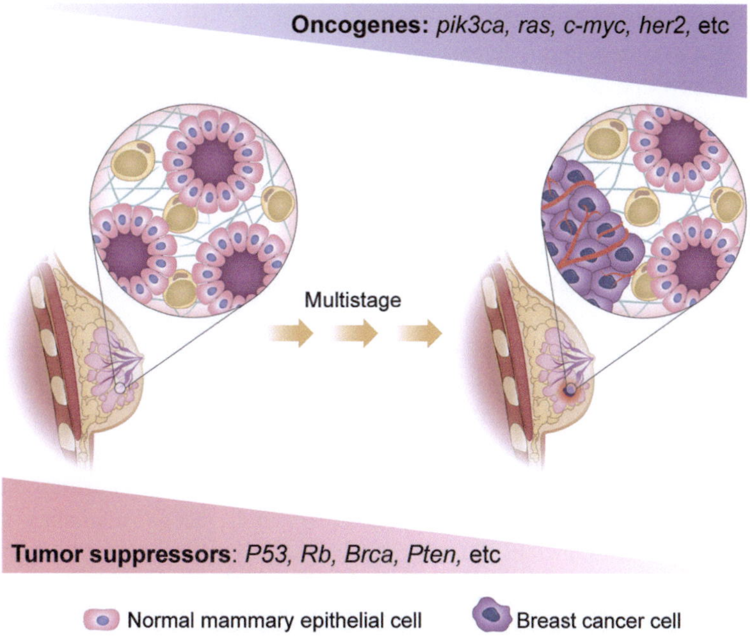

Fig. 1.1 Multistage breast tumorigenesis triggered by coordinated oncogenic mutations, including amplification/overexpression of oncogenes and loss of tumor suppressor genes

domain [11]. Together with HER1 and HER4, HER2 receptor possesses tyrosine kinase activity, structurally attributable to an intracellular tyrosine kinase domain [12]. Up till now, natural ligands for HER2 has not yet been identified, while the other three receptors are able to recognize a variety of ligands such as transforming growth factor α (TGFα), epidermal growth factor (EGF), and heregulins. Even so, HER2 receptor plays a central role in the activation of other family members, serving as a preferred dimerization partner to stabilize the HER2-containing heterodimers [13–15]. Upon HER2 activation, its downstream signal transduction cascades will be set in motion, including Erk1/2 mitogen-activated protein kinases (MAPK), phosphatidylinositol 3-kinase/AKT/mammalian target of rapamycin (PI3K/AKT/mTOR) and Jak/Stat pathways, which ultimately pose a wide range of effects on the cellular biological behaviors of BrCa [16, 17]. Thus, HER2 gene amplification in BrCa leads to enhanced cell growth and proliferation, increased cell motility and invasiveness, accelerated angiogenesis, progressive regional and distant metastases, as well as reduced apoptosis [18].

Considerable clinical data demonstrated that elevated level of HER2 was closely associated with more aggressive phenotypes and, until the advent of HER2-directed therapies, worse disease-free survival was represented by higher recurrence rate and increased mortality [10, 19]. Therefore, abnormal HER2 expression is an independent adverse prognostic factor in both lymph node-negative and lymph node-positive BrCa, paralleling to large tumor size, high histologic grade, DNA aneuploidy, p53 mutation, and deficiency of steroid hormone receptors (estrogen and/or progesterone) as well [20, 21]. Furthermore, a predictive value of HER2 in response to various cancer cures has been long established, indicating a more important role HER2 plays as the therapeutic target in the treatment for BrCa [22].

As a recombinant humanized monoclonal antibody (mAb), trastuzumab (Herceptin) (Genentech Inc. San Francisco, CA, USA; Hoffmann-La Roche Ltd. Basel, Switzerland) is the first genomics-based therapeutic agent and hitherto the only adjuvant treatment approved specifically for patients with HER2-positive early-stage BrCa [23]. Except for interfering the downstream signaling, trastuzumab plays a part in augmenting the antitumor immunity, which will be introduced later in this chapter. Moreover, it markedly impacts the proliferative ability of HER2-overexpressing human breast cancer cells in vitro and in vivo [24–26]. Trastuzumab not only effectively treats HER2-amplified tumors as monotherapy [27] but also improves the clinical outcomes both in the first line and in the adjuvant settings when being administered with and/or sequentially after chemotherapy [28–32]. Endocrine and radiation therapy also benefit from HER2 inactivation in preclinical studies [33, 34]. Unlike trastuzumab, lapatinib (Tykerb/Tyverb) is a small molecule, and dual tyrosine kinase inhibitor (TKI) of HER2 and EGFR, which has been exceptionally approved for specific treatment for patients with advanced HER2-overexpressing malignancies. It functions through reversible inhibition of EGFR/HER2 tyrosine kinase activities and their downstream growth and survival pathways including the MAPK/Erk1/2 and PI3K/Akt pathways [35, 36]. Compared with chemotherapy alone, its combination with lapatinib increases the overall survival rate of patients with advanced BrCa overexpressing HER2 [37]. Generally, the anti-HER2 antibody trastuzumab maintains remission for 1–2 years before resistance develops [38, 39]. Preclinical experiments indicated the lapatinib sensitivity of HER2-enriched and trastuzumab-resistant BrCa cells, which included those with truncated HER2 receptors [40]. In addition, lapatinib administration resulted in enhanced apoptotic effect of the anti-HER2 antibodies [36, 41, 42], suggesting a synergistic interaction between lapatinib and trastuzumab in HER2-positive breast cancer cells. The lapatinib-alone arm of the NeoALTTO trial conferred a supportive conclusion that a combined administration of trastuzumab and lapatinib might provide superior efficacy to monotherapy of either agent, with manageable toxic effects [43].

1.2.2 PI3K Catalytic Subunit-α, PIK3CA

The identification of PI3K pathway as major determinant of trastuzumab resistance suggests that PIK3CA mutation may predict sensitivity to trastuzumab in HER2-positive BrCa [44–49]. With PIK3CA being one of the most commonly mutated genes, the amplification and/or activating mutations of PIK3CA not merely occur in approximately 20% of all breast tumors, more frequently in luminal tumors and in up to 40% of cancers with HER2-positive subtype, but also create tracks in triple-negative breast cancer (TNBC) [50–54]. PIK3CA gene encodes the p110α catalytic subunit of PI3K, which induces activation of the renowned PI3K/AKT pathway in 70% of breast cancers [55]. Various mouse models demonstrated that oncogenic PIK3CA H1047R, the most recurrent mutation, contributed to constitutive PI3K signaling and heterogeneous mammary tumorigenesis [50, 56–58]. More translational studies are needed to characterize signaling pathways about PI3K.

1.2.3 Tumor Protein p53, TP53

Different from PIK3CA that controls oncogenic signaling, another frequently mutated gene TP53 functions as a tumor suppressor gene by encoding p53, a transcription factor (TF) controlling multiple tumor suppressor pathways in BrCa [59]. TP53 mutations occur in at least 80% basal-like BrCa and in 10–30% luminal (ER+) subtypes [60]. Germ line loss of p53 in humans and mice is predisposed to spontaneous cancer formation [61, 62].The arising and development of genetically engineered mouse models (GEMMs) enable deeper in vivo exploration for the cell-of-origin in mammary tumors, through oncogenic PIK3CA H1047R expression and/or somatic p53 deletion [55, 63].

1.2.4 Breast Cancer-Associated Gene 1/2, BRCA1/2

TP53-inactivating mutations occur more often in BRCA1/2-associated breast cancers than in sporadic cases [64]. As the most prominent BrCa genes, BRCA1 and BRCA2 usually cooperate with TP53 in oncogenesis since their mutants cause damage to the intrinsic tumor suppressive abilities. BRCA1 mutation predisposes women carriers to breast cancer, with about 80% risk in their lifetime [65]. BRCA1-related breast tumors are mostly high-grade invasive ductal carcinomas (IDCs) referred to as "triple-negative" BrCa, featuring a lack of expression of ER, PR, and HER2 [66]. Other than ER/HER2-positive BrCa, there is still no tailored therapy available for TNBC. The interaction between BRCA1 and proteins involved in DNA repair confers a role of BRCA1 in the maintenance of genomic stability, which is further demonstrated by the highly sensitive DNA damage response within BRCA1-deficient cells [67–69]. Poly (ADP-ribose) polymerase-1 (PARP1), a key molecule in the damage repair of DNA single-strand breaks, can be a potential therapeutic target since BRCA1/2-deficient cells showed high sensitivity to PARP1 inhibitors [70]. BRCA1 conditional mouse models helped assess the preclinical effects of PARP inhibition in BRCA1-deficient tumors consistently, while PARP inhibitor olaparib displayed inhibitive activity in patients with BRCA-associated malignancies [71]. More researches implied BRCA1's roles as transcriptional regulator, cell cycle checkpoint, as well as assistant for X-chromosome inactivation in cancer evolution [72–74]. Gene expression profiling of BRCA1-mutated breast tumors unveiled close similarity to an undifferentiated basal-like phenotype [75]. Subsequent studies based on genetic predisposition of progenitor cell transformation and BRCA1/p53-deficient transgenic mouse models substantiated the involvement of BRCA1 in mammary epithelial differentiation and luminal progenitor expansion, further indicating luminal progenitor populations as candidate target for basal tumor arising from BRCA1-mutated background [68, 76–79].

1.3 Receptor Activation Triggering Growth Signal Line

Tumor masses intuitively root in relentless cell multiplication under excessive growth-promoting signals from surrounding microenvironment. Long established is the role of growth factors (GFs)-driven signaling in human cancer pathogenesis [80]. Since it takes two to tango, a diverse array of growth factor receptors (GFRs) is mechanically required.

1.3.1 Epidermal Growth Factor Receptor, EGFR

The HER/ErbB family consisting of four homologous transmembrane receptors functions in the regulation of growth factor cellular signaling [81]. Identified in 1978 as the earliest found of the four [82], EGFR was shown to be linked to a transforming viral oncogene a few years later [83]. Nowadays EGFR is known as a typical receptor tyrosine kinase (RTK) with a cytoplasmic tyrosine kinase-containing domain and an extracellular ligand-binding region joined and anchored by a single hydrophobic membrane spanning portion [84, 85].

The EGF-family of peptides binds the ErbB receptors, and due to receptor specificity, EGFR intensely recognizes a specific set of ligands—EGF, TGF-α, amphiregulin, betacellulin, heparin-binding EGF, or epiregulin [41, 42]. Upon ligand binding, the EGFR ectodomain undergoes conformational changes, triggering the formation of homodimeric (with itself) or heterodimeric (preferably with HER2) receptors followed by reciprocal intracellular transphosphorylation on tyrosine residues of each receptor [86]. Phosphorylated tyrosines recruit numerous signal transducers and adaptor molecules from the cytoplasm in shape of a linked complex, which in turn activates the RAS protein responsible for a phosphorylation cascade. A multitude of downstream signaling pathways through MAPK and PI3K thus proceed and eventually exercise a broad range of influences on processes crucial to cancer progression, including prolonged cell survival, runaway proliferation, increased motility along with invasion, and metastasis [84, 87]. In support of this, the aberrant expression of EGFR and/or elevated EGFR-mediated signaling are encountered in a number of solid tumor types, somehow conferring a malignant phenotype and accordingly poor survival [88]. Notably in BrCa, EGFR is usually detectable in IDCs as well as triple-negative malignancy. The inverse correlation between EGFR expression and ER status also substantiates an unfavorable prognosis [89].

Being the first signal-generating protein and proto-oncogene makes EGFR an ideal target in cancer therapeutics. Compared with HER2-directed therapy with trastuzumab, clinical trials of drugs against EGFR in BrCa have long way to go. Clinical benefits have been shown in non-small cell lung cancer (NSCLC) patients treated with EGFR-targeted TKIs due to the presence of activating EGFR mutations in the tyrosine kinase domain [90–92]. However, rare EGFR mutation is detected in patients with TNBC [93]. Monoclonal antibody cetuximab [94] and small molecule TKIs such as gefitinib [95] and erlotinib [96] are thus far two potential therapeutic strategies for the inhibition of EGFR function. Even though experimental studies provided promising results in breast carcinoma cell lines [97, 98], failures predominated in early phase clinical trials which had been intended for encouraging antitumor activity of EGFR-targeted agents in BrCa, TNBC in particular [99]. Despite that, concomitant utilization of EGFR-directed drugs and chemotherapeutic agents seems to hold promise for BrCa cure owing to improved antitumor efficacy, as in the cases of gefitinib/docetaxel and cetuximab/taxanes combination therapies [100, 101]. It is worth noting that using trastuzumab in partnership with gefitinib might attempt dual blockade of both HER2 and EGFR in MBC [102]. Nonetheless, therapeutic benefit from lapatinib, the dual HER2/EGFR inhibitor, turned out to be irrelevant with EGFR expression in TNBC [103]. Of note, growing interest lies in the potential role of EGFR in endocrine-refractory BrCa, which is most likely to involve ligand-independent activation of ERα [104].

1.3.2 Estrogen Receptor-α, ERα

Similar to GFs binding of EGFR, ER (short for ERα in this chapter) is generally responsive to estrogen (E2)—a steroid hormone that stimulates breast cancer cell growth and proliferation. Although ER is detectable in plasma membrane, cytoplasm, and nucleus, the classical genomic action of ER takes place in nucleus as a transcription factor and a signal transducer as well as a nuclear receptor for E2 and its analogues [105]. Derailed ER signaling network in BrCa will be displayed in the following part. The single largest luminal BrCa subtype has been long defined by the expression levels of ER and/or PR, which predict an ER-targeted hormone therapy of choice [106]. For ER-positive BrCa in premenopausal women, endocrine therapies have been grouped into two aspects and widely applied. Tamoxifen, the first molecularly targeted drug, exerts antiestrogen effect through competitive inhibition of ER. Fulvestrant selectively downregulates ER expression. On the other hand, aromatase inhibitors (such as letrozole and exemestane) effectively interrupt E2 biosynthesis, exhibiting a relative clinical superiority in recent years. Even so, therapeutic resistance occurs, and many ER+ tumors relapse decades after apparently successful hormone interventions [107]. The detailed mechanisms of endocrine resistance in ER+ BrCa are under investigation, with emerging cues such as ER mutations and GF signaling [60, 106]. More in-depth work for improved therapeutics by combinatorial targeting of ER and either HER2 or EGFR signaling needs to be done to address this unmet need [108–110].

1.3.3 Integrin

Unlike EGFR and ER sensing external soluble growth signals, another type of microenvironmental sensors is ECM receptor, which accounts for cell-ECM interaction. Prominent among these are integrins, the heterodimers composed of α and β subunits. As transmembrane cell surface receptors, they sense and link ECM to cytoskeleton—the scaffolding that defines cell shape, resulting in the latter's remodeling and wide-ranging cellular responses through affected signaling pathways involving cell polarity, survival, and migration. Structurally, their short cytoplasmic tails are associated with focal adhesion kinase (FAK), a non-receptor tyrosine kinase whose activation allows the assembly of large multi-protein machines at the plasma membrane [111]. And intriguingly, an unusual "inside-out" signaling fashion coexists with the aforementioned "outside-in" event. Integrins responsive to intracellular signals enable cells to anchor the ECM in specific areas called focal adhesions, thereby establishing adhesive interactions with their tissues [112, 113]. The ECM-integrin axis participates in the regulation of many aspects of cellular behaviors and mammary gland biology [114]; any deregulation of integrin pathway can fuel breast neoplasia and progression. Altered integrin expression was observed in normal, hyperplastic, and malignant human breasts [115]. Besides prognostic significance [116], integrins also harbor implication in therapeutic resistance of BrCa [117–119]. Cross-talk between integrins and RTKs, notably EGFR, is a crucial event in BrCa pathology and warrants consideration in the therapeutic field [120]. To date, an α5β1-integrin inhibitor (ATN-161) has entered phase II trials, partnering with chemotherapy [121]. A monoclonal anti-β1 antibody is also under development.

1.4 Cytoplasmic Signaling Circuitry Programing Malignant Hallmarks

As indicated previously, either cell surface or nuclear receptors upon ligand binding, such as E2-ER, GF-EGFR, and ECM-integrin, are coupled to a set of integrator and effector proteins that process the ligand-emitting signals and drive a complex network of signal transduction pathways underlying mammary gland development. Deregulation of this cytoplasmic signaling circuitry usually brings about cellular misbehavior and developmental disorder, or, worse, breast tumorigenesis and progression. In this section we take MAPK, PI3K, and ER pathways as prime examples.

RAS/RAF/MEK/ERK Signaling Of significance are RAS small GTPase-proteins, the binary switches of either inactive (GDP-bound) or active (GTP-bound) form which conducts a series of downstream phosphorylation events through Raf, MEK, and MAPK (Erk1/2) [122, 123]. Although only 2% human breast cancers harbor RAS mutations [124], multiple RTKs (such as EGFR, insulin-like growth factor 1 receptor [IGF1R], and platelet-derived growth factor receptor [PDGFR]) associated with BrCa can maintain RAS in a constitutively activated state, leading to neoplastic cell proliferation [89]. The MAPK cascade has shown signs of hyperactivation in TNBC patients, providing a rationale for components of this pathway as hot therapeutic targets [125, 126]. Unfortunately, no RAS-directed agent for BrCa has yet displayed therapeutic effectiveness, and clinical progression with MEK inhibitors was curtailed by toxicity and potency [127].

PI3K/AKT/mTOR Pathway Most frequently mutated in BrCa, PI3K/AKT/mTOR signaling pathway is a pivotal intracellular signaling system participating in multiple cellular processes including growth, proliferation, and apoptosis [128, 129]. Genetic alterations in PI3K/AKT/mTOR pathway contribute to its robust activation, such as oncogenic mutations in RTK HER2, p110α PI3K-encoded PIK3CA, PI3K activator K-RAS and effectors AKT and mTOR, as well as loss of tumor suppressors PTEN and INPP4B [130]. The cancer-related class IA PI3Ks are heterodimers with p110, a catalytic subunit sequestered by a regulatory subunit p85. The GF binding of cognate RTK somehow recruits p85, from which p110 is released and thus free to phosphorylate the lipid phosphatidylinositol 4,5-bisphosphate (PIP2) to the second messenger phosphatidylinositol 3,4,5-trisphosphate (PIP3). Downstream PIP3 continues to be phosphorylation cascades of phosphoinositide-dependent kinase 1 (PDK1) and AKT, causing the indirect activation of mTOR/Raptor (TORC1) complex that regulates protein synthesis and cell growth. Understanding the functional roles of these centrally located effectors has rationalized the development of anticancer drugs against PI3K/AKT/mTOR pathway. Clinical trials underway involve pan-PI3K inhibitors (buparlisib, pictilisib), dual PI3K/mTOR inhibitor (GDC-0980), mTOR inhibitors (everolimus, ridaforolimus), and AKT inhibitors (MK2206, AZD5363) [131]. Emerging combinatorial strategies concerning PI3K pathway inhibition and endocrine therapy in ER+ BrCa patients are on account of the dynamic interplay between PI3K pathway and ER activation, suggestive of another regulatory role of PI3K in endocrine resistance [132]. Altered PI3K pathway has been conferred as molecular predictor of endocrine responsiveness associated with patient outcome [133–135].

ER Pathway In the classical estrogenic signaling model, E2 binding leads to dimerization and nuclear localization of ER. Through interaction with coactivators (e.g., amplified in breast cancer 1 [AIB1/SRC3]) and corepressors (e.g., nuclear receptor corepressor 1 [NCOR1]) to a greater or lesser extent, respectively, the ER complex further recruits histone acetyl transferase (HAT) such as p300 to activate transcription of different gene sets controlling cell growth and cell cycle progression [105]. Alternatively, PI3K downstream kinases and MAPK can phosphorylate ER at key positions (e.g., S167 and S118) [136, 137] and result in ligand-independent activation of ER that might drive AI-resistant cell proliferation. Such genomic effect can either be estrogen response element (ERE)-dependent (direct binding of ER) or ERE-independent (ER interacting with other TFs like activation protein 1 [AP1]). In addition, ER can exert its growth-promoting functions through non-genomic action in the plasma membrane upon activation of EGFR and IGF1R [138, 139]. Endocrine therapy has been practiced for over a century and continues to be the cornerstone of treatment for ER+ BrCa patients. Progress in unraveling the interlinked signaling transduction upstream and downstream ER will pave the way for dealing with the de novo and acquired endocrine resistance.

Other crucial pathways and regulators, such as Wnt/β-catenin signaling, Notch signaling, Hedgehog signaling, and NK-κB, contribute in various ways to the wicked programing of BrCa. Intricate as the signaling circuitry is, listing will be paused here.

1.5 Deranged Cell Cycle Control Sounds Alarm of Oncocyte Multiplication

As cancers fundamentally feature hallmarks of sustaining proliferative signaling and evading growth suppressors, they are different from normal tissues with strict self-control in the cell growth-and-division cycle. They derange them – both the control and the cell cycle. Targeting disordered cell cycle in BrCa logically holds promise for therapeutic benefit.

The mammalian cell growth-and-division cycle proceeds according to four successive phases: G1 (the first gap phase between M and S phases) → S (DNA synthesis) → G2 (the second gap phase between S and M phases) → M (mitosis). By this way, a single cell can be divided into two daughter cells. Generally, most cells rest in a quiescent, nonproliferating state (G0), which exits from the active cell cycle in early G1. Extrinsic mitogenic stimulations, including GFs, hormone and cell–cell contacts, instigate or drive cell cycle progression from G0 or G1 phase into S and the next M phase via regulation of different cyclin-CDK (cyclin-dependent kinase) axes, notably cyclin D1-CDK4/6 axis in BrCa [140]. Figure 1.2 delineates the overall cell cycle progression governed by key regulators.

Of three different D-type cyclins, cyclin D1 is a nuclear protein encoded by *CCDN1* gene, an E2-responsive gene with oncogenic potential and prognostic value [141]. Physical binding of any D-type cyclin (D1, D2, or D3) to CDK4 or the highly homologous CDK6 leads to enzymatic activation of CDK4/6, which further phosphorylates the retinoblastoma tumor suppressor protein (Rb), resulting in inactivation of its antiproliferative function in late G1. With the E2F TFs released from the hyperphosphorylated Rb (pRb), DNA replication-related genes required for the G1/S phase transition are thus

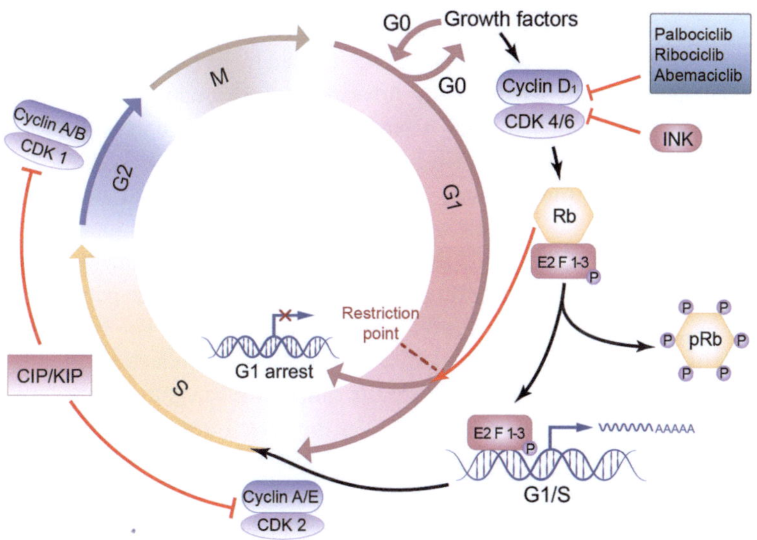

Fig. 1.2 Schematic for the cell growth-and-division cycle regulation via cyclin-CDK-Rb axes. Palbociclib and ribociclib are two CDK4/6 inhibitors with FDA approvals, while abemaciclib is now under evaluation in clinical trials. See text for detailed description

transcriptionally activated [142, 143]. The so-called restriction point in the cell cycle determining cell decision independent of extracellular mitogens—to cycle, or not to cycle—corresponds to the point at which Rb becomes fully phosphorylated, and the late G1 commences entry into S phase [144]. After that, other progressively activated CDKs, such as CDK 1 and CDK 2, in complexes with cyclins E, A, and B, propel cells to enter mitosis from S and G2 phases [144]. Otherwise, the holoenzyme function of CDK4/6 in commitment to DNA synthesis can be attenuated by specific CDK4/6 inhibitors of INK4 proteins such as $p16^{INK4a}$ (so named for a 16-kDa polypeptide encoded by the *INK4a* gene), as well as the Cip/Kip family of universal inhibitors, including $p21^{Cip}$ and $p27^{Kip}$, respectively [145, 146]. The $p16^{INK4a}$-mediated inhibition of CDK4/6 kinase activity subsequently blocks phosphorylation toward the functional Rb, resulting in cell cycle arrest in G1 phase or withdrawal toward the noncycling G0 state, even worse, cell senescence. What's more, p53 serves as negative regulator of cell cycle by activating p21 and, together with Rb, predominates in triggering cell apoptosis and/or senescence [60].

Based on the central roles of cyclin D-CDK4/6-Rb axis in cell cycle control, plausible reasoning exists for their deregulation in breast neoplasia and antiestrogen resistance. High *CCDN1* amplification and/or cyclin D1 overexpression in association with poor prognosis occur in 38% of HER2-enriched, 58% of luminal B, and 29% of luminal A subtypes [50] .Gains in CDK4 are relatively less common – 24%, 25%, and 14%, respectively [147]. It is more likely that basal-like BrCa harbors loss of Rb or lower pRb expression, while RB-pathway disruption in ER+ BrCa confers a worse clinical outcome [148]. Preclinical evidences have demonstrated the hyperactivation of cyclin D1-CDK4/6 pathway orchestrates tumor escape from senescence and contributes to the initiation, maintenance, and development of HER2-driven BrCa, while cyclin D1 or CDK4/6 deficiency retards tumor cell growth [149–151].

The fact that cyclin D1 serves as both a target of E2 signaling [152] and an independent activator of ER [153], whereby its upregulation is mostly observed in ER+ BrCa, heralds clinical trials targeting the cyclin D1-CDK4/6 axis, which potentiates improved efficacy of hormone therapy or reversion of endocrine resistance [154]. Studies in recent years have witnessed a therapeutic breakthrough brought by small molecule CDK4/6 inhibitor (CDK4/6i). Currently, numerous clinical trials are ongoing or planned majorly for three CDK4/6 inhibitors, of which palbociclib (PD0332991) is the most clinically advanced one. In 2015 and 2016, this first-in-class

CDK4/6i gained two separate FDA (the US Food and Drug Administration) approvals to be used in combination with letrozole, or fulvestrant, for treatment of patients with hormone receptor (HR)-positive, HER2-negative MBC prior to, or following endocrine therapy, respectively [155, 156]. Palbociclib is an orally bioavailable, reversible pyridopyrimidine [157] that targets CDK4/6 with high specificity, restoring Rb functionality through dephosphorylation and ultimately arresting tumor cell growth [158]. Ribociclib (LEE011) and abemaciclib (LY2835219) are the other two highly selective CDK4/6 inhibitors in the later stages of clinical development. Besides single-agent activity, LEE011 is undergoing evaluation in phase I/II trials combined with hormonal therapy and PI3K/mTOR-targeted therapies. Such triplet therapy can hitherto be exemplified by LEE011/letrozole/BYL719 or LEE011/exemestane/everolimus in advanced ER+ BrCa [159, 160]. An ongoing phase II study of abemaciclib as monotherapy has shown promising efficacy in patients with HR+/HER2- MBC after at least two prior lines of chemotherapy [161].

Overall, FDA approvals of CDK4/6i combined with endocrine therapy for BrCa treatment highlight long-sought success. Given the success yet confining to HR+/HER2- advanced malignancies, translational investigations concerning the survival benefit from these or other CDK inhibitors, with or without combinatorial therapies, in HER2+ BrCa patients of different clinical settings, will be the next challenge and potential progress.

1.6 Multistep Cancer Progression: From Regional Growth to Ectopic Colonization

1.6.1 Angiogenesis: Building Up Their Own Granary

The rapidly growing tumor tissues, once exceeding 0.5 mm in diameter or located 0.2 mm away from blood vessels, are in great need of oxygen and nutrients for local growth and distant metastases [162]. Therefore, angiogenesis is happening (Fig. 1.3). This process of new blood vessel formation supports malignant tissue requirements through vascular remodeling and subsequent neovascularization carrying blood [163]. As an essential but complex step in cancer development, angiogenesis is regulated by various microenvironmental stimuli, particularly the fine-tuned balance between pro- and anti-angiogenic factors. The "angiogenic switch" turns on when the production of positive regulators precedes the negative one, initiating a set of angiogenic responses that ensure tumor cell survival and growth as well as adequate access to circulation for future metastatic seeding [164].

Aberrant expression of pro-angiogenic factors drives a chaotic growth of immature tumor blood vessels, in parallel with disorderly tumor tissues [165]. Of predominant role is vascular endothelial growth factor (VEGF)-A, which binds with high affinity to its bona fide RTKs that are almost exclusively expressed on vascular endothelial cells (ECs) – VEGFR1/Flt1 and VEGFR2/Flk1/KDR [166]. In case of pathologic conditions such as hypoxia, glucose deprivation, or oncogenic transformation, VEGF-A can be upregulated and secreted by tumor cells and/or tumor-associated macrophages (TAMs), triggering the "angiogenic switch" by directly acting on ECs [167]. It is worth noting that VEGF-A is sequestered by ECM in an inactive form until stromal cell-derived matrix metalloproteinase 9 (MMP9) mediates the proteolytical degradation of ECM components and then allows its bioavailability to VEGFR. With downstream signaling cascades elicited, bioactive VEGF-A not only stimulates EC sprouting, proliferation, and migration but also represses apoptosis and increases capillary permeability, constructing leaky and tortuous newborn blood vessels in and around the tumor, thus providing routes for malignant cell escape and dissemination [166, 168]. Either high VEGF expression or extensive TAM infiltration has been correlated with a greater likelihood of metastatic disease and poor prognosis in BrCa [169, 170]. Other known stromal components that coordinate angiogenic response with tumor metastasis involve basic fibroblast growth factor

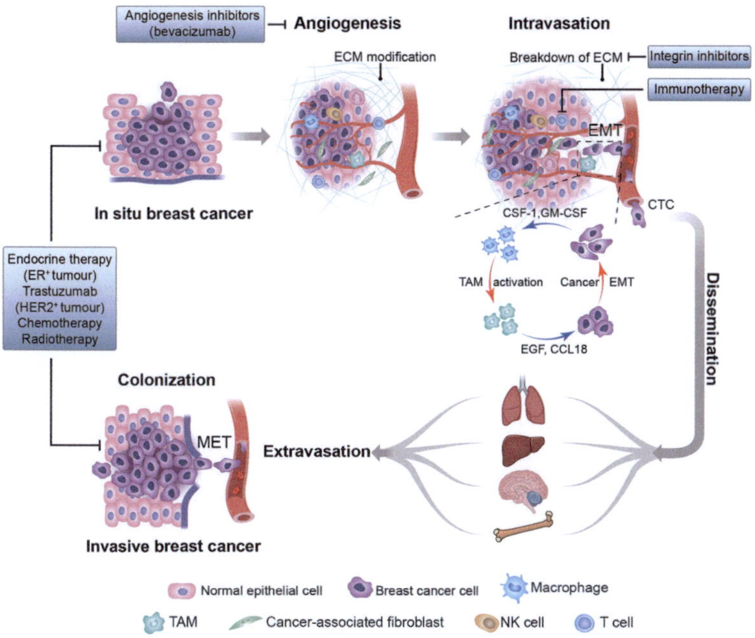

Fig. 1.3 The multistep cascade of breast cancer metastasis, depicting current and potential antitumor therapeutics. The process of angiogenesis is fundamental for the regional tumor growth and subsequent metastatic event. Breast cancer cells at the primary site initially breach the base membrane and ambient stroma to achieve localized invasion. Thereafter, they intravasate into the circulatory system and disseminate at secondary organs, including the lung, liver, brain, and bone. These travelers are also termed as tumor circulating cells. Once settled down at the new environments, they extravasate from the vessels and manage to colonize the tissue, thereby forming macroscopic metastases. Underlying the cancer cell, invasiveness and colonization are two phenotype transdifferentiation programs—epithelial-to-mesenchymal transition and mesenchymal-to-epithelial transition, respectively. Among the stromal cells, tumor-associated macrophages activated by cancer cells can reciprocally induce their epithelial-to-mesenchymal transition, which forms a positive feedback loop in accelerating cancer metastasis. Currently, endocrine therapy is the mainstay for treatment of estrogen receptor-positive (ER+) cancers while trastuzumab corresponds to HER2-positive (HER2+) cancers. Despite the conventional therapies, bevacizumab for angiogenesis blockade, integrin inhibitors to preclude ECM breakdown, and immunotherapy for the boosting of T-cell function also show therapeutic promise in either primary or advanced breast cancers (*Abbreviations*: *ECM* extracellular matrix, *EMT* epithelial-to-mesenchymal transition, *CTC* circulating tumor cell, *MET* mesenchymal-to-epithelial transition, *TAM* tumor-associated macrophage, *NK* natural killer)

(bFGF) and thymidine phosphorylase (TP), ECM-interacting integrins, as well as cancer-associated fibroblasts (CAFs) that deposit ECM proteins and enzymes for remodeling [171, 172]. Such functionalities have rendered them as attractive targets for developing anti-angiogenic drugs. Bevacizumab (Avastin), a humanized mAb against VEGF-A, has been used in partnership with chemotherapy in MBC and exhibited clinical benefit [173]. Sunitinib malate (SU11248) and sorafenib (BAY 43–9006), two TKIs mainly targeting VEGFR-2, are now being evaluated in phase II trials for MBC treatment [174, 175]. As inhibitors for FGFR, integrins and MMPs are also respective therapeutic candidates under development.

1.6.2 Invasion and Metastasis: Breaking Through Local Barriers and Moving Out

With tumor cells exiting the primary habitation and circulating in the blood or lymphatic stream, they are prompted to seed metastases to appropriate secondary organs within the body. BrCa cells preferably metastasize to the lungs, liver, brain, and bone [176]. Underlying the metastatic cascade is a complex succession of cell-biological events – whereby tumor cells invade the basal membrane and surrounding stroma and then intravasate into the bloodstream or lymphatics for dissemination; a minority of circulation survivors go on to extravasate and, with colonization developed in the new environments, macroscopic tumors will arise at distal ectopic sites [177] (Fig. 1.3). For each step, cancer cells closely interact with their surroundings comprising the ECM and stromal cells. The bi-directional interaction between primary tumor and TAMs plays a critically important role in tumor cell invasion and migration. Activated TAMs produce EGF that triggers EGFR signaling in cancer cells, thus promoting the local invasion of mammary carcinoma [178]. Similarly, secretion of IL-4 by breast cancer cells induces cathepsin protease activity in TAMs, further augmenting cancer cell invasiveness [179]. The intravasation of breast cancer cells can be enhanced by perivascular TAMs through EGF secretion. Reciprocally, tumor cell-derived CSF1 (colony-stimulating factor-1) recruits macrophage displaying CSF1 receptor (CSF1R) on their surface, whose activation stimulates EGF expression [180–182]. Chemokine CCL2, along with receptor CCR2, also merits special consideration in amplifying TAMs recruitment and cancer metastasis [183].

Underlying the acquisition of invasive phenotypes is a transdifferentiation program in cancerous epithelial cells, referred to as epithelial-mesenchymal transition (EMT) [184]. Conducted by EMT-inducing Snail-Slug TFs [185], epithelial cells undergo morphological alternation toward a mesenchymal appearance with repressed expression of epithelial-associated genes and increased mesenchymal-gene levels, in parallel with loss of cell polarity and enhanced cell motility and invasiveness, respectively. Cancer EMT process also interacts with TAM activation, whereby a positive-feedback loop involving granulocyte-macrophage colony-stimulating factor (GM-CSF) from mesenchymal-like cancer cells and TAM-derived CCL18 governs BrCa metastasis [186, 187] (Fig. 1.3). Our studies found that cancer cells undergone EMT secrete GM-CSF to activate macrophages to a TAM-like phenotype. Meanwhile, abundant CCL18 from TAMs binds to its functional receptor PITPNM3 and ignites intracellular NF-kB pathway, contributing to EMT induction in breast cancer cells [186]. Earlier work from our group has shown CCL18-PITPNM3 interaction triggers integrin clustering and fibronectin adherence of cancer cells via FAK activation, thereby enhancing their adhesion to ECM and promoting the invasiveness of breast cancer cells [187]. Collectively, cancer cell invasion is not only attributed to weakened cell-to-cell adhesion by EMT but also on account of the enhancement of integrin-mediated tumor cell adherence to the ECM.

Of note, components of the ECM may constitute a link between tumor cells and TAMs. Local growth and invasion of solid tumors as well as metastases greatly depend on the controlled deposition of ECM components, such as various types of collagens, and their degradation by virtue of TAM-released MMPs and the urokinase plasminogen activator (uPa) system [188, 189]. With the ECM components degraded, the polarity-deficient cancer cells are able to migrate freely along the gap formed by loose cell-ECM attachment. CAFs, the majority of stromal cells within tumor masses, act as pathfinders through the ECM. Chemokine receptors help direct the neoplastic migrants to potential habitats [190] while the chemokine gradient together with a diversity of bone marrow-derived stromal cells manage to establish a "premetastatic niche" to facilitate organ-specific colonization [191, 192]. It is postulated that cancer stem cells (CSCs; also named tumor-initiating cells [TILs]), despite rarity, are responsible for initial metastatic seeding as a pioneering force among the disseminated cancer cells, under the control of EMT program [185, 193]. Contrasting EMT here is another process termed mesenchymal-to-epithelial transition (MET). Tumor cells undergoing MET in new colonies regain an epithelial phenotype and thrive into masses characteristic of primary tumor. Translating the knowledge of metastatic biology into tools for clinical decision-making is a big project. However, therapeutically targeting this multistage cascade, at either cellular or molecular level, offers multiple opportunities for BrCa intervention.

There is another burgeoning focus on circulating tumor cells (CTCs) and circulating tumor DNA (ctDNA), which might carry useful information from primary to metastatic tumor deposits and thus serve as clinically useful circulating biomarkers [194]. Surrounded by billions of hematopoietic cells in the bloodstream, CTCs represent very rare but heterogeneous cell populations of prognostic and predictive value [191]. Similar to CSCs, CTCs are also products of EMT and might have a role in CSC enrichment due to their phenotypic resemblance [195]. Detection of CTCs from blood samples and their molecular characterization based on genomic profiling opens new avenues for early diagnosis, prognostic evaluation, as well as clinical management of patients with either primary or metastatic BrCa. Several clinical trials are ongoing to assess CTC status in BrCa, irrespective of molecular subtype, under certain therapeutic intervention in the metastatic setting [196–198].

1.7 Kingdom Defense Between Intrinsic Immunity and Extraneous Cancer: Which Outwits?

In general, nascent tumors are held in severe immunological control and cannot flourish without going through immunoediting, which is an end product of complex interplay between the immune system and the neoplastic cells. The first phase is immunosurveillance, at which cancer cells undergo immune recognition and elimination by adaptive immunity; the second battle with a silent state follows, in the name of equilibrium; lastly, the evil play some tricks, manage to escape from immune destruction, and even create an immunosuppressive but tumor-promoting environment [199]. Although the field of tumor immunology remains in such great flux that awaits further investigation, a growing body of advances has sent immunotherapy to the forefront of cancer cure.

The boosting of antitumor immune function necessitates a thorough understanding of innate and adaptive immunity. The nonspecific innate immune system comprised of natural killer (NK) cells, dendritic cells (DCs), macrophages, granulocytes as well as complements is responsible for a quick attack against tumor cells in their first encounter. Contrarily, adaptive cellular or humoral immunity executed by T (CD4 and CD8) or B lymphocytes develops slowly until specific antigens derived from tumors are presented by mature DCs. The presence of tumor-associated antigens (TAAs) and tumor-specific antigens (TSAs), which are essentially non-mutated and mutated proteins (such as HER2 and mutant p53, respectively), leads to the DC maturation and subsequent activation of naïve T cells in the lymph nodes [199]. Afterward, activated CD8+ cytotoxic T lymphocytes (CTLs), partially attributable to the expansion of IFNγ-producing CD4+ type 1 T helper cells (Th1), migrate to infiltrate the tumoral residence for righteous killing. A correlation has been reported between the extensive tumor-infiltrating lymphocytes (TILs) and good prognosis in patients with early-stage TNBC and HER2+ BrCa [200–202]. Additionally, immune-related gene signatures enriched for CD8+ T-cell responses have been detected via unsupervised gene expression profiling of cancer-associated stroma, predictive of a favorable clinical outcome in BrCa [203].

However, it is still a persistent battle because tumor cells can develop their own ways to evade immune destruction, including (but not limited to) (1) reduced immunogenicity and antigen loss that cause T-cell anergy, (2) generation of excessive immunosuppressive factors like TGF-β which hinder DC maturation and T-cell activation, (3) upregulation of PD-L1 (programmed cell death 1 ligand 1, also known as CD274) and costimulatory molecules CD80 and CD86 for the enhancement of immune checkpoint effects mediated by PD1 (programmed cell death 1) and CTLA4 (cytotoxic T-lymphocyte-associated protein 4, also known as CD152), respectively, (4) production of regulatory T cells (Tregs) that alleviate tumor-reactive T-cell responses, as well as (5) recruitment of myeloid-derived suppressor cells (MDSCs)/CAFs and TAMs which are further educated to establish an inflammatory microenvironment favoring immune escape and tumoral growth. From the therapeutic view, all these are candidate vulnerabilities to be targeted to augment the antitumor immune response, and translational research behind them sheds new light on cancer immunotherapy.

As mentioned previously, trastuzumab is a HER2-directed mAb that blocks pro-tumor HER2 signaling. However, emerging evidence has uncovered its multipronged immunostimulatory effects in either innate or adaptive immunity. Trastuzumab locates tumor cells through its antigen-binding fragment (Fab) binding of HER2, while its constant fragment (Fcγ) engages Fcγ receptors (FcγR) displayed by NK cells or monocytes/macrophages, thereby triggering another wave of immune attack mediated by antibody-dependent cellular cytotoxicity (ADCC) or phagocytosis (ADCP) [204]. Moreover, trastuzumab might assist DCs to take up breast cancer-associated antigens (BCAAs) from immunogenic cell death (ICD) and to prime HER2-targeting CD4+ and/or CD8+ T lymphocytes. Besides, the combination of anti-HER2 therapy and checkpoint blockade with anti-PD1 or anti-CTLA4 mAb showed synergy and constituted a valuable therapeutic goal in preclinical models [205, 206]. A phase Ib study also substantiated the therapeutic efficacy of PD-1 blocker pembrolizumab in approximately 20% of advanced TNBC patients [207]. Passive immunotherapy involving HER2/neu (E75) vaccine alongside GM-CSF has exhibited survival benefit in BrCa patients with low expression of HER2 and high risk for relapse [208]. Another promising immunotherapeutic strategy is adoptive transfer of activated tumor-specific T cells, with chimeric antigen receptors (CAR) engineered T-cells occupying an increasing niche. Despite of the robust therapeutic efficacy of CAR-T cells in CD19+ B-cell malignancies [209, 210], no successful treatment has been reported in solid tumors including BrCa, possibly owing to the corresponding TAAs shortage and inefficient T cell homing to tumor sites as well as the tough immunosuppressive tumor microenvironment.

1.8 Noncoding RNAs: A New Frontier of Translational Research?

Another newborn field concerns the epigenetic regulation of BrCa pathology, which in essence causes heritable changes in gene activity without altering the DNA sequence [211]. Despite that DNA methylation and histone modification are classic mechanisms of action in epigenetic control of gene expression [212], the role of regulatory noncoding RNAs (ncRNAs) has emerged as a major focus of epigenetics research. Our previous publication has reviewed the current knowledge of the ncRNA field, including but not limited to the explicit classification, biogenesis pathway, and functional mechanisms of microRNAs (miRNAs), long ncRNAs (lncRNAs), and circular RNAs (circRNAs) in a diverse array of human cancers. A brief summary of the representative ncRNAs regulating BrCa biology is given in Table 1.1.

The continued identification of ncRNAs involving breast tumorigenesis and progression opens new avenues for developing novel diagnostic and prognostic tools or targetable objects in BrCa treatment and acquired drug resistance. Although ncRNA-based antitumor therapeutics stay in the early stage of development, designing BrCa-available ncRNA-related drugs could draw lesson from the first miRNA agent MRX34 (Mirna Therapeutics, TX, USA; mimic of miR-34), which is currently undergoing trial evaluation for hepatocellular carcinomas [227].

1.9 Concluding Remarks and Future Perspectives

To sum up, tremendous progresses have been made to understand the biological hallmarks of breast malignancy and the regulatory mechanism at multiple aspects underlying oncogenesis and development. More in-depth basic investigations are inspired when research data are successively translated into clinically meaningful tests. With either significantly improved efficacy or mini-

Table 1.1 Representative breast cancer-associated noncoding RNAs

ncRNA		Expression pattern/ functional role	Molecular mode of action	Functional mechanism	References
miRNA	miR-17-92 cluster	UP/OG	Target E2F1	Promote proliferation	[213]
	miR-21	UP/OG	Target PDCD4	Promote growth and metastasis	[214]
	miR-155	UP/OG	Target TRF1	Induce genomic instability	[215]
	let-7/miR98 family	DOWN/TS	Target RAS and HMGA2	Suppress the stemness of breast CSCs (BCSCs), and metastasis	[216, 217]
	miR-34	DOWN/TS	TP53 target; target Fra-1	Promote apoptosis, invasion, and metastasis	[218, 219]
lncRNA	HOTAIR	UP/OG	Scaffold PRC2 and LSD1	Promote invasion and metastasis	[220]
	PANDA	UP/OG	Decoy NF-YA	Promote survival and chemoresistance	[221]
	GAS5	DOWN/TS	Decoy glucocorticoid receptor	Promote apoptosis	[222]
	MALAT1	DOWN/TS	Guide HuR	Suppress EMT and metastasis	[223]
	NKILA	DOWN/TS	Stabilize NF-κB/IκB complex	Suppress cancer-associated inflammation and metastasis	[224]
circRNA	Exosomal circRNAs	UP	Unclear	Circulating diagnostic biomarkers	[225]
piRNA	piR-932	Up	Interact with PIWIL2	Promote BCSCs stemness and metastasis	[226]

UP upregulation, *DOWN* downregulation, *OG* oncogene, *TS* tumor suppressor

mized side effects, more and more novel and effective strategies for prevention and/or treatment would be designed for the malignancy. Though shadowed by challenges, the dawn of precision medicine has emerged, especially for BrCa. Details of the susceptibility of mammary stem cells to malignant transformation, and the relationship between luminal progenitor cells and BCSCs, have been presented in our previous review [228]. Identifying the cellular origin along with key molecule drivers in the cancer onset holds therapeutic promise in early-stage BrCa.

Another unsolved issue concerns the intratumoral and intertumoral heterogeneity of breast tumor, which gives a big strike to the conventional "one-size-fits-all" pattern that treats patients as a homogeneous population in clinical trials or even clinical decision-making. Fortunately, the rapid development of next-generation DNA sequencing technologies has accelerated the pace in demystifying the molecular landscapes of cancerous breast. The genomics-based medicine lays a foundation for precision cancer medicine in a patient-by-patient manner. Moreover, the tumor microenvironment with component complexity remains as a gold mine not yet fully tapped. How this supportive niche facilitates BrCa progression and what kind of responses it makes when coming across the routine antitumor therapies remain to be explored. To bridge the gap between in vitro discovery (bench) and their clinical practice (bedside), well-established preclinical tumor model systems are urgently needed. In addition to GEMMs, the arising of orthotopic patient-derived xenograft (PDX) models helps address the intertumoral heterogeneity and better reflects the biological behavior of human tumors and therapeutic response [229]. Last but not the least, the rational therapeutic combinations and technical optimization, together with more cutting-edge approaches including CTC detection, CAR-T cell immunology, and CRISPR/Cas9 system for RNA interference, despite technically challenging, provide new hope for the treatment and cure of patients with advanced breast cancer.

Conflict of Interest No potential conflicts of interest were disclosed.

References

1. Herschkowitz JI, Simin K, Weigman VJ, Mikaelian I, Usary J, Hu Z, Rasmussen KE, Jones LP, Assefnia S, Chandrasekharan S (2007) Identification of conserved gene expression features between murine mammary carcinoma models and human breast tumors. Genome Biol 8(5):R76
2. Prat A, Perou CM (2011) Deconstructing the molecular portraits of breast cancer. Mol Oncol 5(1):5–23
3. Sotiriou C, Neo S-Y, McShane LM, Korn EL, Long PM, Jazaeri A, Martiat P, Fox SB, Harris AL, Liu ET (2003) Breast cancer classification and prognosis based on gene expression profiles from a population-based study. Proc Natl Acad Sci 100(18):10393–10398
4. Kandoth C, McLellan MD, Vandin F, Ye K, Niu B, Lu C, Xie M, Zhang Q, McMichael JF, Wyczalkowski MA (2013) Mutational landscape and significance across 12 major cancer types. Nature 502(7471):333–339
5. Hudson TJ, Anderson W, Aretz A, Barker AD, Bell C, Bernabé RR, Bhan M, Calvo F, Eerola I, Gerhard DS (2010) International network of cancer genome projects. Nature 464(7291):993–998
6. Paik S, Shak S, Tang G, Kim C, Baker J, Cronin M, Baehner FL, Walker MG, Watson D, Park T (2004) A multigene assay to predict recurrence of tamoxifen-treated, node-negative breast cancer. N Engl J Med 351(27):2817–2826
7. Fidler IJ, Poste G (2008) The "seed and soil" hypothesis revisited. Lancet Oncol 9(8):808
8. Massagué J, Obenauf AC (2016) Metastatic colonization by circulating tumour cells. Nature 529(7586):298–306
9. Wolff AC, Hammond MEH, Schwartz JN, Hagerty KL, Allred DC, Cote RJ, Dowsett M, Fitzgibbons PL, Hanna WM, Langer A (2006) American Society of Clinical Oncology/College of American Pathologists guideline recommendations for human epidermal growth factor receptor 2 testing in breast cancer. J Clin Oncol 25(1):118–145
10. Ross JS, Slodkowska EA, Symmans WF, Pusztai L, Ravdin PM, Hortobagyi GN (2009) The HER-2 receptor and breast cancer: ten years of targeted anti–HER-2 therapy and personalized medicine. Oncologist 14(4):320–368
11. Nielsen DL, Andersson M, Kamby C (2009) HER2-targeted therapy in breast cancer. Monoclonal antibodies and tyrosine kinase inhibitors. Cancer Treat Rev 35(2):121–136
12. Gschwind A, Fischer OM, Ullrich A (2004) The discovery of receptor tyrosine kinases: targets for cancer therapy. Nat Rev Cancer 4(5):361–370
13. Graus-Porta D, Beerli RR, Daly JM, Hynes NE (1997) ErbB-2, the preferred heterodimerization partner of all ErbB receptors, is a mediator of lateral signaling. EMBO J 16(7):1647–1655
14. Tzahar E, Waterman H, Chen X, Levkowitz G, Karunagaran D, Lavi S, Ratzkin BJ, Yarden Y (1996) A hierarchical network of interreceptor interactions determines signal transduction by Neu differentiation factor/neuregulin and epidermal growth factor. Mol Cell Biol 16(10):5276–5287
15. Roskoski R (2004) The ErbB/HER receptor protein-tyrosine kinases and cancer. Biochem Biophys Res Commun 319(1):1–11
16. Nahta R, Esteva FJ (2006) Herceptin: mechanisms of action and resistance. Cancer Lett 232(2):123–138
17. Yeon CH, Pegram MD (2005) Anti-erbB-2 antibody trastuzumab in the treatment of HER2-amplified breast cancer. Investig New Drugs 23(5):391–409
18. Moasser MM (2007) The oncogene HER2: its signaling and transforming functions and its role in human cancer pathogenesis. Oncogene 26(45):6469–6487
19. Chia S, Norris B, Speers C, Cheang M, Gilks B, Gown AM, Huntsman D, Olivotto IA, Nielsen TO, Gelmon K (2008) Human epidermal growth factor receptor 2 overexpression as a prognostic factor in a large tissue microarray series of node-negative breast cancers. J Clin Oncol 26(35):5697–5704
20. Piccart M, Lohrisch C, Di Leo A, Larsimont D (2001) The predictive value of HER2 in breast cancer. Oncology 61(Suppl. 2):73–82
21. Yarden Y (2001) Biology of HER2 and its importance in breast cancer. Oncology 61(Suppl. 2):1–13
22. Ross JS, Fletcher JA (1998) The HER-2/neu oncogene in breast cancer: prognostic factor, predictive factor, and target for therapy. Stem Cells 16(6):413–428
23. Arteaga CL, Sliwkowski MX, Osborne CK, Perez EA, Puglisi F, Gianni L (2012) Treatment of HER2-positive breast cancer: current status and future perspectives. Nat Rev Clin Oncol 9(1):16–32

24. Carter P, Presta L, Gorman CM, Ridgway J, Henner D, Wong W, Rowland AM, Kotts C, Carver ME, Shepard HM (1992) Humanization of an anti-p185HER2 antibody for human cancer therapy. Proc Natl Acad Sci 89(10):4285–4289
25. Tokuda Y, Ohnishi Y, Shimamura K, Iwasawa M, Yoshimura M, Ueyama Y, Tamaoki N, Tajima T, Mitomi T (1996) In vitro and in vivo anti-tumour effects of a humanised monoclonal antibody against c-erbB-2 product. Brit J Cancer 73(11):1362
26. Baselga J, Norton L, Albanell J, Kim Y-M, Mendelsohn J (1998) Recombinant humanized anti-HER2 antibody (Herceptin™) enhances the antitumor activity of paclitaxel and doxorubicin against HER2/neu overexpressing human breast cancer xenografts. Cancer Res 58(13):2825–2831
27. Vogel C, Cobleigh M, Tripathy D, Gutheil J, Harris L, Fehrenbacher L, Slamon D, Murphy M, Novotny W, Burchmore M (2001) First-line, single-agent Herceptin®(trastuzumab) in metastatic breast cancer: a preliminary report. Eur J Cancer 37:25–29
28. Marty M, Cognetti F, Maraninchi D, Snyder R, Mauriac L, Tubiana-Hulin M, Chan S, Grimes D, Antón A, Lluch A (2005) Randomized phase II trial of the efficacy and safety of trastuzumab combined with docetaxel in patients with human epidermal growth factor receptor 2–positive metastatic breast cancer administered as first-line treatment: the M77001 study group. J Clin Oncol 23(19):4265–4274
29. Romond EH, Perez EA, Bryant J, Suman VJ, Geyer CE Jr, Davidson NE, Tan-Chiu E, Martino S, Paik S, Kaufman PA (2005) Trastuzumab plus adjuvant chemotherapy for operable HER2-positive breast cancer. N Engl J Med 353(16):1673–1684
30. Piccart-Gebhart MJ, Procter M, Leyland-Jones B, Goldhirsch A, Untch M, Smith I, Gianni L, Baselga J, Bell R, Jackisch C (2005) Trastuzumab after adjuvant chemotherapy in HER2-positive breast cancer. N Engl J Med 353(16):1659–1672
31. Joensuu H, Kellokumpu-Lehtinen P-L, Bono P, Alanko T, Kataja V, Asola R, Utriainen T, Kokko R, Hemminki A, Tarkkanen M (2006) Adjuvant docetaxel or vinorelbine with or without trastuzumab for breast cancer. N Engl J Med 354(8):809–820
32. Smith I, Procter M, Gelber RD, Guillaume S, Feyereislova A, Dowsett M, Goldhirsch A, Untch M, Mariani G, Baselga J (2007) 2-year follow-up of trastuzumab after adjuvant chemotherapy in HER2-positive breast cancer: a randomised controlled trial. Lancet 369(9555):29–36
33. Kurokawa H, Lenferink AE, Simpson JF, Pisacane PI, Sliwkowski MX, Forbes JT, Arteaga CL (2000) Inhibition of HER2/neu (erbB-2) and mitogen-activated protein kinases enhances tamoxifen action against HER2-overexpressing, tamoxifen-resistant breast cancer cells. Cancer Res 60(20):5887–5894
34. Liang K, Lu Y, Jin W, Ang KK, Milas L, Fan Z (2003) Sensitization of breast cancer cells to radiation by trastuzumab. Mol Cancer Ther 2(11):1113–1120
35. Spector NL, Xia W, Burris H III, Hurwitz H, Dees EC, Dowlati A, O'neil B, Overmoyer B, Marcom PK, Blackwell KL (2005) Study of the biologic effects of lapatinib, a reversible inhibitor of ErbB1 and ErbB2 tyrosine kinases, on tumor growth and survival pathways in patients with advanced malignancies. J Clin Oncol 23(11):2502–2512
36. Konecny GE, Pegram MD, Venkatesan N, Finn R, Yang G, Rahmeh M, Untch M, Rusnak DW, Spehar G, Mullin RJ (2006) Activity of the dual kinase inhibitor lapatinib (GW572016) against HER-2-overexpressing and trastuzumab-treated breast cancer cells. Cancer Res 66(3):1630–1639
37. Geyer CE, Forster J, Lindquist D, Chan S, Romieu CG, Pienkowski T, Jagiello-Gruszfeld A, Crown J, Chan A, Kaufman B (2006) Lapatinib plus capecitabine for HER2-positive advanced breast cancer. N Engl J Med 355(26):2733–2743
38. Ross J, Gray G (2002) Targeted therapy for cancer: the HER-2/neu and Herceptin story. Clinical leadership & management review: the journal of CLMA 17(6):333–340
39. Slamon DJ, Leyland-Jones B, Shak S, Fuchs H, Paton V, Bajamonde A, Fleming T, Eiermann W, Wolter J, Pegram M (2001) Use of chemotherapy plus a monoclonal antibody against HER2 for metastatic breast cancer that overexpresses HER2. N Engl J Med 344(11):783–792
40. Xia W, Liu L-H, Ho P, Spector NL (2004) Truncated ErbB2 receptor (p95ErbB2) is regulated by heregulin through heterodimer formation with ErbB3 yet remains sensitive to the dual EGFR/ErbB2 kinase inhibitor GW572016. Oncogene 23(3):646–653
41. Xia W, Gerard CM, Liu L, Baudson NM, Ory TL, Spector NL (2005) Combining lapatinib (GW572016), a small molecule inhibitor of ErbB1 and ErbB2 tyrosine kinases, with therapeutic anti-ErbB2 antibodies enhances apoptosis of ErbB2-overexpressing breast cancer cells. Oncogene 24(41):6213–6221
42. O'Donovan N, Byrne AT, O'Connor AE, McGee S, Gallagher WM, Crown J (2011) Synergistic interaction between trastuzumab and EGFR/HER-2 tyrosine kinase inhibitors in HER-2 positive breast cancer cells. Investig New Drugs 29(5):752–759
43. Baselga J, Bradbury I, Eidtmann H (2012) First results of the NeoaLTTO trial (BIG 01-06/EGF 106903): Lapatinib with Trastuzumab for HER2-positive early breast cancer (NeoaLTTO): a randomized open label multicenter phase 3 trial. Lancet (London, England) 379(9816):633–640
44. Berns K, Horlings HM, Hennessy BT, Madiredjo M, Hijmans EM, Beelen K, Linn SC, Gonzalez-Angulo AM, Stemke-Hale K, Hauptmann M (2007) A functional genetic approach identifies the PI3K pathway as a major determinant of trastuzumab resistance in breast cancer. Cancer Cell 12(4):395–402

45. Nagata Y, Lan K-H, Zhou X, Tan M, Esteva FJ, Sahin AA, Klos KS, Li P, Monia BP, Nguyen NT (2004) PTEN activation contributes to tumor inhibition by trastuzumab, and loss of PTEN predicts trastuzumab resistance in patients. Cancer Cell 6(2):117–127
46. Wang Y, Liu Y, Du Y, Yin W, Lu J (2013) The predictive role of phosphatase and tensin homolog (PTEN) loss, phosphoinositol-3 (PI3) kinase (PIK3CA) mutation, and PI3K pathway activation in sensitivity to trastuzumab in HER2-positive breast cancer: a meta-analysis. Curr Med Res Opin 29(6):633–642
47. Rexer BN, Arteaga CL (2012) Intrinsic and acquired resistance to HER2-targeted therapies in HER2 gene-amplified breast cancer: mechanisms and clinical implications. Critical Rev Oncogen 17(1): 1–16
48. Cizkova M, Susini A, Vacher S, Cizeron-Clairac G, Andrieu C, Driouch K, Fourme E, Lidereau R, Bièche I (2012) PIK3CA mutation impact on survival in breast cancer patients and in ERα, PR and ERBB2-based subgroups. Breast Cancer Res 14(1):R28
49. Cizkova M, Dujaric M, Lehmann-Che J, Scott V, Tembo O, Asselain B, Pierga J, Marty M, De Cremoux P, Spyratos F (2013) Outcome impact of PIK3CA mutations in HER2-positive breast cancer patients treated with trastuzumab. Brit J Cancer 108(9):1807–1809
50. Network CGA (2012) Comprehensive molecular portraits of human breast tumors. Nature 490(7418):61
51. Samuels Y, Wang Z, Bardelli A, Silliman N, Ptak J, Szabo S, Yan H, Gazdar A, Powell SM, Riggins GJ (2004) High frequency of mutations of the PIK3CA gene in human cancers. Science 304(5670):554–554
52. Janiszewska M, Liu L, Almendro V, Kuang Y, Paweletz C, Sakr RA, Weigelt B, Hanker AB, Chandarlapaty S, King TA (2015) In situ single-cell analysis identifies heterogeneity for PIK3CA mutation and HER2 amplification in HER2-positive breast cancer. Nature Publishing Group
53. Shah SP, Roth A, Goya R, Oloumi A, Ha G, Zhao Y, Turashvili G, Ding J, Tse K, Haffari G (2012) The clonal and mutational evolution spectrum of primary triple-negative breast cancers. Nature 486(7403):395–399
54. Tilch E, Seidens T, Cocciardi S, Reid L, Byrne D, Simpson P, Vargas A, Cummings M, Fox S, Lakhani S (2014) Mutations in EGFR, BRAF and RAS are rare in triple-negative and basal-like breast cancers from Caucasian women. Breast Cancer Res Treat 143(2):385–392
55. Koren S, Reavie L, Couto JP, De Silva D, Stadler MB, Roloff T, Britschgi A, Eichlisberger T, Kohler H, Aina O (2015) PIK3CAH1047R induces multipotency and multi-lineage mammary tumours. Nature 525(7567):114–118
56. Koren S, Bentires-Alj M (2013) Mouse models of PIK3CA mutations: one mutation initiates heterogeneous mammary tumors. FEBS J 280(12):2758–2765
57. Meyer DS, Brinkhaus H, Müller U, Müller M, Cardiff RD, Bentires-Alj M (2011) Luminal expression of PIK3CA mutant H1047R in the mammary gland induces heterogeneous tumors. Cancer Res 71(13):4344–4351
58. Liu P, Cheng H, Santiago S, Raeder M, Zhang F, Isabella A, Yang J, Semaan DJ, Chen C, Fox EA (2011) Oncogenic PIK3CA-driven mammary tumors frequently recur via PI3K pathway-dependent and PI3K pathway-independent mechanisms. Nat Med 17(9):1116–1120
59. Sørlie T, Perou CM, Tibshirani R, Aas T, Geisler S, Johnsen H, Hastie T, Eisen MB, Van De Rijn M, Jeffrey SS (2001) Gene expression patterns of breast carcinomas distinguish tumor subclasses with clinical implications. Proc Natl Acad Sci 98(19):10869–10874
60. Ma CX, Reinert T, Chmielewska I, Ellis MJ (2015) Mechanisms of aromatase inhibitor resistance. Nat Rev Cancer 15(5):261–275. doi:10.1038/nrc3920
61. Donehower LA, Harvey M (1992) Mice deficient for p53 are developmentally normal but susceptible to spontaneous tumours. Nature 356(6366):215
62. Kuperwasser C, Hurlbut GD, Kittrell FS, Dickinson ES, Laucirica R, Medina D, Naber SP, Jerry DJ (2000) Development of spontaneous mammary tumors in BALB/c p53 heterozygous mice: a model for li-Fraumeni syndrome. Am J Pathol 157(6):2151–2159
63. Van Keymeulen A, Lee MY, Ousset M, Brohée S, Rorive S, Giraddi RR, Wuidart A, Bouvencourt G, Dubois C, Salmon I (2015) Reactivation of multipotency by oncogenic PIK3CA induces breast tumour heterogeneity. Nature 525(7567):119–123
64. Greenblatt MS, Chappuis PO, Bond JP, Hamel N, Foulkes WD (2001) TP53 mutations in breast cancer associated with BRCA1 or BRCA2 germ-line mutations. Cancer Res 61(10):4092–4097
65. Drost R, Jonkers J (2009) Preclinical mouse models for BRCA1-associated breast cancer. Brit J Cancer 101(10):1651–1657
66. Johannsson O, Idvall I, Anderson C, Borg Å, Barkardottir R, Egilsson V, Olsson H (1997) Tumour biological features of BRCA1-induced breast and ovarian cancer. Eur J Cancer 33(3):362–371
67. Scully R, Chen J, Plug A, Xiao Y, Weaver D, Feunteun J, Ashley T, Livingston DM (1997) Association of BRCA1 with Rad51 in mitotic and meiotic cells. Cell 88(2):265–275
68. Moynahan ME, Chiu JW, Koller BH, Jasin M (1999) Brca1 controls homology-directed DNA repair. Mol Cell 4(4):511–518
69. Kennedy RD, Quinn JE, Mullan PB, Johnston PG, Harkin DP (2004) The role of BRCA1 in the cellular response to chemotherapy. J Natl Cancer Inst 96(22):1659–1668
70. Farmer H, McCabe N, Lord CJ, Tutt AN, Johnson DA, Richardson TB, Santarosa M, Dillon KJ, Hickson I, Knights C (2005) Targeting the DNA repair defect in BRCA mutant cells as a therapeutic strategy. Nature 434(7035):917–921

71. Fong PC, Boss DS, Yap TA, Tutt A, Wu P, Mergui-Roelvink M, Mortimer P, Swaisland H, Lau A, O'connor MJ (2009) Inhibition of poly (ADP-ribose) polymerase in tumors from BRCA mutation carriers. N Engl J Med 361(2):123–134
72. Deng C-X (2006) BRCA1: cell cycle checkpoint, genetic instability, DNA damage response and cancer evolution. Nucleic Acids Res 34(5):1416–1426
73. Mullan P, Quinn J, Harkin D (2006) The role of BRCA1 in transcriptional regulation and cell cycle control. Oncogene 25(43):5854–5863
74. Ganesan S, Silver DP, Greenberg RA, Avni D, Drapkin R, Miron A, Mok SC, Randrianarison V, Brodie S, Salstrom J (2002) BRCA1 supports XIST RNA concentration on the inactive X chromosome. Cell 111(3):393–405
75. Sørlie T, Tibshirani R, Parker J, Hastie T, Marron J, Nobel A, Deng S, Johnsen H, Pesich R, Geisler S (2003) Repeated observation of breast tumor subtypes in independent gene expression data sets. Proc Natl Acad Sci 100(14):8418–8423
76. Lim E, Vaillant F, Wu D, Forrest NC, Pal B, Hart AH, Asselin-Labat ML, Gyorki DE, Ward T, Partanen A, Feleppa F, Huschtscha LI, Thorne HJ, Fox SB, Yan M, French JD, Brown MA, Smyth GK, Visvader JE, Lindeman GJ (2009) Aberrant luminal progenitors as the candidate target population for basal tumor development in BRCA1 mutation carriers. Nat Med 15(8):907–913. doi:10.1038/nm.2000
77. Proia TA, Keller PJ, Gupta PB, Klebba I, Jones AD, Sedic M, Gilmore H, Tung N, Naber SP, Schnitt S (2011) Genetic predisposition directs breast cancer phenotype by dictating progenitor cell fate. Cell Stem Cell 8(2):149–163
78. Molyneux G, Geyer FC, Magnay F-A, McCarthy A, Kendrick H, Natrajan R, MacKay A, Grigoriadis A, Tutt A, Ashworth A (2010) BRCA1 basal-like breast cancers originate from luminal epithelial progenitors and not from basal stem cells. Cell Stem Cell 7(3):403–417
79. Bai F, Smith M, Chan H, Pei X (2013) Germline mutation of Brca1 alters the fate of mammary luminal cells and causes luminal-to-basal mammary tumor transformation. Oncogene 32(22):2715–2725
80. Normanno N, De Luca A, Bianco C, Strizzi L, Mancino M, Maiello MR, Carotenuto A, De Feo G, Caponigro F, Salomon DS (2006) Epidermal growth factor receptor (EGFR) signaling in cancer. Gene 366(1):2–16
81. Hynes NE, Lane HA (2005) ERBB receptors and cancer: the complexity of targeted inhibitors. Nat Rev Cancer 5(5):341–354
82. Harris RC, Chung E, Coffey RJ (2004) EGF receptor ligands. The EGF receptor family biologic mechanisms and role in cancer Elsevier, California, pp 3–14
83. Downward J, Yarden Y, Mayes E, Scrace G, Totty N, Stockwell P, Ullrich A, Schlessinger J, Waterfield M (1984) Close similarity of epidermal growth factor receptor and v-erb-B oncogene protein sequences. Nature 307(5951):521–527
84. Yarden Y, Sliwkowski MX (2001) Untangling the ErbB signalling network. Nat Rev Mol Cell Biol 2(2):127–137
85. Arteaga CL (2001) The epidermal growth factor receptor: from mutant oncogene in nonhuman cancers to therapeutic target in human neoplasia. J Clin Oncol 19 (Suppl 1):32s-40s
86. Hynes NE, MacDonald G (2009) ErbB receptors and signaling pathways in cancer. Curr Opin Cell Biol 21(2):177–184
87. Jorissen RN, Walker F, Pouliot N, Garrett TP, Ward CW, Burgess AW (2003) Epidermal growth factor receptor: mechanisms of activation and signalling. Exp Cell Res 284(1):31–53
88. Nicholson R, Gee J, Harper M (2001) EGFR and cancer prognosis. Eur J Cancer 37:9–15
89. Bange J, Zwick E, Ullrich A (2001) Molecular targets for breast cancer therapy and prevention. Nat Med 7(5):548
90. Lynch TJ, Bell DW, Sordella R, Gurubhagavatula S, Okimoto RA, Brannigan BW, Harris PL, Haserlat SM, Supko JG, Haluska FG (2004) Activating mutations in the epidermal growth factor receptor underlying responsiveness of non–small-cell lung cancer to gefitinib. N Engl J Med 350(21):2129–2139
91. Paez JG, Jänne PA, Lee JC, Tracy S, Greulich H, Gabriel S, Herman P, Kaye FJ, Lindeman N, Boggon TJ (2004) EGFR mutations in lung cancer: correlation with clinical response to gefitinib therapy. Science 304(5676):1497–1500
92. Pao W, Miller V, Zakowski M, Doherty J, Politi K, Sarkaria I, Singh B, Heelan R, Rusch V, Fulton L (2004) EGF receptor gene mutations are common in lung cancers from "never smokers" and are associated with sensitivity of tumors to gefitinib and erlotinib. Proc Natl Acad Sci U S A 101(36):13306–13311
93. Ali S, Wang K, Johnson A, Rodriguez A, Elvin J, Vergilio J, Suh J, Chumsri S, Morosini D, Yelensky R (2016) Abstract P6-03-02: EGFR genomic alterations in 5605 cases of refractory and metastatic breast cancer. AACR
94. Thomas SM, Grandis JR (2004) Pharmacokinetic and pharmacodynamic properties of EGFR inhibitors under clinical investigation. Cancer Treat Rev 30(3):255–268
95. Ranson M, Hammond LA, Ferry D, Kris M, Tullo A, Murray PI, Miller V, Averbuch S, Ochs J, Morris C (2002) ZD1839, a selective oral epidermal growth factor receptor–tyrosine kinase inhibitor, is well tolerated and active in patients with solid, malignant tumors: results of a phase I trial. J Clin Oncol 20(9):2240–2250
96. Hidalgo M, Siu LL, Nemunaitis J, Rizzo J, Hammond LA, Takimoto C, Eckhardt SG, Tolcher A, Britten CD, Denis L (2001) Phase I and pharmacologic study of OSI-774, an epidermal growth factor receptor tyrosine kinase inhibitor, in patients with advanced solid malignancies. J Clin Oncol 19(13):3267–3279

97. Hong WK, Ullrich A (2000) The role of EGFR in solid tumors and implications for therapy. Oncol Biother 1(1):1–29
98. Ciardiello F, Caputo R, Bianco R, Damiano V, Pomatico G, De Placido S, Bianco AR, Tortora G (2000) Antitumor effect and potentiation of cytotoxic drugs activity in human cancer cells by ZD-1839 (Iressa), an epidermal growth factor receptor-selective tyrosine kinase inhibitor. Clin Cancer Res 6(5):2053–2063
99. Baselga J, Albanell J, Ruiz A, Lluch A, Gascón P, Guillém V, González S, Sauleda S, Marimón I, Tabernero JM (2005) Phase II and tumor pharmacodynamic study of gefitinib in patients with advanced breast cancer. J Clin Oncol 23(23):5323–5333
100. Ciardiello F, Troiani T, Caputo F, De Laurentiis M, Palmieri G, Colantuoni G, Diadema M, De Placido S, Bianco A (2004) A phase II study of gefitinib combined with docetaxel as first-line treatment in patients with advanced breast cancer. J Clin Oncol 22:58s
101. Carey LA, Rugo HS, Marcom PK, Mayer EL, Esteva FJ, Ma CX, Liu MC, Storniolo AM, Rimawi MF, Forero-Torres A (2012) TBCRC 001: randomized phase II study of cetuximab in combination with carboplatin in stage IV triple-negative breast cancer. J Clin Oncol 30(21):2615–2623
102. Moulder SL, Arteaga CL (2003) A phase I/II trial of trastuzumab and gefitinib in patients with metastatic breast cancer that overexpresses HER2/neu (ErbB-2). Clin Breast Cancer 4(2):142–145
103. Finn RS, Press MF, Dering J, Arbushites M, Koehler M, Oliva C, Williams LS, Di Leo A (2009) Estrogen receptor, progesterone receptor, human epidermal growth factor receptor 2 (HER2), and epidermal growth factor receptor expression and benefit from lapatinib in a randomized trial of paclitaxel with lapatinib or placebo as first-line treatment in HER2-negative or unknown metastatic breast cancer. J Clin Oncol 27(24):3908–3915
104. Musgrove EA, Sutherland RL (2009) Biological determinants of endocrine resistance in breast cancer. Nat Rev Cancer 9(9):631–643
105. Manavathi B, Dey O, Gajulapalli VNR, Bhatia RS, Bugide S, Kumar R (2012) Derailed estrogen signaling and breast cancer: an authentic couple. Endocr Rev 34(1):1–32
106. Clarke R, Tyson JJ, Dixon JM (2015) Endocrine resistance in breast cancer–an overview and update. Mol Cell Endocrinol 418:220–234
107. Clarke R, Leonessa F, Welch JN, Skaar TC (2001) Cellular and molecular pharmacology of antiestrogen action and resistance. Pharmacol Rev 53(1):25–72
108. Shou J, Massarweh S, Osborne CK, Wakeling AE, Ali S, Weiss H, Schiff R (2004) Mechanisms of tamoxifen resistance: increased estrogen receptor-HER2/neu cross-talk in ER/HER2–positive breast cancer. J Natl Cancer Inst 96(12):926–935
109. Nicholson RI, Hutcheson IR, Knowlden JM, Jones HE, Harper ME, Jordan N, Hiscox SE, Barrow D, Gee JM (2004) Nonendocrine pathways and endocrine resistance. Clin Cancer Res 10(1):346s–354s
110. Okubo S, Kurebayashi J, Otsuki T, Yamamoto Y, Tanaka K, Sonoo H (2004) Additive antitumour effect of the epidermal growth factor receptor tyrosine kinase inhibitor gefitinib (Iressa, ZD1839) and the antioestrogen fulvestrant (Faslodex, ICI 182,780) in breast cancer cells. Brit J Cancer 90(1):236–244
111. Streuli CH, Akhtar N (2009) Signal co-operation between integrins and other receptor systems. Biochem J 418(3):491–506
112. Hynes RO (1992) Integrins: versatility, modulation, and signaling in cell adhesion. Cell 69(1):11–25
113. Ginsberg MH, Du X, Plow EF (1992) Inside-out integrin signalling. Curr Opin Cell Biol 4(5):766–771
114. Glukhova MA, Streuli CH (2013) How integrins control breast biology. Curr Opin Cell Biol 25(5):633–641
115. Koukoulis G, Virtanen I, Korhonen M, Laitinen L, Quaranta V, Gould V (1991) Immunohistochemical localization of integrins in the normal, hyperplastic, and neoplastic breast. Correlations with their functions as receptors and cell adhesion molecules. Am J Pathol 139(4):787
116. Lu S, Simin K, Khan A, Mercurio AM (2008) Analysis of integrin β4 expression in human breast cancer: association with basal-like tumors and prognostic significance. Clin Cancer Res 14(4):1050–1058
117. Pontiggia O, Sampayo R, Raffo D, Motter A, Xu R, Bissell MJ, de Kier Joffé EB, Simian M (2012) The tumor microenvironment modulates tamoxifen resistance in breast cancer: a role for soluble stromal factors and fibronectin through β1 integrin. Breast Cancer Res Treat 133(2):459–471
118. Lesniak D, Xu Y, Deschenes J, Lai R, Thoms J, Murray D, Gosh S, Mackey JR, Sabri S, Abdulkarim B (2009) β1-integrin circumvents the Antiproliferative effects of Trastuzumab in human epidermal growth factor receptor-2–positive breast cancer. Cancer Res 69(22):8620–8628
119. Jahangiri A, Aghi MK, Carbonell WS (2014) β1 integrin: critical path to antiangiogenic therapy resistance and beyond. Cancer Res 74(1):3–7
120. Nisticò P, Di Modugno F, Spada S, Bissell MJ (2014) β1 and β4 integrins: from breast development to clinical practice. Breast Cancer Res 16(5):459
121. Thundimadathil J (2012) Cancer treatment using peptides: current therapies and future prospects. J Amino Acids 2012:1–13
122. Roberts PJ, Der CJ (2007) Targeting the Raf-MEK-ERK mitogen-activated protein kinase cascade for the treatment of cancer. Oncogene 26(22):3291–3310
123. Samatar AA, Poulikakos PI (2014) Targeting RAS–ERK signalling in cancer: promises and challenges. Nat Rev Drug Discov 13(12):928–942

124. Rowinsky EK, Windle JJ, Von Hoff DD (1999) Ras protein farnesyltransferase: a strategic target for anticancer therapeutic development. J Clin Oncol 17(11):3631–3652
125. Craig DW, O'Shaughnessy JA, Kiefer JA, Aldrich J, Sinari S, Moses TM, Wong S, Dinh J, Christoforides A, Blum JL (2013) Genome and transcriptome sequencing in prospective metastatic triple-negative breast cancer uncovers therapeutic vulnerabilities. Mol Cancer Ther 12(1):104–116
126. Caunt CJ, Sale MJ, Smith PD, Cook SJ (2015) MEK1 and MEK2 inhibitors and cancer therapy: the long and winding road. Nat Rev Cancer 15(10):577–592
127. Rinehart J, Adjei AA, LoRusso PM, Waterhouse D, Hecht JR, Natale RB, Hamid O, Varterasian M, Asbury P, Kaldjian EP (2004) Multicenter phase II study of the oral MEK inhibitor, CI-1040, in patients with advanced non-small-cell lung, breast, colon, and pancreatic cancer. J Clin Oncol 22(22):4456–4462
128. Ihle NT, Lemos R, Wipf P, Yacoub A, Mitchell C, Siwak D, Mills GB, Dent P, Kirkpatrick DL, Powis G (2009) Mutations in the phosphatidylinositol-3-kinase pathway predict for antitumor activity of the inhibitor PX-866 whereas oncogenic Ras is a dominant predictor for resistance. Cancer Res 69(1):143–150
129. Engelman JA (2009) Targeting PI3K signalling in cancer: opportunities, challenges and limitations. Nat Rev Cancer 9(8):550–562
130. Thorpe LM, Yuzugullu H, Zhao JJ (2015) PI3K in cancer: divergent roles of isoforms, modes of activation and therapeutic targeting. Nat Rev Cancer 15(1):7–24
131. Gil EMC (2014) Targeting the PI3K/AKT/mTOR pathway in estrogen receptor-positive breast cancer. Cancer Treat Rev 40(7):862–871
132. Miller TW, Balko JM, Arteaga CL (2011) Phosphatidylinositol 3-kinase and antiestrogen resistance in breast cancer. J Clin Oncol 29(33):4452–4461
133. Karlsson E, Pérez-Tenorio G, Amin R, Bostner J, Skoog L, Fornander T, Sgroi DC, Nordenskjöld B, Hallbeck A-L, Stål O (2013) The mTOR effectors 4EBP1 and S6K2 are frequently coexpressed, and associated with a poor prognosis and endocrine resistance in breast cancer: a retrospective study including patients from the randomised Stockholm tamoxifen trials. Breast Cancer Res 15(5):R96
134. Bostner J, Karlsson E, Pandiyan MJ, Westman H, Skoog L, Fornander T, Nordenskjöld B, Stål O (2013) Activation of Akt, mTOR, and the estrogen receptor as a signature to predict tamoxifen treatment benefit. Breast Cancer Res Treat 137(2):397–406
135. Beelen K, Opdam M, Severson TM, Koornstra RH, Vincent AD, Wesseling J, Muris JJ, Berns EM, Vermorken JB, van Diest PJ (2014) Phosphorylated p-70S6K predicts tamoxifen resistance in postmenopausal breast cancer patients randomized between adjuvant tamoxifen versus no systemic treatment. Breast Cancer Res 16(1):3362
136. Campbell RA, Bhat-Nakshatri P, Patel NM, Constantinidou D, Ali S, Nakshatri H (2001) Phosphatidylinositol 3-kinase/AKT-mediated activation of estrogen receptor α a new model for anti-estrogen resistance. J Biol Chem 276(13):9817–9824
137. Chen D, Washbrook E, Sarwar N, Bates GJ, Pace PE, Thirunuvakkarasu V, Taylor J, Epstein RJ, Fuller-Pace FV, Egly J-M (2002) Phosphorylation of human estrogen receptor [alpha] at serine 118 by two distinct signal transduction pathways revealed by phosphorylation-specific antisera. Oncogene 21(32):4921
138. Fan P, Wang J, Santen RJ, Yue W (2007) Long-term treatment with tamoxifen facilitates translocation of estrogen receptor α out of the nucleus and enhances its interaction with EGFR in MCF-7 breast cancer cells. Cancer Res 67(3):1352–1360
139. Song RX, Barnes CJ, Zhang Z, Bao Y, Kumar R, Santen RJ (2004) The role of Shc and insulin-like growth factor 1 receptor in mediating the translocation of estrogen receptor α to the plasma membrane. Proc Natl Acad Sci 101(7):2076–2081
140. Lim S, Kaldis P (2013) Cdks, cyclins and CKIs: roles beyond cell cycle regulation. Development 140(15):3079–3093
141. Roy PG, Pratt N, Purdie CA, Baker L, Ashfield A, Quinlan P, Thompson AM (2010) High CCND1 amplification identifies a group of poor prognosis women with estrogen receptor positive breast cancer. Int J Cancer 127(2):355–360
142. Nevins JR (1998) Toward an understanding of the functional complexity of the E2F and retinoblastoma families. Cell growth & differentiation : the molecular biology journal of the American Association for Cancer Research 9(8):585–593
143. Burkhart DL, Sage J (2008) Cellular mechanisms of tumour suppression by the retinoblastoma gene. Nat Rev Cancer 8(9):671–682. doi:10.1038/nrc2399
144. Sherr CJ, Beach D, Shapiro GI (2016) Targeting CDK4 and CDK6: from discovery to therapy. Cancer Discov 6(4):353–367
145. Sherr CJ, Roberts JM (1999) CDK inhibitors: positive and negative regulators of G1-phase progression. Genes Dev 13(12):1501–1512
146. Blain S (2008) Switching cyclin D-Cdk4 kinase activity on and off. Cell Cycle 7(7):892–898
147. Murphy CG, Dickler MN (2015) The role of CDK4/6 inhibition in breast cancer. Oncologist 20(5):483–490
148. Ertel A, Dean JL, Rui H, Liu C, Witkiewicz AK, Knudsen KE, Knudsen ES (2010) RB-pathway disruption in breast cancer: differential association with disease subtypes, disease-specific prognosis and therapeutic response. Cell Cycle 9(20):4153–4163. doi:10.4161/cc.9.20.13454
149. Landis MW, Pawlyk BS, Li T, Sicinski P, Hinds PW (2006) Cyclin D1-dependent kinase activity in murine development and mammary tumorigenesis. Cancer Cell 9(1):13–22. doi:10.1016/j.ccr.2005.12.019

150. Yu Q, Geng Y, Sicinski P (2001) Specific protection against breast cancers by cyclin D1 ablation. Nature 411(6841):1017–1021. doi:10.1038/35082500
151. Yu Q, Sicinska E, Geng Y, Ahnstrom M, Zagozdzon A, Kong Y, Gardner H, Kiyokawa H, Harris LN, Stal O, Sicinski P (2006) Requirement for CDK4 kinase function in breast cancer. Cancer Cell 9(1):23–32. doi:10.1016/j.ccr.2005.12.012
152. Sabbah M, Courilleau D, Mester J, Redeuilh G (1999) Estrogen induction of the cyclin D1 promoter: involvement of a cAMP response-like element. Proc Natl Acad Sci U S A 96(20):11217–11222
153. Zwijsen RM, Wientjens E, Klompmaker R, van der Sman J, Bernards R, Michalides RJ (1997) CDK-independent activation of estrogen receptor by cyclin D1. Cell 88(3):405–415
154. Dean JL, Thangavel C, McClendon AK, Reed CA, Knudsen ES (2010) Therapeutic CDK4/6 inhibition in breast cancer: key mechanisms of response and failure. Oncogene 29(28):4018–4032. doi:10.1038/onc.2010.154
155. Beaver JA, Amiri-Kordestani L, Charlab R, Chen W, Palmby T, Tilley A, Zirkelbach JF, Yu J, Liu Q, Zhao L (2015) FDA approval: Palbociclib for the treatment of postmenopausal patients with estrogen receptor–positive, HER2-negative metastatic breast cancer. Clin Cancer Res 21(21):4760–4766
156. Walker AJ, Wedam S, Amiri-Kordestani L, Bloomquist E, Tang S, Sridhara R, Chen W, Palmby TR, Zirkelbach JF, Fu W (2016) FDA approval of Palbociclib in combination with Fulvestrant for the treatment of hormone receptor–positive, HER2-negative metastatic breast cancer. Clin Cancer Res 22(20):4968–4972
157. Flaherty KT, Lorusso PM, Demichele A, Abramson VG, Courtney R, Randolph SS, Shaik MN, Wilner KD, O'Dwyer PJ, Schwartz GK (2012) Phase I, dose-escalation trial of the oral cyclin-dependent kinase 4/6 inhibitor PD 0332991, administered using a 21-day schedule in patients with advanced cancer. Clinical cancer research : an official journal of the American Association for Cancer Research 18(2):568–576. doi:10.1158/1078-0432.ccr-11-0509
158. Fry DW, Harvey PJ, Keller PR, Elliott WL, Meade M, Trachet E, Albassam M, Zheng X, Leopold WR, Pryer NK, Toogood PL (2004) Specific inhibition of cyclin-dependent kinase 4/6 by PD 0332991 and associated antitumor activity in human tumor xenografts. Mol Cancer Ther 3(11):1427–1438
159. Munster PN, Hamilton EP, Franklin C, Bhansali S, Wan K, Hewes B, Juric D (2014) Phase lb study of LEE011 and BYL719 in combination with letrozole in estrogen receptor-positive, HER2-negative breast cancer (ER+, HER2− BC). American Society of Clinical Oncology
160. Bardia A, Modi S, Chavez-Mac Gregor M, Kittaneh M, Marino AJ, Matano A, Bhansali S, Hewes B, Cortes J (2014) Phase Ib/II study of LEE011, everolimus, and exemestane in postmenopausal women with ER+/HER2-metastatic breast cancer. American Society of Clinical Oncology
161. Dickler M, Tolaney S, Rugo H, Cortes J, Dieras V, Patt D, Wildiers H, Frenzel M, Koustenis A, Baselga J (2016) MONARCH1: results from a phase II study of abemaciclib, a CDK4 and CDK6 inhibitor, as monotherapy, in patients with HR+/HER2-breast cancer, after chemotherapy for advanced disease. J Clin Oncol 34(Suppl; abstr 510)
162. Folkman J (1995) Angiogenesis in cancer, vascular, rheumatoid and other disease. Nat Med 1(1):27–30
163. Carmeliet P, Jain RK (2011) Molecular mechanisms and clinical applications of angiogenesis. Nature 473(7347):298–307
164. Ribatti D, Nico B, Crivellato E, Roccaro A, Vacca A (2007) The history of the angiogenic switch concept. Leukemia 21(1):44–52
165. Yarden Y, Baselga J, Miles D (2004) Molecular approach to breast cancer treatment. In: Seminars in oncology. Elsevier, Amsterdam, pp 6–13
166. Carmeliet P (2005) VEGF as a key mediator of angiogenesis in cancer. Oncology 69(Suppl. 3):4–10
167. Baeriswyl V, Christofori G (2009) The angiogenic switch in carcinogenesis. In: Seminars in cancer biology, vol 5. Elsevier, Amsterdam, pp 329–337
168. Bergers G, Benjamin LE (2003) Tumorigenesis and the angiogenic switch. Nat Rev Cancer 3(6):401–410
169. Poon RT-P, Fan S-T, Wong J (2001) Clinical implications of circulating angiogenic factors in cancer patients. J Clin Oncol 19(4):1207–1225
170. Leek RD, Lewis CE, Whitehouse R, Greenall M, Clarke J, Harris AL (1996) Association of macrophage infiltration with angiogenesis and prognosis in invasive breast carcinoma. Cancer Res 56(20):4625–4629
171. Riabov V, Gudima A, Wang N, Mickley A, Orekhov A, Kzhyshkowska J (2014) Role of tumor associated macrophages in tumor angiogenesis and lymphangiogenesis. The regulation of angiogenesis by tissue cell-macrophage interactions: 63
172. Weis SM, Cheresh DA (2011) Tumor angiogenesis: molecular pathways and therapeutic targets. Nat Med 17(11):1359–1370
173. Gasparini G, Longo R, Toi M, Ferrara N (2005) Angiogenic inhibitors: a new therapeutic strategy in oncology. Nat Clin Pract Oncol 2(11):562–577
174. Miller K, Burstein H, Elias A, Rugo H, Cobleigh M, Pegram M, Eisenberg P, Collier M, Adams B, Baum C (2005) Phase II study of SU11248, a multitargeted receptor tyrosine kinase inhibitor (TKI), in patients (pts) with previously treated metastatic breast cancer (MBC). J Clin Oncol 23(90160):563–563
175. Bianchi G, Loibl S, Zamagni C, Ardizzoni A, Raab G, Siena S, Wolf C, Westermeier T, Bergamini L, Gianni L (2005) A phase II multicentre uncontrolled trial of sorafenib (BAY 43–9006) in patients with metastatic breast cancer. In: EJC Supplements, vol 2. Pergamon-Elsevier Science Ltd, Oxford, pp 78–78

176. Eckhardt BL, Francis PA, Parker BS, Anderson RL (2012) Strategies for the discovery and development of therapies for metastatic breast cancer. Nat Rev Drug Discov 11(6):479–497. doi:10.1038/nrd2372
177. Valastyan S, Weinberg RA (2011) Tumor metastasis: molecular insights and evolving paradigms. Cell 147(2):275–292
178. DeNardo DG, Barreto JB, Andreu P, Vasquez L, Tawfik D, Kolhatkar N, Coussens LM (2009) CD4+ T cells regulate pulmonary metastasis of mammary carcinomas by enhancing protumor properties of macrophages. Cancer Cell 16(2):91–102
179. Gocheva V, Wang H-W, Gadea BB, Shree T, Hunter KE, Garfall AL, Berman T, Joyce JA (2010) IL-4 induces cathepsin protease activity in tumor-associated macrophages to promote cancer growth and invasion. Genes Dev 24(3):241–255
180. Condeelis J, Pollard JW (2006) Macrophages: obligate partners for tumor cell migration, invasion, and metastasis. Cell 124(2):263–266
181. Wyckoff J, Wang W, Lin EY, Wang Y, Pixley F, Stanley ER, Graf T, Pollard JW, Segall J, Condeelis J (2004) A paracrine loop between tumor cells and macrophages is required for tumor cell migration in mammary tumors. Cancer Res 64(19):7022–7029
182. Wyckoff JB, Wang Y, Lin EY, J-f L, Goswami S, Stanley ER, Segall JE, Pollard JW, Condeelis J (2007) Direct visualization of macrophage-assisted tumor cell intravasation in mammary tumors. Cancer Res 67(6):2649–2656
183. Qian B-Z, Li J, Zhang H, Kitamura T, Zhang J, Campion LR, Kaiser EA, Snyder LA, Pollard JW (2011) CCL2 recruits inflammatory monocytes to facilitate breast-tumour metastasis. Nature 475(7355):222–225
184. Yang J, Weinberg RA (2008) Epithelial-mesenchymal transition: at the crossroads of development and tumor metastasis. Dev Cell 14(6):818–829
185. Ye X, Tam WL, Shibue T, Kaygusuz Y, Reinhardt F, Ng Eaton E, Weinberg RA (2015) Distinct EMT programs control normal mammary stem cells and tumour-initiating cells. Nature 525(7568):256–260. doi:10.1038/nature14897
186. Su S, Liu Q, Chen J, Chen J, Chen F, He C, Huang D, Wu W, Lin L, Huang W (2014) A positive feedback loop between mesenchymal-like cancer cells and macrophages is essential to breast cancer metastasis. Cancer Cell 25(5):605–620
187. Chen J, Yao Y, Gong C, Yu F, Su S, Chen J, Liu B, Deng H, Wang F, Lin L (2011) CCL18 from tumor-associated macrophages promotes breast cancer metastasis via PITPNM3. Cancer Cell 19(4):541–555
188. Egeblad M, Werb Z (2002) New functions for the matrix metalloproteinases in cancer progression. Nat Rev Cancer 2(3):161–174
189. Dano K, Behrendt N, Hoyer-Hansen G, Johnsen M, Lund LR, Ploug M, Romer J (2005) Plasminogen activation and cancer. Thromb Haemost 93(4):676–681
190. Müller A, Homey B, Soto H, Ge N, Catron D, Buchanan ME, McClanahan T, Murphy E, Yuan W, Wagner SN (2001) Involvement of chemokine receptors in breast cancer metastasis. Nature 410(6824):50–56
191. Mego M, Mani SA, Cristofanilli M (2010) Molecular mechanisms of metastasis in breast cancer—clinical applications. Nat Rev Clin Oncol 7(12):693–701
192. Psaila B, Lyden D (2009) The metastatic niche: adapting the foreign soil. Nat Rev Cancer 9(4):285–293
193. Malanchi I, Santamaria-Martinez A, Susanto E, Peng H, Lehr HA, Delaloye JF, Huelsken J (2012) Interactions between cancer stem cells and their niche govern metastatic colonization. Nature 481(7379):85–89. doi:10.1038/nature10694
194. Alix-Panabieres C, Schwarzenbach H, Pantel K (2012) Circulating tumor cells and circulating tumor DNA. Annu Rev Med 63:199–215. doi:10.1146/annurev-med-062310-094219
195. Aktas B, Tewes M, Fehm T, Hauch S, Kimmig R, Kasimir-Bauer S (2009) Stem cell and epithelial-mesenchymal transition markers are frequently overexpressed in circulating tumor cells of metastatic breast cancer patients. Breast Cancer Res 11(4):R46
196. De Mattos-Arruda L, Cortes J, Santarpia L, Vivancos A, Tabernero J, Reis-Filho JS, Seoane J (2013) Circulating tumour cells and cell-free DNA as tools for managing breast cancer. Nat Rev Clin Oncol 10(7):377–389. doi:10.1038/nrclinonc.2013.80
197. Hagenbeck C, Melcher CA, Janni JW, Schneeweiss A, Fasching PA, Aktas B, Pantel K, Solomayer E-F, Ortmann U, Jaeger BAS (2012) DETECT III: a multicenter, randomized, phase III study to compare standard therapy alone versus standard therapy plus lapatinib in patients (pts) with initially HER2-negative metastatic breast cancer but with HER2-positive circulating tumor cells (CTC). American Society of Clinical Oncology
198. Bidard F-C, Fehm T, Ignatiadis M, Smerage JB, Alix-Panabières C, Janni W, Messina C, Paoletti C, Müller V, Hayes DF (2013) Clinical application of circulating tumor cells in breast cancer: overview of the current interventional trials. Cancer Metastasis Rev 32(1–2):179–188
199. Schreiber RD, Old LJ, Smyth MJ (2011) Cancer immunoediting: integrating immunity's roles in cancer suppression and promotion. Science 331(6024):1565–1570
200. Rody A, Holtrich U, Pusztai L, Liedtke C, Gaetje R, Ruckhaeberle E, Solbach C, Hanker L, Ahr A, Metzler D (2009) T-cell metagene predicts a favorable prognosis in estrogen receptor-negative and HER2-positive breast cancers. Breast Cancer Res 11(2):R15
201. DeNardo DG, Coussens LM (2007) Inflammation and breast cancer. Balancing immune response: crosstalk between adaptive and innate immune cells during breast cancer progression. Breast Cancer Res 9(4):212

202. Bianchini G, Qi Y, Alvarez RH, Iwamoto T, Coutant C, Ibrahim NK, Valero V, Cristofanilli M, Green MC, Radvanyi L (2010) Molecular anatomy of breast cancer stroma and its prognostic value in estrogen receptor–positive and–negative cancers. J Clin Oncol 28(28):4316–4323
203. Finak G, Bertos N, Pepin F, Sadekova S, Souleimanova M, Zhao H, Chen H, Omeroglu G, Meterissian S, Omeroglu A (2008) Stromal gene expression predicts clinical outcome in breast cancer. Nat Med 14(5):518–527
204. Bianchini G, Gianni L (2014) The immune system and response to HER2-targeted treatment in breast cancer. Lancet Oncol 15(2):e58–e68
205. Stagg J, Loi S, Divisekera U, Ngiow SF, Duret H, Yagita H, Teng MW, Smyth MJ (2011) Anti–ErbB-2 mAb therapy requires type I and II interferons and synergizes with anti–PD-1 or anti-CD137 mAb therapy. Proc Natl Acad Sci 108(17):7142–7147
206. Wang Q, Li S-H, Wang H, Xiao Y, Sahin O, Brady SW, Li P, Ge H, Jaffee EM, Muller WJ (2012) Concomitant targeting of tumor cells and induction of T-cell response synergizes to effectively inhibit trastuzumab-resistant breast cancer. Cancer Res 72(17):4417–4428
207. Nanda R, Chow LQ, Dees EC, Berger R, Gupta S, Geva R, Pusztai L, Dolled-Filhart M, Emancipator K, Gonzalez EJ (2015) Abstract S1–09: a phase Ib study of pembrolizumab (MK-3475) in patients with advanced triple-negative breast cancer. AACR
208. Mittendorf EA, Clifton GT, Holmes JP, Clive KS, Patil R, Benavides LC, Gates JD, Sears AK, Stojadinovic A, Ponniah S (2012) Clinical trial results of the HER-2/neu (E75) vaccine to prevent breast cancer recurrence in high-risk patients. Cancer 118(10):2594–2602
209. Kochenderfer JN, Dudley ME, Kassim SH, Somerville RP, Carpenter RO, Stetler-Stevenson M, Yang JC, Phan GQ, Hughes MS, Sherry RM (2014) Chemotherapy-refractory diffuse large B-cell lymphoma and indolent B-cell malignancies can be effectively treated with autologous T cells expressing an anti-CD19 chimeric antigen receptor. J Clin Oncol 33(6):540–549
210. Lee DW, Kochenderfer JN, Stetler-Stevenson M, Cui YK, Delbrook C, Feldman SA, Fry TJ, Orentas R, Sabatino M, Shah NN (2015) T cells expressing CD19 chimeric antigen receptors for acute lymphoblastic leukaemia in children and young adults: a phase 1 dose-escalation trial. Lancet 385(9967):517–528
211. Brookes E, Shi Y (2014) Diverse epigenetic mechanisms of human disease. Annu Rev Genet 48:237–268
212. Bjornsson HT, Fallin MD, Feinberg AP (2004) An integrated epigenetic and genetic approach to common human disease. Trends Genet 20(8):350–358
213. Woods K, Thomson JM, Hammond SM (2007) Direct regulation of an oncogenic micro-RNA cluster by E2F transcription factors. J Biol Chem 282(4):2130–2134
214. Frankel LB, Christoffersen NR, Jacobsen A, Lindow M, Krogh A, Lund AH (2008) Programmed cell death 4 (PDCD4) is an important functional target of the microRNA miR-21 in breast cancer cells. J Biol Chem 283(2):1026–1033
215. Dinami R, Ercolani C, Petti E, Piazza S, Ciani Y, Sestito R, Sacconi A, Biagioni F, le Sage C, Agami R (2014) miR-155 drives telomere fragility in human breast cancer by targeting TRF1. Cancer Res 74(15):4145–4156
216. Yu F, Yao H, Zhu P, Zhang X, Pan Q, Gong C, Huang Y, Hu X, Su F, Lieberman J (2007) Let-7 regulates self renewal and tumorigenicity of breast cancer cells. Cell 131(6):1109–1123
217. Lin X, Chen L, Yao Y, Zhao R, Cui X, Chen J, Hou K, Zhang M, Su F, Chen J (2015) CCL18-mediated down-regulation of miR98 and miR27b promotes breast cancer metastasis. Oncotarget 6(24):20485
218. Kato M, Paranjape T, Ullrich R, Nallur S, Gillespie E, Keane K, Esquela-Kerscher A, Weidhaas J, Slack F (2009) The mir-34 microRNA is required for the DNA damage response in vivo in C. elegans and in vitro in human breast cancer cells. Oncogene 28(25):2419–2424
219. Yang S, Li Y, Gao J, Zhang T, Li S, Luo A, Chen H, Ding F, Wang X, Liu Z (2013) MicroRNA-34 suppresses breast cancer invasion and metastasis by directly targeting Fra-1. Oncogene 32(36):4294–4303
220. Gupta RA, Shah N, Wang KC, Kim J, Horlings HM, Wong DJ, Tsai M-C, Hung T, Argani P, Rinn JL (2010) Long non-coding RNA HOTAIR reprograms chromatin state to promote cancer metastasis. Nature 464(7291):1071–1076
221. Hung T, Wang Y, Lin MF, Koegel AK, Kotake Y, Grant GD, Horlings HM, Shah N, Umbricht C, Wang P (2011) Extensive and coordinated transcription of noncoding RNAs within cell-cycle promoters. Nat Genet 43(7):621–629
222. Mourtada-Maarabouni M, Pickard M, Hedge V, Farzaneh F, Williams G (2009) GAS5, a non-protein-coding RNA, controls apoptosis and is downregulated in breast cancer. Oncogene 28(2):195–208
223. Latorre E, Carelli S, Raimondi I, D'Agostino V, Castiglioni I, Zucal C, Moro G, Luciani A, Ghilardi G, Monti E (2016) The ribonucleic complex HuR-MALAT1 represses CD133 expression and suppresses epithelial–mesenchymal transition in breast cancer. Cancer Res 76(9):2626–2636
224. Su F, Li D, Zeng M, Song E (2015) A cytoplasmic NF-kB interacting long noncoding RNA blocks IkB phosphorylation and suppresses breast cancer metastasis. Cancer Cell 27:370–381
225. Li Y, Zheng Q, Bao C, Li S, Guo W, Zhao J, Chen D, Gu J, He X, Huang S (2015) Circular RNA is enriched and stable in exosomes: a promising biomarker for cancer diagnosis. Cell Res 25(8):981

226. Zhang H, Ren Y, Xu H, Pang D, Duan C, Liu C (2013) The expression of stem cell protein Piwil2 and piR-932 in breast cancer. Surg Oncol 22(4):217–223. doi:10.1016/j.suronc.2013.07.001
227. Pichler M, Calin GA (2015) MicroRNAs in cancer: from developmental genes in worms to their clinical application in patients. Brit J Cancer 113(4):569–573. doi:10.1038/bjc.2015.253
228. Chen X, Liu Q, Song E (2017) Mammary stem cells: angels or demons in mammary gland? Sig Transduct Target Therapy 2:16038
229. Tentler JJ, Tan AC, Weekes CD, Jimeno A, Leong S, Pitts TM, Arcaroli JJ, Messersmith WA, Eckhardt SG (2012) Patient-derived tumour xenografts as models for oncology drug development. Nat Rev Clin Oncol 9(6):338–350

2 Biomarker Studies in Early Detection and Prognosis of Breast Cancer

Gang Li, Jing Hu, and Guohong Hu

Abstract

Breast cancer is characterized with enormous heterogeneity, which represents the major hurdle for accurate diagnosis and curative therapy. It is generally believed that genome unstability and molecular evolvability underlie the robustness of cancer cells in hostile microenvironment and their resilience to therapeutic intervention. Conventional histopathological classification of breast cancer falls short of providing sufficient prognostic and predictive power, and thus biomarkers indicative of tumor intrinsic features at molecular levels have been actively pursued in biomedical researches. Currently, a number of molecular biomarkers are being used in standard clinical practice, including the hormone receptors for breast cancer subtyping and several genes involved in genome maintenance for prediction of breast cancer susceptibility. In addition, a number of biomarkers of single genes or multigene signatures have been approved for clinical use for breast cancer prognosis. A growing body of molecular biomarkers are being studied and tested to facilitate disease diagnosis and management, especially for breast cancer early detection, accurate prediction of metastatic behaviors, and selection of therapy. However, most of them are still at the preclinical stages. Finally, biomarkers of noninvasive protocols, such as serological molecules, have advantages in detection convenience over other biomarker types and therefore are of particular interest in translational and clinical development to improve diagnosis, prognosis, and treatment.

G. Li • J. Hu
Chinese Academy of Sciences, Shanghai Institutes for Biological Sciences, University of Chinese Academy of Sciences, Shanghai, China

G. Hu (✉)
Chinese Academy of Sciences, Shanghai Institutes for Biological Sciences, University of Chinese Academy of Sciences, Shanghai, China

Shanghai Jiao-Tong University School of Medicine, Shanghai 200025, China
e-mail: ghhu@sibs.ac.cn

E. Song, H. Hu (eds.), *Translational Research in Breast Cancer*, Advances in Experimental Medicine and Biology 1026, https://doi.org/10.1007/978-981-10-6020-5_2

Keywords
Molecular biomarkers • Breast cancer • Gang prognosis • Genetic susceptibility • Molecular subtyping • Early detection • Serological biomarkers

2.1 Introduction

Breast cancer is the worldwide leading cause of cancer-related deaths in women [1]. Incommensurate with this urgency, there is still a lack of efficient tools for preventive intervention, therapeutic effect evaluation, and prognosis prediction [2]. The TNM staging system classifies cancer progression by assessing the tumor size, lymphatic involvement, and distant metastasis, which is of great prognostic value. However, in order to achieve personalized or precise treatments, more accurate systems indicative of molecular characteristics of tumors are needed. The refinement of breast cancer biomarker studies will fill up such a gap.

In oncology, biomarkers are referred to as any measurable indicators that demonstrate the presence of malignancy or malignant potentials or predict tumor behaviors, prognosis, or responses to treatments [2]. A number of biomarkers reflecting the molecular status or activity within the tumors have been used for clinical diagnosis of breast cancer, while others possess promising potentials. In this chapter, we will summarize these biomarkers. Although non-molecular biomarkers, including anatomical, pathological, environmental, and lifestyle factors could also affect cancer occurrence and progression, we mainly focus on molecular biomarkers here.

2.2 Biomarkers for Breast Cancer Genetic Susceptibility

The familial clustering of breast cancer occurrence indicates that, despite shared lifestyle or environmental exposure, inherited factors exist to render some people more vulnerable to breast cancer [3]. As early as the 1990s, two major breast cancer suppressor genes *BRCA1* and *BRCA2* were identified [4]. Loss-of-function heterozygous germline mutations in either *BRCA1* or *BRCA2* confer high risk of breast cancer, with the morbidity rate reported to be >80% by the age of 70 in female carriers [5]. Although playing different roles, *BRCA1* and *BRCA2* are both involved in DNA repair by homologous recombination [6]. When *BRCA1*/*BRCA2* is defective, homologous recombination is not functional, and cells are directed toward the error-prone nonhomologous end joining for double-strand DNA breaks, which leads to genetic instability and tumorigenic gene mutations like those in *TP53* or *MYC* [7]. Breast tumors carrying *BRCA1* mutants are linked to basal-like and triple-negative phenotypes [7], but those with *BRCA2* mutations are generally of the luminal subtype [8]. Nevertheless, mutations in both genes often render tumors hypersensitive to certain drugs that block progression of DNA replication forks, such as platinum compounds and olaparib [9], which provides a rational for strategized chemotherapy. In addition to breast cancer, mutations of *BRCA1*/*BRCA2* affect the occurrence of ovarian and other cancers [5, 10]. Like those in *BRCA1* and *BRCA2*, mutations in *TP53* gene also affect breast cancer occurrence with high penetrance. Patients harboring germline *TP53* mutations may suffer from Li-Fraumeni syndrome, which is multi-cancer predisposing, and are predisposed to developing breast cancer [11]. In addition, mutations of other well-studied tumor suppressor genes, including *CDH1* [12], *PTEN* [13], and *STK11* [14], lead to increased risk of breast cancer. Although these breast cancer susceptibility genes are highly penetrant, they only account for a small fraction of familial risk of breast cancer [15]. Moderate- or low-penetrance genes or loci are also important for the studies on breast cancer susceptibility.

Investigations into the biological pathways related with *BRCA1* and *BRCA2* have discovered several moderate-penetrance genes, including *PALB2* [16, 17], *ATM* [18], *BRIP1* [19], and

CHEK2 [20]. *PALB2*, abbreviation of "partner and localizer of *BRCA2*," is implicated in the localization and stability of *BRCA2* whose dysfunction may affect double-strand break repair [21]. ATM is a checkpoint kinase involved in the recognition and repair of DNA double-strand break [22]. BRIP1 is a partner of BRCA1 [19], while *CHEK2* could phosphorylate p53 and *BRCA1* to regulate their activities [23]. Compared with mutated *BRCA1* or *BRCA2*, mutations of these genes confer two- to threefold higher risk of breast cancer [24]. There is also evidence showing *RAD51C* [25] and *BARD1* [26] to be breast cancer susceptibility genes.

However, the high- or moderate-penetrance genetic mutations are far away from accounting for the majority of breast cancer cases. Therefore, it is reasonable to assume that a combination of multiple genetic variants of relatively low penetrance might be a good way to predict breast cancer susceptibility. Genome-wide association studies (GWAS), especially those looking for cancer-predisposing single-nucleotide polymorphisms (SNPs), have identified many genetic variants of breast cancer susceptibility loci, some of which lie in plausible gene regions such as *PTHLH*, *NRIP1* [27], *ESR1*, *LSP1* [28], *FGFR2*, *TNRC9*, *MAP3K1* [29], *TGFBR2*, *MYC*, and *TET2* [30], while others are poorly investigated. These loci are associated with small increases in breast cancer risk. Additional work is needed to determine to what extend these loci could be used for breast cancer risk assessment or prediction.

2.3 Biomarkers of Breast Cancer Molecular Subtypes

According to the intrinsic molecular profiles, breast cancers can be classified into various molecular subtypes. Initially, five subtypes (luminal A, luminal B, HER2-enriched, basal-like, and normal-like) of breast cancer were identified [31]. Before long, other less common subtypes were found, which included claudin-low and molecular apocrine (MA) types. These subtypes are associated with distinct molecular alterations, clinical outcome, and responses to treatments. Therefore, the genes of breast cancer subtyping can be used as important biomarkers to guide clinical treatment and predict prognosis.

Both luminal A and luminal B tumors are estrogen receptor (ER)-positive but distinguished by proliferation and hormone-regulated pathways [32]. The steroid hormone estradiol is a growth stimulus of ER-positive tumors, making ER as an ideal endocrine therapies target. Tamoxifen is currently used for treatment of both early and advanced ER-positive breast cancers, which reduces annual breast cancer death rate by 31% in ER-positive patients [33]. Luminal A tumors have higher expression of hormone-related genes such as progesterone receptor (PR) and *FOXA1*, while luminal B tumors tend to demonstrate upregulation of proliferation-related genes such as *MKI67*, *FGFR1* [34], and *AURKA* [32]. Patients of luminal B subtype have worse distant recurrence-free survival compared to those of luminal A subtype. However, luminal B tumors are more sensitive to chemotherapeutic agents such as anthracycline and taxane [35].

The HER2 (human epidermal growth factor receptor 2)-enriched subtype is characterized by genetic amplification/high expression of HER2 and upregulation of proliferation-related genes such as *GRB7*. This group of breast cancers harbors the highest number of mutations across the genome [32]. Patients with this subtype of breast cancer benefit the most from HER2-targeting treatments such as therapies with anti-HER2 monoclonal antibody trastuzumab and tyrosine kinase inhibitor lapatinib.

The basal-like subtype is characterized by high expression of keratins, markers often expressed by the basal layer of skin, and proliferation-related genes. The majority of basal-like breast cancers are ER-, PR-, and HER2-negative. Therefore, they are called as triple-negative breast cancer (TNBC). Patients with basal-like breast cancer usually do not benefit from endocrine therapy or HER2-targeting treatments [36], which leaves chemotherapy to be the only option. Nevertheless, even after chemotherapy, patients with TNBC may have worse outcomes than those with other subtypes [36]. Recent studies have shown that the basal-like breast cancers or TNBCs are complicated and heterogeneous and can be further divided into multiple subtypes with distinct clinical characteristics. For example, some TNBCs overexpress EGFR and may benefit from anti-EGFR therapy [36]. More studies of molecular subtyping in basal-like tumor are needed. And detailed characterization of this type of breast cancer will help decide better treatment methods.

The claudin-low subtype is characterized by high expression of mesenchymal markers and displays the least differentiation. This kind of tumors lacks the expression of tight junction-related genes such as claudin 3 and E-cadherin and is not sensitive to either hormone therapies or chemotherapy [37]. MA tumors are ER-negative and characterized by androgen receptor (AR) and AR-related gene expression, which makes them sensitive to AR-targeted therapy.

In all, assessment of the abovementioned biomarkers, including fluorescence in situ hybridization (FISH) staining of HER2 DNA copy numbers; immunostaining of hormone receptors ER, PR, and AR as well as the proliferation marker MKI67; and transcriptional analysis of other related genes, is very useful for subtyping, prognosis, and strategizing treatment of breast cancer. In order to improve targeted therapeutic approaches from a molecular subtyping perspective, more studies are needed to further uncover the heterogeneity of each cancer subtype and the molecular underpinning of distinct tumor behaviors among different subtypes.

2.4 Biomarkers for Breast Cancer Prognosis

Molecular biomarkers usually interrogate the intrinsic properties of tumor cells and thus are more promising than conventional anatomical or pathological parameters in predicting disease development. Prognostic biomarkers, often those of single-gene or multigene expression analysis, are highly desirable for personalized treatment of patients. Single-gene biomarkers used for prognosis include uPA and PAI-1, whose low protein levels measured by enzyme-linked immunosorbent assay (ELISA) are associated with lower risk of cancer recurrence. In addition, higher expression of proliferation-associated genes such as Ki67, cyclin D, cyclin E, p27, and p21, which is measured by immunohistochemistry (IHC) or DNA content and S phase and determined by flow cytometry-based parameters, have a certain degree of prognostic and predictive value as uncontrolled proliferation is a hallmark of cancer [38].

As tumor cells are extremely heterogeneous and a single biomarker usually falls short of accurate prediction of cancer progression, multiparameter gene signatures are more appealing for cancer prognosis. With the rapid advancement of genomic technology, numerous gene expression signatures have been reported with the capability of predicting the risk of metastasis, recurrence, and drug resistance in breast cancer. Some of them have been applied and commercialized for clinical use. Onco*type* DX, a 21-gene signature consisting of 16 cancer-related genes (such as Ki67, CCNB1, MMP11, HER2, BCL2, ER) and five reference genes, could be used on patients with early-stage hormone receptor-positive breast cancer to predict disease recurrence and the likely benefits from chemotherapy. Oncotype test has been in use in America and Europe since 2004 with exemption from the standard review of US Food and Drug Administration (FDA). MammaPrint is the first multiparameter genetic test approved by FDA to predict breast cancer recurrence. It is a 70-gene RNA expression profile largely consisting of genes involved in proliferation, metastasis, angiogenesis, and stromal integrity and is fully commercialized as a microarray-based assay for prognosis of patients under the age of 61 years with ER-positive or ER-negative and lymph node-negative breast cancer. MammaPrint could identify groups of patients with very good or very poor prognosis. Rotterdam Signature, a 76-gene microarray assay, is another multigene profile being commercialized and not overlapping with either MammaPrint or Onco*type* DX. Rotterdam Signature could be used to predict the development of distant metastasis within 5 years in patients with lymph node-negative breast cancer, regardless of age, tumor size, and grade [39].

2.5 Biomarkers for Breast Cancer Early Detection

Early detection is among the most critical but challenging tasks of cancer management. If tumors could be detected and diagnosed before cancer cell dissemination, the mortality rate is expected to dramatically decline. So far, there are no reliable methods or technologies to screen tumors before manifestation of clinical symptoms. The mammography is useful but inadequate for its relatively low resolution and sensitivity. Tumor screening at molecular level with biomarkers might be a better choice for breast cancer early detection in the future. Practically, early detection of cancer by biomarker tests is feasible only when it is detected noninvasively, such as being detected in the body fluids of the patients. Numerous studies have showed the possibility to differentiate cancer samples, including those at early stages, from normal controls, by using biomarkers in patient circulating system. For example, serum proteomics, in conjunction with bioinformatics, could be used to detect early-stage breast cancer [40]. RS/DJ-1, a regulator of *PTEN* [41], was shown as a circulating antigen specifically found in sera from 37% of newly diagnosed breast cancer patients, but not in those from healthy controls [42]. Recently, more pilot studies have been performed to show that circulating miRNAs, as well as circulating tumor DNA and exosomes, might be specific and sensitive biomarkers for breast cancer early detection [43], which will be discussed in detail below. However, studies on all these biomarkers are still in preclinical stage. In addition, a study searching for biomarkers in sera taken before diagnosis for cancer screening, which is the experimental design for true cancer early detection, failed to yield positive candidates [44]. It indicates the difficulty in finding such biomarkers in clinical practice. With the advancement of biomedical technologies, detection of biomarkers would become more sensitive and accurate. Nevertheless, because of the enormous heterogeneity of breast cancer, synergistic analysis of multiple biomarkers with integration of information from multiple molecular levels should be applied to achieve early cancer detection before symptom manifestation.

2.6 Biomarkers for Metastasis Organotropism

Breast cancer is a metastatic disease that colonizes in a variety of vital organs such as the bone, lung, brain, and liver to cause mortality. Metastasis of breast cancer is a nonrandom process, as cancer cells within the same tumor prefer different target organs for metastatic colonization. Metastasis to different organs often results in quite distinct clinical outcome. Thus, biomarkers for organotropism have been intensively studied to predict the organ preference of metastatic cells.

Through selecting subpopulations with elevated bone metastasis, Kang et al. have identified a set of 102 genes that are either up- (including *CXCR4*, *FGF5*, *CTGF*, *IL11*, *MMP1*, follistatin, ADAMTS1 proteoglycan-1) or downregulated (including fibronectin, serpin A1, cathepsin B, *DLC1*) in osteolytic bone metastasis [45]. Similarly, gene sets mediating metastasis to the lung [46] and brain [47] have also been identified. Some of these genes affect the overall metastatic ability, while others play organ-specific roles in metastatic colonization. Before long, many genes engaged in organ-specific metastasis have been shown to mediate interaction of cancer cells and microenvironment of target organs. For example, deleted in liver cancer 1 (*DLC1*) was shown to regulate the secretion of parathyroid hormone-like hormone (PTHLH) of cancer cells in a Rho-ROCK-dependent manner and suppress PTHLH-induced osteoclast maturation during bone metastasis of breast cancer [48]. Genes involved in the survival and activation of osteoclasts, such as BMP*s*, Wnt [49], CTFG [50], IL1, IL6, CSF-1, and TNFα [51], are also mediators of breast cancer osteolytic bone metastasis.

Global secretome analysis has also identified new mediators and biomarkers of breast cancer bone metastasis such as CST1/2/4/6, PLAT, PLAU, PLOD2, and COL6A1 [52].

Despite genes implicated in proliferation, survival, invasion, or epithelial-mesenchymal transition [53], cross talks between cancer cell and lung microenvironment are also mediators of lung metastasis [46]. The normal lung cellular elements could express considerable amount of BMP to induce dormancy of cancer cells, which is quite different from those in the bone and brain. To overcome this organ-specific anti-metastatic pathway, breast cancer cells express Coco to block BMP signals to undergo reactivation [54]. Neutrophiles [55] and macrophage [56] in the lung, cancer-associated fibroblasts, and mesenchymal stem cells are also shown to affect breast cancer lung metastasis, which suggests that targeting noncancer cell component may be an alternative choice in future cancer therapy.

The interaction between cancer cells and brain-resident cells also helps metastatic colonization in the brain. For example, astrocytes, the most abundant cell type in the brain, could deliver exosomal miRNAs that silence PTEN expression [57] or transfer cGAMP through gap junctions [57] into cancer cells and thus support tumoral growth and chemoresistance in the brain. Therefore, these astrocyte-derived molecules can be used as biomarkers to predict the risk of brain metastasis of breast cancer.

In addition, Hoshino et al. showed that tumor exosomes can be delivered into various distant organs prior to cancer cell dissemination and prepare the pre-metastatic niche in the bone, lung, and liver. Interestingly, exosomes containing different integrin proteins displayed distinct preference for metastasis organs [58], which indicates that exosomal integrins are candidate biomarkers to predict metastasis to multiple organs. Furthermore, a set of miRNAs, including miR-200 family [59], miR-21, miR-155, miR-34a, let-7 [59], miR-31 [60], miR-9 [61], and miR-10b [62], have been shown to mediate the metastasis of breast cancer. These fundamental findings have profound implications, but to what extent they may improve the clinical treatments and benefit patients needs validation in both preclinical and clinical settings.

2.7 Serum Biomarkers

Serum biomarker is a particular class of biomarkers showing great clinical potentials. They can be measured noninvasively and are highly sensitive due to the development of technology such as deep sequencing. There are serum biomarkers at DNA, RNA, protein, as well as cellular levels, all of which display advantages for breast cancer early detection, monitoring, and prognosis over biomarkers that require invasive procedures. Therefore, we will discuss serum biomarkers in this section. Although noninvasive biomarkers could also be detected in other types of body fluids like nipple aspirate fluid [63], urine [64], and ductal lavage fluid [65], we mainly focus on serum here.

2.7.1 Circulating Proteins

The carcinoma antigen 15-3 (CA 15-3) and CA 27-29 assays are clinically used to monitor response to tumor treatments and predict cancer recurrence. Both biomarkers are different epitopes of the MUC1 gene and among the most widely used clinical serum biomarkers for breast cancer. As a transmembrane glycoprotein, MUC-1 is implicated in cell adhesion, immunity, and metastasis [66, 67]. CA 15-3 increases either before or during recurrence in ~70% of cases [68, 69] and decreases after chemotherapy [68], which makes it a good biomarker for monitoring patients with metastasis during active therapy. CA 15-3 and CA 27-29 should be used in conjunction with disease history, diagnostic imaging, and physical examination [39] and are not recommended for screening, diagnosis, and staging considering the low sensitivity of these markers [70]. So is the case with carcinoembryonic antigen (CEA), a set of highly related glycoproteins involved in cell adhesion [39].

2.7.2 Circulating Tumor Cells

Breast cancer cells invade into blood vascular system before metastasizing to distal organs. Enumeration of circulating tumor cells demonstrates great clinical significance as many studies show that the increment in circulating tumor cells is associated with decreased progression-free survival or overall survival in patients [71]. Moreover, circulating tumor cell measurement could be used to predict therapy efficacy and resistance [33]. The CellSearch Assay, a system that distinguishes cancer cells from other circulating cells by cytokeratin and CD45 staining, has received the FDA approval. However, very low numbers of circulating tumor cells are found in many patients with the CellSearch system [72], which may lower the sensitivity and thus limit clinical application of this method.

2.7.3 Circulating Tumor DNA

It is known that somatic genetic alterations, including single-base substitutions, insertions, deletions, or translocations, occur in all kinds of cancers [73]. Cell-free tumor DNA or circulating tumor DNA (ctDNA) refer to DNA fragments carrying tumor-specific sequence mutations found in serum and other body fluids of the patients. The development of sequencing technologies enables identification of mutations in certain genes such as *TP53* and *PIK3CA* in serum of cancer patients [72], which also allows for cancer detection or monitoring of therapy responses with ctDNA. Notably, the alteration of ctDNA could be detected even in patients without detectable circulating tumor cells [73] or elevates protein markers such as CA 15-3 [72], which highlights its superior sensitivity and greater clinical potentials. In addition to genetic mutations, epigenetic alterations could also be assayed in ctDNA. Silence of tumor suppressor genes due to hypermethylation at the promoter regions, modification in chromatin structures, and oncogene activation resulting from histone acetylation and DNA conformation changes could promote breast cancer progression. Epigenetic silencing of BRCA1, GSTP1 [74], DOK7 [75], and RASSF1A is frequently observed. For example, RASSF1A is hypermethylated in 60–77% of breast cancers, which is rarely observed in normal tissues [76]. Many studies suggest that aberrant DNA methylation in breast tumor sample could also be detected in body fluids [77], making it an alternative candidate as circulating biomarkers.

2.7.4 Circulating microRNAs

Many microRNAs have been found to play key roles in cancer initiation, progression and metastasis [78, 79]. Before long, studies showed that miRNAs are also detectable in the plasma of patients. MiRNAs are stable in serum and display distinct expression patterns in cancer and normal individuals [80], which makes them good biomarkers for breast cancer detection or prognosis. The miRNAs reported to be significantly upregulated in plasma of breast cancer patients include miR-141, miR-200a/b/c, miR-203, mimR-210, miR-375, miR-801 [81], miR-10 [82], miR-155 [82], miR-191 [82], miR-382 [82], miR-451 [83], miR-21 [84], miR-148b, miR-376c, miR-409-3p, and miR-801 [85], while miR-181a [86], miR-768-3p [81], miR-145 [83], and miR-92a [84] were found to be downregulated when compared to healthy controls. Interestingly, miR-155 was highly expressed in sera of PR-positive breast cancer patients than that of PR-negative patients [87]. The expression level of miR-125b was associated with chemotherapeutic resistance of breast cancer [88]. Some of these miRNAs have showed potentials with enough sensitivity for early detection and/or prognosis of breast cancer in preliminary studies but need to be further validated by researches with larger clinical cohorts and more stringent procedure before being determined as clinically applicable. Interestingly, some miRNAs demonstrate different expression patterns in patient serum and tumor tissues, a phenomenon worthy of further investigation [85].

Exosomes are small, membrane-bound vesicles which contain a broad range of molecules including proteins, mRNAs, miRNAs, and other kinds of noncoding RNAs. Exosomes mediate the cross talk between cancer cells and normal or cancer-related stromal cells to regulate tumor growth and metastasis [89]. For example, astrocyte-derived exosomes could transfer *PTEN*-targeting miRNAs to tumor cells to suppress *PTEN* expression and promote metastasis [57]. Tumor-derived exosomes were also found to be present in peripheral circulation [90], which bestows it with the potential to be an ideal biomarker for various types of cancers [90]. In order to elucidate the clinical values of exosomes, it is crucially important to profile the contents of exosomes derived from different cancer subtypes or patients receiving distinct treatments.

2.8 Conclusion

In summary, all kinds of biomarkers have been intensively studied for detection, subtyping, and prognosis of breast cancer patients. Many of them have shown great clinical values, while others possess huge potentials. Better clinical application of these biomarkers may be achieved by refinement of biomarker studies and elucidation of the underlying mechanisms on these biomarkers' association with disease status and progression. Furthermore, it is important not only to find new reliable biomarkers but also to define the standard operating protocols for analysis of existing biomarkers in order to offer accurate and reproducible information for clinical consideration. For example, to analyze the expression of a gene, several methods could be used, including IHC, RT-PCR, and ELISA. These expression assays are distinct in terms of sensitivity and resolution and reflect the gene activity at different levels. For clinical application of a particular biomarker, different aspects including sample types, analysis methods and protocols, reagents, and cutoff values for the biomarker status should all be defined and standardized. From this point of view, we still have a long way to go before realizing clinical translation of biomarkers currently identified with promising potentials for breast cancer diagnosis and treatments.

References

1. Media Centre-Cancer (Fact Sheet) (2017) World Health Organization http://www.who.int/mediacentre/factsheets/fs297/en/. Accessed Feb 1 2017
2. Hinestrosa MC, Dickersin K, Klein P, Mayer M, Noss K, Slamon D, Sledge G, Visco FM (2007) Shaping the future of biomarker research in breast cancer to ensure clinical relevance. Nat Rev Cancer 7(4):309–315. doi:10.1038/nrc2113
3. Lynch HT, Krush AJ (1971) Carcinoma of the breast and ovary in three families. Surg Gynecol Obstet 133(4):644–648
4. Hall JM, Lee MK, Newman B, Morrow JE, Anderson LA, Huey B, King MC (1990) Linkage of early-onset familial breast cancer to chromosome 17q21. Science 250(4988):1684–1689
5. Ford D, Easton DF, Stratton M, Narod S, Goldgar D, Devilee P, Bishop DT, Weber B, Lenoir G, Chang-Claude J, Sobol H, Teare MD, Struewing J, Arason A, Scherneck S, Peto J, Rebbeck TR, Tonin P, Neuhausen S, Barkardottir R, Eyfjord J, Lynch H, Ponder BA, Gayther SA, Zelada-Hedman M et al (1998) Genetic heterogeneity and penetrance analysis of the BRCA1 and BRCA2 genes in breast cancer families. The breast cancer linkage consortium. Am J Hum Genet 62(3):676–689
6. Lord CJ, Ashworth A (2016) BRCAness revisited. Nat Rev Cancer 16(2):110–120. doi:10.1038/nrc.2015.21
7. Turner N, Tutt A, Ashworth A (2004) Hallmarks of 'BRCAness' in sporadic cancers. Nat Rev Cancer 4(10):814–819. doi:10.1038/nrc1457
8. Prakash R, Zhang Y, Feng W, Jasin M (2015) Homologous recombination and human health: the roles of BRCA1, BRCA2, and associated proteins. Cold Spring Harb Perspect Biol 7(4):a016600. doi:10.1101/cshperspect.a016600
9. Fong PC, Boss DS, Yap TA, Tutt A, Wu P, Mergui-Roelvink M, Mortimer P, Swaisland H, Lau A, O'Connor MJ, Ashworth A, Carmichael J, Kaye SB, Schellens JH, de Bono JS (2009) Inhibition of poly(ADP-ribose) polymerase in tumors from BRCA mutation carriers. N Engl J Med 361(2):123–134. doi:10.1056/NEJMoa0900212
10. Couch FJ, Nathanson KL, Offit K (2014) Two decades after BRCA: setting paradigms in personalized cancer care and prevention. Science 343(6178):1466–1470. doi:10.1126/science.1251827
11. Hisada M, Garber JE, Fung CY, Fraumeni JF Jr, Li FP (1998) Multiple primary cancers in families with li-Fraumeni syndrome. J Natl Cancer Inst 90(8):606–611
12. Kaurah P, MacMillan A, Boyd N, Senz J, De Luca A, Chun N, Suriano G, Zaor S, Van Manen L, Gilpin C, Nikkel S, Connolly-Wilson M, Weissman S, Rubinstein WS, Sebold C, Greenstein R, Stroop J, Yim D, Panzini B, McKinnon W, Greenblatt M, Wirtzfeld D, Fontaine D, Coit D, Yoon S, Chung D, Lauwers G, Pizzuti A, Vaccaro C, Redal MA, Oliveira C, Tischkowitz M, Olschwang S, Gallinger S, Lynch

H, Green J, Ford J, Pharoah P, Fernandez B, Huntsman D (2007) Founder and recurrent CDH1 mutations in families with hereditary diffuse gastric cancer. JAMA 297(21):2360–2372. doi:10.1001/jama.297.21.2360

13. Marsh DJ, Kum JB, Lunetta KL, Bennett MJ, Gorlin RJ, Ahmed SF, Bodurtha J, Crowe C, Curtis MA, Dasouki M, Dunn T, Feit H, Geraghty MT, Graham JM Jr, Hodgson SV, Hunter A, Korf BR, Manchester D, Miesfeldt S, Murday VA, Nathanson KL, Parisi M, Pober B, Romano C, Eng C et al (1999) PTEN mutation spectrum and genotype-phenotype correlations in Bannayan-Riley-Ruvalcaba syndrome suggest a single entity with Cowden syndrome. Hum Mol Genet 8(8):1461–1472
14. Hearle N, Schumacher V, Menko FH, Olschwang S, Boardman LA, Gille JJ, Keller JJ, Westerman AM, Scott RJ, Lim W, Trimbath JD, Giardiello FM, Gruber SB, Offerhaus GJ, de Rooij FW, Wilson JH, Hansmann A, Moslein G, Royer-Pokora B, Vogel T, Phillips RK, Spigelman AD, Houlston RS (2006) Frequency and spectrum of cancers in the Peutz-Jeghers syndrome. Clin Cancer Res 12(10):3209–3215. doi:10.1158/1078-0432.CCR-06-0083
15. Thompson D, Easton D (2004) The genetic epidemiology of breast cancer genes. J Mammary Gland Biol Neoplasia 9(3):221–236. doi:10.1023/B:JOMG.0000048770.90334.3b
16. Rahman N, Seal S, Thompson D, Kelly P, Renwick A, Elliott A, Reid S, Spanova K, Barfoot R, Chagtai T, Jayatilake H, McGuffog L, Hanks S, Evans DG, Eccles D, Breast Cancer Susceptibility C, Easton DF, Stratton MR (2007) PALB2, which encodes a BRCA2-interacting protein, is a breast cancer susceptibility gene. Nat Genet 39(2):165–167. doi:10.1038/ng1959
17. Erkko H, Xia B, Nikkila J, Schleutker J, Syrjakoski K, Mannermaa A, Kallioniemi A, Pylkas K, Karppinen SM, Rapakko K, Miron A, Sheng Q, Li G, Mattila H, Bell DW, Haber DA, Grip M, Reiman M, Jukkola-Vuorinen A, Mustonen A, Kere J, Aaltonen LA, Kosma VM, Kataja V, Soini Y, Drapkin RI, Livingston DM, Winqvist R (2007) A recurrent mutation in PALB2 in Finnish cancer families. Nature 446(7133):316–319. doi:10.1038/nature05609
18. Thompson D, Duedal S, Kirner J, McGuffog L, Last J, Reiman A, Byrd P, Taylor M, Easton DF (2005) Cancer risks and mortality in heterozygous ATM mutation carriers. J Natl Cancer Inst 97(11):813–822. doi:10.1093/jnci/dji141
19. Seal S, Thompson D, Renwick A, Elliott A, Kelly P, Barfoot R, Chagtai T, Jayatilake H, Ahmed M, Spanova K, North B, McGuffog L, Evans DG, Eccles D, Breast Cancer Susceptibility C, Easton DF, Stratton MR, Rahman N (2006) Truncating mutations in the Fanconi anemia J gene BRIP1 are low-penetrance breast cancer susceptibility alleles. Nat Genet 38(11):1239–1241. doi:10.1038/ng1902
20. Meijers-Heijboer H, van den Ouweland A, Klijn J, Wasielewski M, de Snoo A, Oldenburg R, Hollestelle A, Houben M, Crepin E, van Veghel-Plandsoen M, Elstrodt F, van Duijn C, Bartels C, Meijers C, Schutte M, McGuffog L, Thompson D, Easton D, Sodha N, Seal S, Barfoot R, Mangion J, Chang-Claude J, Eccles D, Eeles R, Evans DG, Houlston R, Murday V, Narod S, Peretz T, Peto J, Phelan C, Zhang HX, Szabo C, Devilee P, Goldgar D, Futreal PA, Nathanson KL, Weber B, Rahman N, Stratton MR, Consortium CH-BC (2002) Low-penetrance susceptibility to breast cancer due to CHEK2(*)1100delC in noncarriers of BRCA1 or BRCA2 mutations. Nat Genet 31(1):55–59. doi:10.1038/ng879
21. Xia B, Sheng Q, Nakanishi K, Ohashi A, Wu J, Christ N, Liu X, Jasin M, Couch FJ, Livingston DM (2006) Control of BRCA2 cellular and clinical functions by a nuclear partner, PALB2. Mol Cell 22(6):719–729. doi:10.1016/j.molcel.2006.05.022
22. Savitsky K, Bar-Shira A, Gilad S, Rotman G, Ziv Y, Vanagaite L, Tagle DA, Smith S, Uziel T, Sfez S, Ashkenazi M, Pecker I, Frydman M, Harnik R, Patanjali SR, Simmons A, Clines GA, Sartiel A, Gatti RA, Chessa L, Sanal O, Lavin MF, Jaspers NG, Taylor AM, Arlett CF, Miki T, Weissman SM, Lovett M, Collins FS, Shiloh Y (1995) A single ataxia telangiectasia gene with a product similar to PI-3 kinase. Science 268(5218):1749–1753
23. Ahn J, Urist M, Prives C (2004) The Chk2 protein kinase. DNA Repair 3(8–9):1039–1047. doi:10.1016/j.dnarep.2004.03.033
24. Stratton MR, Rahman N (2008) The emerging landscape of breast cancer susceptibility. Nat Genet 40(1):17–22. doi:10.1038/ng.2007.53
25. Meindl A, Hellebrand H, Wiek C, Erven V, Wappenschmidt B, Niederacher D, Freund M, Lichtner P, Hartmann L, Schaal H, Ramser J, Honisch E, Kubisch C, Wichmann HE, Kast K, Deissler H, Engel C, Muller-Myhsok B, Neveling K, Kiechle M, Mathew CG, Schindler D, Schmutzler RK, Hanenberg H (2010) Germline mutations in breast and ovarian cancer pedigrees establish RAD51C as a human cancer susceptibility gene. Nat Genet 42(5):410–414. doi:10.1038/ng.569
26. Karppinen SM, Heikkinen K, Rapakko K, Winqvist R (2004) Mutation screening of the BARD1 gene: evidence for involvement of the Cys557Ser allele in hereditary susceptibility to breast cancer. J Med Genet 41(9):e114. doi:10.1136/jmg.2004.020669
27. Ghoussaini M, Fletcher O, Michailidou K, Turnbull C, Schmidt MK, Dicks E, Dennis J, Wang Q, Humphreys MK, Luccarini C, Baynes C, Conroy D, Maranian M, Ahmed S, Driver K, Johnson N, Orr N, dos Santos SI, Waisfisz Q, Meijers-Heijboer H, Uitterlinden AG, Rivadeneira F, Netherlands Collaborative Group on Hereditary B, Ovarian C, Hall P, Czene K, Irwanto A, Liu J, Nevanlinna H, Aittomaki K, Blomqvist C, Meindl A, Schmutzler RK, Muller-Myhsok B, Lichtner P, Chang-Claude J, Hein R, Nickels S, Flesch-Janys D, Tsimiklis H, Makalic E, Schmidt D, Bui M, Hopper JL, Apicella C, Park DJ, Southey

M, Hunter DJ, Chanock SJ, Broeks A, Verhoef S, Hogervorst FB, Fasching PA, Lux MP, Beckmann MW, Ekici AB, Sawyer E, Tomlinson I, Kerin M, Marme F, Schneeweiss A, Sohn C, Burwinkel B, Guenel P, Truong T, Cordina-Duverger E, Menegaux F, Bojesen SE, Nordestgaard BG, Nielsen SF, Flyger H, Milne RL, Alonso MR, Gonzalez-Neira A, Benitez J, Anton-Culver H, Ziogas A, Bernstein L, Dur CC, Brenner H, Muller H, Arndt V, Stegmaier C, Familial Breast Cancer S, Justenhoven C, Brauch H, Bruning T, Gene Environment Interaction of Breast Cancer in Germany N, Wang-Gohrke S, Eilber U, Dork T, Schurmann P, Bremer M, Hillemanns P, Bogdanova NV, Antonenkova NN, Rogov YI, Karstens JH, Bermisheva M, Prokofieva D, Khusnutdinova E, Lindblom A, Margolin S, Mannermaa A, Kataja V, Kosma VM, Hartikainen JM, Lambrechts D, Yesilyurt BT, Floris G, Leunen K, Manoukian S, Bonanni B, Fortuzzi S, Peterlongo P, Couch FJ, Wang X, Stevens K, Lee A, Giles GG, Baglietto L, Severi G, McLean C, Alnaes GG, Kristensen V, Borrensen-Dale AL, John EM, Miron A, Winqvist R, Pylkas K, Jukkola-Vuorinen A, Kauppila S, Andrulis IL, Glendon G, Mulligan AM, Devilee P, van Asperen CJ, Tollenaar RA, Seynaeve C, Figueroa JD, Garcia-Closas M, Brinton L, Lissowska J, Hooning MJ, Hollestelle A, Oldenburg RA, van den Ouweland AM, Cox A, Reed MW, Shah M, Jakubowska A, Lubinski J, Jaworska K, Durda K, Jones M, Schoemaker M, Ashworth A, Swerdlow A, Beesley J, Chen X, kConFab I, Australian Ovarian Cancer Study G, Muir KR, Lophatananon A, Rattanamongkongul S, Chaiwerawattana A, Kang D, Yoo KY, Noh DY, Shen CY, Yu JC, Wu PE, Hsiung CN, Perkins A, Swann R, Velentzis L, Eccles DM, Tapper WJ, Gerty SM, Graham NJ, Ponder BA, Chenevix-Trench G, Pharoah PD, Lathrop M, Dunning AM, Rahman N, Peto J, Easton DF (2012) Genome-wide association analysis identifies three new breast cancer susceptibility loci. Nat Genet 44(3):312–318. doi:10.1038/ng.1049

28. Turnbull C, Ahmed S, Morrison J, Pernet D, Renwick A, Maranian M, Seal S, Ghoussaini M, Hines S, Healey CS, Hughes D, Warren-Perry M, Tapper W, Eccles D, Evans DG, Breast Cancer Susceptibility C, Hooning M, Schutte M, van den Ouweland A, Houlston R, Ross G, Langford C, Pharoah PD, Stratton MR, Dunning AM, Rahman N, Easton DF (2010) Genome-wide association study identifies five new breast cancer susceptibility loci. Nat Genet 42(6):504–507. doi:10.1038/ng.586

29. Easton DF, Pooley KA, Dunning AM, Pharoah PD, Thompson D, Ballinger DG, Struewing JP, Morrison J, Field H, Luben R, Wareham N, Ahmed S, Healey CS, Bowman R, collaborators S, Meyer KB, Haiman CA, Kolonel LK, Henderson BE, Le Marchand L, Brennan P, Sangrajrang S, Gaborieau V, Odefrey F, Shen CY, Wu PE, Wang HC, Eccles D, Evans DG, Peto J, Fletcher O, Johnson N, Seal S, Stratton MR, Rahman N, Chenevix-Trench G, Bojesen SE, Nordestgaard BG, Axelsson CK, Garcia-Closas M, Brinton L, Chanock S, Lissowska J, Peplonska B, Nevanlinna H, Fagerholm R, Eerola H, Kang D, Yoo KY, Noh DY, Ahn SH, Hunter DJ, Hankinson SE, Cox DG, Hall P, Wedren S, Liu J, Low YL, Bogdanova N, Schurmann P, Dork T, Tollenaar RA, Jacobi CE, Devilee P, Klijn JG, Sigurdson AJ, Doody MM, Alexander BH, Zhang J, Cox A, Brock IW, MacPherson G, Reed MW, Couch FJ, Goode EL, Olson JE, Meijers-Heijboer H, van den Ouweland A, Uitterlinden A, Rivadeneira F, Milne RL, Ribas G, Gonzalez-Neira A, Benitez J, Hopper JL, McCredie M, Southey M, Giles GG, Schroen C, Justenhoven C, Brauch H, Hamann U, Ko YD, Spurdle AB, Beesley J, Chen X, kConFab, Group AM, Mannermaa A, Kosma VM, Kataja V, Hartikainen J, Day NE, Cox DR, Ponder BA (2007) Genome-wide association study identifies novel breast cancer susceptibility loci. Nature 447(7148):1087–1093. doi:10.1038/nature05887

30. Michailidou K, Hall P, Gonzalez-Neira A, Ghoussaini M, Dennis J, Milne RL, Schmidt MK, Chang-Claude J, Bojesen SE, Bolla MK, Wang Q, Dicks E, Lee A, Turnbull C, Rahman N, Breast, Ovarian Cancer Susceptibility C, Fletcher O, Peto J, Gibson L, Dos Santos Silva I, Nevanlinna H, Muranen TA, Aittomaki K, Blomqvist C, Czene K, Irwanto A, Liu J, Waisfisz Q, Meijers-Heijboer H, Adank M, Hereditary B, Ovarian Cancer Research Group N, van der Luijt RB, Hein R, Dahmen N, Beckman L, Meindl A, Schmutzler RK, Muller-Myhsok B, Lichtner P, Hopper JL, Southey MC, Makalic E, Schmidt DF, Uitterlinden AG, Hofman A, Hunter DJ, Chanock SJ, Vincent D, Bacot F, Tessier DC, Canisius S, Wessels LF, Haiman CA, Shah M, Luben R, Brown J, Luccarini C, Schoof N, Humphreys K, Li J, Nordestgaard BG, Nielsen SF, Flyger H, Couch FJ, Wang X, Vachon C, Stevens KN, Lambrechts D, Moisse M, Paridaens R, Christiaens MR, Rudolph A, Nickels S, Flesch-Janys D, Johnson N, Aitken Z, Aaltonen K, Heikkinen T, Broeks A, Veer LJ, van der Schoot CE, Guenel P, Truong T, Laurent-Puig P, Menegaux F, Marme F, Schneeweiss A, Sohn C, Burwinkel B, Zamora MP, Perez JI, Pita G, Alonso MR, Cox A, Brock IW, Cross SS, Reed MW, Sawyer EJ, Tomlinson I, Kerin MJ, Miller N, Henderson BE, Schumacher F, Le Marchand L, Andrulis IL, Knight JA, Glendon G, Mulligan AM, kConFab I, Australian Ovarian Cancer Study G, Lindblom A, Margolin S, Hooning MJ, Hollestelle A, van den Ouweland AM, Jager A, Bui QM, Stone J, Dite GS, Apicella C, Tsimiklis H, Giles GG, Severi G, Baglietto L, Fasching PA, Haeberle L, Ekici AB, Beckmann MW, Brenner H, Muller H, Arndt V, Stegmaier C, Swerdlow A, Ashworth A, Orr N, Jones M, Figueroa J, Lissowska J, Brinton L, Goldberg MS, Labreche F, Dumont M, Winqvist R, Pylkas K, Jukkola-Vuorinen A, Grip M, Brauch H, Hamann U, Bruning T, Network G, Radice P, Peterlongo P, Manoukian S, Bonanni B, Devilee P, Tollenaar RA, Seynaeve C, van Asperen CJ, Jakubowska A, Lubinski J, Jaworska K, Durda K, Mannermaa A, Kataja V, Kosma VM,

Hartikainen JM, Bogdanova NV, Antonenkova NN, Dork T, Kristensen VN, Anton-Culver H, Slager S, Toland AE, Edge S, Fostira F, Kang D, Yoo KY, Noh DY, Matsuo K, Ito H, Iwata H, Sueta A, Wu AH, Tseng CC, Van Den Berg D, Stram DO, Shu XO, Lu W, Gao YT, Cai H, Teo SH, Yip CH, Phuah SY, Cornes BK, Hartman M, Miao H, Lim WY, Sng JH, Muir K, Lophatananon A, Stewart-Brown S, Siriwanarangsan P, Shen CY, Hsiung CN, Wu PE, Ding SL, Sangrajrang S, Gaborieau V, Brennan P, McKay J, Blot WJ, Signorello LB, Cai Q, Zheng W, Deming-Halverson S, Shrubsole M, Long J, Simard J, Garcia-Closas M, Pharoah PD, Chenevix-Trench G, Dunning AM, Benitez J, Easton DF (2013) Large-scale genotyping identifies 41 new loci associated with breast cancer risk. Nat Genet 45(4):353–361., 361e351-352. doi:10.1038/ng.2563

31. Sorlie T, Tibshirani R, Parker J, Hastie T, Marron JS, Nobel A, Deng S, Johnsen H, Pesich R, Geisler S, Demeter J, Perou CM, Lonning PE, Brown PO, Borresen-Dale AL, Botstein D (2003) Repeated observation of breast tumor subtypes in independent gene expression data sets. Proc Natl Acad Sci U S A 100(14):8418–8423. doi:10.1073/pnas.0932692100
32. Prat A, Pineda E, Adamo B, Galvan P, Fernandez A, Gaba L, Diez M, Viladot M, Arance A, Munoz M (2015) Clinical implications of the intrinsic molecular subtypes of breast cancer. Breast 24(Suppl 2):S26–S35. doi:10.1016/j.breast.2015.07.008
33. Weigel MT, Dowsett M (2010) Current and emerging biomarkers in breast cancer: prognosis and prediction. Endocr Relat Cancer 17(4):R245–R262. doi:10.1677/ERC-10-0136
34. Reis-Filho JS, Pusztai L (2011) Gene expression profiling in breast cancer: classification, prognostication, and prediction. Lancet 378(9805):1812–1823. doi:10.1016/S0140-6736(11)61539-0
35. Lonning PE (2012) Poor-prognosis estrogen receptor-positive disease: present and future clinical solutions. Therapeutic advances in medical oncology 4(3):127–137. doi:10.1177/1758834012439338
36. Foulkes WD, Smith IE, Reis-Filho JS (2010) Triple-negative breast cancer. N Engl J Med 363(20):1938–1948. doi:10.1056/NEJMra1001389
37. Sabatier R, Finetti P, Guille A, Adelaide J, Chaffanet M, Viens P, Birnbaum D, Bertucci F (2014) Claudin-low breast cancers: clinical, pathological, molecular and prognostic characterization. Mol Cancer 13:228. doi:10.1186/1476-4598-13-228
38. Harris L, Fritsche H, Mennel R, Norton L, Ravdin P, Taube S, Somerfield MR, Hayes DF, Bast RC, Jr., American Society of Clinical O (2007) American Society of Clinical Oncology 2007 update of recommendations for the use of tumor markers in breast cancer. Journal of clinical oncology : official journal of the American Society of Clinical Oncology 25(33):5287–5312. doi:10.1200/JCO.2007.14.2364
39. Kyle RA, Yee GC, Somerfield MR, Flynn PJ, Halabi S, Jagannath S, Orlowski RZ, Roodman DG, Twilde P, Anderson K, American Society of Clinical O (2007) American Society of Clinical Oncology 2007 clinical practice guideline update on the role of bisphosphonates in multiple myeloma. Journal of clinical oncology : official journal of the American Society of Clinical Oncology 25(17):2464–2472. doi:10.1200/JCO.2007.12.1269
40. Li J, Zhang Z, Rosenzweig J, Wang YY, Chan DW (2002) Proteomics and bioinformatics approaches for identification of serum biomarkers to detect breast cancer. Clin Chem 48(8):1296–1304
41. Kim RH, Peters M, Jang Y, Shi W, Pintilie M, Fletcher GC, DeLuca C, Liepa J, Zhou L, Snow B, Binari RC, Manoukian AS, Bray MR, Liu FF, Tsao MS, Mak TW (2005) DJ-1, a novel regulator of the tumor suppressor PTEN. Cancer Cell 7(3):263–273. doi:10.1016/j.ccr.2005.02.010
42. Le Naour F, Misek DE, Krause MC, Deneux L, Giordano TJ, Scholl S, Hanash SM (2001) Proteomics-based identification of RS/DJ-1 as a novel circulating tumor antigen in breast cancer. Clin Cancer Res 7(11):3328–3335
43. Zhao H, Shen J, Medico L, Wang D, Ambrosone CB, Liu S (2010) A pilot study of circulating miRNAs as potential biomarkers of early stage breast cancer. PLoS One 5(10):e13735. doi:10.1371/journal.pone.0013735
44. Kazarian A, Blyuss O, Metodieva G, Gentry-Maharaj A, Ryan A, Kiseleva EM, Prytomanova OM, Jacobs IJ, Widschwendter M, Menon U, Timms JF (2017) Testing breast cancer serum biomarkers for early detection and prognosis in pre-diagnosis samples. Br J Cancer 116(4):501–508. doi:10.1038/bjc.2016.433
45. Kang Y, Siegel PM, Shu W, Drobnjak M, Kakonen SM, Cordon-Cardo C, Guise TA, Massague J (2003) A multigenic program mediating breast cancer metastasis to bone. Cancer Cell 3(6):537–549
46. Minn AJ, Gupta GP, Siegel PM, Bos PD, Shu W, Giri DD, Viale A, Olshen AB, Gerald WL, Massague J (2005) Genes that mediate breast cancer metastasis to lung. Nature 436(7050):518–524
47. Bos PD, Zhang XH, Nadal C, Shu W, Gomis RR, Nguyen DX, Minn AJ, van de Vijver MJ, Gerald WL, Foekens JA, Massague J (2009) Genes that mediate breast cancer metastasis to the brain. Nature 459(7249):1005–1009. doi:10.1038/nature08021
48. Wang Y, Lei R, Zhuang X, Zhang N, Pan H, Li G, Hu J, Pan X, Tao Q, Fu D, Xiao J, Chin YE, Kang Y, Yang Q, Hu G (2014) DLC1-dependent parathyroid hormone-like hormone inhibition suppresses breast cancer bone metastasis. J Clin Invest 124(4):1646–1659. doi:10.1172/JCI71812
49. Harada S, Rodan GA (2003) Control of osteoblast function and regulation of bone mass. Nature 423(6937):349–355. doi:10.1038/nature01660
50. Logothetis CJ, Lin SH (2005) Osteoblasts in prostate cancer metastasis to bone. Nat Rev Cancer 5(1):21–28. doi:10.1038/nrc1528
51. Rucci N, Sanita P, Delle Monache S, Alesse E, Angelucci A (2014) Molecular pathogenesis of bone metastases in breast cancer: proven and emerging

therapeutic targets. World journal of clinical oncology 5(3):335–347. doi:10.5306/wjco.v5.i3.335
52. Blanco MA, LeRoy G, Khan Z, Aleckovic M, Zee BM, Garcia BA, Kang Y (2012) Global secretome analysis identifies novel mediators of bone metastasis. Cell Res 22(9):1339–1355. doi:10.1038/cr.2012.89
53. DiMeo TA, Anderson K, Phadke P, Fan C, Perou CM, Naber S, Kuperwasser C (2009) A novel lung metastasis signature links Wnt signaling with cancer cell self-renewal and epithelial-mesenchymal transition in basal-like breast cancer. Cancer Res 69(13):5364–5373. doi:10.1158/0008-5472.CAN-08-4135
54. Gao H, Chakraborty G, Lee-Lim AP, Mo Q, Decker M, Vonica A, Shen R, Brogi E, Brivanlou AH, Giancotti FG (2012) The BMP inhibitor coco reactivates breast cancer cells at lung metastatic sites. Cell 150(4):764–779. doi:10.1016/j.cell.2012.06.035
55. Wculek SK, Malanchi I (2015) Neutrophils support lung colonization of metastasis-initiating breast cancer cells. Nature 528(7582):413–417. doi:10.1038/nature16140
56. Chen Q, Zhang XH, Massague J (2011) Macrophage binding to receptor VCAM-1 transmits survival signals in breast cancer cells that invade the lungs. Cancer Cell 20(4):538–549. doi:10.1016/j.ccr.2011.08.025
57. Zhang L, Zhang S, Yao J, Lowery FJ, Zhang Q, Huang WC, Li P, Li M, Wang X, Zhang C, Wang H, Ellis K, Cheerathodi M, McCarty JH, Palmieri D, Saunus J, Lakhani S, Huang S, Sahin AA, Aldape KD, Steeg PS, Yu D (2015) Microenvironment-induced PTEN loss by exosomal microRNA primes brain metastasis outgrowth. Nature 527(7576):100–104. doi:10.1038/nature15376
58. Hoshino A, Costa-Silva B, Shen TL, Rodrigues G, Hashimoto A, Mark MT, Molina H, Kohsaka S, Di Giannatale A, Ceder S, Singh S, Williams C, Soplop N, Uryu K, Pharmer L, King T, Bojmar L, Davies AE, Ararso Y, Zhang T, Zhang H, Hernandez J, Weiss JM, Dumont-Cole VD, Kramer K, Wexler LH, Narendran A, Schwartz GK, Healey JH, Sandstrom P, Labori KJ, Kure EH, Grandgenett PM, Hollingsworth MA, de Sousa M, Kaur S, Jain M, Mallya K, Batra SK, Jarnagin WR, Brady MS, Fodstad O, Muller V, Pantel K, Minn AJ, Bissell MJ, Garcia BA, Kang Y, Rajasekhar VK, Ghajar CM, Matei I, Peinado H, Bromberg J, Lyden D (2015) Tumour exosome integrins determine organotropic metastasis. Nature 527 (7578):329–+. doi:10.1038/nature15756
59. Korpal M, Kang Y (2008) The emerging role of miR-200 family of microRNAs in epithelial-mesenchymal transition and cancer metastasis. RNA Biol 5(3):115–119
60. Valastyan S, Reinhardt F, Benaich N, Calogrias D, Szasz AM, Wang ZC, Brock JE, Richardson AL, Weinberg RA (2009) A pleiotropically acting microRNA, miR-31, inhibits breast cancer metastasis. Cell 137(6):1032–1046. doi:10.1016/j.cell.2009.03.047
61. Ma L, Young J, Prabhala H, Pan E, Mestdagh P, Muth D, Teruya-Feldstein J, Reinhardt F, Onder TT, Valastyan S, Westermann F, Speleman F, Vandesompele J, Weinberg RA (2010) miR-9, a MYC/MYCN-activated microRNA, regulates E-cadherin and cancer metastasis. Nat Cell Biol 12(3):247–256. doi:10.1038/ncb2024
62. Ma L, Reinhardt F, Pan E, Soutschek J, Bhat B, Marcusson EG, Teruya-Feldstein J, Bell GW, Weinberg RA (2010) Therapeutic silencing of miR-10b inhibits metastasis in a mouse mammary tumor model. Nat Biotechnol 28(4):341–347. doi:10.1038/nbt.1618
63. Sauter ER, Zhu W, Fan XJ, Wassell RP, Chervoneva I, Du Bois GC (2002) Proteomic analysis of nipple aspirate fluid to detect biologic markers of breast cancer. Br J Cancer 86(9):1440–1443. doi:10.1038/sj.bjc.6600285
64. Kennedy S (2001) Proteomic profiling from human samples: the body fluid alternative. Toxicol Lett 120(1–3):379–384
65. Evron E, Dooley WC, Umbricht CB, Rosenthal D, Sacchi N, Gabrielson E, Soito AB, Hung DT, Ljung B, Davidson NE, Sukumar S (2001) Detection of breast cancer cells in ductal lavage fluid by methylation-specific PCR. Lancet 357(9265):1335–1336
66. Duffy MJ, Shering S, Sherry F, McDermott E, O'Higgins N (2000) CA 15-3: a prognostic marker in breast cancer. Int J Biol Markers 15(4):330–333
67. Duffy MJ (1999) CA 15-3 and related mucins as circulating markers in breast cancer. Ann Clin Biochem 36:579–586
68. Duffy MJ, Duggan C, Keane R, Hill AD, McDermott E, Crown J, O'Higgins N (2004) High preoperative CA 15-3 concentrations predict adverse outcome in node-negative and node-positive breast cancer: study of 600 patients with histologically confirmed breast cancer. Clin Chem 50(3):559–563. doi:10.1373/clinchem.2003.025288
69. Gumireddy K, Li A, Gimotty PA, Klein-Szanto AJ, Showe LC, Katsaros D, Coukos G, Zhang L, Huang Q (2009) KLF17 is a negative regulator of epithelial-mesenchymal transition and metastasis in breast cancer. Nat Cell Biol 11(11):1297–1304. doi:10.1038/ncb1974
70. Uehara M, Kinoshita T, Hojo T, Akashi-Tanaka S, Iwamoto E, Fukutomi T (2008) Long-term prognostic study of carcinoembryonic antigen (CEA) and carbohydrate antigen 15-3 (CA 15-3) in breast cancer. Int J Clin Oncol 13(5):447–451. doi:10.1007/s10147-008-0773-3
71. Cristofanilli M, Hayes DF, Budd GT, Ellis MJ, Stopeck A, Reuben JM, Doyle GV, Matera J, Allard WJ, Miller MC, Fritsche HA, Hortobagyi GN, Terstappen LW (2005) Circulating tumor cells: a novel prognostic factor for newly diagnosed metastatic breast cancer. J Clin Oncol 23(7):1420–1430. doi:10.1200/JCO.2005.08.140

72. Dawson SJ, Tsui DW, Murtaza M, Biggs H, Rueda OM, Chin SF, Dunning MJ, Gale D, Forshew T, Mahler-Araujo B, Rajan S, Humphray S, Becq J, Halsall D, Wallis M, Bentley D, Caldas C, Rosenfeld N (2013) Analysis of circulating tumor DNA to monitor metastatic breast cancer. N Engl J Med 368(13):1199–1209. doi:10.1056/NEJMoa1213261
73. Bettegowda C, Sausen M, Leary RJ, Kinde I, Wang Y, Agrawal N, Bartlett BR, Wang H, Luber B, Alani RM, Antonarakis ES, Azad NS, Bardelli A, Brem H, Cameron JL, Lee CC, Fecher LA, Gallia GL, Gibbs P, Le D, Giuntoli RL, Goggins M, Hogarty MD, Holdhoff M, Hong SM, Jiao Y, Juhl HH, Kim JJ, Siravegna G, Laheru DA, Lauricella C, Lim M, Lipson EJ, Marie SK, Netto GJ, Oliner KS, Olivi A, Olsson L, Riggins GJ, Sartore-Bianchi A, Schmidt K, Shih l M, Oba-Shinjo SM, Siena S, Theodorescu D, Tie J, Harkins TT, Veronese S, Wang TL, Weingart JD, Wolfgang CL, Wood LD, Xing D, Hruban RH, Wu J, Allen PJ, Schmidt CM, Choti MA, Velculescu VE, Kinzler KW, Vogelstein B, Papadopoulos N, Diaz LA, Jr. (2014) Detection of circulating tumor DNA in early- and late-stage human malignancies. Science translational medicine 6(224):224ra224. doi:10.1126/scitranslmed.3007094
74. Esteller M, Corn PG, Urena JM, Gabrielson E, Baylin SB, Herman JG (1998) Inactivation of glutathione S-transferase P1 gene by promoter hypermethylation in human neoplasia. Cancer Res 58(20):4515–4518
75. Heyn H, Carmona FJ, Gomez A, Ferreira HJ, Bell JT, Sayols S, Ward K, Stefansson OA, Moran S, Sandoval J, Eyfjord JE, Spector TD, Esteller M (2013) DNA methylation profiling in breast cancer discordant identical twins identifies DOK7 as novel epigenetic biomarker. Carcinogenesis 34(1):102–108. doi:10.1093/carcin/bgs321
76. Dworkin AM, Huang TH, Toland AE (2009) Epigenetic alterations in the breast: implications for breast cancer detection, prognosis and treatment. Semin Cancer Biol 19(3):165–171. doi:10.1016/j.semcancer.2009.02.007
77. Radpour R, Barekati Z, Kohler C, Lv Q, Burki N, Diesch C, Bitzer J, Zheng H, Schmid S, Zhong XY (2011) Hypermethylation of tumor suppressor genes involved in critical regulatory pathways for developing a blood-based test in breast cancer. PLoS One 6(1):e16080. doi:10.1371/journal.pone.0016080
78. Calin GA, Croce CM (2006) MicroRNA signatures in human cancers. Nat Rev Cancer 6(11):857–866. doi:10.1038/nrc1997
79. He L, He X, Lim LP, de Stanchina E, Xuan Z, Liang Y, Xue W, Zender L, Magnus J, Ridzon D, Jackson AL, Linsley PS, Chen C, Lowe SW, Cleary MA, Hannon GJ (2007) A microRNA component of the p53 tumour suppressor network. Nature 447(7148):1130–1134
80. Mitchell PS, Parkin RK, Kroh EM, Fritz BR, Wyman SK, Pogosova-Agadjanyan EL, Peterson A, Noteboom J, O'Briant KC, Allen A, Lin DW, Urban N, Drescher CW, Knudsen BS, Stirewalt DL, Gentleman R, Vessella RL, Nelson PS, Martin DB, Tewari M (2008) Circulating microRNAs as stable blood-based markers for cancer detection. Proc Natl Acad Sci U S A 105(30):10513–10518. doi:10.1073/pnas.0804549105
81. Madhavan D, Zucknick M, Wallwiener M, Cuk K, Modugno C, Scharpff M, Schott S, Heil J, Turchinovich A, Yang R, Benner A, Riethdorf S, Trumpp A, Sohn C, Pantel K, Schneeweiss A, Burwinkel B (2012) Circulating miRNAs as surrogate markers for circulating tumor cells and prognostic markers in metastatic breast cancer. Clin Cancer Res 18(21):5972–5982. doi:10.1158/1078-0432.CCR-12-1407
82. Mar-Aguilar F, Mendoza-Ramirez JA, Malagon-Santiago I, Espino-Silva PK, Santuario-Facio SK, Ruiz-Flores P, Rodriguez-Padilla C, Resendez-Perez D (2013) Serum circulating microRNA profiling for identification of potential breast cancer biomarkers. Dis Markers 34(3):163–169. doi:10.3233/DMA-120957
83. Ng EK, Li R, Shin VY, Jin HC, Leung CP, Ma ES, Pang R, Chua D, Chu KM, Law WL, Law SY, Poon RT, Kwong A (2013) Circulating microRNAs as specific biomarkers for breast cancer detection. PLoS One 8(1):e53141. doi:10.1371/journal.pone.0053141
84. Si H, Sun X, Chen Y, Cao Y, Chen S, Wang H, Hu C (2013) Circulating microRNA-92a and microRNA-21 as novel minimally invasive biomarkers for primary breast cancer. J Cancer Res Clin Oncol 139(2):223–229. doi:10.1007/s00432-012-1315-y
85. Cuk K, Zucknick M, Heil J, Madhavan D, Schott S, Turchinovich A, Arlt D, Rath M, Sohn C, Benner A, Junkermann H, Schneeweiss A, Burwinkel B (2013) Circulating microRNAs in plasma as early detection markers for breast cancer. Int J Cancer 132(7):1602–1612. doi:10.1002/ijc.27799
86. Guo LJ, Zhang QY (2012) Decreased serum miR-181a is a potential new tool for breast cancer screening. Int J Mol Med 30(3):680–686. doi:10.3892/ijmm.2012.1021
87. Zhu W, Qin W, Atasoy U, Sauter ER (2009) Circulating microRNAs in breast cancer and healthy subjects. BMC Res Notes 2:89. doi:10.1186/1756-0500-2-89
88. Wang H, Tan G, Dong L, Cheng L, Li K, Wang Z, Luo H (2012) Circulating MiR-125b as a marker predicting chemoresistance in breast cancer. PLoS One 7(4):e34210. doi:10.1371/journal.pone.0034210
89. Alderton GK (2012) Metastasis. Exosomes drive premetastatic niche formation. Nat Rev Cancer 12(7):447. doi:10.1038/nrc3304
90. Taylor DD, Gercel-Taylor C (2008) MicroRNA signatures of tumor-derived exosomes as diagnostic biomarkers of ovarian cancer. Gynecol Oncol 110(1):13–21. doi:10.1016/j.ygyno.2008.04.033

The Preventive Intervention of Hereditary Breast Cancer

3

Ayong Cao, Liang Huang, and Zhimin Shao

Abstract

Approximately 5–10% of breast cancer is considered to be hereditary. Familial breast cancers exhibit a dominant hereditary pattern, which typically have an early age of onset and are accompanied by symptoms of ovarian cancer, bilateral breast cancer, or male breast cancer. BRCA gene mutation carriers should be regarded as high-risk groups for breast cancer, which necessitates early examination of breast cancer. Studies have built up kinds of predictive models and recommended that female BRCA mutation carriers should receive breast self-test training and take monthly breast self-examination. Familial or hereditary breast cancer family members are high-risk groups, and their risks of breast cancer can be reduced by chemoprevention, including dietary composition adjustment and application of endocrine drugs. In recent years, large-scale clinical trials have shown the important role of chemoprevention in reducing the occurrence of hereditary breast cancer. Prophylactic mastectomy is also suitable for healthy women with high breast cancer risk factors. It can reduce the incidence rate of breast cancer in high-risk women by 90% and decrease the breast cancer mortality rate in medium-risk and high-risk women by 100% and 81%, respectively.

Keywords

Hereditary breast cancer • BRCA-1/2 • Genetic counseling • Chemoprevention • Prophylactic mastectomy

A. Cao • L. Huang • Z. Shao (✉)
Department of Breast Surgery, Shanghai Cancer Center/Cancer Institute, Fudan University, No.270 Dong'an Road, Shanghai, China
e-mail: zhimingshao@fudan.edu.cn; zhimingshao@yahoo.com

E. Song, H. Hu (eds.), *Translational Research in Breast Cancer*, Advances in Experimental Medicine and Biology 1026, https://doi.org/10.1007/978-981-10-6020-5_3

3.1 Definition of Hereditary Breast Cancer and Its Characteristics and Epidemics

At present, approximately 5–10% of breast cancer is considered to be hereditary. Familial breast cancers exhibit a dominant hereditary pattern, which typically have an early age of onset and are accompanied by symptoms of ovarian cancer, bilateral breast cancer, or male breast cancer [1]. Familial breast cancer usually shows familial aggregation. It is generally believed that the occurrence of breast cancer in two blood-related family members indicates familial breast cancer. In Japan, in addition to the proband, familial breast cancer is defined when there are two or more first-degree relatives with breast cancer in one family, among whom at least one patient meets one of the following conditions: (1) being less than 40 years of age at the time of onset, (2) suffering sequential or simultaneous development of bilateral breast cancer, (3) or enduring sequential or simultaneous development of malignancies outside the breast.

Breast cancer with clear genetic factors is called hereditary breast cancer. This type of breast cancer accounts for 5–10% of the entire breast cancer population. Many genes are associated with hereditary breast cancer, among which the *BRCA-1* and *BRCA-2* genes have the strongest correlation with hereditary breast cancer, followed by *p53*, *PTEN*, *ATM*, and others. Most hereditary breast cancer cases show familial aggregation, whereas a small proportion is sporadic.

3.2 Hereditary Breast Cancer Syndrome

3.2.1 BRCA-1- and BRCA-2-Associated Breast Cancer

Although gene mutations in *BRCA-1* and *BRCA-2* lead to an increased risk of breast cancer and ovarian cancer, these two genes have incomplete penetrance. It has been reported that the cumulative risk of breast cancer for carriers of *BRCA-1* and *BRCA-2* mutations is 45–87% at the age of 70. This penetrance is also affected by gene mutation types, exogenous factors, and lifestyle. Early studies have shown that most *BRCA-1* and *BRCA-2* mutations are associated with hereditary breast cancer, but increasing studies suggest that the two genes can only account for approximately 25% to 28% of hereditary breast cancer risk. However, we cannot exclude the possibility that current detection methods are not sufficient to identify other new mutations of *BRCA-1/2*. *BRCA-1* mutation carriers also have increased risk of developing other malignancies, such as ovarian cancer and fallopian tube cancer, and *BRCA-2* mutation carriers are at increased risk of developing male breast cancer, prostate cancer, pancreatic cancer, gastrointestinal cancer, and malignant melanoma.

BRCA-1 and *BRCA-2* both have complex genomic structures. *BRCA-1* contains 24 exons and encodes a protein consisting of 1863 amino acids, while *BRCA-2* contains 27 exons and encodes a larger protein (3418 amino acids). The first exons of both genes do not encode proteins, and the 11th exons of both genes have the longest sequence among all exons [2, 3]. *BRCA-1* and *BRCA-2* act as tumor suppressor genes, and the encoded proteins play an important role in maintaining genomic integrity through DNA damage signal transduction and injury repair. *BRCA-1* and *BRCA-2* are involved in RAD51-mediated homologous recombination (HR). In the event of DNA double-stranded damage, BRCA-1 assists in the accurate localization of RAD51 to DNA damage, followed by phosphorylation, while BRCA-2 forms a complex with RAD51 to function downstream of the pathway. In the absence of *BRCA-1* and *BRCA-2*, HR-mediated repair of DNA cannot be performed, and DNA damage repair is conducted through the more error-prone nonhomologous end-joining (NHEJ) pathway, resulting in genome instability.

Until now, the Breast Cancer Information Core (BIC) database has included 1790 genetic variants of the *BRCA-1* gene and 2000 genetic variants of the *BRCA-2* gene [4], of which approximately 53%–55% of the variants are detected only in a single family. These mutation sites are scattered throughout the coding region of the genes. The most common pathogenic mutations are small deletions, insertions or frameshift mutations, while splice site mutations and large fragments of gene rearrangements are also observed in both genes [5]. Other forms of mutations, such as missense mutation, silent mutations, and genetic polymorphisms, are relatively common, but it remains challenging to clinically interpret the pathogenic potential of these loci. In the *BRCA-1* and *BRCA-2* genes, approximately 1800 significant sequence variants are considered to be unclassified variants (UVs). To assess the clinical significance of these rare mutations, the Evidence-based Network for the Interpretation of Germline Mutant Alleles (ENIGMA) was established in 2009 to collect globally valid data and resources in order to better understand the clinical significance of these UVs [6].

Germline mutations in the *BRCA-1* and *BRCA-2* genes conform to the first strike in the classical Knudson's "two-strike" theory, whereas mutations of the second inactivation system usually involve the loss of wild-type loci, i.e., loss of heterozygosity (LOH). LOH is present in tumor tissues of the vast majorities of mutation carriers [7, 8]. The mechanism of the other inactivation system is epigenetic silencing caused by promoter methylation, which has an incidence rate of approximately 9–13% in sporadic breast cancer and an incidence rate of up to 42% in non-*BRCA-1/2* hereditary breast cancer [9–11]. In contrast, *BRCA-1* gene promoter methylation is very rare in *BRCA-1* and *BRCA-2* mutation carriers. In the case of *BRCA-2*, almost no promoter methylation is observed in sporadic breast cancer or familial breast cancer [12].

3.2.2 Li-Fraumeni Syndrome

Li-Fraumeni syndrome is a rare autosomal recessive disease that was first reported in 1969 and was named after the two physicians who first discovered it. It is a syndrome of malignancies with familial aggregation, including breast cancer, soft tissue sarcoma, osteosarcoma, brain tumors, leukemia, and adrenal cortical malignancies. Familial aggregation analysis confirmed that the penetrance of carriers is 50% and 90% at the ages of 30 and 40 years, respectively. The proportion of breast cancer is very high in Li-Fraumeni syndrome, and mutations of the tumor suppressor gene *p53* were found to be closely related with this syndrome [13, 14]. A total of 50%–70% of families carrying Li-Fraumeni syndrome have mutations in the *p53* gene, whereas the positive rate of *p53* gene mutation is 1% in breast cancer patients, with an onset age <40 years.

3.2.3 Ataxia Telangiectasia

Ataxia telangiectasia is an autosomal recessive hereditary disease, with which patients display eyelid telangiectasia, cerebellar ataxia, immunodeficiency, and susceptibility to leukemia, lymphoma, and other diseases. The susceptibility gene of this disease is *ATM*, which is located on human chromosome 11q. The *ATM* gene mutation rate in the population is 1%, and this disease is closely related to breast cancer [15]. It has been shown that the risk of breast cancer in heterozygotic carriers of *ATM* mutations is at least four times that of noncarriers and that the risk of breast cancer in heterozygotic carriers of *ATM* mutations can be increased after exposure to radiation. Other studies have shown that the risk of breast cancer in heterozygotic carriers of *ATM* mutations does not increase. The clinical guidance value of ATM still requires further investigation [16].

3.2.4 Cowden Syndrome

Cowden syndrome is a rare autosomal dominant genetic disease, whose clinical manifestations include multiple hamartoma-like lesions, early-onset breast cancer, and thyroid cancer. Hamartoma-like lesions often occur in the skin, oral mucosa, mammary gland, and intestinal tract, including papilloma of the lips and the oral mucosa and limb horny warts. Most Cowden syndrome patients have skin lesions at age 20 and often do not have a family history of breast cancer. Furthermore, 75% of female Cowden syndrome patients are associated with benign breast diseases, such as ductal hyperplasia, intraductal papilloma, breast disease, fibroadenoma, and cystic fibrosis. Approximately 10% of patients with Cowden syndrome are associated with thyroid cancer. The *PTEN/MMAC1/TEP1* gene is the Cowden syndrome susceptibility gene, which is located on human chromosome 10q22–23. This gene was successfully cloned in 1997 [17].

3.2.5 Peutz-Jeghers Syndrome

Peutz-Jeghers syndrome is an autochromosomal dominant hereditary disease, whose common lesions include gastrointestinal hamartoma-like polyps and melanin deposition in the skin mucosa, the latter of which is commonly found in the oral mucosa, lips, fingers, and toes. Some Peutz-Jeghers syndrome patients also have breast cancer, and the average age of onset is 39 years old. The *STK11* gene located on human chromosome 19p13 is closely related to the occurrence of this disease [18].

3.2.6 Muir-Torre Syndrome

Muir-Torre syndrome is a variant of hereditary nonpolyposis colorectal cancer (also known as Lynch syndrome type II) and an autosomal dominant hereditary disease. Its manifestations include multiple sebaceous tumors and skin tumors, such as keratoacanthoma and basal cell carcinoma, accompanied by cancers in the small intestine, large intestine, throat, stomach, endometrium, kidney, bladder, ovary, and breast. Female patients present an increased risk of postmenopausal breast cancer. The *MLH-1* and *MSH-2* genes are associated with this disease [19].

3.3 Pathological Characteristics of *BRCA* Mutation-Associated Breast Cancer

Most *BRCA-1* and *BRCA-2* mutation-associated breast cancers are invasive ductal carcinoma (>80%). There are significant differences in the pathologic features of *BRCA-1* and *BRCA-2* mutation breast cancers, which are also different from sporadic breast cancer. A recent study of 4325 cases of *BRCA-1* gene mutation carriers and 2568 cases of *BRCA-2* gene mutation carriers found that the estrogen receptor (ER)-negative rate of *BRCA-1* mutation breast cancer reached 78%, whereas the ER-negative rate of *BRCA-2* mutation breast cancer was 23%. Even after adjustment for factors, such as age of onset, the correlation between *BRCA-1* gene mutations and the ER-negative rate in *BRCA-1* mutation breast cancer was significantly higher than that in sporadic breast cancer. The HER2/neu amplification ratio was only approximately 10% in the gene mutation carriers, which was significantly lower than that in sporadic breast cancer. As a result, the proportion of triple-negative breast cancer (ER, progesterone receptor (PR), and neu are all negative) in *BRCA-1* gene mutation-associated breast cancer was as high as 69%, which was significantly higher than that in *BRCA-2* gene mutation-associated breast cancer (16%) [20]. In addition, it has been reported that the *p53* gene mutation rate and the positive immunohistochemical detection rate of *p53* are significantly increased in *BRCA-1* mutation breast cancer compared with sporadic breast cancer, and its germline mutant form is also significantly different in *BRCA-1* mutation breast cancer compared to sporadic breast cancer. Based on the gene expression profile of breast cancer, it has been found that most *BRCA-1* mutation breast cancers show a basal-like/myoepithelial phenotype and express several basal markers, such as CK5/CK6, CK14, caveolin, vimentin, laminin, p-cadherin, osteonectin, and EGFR [21, 22].

It has also been reported that the proportion of medullary carcinoma in *BRCA-1* mutation-associated breast cancer is higher than that in sporadic breast cancer (9% and 2%, respectively). The probability of *BRCA-1* mutation in myeloid carcinoma is as high as 11% [23]. Medullary carcinoma is characterized by low differentiation, high grade, and vascular invasion, but the prognosis of medullary carcinoma is usually better, which may be related to the low incidence of lymph node metastasis [24]. Compared with sporadic breast cancer, *BRCA-1* mutation-associated breast cancer has a higher nuclear grade and division index and is more prone to necrosis and lymphatic invasion. All of the above findings indicate that *BRCA-1* mutation-associated breast cancer has a more aggressive biological behavior and poorer prognosis than sporadic breast cancer.

In contrast to *BRCA-1* mutation-associated breast cancers, the pathologic features of *BRCA-2* mutation-associated breast cancers are more similar to those of sporadic breast cancer. The positive rate of ER in *BRCA-2* mutation-associated breast cancers is significantly higher than that in *BRCA-1* mutation-associated breast cancers, and the positive expression rate of ER in *BRCA-2* mutation-associated breast cancers decreases with increased age of onset [25]. Several studies have reported that *BRCA-2* mutation-associated breast cancers show no amplification or low-level amplification of HER2/neu. Recent studies have reported that the expression levels of fibroblast growth factor 1 (FGF1) and fibroblast growth factor receptor 2 (FGFR2) are significantly higher in *BRCA-2* mutation-associated breast cancers than in *BRCA-1* mutation-associated breast cancers [26].

In terms of gene expression profiles, most BRCA-2 mutation-associated breast cancers are of the luminal phenotypes that express ER, PR, CK8, and CK18 [27]. The positive rate of the p53 gene is similar in BRCA-1 mutation-associated breast cancers and BRCA-2 mutation-associated breast cancers [28], whereas caveolin1 is not expressed in BRCA-2 mutation-associated breast cancers, which differs significantly from BRCA-1 mutation-associated breast cancers [29]. In addition, the expression rates of cyclin D1, BAX, and BCL2 in BRCA-2 mutation-associated breast cancers are also significantly higher than those in BRCA-1 mutation-associated breast cancers.

3.4 Gene Mutation Risk Prediction and Prognosis

Gail Model In 1989, Gail et al. published a statistical analysis model using case-control data to predict the risk of invasive breast cancer and in situ carcinoma in white women who underwent mammography each year (Gail-1) [30]. The tools of this assessment model utilize breast cancer risk factors, including breast cancer history, age, menarche age, primiparous age, number of breast cancer patients in first-degree relatives, breast biopsy result, and race. In 1990, Costantino and Gail et al. [31] reported a modified Gail model validation study. The 5-year follow-up study showed that the ratio of the estimated Gail-2 absolute risks to the actual incidence of breast cancer (E/O) was 1.03 (0.88–1.21), better than the Gail-1 E/O value (0.84 (0.73–0.97)). In 2001, the US Food and Drug Administration approved preventative tamoxifen treatment for women with Gail 5-year risk forecasts ≥1.66% and over 35 years of age [32]. The US National Cancer Institute later published the Breast Cancer Risk Prediction Tool (http: //www.cancer.gov/bcrisktool) online to allow self-assessment in US women.

Data of the Gail model were collected from the US white patients, but the model has been validated in many different populations. Rockhill reported a Gail risk analysis of 82,109 women. The follow-up results showed that the ratio of the expected number of breast cancer cases to the actual observed number was 0.94 (95% CI 0.89 to 0.99), confirming the prediction efficacy of the Gail model. Currently, the research population has been extended from Caucasian to African-American, Hispanic, Asian American, American Indian, and Alaska Native. Race or ethnicity can affect the risk calculation of breast cancer. Whether this tool is suitable for assessing women living outside the United States still needs to be investigated.

BRCAPro Model The BRCAPro model is the most widely used gene model. This model is based on the screening of *BRCA-1/2* gene mutation carriers using Bayes theorem as well as the parameters of the conditions of breast cancer and ovarian cancer in first-degree and second-degree relatives and the age of onset of the diseased family members [33, 34]. Berry et al. [35] subsequently used the BRCAPro model to predict the risk of breast cancer and ovarian cancer in 301 individuals who underwent *BRCA-1/2* mutation detection and later confirmed the feasibility of the BRCAPro model. At present, the model has been continuously updated based on published literature and data and been combined with reports of the penetrance of *BRCA-1 and BRCA-2* gene mutations.

Myriad Model The Myriad model [36] uses information from 10,000 cases having undergone routine *BRCA-1/2* mutation detection (7461 cases) or detection of three ancestor mutations of Ashkenazi Jewish origin, which include family history, age of onset of breast or ovarian cancer, and presence or absence of invasive cancer, to establish a model predicting the possibility of carrying mutations. This model found that women with family history and of Jewish descent were more likely to carry mutations, while high-risk African populations and other non-Jewish and European descent populations had similar rates of mutation. In addition, the pathological type had a big impact on the possibility of carrying mutations, as the mutation rate in in situ ductal carcinoma patients aged <50 years was significantly lower than that in patients with invasive ductal carcinoma. The model has been published on Myriad's website and is regularly updated based on the expansion of the sample size.

Penn II Model The Couch (Penn I) model was established in 1997 and initially used to predict only the likelihood of *BRCA-1* mutations among 169 breast cancer families. This model was then updated to obtain a new online Penn II model for predicting the likelihood of *BRCA-1/2* mutations. It uses logistic regression analysis to predict the likelihood of *BRCA-1/2* mutation in individuals or families; incorporates specific clinical features that are ignored by most other models, such as bilateral breast cancer and the presence or absence of prostate cancer and pancreatic cancer; and includes a risk assessment of third-degree relatives. Lindner NM et al. validated the LAMBDA, BRCAPro, modified Penn I (Couch), Myriad II, and Penn II models in 285 probands from different breast cancer families, and the results showed that the Penn II model had the best prediction efficacy [37].

Rao Nanyan et al. from the Fudan University Shanghai Cancer Center used breast cancer-related information from 212 patients receiving *BRCA-1/2* gene mutation detection in the multicenter database to validate the Penn, Myriad, and BRCAPro prediction models originally used in Western populations. The results showed that the prediction efficacies of the three models were similar, as the area under the curve (AUC) values of the receiver operating characteristic (ROC) curve were approximately 0.7. The three models predicted higher ROC values and positive likelihood ratios in 66 high-risk families. If the cutoff point was set at 10%, the BRCAPro model had the maximum prediction value for *BRCA* mutations [38].

As the current breast cancer prediction models are based on Western populations, their predictive efficacy for Chinese population remains unclear. Therefore, it is urgently important to establish a gene mutation prediction model suitable for the Chinese population. Among the sample population, nonparametric analysis showed that genetic mutation carriers were related to a family history of ovarian cancer and gastric cancer, as well as the age of breast cancer onset in the family. Hu Zhen et al. from the Fudan University Shanghai Cancer Center established a prediction model for the Chinese population based on these related factors and used samples from another independent cohort to verify this model compared with the Western population model. The results showed that when using this model to draw the ROC curve, the AUC was above 0.8, whereas the AUC of the BRCAPro model was similar to that of the previous study at 0.7. This model has been proved to be suitable for prediction in the Chinese population, but it is not perfect. By obtaining more comprehensive clinical and pedigree information of the cases and by appropriately adding more cases, we can further improve the predictive ability and accuracy of this model [39].

3.5 Genetic Counseling and Management of Gene Mutation Carriers and the Current Status of Genetic Testing

Genetic counseling is a process of advising patients with genetic diseases or relatives who are at risk of having the disease, the probability of onset or heredity, and the method of prevention or mitigation. Genetic counseling is a means of preventing genetic diseases, which must be based on an accurate diagnosis. We should focus on the proband, that is, the first patient in a family, conduct a meticulous family survey, perform a pedigree analysis, and estimate the hereditary forms and possibility of disease occurrence in offspring.

Genetic counseling on BRCA mutation testing can be carried out by trained professionals. The processes of genetic counseling mainly include risk assessment of potentially detrimental BRCA mutations, education of possible outcomes of the test, identification of family members who should receive gene mutation detection, list of several interventions aimed at disease screening and reducing the risk of disease or use of surgical controls for people in need, and explanation of the results of genetic testing. The presence of the following family history factors could lead to an increased probability of carrying potentially deleterious BRCA mutations. They are the detection of breast cancer before 50 years of age, bilateral breast cancer, breast cancer and ovarian cancer family history, at least one male family member with breast cancer, multiple cases of breast cancer in the family, at least one BRCA-related primary cancer patient in the family, and Nordic Jewish descent [40]. In practice, genetic counseling staff can use a number of family risk stratification tools, such as the Ontario Family History Assessment Tool, the Manchester Scoring System, the Pedigree Assessment Tool, and FHS-7, to determine whether or not to conduct advanced genetic counseling and which patients are suitable for the detection of *BRCA* gene mutations [41].

BRCA gene testing is generally recommended for women after 18 years of age. There is ample evidence suggesting that existing gene sequencing techniques can accurately detect *BRCA* mutations. When a person in the family suffers from breast cancer or has a family history of the disease, which suggests the existence of cancer susceptibility factors, he or she can visit the clinic to accept genetic counseling and seek explanation of questions related to genetic testing. If the consultants decide that the test results can help determine a treatment strategy, the person receiving consultation needs to receive further testing for *BRCA* mutations. It is necessary to fully inform the patient of the potential issues caused by genetic testing, with the premise that the patient has the intention to take the test. The approach of mutation analysis relies on the family history of the patient. A family member with a definite genetic mutation in the family or a patient derived from a racial group carrying a particular gene mutation (such as a Caucasian Jewish woman) may first have these specific mutation sites detected [42]. Patients without these characteristics need to have the whole gene sequence tested. Among these patients, if possible, genetic testing should first be conducted for those with relatives who have breast cancer or ovarian cancer, so as to determine whether these diseased family members may carry pathogenic mutations.

If the specific mutation sites of the *BRCA* gene are known, gene detection is easy and highly sensitive and specific. However, how to interpret the test results is a complex problem and requires subsequent genetic counseling. The results of genetic testing are generally expressed as positive mutations (i.e., potentially deleterious mutations), UVs, uncertain negative results, and definite negative results. If negative test results are obtained for women with relatives who have known *BRCA* mutations (definite negative), the results need to be confirmed again, and the possible risks should be assessed. Some studies have shown that women who are clinically negative for *BRCA* mutations also have the same risk of developing breast cancer [43–46]. However, a meta-analysis suggested no increase in risk (8). Uncertain negative results indicate that either no potentially harmful mutations have been detected or other members of the family have not yet been tested and that no mutations have been found in the family members who have been tested. Studies have suggested that women who have uncertain negative testing results may also have an increased risk of developing breast cancer [47].

Gene mutation detection is conducive to detecting high-risk groups for hereditary breast cancer as early as possible and to taking preventive measures as early as possible to improve its prognosis [48]. In 1996, commercial use of genetic testing first appeared in the United States. Since then, it has become a routine diagnostic approach in developed countries in Europe and America to conduct genetic testing targeting populations with hereditary breast cancer and to apply the test results to guide clinical treatment decision-making and high-risk population screening and monitoring. Most breast cancer medical centers have established a multidisciplinary team for the diagnosis and treatment of hereditary breast cancer, which includes experts of tumor surgery, plastic surgery, oncology, psychiatry and psychology, tumor molecular biology, and genetics to provide professional diagnosis and treatment services for hereditary breast cancer patients and their relatives.

Previously, the internationally renowned movie star Angelina Jolie received a bilateral mastectomy to reduce the risk of hereditary breast cancer. This brave behavior has allowed genetic testing for prevention of hereditary breast cancer to reenter the public spotlight. Clinical researches of more than 30 years in Europe and the United States found that regular health screening, drug prevention, and preventive surgery for *BRCA-1/2* mutant gene carriers could reduce the risk of breast cancer by as much as 90% [49]. Through this test, subjects can understand their own risk of breast cancer and family genetic background. It can also help future generations to carry out early prevention and intervention of breast cancer.

From a technical point of view, it is not difficult to carry out accurate detection of *BRCA-1/2* mutations and other susceptible gene mutations. However, due to a lack of standardization for genetic testing techniques in China, professionals of data interpretation and genetic counseling are scarce. Many psychological and ethical issues involved in genetic testing cannot be solved, and research data on the prevention, treatment, and intervention in gene mutation carriers are still lacking. Breast cancer-related genetic testing in China has not been approved by the government to become a routine procedure in clinical diagnosis and treatment. Since many institutes in China carry out genetic testing as part of scientific research projects [50], their test results cannot be directly applied to treatment development. However, based on years of being practiced outside China, breast cancer genetic testing is bound to provide a great help for the prevention and treatment of breast cancer. We believe that in the near future, with the development of detection technology, accumulation of more complete and detailed research data, and establishment of professional teams for hereditary breast cancer treatment, breast cancer genetic testing could be recommended as a standard diagnosis and treatment procedure to meet the needs of the specific patient groups in China.

3.6 Disease Screening and Follow-Up of Patients with Hereditary Breast Cancer

BRCA gene mutation carriers should be regarded as high-risk groups for breast cancer, which necessitates early examination of breast cancer. Studies have recommended that female *BRCA* mutation carriers should receive breast self-test training and take monthly breast self-examination from the age of 18. Clinical breast examination for every 6 months is also recommended for those at the age of 25 or above [51]. However, a study enrolling 236 Caucasian women with *BRCA* mutations suggests that clinical examinations are not particularly sensitive and can detect only 9.1% of breast cancer patients [52]. Examination with molybdenum target screening can significantly improve the detection rate of breast cancer, as the detection sensitivity is increased by 45% and the specificity can reach 99.8%.

Over the past few years, there has been much debate about the radioactive hazards of molybdenum target screening and its decrease of sensitivity caused by dense mammary glands in young women. Therefore, the combined application of ultrasound and magnetic resonance imaging (MRI) is promoted as an alternative screening tool. MRI can diagnose breast cancer that has been missed by ultrasonography and mammography. Warner et al. reported that in 236 female Caucasian *BRCA* mutation carriers, the combination of MRI, ultrasound, and molybdenum target screening could improve the detection sensitivity from 65% to 95%. A recent study of large-scale screening with a longer duration enrolled a total of 496 women. The subjects received 12 years of continuous screening, and 57 cases of breast cancer were identified, 97% of which were early cases at stage 0 or 1. This study and other evidence have promoted updates of the National Comprehensive Cancer Network (NCCN) guidelines, which suggest that female *BRCA* mutation carriers should start to receive molybdenum target and MRI screening starting from the age of 25 [53]. There are also controversies regarding the recommendations of the NCCN guidelines, such as whether exposure to radiation is inherently risky. In the general population, premature exposure to radiation is a recognized risk factor for breast cancer. As the *BRCA* gene itself is a key member of the DNA double-strand break repair pathway [54–56], *BRCA* mutation carriers lack the capacity to repair DNA damage caused by radiation, which subsequently results in susceptibility to breast cancer [57, 58].

Other studies have suggested that the radiation dose of the molybdenum target screening is approximately 0.004 Gy per examination. Therefore, the application of diagnostic molybdenum target screening in people younger than 30 years may increase the risk of radiation-induced breast cancer. Overall, these findings caution about the use of molybdenum target screening in young populations.

The disease characteristics and occurrence peaks of breast cancer in China are not exactly the same as those in Europe and America. Considering the factors of earlier breast cancer occurrence peak, larger premenopausal proportion, and denser breast tissues among Chinese patients, molybdenum target-based screening methods are not fully applicable in China. On the other hand, ultrasound examination is nonradioactive and noninvasive. Its cost is lower than the molybdenum target examination, and its image of the dense mammary gland is clearer. It could find lesions ≥2 mm and easily identify cysts. Therefore, ultrasound can be used as an auxiliary means of breast screening. *The Guidelines and Norms on the Management of Breast Cancer Diagnosis (2013 version)* from the Chinese Anti-Cancer Association clearly states that ultrasound can be used as a combined examination method for mammography screening or a supplementary examination method for BI-RADS 0 level patients as revealed by mammography screening.

3.7 Prevention of Hereditary Breast Cancer

3.7.1 Chemoprevention

Sporn and Newton [59] first proposed the concept of tumor chemoprevention in 1979, that is, the use of drugs to prevent or reverse the process of cancer, thus preventing the occurrence of cancer and reversing precancerous lesions. In the evolution of cancer, tissues and cells first differentiate, i.e., the development of precancerous lesions, which then develop into invasive tumors after a long period of time. This process provides a temporal possibility for chemical interventions and allows drugs to have sufficient time to reverse the abnormal differentiation of the cells.

Familial or hereditary breast cancer family members are high-risk groups, and their risks of breast cancer can be reduced by chemoprevention, including dietary composition adjustment and application of endocrine drugs. In recent years, large-scale clinical trials have shown the important role of chemoprevention in reducing the occurrence of hereditary breast cancer.

Selective Estrogenic Receptor Modulators (SERMs) SERMs are a class of drugs acting on ERs, which are tissue-specific, and can be divided into estrogen-like drugs and estrogen antagonists, represented by tamoxifen and raloxifene. There are three representative large-scale clinical trials demonstrating the preventive effects of these two drugs on breast cancer in high-risk groups: the National Surgical Adjuvant Breast and Bowel Project (NSABP-P1) trial, the International Breast Cancer Intervention Study (IBIS)-1 trial, and the NSABP Study of Tamoxifen and Raloxifene (STAR) trial. The results of the NSABP-P1 trial showed that tamoxifen reduced the risk of invasive breast cancer by 49%, lowered the risk of noninvasive breast cancer by 50%, and decreased the recurrence rate of ER-positive breast cancer by 69%, whereas it had no effect on ER-negative breast cancer [60]. The IBIS-1 study began in 1992 and continued into mid-2001, which identified 7154 women as having increased risk of breast cancer. Patients were randomized to the tamoxifen and placebo treatment groups, and the treatment lasted for 5 years, with an average follow-up of 16 years. It is worth mentioning that nearly half of the enrolled individual subjects were simultaneously using hormone replacement therapy during this study. The results showed that the use of tamoxifen for 5 years reduced the incidence of breast cancer by 29%. In addition, subgroup analysis of this study showed that the prevention effect provided by tamoxifen was reduced by the use of hormone replacement therapy [61]. The STAR was designed to study the chemoprevention effects of raloxifene on breast cancer in postmenopausal women with high risk factors. The results showed that the incidence rates of invasive breast cancer in both groups were similar, while the incidence of noninvasive breast cancer in the tamoxifen group was slightly lower than that in the raloxifene group (RR = 1.40), which was not statistically significant. However, in the case of drug safety, the risks of endometrial cancer and thromboembolism in the raloxifene group were significantly lower than those in the tamoxifen group (RR = 0.62 and 0.70, respectively). The results of the STAR study suggested that raloxifene and tamoxifen were similar in their efficacy of breast cancer prevention, but raloxifene was superior in terms of having fewer side effects [62, 63].

Jordan reported the effect of raloxifene on the incidence of breast cancer in 10,553 osteoporotic women. The results showed that the new incidence rate of breast cancer was 1.7‰ and that of the placebo group was 3.7‰. The relative risk ratio was 0.46, and the incidence rate of breast cancer was reduced by 54%. Raloxifene had a significant effect on ER-positive breast cancer, with a 70% reduction in the incidence rate, and had no effect on ER-negative tumors [64].

SERMs mainly reduce the incidence of ER-positive breast cancer. Among *BRCA*-associated breast cancer patients, *BRCA-2* mutation carriers predominantly have ER-positive breast cancer (75%), whereas *BRCA-1* mutation carriers tend to develop ER-negative breast cancer (80%). Therefore, prevention of breast cancer by blocking estrogen should be more effective in *BRCA-2* mutation carriers than in *BRCA-1* mutation carriers. Several clinical studies support this hypothesis. In the NSABP-P1 study, among 19 mutation carriers, tamoxifen-mediated chemoprevention could reduce the risk of breast cancer by 50% in the *BRCA-2* mutation carriers but was ineffective in *BRCA-1* mutation carriers (RR = 1.67) [60]. However, due to the low number of mutation cases, there is not sufficient evidence. Thus far, the efficacy of applying tamoxifen and other similar drugs to prevent

BRCA-related breast cancer has not been fully confirmed.

In addition, other studies have suggested that tamoxifen can significantly reduce the incidence of contralateral breast cancer in *BRCA-1* and *BRCA-2*-associated breast cancers after surgery and lower the postsurgical occurrence of ipsilateral breast cancer after breast-conserving therapy. A study explored the synergistic effect of prophylactic oophorectomy and tamoxifen on *BRCA* mutation-associated breast cancer patients and suggested that combination therapy had a greater effect on reducing the incidence of contralateral breast cancer than any single approach.

Aromatase Inhibitors Aromatase inhibitors can prevent the conversion of androgen to estrogen. They act as standard endocrine therapy medication for postmenopausal early-stage breast cancer patients with positive ER scores and can significantly improve the comprehensive treatment effect of breast cancer. As a new drug for breast cancer prevention, aromatase inhibitors can reduce the incidence of invasive breast cancer in postmenopausal women and prevent contralateral breast cancer in the early stage [65].

The IBIS-II prevention test evaluated the preventive efficacy of anastrozole, in comparison with placebo, in postmenopausal women who did not have breast cancer but had a high risk of developing it. A total of 1920 high-risk women were treated with anastrozole for 5 years, and 1944 cases in the control group were treated with placebo. The results showed that anastrozole could reduce the incidence rate of breast cancer by 53% for postmenopausal subjects. The 7-year cumulative incidence rate of breast cancer was 2.8% in the anastrozole treatment group and 5.6% in the placebo group. Meanwhile, the side effects caused by the treatment did not increase significantly [66].

MAP.3 is a randomized, double-blind, placebo-controlled, multicenter, international phase III clinical trial, which evaluates the effect of exemestane on postmenopausal women for breast cancer prevention. The participating countries included the United States, Canada, France, and Spain. From 2004 to 2010, 4560 postmenopausal women with high risk of breast cancer were enrolled in the group. The enrolled subjects had at least one of the following high-risk factors: age greater than or equal to 60 years; Gail risk score greater than 1.66%; ductal atypical hyperplasia, lobular atypical hyperplasia, or in situ lobular carcinoma indicated by previous breast biopsy; or previous incidence of in situ ductal carcinoma followed by total mastectomy. The MAP.3 test results showed that exemestane not only reduced the incidence rate of invasive breast cancer by 65% but also reduced the incidence rate of DCIS, ADH, ALH, and LCIS. Reduction of these precancerous lesions is likely to be transformed into a more pronounced decrease in invasive breast cancer during longer-term follow-up. It suggests that exemestane has an effect on the prevention of breast cancer. In addition, this study did not find serious safety problems (including increased risk of osteoporosis, cardiovascular events, or occurrence of other tumors) during the 3-year follow-up [67].

However, current clinical studies on aromatase inhibitor-mediated breast cancer prevention do not specifically limit the subject population to the *BRCA* mutation carriers. Whether anastrozole or exemestane can prevent the occurrence of *BRCA*-related breast cancer is unclear.

Other Nonselective ER Regulators Other drugs, such as N-(4-hydroxyphenyl) retinamide (4-HPR; fenretinide) and soy products, can play a preventive role in an ER-independent manner. Long-term use of 4-HPR can reduce the incidence rate of contralateral breast cancer and the recurrence rate of ipsilateral breast cancer in premenopausal early-stage breast cancer patients [68]. However, rigorous designs of large prospective clinical trials to support the long-term use of these drugs are still lacking.

In summary, in recent years, the field of chemical prevention of breast cancer has become very active. New drugs continue to emerge, mechanistic investigations have been increased, and combined chemoprevention has been reported, which all provide a good prospect for breast cancer prevention in high-risk groups. However, many problems still need to be further resolved. We should perform a more comprehensive and in-depth mechanistic investigation on chemoprevention for breast cancer, further investigate the selection of drug dose for breast cancer prevention, explore how to reduce its toxic side effects in long-term use, and develop gene-specific drugs with precise targeting and fewer toxic side effects. Based on current understanding of tumor pathogenesis, the combined approaches of prevention therapies acting on multiple targets, orderly replacement of different drugs, and procedures designed to prevent individual tumor will further improve the chemoprevention of breast cancer and ultimately achieve the goal of breast cancer prevention and treatment.

3.7.2 Prophylactic Mastectomy

Prophylactic mastectomy is suitable for healthy women with high breast cancer risk factors. It can reduce the incidence rate of breast cancer in high-risk women by 90% and decrease the breast cancer mortality rate in medium-risk and high-risk women by 100% and 81%, respectively. Therefore, prophylactic mastectomy can significantly reduce the risk of familial breast cancer. However, subcutaneous mastectomy may have poor treatment efficacy due to residual breast tissues under the nipple and areola [69]. The 2010 PROSE study was a large, prospective, multicenter study that enrolled 2484 female cases with *BRCA-1/2* mutations. The cases came from 22 research centers in Europe and North America from 1974 to 2008. In the study, 247 women underwent prophylactic bilateral mastectomy, and no breast cancer occurred later. Meanwhile, 7% of women in the control group who did not receive prophylactic surgery had breast cancer [70].

3.7.3 Prophylactic Oophorectomy

In addition to breast cancer, *BRCA* mutation carriers also have increased risks of developing ovarian cancer. Prophylactic oophorectomy can reduce the risk of both ovarian cancer and breast cancer [71].

In one study, compared with women who did not receive risk-reducing salpingo-oophorectomy (RRSO), women who received RRSO showed significant reductions in mortality risk due to various factors, breast cancer-specific mortality, and ovarian cancer-specific mortality [70]. Rebbeck et al. performed a meta-analysis of ten studies and found that RRSO significantly reduced the risk of breast cancer in female carriers of various mutation subtypes. In *BRCA-1/2* mutation carriers, the HR value was 0.49, while similar efficacy was observed in *BRCA-1* and *BRCA-2* mutation carriers [72]. Similarly, RRSO can significantly reduce the risk of ovarian cancer in *BRCA-1/2* mutation carriers. However, there is no sufficient evidence to confirm whether there is a consistent outcome among different populations carrying *BRCA-1* and *BRCA-2* mutations.

Breast cancer in *BRCA-2* mutation carriers is predominantly ER-positive (75%), whereas *BRCA-1* mutation carriers tend to develop ER-negative breast cancer (80%). Thus, prevention of breast cancer by blocking estrogen should be more effective in *BRCA-2* mutation carriers than in *BRCA-1* mutation carriers. Several clinical studies support this hypothesis. In the NSABP-P1 study, the risk of breast cancer was reduced by 50% in the 19 *BRCA-2* mutation carriers by tamoxifen-mediated chemoprevention, but this type of prevention was ineffective for *BRCA-1* mutation carriers. However, due to the limited number of mutation carriers, the evidence was not sufficient. In a collaborative study between the PROSE study group and Memorial Sloan Kettering Cancer Center (MSKCC), RRSO reduced the incidence rate of breast cancer by 72% in *BRCA-2* mutation carriers, whereas that for *BRCA-1* mutation carriers was only reduced by 49% [73]. Therefore, it can be inferred that RRSO can reduce the risk of ER-positive cancer rather than ER-negative breast cancer. The results published by the PROSE team also suggest that RRSO reduces the risk of breast cancer by 64% and 37%, respectively, in *BRCA-2* and *BRCA-1* mutation carriers [70]. Therefore, the role of

RRSO in reducing the incidence of breast cancer should be mainly reflected in the *BRCA-2* mutation carriers. Of the 120 *BRCA-2* mutation carriers who received RRSO, no breast cancer-related mortality occurred, and 6 of the *BRCA-2* mutation carriers without oophorectomy died. In the *BRCA-1* mutation carriers, the specific mortality rate of breast cancer did not decrease significantly.

NCCN guidelines recommend that RRSO be performed in *BRCA* mutation carriers between the ages of 35 and 40 years [74]. In most women, this early arrival of menopause can be tolerated, although the decline in quality of life requires the application of exogenous hormone replacement therapy. This has led some clinicians to advocate the simultaneous removal of the uterus at the time of RRSO to facilitate the implementation of a single estrogen treatment [75]. However, a report by the PROSE study showed that any type of estrogen replacement therapy used after implementing RRSO could reduce the risk of breast cancer [76].

3.8 Existing Issues and Future Research Directions and Prospects

At present, genetic counseling, screening, and early intervention for hereditary breast cancer and its high-risk population have received increasing attention from oncologists. However, we need to maintain an objective and calm mind, as most clinical practices are carried out in developed countries. A large number of problems are yet to be solved, such as the high cost of genetic testing, the identification of large quantities of variants with "uncertain significance" in the test results, the psychological burden on patients, as well as the unnecessary prophylactic surgery, chemical prevention, or intensive screening. In addition, prophylactic mastectomy may lead to more physical damage, loss of breast function, and impaired sexual life. Therefore, a careful evaluation of intervention therapies and their short- and long-term impacts on the quality of life is essential to treatment implementation for *BRCA* mutation carriers.

References

1. Honrado E, Benitez J, Palacios J (2005) The molecular pathology of hereditary breast cancer: genetic testing and therapeutic implications. Mod Pathol 18(10):1305–1320. doi:10.1038/modpathol.3800453
2. Smith TM, Lee MK, Szabo CI, Jerome N, McEuen M, Taylor M, Hood L, King MC (1996) Complete genomic sequence and analysis of 117 kb of human DNA containing the gene BRCA1. Genome Res 6(11):1029–1049
3. Wooster R, Bignell G, Lancaster J, Swift S, Seal S, Mangion J, Collins N, Gregory S, Gumbs C, Micklem G (1995) Identification of the breast cancer susceptibility gene BRCA2. Nature 378(6559):789–792. doi:10.1038/378789a0
4. Database TBCIC (2012) Available at http://research.nhgri.nih.gov/projects/bic/
5. Thomassen M, Gerdes AM, Cruger D, Jensen PK, Kruse TA (2006) Low frequency of large genomic rearrangements of BRCA1 and BRCA2 in western Denmark. Cancer Genet Cytogenet 168(2):168–171. doi:10.1016/j.cancergencyto.2005.12.016
6. Spearman AD, Sweet K, Zhou XP, McLennan J, Couch FJ, Toland AE (2008) Clinically applicable models to characterize BRCA1 and BRCA2 variants of uncertain significance. J Clin Oncol 26(33):5393–5400. doi:10.1200/jco.2008.17.8228
7. Collins N, McManus R, Wooster R, Mangion J, Seal S, Lakhani SR, Ormiston W, Daly PA, Ford D, Easton DF et al (1995) Consistent loss of the wild type allele in breast cancers from a family linked to the BRCA2 gene on chromosome 13q12-13. Oncogene 10(8):1673–1675
8. Cornelis RS, Neuhausen SL, Johansson O, Arason A, Kelsell D, Ponder BA, Tonin P, Hamann U, Lindblom A, Lalle P et al (1995) High allele loss rates at 17q12-q21 in breast and ovarian tumors from BRCAl-linked families. The breast cancer linkage consortium. Genes Chromosomes Cancer 13(3):203–210
9. Larsen MJ, Thomassen M, Tan Q, Laenkholm AV, Bak M, Sorensen KP, Andersen MK, Kruse TA, Gerdes AM (2014) RNA profiling reveals familial aggregation of molecular subtypes in non-BRCA1/2 breast cancer families. BMC Med Genet 7:9. doi:10.1186/1755-8794-7-9
10. Honrado E, Osorio A, Milne RL, Paz MF, Melchor L, Cascon A, Urioste M, Cazorla A, Diez O, Lerma E, Esteller M, Palacios J, Benitez J (2007) Immunohistochemical classification of non-BRCA1/2 tumors identifies different groups that demonstrate the heterogeneity of BRCAX families. Modern Pathol 20(12):1298–1306. doi:10.1038/modpathol.3800969

11. Esteller M, Silva JM, Dominguez G, Bonilla F, Matias-Guiu X, Lerma E, Bussaglia E, Prat J, Harkes IC, Repasky EA, Gabrielson E, Schutte M, Baylin SB, Herman JG (2000) Promoter hypermethylation and BRCA1 inactivation in sporadic breast and ovarian tumors. J Natl Cancer Inst 92(7):564–569
12. Collins N, Wooster R, Stratton MR (1997) Absence of methylation of CpG dinucleotides within the promoter of the breast cancer susceptibility gene BRCA2 in normal tissues and in breast and ovarian cancers. Br J Cancer 76(9):1150–1156
13. Birch JM, Blair V, Kelsey AM, Evans DG, Harris M, Tricker KJ, Varley JM (1998) Cancer phenotype correlates with constitutional TP53 genotype in families with the li-Fraumeni syndrome. Oncogene 17(9):1061–1068. doi:10.1038/sj.onc.1202033
14. Malkin D, Li FP, Strong LC, Fraumeni JF Jr, Nelson CE, Kim DH, Kassel J, Gryka MA, Bischoff FZ, Tainsky MA et al (1990) Germ line p53 mutations in a familial syndrome of breast cancer, sarcomas, and other neoplasms. Science (New York, NY) 250(4985):1233–1238
15. Ding SL, Sheu LF, Yu JC, Yang TL, Chen BF, Leu FJ, Shen CY (2004) Abnormality of the DNA double-strand-break checkpoint/repair genes, ATM, BRCA1 and TP53, in breast cancer is related to tumour grade. Br J Cancer 90(10):1995–2001. doi:10.1038/sj.bjc.6601804
16. Khanna KK, Chenevix-Trench G (2004) ATM and genome maintenance: defining its role in breast cancer susceptibility. J Mammary Gland Biol Neoplasia 9(3):247–262. doi:10.1023/B:JOMG.0000048772.92326.a1
17. Li DM, Sun H (1997) TEP1, encoded by a candidate tumor suppressor locus, is a novel protein tyrosine phosphatase regulated by transforming growth factor beta. Cancer Res 57(11):2124–2129
18. Nakanishi C, Yamaguchi T, Iijima T, Saji S, Toi M, Mori T, Miyaki M (2004) Germline mutation of the LKB1/STK11 gene with loss of the normal allele in an aggressive breast cancer of Peutz-Jeghers syndrome. Oncology 67(5–6):476–479. doi:10.1159/000082933
19. Thiffault I, Hamel N, Pal T, McVety S, Marcus VA, Farber D, Cowie S, Deschenes J, Meschino W, Odefrey F, Goldgar D, Graham T, Narod S, Watters AK, MacNamara E, Du Sart D, Chong G, Foulkes WD (2004) Germline truncating mutations in both MSH2 and BRCA2 in a single kindred. Br J Cancer 90(2):483–491. doi:10.1038/sj.bjc.6601424
20. Mavaddat N, Barrowdale D, Andrulis IL, Domchek SM, Eccles D, Nevanlinna H, Ramus SJ, Spurdle A, Robson M, Sherman M, Mulligan AM, Couch FJ, Engel C, McGuffog L, Healey S, Sinilnikova OM, Southey MC, Terry MB, Goldgar D, O'Malley F, John EM, Janavicius R, Tihomirova L, Hansen TV, Nielsen FC, Osorio A, Stavropoulou A, Benitez J, Manoukian S, Peissel B, Barile M, Volorio S, Pasini B, Dolcetti R, Putignano AL, Ottini L, Radice P, Hamann U, Rashid MU, Hogervorst FB, Kriege M, van der Luijt RB, Hebon PS, Frost D, Evans DG, Brewer C, Walker L, Rogers MT, Side LE, Houghton C, Embrace WJ, Godwin AK, Schmutzler RK, Wappenschmidt B, Meindl A, Kast K, Arnold N, Niederacher D, Sutter C, Deissler H, Gadzicki D, Preisler-Adams S, Varon-Mateeva R, Schonbuchner I, Gevensleben H, Stoppa-Lyonnet D, Belotti M, Barjhoux L, Collaborators GS, Isaacs C, Peshkin BN, Caldes T, de la Hoya M, Canadas C, Heikkinen T, Heikkila P, Aittomaki K, Blanco I, Lazaro C, Brunet J, Agnarsson BA, Arason A, Barkardottir RB, Dumont M, Simard J, Montagna M, Agata S, D'Andrea E, Yan M, Fox S, kConFab I, Rebbeck TR, Rubinstein W, Tung N, Garber JE, Wang X, Fredericksen Z, Pankratz VS, Lindor NM, Szabo C, Offit K, Sakr R, Gaudet MM, Singer CF, Tea MK, Rappaport C, Mai PL, Greene MH, Sokolenko A, Imyanitov E, Toland AE, Senter L, Sweet K, Thomassen M, Gerdes AM, Kruse T, Caligo M, Aretini P, Rantala J, von Wachenfeld A, Henriksson K, Collaborators S-B, Steele L, Neuhausen SL, Nussbaum R, Beattie M, Odunsi K, Sucheston L, Gayther SA, Nathanson K, Gross J, Walsh C, Karlan B, Chenevix-Trench G, Easton DF, Antoniou AC, Consortium of Investigators of Modifiers of B (2012) Pathology of breast and ovarian cancers among BRCA1 and BRCA2 mutation carriers: results from the consortium of investigators of modifiers of BRCA1/2 (CIMBA). Cancer Epidemiol Biomark Prev 21(1):134–147. doi:10.1158/1055-9965.EPI-11-0775
21. van der Groep P, van der Wall E, van Diest PJ (2011) Pathology of hereditary breast cancer. Cell Oncol (Dordr) 34(2):71–88. doi:10.1007/s13402-011-0010-3
22. Lakhani SR, Reis-Filho JS, Fulford L, Penault-Llorca F, van der Vijver M, Parry S, Bishop T, Benitez J, Rivas C, Bignon YJ, Chang-Claude J, Hamann U, Cornelisse CJ, Devilee P, Beckmann MW, Nestle-Kramling C, Daly PA, Haites N, Varley J, Lalloo F, Evans G, Maugard C, Meijers-Heijboer H, Klijn JG, Olah E, Gusterson BA, Pilotti S, Radice P, Scherneck S, Sobol H, Jacquemier J, Wagner T, Peto J, Stratton MR, McGuffog L, Easton DF, Breast Cancer Linkage C (2005) Prediction of BRCA1 status in patients with breast cancer using estrogen receptor and basal phenotype. Clin Cancer Res 11(14):5175–5180. doi:10.1158/1078-0432.CCR-04-2424
23. Eisinger F, Jacquemier J, Charpin C, Stoppa-Lyonnet D, Bressac-de Paillerets B, Peyrat JP, Lonny M, Guinebretiere JM, Sauvan R, Noguchi T, Birnbaum D, Sobol H (1998) Mutations at BRCA1: the medullary breast carcinoma revisited. Cancer Res 58(8):1588–1592
24. Jensen ML, Kiaer H, Andersen J, Jensen V, Melsen F (1997) Prognostic comparison of three classifications for medullary carcinomas of the breast. Histopathology 30(6):523–532
25. Foulkes WD, Brunet JS, Stefansson IM, Straume O, Chappuis PO, Begin LR, Hamel N, Goffin JR, Wong

N, Trudel M, Kapusta L, Porter P, Akslen LA (2004) The prognostic implication of the basal-like (cyclin E high/p27 low/p53+/glomeruloid-microvascular-proliferation+) phenotype of BRCA1-related breast cancer. Cancer Res 64(3):830–835

26. Bane AL, Pinnaduwage D, Colby S, Reedijk M, Egan SE, Bull SB, O'Malley FP, Andrulis IL (2009) Expression profiling of familial breast cancers demonstrates higher expression of FGFR2 in BRCA2-associated tumors. Breast Cancer Res Treat 117(1):183–191. doi:10.1007/s10549-008-0087-1
27. Lakhani SR, Van De Vijver MJ, Jacquemier J, Anderson TJ, Osin PP, McGuffog L, Easton DF (2002) The pathology of familial breast cancer: predictive value of immunohistochemical markers estrogen receptor, progesterone receptor, HER-2, and p53 in patients with mutations in BRCA1 and BRCA2. J Clin Oncol 20(9):2310–2318. doi:10.1200/JCO.2002.09.023
28. Palacios J, Honrado E, Osorio A, Cazorla A, Sarrio D, Barroso A, Rodriguez S, Cigudosa JC, Diez O, Alonso C, Lerma E, Dopazo J, Rivas C, Benitez J (2005) Phenotypic characterization of BRCA1 and BRCA2 tumors based in a tissue microarray study with 37 immunohistochemical markers. Breast Cancer Res Treat 90(1):5–14. doi:10.1007/s10549-004-1536-0
29. Pinilla SM, Honrado E, Hardisson D, Benitez J, Palacios J (2006) Caveolin-1 expression is associated with a basal-like phenotype in sporadic and hereditary breast cancer. Breast Cancer Res Treat 99(1):85–90. doi:10.1007/s10549-006-9184-1
30. Gail MH, Brinton LA, Byar DP, Corle DK, Green SB, Schairer C, Mulvihill JJ (1989) Projecting individualized probabilities of developing breast cancer for white females who are being examined annually. J Natl Cancer Inst 81(24):1879–1886
31. Costantino JP, Gail MH, Pee D, Anderson S, Redmond CK, Benichou J, Wieand HS (1999) Validation studies for models projecting the risk of invasive and total breast cancer incidence. J Natl Cancer Inst 91(18):1541–1548
32. Rockhill B, Spiegelman D, Byrne C, Hunter DJ, Colditz GA (2001) Validation of the Gail et al. model of breast cancer risk prediction and implications for chemoprevention. J Natl Cancer Inst 93(5):358–366
33. Berry DA, Parmigiani G, Sanchez J, Schildkraut J, Winer E (1997) Probability of carrying a mutation of breast-ovarian cancer gene BRCA1 based on family history. J Natl Cancer Inst 89(3):227–238
34. Parmigiani G, Berry D, Aguilar O (1998) Determining carrier probabilities for breast cancer-susceptibility genes BRCA1 and BRCA2. Am J Hum Genet 62(1):145–158
35. Berry DA, Iversen ES Jr, Gudbjartsson DF, Hiller EH, Garber JE, Peshkin BN, Lerman C, Watson P, Lynch HT, Hilsenbeck SG, Rubinstein WS, Hughes KS, Parmigiani G (2002) BRCAPRO validation, sensitivity of genetic testing of BRCA1/BRCA2, and prevalence of other breast cancer susceptibility genes. J Clin Oncol 20(11):2701–2712. doi:10.1200/JCO.2002.05.121
36. Frank TS, Deffenbaugh AM, Reid JE, Hulick M, Ward BE, Lingenfelter B, Gumpper KL, Scholl T, Tavtigian SV, Pruss DR, Critchfield GC (2002) Clinical characteristics of individuals with germline mutations in BRCA1 and BRCA2: analysis of 10,000 individuals. J Clin Oncol 20(6):1480–1490. doi:10.1200/JCO.2002.20.6.1480
37. Lindor NM, Johnson KJ, Harvey H, Shane Pankratz V, Domchek SM, Hunt K, Wilson M, Cathie Smith M, Couch F (2010) Predicting BRCA1 and BRCA2 gene mutation carriers: comparison of PENN II model to previous study. Familial Cancer 9(4):495–502. doi:10.1007/s10689-010-9348-3
38. Rao NY, Hu Z, Yu JM, Li WF, Zhang B, Su FX, Wu J, Shen ZZ, Huang W, Shao ZM (2009) Evaluating the performance of models for predicting the BRCA germline mutations in Han Chinese familial breast cancer patients. Breast Cancer Res Treat 116(3):563–570. doi:10.1007/s10549-008-0181-4
39. Rao NY, Hu Z, Li WF, Huang J, Ma ZL, Zhang B, Su FX, Zhou J, Di GH, Shen KW, Wu J, Lu JS, Luo JM, Yuan WT, Shen ZZ, Huang W, Shao ZM (2009) Models for predicting BRCA1 and BRCA2 mutations in Han Chinese familial breast and/or ovarian cancer patients. Breast Cancer Res Treat 113(3):467–477. doi:10.1007/s10549-008-9965-9
40. Moyer VA, Force USPST (2014) Risk assessment, genetic counseling, and genetic testing for BRCA-related cancer in women: U.S. preventive services task force recommendation statement. Ann Intern Med 160(4):271–281. doi:10.7326/M13-2747
41. Antoniou AC, Hardy R, Walker L, Evans DG, Shenton A, Eeles R, Shanley S, Pichert G, Izatt L, Rose S, Douglas F, Eccles D, Morrison PJ, Scott J, Zimmern RL, Easton DF, Pharoah PD (2008) Predicting the likelihood of carrying a BRCA1 or BRCA2 mutation: validation of BOADICEA, BRCAPRO, IBIS, myriad and the Manchester scoring system using data from UK genetics clinics. J Med Genet 45(7):425–431. doi:10.1136/jmg.2007.056556
42. Hartge P, Struewing JP, Wacholder S, Brody LC, Tucker MA (1999) The prevalence of common BRCA1 and BRCA2 mutations among Ashkenazi Jews. Am J Hum Genet 64(4):963–970
43. Gronwald J, Cybulski C, Lubinski J, Narod SA (2007) Phenocopies in breast cancer 1 (BRCA1) families: implications for genetic counselling. J Med Genet 44(4):e76. doi:10.1136/jmg.2006.048462
44. Rowan E, Poll A, Narod SA (2007) A prospective study of breast cancer risk in relatives of BRCA1/BRCA2 mutation carriers. J Med Genet 44(8):e89; author reply e88
45. Smith A, Moran A, Boyd MC, Bulman M, Shenton A, Smith L, Iddenden R, Woodward ER, Lalloo F, Maher ER, Evans DG (2007) Phenocopies in BRCA1 and BRCA2 families: evidence for modifier genes and implications for screening. J Med Genet 44(1):10–15. doi:10.1136/jmg.2006.043091

46. Vos JR, de Bock GH, Teixeira N, van der Kolk DM, Jansen L, Mourits MJ, Oosterwijk JC (2013) Proven non-carriers in BRCA families have an earlier age of onset of breast cancer. Eur J Cancer 49(9):2101–2106. doi:10.1016/j.ejca.2013.02.018
47. Nelson HD, Fu R, Goddard K, Mitchell JP, Okinaka-Hu L, Pappas M, Zakher B (2013) In: risk assessment, genetic counseling, and genetic testing for BRCA-related cancer: systematic review to update the U.S. preventive services task force recommendation. U.S. preventive services task force evidence syntheses, formerly systematic evidence reviews. Rockville (MD)
48. Weir HK, Thun MJ, Hankey BF, Ries LA, Howe HL, Wingo PA, Jemal A, Ward E, Anderson RN, Edwards BK (2003) Annual report to the nation on the status of cancer, 1975-2000, featuring the uses of surveillance data for cancer prevention and control. J Natl Cancer Inst 95(17):1276–1299
49. Trainer AH, Lewis CR, Tucker K, Meiser B, Friedlander M, Ward RL (2010) The role of BRCA mutation testing in determining breast cancer therapy. Nat Rev Clin Oncol 7(12):708–717. doi:10.1038/nrclinonc.2010.175
50. Cao AY, Hu Z, Shao ZM (2010) Mutation screening of breast cancer susceptibility genes in Chinese high-risk families: the results will develop the genetic testing strategy in China. Breast Cancer Res Treat 120(1):271–272. doi:10.1007/s10549-009-0598-4
51. Hereditary Breast and/or Ovarian Cancer Syndrome (2012.) http://www.nccn.org/professionals/physician_gls/pdf/genetics_screening.pdf. Accessed 20 Dec 2012
52. Warner E, Plewes DB, Hill KA, Causer PA, Zubovits JT, Jong RA, Cutrara MR, DeBoer G, Yaffe MJ, Messner SJ, Meschino WS, Piron CA, Narod SA (2004) Surveillance of BRCA1 and BRCA2 mutation carriers with magnetic resonance imaging, ultrasound, mammography, and clinical breast examination. JAMA 292(11):1317–1325. doi:10.1001/jama.292.11.1317
53. Breast Cancer Screening and Diagnosis (2012) http://www.nccn.org/professionals/physician_gls/pdf/breast-screening.pdf. Accessed 20 Dec 2012
54. Nieuwenhuis B, Van Assen-Bolt AJ, Van Waarde-Verhagen MA, Sijmons RH, Van der Hout AH, Bauch T, Streffer C, Kampinga HH (2002) BRCA1 and BRCA2 heterozygosity and repair of X-ray-induced DNA damage. Int J Radiat Biol 78(4):285–295. doi:10.1080/09553000110097974
55. Powell SN, Kachnic LA (2003) Roles of BRCA1 and BRCA2 in homologous recombination, DNA replication fidelity and the cellular response to ionizing radiation. Oncogene 22(37):5784–5791. doi:10.1038/sj.onc.1206678
56. Yoshida K, Miki Y (2004) Role of BRCA1 and BRCA2 as regulators of DNA repair, transcription, and cell cycle in response to DNA damage. Cancer Sci 95(11):866–871
57. Formenti SC, Preston-Martin S, Haffty BG (2000) BRCA1/2 germline mutations: a marker for radioresistance or radiosensitivity? J Clin Oncol 18(5):1159–1160. doi:10.1200/JCO.2000.18.5.1159
58. Xia F, Powell SN (2002) The molecular basis of radiosensitivity and chemosensitivity in the treatment of breast cancer. Semin Radiat Oncol 12(4):296–304. doi:10.1053/srao.2002.35250
59. Sporn MB, Newton DL (1979) Chemoprevention of cancer with retinoids. Fed Proc 38(11):2528–2534
60. King MC, Wieand S, Hale K, Lee M, Walsh T, Owens K, Tait J, Ford L, Dunn BK, Costantino J, Wickerham L, Wolmark N, Fisher B, National Surgical Adjuvant B, Bowel P (2001) Tamoxifen and breast cancer incidence among women with inherited mutations in BRCA1 and BRCA2: National Surgical Adjuvant Breast and bowel project (NSABP-P1) breast cancer prevention trial. JAMA 286(18):2251–2256
61. Mokbel K, Singh-Ranger G (2002) Breast cancer chemoprevention--an update. Curr Med Res Opin 18(6):329–331
62. Vogel VG (2009) The NSABP study of tamoxifen and Raloxifene (STAR) trial. Expert Rev Anticancer Ther 9(1):51–60. doi:10.1586/14737140.9.1.51
63. Vogel VG, Costantino JP, Wickerham DL, Cronin WM, Cecchini RS, Atkins JN, Bevers TB, Fehrenbacher L, Pajon ER, Wade JL, 3rd, Robidoux A, Margolese RG, James J, Runowicz CD, Ganz PA, Reis SE, McCaskill-Stevens W, Ford LG, Jordan VC, Wolmark N, National Surgical Adjuvant B, Bowel P (2010) Update of the National Surgical Adjuvant Breast and Bowel Project Study of Tamoxifen and Raloxifene (STAR) P-2 Trial: preventing breast cancer. Cancer Prev Res (Phila) 3(6):696–706. doi:10.1158/1940-6207.CAPR-10-0076
64. Jordan VC (2007) Beyond raloxifene for the prevention of osteoporosis and breast cancer. Br J Pharmacol 150(1):3–4. doi:10.1038/sj.bjp.0706962
65. Beattie MS, Costantino JP, Cummings SR, Wickerham DL, Vogel VG, Dowsett M, Folkerd EJ, Willett WC, Wolmark N, Hankinson SE (2006) Endogenous sex hormones, breast cancer risk, and tamoxifen response: an ancillary study in the NSABP breast cancer prevention trial (P-1). J Natl Cancer Inst 98(2):110–115. doi:10.1093/jnci/djj011
66. Cuzick J, Sestak I, Forbes JF, Dowsett M, Knox J, Cawthorn S, Saunders C, Roche N, Mansel RE, von Minckwitz G, Bonanni B, Palva T, Howell A, Investigators I-I (2014) Anastrozole for prevention of breast cancer in high-risk postmenopausal women (IBIS-II): an international, double-blind, randomised placebo-controlled trial. Lancet 383(9922):1041–1048. doi:10.1016/S0140-6736(13)62292-8
67. Goss PE, Ingle JN, Ales-Martinez JE, Cheung AM, Chlebowski RT, Wactawski-Wende J, McTiernan A, Robbins J, Johnson KC, Martin LW, Winquist E, Sarto GE, Garber JE, Fabian CJ, Pujol P, Maunsell E, Farmer P, Gelmon KA, Tu D, Richardson H, Investigators NCMS (2011) Exemestane for breast-

cancer prevention in postmenopausal women. N Engl J Med 364(25):2381–2391. doi:10.1056/NEJMoa1103507
68. Vantyghem SA, Wilson SM, Postenka CO, Al-Katib W, Tuck AB, Chambers AF (2005) Dietary genistein reduces metastasis in a postsurgical orthotopic breast cancer model. Cancer Res 65(8):3396–3403. doi:10.1158/0008-5472.CAN-04-4109
69. Lostumbo L, Carbine NE, Wallace J (2010) Prophylactic mastectomy for the prevention of breast cancer. Cochrane Database Syst Rev 11:CD002748. doi:10.1002/14651858.CD002748.pub3
70. Domchek SM, Friebel TM, Singer CF, Evans DG, Lynch HT, Isaacs C, Garber JE, Neuhausen SL, Matloff E, Eeles R, Pichert G, Van t'veer L, Tung N, Weitzel JN, Couch FJ, Rubinstein WS, Ganz PA, Daly MB, Olopade OI, Tomlinson G, Schildkraut J, Blum JL, Rebbeck TR (2010) Association of risk-reducing surgery in BRCA1 or BRCA2 mutation carriers with cancer risk and mortality. JAMA 304(9):967–975. doi:10.1001/jama.2010.1237
71. Rebbeck TR, Lynch HT, Neuhausen SL, Narod SA, Van't Veer L, Garber JE, Evans G, Isaacs C, Daly MB, Matloff E, Olopade OI, Weber BL, Prevention, Observation of Surgical End Points Study G (2002) Prophylactic oophorectomy in carriers of BRCA1 or BRCA2 mutations. N Engl J Med 346 (21):1616–1622. doi:10.1056/NEJMoa012158
72. Rebbeck TR, Kauff ND, Domchek SM (2009) Meta-analysis of risk reduction estimates associated with risk-reducing salpingo-oophorectomy in BRCA1 or BRCA2 mutation carriers. J Natl Cancer Inst 101(2):80–87. doi:10.1093/jnci/djn442
73. Kauff ND, Domchek SM, Friebel TM, Robson ME, Lee J, Garber JE, Isaacs C, Evans DG, Lynch H, Eeles RA, Neuhausen SL, Daly MB, Matloff E, Blum JL, Sabbatini P, Barakat RR, Hudis C, Norton L, Offit K, Rebbeck TR (2008) Risk-reducing salpingo-oophorectomy for the prevention of BRCA1- and BRCA2-associated breast and gynecologic cancer: a multicenter, prospective study. J Clin Oncol 26(8):1331–1337. doi:10.1200/JCO.2007.13.9626
74. Hereditary Breast and/or Ovarian Cancer Syndrome. http://www.nccn.org/professionals/physician_gls/pdf/genetics_screening.pdf. 20 December 2012
75. Euhus D (2012) Managing the breast in patients who test positive for hereditary breast cancer. Ann Surg Oncol 19(6):1738–1744. doi:10.1245/s10434-012-2258-x
76. Rebbeck TR, Friebel T, Wagner T, Lynch HT, Garber JE, Daly MB, Isaacs C, Olopade OI, Neuhausen SL, van't Veer L, Eeles R, Evans DG, Tomlinson G, Matloff E, Narod SA, Eisen A, Domchek S, Armstrong K, Weber BL, Group PS (2005) Effect of short-term hormone replacement therapy on breast cancer risk reduction after bilateral prophylactic oophorectomy in BRCA1 and BRCA2 mutation carriers: the PROSE Study Group. J Clin Oncol 23(31):7804–7810. doi:10.1200/JCO.2004.00.8151

Predicting and Overcoming Chemotherapeutic Resistance in Breast Cancer

4

Kyung-Hee Chun, Jong Hoon Park, and Siting Fan

Abstract

Our understanding of breast cancer and its therapeutic approach has improved greatly due to the advancement of molecular biology in recent years. Clinically, breast cancers are characterized into three basic types based on their immunohistochemical properties. They are triple-negative breast cancer, estrogen receptor (ER) and progesterone receptor (PR)-positive-HR positive breast cancer, and human epidermal growth factor receptor 2 (HER2)-positive breast cancer. Even though these subtypes have been characterized, assessment of a breast cancer's receptor status is still widely used to determine whether or not a targeted therapy could be applied. Moreover, drug resistance is common in all breast cancer types despite the different treatment modalities applied. The development of resistance to different therapeutics is not mutually exclusive. It seems that tumor could be resistant to multiple treatment strategies, such as being both chemoresistant and monoclonal antibody resistant. However, the underlying mechanisms are complicated and need further investigation. In this chapter, we aim to provide a brief review of the different types of breast cancer and their respective treatment strategies. We also review the

K.-H. Chun (✉)
College of Medicine, Yonsei University, Seoul, South Korea
e-mail: KHCHUN@yuhs.ac

J.H. Park (✉)
Department of Biological Science, Sookmyung Women's University, Seoul, South Korea
e-mail: parkjh@sookmyung.ac.kr

S. Fan
Breast Tumor Center, Sun Yat-Sen Memorial Hospital, Sun Yat-Sen University, Guangzhou 510120, China

Guangdong Provincial Key Laboratory of Malignant Tumor Epigenetics and Gene Regulation, Sun Yat-Sen Memorial Hospital, Sun Yat-Sen University, Guangzhou 510120, China

E. Song, H. Hu (eds.), *Translational Research in Breast Cancer*, Advances in Experimental Medicine and Biology 1026, https://doi.org/10.1007/978-981-10-6020-5_4

possible mechanisms of potential drug resistance associated with each treatment type. We believe that a better understanding of the drug resistance mechanisms can lead to a more effective and efficient therapeutic success.

Keywords
Chemotherapeutic resistance • Breast cancer subtype • Molecular mechanism

4.1 Introduction

Prognosis and classification of breast cancer not only depend on the elucidation of tumor morphology but also on the expression levels of three protein biomarkers. These proteins include progesterone receptor (PR), estrogen receptor (ER), and human epidermal growth factor (EGF) receptor 2 (HER2, also known as EGFR2), which are particular therapeutic targets for cancer treatment. Tumors that do not express any of these three markers are generally known as triple-negative breast cancers (TNBCs, ER-/PR-/HER2-). Treatment strategies for this type of cancer are usually conventional, which include radiation therapy and chemotherapy [1].

In the early 2000s, studies on expression profiling indicated that breast cancer was more heterogeneous than originally assumed, with classification of its subtypes reaching beyond the ER/PR/HER2 area [2]. Five breast cancer subtypes with significant transcriptional importance have been investigated, including basal-like breast cancer, HER2-positive (HER2+, also known as Erb-B2 receptor tyrosine kinase 2 positive, ERBB2+) breast cancer, luminal A breast cancer, luminal B breast cancer, and normal-like breast cancer (although the normal-like subtype is currently not considered as originating from breast cancer cells). These subtypes not only correlate with significantly different histological features and disease progression but also affect responses to therapeutic treatments and clinical results. Mapped to ER+/PR+ tumors, luminal A tumors are generally low grade and thus can be treated by hormonal therapies, such as aromatase inhibitors (AIs) and the ER modulator, tamoxifen. Their therapeutic outcomes are usually the best among those of all subtypes. Being more aggressive, luminal B tumors are usually ER+ and possibly involve in either HER2 overexpression or high levels of marker of proliferation Ki-67 (MKi67) [3]. This type of tumors can be treated by hormonal therapy but frequently relapses and shows poor clinical outcomes. Driven by amplification of ERBB2/HER2, HER2+ tumors are intrinsically aggressive but treatable with trastuzumab and pertuzumab, which are monoclonal antibodies, or lapatinib, which is small-molecule kinase inhibitor. Molecularly different from ER+ tumors, basal-like tumors are ER-, which do not often respond to hormonal and/or AI therapies. Therefore, they are mainly treated by chemotherapy and radiotherapy. Even though TNBCs overlap with basal-like cancers with up to 70%, they are histopathologically and clinically different and could not be regarded as synonymous terms. Most patients suffering from triple-negative breast cancer (TNBC) have poor therapeutic outcomes. Compared to patients with other breast cancer subtypes, only 30–45% of TNBC patients achieve a pathological complete response (pCR) and realize similar survival rates. Thus, TNBCs are still a big obstacle for development of aggressive breast cancer treatment [4, 5].

Adjuvant treatments, such as hormonal agents, trastuzumab, and cytotoxic chemotherapy, in sequence and/or in combination, are usually used to avoid disease recurrence from micrometastases for patients with stage I, II, or IIIA or operable stage IIIC breast cancers, who have undergone radical surgery and are currently receiving radiotherapy. Systemic therapies are administered

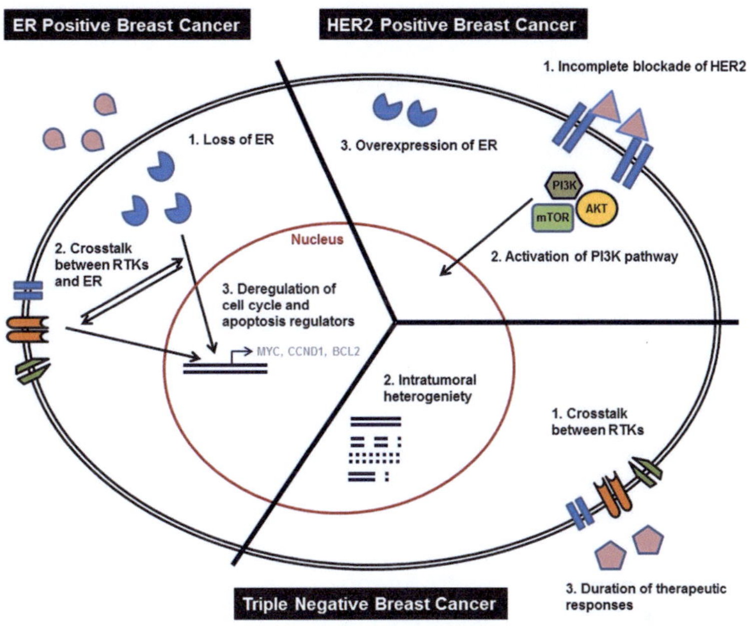

Fig. 4.1 Scheme of molecular mechanisms of chemotherapy for breast cancer

based on the assessment of clinicopathologic features, such as nodal involvement, tumor size, HER2 gene amplification, and hormone receptor status [6, 7]. Systemic chemotherapy or neoadjuvant hormone therapy is applied for treating stage IIIB and inoperable stage IIIC breast cancer, which is often followed by surgery and radiotherapy to downstage locally advanced tumors. Meanwhile, for treatment of stage IV and metastatic breast cancer, hormonal agents, trastuzumab or lapatinib, are usually employed with palliative intent. They are conventional cytotoxic drugs which could be used in sequence as a single agent most of the time or in two-drug combinations sometimes [8].

In this chapter, we will briefly summarize the molecular mechanisms involved in chemotherapeutic resistance in breast cancer (Fig. 4.1) and then focus on novel biomarkers that are currently under investigation for predicting chemoresistance, as well as therapeutic targets to overcome chemoresistance, and the refinement of therapeutic strategies for treating patients with breast cancer.

4.2 Mechanisms of Chemotherapeutic Resistance in Breast Cancer

4.2.1 Estrogen Receptor-Positive Breast Cancer

4.2.1.1 Estrogen Receptor and Its Signaling

Estrogen functions through ERs, a type of DNA-binding protein that stimulates a number of downstream genes related to reproductive and sexual health, metabolism, and bone resorption [9]. ER is a very strong mitogenic receptor that can enhance breast cancer cell proliferation and survival. Specifically, its expression has been observed in about 70% of breast cancers. Increased ER expression in human breast tissue enhances the risk of breast cancer development [10]. Estrogen binding to ER could cause receptor dimerization and translocation to the nucleus in the ER, which promote coregulator binding and modulate ER transcription at particular consensus DNA factors, such as estrogen response elements (EREs), in the target genes' enhancer/promoter region [11]. The recruited types of coregulatory proteins are based on the specificity of ligand, which either ameliorates or suppresses ER transcription by recruiting coactivators or corepressors. ER encoded by distinct genes displays different functions. For example, ERα facilitates breast cancer initiation and progression, while the function of ERβ in breast cancer is under discussion. Studies have recently found that ERβ acts like an antagonist of ERα and impairs the ability of estrogen to promote proliferation. Interestingly, decreased levels of ERβ protein predict resistance to tamoxifen treatment. If not otherwise specified, we will use "ER" to indicate "ERα" in this chapter [12].

4.2.1.2 Chemotherapeutic Agents for Estrogen Receptor-Positive Breast Cancer

The developmental strategies for treating ER+ breast cancer have introduced three therapeutic agents, which are selective estrogen receptor modulators, such as tamoxifen; selective estrogen receptor downregulators, such as fulvestrant; and estrogen synthesis inhibitors, such as AIs like anastrozole, letrozole, and exemestane [13].

For more than three decades, tamoxifen has been used to successfully treat patients with ER+ breast cancer. It hinders the binding of estrogen to ER and activates a distinct receptor conformation, which allows ER to selectively correlate with corepressor complexes instead of coactivators, and leads to the termination of estrogen-activated gene transcription. Based on the specific cell or tissue type, this agent activates different genes and promotes or inhibits ER function accordingly. Furthermore, tamoxifen is capable to induce ER non-genomic, extranuclear pathways. Nevertheless, about half of the patients with metastatic disease have no response to first-line treatment with tamoxifen, which is known as de novo resistance. Further study reveals that a lot of first-time responders eventually stop responding to treatment, which is known as acquired resistance [14, 15].

Selective estrogen receptor downregulators, like fulvestrant, fight against estrogen to bind to ER. Moreover, this class of drugs blocks receptor dimerization and causes degradation of receptor protein, leading to a more complete antiestrogen impact on both genomic and non-genomic ER signaling. Approved by FDA, fulvestrant is used to treat ER+ metastatic breast cancer patients who have endured antiestrogen therapies but still suffer from disease progression [16].

AI suppresses the levels of plasma estrogen in women either by inhibiting or inactivating aromatase. Officially called cytochrome P450 family 19 subfamily A member 1 (CYP19A1), the enzyme aromatase takes main responsibility for the synthesis of estrogens from androgenic substrates. Inactivation of aromatase by AIs results in the inhibition of estrogen synthesis, presenting another therapeutic strategy of ER antagonism. AI decreases circulating estrogen to 1–10% of pretreatment levels in postmenopausal women and would not be efficacious in premenopausal patients without simultaneous ovarian suppression. Currently, the AIs for clinical use are divided into two types: irreversible steroidal inactivators (e.g., exemestane) and reversible nonsteroidal inhibitors (e.g., anastrozole and letrozole) [17].

4.2.1.3 Chemotherapy Resistance Mechanisms in Estrogen Receptor-Positive Breast Cancer

In recent years, several mechanisms describing ER+ breast cancer's resistance to endocrine therapy have been identified.

4.2.1.3.1 Loss of Estrogen Receptor and Transcriptional Machinery Expression and Activity

ER expression, which is considered to be mainly regulated by epigenetic and posttranscriptional mechanisms at non-genomic level, has become the most important biomarker for prediction of endocrine treatment response. However, about 20% of breast cancer patients undergoing endocrine therapy may suffer from ER loss, which could happen over time at multiple levels and by different mechanisms. The decrease of ER expression does not seem to be related to the loss of heterozygosity at the ER gene locus, as the latter has little effect on the former. Furthermore, ER cooperates with regulatory proteins to form a transcription initiation complex, exerting its influence on gene expression. Any changes in the proteins could have great impact on the efficacy of endocrine therapy [18].

4.2.1.3.2 Cross Talk Between Receptor Tyrosine Kinases and Their Downstream Pathways

By either cooperating with ER signaling or bypassing it, receptor tyrosine kinases (RTKs), as well as their downstream signaling pathways, are capable of alternatively accelerating tumor growth. RTKs pathways modulate ER activity to directly negate or repress the inhibitory effects of endocrine therapy. Accumulating evidence indicates that tamoxifen resistance in breast cancer cell lines is caused by augmented expression of epidermal growth factor receptor (EGFR), HER2, and insulin-like growth factor 1 receptor (IGF1R). Moreover, HER2 and/or EGFR overexpression is related to poorer treatment results for patients taking tamoxifen [19]. Meanwhile, the genetic or epigenetic modifications of signal transduction intermediates, such as activating mutations in phosphatidylinositol 3-kinases (PI3Ks), the loss of heterozygosity or methylation of the tumor-suppressor PTEN, and the following activation of AKT serine/threonine kinase 1 (AKT1) usually cause overexpression of RTK signaling. To be specific, ER and its coregulator proteins are phosphorylated by multiple growth factor-dependent and stress-related intracellular kinases, such as p38, p42, and p44 mitogen-activated protein kinases (MAPKs), phosphoinositide 3-kinases (PI3Ks), AKT1, and ribosomal protein S6 kinase A1 (RPS6KA1). The phosphorylation induces stimulation of ER, independent of ligand or in combination with antiestrogens, transforming the classical ER transcriptional pathway to nonclassical nuclear genomic pathway [20].

4.2.1.3.3 Deregulation of Cell Cycle and Apoptosis Regulators

Tumor sensitivity to endocrine treatment can be affected by both positive and negative regulations of cell cycle. For instance, overexpression of the positive regulators, such as MYC proto-oncogene (MYC), cyclin E1 (CCNE1), and cyclin D1 (CCND1), enhances resistance to endocrine therapy either through activation of the cyclin-dependent kinases essential for G1 phase or through the relief of inhibitory effects by the negative cell cycle regulators, cyclin-dependent kinase inhibitor 1A (CDKN1A), and cyclin-dependent kinase inhibitor 1B (CDKN1B, also known as $p27^{kip1}$). Both the decreased expression, stability, or activity of CDKN1A and CDKN1B and the inactivation of tumor suppressor, retinoblastoma transcriptional corepressor 1 (RB1), are involved in resistance to tamoxifen [21, 22], which can be initiated by activation of RTKs and their downstream signaling pathways through modulation of transcription factors, microRNAs (miRNAs), or protein phosphorylation. Plus, overexpression of antiapoptotic molecules (e.g., B-cell lymphoma-extra-large protein) and decreased expression of proapoptotic molecules (e.g., BCL2-interacting killer and caspase-9) can modulate apoptosis mediated by antiestrogen. With cell cycle regulators, RTK and transcription factor signaling directly regulate the activity of the apoptotic/survival molecules through stimulation of the protein complex, nuclear factor kappa-light-chain-enhancer of activated B cells (NF-κB) [23].

4.2.2 HER2-Positive Breast Cancer

4.2.2.1 Epidermal Growth Factor Receptor and Its Signaling

The HER (human epidermal growth factor receptor) family is a group of structurally related receptor tyrosine kinases contributing to the growth, development, and differentiation of a number of organs, such as the breast. The HER family consists of HER1 (more commonly known as epidermal growth factor receptor (EGFR)), HER2, HER3, and HER4 [24]. HER signaling activity is tightly mediated in a ligand-dependent manner, with the binding of the ligand to its corresponding HER causing receptor homo- or heterodimerization, followed by the subsequent activation of the intracellular kinase domain and transphosphorylation of tyrosine residues [25]. No ligand has yet been found in HER2, and thus the latter relies on a ligand-activated partner receptor for cross activation. Nevertheless, among all HERs, HER2 is a better partner for heterodimerization, whereas the HER2-HER3 complex has the greatest transforming ability, despite the kinase domain of HER3 being inactive. When overexpressed, HER2 can form homo- and/or heterodimers capable of initiating downstream signaling in a ligand-independent manner. Thus, the amplification of HER2 and/or overexpression of the HER2 protein in breast cancer is sufficient to induce ligand-independent signaling. The major biological readouts of both ligand-dependent and ligand-independent HER activation are PI3K/AKT1 and MAPK signaling resulting in cell cycle progression, proliferation, and survival [26].

4.2.2.2 Chemotherapeutic Agents for HER2-Positive Breast Cancer

Amplification or overexpression of HER2, which can induce aggressive tumor behavior such as rapid growth and frequent metastasis, has been found in 20% to 25% of breast cancers. Targeting HER2 has been successful in clinical trial, which presents an attractive treatment option for breast cancer [27].

HER2 status can be determined by measuring receptor protein expression at the cell membrane using immunohistochemistry (IHC) or gene amplification by fluorescent in situ hybridization (FISH). All patients with strong HER2 IHC scoring (IHC3+), or with more than six HER2 or HER2/CEP17 gene copy numbers in the nucleus (FISH ratio >2.2), are considered to be HER2+ [28].

Trastuzumab, the humanized monoclonal antibody, is the first therapeutic agent invented for targeting HER2. Although trastuzumab has been shown to interact with the extracellular domain of HER2 to suppress the latter's function, its action mechanism is not completely understood. Trastuzumab may have a stronger inhibition effect on signaling from HER2 homodimers than from HER1-HER2 heterodimers, or HER3. It also downregulates the PI3K/AKT1 signaling pathway, which activates apoptosis in human tumors, and induces antibody-dependent cell-mediated cytotoxicity [29].

Several other drugs that inhibit multiple ERBB receptors are available and show more promise than solitary HER1 inhibitors for disease treatment. Pertuzumab binds to the heterodimerization domain of HER2 and blocks its interaction with HER1 and HER3 [30]. Lapatinib, afatinib, and neratinib are dual kinase inhibitors (HER1/HER2). Lapatinib, an FDA-approved anti-HER2 agent, is a small dual HER1 and HER2 TKI molecule. Lapatinib is used in combination therapy with DNA-damaging substances [31].

As an antibody-drug conjugate, T-DM1 (ado-trastuzumab emtansine) is a first-class therapeutic agent linking its molecules to the antibody trastuzumab and internalizing the compound upon binding. Approved by FDA for its good clinical trial performance in the second- and third-line treatment of HER2+ metastatic breast cancer, T-DM1 is now being studied in first-line settings to treat patients with metastatic or early-stage HER2+ breast cancer [32]. Additionally, potent kinase inhibitors of HER1 include gefitinib and erlotinib [33].

4.2.2.2.1 HER2 Expression and Breast Cancer Subtype

Recent classifications of breast cancer have established five subtypes: luminal A, luminal B, HER2+, basal-like, and claudin-low. Gene expression signatures also confirm the presence of subtypes within HER2+ breast cancer [34]. These subtypes require different management, and virtually every HER2+ breast cancer patient receives chemotherapy alongside anti-HER2 treatment. About twice as many patients gain advantages from trastuzumab when combined with chemotherapy, including higher objective response rates, a longer duration of response, and extended survival.

In the case of ER+/HER2+ disease, there is cross talk between the estrogen and HER signaling pathways, and simultaneous targeting of both pathways could be successful in some patients to prevent resistance to ER and HER2 treatments, even possibly eradicating the need for systemic cytotoxic treatment [27].

4.2.2.2.2 Combination Strategies for HER2-Positive Breast Cancer

The efficacy of different doublet and triplet combinations for HER2+ tumor xenografts has been studied, where the three-drug combination of gefitinib, pertuzumab, and trastuzumab is found to block signaling from all HER1, HER2, and HER3 receptor homo- and heterodimer pairs. It not only eradicates HER2-overexpressing xenografts in mice but also displays higher therapeutic efficiency than any single agent or two-drug regimen such as trastuzumab and pertuzumab. Other findings show that the lapatinib/trastuzumab combination efficiently eradicates HER2-overexpressing xenografts. Despite the low expression levels of HER1 (i.e., EGFR), inhibition of its activity enhances efficacy for treatment purpose. In ER+ tumors, endocrine therapy is also needed to achieve optimal antitumor effects. Even lower drug doses of the lapatinib/trastuzumab combination and intermittent administration of the regimen show effects on eradicating most tumors [35].

The above findings have laid a solid biological foundation for clinical trials combining two anti-HER2 agents (i.e., dual inhibition). Relevant researches have demonstrated enhanced treatment effect and improved outcomes for patient with combined chemotherapy in the metastatic setting. Similar studies in neoadjuvant and adjuvant settings have been carried out, and some of the trials are still ongoing [36].

To verify whether or not the increased inhibition of HER2 pathway would promote treatment efficiency, the combined anti-HER2 therapies were put forward in the neoadjuvant (or preoperative) setting for clinical trials. The studies aimed to investigate the combined therapies' effect after adding another anti-HER2 agent to trastuzumab, namely, lapatinib or pertuzumab, along with chemotherapy. Approved by the US Food and Drug Administration (FDA), pCR was used in all of the trials despite that it was an endpoint with variable definitions in the available literature and some limitations in correlation with outcomes. Similar results were presented in the neoadjuvant dual inhibition trials, indicating that the combination of dual HER2 inhibitors with chemotherapy achieved higher efficacy than trastuzumab plus chemotherapy. The positive results from experiments utilizing pertuzumab/trastuzumab as anti-HER2 therapy allowed the FDA to approve the combination of pertuzumab with trastuzumab and chemotherapy for treatment purpose in the neoadjuvant setting [37].

4.2.2.3 Chemotherapy Resistance Mechanisms in HER2-Positive Breast Cancer

Three major categories have been suggested to include different types of resistance to anti-HER2 therapy with trastuzumab. The first is redundancy within the HER receptor layer, indicating the capacity of signaling pathways to keep on functioning even after partial inhibition by the redundant ligands and receptors that activate alternative dimerization patterns. The second category is reactivation, which implies that pathway signaling at/downstream of the receptor layer or after loss of downstream negative regulation mechanisms is activated to stimulate HER or downstream mutations. The last category is escape by using pre-existing or acquired pathways during resistance, which do not usually drive cancer cells when HER2 is not inhibited [38].

4.2.2.3.1 Incomplete Blockade of HER Receptors

HER receptors will be incompletely blocked if the drug or drugs used for treatment fail to efficiently inhibit signaling from all HER family dimer pairs and fully prevent downstream signaling. By using combinations of targeted agents, clinical studies aim to investigate regimens with higher treatment efficacy which completely block HER2 as the major driver pathway [39].

4.2.2.3.2 Activation of the PI3K Pathway

As an influential downstream signaling pathway, the PI3K/AKT1 pathway is stimulated by HER2 signaling. Preclinical finding shows that constitutive activation of the PI3K/AKT1 pathway by reduced levels of its tumor-suppressor PTEN, or by active mutations in the PI3K catalytic subunit alpha (PI3KCA) gene, may inhibit response to trastuzumab and other anti-HER2 drugs, which is also known as drug resistance. The enrichment or emergence of PI3KCA mutations activates PI3K signaling, which may also cause acquired resistance to lapatinib in experimental models. Low levels but not necessarily complete loss of PTEN initiate the PI3K pathway and decrease treatment efficacy. Preclinical results indicate that adding PI3K/mTOR/AKT1 pathway inhibitors to anti-HER2 treatment can suppress resistance in tumors with PI3KCA mutations. The detailed mechanisms are currently under investigation in clinical trials [40].

4.2.2.3.3 Overexpression of Estrogen Receptor

ER is expressed in about half of HER2+ breast cancer tumors. The ER and the HER pathways positively and negatively regulate each other via complex bidirectional cross talk, where one pathway can become the escape route for the other in targeted therapy. Many preclinical experiments have used a variety of HER2+ breast cancer models to demonstrate the role of ER and ER signaling in evading HER2 inhibition and enhancing drug resistance. Pre-existing or restored ER levels and/or activity is shown to regulate de novo or acquired resistance to intensive anti-HER2 therapy in ER+/HER2+ breast cancer cells. A study that uses samples from a neoadjuvant trial with lapatinib found that ER and B-cell lymphoma 2 (BCL2) levels were increased in a parallel manner after lapatinib treatment. Substantial clinical reports also indicate that the ER pathway provides an escape mechanism from HER2 inhibition.

4.2.2.3.4 Fcγ Receptor Polymorphisms

One of the mechanisms by which trastuzumab is assumed to act on tumor cells is that its interactions with Fcγ receptors (FcγR) on leukocytes induce antibody-dependent cell-mediated cytotoxicity (ADCC). There are indications that Fcγ receptor polymorphisms have an effect on ADCC. Amino acid substitutions in Fc fragment of IgG receptor IIIa (FCGR3A) and IIa (FCGR2A) genes at positions 158 (158 V/V) and 131 (131H/H) mediate the strength of binding to the antibodies [41].

4.2.3 Triple-Negative Breast Cancer

TNBC is defined, immunohistochemically, by the lack of expression of ERs and PRs and lack of overexpression of HER2 [42]. According to recent guidelines, patients should be considered ER-/PR- when IHC shows <1% of cells are positive for hormonal receptors. About 15–20% of breast cancer cases at diagnosis are TNBCs, especially for young female patients (<40 years of age) [43]. TNBCs, generally regarded as invasive ductal carcinomas, have the characteristics of high proliferative capacity, poor differentiation, and a large overall tumor size. Instead of metastasizing to the bone and soft tissues, as most breast cancer subtypes do, TNBC tumors tend to disseminate to the brain and lungs. Furthermore, unlike other breast cancer subtypes, TNBCs establish no correlation between tumor size and positive lymph nodes. The survival rate of TNBC patients is expected to be 70% for the first 5 year, which is 10% lower than other subtypes [44].

Development of identifying breast cancer heterogeneity [2, 3, 45] suggests an alternative method for subgrouping patients. Identifying targetable vulnerabilities within these subgroups is essential to tailor therapeutic approach for improved treatment effects. Meanwhile, intra- and inter-tumor heterogeneity in TNBCs, along with inherent and acquired resistance to therapies, has become a major challenge to develop feasible targeted strategy for TNBCs. To overcome some of these challenges, researchers have started studies of TNBCs based on large-scale gene expression and genome [46–49]. Their studies indicate that TNBCs are heterogeneous and consisted of at least four to six definable molecular subtypes [46, 50]. These subtypes express elements of distinct oncogenic signaling pathways, which imply the existence of potential molecular targets. Based on previous and ongoing researches, single agents may not exert effects against TNBC, while their combination with other therapies shows promise in preventing intrinsic or acquired drug resistance.

4.2.3.1 Treatment Strategies for Triple-Negative Breast Cancer

Due to the lack of high-frequency oncogenic drivers, which have been causally proven to target disease heterogeneity, treatment for TNBC faces a big challenge in clinical practice [46, 48]. As the core of treatment, chemotherapy usually participates in regulating taxanes, anthracyclines, and/or platinum compounds to disrupt the functions of cancer cell. The nature of chemotherapy and whether chemotherapy choices should be distinct among TNBC subtypes are currently under discussion, as most TNBC patients who have received chemotherapy do not achieve a pCR. Studies showed that adding platinum compounds to standard chemotherapy doubled the number of TNBC patients achieving a pCR [51]. Compared to patients with residual disease from other breast cancer subtypes, TNBC patients who do not achieve a PCR display worse results [52]. Therefore, more research is encouraged to investigate effective therapeutic strategies for TNBCs.

4.2.3.2 Classification of Triple-Negative Breast Cancer by Molecular Subtype

About 70% of TNBC tumors are shown to be basal-like by gene expression profiling [53]. Meanwhile, expression of ER/PR or HER2 and at least one other basal molecular marker, such as cytokeratins 5 and 6 (CK5/CK6), CK14, CK17, caveolin 1/2, and EGF receptor (EGFR), has been observed in a significant number of basal-like tumors, indicating that TNBCs distinguish themselves with histopathological features and are represented in various mRNA expression-based subtypes. Being clinically heterogeneous, TNBCs also have variations in morphology, mutational phenotype, as well as signaling profiles between tumors [46].

The meta-analysis of gene expression-profiling data from 21 breast cancer samples suggests seven subclasses for TNBCs (six definable subclasses and an unstable one) [46], including basal-like (BL1 and BL2 of basal or myoepithelial origin), immunomodulatory (IM), mesenchymal stemlike (MSL), mesenchymal-like (M), and luminal androgen receptor expression (LAR). All of the subclasses react differently to neoadjuvant chemotherapy (NAC) and correspond well with pCR rates [54]. After NAC, BL1 tumors display the highest pCR rates (52%) among all of the subclasses. BL1 tumors, which are enriched for cell cycle and DNA damage response (DDR) genes, highly express MKi67 and respond well to antimitotic agents like docetaxel, taxanes like paclitaxel, and DNA-damaging agent cisplatin. Meanwhile, patients of BL2 tumors rarely achieve a pCR, even though the tumors are abundant in metabolic signaling genes, survival-mediated receptor tyrosine kinases (RTKs), and proliferation genes. Epithelial-mesenchymal transition (EMT) markers and growth factor signaling pathway elements are highly expressed in the M and MSL subclasses, the cell lines of which display sensitivity to sarcoma family kinase (SRC) and PI3K/mammalian target of rapamycin (mTOR) inhibitors [46]. pCR rates are moderate (23–31%) in MSL tumors, which express decreased levels of proliferation-related genes with a low mitotic index. Moreover, these tumor types and their cell lines are featured with consistent upregulation of transforming growth factor beta (TGFB) receptor type III (TGFBRIII), which promotes tumor growth in vivo and the migration and invasion of MSL cell lines in vitro [55]. In LAR tumors, an abundance of hormone-regulated signaling pathways has been observed. These pathways, such as steroid synthesis and androgen receptor (AR) signaling, generally induce PI3KCA activating mutations and demonstrate low levels of responses to chemotherapy (10% pCR rates). In preclinical models, this tumor subtype could be treated with the combined regimen of antiandrogen and bicalutamide targeting AR and PI3K [56]. The IM subclass shows moderate pCR rates like M tumors and is enriched for immune response-mediated cell signaling with antigen presentation and T-cell functions.

The main characteristics of TNBCs involve mutations in or loss of tumor protein p53 (TP53) (in ~85% of TNBCs), mutation or loss of RB1 (20%), amplification of myeloid cell leukemia 1 (MCL1) (54%), and amplification of v-MYC avian myelocytomatosis viral oncogene homolog (MYCLK1) (35%) [57]. Moreover, these tumor subtypes display mutation or loss of PTEN (35%), mutations in PI3KCA (7%), and loss of inositol polyphosphate-4-phosphatase type II B (INPP4B) (30%). Other features of TNBCs include widespread chromosomal instability with frequent gains on chromosomal arms 1q, 3q, 8q and 12q, as well as loss on 4q, 5q, and 8p. Regardless of familial history of breast cancer, TNBCs are also related to germline dysfunction involving breast cancer 1 and 2 and DNA repair-associated (BRCA1 and BRCA2) genes [58, 59]. Additionally, other breast cancer predisposition genes participating in homologous recombination (HR) have been found in a small number of TNBCs patients. Such genes are comprised of BRCA1-associated RING domain 1 (BARD1), partner and localizer of BRCA2 (PALB2), and BRCA1-interacting protein C-terminal helicase 1 (BRIP1) [1].

4.2.3.3 Biomarkers for Triple-Negative Breast Cancer-Targeted Therapy

4.2.3.3.1 BRCA1 Status for Targeted Triple-Negative Breast Cancer Therapy

Mutation of BRCA1 is rare in sporadic TNBC. Nevertheless, some TNBC patients still show a BRCA1 mutation carrier-like phenotype [60, 61]. According to the seminal reports, both BRCA1- and BRCA2-deficient cell lines were highly sensitive to poly (ADP-ribose) polymerase 1 (PARP1) inhibition, which triggered apoptosis and instability. Either tumors or cell lines with mutations in or inactivation of BRCA1 or BRCA2 have impaired HR and thus rely on mechanisms involving PARP1 for DNA damage repair, such as the alternative nonhomologous end-joining (alt-NHEJ) and base excision repair (BER) pathways. As the cells are unable to repair their DNAs, they will be forced to undergo apoptosis if PARP1 is inhibited in BRCA defective tumors [62–64]. While the simultaneous loss of PARP1 and BRCA1 genes results in cell death, deletion of either of the genes individually does not affect cell viability. Increasing attention has been drawn to PARP1 inhibition as a promising therapeutic strategy for treating cancers with BRCA1 mutations via synthetic lethality.

Several early-phase clinical trials have investigated the efficacy of PARP1 inhibitors in the treatment of TNBC patients. Initial results in metastatic TNBC patients compared chemotherapy alone with the PARP1 inhibitor iniparib in combination with chemotherapy and displayed a significant improvement with the median overall survival (OS) being 7.7 vs 12.3 months and the progression-free survival (PFS) being 3.6 vs 5.9 months. Meanwhile, in a recent phase III clinical trial, the results were more modest with a PFS of 4.6 vs 5.6 months [65, 66]. In a recent phase I multicenter trial, 37% of the patients with metastatic TNBCs partially responded to the combined regimen of olaparib and paclitaxel, which was administered either as first- or second-line treatment. The trial results confirmed a moderate response in this cohort following PARP inhibition [67]. Meanwhile, olaparib treatment for patients with non-BRCA-associated TNBCs observed no objective responses [68] in phase II studies, perhaps due to the small sample size used in the study (26 breast cancer patients) or the selection of patients who have been intensively treated with chemotherapy. Other ongoing projects include phase I and II clinical trials of PARP inhibitors, which stratify TNBC patients based on HR status and aim to achieve more favorable response rates.

4.2.3.3.2 Control of the Cell Cycle and DNA-Damage Response in Triple-Negative Breast Cancer

With several pharmaceutical companies exploring inhibitors of DNA damage checkpoint kinases, such as ataxia-telangiectasia (AT) mutated serine/threonine kinase (ATM), checkpoint kinase 1/2 (CHK1/CHK2), and ataxia-telangiectasia and Rad3-related Ser/Thr kinase (ATR), targeting DNA-damage-induced cell cycle arrest has become a major focus for chemotherapeutic research. Cell cycle arrest gives cancer cells time to repair their DNA and is therefore a survival mechanism in the latter. Cell cycle control interference could cause "inappropriate" cell cycle progression, which accumulates DNA damage and triggers cancer cell death. Thus, abrogation of checkpoints could activate apoptotic cascades before DNA repair is complete. TNBCs rely on CHK1 to arrest cell cycle progression. Despite that TNBCs have no amplification or mutation of CHK1/CHK2, which has been observed in the oncogenic targets discussed above, CHK1 inhibition presents a promising strategy for treating aggressive breast tumors with mutated TP53 [69].

Cell cycle promoters, such as cyclin-dependent kinases (CDKs), enhance progression through the cell cycle under activation of cyclins. Directly targeting the promoters presents an alternative approach to target cell cycle checkpoints. Naturally occurring CDK inhibitors could efficiently inhibit CDKs. However, tumorigenesis may induce overexpression of cyclins or inactivation of CDK inhibitors (such as the INK4 class of inhibitors, notably p16), which leads to uncontrolled proliferation. CDKs are not mutated or amplified in TNBCs as they are in the oncogenic targets discussed above, which is similar to CHK1/CHK2. Nevertheless, the targeting of CDKs may promote apoptosis by limiting cycle progression of tumor cells [1].

CHK1, which belongs to the Ser/Thr protein kinase family, phosphorylates the protein phosphatase, cell division cycle 25A (CDC25A), to delay cell cycle progression in response to double-strand breaks (DSBs). Checkpoint-mediated cell cycle arrest requires CHK1 to respond to DNA damage or the presence of unreplicated DNA. As expected, CHK1 is overexpressed in TNBC cells, which are fast-dividing and genomically unstable. Studies showed that the CHK1 inhibitor UCN-01 [70] enhanced cisplatin sensitivity and evaded the DNA-damage-dependent G2 checkpoint induced by cisplatin treatment [71]. Other phase I and II clinical trials are currently investigating AZD7762 (AstraZeneca), PF-477736 (Pfizer), CHK1 inhibitors, SCH900776 (Schering-Plough), and LY2606368 (Eli Lilly), either as a single agent or in combination with other chemotherapies. By far, TP53 mutation has been the most common incident in TNBCs (~85%) [48, 57]. In particular, promising preclinical results have been shown in CHK1 inhibitors (UCN-01 or AZD7762) describing the sensitizing of TNBC xenografts with TP53 mutations to chemotherapy [72].

The combined treatment of CHK2 inhibitor LY2606368 with chemotherapy is currently being investigated in phase I trials for patients with advanced or metastatic solid tumors and in phase II trials for patients with germline BRCA1/BRCA2 mutations (including TNBCs) [73]. The inhibitor of WEE1 G2 checkpoint kinase was found to contribute to the increase of cell death in TNBC cells with TP53 mutations [74].

Several CDK inhibitors, such as UCN-01, have been developed to inactivate CDK2 by dephosphorylation and inhibit CHK1 [75], delaying cell cycle progression into S phase. In 35% of TNBCs, selective inhibition of CDK1 and CDK2 is amplified, which could be synthetically lethal with MYC [76]. Inhibition of CDK1 or CDK2 promotes apoptosis through BCL2-like 11 (BCL2L11) activation in TNBC xenografts, indicating the effect of CDK1/CDK2 inhibition on treating MYC-driven TNBC [76]. CDK inhibitors make breast tumor cells susceptible to PARP inhibitors [77]. The inhibition of CDK1 sensitizes wild-type BRCA1 cancer cells to PARP1 inhibition, presenting a feasible strategy to extend the utility of PARP inhibitors for BRCA1-/BRCA2-proficient cells.

CDK4/CDK6 inhibition, together with PI3KCA inhibition, has recently displayed promising benefits for breast tumor treatment in several PI3KCA-mutant xenograft tumor models [78]. The RB1 tumor suppressor, which is absent in 20% of TNBCs [57, 79], decides the sensitivity to CDK4/CDK6 inhibition [80]. Moreover, two preclinical studies on CDK4/CDK6 inhibition have recently demonstrated that the growth of RB+ TNBC cells is retarded by the induction of G1 arrest, while RB- cells are completely resistant [81, 82].

4.2.3.3.3 Targeting Receptor Tyrosine Kinases in Triple-Negative Breast Cancer

About 80% of TNBCs display certain levels of constitutive activation for members of the EGFR/BRAF proto-oncogene, Ser/Thr kinase (BRAF) signaling pathway in large-scale genomic analyses [48, 57, 83–85]. Despite that EGFR overexpression in TNBC plays a significant role in poor OS [86], EGFR inhibitors present limited effects during preclinical and clinical trials [87]. AKT1 and HER3 signaling pathways have been further proved to mediate signaling pathway activation as feedback loops to induce acquired resistance [86]. Furthermore, modest results were observed in two independent phase II clinical trials that aimed to evaluate the efficacy of cetuximab combined treatment with either cisplatin or carboplatin [88, 89]. The findings indicate that EGFR-targeted therapy in an appropriate drug combination could still be a feasible option for cancer treatment.

PDGF-A, PDGF-B, PDGF-C, and PDGF-D comprise the platelet-derived growth factor (PDGF) family and bind either as homo- or heterodimers to one of the two RTKs, PDGF receptor alpha (PDGFR-A) or PDGF receptor beta (PDGFR-B), to mediate cell migration, proliferation, and survival [90, 91]. The observation of PDGF overexpression in breast cancer predicts the diagnosis of an advanced tumor stage, as well as a malignant phenotype and a poor OS rate [91, 92]. Approved by the USFDA for treating chronic myeloid leukemia (CML) [93], inhibitors, such as imatinib, target the phosphorylation of RTKs, including PDGFR-B and KIT proto-oncogene receptor tyrosine kinase (KIT). Studies have been carried out focusing on the monoclonal antibody, bevacizumab, which specifically targets vascular endothelial growth factor A (VEGFA), and the small-molecule kinase inhibitor, sunitinib, which inhibits members of both the VEGF and PDGF family, in the context of tumor suppression in TNBC [94]. In TNBC xenograft models, sunitinib decreased a large amount of tumor volume through the inhibition of angiogenesis, which was induced by VEGFA signaling [94]. Even though no changes in OS were observed, the addition of bevacizumab to paclitaxel chemotherapy was found to double both the response rates and duration of PFS [95, 96].

In TNBC cells, the physical interaction between hepatocyte growth factor receptor (HGFR), also known as proto-oncogene c-Met, and AXL receptor tyrosine kinase (AXL) demonstrates a significant signaling cross talk. Besides, AXL, which composes a complex with other HER family members, as well as with HGFR and PDGFR in TNBC cells, significantly diversifies EGFR signaling and limits the response to EGFR-targeted inhibitors [97]. Such highlights a widespread role of these RTKs in TNBCs [97]. Moreover, the amplification of fibroblast growth factor receptor 2 (FGFR2) is observed in 2–4% of TNBCs [98, 99]. In xenografts, FGFR inhibition interferes with FGF signaling and significantly impairs tumor formation, indicating that FGFR2 may be a potential target for TNBC treatment [98].

4.2.3.3.4 Targeting Oncogenic Signaling in Triple-Negative Breast Cancer

Recent research found that mutated rapidly accelerated sarcoma (RAS) and rapidly accelerated fibrosarcoma (RAF) family genes are responsible for 2% of breast cancers [57, 100]. During metastasis in TNBC, copy number changes constitute the main mechanisms involved in RAS and RAF activation [101], and aberrant MAPK activity is implicated in TNBC development and progression [102]. Furthermore, negative regulators, such as neurofibromin (NF1) and dual-specificity phosphatases (DUSPs), regulate the activation of RAS-mediated signaling. Somatic mutations in NF1 [57, 85, 103] and promoter methylation of DUSP4 activate RAS/ERK signaling related to high induced proliferation of tumor cell following NAC in basal breast cancer [104]. Overexpression of mitogen-activated protein kinase kinase 1/2 (MAP2K1/MAP2K2, also known as MEK1/MEK2) and mitogen-activated protein kinase 1 (MAPK1) also contributes to the reduction of PFS and OS. Nevertheless, modest results were showed in phase I clinical trials with MAP2K1/MAP2K2 inhibition in solid tumors [105, 106], because the kinome reprogramed through RTKs in TNBC cell lines as an adaptive resistance mechanism after MAP2K1/MAP2K2 inhibition [107]. In addition, dual activation of MAP2K1 and PI3K pathways was observed in some TNBC patients, indicating the tumor growth inhibition effect through co-inhibition of both pathways [108, 109].

PI3K pathway overactivation caused by PTEN mutation or deletion and loss of heterozygosity at the INPP4B locus is observed in 60% of TNBC patients [57, 48, 110–112], while the pathway overactivation induced by activating PI3KCA mutations is seen in only 8% of TNBCs. Overexpression of RTKs may also contribute to overactivation of this pathway. Likewise, the activation of RAS signaling is regulated by negative feedback mechanisms in the activator of transcription (JAK/STAT) pathway and the PI3K/mechanistic target of rapamycin (mTOR)/AKT1 and Janus family of kinases/signal transducer. Preclinical studies found that the hyperactivation of AKT1 and mTOR was correlated with poor prognosis in TNBC patients, indicating that dual inhibition of these molecules could be a promising therapeutic strategy [112–114]. A significant improvement in PFS following PI3K/mTOR/AKT1 inhibition in combination with chemotherapy in metastatic TNBC has been highlighted in a recently completed phase I trial [115]. In addition, PI3K suppression in TNBC confers sensitivity to PARP inhibition possibly through impairing DNA repair as well as sensitizing both BRCA1-proficient and BRCA1-deficient TNBC patient-derived xenografts (PDXs) [116, 117]. In compliance with the above findings, PARP inhibitors exhibit preclinical activity in tumors with wild-type BRCA1 and loss of PTEN [118]. However, the result needs to be testified in clinical trials.

As supported by several studies, deregulation of JAK/STAT pathway takes up a major role in TNBC [114, 119]. According to molecular-profiling studies, Janus kinase 1 (JAK1) and JAK2 are more abundantly expressed in TNBC patients with residual disease [51], while the constitutive overexpression of STAT3 acts as an oncogenic driver for uncontrolled cell proliferation and angiogenesis [119, 120]. The JAK1/JAK2 inhibitor, ruxolitinib, has been approved for treating myelofibrosis and is currently administered in phase II trials as a single agent or in combination with paclitaxel for TNBC patients.

4.2.3.4 Biomarkers for Development of Immune-Modulating Triple-Negative Breast Cancer Therapeutics

4.2.3.4.1 Tumor-Infiltrating Lymphocytes

Tumor-infiltrating lymphocytes (TILs) determine the efficacy of conventional chemotherapies not only as a prognostic biomarker but also as a predictive factor and thus have gained much attention in the past years [121, 122]. As suggested by gene expression profiling and immunohistochemistry (IHC) staining, augmented TILs predict a better prognosis in TNBCs [123–125]. In IHC studies of cohorts of patients with TNBC undergoing chemotherapy and neoadjuvant therapy, higher numbers of TILs indicated a stronger correlation with prognosis in response to treatment, regardless of the lymphocyte subtype [124, 126–128].

In a recent neoadjuvant GeparSixto trial (NCT01426880), immunological markers, such as indoleamine 2,3-dioxygenase (IDO1), T-cell surface glycoprotein CD8 alpha chain (CD8A), C-C motif chemokine ligand 5 (CCL5), and programmed cell death 1 ligand 1 (PD-L1), were shown to positively correlate with stromal TILs and increased pCR after chemotherapy. Patients with high levels of TILs also achieved a greater pCR rate than patients with low levels of TILs, with a pCR odds ratio of 1.22 per 10% increase in TILs [129]. These findings indicate that the type and the level of immune infiltration may determine the clinical outcome in some TNBC patients, especially those with basal TNBC who may benefit from immune-based therapies. However, clinical evidence is still limited to verify the studies above. An immunomodulatory subtype of TNBC has been recently described with characteristics of elevated expression of genes involved in antigen processing and T-cell functions, indicating that the subtype may benefit from immunotherapy against TNBCs [46].

Tumor antigens in TNBC, which are not shared by normal cells, have been proved to be effective targets for immunotherapy. Cancer/testis (CT) antigens, such as CT antigen 1A (CTAG1A), melanoma-antigen (MAGE) family member A3 (MAGEA3), and MAGEA4, are expressed in testicular germ cells but not in somatic cells. They are also present in a substantial proportion of TNBC cells [130, 131]. Therapeutic tumor vaccines targeting MAGEA3 have been tested in clinical experiments for treatment of melanoma and non-small cell lung cancer (NSCLC). Due to failure of identifying a subpopulation of MAGEA3+ NSCLC patients that may benefit from this treatment, GlaxoSmithKline (GSK) has recently stopped its phase III trial of a MAGEA3 cancer immunotherapeutic for NSCLC patients. Therapies targeting MAGEA3 in TNBCs are still under preclinical investigation [132]. Mesothelin, a cell surface glycoprotein present on mesothelial cells, is expressed in 34% of TNBCs, while glycosylated mucin-1, a cell surface-associated peptide covalently linked to a toll-like receptor (TLR) agonist, generates a potent therapeutic antitumor response [133].

The potential effectiveness of T-cell responses in patients with breast cancer is reflected by the presence of TIL-induced upregulation of genes as immunosuppressive markers, such as PD-L1, programmed cell death protein 1 (PDCD1), cytotoxic T-lymphocyte protein 4 (CTLA-4), and IDO1. Moreover, upregulation of these genes was found to be predictive of a better clinical outcome with chemotherapy in the treatment of TNBCs [129] and basal-like breast cancers [134], indicating that combined immunotherapies targeting the above immunosuppressive pathways with chemotherapies could be a successful approach for treating TNBC.

4.2.3.4.2 Programmed Cell Death 1 Ligand 1 Expression

In TNBC, overexpression of PD-L1 was reported in 20% of patients [135], which positively correlated with the number of TILs. A retrospectively validated IHC measurement of 636 breast cancer samples further supported the above finding by demonstrating a positive correlation among the levels of PD-L1, TILs, and RFS. Higher levels of PD-L1 were also shown to be intensively related to higher TILs and better prognosis in the neoadjuvant treatment of TNBC [136, 137]. Moreover, phase I clinical trials with PD-L1-specific antibodies (Merck's pembrolizumab, formerly known as MK-3475, and Roche's MPDL3280A) displayed promising results, which were presented at the 2014 San Antonio Breast Cancer Symposium. However, further studies are required to determine whether or not PD-L1 is a predictive biomarker for the response to anti-PD-L1 immunotherapy.

4.2.3.4.3 Cluster of Differentiation 73 and Other Immune Modulators

Overexpression of cluster of differentiation 73 (CD73) indicates a worse prognosis and resistance to chemotherapy in TNBC [138]. Decreased metastatic burden for CD73-positive tumors instead of CD73-negative tumor cells was shown in TNBC mouse models after anti-CD73 therapy or administration of adenosine receptor inhibitors, which suggested that CD73 could be a potential biomarker for CD73-targeted therapy. Notably, immunotherapies targeting the CD73-adenosine pathway can be efficient treatment methods when combined with other immunotherapy approaches such as anti-PD-1 and anti-cytotoxic T-lymphocyte-associated protein 4 (CTLA-4) [139].

4.2.3.5 Mechanisms of Resistance to Targeted Therapies in Triple-Negative Breast Cancer

4.2.3.5.1 Cross Talk Between Receptor Tyrosine Kinase Pathways

Numerous preclinical and clinical studies indicate that the resistance of TNBC to single-agent inhibitor therapies [140] owes much to the tumor cells' cross-resistance mechanisms [107, 141]. For instance, TNBC cells increase the expression and activation of the RTKs AXL and PDGFR-B and thus bypass the action of the MAP2K1/MAP2K2 inhibitor, AZD6244 [107]. In the allosteric binding pocket of MAP2K1, resistance to

MAP2K1/MAP2K2 inhibitors is also related to the acquired mutations [141]. In the case of RTK pathway inhibitors, evidence from in vitro and in vivo models [142, 143] indicates that pathway redundancies and reprogramming of the kinome effectively bypass targeted inhibition and therefore promote resistance, which could be caused by alterations in pathway cross talk or feedback inhibition involving the acquisition of survival-enhancing mutations [144].

4.2.3.5.2 Intratumoral Heterogeneity of Triple-Negative Breast Cancers

Another example of acquired resistance involves research reporting that the intratumoral heterogeneity of TNBCs renders abundant expression of a specific population of tumor cells ($CD44^{high}$/$CD24^{low}$ stem cell-like subpopulation), which then induces tumor recurrence and resistance after treatment, as recently demonstrated in a report on X-box binding protein 1 (XBP1) [145]. According to the report, an increase of XBP1 splicing (i.e., activation) was observed in the $CD44^{high}$/$CD24^{low}$ population of TNBC. Following chemotherapy, XBP1 depletion reduced the induction of the $CD44^{high}$/$CD24^{low}$ stem cell-like subpopulation and significantly inhibited relapse in TNBC xenograft tumors. The study implies that XBP1 promotes the acquisition of cancer stem cell-like properties and contributes to the acquired resistance to chemotherapy in tumor cells.

4.2.3.5.3 Duration of Therapeutic Responses

Preclinical research in a genetically engineered mouse model with BRCA1 deficiency explored the efficacy of PARP inhibition and found that resistance could still occur within 1 year after treatment [61, 146]. The underlying mechanisms remain to be further investigated, in spite of the significant role played by the restoration of the BRCA expression and HR activity through genetic reversion to the original mutation and upregulation of multidrug resistance protein 1 (MRP1)-mediated drug efflux [146]. Preclinical evidence further testified the other mechanisms of resistance to PARP inhibitor (olaparib) in mammary tumor models of BRCA1-deficient mouse, which involves the rewiring of the DNA damage response because of loss of tumor protein p53 binding protein 1 (TP53BP1). However, this finding has not yet been validated in clinical research on human tumors [147]. Meanwhile, HR deficiency in ovarian cancer cell lines has been found to promote expression of DNA polymerase theta (Polθ) and induce resistance to PARP inhibition and DNA-damaging agents [148].

Tumor resistance still poses a big challenge for establishing durable responses to treatment, because tumor cells adapt their signaling circuitry to take advantage of redundancies and feedback mechanisms. Mechanisms of drug resistance are compared to a map of transportation or subway. Even if a commuter line is blocked, passengers could still find another route to reach their destinations, which will have repercussions throughout the network [149]. In order to prevent pathway rewiring and upregulation of compensatory pathways, systems-biology models [150–152] suggest that multiple "hubs" within oncogenic pathways should be simultaneously inhibited. Oncogenes and other disease-associated genes are generally seen as part of larger networks. It could be a good strategy to target neighboring proteins in the network that break signaling but do not trigger drug resistance [153]. An augmenting interest also focuses on the use of single inhibitors to target several molecules simultaneously, which could help overcome the issues caused by intra- and inter-tumor heterogeneity of TNBC. In spite of the lack of perfect models for human, which is due to the inadequacy of PDXs to activate antitumor immunity, TNBC immunotherapy still shows promise in improving clinical outcomes within the TNBC therapeutic armamentarium [1].

4.2.4 Metastatic Breast Cancer

Targeted therapies improve treatment for advanced breast cancer and prolong patient survival. However, because of drug resistance mechanisms, these tumors frequently relapse [154]. Some studies have shown that 30–50% of ER + breast cancer cases do not respond to tamoxifen therapy [154, 155]. HER2-overexpressing breast cancer patients can develop metastases, where only 11–34% of metastatic tumors respond to trastuzumab monotherapy [154, 156]. Although TNBC patients show ~39% response to chemotherapy, the presence of residual disease is related to an early risk of relapse [52, 157].

Metastatic disease remains the leading cause of death in breast cancer patients, whereas personalized medicine for metastatic disease presents greater hurdles than nonmetastatic disease. Metastasis is described as the migration of tumor cells from the primary tumor, after which are intravasation, survival, extravasation of the circulatory system, and progressive colonization of a distant site. Even though only 5–10% of newly diagnosed breast cancer patients display cancer metastasizing to distant body parts, there remains a high risk for patients with localized primary disease to develop metastatic disease after successful primary tumor resection and a voisin therapy [158–160]. While these findings, coupled with the fact that distant recurrent diseases are viewed as being incurable, imply the high clinical burdens of metastatic breast cancer, better strategies for clinical intervention are being explored to meet the urgent demand [158].

4.2.4.1 Tumor Invasion from Early Preneoplastic Lesions

The development of clinically detectable distant metastases involves multiple genetic and epigenetic alterations, affects both tumor cells and the surrounding stroma, and allows seeding of metastases at distant sites. During the process, primary tumor cells must first invade and escape from complex physical barriers at the primary site, which include the extracellular matrix, basement membrane, and vasculature. The cells then intravasate into the lymphatic or vascular system, exit to infiltrate distant organs, and continue to proliferate in this foreign milieu. Both the cell of origin and the oncogenic alterations in the tumor are considered to influence the metastatic propensity of a tumor cell [154, 160].

With genes being divided into four main categories, a tumor's ability to metastasize may depend on the type of oncogenic driver mutation and associated genes [158]. The first category is called metastasis initiation genes, which allow aggressive cells to invade the surrounding tissue, attract a supportive stroma [154], facilitate the dispersion of cancer cells and may also infiltrate distant metastatic niches. This step involves several genes participating in the epithelial-to-mesenchymal transition (EMT; e.g., TWIST1, SNAI1, and SNAI2), extracellular matrix degradation (e.g., matrix metalloproteinases, MMPs), hypoxia (e.g., HIF1A), and angiogenesis (e.g., VEGFA) [161–163]. Expression of these genes from the first category, along with their target genes, is prognostic of poor outcomes in primary tumors [164].

The second category is composed of metastasis progression genes, which work with each other in tumor cells to promote extravasation, survival, and reinitiating of tumor growth in the invaded parenchyma based on specific tissues [164]. Repeated identifications of gene expression signatures in primary tumors with relapse potential have proved the expression of metastasis progression genes, which include PTGS2, EREG, LOX, ANGPLTL4, and CLDN2 [165–167].

Metastasis virulence genes, which endow disseminated cells with the ability to overtly colonize distant sites, comprise the third group. Between dissemination and colonization, tumor cells must endure essential alterations to proliferate and survive in the foreign tissue, which is known as a latency period (metastatic dormancy period) [164]. For instance, essential osteoclast mobilizing factors, including interleukin-11 (IL-11), vascular cell adhesion protein 1 (VCAM-1), and parathyroid hormone-related protein (PTHrP), play a crucial role in establishing osteolytic metastases [166, 168].

The fourth category concerns metastasis-suppressor genes that prolong metastatic latency and prevent metastatic cells from reinitiating growth upon infiltrating distant organs. Among this group of genes, cystatin E/M (CST6) has been found to inhibit breast cancer bone metastases [169]. Retinoic acid receptor responder protein 3 (RARRES3) has been recently identified to potentially suppress breast cancer lung metastasis. And metastasis-suppressor KiSS-1 (KISS1) functions as a metastatic suppressor in breast cancer and other malignancies [170].

4.2.4.2 Bone Metastasis

Breast cancer cells preferentially cause osteolytic lesions in the bone, the most common distant metastatic site for breast cancer and the third most prevalent site of cancer metastases in general. Bone metastasis is often related to severe pain and other comorbidities [166]. Protransforming growth factor alpha (TGFα) induces osteoclasts to secrete PTHrP, tumor necrosis factor alpha (TNFα), and cytokines (including interleukins 1, 6, 8, and 11) and prompts osteoblasts to release the receptor activator of NF-κB (RANK) ligand (RANKL) (also known as the tumor necrosis factor ligand superfamily member 11, TNFSF11), all of which stimulate osteoclast differentiation [166, 171–173]. Osteoclasts in turn demineralize the bone, which releases growth factors such as bone morphogenetic proteins, IGF1 and TGFα from the bone matrix to support cancer cell proliferation. Other important proteins that mediate the specific functions crucial for cancer cells localization and colonization to the bone include cytokines (C-X-C motif chemokine receptor 4 (CXCR4), CXC ligand 12, and TGFα), VCAM-1, NF-κB, Jagged1, SRC, osteopontin, MMP1, integrin avb3, cadherins (e.g., cadherin 11), and adrenomedullin[174–177]. Further investigation demonstrated that miRNAs miR-141, miR-219, and miR-34a were expressed in a xenograft mouse model to prohibit bone metastases. MiR-16 and miR-378 expressions were also shown to correlate with bone metastasis burden, suggesting mRNAs' role as potential therapeutic targets and clinical biomarkers for bone metastases [178].

4.2.4.3 Brain Metastases

Compared with other subtypes, HER2+ breast cancer may dramatically increase the risk of brain metastases. Therefore, development of effective therapies controlling extracranial metastatic HER2+ breast cancer becomes even more important [179]. Endothelial cells, astrocytes, and pericytes compose the blood-brain barrier (BBB), which allows penetration of tumor cells to seed metastases in the brain. Cathepsin S (CTSS) expression in experimental xenografts was shown to facilitate transmigration of the BBB through cleavage of tight junction proteins. An inverse correlation between CTSS expression and brain metastasis-free survival was also observed in primary breast tumors [165, 180]. High levels of anti-plasminogen activator serpins, such as neuroserpin (SERPINI1) and serpin B2 (SERPINB2), are expressed in brain metastatic breast cancer cells [165, 181, 182], which inhibit the generation of plasmin and the following suppression of metastasis formation. In the brain microenvironment, prostaglandin G/H synthase 2 (PTGS2), EGFR ligand HBEGF, the α2, 6-sialyltransferase ST6GALNAC5, neural cell adhesion molecule L1 (L1CAM), SERPINI1, and plasminogen (PLG) work with astrocytes, pericytes, and other cell types as mediators of cancer cell passage through the BBB, collectively enhancing vascular co-option and survival of breast cancer cells in the brain [165, 181, 182].

4.2.4.4 Lung Metastases

The development of lung metastases is related to multiple gene-encoding cytokines or their secreted products supporting transendothelial migration from circulation into the lung parenchyma [183]. RARRES3 downregulation facilitated the adhesion of the tumor cells to the lung parenchyma [184]. Thus, tenascin C (TNC) overexpression and retinoic acid receptor responder 3 (RARRES3) downregulation may serve as potential biomarkers for the identification of patients at high risk of lung relapse. Other noteworthy factors that function to mediate lung metastasis include TGFα, epiregulin, PTGS2, MMP1 and MMP2, angiopoietin-related protein 3 (ANGPTL4), DNA-binding protein inhibitor ID-1 (ID1), protein lysine-rich CEACAM1 co-isolated (LYRIC), and VCAM-1 [183, 185, 186].

4.2.4.5 Liver Metastases

Similar to metastases of the bone, brain, and lungs, liver metastases of breast cancer are closely linked to breast cancer cell-secreted chemokines and their cognate receptors in a number of studies. The levels of C-C motif chemokine ligand 9 (CCL9) and CX3CL1 were strongly and equivalently elevated in liver metastatic breast cancer cells in mice [187]. The tight junction protein, claudin 2 (CLDN2), was also found to be significantly expressed in both breast cancer liver metastases and primary tumors with an increased predilection to metastasize to the liver. Moreover, the development of breast cancer liver metastases involves participation of CXCR4 and its ligand (CXCL12, also known as SDF1), cadherins, integrins, and other claudins [187, 188]. In all, the complex interactions between dispersed cells and specific stromal components in metastatic niches significantly contribute to the apparent colonization of specific organs [158]. Tumor cell-derived proteases and their regulators majorly undergo stage-specific expression changes during metastatic seeding and outgrowth in different organs, while stroma-derived genes are primarily regulated in a tissue-specific manner.

4.2.4.6 Therapeutic Approach for Metastatic Disease

4.2.4.6.1 HER2-Targeted Therapies

HER2 is overexpressed in animal models to promote metastasis to the lymph nodes, bone, lung, and brain [189]. Several pathways have been found in studies on HER2 metastasis, which include a bidirectional interaction with the TGFB-Smad pathway, an increase in expression and stability of the homing chemokine receptor CXCR4[190], an activation of p60-SRC with successive phosphorylation of focal adhesion kinase 1 (FADK 1) at tyr861 and activation of p120/RAC1/CDC42 [191], and an increase in angiogenesis through upregulation of VEGFA [192] and angiopoietin-2. It is of great importance to determine whether inhibition of these pathways combined with HER2 therapy provides better strategies for preventing metastasis or lesion shrinkages [193].

4.2.4.6.2 Estrogen Receptor-Targeted Therapies

ER+ tumors tend to metastasize to the bone and often metastasize late [194]. A recent long-term follow-up study on XXXX patients receiving tamoxifen therapy for 5 years has showed that metastatic relapses appear over the 10 years after treatment and do not level off at that time point, implying a consistent break from dormancy. ER signaling and the epithelial-mesenchymal transition were also found to be connected in previous research. The various options of endocrine therapy provide a great opportunity for selecting efficient combination between therapeutic agents and optimal sequence to overcome tumor recurrence or relapse. Current researches focus on verifying combinations of growth factor and PI3K pathway targeting agents, such as everolimus and gefitinib [195].

4.2.4.6.3 Bisphosphonates and RANK Ligand Antibodies

Bone metastatic breast cancer patients have different therapeutic options, such as bisphosphonates and denosumab, which is generally combined with endocrine or HER2-directed therapies. The process of bone metastasis is often referred to as a vicious cycle [166].

Both TGFB and PTHrP have a critical role in osteolysis. Tumor cells produce PTHrP and stimulate osteoblasts and osteoclasts via the RANKL pathway, which often leads to bone resorption. As a human monoclonal antibody against RANKL, denosumab was found to reduce the risk of developing multiple skeletal-related events (i.e., time to first and subsequent events) by 23% in a phase III trial, which was more significant than the bisphosphonate, zoledronic acid [196].

4.2.4.7 Resistance Reasons for Metastatic Breast Cancer Therapy

4.2.4.7.1 Intratumoral Heterogeneity

Heterogeneity in primary tumors has been demonstrated by recent researches in metastases of relevant breast cancer at morphological, molecular, and genomic levels. Such heterogeneity may determine response to anticancer therapy. According to a study classifying traditional markers such as HER2, ER, and PR, 5–22%, 13–33%, and 31–32% discordance occurs, respectively, between the primary tumor and distant metastases [197].

The poor correlation between p-AKT immunohistochemical levels and p4EBP1 expression shows a common dissension between primary tumors and metastases for other therapeutic target. According to Wu and his colleagues [198], a diverse heterogeneity between primary breast carcinomas and their paired metastases was observed even among different metastatic breast cancer cells from the same patient. ER and PR were found to be downregulated, and PTGS2, HGFR, EGFR, and mesothelin were overexpressed in metastatic lesions compared with primary lesions. In the primary breast cancer and some metastatic breast cancers, the therapeutic targets were identified to be different from targets found in all metastatic sites [198].

As shown by a recent mutational profiling, multiple genetic conversions, which are not unique to metastases, frequently occur in the secondary lesions of breast cancer metastases and primary tumors. Take the mutations in the tumor-suppressor gene TP53, which are usually present in 25% of primary breast cancers, as an example. They were abundantly observed in a series of 23 brain metastases of breast cancer (87%), with a more complex and superior mutation processes of TP53, such as frameshift, splice, and nonsense mutations, as well as in-frame insertions and deletions [199]. Previous studies have described the enhanced expansion of MYC in systemic metastases compared to primary breast cancers. Because allelic imbalance is more frequently observed in brain metastases than in primary breast tumors, mutation or loss of the tumor-suppressor PTEN is more commonly found in the former [200].

4.2.4.7.2 Genomic Instability

Heterogeneity is likely to be driven by the precariousness of metastatic breast cancer cells, and dissimilar clonal evolution could provide molecular contention during disease relapse. Tumor evolution could start early, as a broad range of clonal genomic heterogeneity was demonstrated by analysis comparing primary tumor and concurrent lymph node infiltrates through the utilization of comparative genomic hybridization [201]. Malfunction of the DNA break repair system, mitotic chromosome transmission, or the spindle mitotic checkpoint results in chromosomal lesions, which are hallmarks of genomic instability. Subsequently, high mutation rates and chromosomal rearrangements (deletions, duplications, and amplifications) could promote cancer progression in tumor-suppressor genes interference, fusion proteins formation, enzymes activation, or oncogenes amplification [202].

4.2.4.7.3 Tumor Subpopulations

Cancer stem cells or cancer-initiating cells compose one of the tumor subpopulations. Along with molecular heterogeneity, multiple functions are assumed for subpopulations of tumor cells. The propagation of tumor cells with stem cell-like properties and proliferative capacities could contribute to metastases at distant sites [162]. Stem cell-like properties not only present the potential to metastasize but also induce chemotherapeutic resistance [203].

Another tumor subpopulation is dormant tumor cells. Metastasis could occur shortly after primary tumor development for some breast cancer patients, or appear years or even decades after initial treatment for some others. Tumor cells that remain potential for an extended period of time are termed dormant. Dormancy can be heterogeneous, with a balance of proliferation, apoptosis, cell cycle quiescence, and/or antiangiogenic mechanisms [204]. Dormant breast cancers exhibited doxorubicin defiance in an experimental study [205]. They also present the potential to survive from the initial chemotherapy and could be awakened years later.

4.2.4.7.4 Microenvironmental Influences

Tumor microenvironment is a primary contributor of tumor metastasis, comprising fibroblasts, vasculature, immune, and inflammatory cells and the extracellular matrix. It is also known as niche and takes part in complementary interactions between cancers cells and their surrounding factors [206]. Microenvironment is by no means stationary, as it could be altered by cancer cells and infiltrated by immune and other circulating cells, a property of which may stimulate tumor progression [193]. Even in experimental models, there is a lack of thorough profiles of the metastatic microenvironment by organ site through time. The urgency for a complete portrait of microenvironment indicates its importance for chemotherapeutic resistance [207].

4.2.5 Breast Cancer Stem Cells (Tumor-Initiating Cells)

Although adjuvant therapy plays a critical part in the management of early breast cancer, regional relapse still occurs [208]. Currently, about 40% of all breast cancer patients suffer from recurrences, with 10–20% of them being local and 60–70% being distant metastases [209]. The deficiency of treatment with adjuvant therapies for breast cancer patients is considered to be caused by a borderline classification of the disease, which may lead to some but not all recurrences. Local and metastatic recurrence after the surgical treatment for primary tumor could be caused by surgery residues of tumor cells or early micrometastases resistant to supplementary treatments. Moreover, the resistance to neoadjuvant systemic therapy may explain cancer recurrence and proliferation in locally advanced breast tumor cells. Even though these therapies may fail to cure most solid tumors due to disease recurrence and local metastasis, an intense interest has been triggered in the controversial cancer stem cell (CSC) model, where a therapy-resistant subpopulation of cells has been found to have the capacity to regenerate cancer cells. The CSC hypothesis defines "a small subset of cells within a cancer that composes a reservoir of self-sustaining cells with the restricted ability to self-renew and to cause the heterogeneous lineages of cancer cells that comprise the tumor." Preclinical results based on the cell lines and cancer models sustain that tumor-initiating cells derived from breast cancer are comparatively defiant to chemotherapy and radiation. The association between the CSC hypothesis and the normal breast epithelial hierarchy supports further investigation on the histogenesis of breast cancer, including the normal cellular origin of a certain breast cancer subtype. However, the importance of understanding tumor cell origin for ameliorating breast cancer outcomes is yet to be decided [208].

4.2.5.1 Biomarkers for Breast Cancer Stem Cells

Studies have identified several biomarkers for breast cancer stem cells (BCSCs), including HER2, integrin alpha-6 (ITGA6), epithelial cell adhesion molecule (EpCAM), acetaldehyde dehydrogenase activity, and phosphatase and PTEN. The first biomarkers for human BCSCs were put forward by Al-Hajj and his colleagues in 2003. The authors isolated CD44+/CD24−/^{low}Lin- cells from primary human/xenograft breast cancers by using two cell surface markers, the glycoproteins CD44 and CD24, and demonstrated that this cell population was enriched in BCSCs and more tumorigenic than the others. Currently, the EpCAM+/CD44+/CD24- putative BCSCs have been extensively investigated. Based on the enhanced tumorigenicity observed in a human breast cancer metastasis xenograft model, this surface marker expression profile has been proposed as a CSC phenotype and become the study focus for many researchers [208].

Surrogate markers for CD44C/CD24K cells, such as protein C receptor (PROCR) and ganglioside GD2 (a glycosphingolipid), have been suggested as cell surface receptors specifically expressed in CD44+ cells [210]. Nearly all GD2C cells are CD44+/CD24-, and the knockdown of GD3 synthase, an enzyme involved in the synthesis of GD2 in GD2+ cells, could decrease CSC properties. These findings suggest that GD2 and CD44 could be phenotypic indicators for the underlying mechanisms stimulating CSC activity.

Recent research has also identified integrins to be human BCSC biomarkers. For example,

ITGA6 marks CSCs in ERK and TNBC xenografts [211].

Distinct groups of cell surface proteins have been found to be enriched in CSCs in different types of mouse mammary tumors, such as the stem combinations of cell antigen 1 (Sca1) and CD24 (Sca1+/CD24+) and the b3 integrin, CD61, combined with CD49f (CD49fhi/CD61hi), in mouse mammary tumor virus (MMTV)-Neu tumors; the CD24+/CD49f+ combination in MMTV-PyMT tumors; the CD61+/CD29lo/CD24+ combination; the thymocyte antigen 1 (Thy1)C/CD24C combination; the EpCAM^{low}/CD49fhi combination; and the CD24hi/CD49fhi combination in MMTV-Wnt1 tumors. Moreover, TICs were found to be enriched in CD29hi/CD24+ cells from both p53-null and BRCA1-deficient tumors [212].

The employment of non-cell surface markers as biomarkers for CSCs has proved to be successful. For instance, in mouse mammary glands and human breast cancers, stem cells could be pinpointed by referring to the activity of aldehyde dehydrogenase (ALDH). The limited overlap between the CD44C/CD24K and ALDH1C populations implies the existence of different breast TIC groups even within individual cancers. Thus, ALDH activity may only be applicable for certain breast cancer subtypes, which is also the case for CD44+/CD24−/^{low}Lin- markers [213].

4.2.5.2 Therapeutic Cancer Stem Cell Approach in Breast Cancer

Cells with rapidly dividing properties are easily targeted by cytotoxic drugs. A self-renewing, long-lived, and relatively quiescent CSC population may be more resistant to therapy. Moreover, the survival of these cells could be affected by post-cancer treatment recurrence. A delicate balance exists between DNA replication and repair in cell proliferation, self-renewal, and quiescence to maintain stem cell. The levels of DNA repair in human embryonic and adult stem cells were found to be elevated in comparison with XXXX, which provided an approach to enhance survival [214].

4.2.5.2.1 DNA Repair and Checkpoint Inhibitors

Characterization has been made on nonhomologous end-joining inhibitors and the BER and homologous recombination DNA repair pathway. The mechanism and effect of checkpoint abrogation on anticancer treatments, especially the G2 checkpoint, were also evaluated. Inhibition of the checkpoint kinases, particularly CHK1, may shed light on treatment of BCSCs, where prolonged G2 arrest has been observed. Moreover, glioblastoma CSC sensitivity could be restored by these checkpoint inhibitors under exposure to ionizing radiation [215].

4.2.5.2.2 Targeting Cellular Signaling in Breast Cancer Stem Cells

Microenvironment and the developmental signaling pathways, such as Hedgehog, Notch, and Wnt, are associated with the renewal and differentiation of tumor cells. Based on the above mechanisms, targeting CSCs is suggested by many studies as an efficient way to tackle cancer [216].

PTEN/PI3K/AKT1/b-catenin pathway has a significant impact on regulating breast cancer stem/progenitor cells. Small-molecule inhibitors targeting components of RAS/PI3K/PTEN/mTOR, the Ca^{2+}/calmodulin-dependent protein kinase, and RAS/RAF/MAPK/ERK pathways could synergize with each other and induce death in drug-resistant breast cancer cells after conventional therapy [208].

While the mechanisms remain obscure in regard to the effect of the tamoxifen analogue, N, N-diethyl-2-[4-(phenylmethyl)phenoxy]ethanamine (DPPE; tesmilifene), treatment with DPPE alone reduces mammosphere formation and viability of CD44+/CD24- breast cancer cells. Furthermore, DPPE works with doxorubicin to achieve complete eradication of tumor cells. EGFR signaling is associated with the stemness of human breast cancer cells in a positive manner. Inhibition of EGFR signaling disrupts mammosphere formation. Unlike chemotherapy, lapatinib (an EGFR1/HER2 tyrosine kinase inhibitor) does not result in an increased expression of

CD44+/CD24−/Lin- BCSCs. An anti-EGFR monoclonal antibody has been shown to disrupt mammosphere formation and reduce the number of CD44þ+/CD24+ cells in mammospheres [208].

Despite the unclarified role of heat shock protein-90 (HSP90) in CSCs, it remains as a putative therapeutic target. The application of HSP90 inhibitor, 17-N-allylamino-17-demethoxygeldanamycin, proved to be efficient in removing CSC by studies utilizing tumorigenic glioma stem cells [124]. Even though the effect of HSP90 inhibitors on BCSCs has not been explored, the ectopic expression of engrailed homeobox protein 1 (En1) shows close association with a stem cell phenotype inhibited by 17-N-allylamino-17-demethoxygeldanamycin (17-AAG) [208].

4.2.5.3 Resistance to Chemotherapy in Breast Cancer Stem Cells

De novo chemoresistant BCSCs could be a key contributor to disease relapse. This assumption is supported by two important pieces of evidence: (1) chemotherapy treatment increases the number of cells expressing markers of breast cancer TICs, and (2) tumors enriched in markers of breast TICs are comparatively resistant to chemotherapy [211]. Following chemotherapy, residual breast cancers and cancer cell lines were found to increase $CD44^{high}/CD24^{low}$ subpopulations, MSFE and tumor-initiating efficiency, and expression of TIC signature genes [211].

Chemotherapy drugs usually need cell division to efficiently induce apoptosis, which gives rise to the hypothesis that the quiescent property of BCSCs may cause chemoresistance [211]. The slow cycling nature of BCSCs is supported by several findings: (1) markers from the gene expression signature derived from quiescent normal human mammary stem cells can be used to isolate BCSCs, suggesting that BCSCs are correspondingly quiescent; (2) the dye retention side population (SP) is enriched for CSCs in breast cancer cell lines, showing increased expression of negative cell cycle regulators [211]; (3) CD44+/CD24−/low cells isolated from breast cancer cell lines are slow cycling and resistant to chemotherapies [211].

The increased number of ATP-binding cassette (ABC) transporters, which actively pump drugs out of cells of the cancer bulk, could also stimulate chemoresistance to BCSCs [211]. The ATP-binding cassette subfamily G member 2 (ABCG2) involving in multidrug resistance in vitro (also known as breast cancer resistance protein (BCRP)) was first identified in a multidrug-resistant sublime of MCF7. Likewise, the increased expression of another well-characterized ABC transporter, the multidrug resistance protein 1 (MRP1, also referred to as P-glycoprotein (Pgp)), correlates with doxorubicin resistance in multiple breast cancer cell lines [211]. In breast cancer models, high BCRP expression corresponds with high levels of HER2 expression, lymph node metastasis, and an advanced stage of breast cancer. MRP1 is increasingly expressed after exposure to chemotherapy, indicating that cancer cells with MRP1 expression may be resistant to chemotherapy [211].

4.3 Prediction Biomarkers for Overcoming Chemotherapeutic Resistance in Breast Cancer

4.3.1 Estrogen Receptor-Positive Breast Cancer

Biomarkers that predict development of broad endocrine resistance and resistance to a specific agent are essential for effective treatment of estrogen receptor alpha (ERα)+ breast cancer [217]. They could also assist in selecting patients who would benefit the most from additional regimen without targeting ERα. Biomarkers for different biological mechanisms could be applicable to different patient subtypes in predicting either broad endocrine resistance or resistance to a specific agent [217]. ERα activity is reliant on coactivating proteins. Therefore, modified coactivator expression is likely to cause endocrine resistance [218].

4.3.1.1 Lack of Hormone Receptor

4.3.1.1.1 Estrogen Receptor

ERα (identified by immunohistochemical staining and also known as ESR1) is the only clinically used biomarker to evaluate the response to endocrine therapy. While endocrine therapy is not effective for complete ERα- disease, patients with the highest ERα expression improve slightly better than those with low receptor expression during tamoxifen therapy. In 2010, a genomic index for endocrine therapy sensitivity was reported, which was based on the mean expression level of genes sharing both positive and negative relations with ER [219].

In addition, receptor sensitivity to antiestrogen therapy can be affected by posttranslational modifications of the receptor behavior mostly during phosphorylation events [217]. Phosphorylation of ERα at Ser118 can be compromised by MAPK1 or MAPK3 in an estradiol-independent manner, whereas CDK7 mediates estradiol-induced phosphorylation at Ser118 [220]. Phosphorylation at Ser305 by protein kinase A (PKA, subtypes not specified) or Ser/Thr protein kinase PAK 1 (PAK1) results in a conformational change in the receptor after tamoxifen binding, which nullifies tamoxifen's antagonistic effects.

Even though the expression of ERα is related with the response to endocrine therapy, the mechanisms of ERβ effect is not yet identified. As shown by in vitro studies, ERβ sensitizes breast cancer cells to the antiestrogenic actions of endoxifen (a tamoxifen metabolite). Clinical data also implies a favorable outcome after tamoxifen therapy in patients with high expression levels of ERβ in their cancer cells [221]. However, further research is required to confirm the finding.

4.3.1.1.2 Progesterone Receptor

Since expression of the progesterone receptor (PR) gene (*PGR*) is highly dependent on an unaltered and fully functional ERα signaling pathway, a lack of PR expression can cause insensitivity to antiestrogenic treatment [217]. Low PR expression may be representative of downregulation of *PGR* due to excessive growth factor receptor signaling (caused by overexpression of HER2), a mechanism that can also cause endocrine resistance [217]. However, a meta-analysis reported no difference in tamoxifen benefit between patients with PR+ and PR- tumors [222], casting doubt on the potentially involved mechanisms.

4.3.1.2 Activation of Growth Factor Pathways

Growth factor receptor pathways can excite growth independent of ER, or affect the endocrine therapy response via cross talk with ER [73]. In vitro trigger of these growth factor receptor pathways by overexpression of different growth factor receptors, such as EGFR (HER1), HER2, IGF1R, and fibroblast growth factor receptor 1 (FGFR1), was found to result in tamoxifen resistance [217].

4.3.1.2.1 Epidermal Growth Factor Receptors

Analysis of all three EGFRs (HER1, HER2, and HER3) in the tamoxifen and exemestane adjuvant multinational (TEAM) trial revealed that only patients with HER1–HER3-negative tumors had a decreased risk of relapse when receiving exemestane versus tamoxifen. High EGFR protein expression, but not HER2 amplification, has been associated with a reduced benefit from adjuvant tamoxifen [217]. Overexpression of EGFRs (EGFR and receptor tyrosine-protein kinase (ERBB2)) activates intracellular signaling cascades, including MAPK and PI3K, fosters cellular proliferation, and induces a worsened and shortened response to tamoxifen in ER+ breast cancer [218]. Patients with HER1–HER3-positive tumors had poorer prognosis in both treatment arms, suggesting the relative resistance to both endocrine therapies in these women. HER2 status was not a predictive factor for the differential benefit from anastrozole versus tamoxifen in the Arimidex, tamoxifen, alone or in combination (ATAC) trial [223].

4.3.1.2.2 Insulin-Like Growth Factor 1 Receptor

As insulin growth factor 1 receptor (IGF1R) is an ER-dependent gene, its expression likely affects disease prognosis. Inhibitors of IGF1R help to stimulate tumor growth by inhibiting apoptosis and inducing metastasis and angiogenesis [217]. Preclinical trials suggest that the increased activity of IGF1R signaling pathways may promote resistance to endocrine therapy, whereas blocking or inhibiting these pathways may increase the efficacy of second-line hormonal therapies for breast cancer [224]. Although the total IGF1R level might reflect ERα activity, phosphorylated IGF1R reflected active IGF1R signaling and was indicative of poor survival in a series of patients with primary breast cancer treated with different endocrine therapies [225]. The exact prognostic and predictive value of IGF1R and its phosphorylated counterpart requires further investigation.

4.3.1.3 Mutation of PI3KCA and AKT Activation

Besides the upregulation of growth factor receptors, altered activity of the involved kinase could dysregulate growth factor pathways to encourage cell proliferation in an ER-independent manner [217]. Mutations in PI3KCA are associated with increased PI3K activity in vitro and occur in approximately 30% of breast cancers, primarily in exon 9 that encodes the helical domain and exon 20 that encodes the kinase domain [226]. PI3KCA mutation with AKT activation (defined as positive for AKT1, AKT2, or phosphorylated AKT1) did not benefit from tamoxifen, whereas those without PI3KCA mutations and/or lack of AKT activation did benefit. A PI3KCA gene signature (derived from exon 20 mutation analysis) was shown to be a predictor of favorable outcome after treatment with adjuvant tamoxifen and was associated with the relatively low expression of downstream activated proteins [227]. Therefore, the existence of PI3KCA mutation itself is not an adequate predictor of benefit from endocrine therapy. Notably, the activation status of the PI3K and MAPK pathways might affect the response [217].

Furthermore, while in vitro knockdown of PTEN (an inhibitor of the AKT1/mTOR pathway) leads to hormone-independent growth and resistance to endocrine therapy, a direct relationship between PTEN expression and endocrine therapy response in patients has not been verified [217].

4.3.1.4 Overexpression of mTOR

The PI3K/AKT/mTOR pathway is activated by multiple growth factors, which control the regulation of cellular growth and proliferation at the translational level [228]. In breast cancer, the hyperactivation of this pathway involves different activities, which include stimulation of tyrosine kinase receptors (IGFR, RBB, and FGFR) and gene alterations like mutations in PI3K or AKT or PTEN loss [229]. Such modifications trigger a cascade of consecutive events, which cause the phosphorylation of transcription factors and the subsequent increase in cellular survival, proliferation, and angiogenesis [229]. Hormone-treated patients displaying hyperactivation of the PI3K/AKT/mTOR pathway develop a more aggressive phenotype, with a decrease in PFS and OS due to resistance to antiestrogenic therapy [13]. Although PI3K inhibitors are still in the early stages of development, mTOR inhibitors combined with hormonal therapy have already been tested in the treatment of breast cancer [230].

4.3.1.5 Deregulation of Cell Cycle Regulators

Cell cycle regulators are sensitive to the effects caused by tamoxifen and aromatase inhibitors. The proteins involved in cell cycle advancement usually represent poor prognosis, for they could be interpreted as biomarkers for cell proliferation. Meanwhile, there are other biomarkers exhibiting independent predictive value for endocrine therapy response in breast cancer patients [217].

Among premenopausal patients casually assigned to receive adjuvant tamoxifen or no systemic treatment, patients with CCND1-amplified tumors had an adverse response to tamoxifen, whereas patients with tumors without CCND1 amplification benefited from tamoxifen. The finding suggests that cyclin D1 directly affects tamoxifen response [231]. Cyclin A, cyclin E1, and CDK2- are associated with poor outcome in endocrine-treated patients, which is most likely a prognostic effect caused by the markers accompanied with cell proliferation [21]. High expression of cyclin-dependent kinase inhibitor 1B (CDKN1B) is the exception and has been found to be an independent predictive factor for benefit from tamoxifen in premenopausal patients [217].

Homeobox protein Hox-B13 (HOXB13) is a member of the homeobox gene family, a group of genes that encode transcriptional regulators of cell growth and differentiation [232]. Although HOXB13 levels in ERα+ breast cancer were originally identified as a prognostic biomarker, high HOXB13 levels were also shown to be predictive for lack of tamoxifen benefit in postmenopausal patients [232].

4.3.1.6 Drug Metabolism Resistance

Tamoxifen commands metabolic activation via the cytochrome P450-mediated pathway to amplify its affinity for ER. Endoxifen is currently considered as the most potent tamoxifen metabolite. Direct measurement of endoxifen levels might be a factual predictor of tamoxifen benefit. In a series of tamoxifen-treated patients with primary breast cancer, a significant reduction in recurrence rate was shown for patients with high endoxifen levels [233].

High tumor levels of estradiol 17β-dehydrogenase type 1 (17β-HSD1, also known as 17β-hydroxysteroid dehydrogenase type 1), the enzyme responsible for the conversion of estrone into the more potent compound estradiol, were estimated for lack of tamoxifen benefit in premenopausal patients [234]. Additionally, 17β-HSD2 catalyzes the oxidation of estradiol to estrone. In postmenopausal patients, a high 17β-HSD1 to 17β-HSD2 ratio predicted a lack of tamoxifen benefit [235].

4.3.2 HER2-Positive Breast Cancer

Half of the patients express primary resistance to trastuzumab [236], and a large portion acquire resistance before the end of the first year of treatment in metastatic settings. In metastatic breast cancer, 15–26% overall response rates were achieved with trastuzumab monotherapy. Half of the patients expressed primary resistance to trastuzumab and a large portion acquired resistance before the end of the first year of treatment in the metastatic setting [27]. Attempts were made to develop a comprehensive classification systems aiding patient selection.

There is a wide-ranging set of potential biomarkers capable of stratifying patients by their response to trastuzumab. These include HER2 amplification, impaired access to the binding site (p95HER2, Δ16HER2, MUC4), augmented signaling through other ERBB family receptors (HER1, HER3, HER4) and their ligands [27], activation of HER2 targets by alternate heterodimers (ephrin type-A receptor 2 (EphA2), IGF1R, growth differentiation factor 15 (GDF15), and the cleaved form of the mucin-1 protein (MUC1*)), signaling triggered by downstream members (PI3KCA, PTEN, SRC, mTOR), altered expression of cell cycle and apoptotic regulators (CDKs, CDKN1B, BCL2), hormone receptor status, resistance to antibody-dependent cellular cytotoxicity (FcγR), and altered miRNA expression signatures [237].

Multigenic molecular profile analyses have revealed further genes not directly associated with the classical oncogenic pathways. Although numerous biomarkers have shown promise in preclinical studies, many have delivered controversial results when evaluated in clinical trials [238]. One of the keys for targeting ERBB2 will be to consider the entire ERBB family and downstream associated pathways responsible for the malignant transformation.

4.3.2.1 HER2 as a Biomarker for Therapeutic Resistance

4.3.2.1.1 HER2 Gene Amplification, Protein Expression, and Quantification

There is evidence that both locally advanced and metastatic breast cancer patients with high gene amplification levels (such as a HER2 gene copy number over 10 or HER2/CEP17 ratio over four) derive greater benefits from trastuzumab treatment. However, disease progression associated with HER2 expression is complex. According to more recent studies, low, intermediate, and high HER2 expression corresponds to differences in trastuzumab sensitivity [27]. Time to progression is short when HER2 expression is low, but increases significantly at intermediate HER2 levels [27, 239]. Nevertheless, time to progression abruptly decreases when HER2 is too high, such as being fivefold or more of the intermediate values. HER2 amplification measured by FISH, but not IHC, was related to pCR, suggesting that FISH may be superior to IHC at predicting overall survival in neoadjuvant settings.

Moreover, the levels of HER2 measured by HER mark predict the response to trastuzumab and trastuzumab combined with lapatinib, when HER2 expression is above the median, irrespective of hormone receptor status. A quantitative HER2 analysis, such as HER mark (an application from the VeraTag platform), may serve as a better predictor of trastuzumab's efficacy. HER mark allows simultaneous measurement of the total HER2 protein levels and the levels of functional HER2 homodimers on the surface of the cells [27]. HER2 expression assessed by HER mark and IHC values correlate. However, HER mark provides more quantifiable values [240].

Conversely, in a subset of HER2+ tumors, the HER2 gene is amplified without overexpression of the protein, and these patients are expected to gain significantly less from trastuzumab therapy [241].

Altogether, precise, quantitative HER2 assessment emerges as one of the most promising candidates for the accurate prediction of trastuzumab efficiency.

4.3.2.1.2 HER2 Homodimers

HER2 oligomer status has been associated with trastuzumab sensitivity [29], as HER2 heterodimers activate the PI3K pathway [242]. Meanwhile, HER2 overexpression can lead to HER2 homodimerization that activates the RAS/ERK signaling pathway. HER2+ metastatic breast cancer patients with high levels of HER2 homodimers experience longer time to progression, which prolongs overall survival [243]. The level of HER2 homodimers correlates with longer survival after trastuzumab treatment in metastatic patients, and patients with the highest homodimer expression seem to benefit less from concurrent chemotherapy than patients expressing lower HER2 homodimer levels.

4.3.2.1.3 Intratumor Heterogeneity and Selection of a HER2-Negative State

Primary tumors and metastatic lesions do not always retain HER2 positivity after neoadjuvant trastuzumab treatment [244]. In up to one-third of patients where pCR was not achieved, the residual lesions transformed to HER2- status associated with poor survival. One-fourth of patients diagnosed with HER2+ disease had HER2-negative metastatic lesions. The presence of such discordance was associated with shorter overall survival and suggested the administration of chemotherapy, but not with trastuzumab treatment.

4.3.2.1.4 HER2 Extracellular Domain in the Serum

Elevated serum HER2 extracellular domain levels (HER2/ECD) signal poor prognosis. Although the evidence is somewhat controversial, a decrease in HER2/ECD is associated with the magnitude of response to anti-HER2 therapy [30]. Increased objective response rate and delayed disease progression were observed in patients who achieved a significant decrease in ECD levels [245]. Similarly, longer progression-free survival characterized metastatic breast cancer patients with constantly low serum HER2/ECD levels, or those who achieved low serum ECD levels during treatment. The role of HER2/ECD is even more controversial in predicting survival, as high serum HER2/ECD levels indicate a significantly higher objective response rate but do not predict survival [246].

Furthermore, evidence suggests the applicability of HER2/ECD levels for detecting breast cancer progression or regression. Patients with HER2+ disease had significantly higher HER2/

ECD levels (24%) than other conventional tumor markers (7.4–12.9%) [27]. In particular, patients with stage 3, stage 4, and recurrent disease showed significantly higher HER2/ECD levels than other tumor markers. In patients with recurrent disease, recurrence was associated with the elevation of serum HER2/ECD levels from the baseline. The above findings suggest that elevated serum HER2/ECD levels signal the emergence of HER2+ metastatic disease [247].

4.3.2.1.5 Impaired Access to the HER2 Binding Site

p95HER2 comprises a series of carboxy-terminal HER2 receptor fragments. As a result of proteolytic shedding likely initiated by ADAM10 metalloproteases or the alternative initiation of translation, the HER2 receptor loses the fourth extracellular domain that binds trastuzumab [248]. The p95HER2 fragment with a size of 100–115 kDa, called 611CTF (carboxy-terminal fragment), is generated by alternative initiation of translation and is highly oncogenic due to constitutive homodimerization.

In about one-third of HER2+ tumors [249], high p95HER2 was found to be generally related to worse prognosis and was therefore considered to be a marker of trastuzumab resistance [250]. In metastatic breast cancer patients treated with trastuzumab, high p95HER2 expression was associated with shorter progression-free and overall survival. Elevated levels of p95HER2 were also predictive of poor prognosis and significantly correlated with low progression-free and overall survival.

Δ16HER2 is a HER2 splice variant generated by exon 16 skipping, causing a conformational change in the extracellular domain of the receptor that leads to stable homodimerization [27]. The relationship between Δ16HER2 expression and trastuzumab sensitivity is not fully understood [27]. However, it is clear that the Δ16HER2 splice variant mediates a strong oncogenic potential in tumor cells [27], where Δ16HER2 is present in 89% of HER2+, node-positive primary breast cancer, increasing the heterogeneity of the disease [251].

4.3.2.2 Augmented Signaling in ERBB Family Receptors and Their Ligands

4.3.2.2.1 HER1

The integrated signaling network linking all ERBB RTKs allows for the escaping of the effects of a single receptor blockade via alternative signaling routes [27]. HER1 was identified by co-expression as a potential independent negative predictor of pCR [252]. Trastuzumab-resistant BT474 cells not only expressed higher amounts of HER1 and its ligands but also displayed increased levels of HER1:HER2 heterodimer formation [253]. In a larger study with 155 trastuzumab-treated patients in either metastatic or adjuvant/neoadjuvant settings, 15% of tumors overexpressed HER1, and the patients also experienced decreased overall survival. Targeting HER1 with the tyrosine kinase inhibitor (TKI), gefitinib, eliminated in vitro HER2 phosphorylation and suppressed HER2-driven proliferation in HER2+ breast cancer cell lines [254].

4.3.2.2.2 HER3

Although trastuzumab blocks ligand-independent formation of HER2 homodimers and HER2:HER3 heterodimers, HER2:HER3 heterodimers have the most potent carcinogenic effect [27]. Growth factor-induced HER2:HER3 heterodimers initiate AKT//mTOR signaling, providing an escape route from the effect of trastuzumab. Inhibition of PI3K abolished AKT activation. However, it initiated a compensatory ERK activation by increasing HER3 expression [27]. Consequently, combined anti-HER2 and anti-MEK treatments will be required to prevent this alternative signaling. Excess TGFB stimulates the PI3K/AKT pathway through HER3 activation [255]. Clinical studies have attributed predictive potential to HER3 expression in order to stratify patients by prospective treatment response [27]. Trastuzumab-treated HER2+ metastatic breast cancer patients with high HER3 expression measured by the VeraTag assay were associated with shorter progression-free and overall survival than patients with low HER3 expression. In patients treated with adjuvant trastuzumab, the presence of HER2/HER3 heterodimers and loss of p21 expression were associated with poor survival prognosis.

Long-term anti-HER2 treatment may induce the reprogramming of tumor cells to activate alternate signaling pathways, even in the absence of visible effects [27]. Extended trastuzumab treatment modulated the ERBB signaling network and increased HER1 and HER3 expression in HER2+ trastuzumab-resistant cell lines [256]. In 125 metastatic breast cancer patients, retrospective analysis of HER3 expression, using IHC, revealed an association between HER3 negativity and better progression-free survival after taxane and trastuzumab treatment [257].

4.3.2.2.3 HER4

The role of HER4 in trastuzumab resistance is rather controversial. Multiple reports attribute antiproliferative effects to HER4 [258]. In a neoadjuvant setting, HER2+ patients with HER4 co-expression measured by IHC experienced a significant delay in the development of metastases and improved progression-free survival after adjuvant therapy. These findings suggest the use of HER4 as a potential biomarker to predict clinical outcomes. However, the type of HER4 isoforms and location of HER4 seem to be crucial for its prognostic power [27]. Cytoplasmic HER4 has been linked with increased survival, while nuclear translocation of HER4 has been associated with trastuzumab resistance during monotherapy and presented as a poor prognostic factor in HER2+ breast cancer [27]. Additionally, anti-HER2 treatment induced HER4 overexpression, cleavage, and nuclear translocation.

4.3.2.3 Activation of Growth Factor Receptors, Their Ligands, and Membrane Proteins

4.3.2.3.1 Hepatocyte Growth Factor Receptor

In HER2+ breast cancer, receptor tyrosine kinase HGFR is frequently overexpressed and predicts poor prognosis with trastuzumab [259]. In HER2-overexpressing cell cultures, trastuzumab treatment results in rapid HGFR upregulation [259]. In 130 trastuzumab-treated metastatic breast cancer patients, HGFR and its ligand, hepatocyte growth factor (HGF), had a strong correlation with HGF FISH positivity, which correlates with a higher failure rate to trastuzumab administration and a shorter time to progression [260].

4.3.2.3.2 Ephrin Type-A Receptor 2

Overexpression of EphA2 was observed in trastuzumab-resistant cell lines, where in vivo inhibition of EphA2 expression restored sensitivity to anti-HER2 therapy. In a large cohort of human breast cancer samples, elevated levels of EphA2 correlated with worse disease-free and overall survival [261].

4.3.2.3.3 Insulin-Like Growth Factor-I Receptor

Overexpression of IGF1R results in amplified trastuzumab resistance and is associated with shorter progression-free survival in patients treated with adjuvant trastuzumab. Stimulation of 1GF1R by IGF1 results in increased HER2 phosphorylation. HER2 also participates in distinctive protein-protein interactions, such as hetero-trimerization with IGFIR and HER3. Co-targeting HER2 and IGF1R has been suggested as a viable strategy to improve trastuzumab responsiveness [262].

4.3.2.3.4 Growth Differentiation Factor 15

GDF15 has been reported to be secreted by adipocytes. In HER2-overexpressing cell cultures, TGFB-related GDF15 induced phosphorylation of HER2 and reduced trastuzumab sensitivity [263]. Resistance to trastuzumab was associated with increased GDF15 expression, and GDF15 knockdown restored trastuzumab sensitivity in resistant cells. Inhibition of either TGFB receptor or SRC blocked GDF15-mediated trastuzumab resistance, supporting GDF15-mediated TGFB/SRC/HER2 cross talk signaling as a novel mechanism of trastuzumab resistance [263].

4.3.2.3.5 Mucin-1

MUC1*, the transmembrane cleavage product of MUC1, functions as a growth factor receptor in cancer cells [264]. In a cell line induced to be resistant to trastuzumab, MUC1 expression increased dramatically. Moreover, MUC1-overexpressing cells were shown to acquire resistance to standard chemotherapeutic agents. Treatment with MUC1 antagonist sensitized intrinsically trastuzumab-resistant HER2-overexpressing breast cancer cell lines to trastuzumab, with resistance to chemotherapeutic drugs also reversed. Treatment with the MUC1 inhibitor, GO-203, reduced p-HER2 levels and exerted synergistic effects with trastuzumab treatment [27].

4.3.2.3.6 Mucin-4

MUC4, a membrane-associated high molecular weight glycoprotein that can mask membrane proteins, interacts with and activates HER2. Overexpression of MUC4 reduces binding of anti-HER2 antibodies, including trastuzumab, although it does not affect HER2 expression levels [27]. MUC4 silencing by RNA interference increases trastuzumab binding. Mucins, including MUC4, were upregulated in xenograft models of ER+/HER2+ breast cancer that developed resistance against trastuzumab/lapatinib and tamoxifen/estrogen deprivation treatment[27]. These model tumors shifted their molecular phenotype toward being more ER-negative/HER2+ following mucin upregulation [27].

4.3.2.4 Activation of PI3K/AKT/mTOR Signaling Pathway

This pathway can be activated via HER3 and HER2 overexpression, low PTEN expression, PI3KCA mutations, AKT1 mutations, and altered INPP4B expression [27].

Mutations in the PI3KCA gene encoding the p110α catalytic subunit of PI3K are mainly found in HER2+ and/or ER+ breast cancers, with PI3KCA amplification mainly found in TNBCs. Both in vitro and in vivo studies indicate that mutations in PI3KCA confer resistance to trastuzumab. PTEN loss has been observed in HER2+ breast cancer patients and linked to trastuzumab resistance, poor prognosis, and decreased survival in a number of studies [257]. Additionally, metastatic breast cancer patients with high PTEN expression experienced longer progression-free and overall survival than patients with PTEN expression below 20, as measured by IHC [257]. In about 50% of HER2+ breast cancer patients, the PI3K/AKT signaling pathway is activated by low PTEN and/or the presence of PI3KCA mutations. A cell culture-based large-scale genetic screen using RNA interference identified PTEN loss and PI3KCA mutations as key biomarkers linked to trastuzumab resistance, whereas low PTEN expression was predictive of resistance against anti-HER2 therapy [257]. Additionally, when combined with PI3KCA mutation, PTEN was capable of identifying twice as many patients for increased risk of progression [27]. In another study, PTEN loss and PI3KCA mutations were not associated with response to trastuzumab and clinical outcome at all [27]. Based on current evidence, PI3KCA mutation and/or PTEN loss should not exclude HER2+ patients from potentially beneficial trastuzumab treatment.

PTEN directly and specifically dephosphorylates the non-receptor tyrosine kinase, SRC [265]. Loss of PTEN results in high amounts of active, phosphorylated SRC that facilitates HER1 and HER3 expression in vitro [27]. A single retrospective study investigated the relation between SRC expression and sensitivity to trastuzumab [27]. High levels of active SCR in tumors after trastuzumab treatment were associated with lower clinical response rates and lower overall survival, thereby contributing to trastuzumab resistance [266].

The GeparQuattro trial explored biomarkers for trastuzumab resistance in 153 HER2+ breast cancer patients using IHC and identified multiple markers related to mTOR pathway activation, such as p4EBP1, and the stem cell marker, ALDH1.

4.3.2.5 Deregulation of Cell Cycle and Apoptosis Regulator

In resistant cell culture models of HER2+ breast cancer, CDK4/CDK6 inhibition reduced cell proliferation, which was recapitulated by a finding in mouse xenografts. CDK4/CDK6 inhibition and small-molecule TKIs, such as lapatinib, had additive effects on cell growth suppression. CDK4/CDK6 inhibitors also exerted complementary action on residual clones remaining after T-DM1 therapy [267].

In HER2-amplified trastuzumab-sensitive cells, treatment with XXXX downregulated AKT1 expression, generating an increased expression of CDKN1B that caused downregulation of cyclin D1 and thus promoting cell cycle arrest [268]. In contrast, trastuzumab-resistant HER2+ cells displayed an increased proliferation rate accompanied by CDKN1B downregulation and an increase in the production of CDK2. Therefore, CDKN1B downregulation is considered as a biomarker for trastuzumab resistance, and increased CDKN1B has been shown to restore trastuzumab sensitivity [27]. Knockdown of protein phosphatase 1H (PPM1H) in HER2-overexpressing breast cancer cell culture promoted proliferation in the presence of trastuzumab. In situ hybridization on formalin-fixed paraffin-embedded (FFPE) samples revealed a trend between low tumor PPM1H expression and worse patient outcome [269]. CDKN1B blocks the activity of cyclin E/CDK2 complexes. In trastuzumab-resistant HER2+ cells, cyclin E was overexpressed in comparison with that in sensitive cells. In patients receiving first-line trastuzumab therapy, cyclin E overexpression was related to lower clinical benefit rates and shorter progression-free survival in contrast to cyclin E-negative patients (6 and 14 months, respectively) [270].

HER2-overexpressing BT474 breast cancer cells with acquired resistance to trastuzumab have increased expression of the antiapoptotic protein, BCL2, and increased sensitivity to BCL2 inhibitor ABT-737, compared to the parental cells. Concurrent treatment of the resistant cells with ABT-737 increased trastuzumab-mediated growth inhibition. Inhibition of PI3K and the IκB kinase complex (IKK) resulted in the downregulation of BCL2 and increased sensitivity to trastuzumab in resistant cells [271].

4.3.2.6 Overexpression of Estrogen Receptor

Failure of anti-HER2 agents is due to the alternative escape mechanisms offered by ER-related pathways in HER2+/ER+ breast cancer. Meta-analysis of 12 international trials has revealed that higher pCR rates are associated with ER-negative status, regardless of the treatment regimen. There is an inverse relationship between HER2 and ER, which can be explained by the negative control HER2 exerts on ER expression [27].

ER can activate HER signaling members, for example, by increasing growth factor expression of other RTKs, such as TGFα and IGF1. ER can downregulate HER1 and HER2, and upregulate IGF1R, opening alternative pathways to escape the anti-HER blockade. ER+ tumors that overexpress HER2 show both de novo and acquired resistance to endocrine therapy [234]. Targeting these pathways together allows simultaneous blocking of HER2- and ER-mediated signaling.

4.3.2.7 Role of miRNA in HER2-Targeted Therapy

In HER2+ human cell lines with acquired resistance to trastuzumab, upregulated miR-21 was accompanied by reduced expression of PTEN. Blocking miR-21 both in vitro and in vivo increased PTEN production, decreased cell proliferation, and restored trastuzumab sensitivity to the tumor cells. Higher miR-21 expression levels were associated with a decreased response to trastuzumab in pretreatment tumor biopsies, whereas trastuzumab treatment further increased miR-21 expression [272].

Trastuzumab treatment also induced miR-26a and miR-30b expression in HER2+ breast cancer cells. As revealed by cell cycle analysis, miRNAs induced G1 arrest and apoptosis, and miR-30b decreased expression of cyclin E2 posttranscriptionally, suggesting that the therapeutic action of trastuzumab may be mediated by miRNAs as well [273].

4.3.2.8 Increasing Tumor-Infiltrating Lymphocytes

An adjuvant phase III study has found that TILs are predictive of the benefit of trastuzumab and chemotherapy in early-stage HER2+ breast cancer. The neoadjuvant GeparQuattro trial further confirmed these findings, showing that every 10% increase in TILs is associated with higher pCR levels after trastuzumab and chemotherapy treatment [274]. Experimental observations have described that trastuzumab relies at least partially on the role of interferon I and II in mediating antitumor effects. These findings suggest that agents inhibiting T-cell-targeting immunosuppressive mechanisms could also be beneficial for anti-HER2 therapy [27].

4.3.3 Triple-Negative Breast Cancer

Numerous studies have identified that collective signaling molecules, such as RTKs, are often found increased in cancers and could operate TNBC targets [1]. Specifically, RTKs such as EGFRs, FGFR, TGFB receptor (TGFRB), PDGFR, VEGF receptor (VEGFR), HGFR, IGF1R, AXL, and non-receptor tyrosine kinase, SRC, could be likely exploited as targets for TNBCs [1]. Similarly, common signaling intermediates that are regulated by the aforementioned proteins have been suggested as good therapeutic targets [1]. They consist of elements of PI3K/AKT/mTOR pathway, JAK/STAT pathway, and RAF/MEK/ERK pathway, as well as Wnt/β-catenin pathway signaling [1]. Even though these targets are authorized in cell lines and preclinical models, none could be used as a single agent to reverse the aggressiveness of TNBC tumors in clinical trials, which is due to intrinsic and acquired drug resistance [1].

4.3.4 Metastatic Breast Cancer

4.3.4.1 Prognostic and Predictive Signatures for Therapeutic Outcomes

In addition to common histopathological tools, a number of multigene expression signatures have been distinguished for estimating the natural history of metastatic breast cancer.

4.3.4.1.1 Metastatic Biopsies

An actual number of patients have incongruous findings between matched biopsies of primary tumor and metastatic sites. Furthermore, one metastasis may diverge from another within the same individual, which raises the question of how many biopsies are needed. Biopsy of metastatic tissue could conceivably enhance the clinical outcome by identifying what new genetic or molecular pathways are activated, which could lead to converted and ideally more efficacious therapies.

4.3.4.1.2 Circulating Tumor Cells

Detection and bio-characterization of circulating tumor cells (CTCs) in the peripheral blood of patients with progressed breast cancer could suitably serve as a real-time tumor biopsy. CTCs are rare events, occurring at rates as low as one cell per 10^5 to 10^7 peripheral blood mononucleated cells [275], with their detection complicated by significant leukocyte contamination [193]. Detection based on EpCAM or cytokeratin expression in CTCs is potentially complicated by the epithelial-mesenchymal transition, where expression of these proteins is lost [276]. Prospective studies have described that detection of CTCs in metastatic breast cancer can successfully predict progression-free survival and overall survival [277].

Although CTCs are favorable, several issues in the literature need to be resolved before CTCs can be proposed as a part of a personalized medicine regimen [154, 193]. The proportion of CTCs that are metastatically competent, as opposed to emitting tumor cells that are destined to die, is unclear [154, 193]. According to the experimental scholarly, the vast majority of tumor cells that shed into the circulation never form a metastatic lesion [154, 193]. In addition, it is known that the molecular properties of CTCs from the primary tumor do not always match with those from the same blood sample, and heterogeneous CTC subpopulations with disparate hormone receptor and other phenotypes coexist [154, 193].

4.3.4.1.3 microRNAs

Other components of blood, such as conventional cancer markers and miRNAs, may also have predictive ability [154, 193].

4.3.5 Breast Cancer Stem Cells (Tumor-Initiating Cells)

In more than half of the breast tumors, the total and phosphorylated STAT3 has been enhanced, which is associated with poor prognostic outcome and an aggressive phenotype [1]. Preclinical research in mouse models found that STAT3 knockdown slowed down tumor growth and enhanced response to chemotherapeutic agents [278]. Moreover, a stimulation of the IL6/JAK2/STAT3 signaling pathway was observed in CD44+/CD24+ stem cell-like cells of basal-like breast cancer. BSK805, a JAK2 inhibitor for inhibition of this pathway, was found to attenuate xenograft tumor formation and exclude stem cell-like cells, which further suggested the crucial role of JAK2/STAT3 in tumor initiation [1].

4.4 Concluding Remarks and Future Perspectives

Despite improved researches in chemotherapeutic resistance, considerable numbers of patients have faced obstacles of complexities to overcome chemoresistance in breast cancer. The current challenges are understanding the specific molecular mechanism and identifying novel targeted therapies that are effective in chemoresistant breast cancer. A thorough investigation of the molecular mechanisms of tumor resistance could shed light on using anticancer drugs in an effective way and help to predict the response of chemotherapy more accurately.

In some cases of chemotherapy-resistant patients, the molecular mechanism could be changed after treatment. Some cancer cells that survive after chemotherapy can find an escape route through the activation of other signaling pathways. Hence, compensatory signaling pathways exacerbate the challenges faced when treating chemotherapy-resistant patients. Therefore, rational drug combinations and other ingenious strategies are needed to overcome chemotherapy resistance.

Recently, advanced high-throughput techniques such as next-generation sequencing have provided meaningful data that can be used as prognostic markers for chemotherapy-resistant patients. In addition, identifying new multi-targeted drugs can be considered promising therapeutic strategies. Before applying novel drugs to improve prognosis, improved in vivo systems such as patient-derived xenograft (PDX) provide opportunities to predict responses and determine the therapeutic value. Thus, incorporating high-throughput techniques will help to overcome chemotherapeutic resistance in breast cancer.

References

1. Kalimutho M, Parsons K, Mittal D, Lopez JA, Srihari S, Khanna KK (2015) Targeted therapies for triple-negative breast cancer: combating a stubborn disease. Trends Pharmacol Sci 36(12):822–846. doi:10.1016/j.tips.2015.08.009
2. Sorlie T, Perou CM, Tibshirani R, Aas T, Geisler S, Johnsen H, Hastie T, Eisen MB, van de Rijn M, Jeffrey SS, Thorsen T, Quist H, Matese JC, Brown PO, Botstein D, Lonning PE, Borresen-Dale AL (2001) Gene expression patterns of breast carcinomas distinguish tumor subclasses with clinical implications. Proc Natl Acad Sci U S A 98(19):10869–10874. doi:10.1073/pnas.191367098
3. Perou CM, Sorlie T, Eisen MB, van de Rijn M, Jeffrey SS, Rees CA, Pollack JR, Ross DT, Johnsen H, Akslen LA, Fluge O, Pergamenschikov A, Williams C, Zhu SX, Lonning PE, Borresen-Dale AL, Brown PO, Botstein D (2000) Molecular portraits of human breast tumours. Nature 406(6797):747–752. doi:10.1038/35021093
4. Rakha E, Ellis I, Reis-Filho J (2008) Are triple-negative and basal-like breast cancer synonymous? Clin Cancer Res 14(2):618.; author reply 618-619. doi:10.1158/1078-0432.CCR-07-1943
5. Rakha EA, Tan DS, Foulkes WD, Ellis IO, Tutt A, Nielsen TO, Reis-Filho JS (2007) Are triple-negative tumours and basal-like breast cancer synonymous? Breast Cancer Res 9(6):404.; author reply 405. doi:10.1186/bcr1827
6. Goldhirsch A, Ingle JN, Gelber RD, Coates AS, Thurlimann B, Senn HJ, Panel m (2009) Thresholds for therapies: highlights of the St Gallen international expert consensus on the primary therapy of early breast cancer 2009. Annals Oncol 20(8):1319–1329. doi:10.1093/annonc/mdp322
7. Virnig BA, Wang SY, Shamilyan T, Kane RL, Tuttle TM (2010) Ductal carcinoma in situ: risk factors and impact of screening. J Natl Cancer Inst Monogr 2010(41):113–116. doi:10.1093/jncimonographs/lgq024
8. Clarke M, Collins R, Darby S, Davies C, Elphinstone P, Evans V, Godwin J, Gray R, Hicks

C, James S, MacKinnon E, McGale P, McHugh T, Peto R, Taylor C, Wang Y, Early Breast Cancer Trialists' Collaborative G (2005) Effects of radiotherapy and of differences in the extent of surgery for early breast cancer on local recurrence and 15-year survival: an overview of the randomised trials. Lancet 366(9503):2087–2106. doi:10.1016/S0140-6736(05)67887-7
9. Ma L, Mao R, Shen K, Zheng Y, Li Y, Liu J, Ni L (2014) Atractylenolide I-mediated notch pathway inhibition attenuates gastric cancer stem cell traits. Biochem Biophys Res Commun 450(1):353–359. doi:10.1016/j.bbrc.2014.05.110
10. Clark GM, Osborne CK, McGuire WL (1984) Correlations between estrogen receptor, progesterone receptor, and patient characteristics in human breast cancer. J Clin Oncol 2(10):1102–1109. doi:10.1200/JCO.1984.2.10.1102
11. Klinge CM (2001) Estrogen receptor interaction with estrogen response elements. Nucleic Acids Res 29(14):2905–2919
12. Marino M, Galluzzo P, Ascenzi P (2006) Estrogen signaling multiple pathways to impact gene transcription. Curr Genomics 7(8):497–508
13. Dalmau E, Armengol-Alonso A, Munoz M, Segui-Palmer MA (2014) Current status of hormone therapy in patients with hormone receptor positive (HR+) advanced breast cancer. Breast 23(6):710–720. doi:10.1016/j.breast.2014.09.006
14. Jiang Q, Zheng S, Wang G (2013) Development of new estrogen receptor-targeting therapeutic agents for tamoxifen-resistant breast cancer. Future Med Chem 5(9):1023–1035. doi:10.4155/fmc.13.63
15. Musgrove EA, Sutherland RL (2009) Biological determinants of endocrine resistance in breast cancer. Nat Rev Cancer 9(9):631–643. doi:10.1038/nrc2713
16. Bross PF, Cohen MH, Williams GA, Pazdur R (2002) FDA drug approval summaries: fulvestrant. Oncologist 7(6):477–480
17. Lonning PE, Eikesdal HP (2013) Aromatase inhibition 2013: clinical state of the art and questions that remain to be solved. Endocr Relat Cancer 20(4):R183–R201. doi:10.1530/ERC-13-0099
18. Osborne CK, Schiff R (2011) Mechanisms of endocrine resistance in breast cancer. Annu Rev Med 62:233–247. doi:10.1146/annurev-med-070909-182917
19. Arpino G, Wiechmann L, Osborne CK, Schiff R (2008) Crosstalk between the estrogen receptor and the HER tyrosine kinase receptor family: molecular mechanism and clinical implications for endocrine therapy resistance. Endocr Rev 29(2):217–233. doi:10.1210/er.2006-0045
20. Yuan TL, Cantley LC (2008) PI3K pathway alterations in cancer: variations on a theme. Oncogene 27(41):5497–5510. doi:10.1038/onc.2008.245
21. Casimiro MC, Crosariol M, Loro E, Li Z, Pestell RG (2012) Cyclins and cell cycle control in cancer and disease. Genes Cancer 3(11–12):649–657. doi:10.1177/1947601913479022
22. Musgrove EA, Caldon CE, Barraclough J, Stone A, Sutherland RL (2011) Cyclin D as a therapeutic target in cancer. Nat Rev Cancer 11(8):558–572. doi:10.1038/nrc3090
23. Lomonosova E, Chinnadurai G (2008) BH3-only proteins in apoptosis and beyond: an overview. Oncogene 27(Suppl 1):S2–19. doi:10.1038/onc.2009.39
24. Nuciforo P, Radosevic-Robin N, Ng T, Scaltriti M (2015) Quantification of HER family receptors in breast cancer. Breast Cancer Res 17:53. doi:10.1186/s13058-015-0561-8
25. Wieduwilt MJ, Moasser MM (2008) The epidermal growth factor receptor family: biology driving targeted therapeutics. Cell Mol Life Sci 65(10):1566–1584. doi:10.1007/s00018-008-7440-8
26. Creedon H, Byron A, Main J, Hayward L, Klinowska T, Brunton VG (2014) Exploring mechanisms of acquired resistance to HER2 (human epidermal growth factor receptor 2)-targeted therapies in breast cancer. Biochem Soc Trans 42(4):822–830. doi:10.1042/BST20140109
27. Menyhart O, Santarpia L, Gyorffy B (2015) A comprehensive outline of Trastuzumab resistance biomarkers in HER2 overexpressing breast cancer. Curr Cancer Drug Targets 15(8):665–683
28. Hanna WM, Ruschoff J, Bilous M, Coudry RA, Dowsett M, Osamura RY, Penault-Llorca F, van de Vijver M, Viale G (2014) HER2 in situ hybridization in breast cancer: clinical implications of polysomy 17 and genetic heterogeneity. Modern Pathol 27(1):4–18. doi:10.1038/modpathol.2013.103
29. Fiszman GL, Jasnis MA (2011) Molecular mechanisms of Trastuzumab resistance in HER2 overexpressing breast cancer. Int J Breast Cancer 2011:352182. doi:10.4061/2011/352182
30. Tai W, Mahato R, Cheng K (2010) The role of HER2 in cancer therapy and targeted drug delivery. J Control Release 146(3):264–275. doi:10.1016/j.jconrel.2010.04.009
31. Untch M, Loibl S, Bischoff J, Eidtmann H, Kaufmann M, Blohmer JU, Hilfrich J, Strumberg D, Fasching PA, Kreienberg R, Tesch H, Hanusch C, Gerber B, Rezai M, Jackisch C, Huober J, Kuhn T, Nekljudova V, von Minckwitz G, German Breast G, Arbeitsgemeinschaft Gynakologische Onkologie-Breast Study G (2012) Lapatinib versus trastuzumab in combination with neoadjuvant anthracycline-taxane-based chemotherapy (GeparQuinto, GBG 44): a randomised phase 3 trial. Lancet Oncol 13(2):135–144. doi:10.1016/S1470-2045(11)70397-7
32. Oostra DR, Macrae ER (2014) Role of trastuzumab emtansine in the treatment of HER2-positive breast cancer. Breast Cancer 6:103–113. doi:10.2147/BCTT.S67297
33. Dai Q, Ling YH, Lia M, Zou YY, Kroog G, Iwata KK, Perez-Soler R (2005) Enhanced sensitivity to

the HER1/epidermal growth factor receptor tyrosine kinase inhibitor erlotinib hydrochloride in chemotherapy-resistant tumor cell lines. Clin Cancer Res 11(4):1572–1578. doi:10.1158/1078-0432.CCR-04-0993
34. Prat A, Adamo B, Fan C, Peg V, Vidal M, Galvan P, Vivancos A, Nuciforo P, Palmer HG, Dawood S, Rodon J, Ramon y Cajal S, Del Campo JM, Felip E, Tabernero J, Cortes J (2013) Genomic analyses across six cancer types identify basal-like breast cancer as a unique molecular entity. Sci Rep 3:3544. doi:10.1038/srep03544
35. Rimawi MF, Mayer IA, Forero A, Nanda R, Goetz MP, Rodriguez AA, Pavlick AC, Wang T, Hilsenbeck SG, Gutierrez C, Schiff R, Osborne CK, Chang JC (2013) Multicenter phase II study of neoadjuvant lapatinib and trastuzumab with hormonal therapy and without chemotherapy in patients with human epidermal growth factor receptor 2-overexpressing breast cancer: TBCRC 006. J Clin Oncol 31(14):1726–1731. doi:10.1200/JCO.2012.44.8027
36. Zeichner SB, Terawaki H, Gogineni K (2016) A review of systemic treatment in metastatic triple-negative breast cancer. Breast Cancer Basic Clin Res 10:25–36. doi:10.4137/BCBCR.S32783
37. Advani P, Cornell L, Chumsri S, Moreno-Aspitia A (2015) Dual HER2 blockade in the neoadjuvant and adjuvant treatment of HER2-positive breast cancer. Breast Cancer 7:321–335. doi:10.2147/BCTT.S90627
38. Rimawi MF, De Angelis C, Schiff R (2015) Resistance to anti-HER2 therapies in breast cancer. American Society of Clinical Oncology educational book American Society of Clinical Oncology Meeting: e157–164. doi:10.14694/EdBook_AM.2015.35.e157
39. Nahta R (2012) Molecular mechanisms of Trastuzumab-based treatment in HER2-overexpressing breast cancer. ISRN Oncol 2012:428062. doi:10.5402/2012/428062
40. Khan KH, Yap TA, Yan L, Cunningham D (2013) Targeting the PI3K-AKT-mTOR signaling network in cancer. Chin J Cancer 32(5):253–265. doi:10.5732/cjc.013.10057
41. Pohlmann PR, Mayer IA, Mernaugh R (2009) Resistance to Trastuzumab in breast cancer. Clin Cancer Res 15(24):7479–7491. doi:10.1158/1078-0432.CCR-09-0636
42. Carey L, Winer E, Viale G, Cameron D, Gianni L (2010) Triple-negative breast cancer: disease entity or title of convenience? Nat Rev Clin Oncol 7(12):683–692. doi:10.1038/nrclinonc.2010.154
43. Bauer KR, Brown M, Cress RD, Parise CA, Caggiano V (2007) Descriptive analysis of estrogen receptor (ER)-negative, progesterone receptor (PR)-negative, and HER2-negative invasive breast cancer, the so-called triple-negative phenotype: a population-based study from the California cancer registry. Cancer 109(9):1721–1728. doi:10.1002/cncr.22618
44. Brouckaert O, Wildiers H, Floris G, Neven P (2012) Update on triple-negative breast cancer: prognosis and management strategies. Int J Womens Health 4:511–520. doi:10.2147/IJWH.S18541
45. Ali HR, Rueda OM, Chin SF, Curtis C, Dunning MJ, Aparicio SA, Caldas C (2014) Genome-driven integrated classification of breast cancer validated in over 7,500 samples. Genome Biol 15(8):431. doi:10.1186/s13059-014-0431-1
46. Lehmann BD, Bauer JA, Chen X, Sanders ME, Chakravarthy AB, Shyr Y, Pietenpol JA (2011) Identification of human triple-negative breast cancer subtypes and preclinical models for selection of targeted therapies. J Clin Invest 121(7):2750–2767. doi:10.1172/JCI45014
47. Podo F, Buydens LM, Degani H, Hilhorst R, Klipp E, Gribbestad IS, Van Huffel S, van Laarhoven HW, Luts J, Monleon D, Postma GJ, Schneiderhan-Marra N, Santoro F, Wouters H, Russnes HG, Sorlie T, Tagliabue E, Borresen-Dale AL, Consortium F (2010) Triple-negative breast cancer: present challenges and new perspectives. Mol Oncol 4 (3):209–229. doi:10.1016/j.molonc.2010.04.006
48. Shah SP, Roth A, Goya R, Oloumi A, Ha G, Zhao Y, Turashvili G, Ding J, Tse K, Haffari G, Bashashati A, Prentice LM, Khattra J, Burleigh A, Yap D, Bernard V, McPherson A, Shumansky K, Crisan A, Giuliany R, Heravi-Moussavi A, Rosner J, Lai D, Birol I, Varhol R, Tam A, Dhalla N, Zeng T, Ma K, Chan SK, Griffith M, Moradian A, Cheng SW, Morin GB, Watson P, Gelmon K, Chia S, Chin SF, Curtis C, Rueda OM, Pharoah PD, Damaraju S, Mackey J, Hoon K, Harkins T, Tadigotla V, Sigaroudinia M, Gascard P, Tlsty T, Costello JF, Meyer IM, Eaves CJ, Wasserman WW, Jones S, Huntsman D, Hirst M, Caldas C, Marra MA, Aparicio S (2012) The clonal and mutational evolution spectrum of primary triple-negative breast cancers. Nature 486(7403):395–399. doi:10.1038/nature10933
49. Xu H, Eirew P, Mullaly SC, Aparicio S (2014) The omics of triple-negative breast cancers. Clin Chem 60(1):122–133. doi:10.1373/clinchem.2013.207167
50. Burstein MD, Tsimelzon A, Poage GM, Covington KR, Contreras A, Fuqua SA, Savage MI, Osborne CK, Hilsenbeck SG, Chang JC, Mills GB, Lau CC, Brown PH (2015) Comprehensive genomic analysis identifies novel subtypes and targets of triple-negative breast cancer. Clin Cancer Res 21(7):1688–1698. doi:10.1158/1078-0432.CCR-14-0432
51. Petrelli F, Coinu A, Borgonovo K, Cabiddu M, Ghilardi M, Lonati V, Barni S (2014) The value of platinum agents as neoadjuvant chemotherapy in triple-negative breast cancers: a systematic review and meta-analysis. Breast Cancer Res Treat 144(2):223–232. doi:10.1007/s10549-014-2876-z
52. Liedtke C, Mazouni C, Hess KR, Andre F, Tordai A, Mejia JA, Symmans WF, Gonzalez-Angulo AM, Hennessy B, Green M, Cristofanilli M, Hortobagyi GN, Pusztai L (2008) Response to neoadjuvant therapy and long-term survival in patients

with triple-negative breast cancer. J Clin Oncol 26(8):1275–1281. doi:10.1200/JCO.2007.14.4147
53. Kreike B, van Kouwenhove M, Horlings H, Weigelt B, Peterse H, Bartelink H, van de Vijver MJ (2007) Gene expression profiling and histopathological characterization of triple-negative/basal-like breast carcinomas. Breast Cancer Res 9(5):R65. doi:10.1186/bcr1771
54. Masuda H, Baggerly KA, Wang Y, Zhang Y, Gonzalez-Angulo AM, Meric-Bernstam F, Valero V, Lehmann BD, Pietenpol JA, Hortobagyi GN, Symmans WF, Ueno NT (2013) Differential response to neoadjuvant chemotherapy among 7 triple-negative breast cancer molecular subtypes. Clinical Cancer Res 19(19):5533–5540. doi:10.1158/1078-0432.CCR-13-0799
55. Jovanovic B, Beeler JS, Pickup MW, Chytil A, Gorska AE, Ashby WJ, Lehmann BD, Zijlstra A, Pietenpol JA, Moses HL (2014) Transforming growth factor beta receptor type III is a tumor promoter in mesenchymal-stem like triple negative breast cancer. Breast Cancer Res 16(4):R69. doi:10.1186/bcr3684
56. Lehmann BD, Bauer JA, Schafer JM, Pendleton CS, Tang L, Johnson KC, Chen X, Balko JM, Gomez H, Arteaga CL, Mills GB, Sanders ME, Pietenpol JA (2014) PIK3CA mutations in androgen receptor-positive triple negative breast cancer confer sensitivity to the combination of PI3K and androgen receptor inhibitors. Breast Cancer Res 16(4):406. doi:10.1186/s13058-014-0406-x
57. Cancer Genome Atlas N (2012) Comprehensive molecular portraits of human breast tumours. Nature 490(7418):61–70. doi:10.1038/nature11412
58. Lips EH, Mulder L, Oonk A, van der Kolk LE, Hogervorst FB, Imholz AL, Wesseling J, Rodenhuis S, Nederlof PM (2013) Triple-negative breast cancer: BRCAness and concordance of clinical features with BRCA1-mutation carriers. Br J Cancer 108(10):2172–2177. doi:10.1038/bjc.2013.144
59. Couch FJ, Hart SN, Sharma P, Toland AE, Wang X, Miron P, Olson JE, Godwin AK, Pankratz VS, Olswold C, Slettedahl S, Hallberg E, Guidugli L, Davila JI, Beckmann MW, Janni W, Rack B, Ekici AB, Slamon DJ, Konstantopoulou I, Fostira F, Vratimos A, Fountzilas G, Pelttari LM, Tapper WJ, Durcan L, Cross SS, Pilarski R, Shapiro CL, Klemp J, Yao S, Garber J, Cox A, Brauch H, Ambrosone C, Nevanlinna H, Yannoukakos D, Slager SL, Vachon CM, Eccles DM, Fasching PA (2015) Inherited mutations in 17 breast cancer susceptibility genes among a large triple-negative breast cancer cohort unselected for family history of breast cancer. J Clin Oncol 33(4):304–311. doi:10.1200/JCO.2014.57.1414
60. Audeh MW (2014) Novel treatment strategies in triple-negative breast cancer: specific role of poly(adenosine diphosphate-ribose) polymerase inhibition. Pharmgenom Pers Med 7:307–316. doi:10.2147/PGPM.S39765
61. Anders CK, Winer EP, Ford JM, Dent R, Silver DP, Sledge GW, Carey LA (2010) Poly(ADP-ribose) polymerase inhibition: "targeted" therapy for triple-negative breast cancer. Clin Cancer Res 16(19):4702–4710. doi:10.1158/1078-0432.CCR-10-0939
62. Birkbak NJ, Wang ZC, Kim JY, Eklund AC, Li Q, Tian R, Bowman-Colin C, Li Y, Greene-Colozzi A, Iglehart JD, Tung N, Ryan PD, Garber JE, Silver DP, Szallasi Z, Richardson AL (2012) Telomeric allelic imbalance indicates defective DNA repair and sensitivity to DNA-damaging agents. Cancer Discov 2(4):366–375. doi:10.1158/2159-8290.CD-11-0206
63. Naipal KA, Verkaik NS, Ameziane N, van Deurzen CH, Ter Brugge P, Meijers M, Sieuwerts AM, Martens JW, O'Connor MJ, Vrieling H, Hoeijmakers JH, Jonkers J, Kanaar R, de Winter JP, Vreeswijk MP, Jager A, van Gent DC (2014) Functional ex vivo assay to select homologous recombination-deficient breast tumors for PARP inhibitor treatment. Clin Cancer Res 20(18):4816–4826. doi:10.1158/1078-0432.CCR-14-0571
64. Watkins JA, Irshad S, Grigoriadis A, Tutt AN (2014) Genomic scars as biomarkers of homologous recombination deficiency and drug response in breast and ovarian cancers. Breast Cancer Res 16(3):211. doi:10.1186/bcr3670
65. O'Shaughnessy J, Osborne C, Pippen JE, Yoffe M, Patt D, Rocha C, Koo IC, Sherman BM, Bradley C (2011) Iniparib plus chemotherapy in metastatic triple-negative breast cancer. N Engl J Med 364(3):205–214. doi:10.1056/NEJMoa1011418
66. O'Shaughnessy J, Schwartzberg L, Danso MA, Miller KD, Rugo HS, Neubauer M, Robert N, Hellerstedt B, Saleh M, Richards P, Specht JM, Yardley DA, Carlson RW, Finn RS, Charpentier E, Garcia-Ribas I, Winer EP (2014) Phase III study of iniparib plus gemcitabine and carboplatin versus gemcitabine and carboplatin in patients with metastatic triple-negative breast cancer. J Clin Oncol 32(34):3840–3847. doi:10.1200/JCO.2014.55.2984
67. Dent RA, Lindeman GJ, Clemons M, Wildiers H, Chan A, McCarthy NJ, Singer CF, Lowe ES, Watkins CL, Carmichael J (2013) Phase I trial of the oral PARP inhibitor olaparib in combination with paclitaxel for first- or second-line treatment of patients with metastatic triple-negative breast cancer. Breast Cancer Res 15(5):R88. doi:10.1186/bcr3484
68. Gelmon KA, Tischkowitz M, Mackay H, Swenerton K, Robidoux A, Tonkin K, Hirte H, Huntsman D, Clemons M, Gilks B, Yerushalmi R, Macpherson E, Carmichael J, Oza A (2011) Olaparib in patients with recurrent high-grade serous or poorly differentiated ovarian carcinoma or triple-negative breast cancer: a phase 2, multicentre, open-label, non-randomised study. Lancet Oncol 12(9):852–861. doi:10.1016/S1470-2045(11)70214-5
69. Dent P, Tang Y, Yacoub A, Dai Y, Fisher PB, Grant S (2011) CHK1 inhibitors in combination chemo-

therapy: thinking beyond the cell cycle. Mol Interv 11(2):133–140. doi:10.1124/mi.11.2.11
70. Takahashi I, Kobayashi E, Asano K, Yoshida M, Nakano H (1987) UCN-01, a selective inhibitor of protein kinase C from Streptomyces. J Antibiot (Tokyo) 40(12):1782–1784
71. Bunch RT, Eastman A (1996) Enhancement of cisplatin-induced cytotoxicity by 7-hydroxystaurosporine (UCN-01), a new G2-checkpoint inhibitor. Clin Cancer Res 2(5):791–797
72. Ma CX, Cai S, Li S, Ryan CE, Guo Z, Schaiff WT, Lin L, Hoog J, Goiffon RJ, Prat A, Aft RL, Ellis MJ, Piwnica-Worms H (2012) Targeting Chk1 in p53-deficient triple-negative breast cancer is therapeutically beneficial in human-in-mouse tumor models. J Clin Invest 122(4):1541–1552. doi:10.1172/JCI58765
73. Castedo M, Perfettini JL, Roumier T, Yakushijin K, Horne D, Medema R, Kroemer G (2004) The cell cycle checkpoint kinase Chk2 is a negative regulator of mitotic catastrophe. Oncogene 23(25):4353–4361. doi:10.1038/sj.onc.1207573
74. Aarts M, Sharpe R, Garcia-Murillas I, Gevensleben H, Hurd MS, Shumway SD, Toniatti C, Ashworth A, Turner NC (2012) Forced mitotic entry of S-phase cells as a therapeutic strategy induced by inhibition of WEE1. Cancer Discov 2(6):524–539. doi:10.1158/2159-8290.CD-11-0320
75. Akiyama T, Yoshida T, Tsujita T, Shimizu M, Mizukami T, Okabe M, Akinaga S (1997) G1 phase accumulation induced by UCN-01 is associated with dephosphorylation of Rb and CDK2 proteins as well as induction of CDK inhibitor p21/Cip1/WAF1/Sdi1 in p53-mutated human epidermoid carcinoma A431 cells. Cancer Res 57(8):1495–1501
76. Horiuchi D, Kusdra L, Huskey NE, Chandriani S, Lenburg ME, Gonzalez-Angulo AM, Creasman KJ, Bazarov AV, Smyth JW, Davis SE, Yaswen P, Mills GB, Esserman LJ, Goga A (2012) MYC pathway activation in triple-negative breast cancer is synthetic lethal with CDK inhibition. J Exp Med 209(4):679–696. doi:10.1084/jem.20111512
77. Johnson N, Li YC, Walton ZE, Cheng KA, Li D, Rodig SJ, Moreau LA, Unitt C, Bronson RT, Thomas HD, Newell DR, D'Andrea AD, Curtin NJ, Wong KK, Shapiro GI (2011) Compromised CDK1 activity sensitizes BRCA-proficient cancers to PARP inhibition. Nat Med 17(7):875–882. doi:10.1038/nm.2377
78. Vora SR, Juric D, Kim N, Mino-Kenudson M, Huynh T, Costa C, Lockerman EL, Pollack SF, Liu M, Li X, Lehar J, Wiesmann M, Wartmann M, Chen Y, Cao ZA, Pinzon-Ortiz M, Kim S, Schlegel R, Huang A, Engelman JA (2014) CDK 4/6 inhibitors sensitize PIK3CA mutant breast cancer to PI3K inhibitors. Cancer Cell 26(1):136–149. doi:10.1016/j.ccr.2014.05.020
79. Migliaccio I, Di Leo A, Malorni L (2014) Cyclin-dependent kinase 4/6 inhibitors in breast cancer therapy. Curr Opin Oncol 26(6):568–575. doi:10.1097/CCO.0000000000000129
80. Robinson TJ, Liu JC, Vizeacoumar F, Sun T, Maclean N, Egan SE, Schimmer AD, Datti A, Zacksenhaus E (2013) RB1 status in triple negative breast cancer cells dictates response to radiation treatment and selective therapeutic drugs. PLoS One 8(11):e78641. doi:10.1371/journal.pone.0078641
81. Dean JL, McClendon AK, Hickey TE, Butler LM, Tilley WD, Witkiewicz AK, Knudsen ES (2012) Therapeutic response to CDK4/6 inhibition in breast cancer defined by ex vivo analyses of human tumors. Cell Cycle 11(14):2756–2761. doi:10.4161/cc.21195
82. McClendon AK, Dean JL, Rivadeneira DB, Yu JE, Reed CA, Gao E, Farber JL, Force T, Koch WJ, Knudsen ES (2012) CDK4/6 inhibition antagonizes the cytotoxic response to anthracycline therapy. Cell Cycle 11(14):2747–2755. doi:10.4161/cc.21127
83. Curtis C, Shah SP, Chin SF, Turashvili G, Rueda OM, Dunning MJ, Speed D, Lynch AG, Samarajiwa S, Yuan Y, Graf S, Ha G, Haffari G, Bashashati A, Russell R, McKinney S, Group M, Langerod A, Green A, Provenzano E, Wishart G, Pinder S, Watson P, Markowetz F, Murphy L, Ellis I, Purushotham A, Borresen-Dale AL, Brenton JD, Tavare S, Caldas C, Aparicio S (2012) The genomic and transcriptomic architecture of 2,000 breast tumours reveals novel subgroups. Nature 486(7403):346–352. doi:10.1038/nature10983
84. Banerji S, Cibulskis K, Rangel-Escareno C, Brown KK, Carter SL, Frederick AM, Lawrence MS, Sivachenko AY, Sougnez C, Zou L, Cortes ML, Fernandez-Lopez JC, Peng S, Ardlie KG, Auclair D, Bautista-Pina V, Duke F, Francis J, Jung J, Maffuz-Aziz A, Onofrio RC, Parkin M, Pho NH, Quintanar-Jurado V, Ramos AH, Rebollar-Vega R, Rodriguez-Cuevas S, Romero-Cordoba SL, Schumacher SE, Stransky N, Thompson KM, Uribe-Figueroa L, Baselga J, Beroukhim R, Polyak K, Sgroi DC, Richardson AL, Jimenez-Sanchez G, Lander ES, Gabriel SB, Garraway LA, Golub TR, Melendez-Zajgla J, Toker A, Getz G, Hidalgo-Miranda A, Meyerson M (2012) Sequence analysis of mutations and translocations across breast cancer subtypes. Nature 486(7403):405–409. doi:10.1038/nature11154
85. Stephens PJ, Tarpey PS, Davies H, Van Loo P, Greenman C, Wedge DC, Nik-Zainal S, Martin S, Varela I, Bignell GR, Yates LR, Papaemmanuil E, Beare D, Butler A, Cheverton A, Gamble J, Hinton J, Jia M, Jayakumar A, Jones D, Latimer C, Lau KW, McLaren S, McBride DJ, Menzies A, Mudie L, Raine K, Rad R, Chapman MS, Teague J, Easton D, Langerod A, Oslo Breast Cancer C, Lee MT, Shen CY, Tee BT, Huimin BW, Broeks A, Vargas AC, Turashvili G, Martens J, Fatima A, Miron P, Chin SF, Thomas G, Boyault S, Mariani O, Lakhani SR, van de Vijver M, van't Veer L, Foekens J, Desmedt C,

Sotiriou C, Tutt A, Caldas C, Reis-Filho JS, Aparicio SA, Salomon AV, Borresen-Dale AL, Richardson AL, Campbell PJ, Futreal PA, Stratton MR (2012) The landscape of cancer genes and mutational processes in breast cancer. Nature 486(7403):400–404. doi:10.1038/nature11017

86. Corkery B, Crown J, Clynes M, O'Donovan N (2009) Epidermal growth factor receptor as a potential therapeutic target in triple-negative breast cancer. Ann Oncol 20(5):862–867. doi:10.1093/annonc/mdn710
87. Amin DN, Sergina N, Ahuja D, McMahon M, Blair JA, Wang D, Hann B, Koch KM, Shokat KM, Moasser MM (2010) Resiliency and vulnerability in the HER2-HER3 tumorigenic driver. Sci Transl Med 2(16):16ra17. doi:10.1126/scitranslmed.3000389
88. Carey LA, Rugo HS, Marcom PK, Mayer EL, Esteva FJ, Ma CX, Liu MC, Storniolo AM, Rimawi MF, Forero-Torres A, Wolff AC, Hobday TJ, Ivanova A, Chiu WK, Ferraro M, Burrows E, Bernard PS, Hoadley KA, Perou CM, Winer EP (2012) TBCRC 001: randomized phase II study of cetuximab in combination with carboplatin in stage IV triple-negative breast cancer. J Clin Oncol 30(21):2615–2623. doi:10.1200/JCO.2010.34.5579
89. Baselga J, Gomez P, Greil R, Braga S, Climent MA, Wardley AM, Kaufman B, Stemmer SM, Pego A, Chan A, Goeminne JC, Graas MP, Kennedy MJ, Ciruelos Gil EM, Schneeweiss A, Zubel A, Groos J, Melezinkova H, Awada A (2013) Randomized phase II study of the anti-epidermal growth factor receptor monoclonal antibody cetuximab with cisplatin versus cisplatin alone in patients with metastatic triple-negative breast cancer. J Clin Oncol 31(20):2586–2592. doi:10.1200/JCO.2012.46.2408
90. Carvalho I, Milanezi F, Martins A, Reis RM, Schmitt F (2005) Overexpression of platelet-derived growth factor receptor alpha in breast cancer is associated with tumour progression. Breast Cancer Res 7(5):R788–R795. doi:10.1186/bcr1304
91. Coltrera MD, Wang J, Porter PL, Gown AM (1995) Expression of platelet-derived growth factor B-chain and the platelet-derived growth factor receptor beta subunit in human breast tissue and breast carcinoma. Cancer Res 55(12):2703–2708
92. Weigel MT, Dahmke L, Schem C, Bauerschlag DO, Weber K, Niehoff P, Bauer M, Strauss A, Jonat W, Maass N, Mundhenke C (2010) In vitro effects of imatinib mesylate on radiosensitivity and chemosensitivity of breast cancer cells. BMC Cancer 10:412. doi:10.1186/1471-2407-10-412
93. Smith BD (2011) Imatinib for chronic myeloid leukemia: the impact of its effectiveness and long-term side effects. J Natl Cancer Inst 103(7):527–529. doi:10.1093/jnci/djr073
94. Chinchar E, Makey KL, Gibson J, Chen F, Cole SA, Megason GC, Vijayakumar S, Miele L, Gu JW (2014) Sunitinib significantly suppresses the proliferation, migration, apoptosis resistance, tumor angiogenesis and growth of triple-negative breast cancers but increases breast cancer stem cells. Vasc Cell 6:12. doi:10.1186/2045-824X-6-12
95. Yadav BS, Sharma SC, Chanana P, Jhamb S (2014) Systemic treatment strategies for triple-negative breast cancer. World J Clin Oncol 5(2):125–133. doi:10.5306/wjco.v5.i2.125
96. Kumler I, Christiansen OG, Nielsen DL (2014) A systematic review of bevacizumab efficacy in breast cancer. Cancer Treat Rev 40(8):960–973. doi:10.1016/j.ctrv.2014.05.006
97. Meyer AS, Miller MA, Gertler FB, Lauffenburger DA (2013) The receptor AXL diversifies EGFR signaling and limits the response to EGFR-targeted inhibitors in triple-negative breast cancer cells. Sci Signal 6(287):ra66. doi:10.1126/scisignal.2004155
98. Sharpe R, Pearson A, Herrera-Abreu MT, Johnson D, Mackay A, Welti JC, Natrajan R, Reynolds AR, Reis-Filho JS, Ashworth A, Turner NC (2011) FGFR signaling promotes the growth of triple-negative and basal-like breast cancer cell lines both in vitro and in vivo. Clin Cancer Res 17(16):5275–5286. doi:10.1158/1078-0432.CCR-10-2727
99. Turner N, Lambros MB, Horlings HM, Pearson A, Sharpe R, Natrajan R, Geyer FC, van Kouwenhove M, Kreike B, Mackay A, Ashworth A, van de Vijver MJ, Reis-Filho JS (2010) Integrative molecular profiling of triple negative breast cancers identifies amplicon drivers and potential therapeutic targets. Oncogene 29(14):2013–2023. doi:10.1038/onc.2009.489
100. Cerami E, Gao J, Dogrusoz U, Gross BE, Sumer SO, Aksoy BA, Jacobsen A, Byrne CJ, Heuer ML, Larsson E, Antipin Y, Reva B, Goldberg AP, Sander C, Schultz N (2012) The cBio cancer genomics portal: an open platform for exploring multidimensional cancer genomics data. Cancer Discov 2(5):401–404. doi:10.1158/2159-8290.CD-12-0095
101. Craig DW, O'Shaughnessy JA, Kiefer JA, Aldrich J, Sinari S, Moses TM, Wong S, Dinh J, Christoforides A, Blum JL, Aitelli CL, Osborne CR, Izatt T, Kurdoglu A, Baker A, Koeman J, Barbacioru C, Sakarya O, De La Vega FM, Siddiqui A, Hoang L, Billings PR, Salhia B, Tolcher AW, Trent JM, Mousses S, Von Hoff D, Carpten JD (2013) Genome and transcriptome sequencing in prospective metastatic triple-negative breast cancer uncovers therapeutic vulnerabilities. Mol Cancer Ther 12(1):104–116. doi:10.1158/1535-7163.MCT-12-0781
102. Giltnane JM, Balko JM (2014) Rationale for targeting the Ras/MAPK pathway in triple-negative breast cancer. Discov Med 17(95):275–283
103. Wallace MD, Pfefferle AD, Shen L, McNairn AJ, Cerami EG, Fallon BL, Rinaldi VD, Southard TL, Perou CM, Schimenti JC (2012) Comparative oncogenomics implicates the neurofibromin 1 gene (NF1) as a breast cancer driver. Genetics 192(2):385–396. doi:10.1534/genetics.112.142802

104. Balko JM, Cook RS, Vaught DB, Kuba MG, Miller TW, Bhola NE, Sanders ME, Granja-Ingram NM, Smith JJ, Meszoely IM, Salter J, Dowsett M, Stemke-Hale K, Gonzalez-Angulo AM, Mills GB, Pinto JA, Gomez HL, Arteaga CL (2012) Profiling of residual breast cancers after neoadjuvant chemotherapy identifies DUSP4 deficiency as a mechanism of drug resistance. Nat Med 18(7):1052–1059. doi:10.1038/nm.2795
105. Robert C, Dummer R, Gutzmer R, Lorigan P, Kim KB, Nyakas M, Arance A, Liszkay G, Schadendorf D, Cantarini M, Spencer S, Middleton MR (2013) Selumetinib plus dacarbazine versus placebo plus dacarbazine as first-line treatment for BRAF-mutant metastatic melanoma: a phase 2 double-blind randomised study. Lancet Oncol 14(8):733–740. doi:10.1016/S1470-2045(13)70237-7
106. Kirkwood JM, Bastholt L, Robert C, Sosman J, Larkin J, Hersey P, Middleton M, Cantarini M, Zazulina V, Kemsley K, Dummer R (2012) Phase II, open-label, randomized trial of the MEK1/2 inhibitor selumetinib as monotherapy versus temozolomide in patients with advanced melanoma. Clin Cancer Res 18(2):555–567. doi:10.1158/1078-0432.CCR-11-1491
107. Duncan JS, Whittle MC, Nakamura K, Abell AN, Midland AA, Zawistowski JS, Johnson NL, Granger DA, Jordan NV, Darr DB, Usary J, Kuan PF, Smalley DM, Major B, He X, Hoadley KA, Zhou B, Sharpless NE, Perou CM, Kim WY, Gomez SM, Chen X, Jin J, Frye SV, Earp HS, Graves LM, Johnson GL (2012) Dynamic reprogramming of the kinome in response to targeted MEK inhibition in triple-negative breast cancer. Cell 149(2):307–321. doi:10.1016/j.cell.2012.02.053
108. Hoeflich KP, O'Brien C, Boyd Z, Cavet G, Guerrero S, Jung K, Januario T, Savage H, Punnoose E, Truong T, Zhou W, Berry L, Murray L, Amler L, Belvin M, Friedman LS, Lackner MR (2009) In vivo antitumor activity of MEK and phosphatidylinositol 3-kinase inhibitors in basal-like breast cancer models. Clin Cancer Res 15(14):4649–4664. doi:10.1158/1078-0432.CCR-09-0317
109. Mirzoeva OK, Das D, Heiser LM, Bhattacharya S, Siwak D, Gendelman R, Bayani N, Wang NJ, Neve RM, Guan Y, Hu Z, Knight Z, Feiler HS, Gascard P, Parvin B, Spellman PT, Shokat KM, Wyrobek AJ, Bissell MJ, McCormick F, Kuo WL, Mills GB, Gray JW, Korn WM (2009) Basal subtype and MAPK/ERK kinase (MEK)-phosphoinositide 3-kinase feedback signaling determine susceptibility of breast cancer cells to MEK inhibition. Cancer Res 69(2):565–572. doi:10.1158/0008-5472.CAN-08-3389
110. Fedele CG, Ooms LM, Ho M, Vieusseux J, O'Toole SA, Millar EK, Lopez-Knowles E, Sriratana A, Gurung R, Baglietto L, Giles GG, Bailey CG, Rasko JE, Shields BJ, Price JT, Majerus PW, Sutherland RL, Tiganis T, McLean CA, Mitchell CA (2010) Inositol polyphosphate 4-phosphatase II regulates PI3K/Akt signaling and is lost in human basal-like breast cancers. Proc Natl Acad Sci U S A 107(51):22231–22236. doi:10.1073/pnas.1015245107
111. Gewinner C, Wang ZC, Richardson A, Teruya-Feldstein J, Etemadmoghadam D, Bowtell D, Barretina J, Lin WM, Rameh L, Salmena L, Pandolfi PP, Cantley LC (2009) Evidence that inositol polyphosphate 4-phosphatase type II is a tumor suppressor that inhibits PI3K signaling. Cancer Cell 16(2):115–125. doi:10.1016/j.ccr.2009.06.006
112. Gordon V, Banerji S (2013) Molecular pathways: PI3K pathway targets in triple-negative breast cancers. Clin Cancer Res 19(14):3738–3744. doi:10.1158/1078-0432.CCR-12-0274
113. Yunokawa M, Koizumi F, Kitamura Y, Katanasaka Y, Okamoto N, Kodaira M, Yonemori K, Shimizu C, Ando M, Masutomi K, Yoshida T, Fujiwara Y, Tamura K (2012) Efficacy of everolimus, a novel mTOR inhibitor, against basal-like triple-negative breast cancer cells. Cancer Sci 103(9):1665–1671. doi:10.1111/j.1349-7006.2012.02359.x
114. Montero JC, Esparis-Ogando A, Re-Louhau MF, Seoane S, Abad M, Calero R, Ocana A, Pandiella A (2014) Active kinase profiling, genetic and pharmacological data define mTOR as an important common target in triple-negative breast cancer. Oncogene 33(2):148–156. doi:10.1038/onc.2012.572
115. Ganesan P, Moulder S, Lee JJ, Janku F, Valero V, Zinner RG, Naing A, Fu S, Tsimberidou AM, Hong D, Stephen B, Stephens P, Yelensky R, Meric-Bernstam F, Kurzrock R, Wheler JJ (2014) Triple-negative breast cancer patients treated at MD Anderson Cancer Center in phase I trials: improved outcomes with combination chemotherapy and targeted agents. Mol Cancer Ther 13(12):3175–3184. doi:10.1158/1535-7163.MCT-14-0358
116. Ibrahim YH, Garcia-Garcia C, Serra V, He L, Torres-Lockhart K, Prat A, Anton P, Cozar P, Guzman M, Grueso J, Rodriguez O, Calvo MT, Aura C, Diez O, Rubio IT, Perez J, Rodon J, Cortes J, Ellisen LW, Scaltriti M, Baselga J (2012) PI3K inhibition impairs BRCA1/2 expression and sensitizes BRCA-proficient triple-negative breast cancer to PARP inhibition. Cancer Discov 2(11):1036–1047. doi:10.1158/2159-8290.CD-11-0348
117. Juvekar A, Burga LN, Hu H, Lunsford EP, Ibrahim YH, Balmana J, Rajendran A, Papa A, Spencer K, Lyssiotis CA, Nardella C, Pandolfi PP, Baselga J, Scully R, Asara JM, Cantley LC, Wulf GM (2012) Combining a PI3K inhibitor with a PARP inhibitor provides an effective therapy for BRCA1-related breast cancer. Cancer Discov 2(11):1048–1063. doi:10.1158/2159-8290.CD-11-0336
118. Mendes-Pereira AM, Martin SA, Brough R, McCarthy A, Taylor JR, Kim JS, Waldman T, Lord CJ, Ashworth A (2009) Synthetic lethal targeting of PTEN mutant cells with PARP inhibitors. EMBO Mol Med 1(6–7):315–322. doi:10.1002/emmm.200900041

119. Marotta LL, Almendro V, Marusyk A, Shipitsin M, Schemme J, Walker SR, Bloushtain-Qimron N, Kim JJ, Choudhury SA, Maruyama R, Wu Z, Gonen M, Mulvey LA, Bessarabova MO, Huh SJ, Silver SJ, Kim SY, Park SY, Lee HE, Anderson KS, Richardson AL, Nikolskaya T, Nikolsky Y, Liu XS, Root DE, Hahn WC, Frank DA, Polyak K (2011) The JAK2/STAT3 signaling pathway is required for growth of CD44(+)CD24(−) stem cell-like breast cancer cells in human tumors. J Clin Invest 121(7):2723–2735. doi:10.1172/JCI44745
120. Wei W, Tweardy DJ, Zhang M, Zhang X, Landua J, Petrovic I, Bu W, Roarty K, Hilsenbeck SG, Rosen JM, Lewis MT (2014) STAT3 signaling is activated preferentially in tumor-initiating cells in claudin-low models of human breast cancer. Stem Cells 32(10):2571–2582. doi:10.1002/stem.1752
121. Fridman WH, Galon J, Pages F, Tartour E, Sautes-Fridman C, Kroemer G (2011) Prognostic and predictive impact of intra- and peritumoral immune infiltrates. Cancer Res 71(17):5601–5605. doi:10.1158/0008-5472.CAN-11-1316
122. Matsumoto H, Koo SL, Dent R, Tan PH, Iqbal J (2015) Role of inflammatory infiltrates in triple negative breast cancer. J Clin Pathol 68(7):506–510. doi:10.1136/jclinpath-2015-202944
123. Finak G, Bertos N, Pepin F, Sadekova S, Souleimanova M, Zhao H, Chen H, Omeroglu G, Meterissian S, Omeroglu A, Hallett M, Park M (2008) Stromal gene expression predicts clinical outcome in breast cancer. Nat Med 14(5):518–527. doi:10.1038/nm1764
124. Loi S, Michiels S, Salgado R, Sirtaine N, Jose V, Fumagalli D, Kellokumpu-Lehtinen PL, Bono P, Kataja V, Desmedt C, Piccart MJ, Loibl S, Denkert C, Smyth MJ, Joensuu H, Sotiriou C (2014) Tumor infiltrating lymphocytes are prognostic in triple negative breast cancer and predictive for trastuzumab benefit in early breast cancer: results from the FinHER trial. Ann Oncol 25(8):1544–1550. doi:10.1093/annonc/mdu112
125. Liu S, Lachapelle J, Leung S, Gao D, Foulkes WD, Nielsen TO (2012) CD8+ lymphocyte infiltration is an independent favorable prognostic indicator in basal-like breast cancer. Breast Cancer Res 14(2):R48. doi:10.1186/bcr3148
126. West NR, Milne K, Truong PT, Macpherson N, Nelson BH, Watson PH (2011) Tumor-infiltrating lymphocytes predict response to anthracycline-based chemotherapy in estrogen receptor-negative breast cancer. Breast Cancer Res 13(6):R126. doi:10.1186/bcr3072
127. Loi S, Sirtaine N, Piette F, Salgado R, Viale G, Van Eenoo F, Rouas G, Francis P, Crown JP, Hitre E, de Azambuja E, Quinaux E, Di Leo A, Michiels S, Piccart MJ, Sotiriou C (2013) Prognostic and predictive value of tumor-infiltrating lymphocytes in a phase III randomized adjuvant breast cancer trial in node-positive breast cancer comparing the addition of docetaxel to doxorubicin with doxorubicin-based chemotherapy: BIG 02-98. J Clin Oncol 31(7):860–867. doi:10.1200/JCO.2011.41.0902
128. Garcia-Martinez E, Gil GL, Benito AC, Gonzalez-Billalabeitia E, Conesa MA, Garcia Garcia T, Garcia-Garre E, Vicente V, Ayala de la Pena F (2014) Tumor-infiltrating immune cell profiles and their change after neoadjuvant chemotherapy predict response and prognosis of breast cancer. Breast Cancer Res 16(6):488. doi:10.1186/s13058-014-0488-5
129. Denkert C, von Minckwitz G, Brase JC, Sinn BV, Gade S, Kronenwett R, Pfitzner BM, Salat C, Loi S, Schmitt WD, Schem C, Fisch K, Darb-Esfahani S, Mehta K, Sotiriou C, Wienert S, Klare P, Andre F, Klauschen F, Blohmer JU, Krappmann K, Schmidt M, Tesch H, Kummel S, Sinn P, Jackisch C, Dietel M, Reimer T, Untch M, Loibl S (2015) Tumor-infiltrating lymphocytes and response to neoadjuvant chemotherapy with or without carboplatin in human epidermal growth factor receptor 2-positive and triple-negative primary breast cancers. J Clin Oncol 33(9):983–991. doi:10.1200/JCO.2014.58.1967
130. Cabezon T, Gromova I, Gromov P, Serizawa R, Timmermans Wielenga V, Kroman N, Celis JE, Moreira JM (2013) Proteomic profiling of triple-negative breast carcinomas in combination with a three-tier orthogonal technology approach identifies Mage-A4 as potential therapeutic target in estrogen receptor negative breast cancer. Mol Cell Proteomics 12(2):381–394. doi:10.1074/mcp.M112.019786
131. Curigliano G, Viale G, Ghioni M, Jungbluth AA, Bagnardi V, Spagnoli GC, Neville AM, Nole F, Rotmensz N, Goldhirsch A (2011) Cancer-testis antigen expression in triple-negative breast cancer. Ann Oncol 22(1):98–103. doi:10.1093/annonc/mdq325
132. Singh M, Ramos I, Asafu-Adjei D, Quispe-Tintaya W, Chandra D, Jahangir A, Zang X, Aggarwal BB, Gravekamp C (2013) Curcumin improves the therapeutic efficacy of listeria(at)-Mage-b vaccine in correlation with improved T-cell responses in blood of a triple-negative breast cancer model 4T1. Cancer Med 2(4):571–582. doi:10.1002/cam4.94
133. Parinyanitikul N, Blumenschein GR, Wu Y, Lei X, Chavez-Macgregor M, Smart M, Gonzalez-Angulo AM (2013) Mesothelin expression and survival outcomes in triple receptor negative breast cancer. Clin Breast Cancer 13(5):378–384. doi:10.1016/j.clbc.2013.05.001
134. Jacquemier J, Bertucci F, Finetti P, Esterni B, Charafe-Jauffret E, Thibult ML, Houvenaeghel G, Van den Eynde B, Birnbaum D, Olive D, Xerri L (2012) High expression of indoleamine 2,3-dioxygenase in the tumour is associated with medullary features and favourable outcome in basal-like breast carcinoma. Int J Cancer 130(1):96–104. doi:10.1002/ijc.25979

135. Mittendorf EA, Philips AV, Meric-Bernstam F, Qiao N, Wu Y, Harrington S, Su X, Wang Y, Gonzalez-Angulo AM, Akcakanat A, Chawla A, Curran M, Hwu P, Sharma P, Litton JK, Molldrem JJ, Alatrash G (2014) PD-L1 expression in triple-negative breast cancer. Cancer Immunol Res 2(4):361–370. doi:10.1158/2326-6066.CIR-13-0127
136. Sabatier R, Finetti P, Mamessier E, Adelaide J, Chaffanet M, Ali HR, Viens P, Caldas C, Birnbaum D, Bertucci F (2015) Prognostic and predictive value of PDL1 expression in breast cancer. Oncotarget 6(7):5449–5464. doi:10.18632/oncotarget.3216
137. Wimberly H, Brown JR, Schalper K, Haack H, Silver MR, Nixon C, Bossuyt V, Pusztai L, Lannin DR, Rimm DL (2015) PD-L1 expression correlates with tumor-infiltrating lymphocytes and response to neoadjuvant chemotherapy in breast cancer. Cancer Immunol Res 3(4):326–332. doi:10.1158/2326-6066.CIR-14-0133
138. Loi S, Pommey S, Haibe-Kains B, Beavis PA, Darcy PK, Smyth MJ, Stagg J (2013) CD73 promotes anthracycline resistance and poor prognosis in triple negative breast cancer. Proc Natl Acad Sci U S A 110(27):11091–11096. doi:10.1073/pnas.1222251110
139. Allard B, Pommey S, Smyth MJ, Stagg J (2013) Targeting CD73 enhances the antitumor activity of anti-PD-1 and anti-CTLA-4 mAbs. Clin Cancer Res 19(20):5626–5635. doi:10.1158/1078-0432.CCR-13-0545
140. Graves LM, Duncan JS, Whittle MC, Johnson GL (2013) The dynamic nature of the kinome. Biochem J 450(1):1–8. doi:10.1042/BJ20121456
141. Hatzivassiliou G, Liu B, O'Brien C, Spoerke JM, Hoeflich KP, Haverty PM, Soriano R, Forrest WF, Heldens S, Chen H, Toy K, Ha C, Zhou W, Song K, Friedman LS, Amler LC, Hampton GM, Moffat J, Belvin M, Lackner MR (2012) ERK inhibition overcomes acquired resistance to MEK inhibitors. Mol Cancer Ther 11(5):1143–1154. doi:10.1158/1535-7163.MCT-11-1010
142. Logue JS, Morrison DK (2012) Complexity in the signaling network: insights from the use of targeted inhibitors in cancer therapy. Genes Dev 26(7):641–650. doi:10.1101/gad.186965.112
143. Holohan C, Van Schaeybroeck S, Longley DB, Johnston PG (2013) Cancer drug resistance: an evolving paradigm. Nat Rev Cancer 13(10):714–726. doi:10.1038/nrc3599
144. Yap TA, Omlin A, de Bono JS (2013) Development of therapeutic combinations targeting major cancer signaling pathways. J Clin Oncol 31(12):1592–1605. doi:10.1200/JCO.2011.37.6418
145. Chen X, Iliopoulos D, Zhang Q, Tang Q, Greenblatt MB, Hatziapostolou M, Lim E, Tam WL, Ni M, Chen Y, Mai J, Shen H, Hu DZ, Adoro S, Hu B, Song M, Tan C, Landis MD, Ferrari M, Shin SJ, Brown M, Chang JC, Liu XS, Glimcher LH (2014) XBP1 promotes triple-negative breast cancer by controlling the HIF1alpha pathway. Nature 508(7494):103–107. doi:10.1038/nature13119
146. Rottenberg S, Jaspers JE, Kersbergen A, van der Burg E, Nygren AO, Zander SA, Derksen PW, de Bruin M, Zevenhoven J, Lau A, Boulter R, Cranston A, O'Connor MJ, Martin NM, Borst P, Jonkers J (2008) High sensitivity of BRCA1-deficient mammary tumors to the PARP inhibitor AZD2281 alone and in combination with platinum drugs. Proc Natl Acad Sci U S A 105(44):17079–17084. doi:10.1073/pnas.0806092105
147. Jaspers JE, Kersbergen A, Boon U, Sol W, van Deemter L, Zander SA, Drost R, Wientjens E, Ji J, Aly A, Doroshow JH, Cranston A, Martin NM, Lau A, O'Connor MJ, Ganesan S, Borst P, Jonkers J, Rottenberg S (2013) Loss of 53BP1 causes PARP inhibitor resistance in Brca1-mutated mouse mammary tumors. Cancer Discov 3(1):68–81. doi:10.1158/2159-8290.CD-12-0049
148. Ceccaldi R, Liu JC, Amunugama R, Hajdu I, Primack B, Petalcorin MI, O'Connor KW, Konstantinopoulos PA, Elledge SJ, Boulton SJ, Yusufzai T, D'Andrea AD (2015) Homologous-recombination-deficient tumours are dependent on Poltheta-mediated repair. Nature 518(7538):258–262. doi:10.1038/nature14184
149. Sandmann T, Boutros M (2012) Screens, maps & networks: from genome sequences to personalized medicine. Curr Opin Genet Dev 22(1):36–44. doi:10.1016/j.gde.2012.02.001
150. Mishra SK, Bhowmick SS, Chua H, Zhang F, Zheng J (2015) Computational cell fate modelling for discovery of rewiring in apoptotic network for enhanced cancer drug sensitivity. BMC Syst Biol 9(Suppl 1):S4. doi:10.1186/1752-0509-9-S1-S4
151. Muellner MK, Mair B, Ibrahim Y, Kerzendorfer C, Lechtermann H, Trefzer C, Klepsch F, Muller AC, Leitner E, Macho-Maschler S, Superti-Furga G, Bennett KL, Baselga J, Rix U, Kubicek S, Colinge J, Serra V, Nijman SM (2015) Targeting a cell state common to triple-negative breast cancers. Mol Syst Biol 11(1):789. doi:10.15252/msb.20145664
152. Pearl LH, Schierz AC, Ward SE, Al-Lazikani B, Pearl FM (2015) Therapeutic opportunities within the DNA damage response. Nat Rev Cancer 15(3):166–180. doi:10.1038/nrc3891
153. Barabasi AL, Gulbahce N, Loscalzo J (2011) Network medicine: a network-based approach to human disease. Nat Rev Genet 12(1):56–68. doi:10.1038/nrg2918
154. Trape AP, Gonzalez-Angulo AM (2012) Breast cancer and metastasis: on the way toward individualized therapy. Cancer Genomics Proteomics 9(5):297–310
155. Girault I, Bieche I, Lidereau R (2006) Role of estrogen receptor alpha transcriptional coregulators in tamoxifen resistance in breast cancer. Maturitas 54(4):342–351. doi:10.1016/j.maturitas.2006.06.003

156. Vu T, Claret FX (2012) Trastuzumab: updated mechanisms of action and resistance in breast cancer. Front Oncol 2:62. doi:10.3389/fonc.2012.00062
157. Huober J, von Minckwitz G, Denkert C, Tesch H, Weiss E, Zahm DM, Belau A, Khandan F, Hauschild M, Thomssen C, Hogel B, Darb-Esfahani S, Mehta K, Loibl S (2010) Effect of neoadjuvant anthracycline-taxane-based chemotherapy in different biological breast cancer phenotypes: overall results from the GeparTrio study. Breast Cancer Res Treat 124(1):133–140. doi:10.1007/s10549-010-1103-9
158. Kimbung S, Loman N, Hedenfalk I (2015) Clinical and molecular complexity of breast cancer metastases. Semin Cancer Biol 35:85–95. doi:10.1016/j.semcancer.2015.08.009
159. Cardoso F, Costa A, Norton L, Cameron D, Cufer T, Fallowfield L, Francis P, Gligorov J, Kyriakides S, Lin N, Pagani O, Senkus E, Thomssen C, Aapro M, Bergh J, Di Leo A, El Saghir N, Ganz PA, Gelmon K, Goldhirsch A, Harbeck N, Houssami N, Hudis C, Kaufman B, Leadbeater M, Mayer M, Rodger A, Rugo H, Sacchini V, Sledge G, van't Veer L, Viale G, Krop I, Winer E (2012) 1st international consensus guidelines for advanced breast cancer (ABC 1). Breast 21(3):242–252. doi:10.1016/j.breast.2012.03.003
160. Gupta GP, Massague J (2006) Cancer Metastasis: building a framework. Cell 127(4):679–695. doi:10.1016/j.cell.2006.11.001
161. Vanharanta S, Massague J (2013) Origins of metastatic traits. Cancer Cell 24(4):410–421. doi:10.1016/j.ccr.2013.09.007
162. Valastyan S, Weinberg RA (2011) Tumor metastasis: molecular insights and evolving paradigms. Cell 147(2):275–292. doi:10.1016/j.cell.2011.09.024
163. Kang Y, Pantel K (2013) Tumor cell dissemination: emerging biological insights from animal models and cancer patients. Cancer Cell 23(5):573–581. doi:10.1016/j.ccr.2013.04.017
164. Nguyen DX, Bos PD, Massague J (2009) Metastasis: from dissemination to organ-specific colonization. Nat Rev Cancer 9(4):274–284. doi:10.1038/nrc2622
165. Bos PD, Zhang XH, Nadal C, Shu W, Gomis RR, Nguyen DX, Minn AJ, van de Vijver MJ, Gerald WL, Foekens JA, Massague J (2009) Genes that mediate breast cancer metastasis to the brain. Nature 459(7249):1005–1009. doi:10.1038/nature08021
166. Kang Y, Siegel PM, Shu W, Drobnjak M, Kakonen SM, Cordon-Cardo C, Guise TA, Massague J (2003) A multigenic program mediating breast cancer metastasis to bone. Cancer Cell 3(6):537–549
167. Tabaries S, Dong Z, Annis MG, Omeroglu A, Pepin F, Ouellet V, Russo C, Hassanain M, Metrakos P, Diaz Z, Basik M, Bertos N, Park M, Guettier C, Adam R, Hallett M, Siegel PM (2011) Claudin-2 is selectively enriched in and promotes the formation of breast cancer liver metastases through engagement of integrin complexes. Oncogene 30(11):1318–1328. doi:10.1038/onc.2010.518
168. Lu X, Mu E, Wei Y, Riethdorf S, Yang Q, Yuan M, Yan J, Hua Y, Tiede BJ, Lu X, Haffty BG, Pantel K, Massague J, Kang Y (2011) VCAM-1 promotes osteolytic expansion of indolent bone micrometastasis of breast cancer by engaging alpha4beta1-positive osteoclast progenitors. Cancer Cell 20(6):701–714. doi:10.1016/j.ccr.2011.11.002
169. Jin L, Zhang Y, Li H, Yao L, Fu D, Yao X, Xu LX, Hu X, Hu G (2012) Differential secretome analysis reveals CST6 as a suppressor of breast cancer bone metastasis. Cell Res 22(9):1356–1373. doi:10.1038/cr.2012.90
170. Ohtaki T, Shintani Y, Honda S, Matsumoto H, Hori A, Kanehashi K, Terao Y, Kumano S, Takatsu Y, Masuda Y, Ishibashi Y, Watanabe T, Asada M, Yamada T, Suenaga M, Kitada C, Usuki S, Kurokawa T, Onda H, Nishimura O, Fujino M (2001) Metastasis suppressor gene KiSS-1 encodes peptide ligand of a G-protein-coupled receptor. Nature 411(6837):613–617. doi:10.1038/35079135
171. Yin JJ, Selander K, Chirgwin JM, Dallas M, Grubbs BG, Wieser R, Massague J, Mundy GR, Guise TA (1999) TGF-beta signaling blockade inhibits PTHrP secretion by breast cancer cells and bone metastases development. J Clin Invest 103(2):197–206. doi:10.1172/JCI3523
172. Hirbe AC, Morgan EA, Weilbaecher KN (2010) The CXCR4/SDF-1 chemokine axis: a potential therapeutic target for bone metastases? Curr Pharm Des 16(11):1284–1290
173. Krzeszinski JY, Wan Y (2015) New therapeutic targets for cancer bone metastasis. Trends Pharmacol Sci 36(6):360–373. doi:10.1016/j.tips.2015.04.006
174. Siclari VA, Mohammad KS, Tompkins DR, Davis H, McKenna CR, Peng X, Wessner LL, Niewolna M, Guise TA, Suvannasankha A, Chirgwin JM (2014) Tumor-expressed adrenomedullin accelerates breast cancer bone metastasis. Breast Cancer Res 16(6):458. doi:10.1186/s13058-014-0458-y
175. Tamura D, Hiraga T, Myoui A, Yoshikawa H, Yoneda T (2008) Cadherin-11-mediated interactions with bone marrow stromal/osteoblastic cells support selective colonization of breast cancer cells in bone. Int J Oncol 33(1):17–24
176. Sethi N, Dai X, Winter CG, Kang Y (2011) Tumor-derived JAGGED1 promotes osteolytic bone metastasis of breast cancer by engaging notch signaling in bone cells. Cancer Cell 19(2):192–205. doi:10.1016/j.ccr.2010.12.022
177. Zhang XH, Wang Q, Gerald W, Hudis CA, Norton L, Smid M, Foekens JA, Massague J (2009) Latent bone metastasis in breast cancer tied to Src-dependent survival signals. Cancer Cell 16(1):67–78. doi:10.1016/j.ccr.2009.05.017
178. Ell B, Mercatali L, Ibrahim T, Campbell N, Schwarzenbach H, Pantel K, Amadori D, Kang Y (2013) Tumor-induced osteoclast miRNA changes as regulators and biomarkers of osteolytic bone metas-

tasis. Cancer Cell 24(4):542–556. doi:10.1016/j.ccr.2013.09.008
179. Burstein HJ, Lieberman G, Slamon DJ, Winer EP, Klein P (2005) Isolated central nervous system metastases in patients with HER2-overexpressing advanced breast cancer treated with first-line trastuzumab-based therapy. Ann Oncol 16(11):1772–1777. doi:10.1093/annonc/mdi371
180. Sevenich L, Bowman RL, Mason SD, Quail DF, Rapaport F, Elie BT, Brogi E, Brastianos PK, Hahn WC, Holsinger LJ, Massague J, Leslie CS, Joyce JA (2014) Analysis of tumour- and stroma-supplied proteolytic networks reveals a brain-metastasis-promoting role for cathepsin S. Nat Cell Biol 16(9):876–888. doi:10.1038/ncb3011
181. Valiente M, Obenauf AC, Jin X, Chen Q, Zhang XH, Lee DJ, Chaft JE, Kris MG, Huse JT, Brogi E, Massague J (2014) Serpins promote cancer cell survival and vascular co-option in brain metastasis. Cell 156(5):1002–1016. doi:10.1016/j.cell.2014.01.040
182. Armulik A, Genove G, Betsholtz C (2011) Pericytes: developmental, physiological, and pathological perspectives, problems, and promises. Dev Cell 21(2):193–215. doi:10.1016/j.devcel.2011.07.001
183. Oskarsson T, Acharyya S, Zhang XH, Vanharanta S, Tavazoie SF, Morris PG, Downey RJ, Manova-Todorova K, Brogi E, Massague J (2011) Breast cancer cells produce tenascin C as a metastatic niche component to colonize the lungs. Nat Med 17(7):867–874. doi:10.1038/nm.2379
184. Morales M, Arenas EJ, Urosevic J, Guiu M, Fernandez E, Planet E, Fenwick RB, Fernandez-Ruiz S, Salvatella X, Reverter D, Carracedo A, Massague J, Gomis RR (2014) RARRES3 suppresses breast cancer lung metastasis by regulating adhesion and differentiation. EMBO Mol Med 6(7):865–881. doi:10.15252/emmm.201303675
185. Chen Q, Zhang XH, Massague J (2011) Macrophage binding to receptor VCAM-1 transmits survival signals in breast cancer cells that invade the lungs. Cancer Cell 20(4):538–549. doi:10.1016/j.ccr.2011.08.025
186. Padua D, Zhang XH, Wang Q, Nadal C, Gerald WL, Gomis RR, Massague J (2008) TGFbeta primes breast tumors for lung metastasis seeding through angiopoietin-like 4. Cell 133(1):66–77. doi:10.1016/j.cell.2008.01.046
187. Tabaries S, Ouellet V, Hsu BE, Annis MG, Rose AA, Meunier L, Carmona E, Tam CE, Mes-Masson AM, Siegel PM (2015) Granulocytic immune infiltrates are essential for the efficient formation of breast cancer liver metastases. Breast Cancer Res 17:45. doi:10.1186/s13058-015-0558-3
188. Ma R, Feng Y, Lin S, Chen J, Lin H, Liang X, Zheng H, Cai X (2015) Mechanisms involved in breast cancer liver metastasis. J Transl Med 13:64. doi:10.1186/s12967-015-0425-0
189. Palmieri D, Bronder JL, Herring JM, Yoneda T, Weil RJ, Stark AM, Kurek R, Vega-Valle E, Feigenbaum L, Halverson D, Vortmeyer AO, Steinberg SM, Aldape K, Steeg PS (2007) Her-2 overexpression increases the metastatic outgrowth of breast cancer cells in the brain. Cancer Res 67(9):4190–4198. doi:10.1158/0008-5472.CAN-06-3316
190. Li YM, Pan Y, Wei Y, Cheng X, Zhou BP, Tan M, Zhou X, Xia W, Hortobagyi GN, Yu D, Hung MC (2004) Upregulation of CXCR4 is essential for HER2-mediated tumor metastasis. Cancer Cell 6(5):459–469. doi:10.1016/j.ccr.2004.09.027
191. Vadlamudi RK, Sahin AA, Adam L, Wang RA, Kumar R (2003) Heregulin and HER2 signaling selectively activates c-Src phosphorylation at tyrosine 215. FEBS Lett 543(1–3):76–80
192. Yen L, You XL, Al Moustafa AE, Batist G, Hynes NE, Mader S, Meloche S, Alaoui-Jamali MA (2000) Heregulin selectively upregulates vascular endothelial growth factor secretion in cancer cells and stimulates angiogenesis. Oncogene 19(31):3460–3469. doi:10.1038/sj.onc.1203685
193. Marino N, Woditschka S, Reed LT, Nakayama J, Mayer M, Wetzel M, Steeg PS (2013) Breast cancer metastasis: issues for the personalization of its prevention and treatment. Am J Pathol 183(4):1084–1095. doi:10.1016/j.ajpath.2013.06.012
194. Kim RS, Avivar-Valderas A, Estrada Y, Bragado P, Sosa MS, Aguirre-Ghiso JA, Segall JE (2012) Dormancy signatures and metastasis in estrogen receptor positive and negative breast cancer. PLoS One 7(4):e35569. doi:10.1371/journal.pone.0035569
195. Loi S, Michiels S, Baselga J, Bartlett JM, Singhal SK, Sabine VS, Sims AH, Sahmoud T, Dixon JM, Piccart MJ, Sotiriou C (2013) PIK3CA genotype and a PIK3CA mutation-related gene signature and response to everolimus and letrozole in estrogen receptor positive breast cancer. PLoS One 8(1):e53292. doi:10.1371/journal.pone.0053292
196. Stopeck AT, Lipton A, Body JJ, Steger GG, Tonkin K, de Boer RH, Lichinitser M, Fujiwara Y, Yardley DA, Viniegra M, Fan M, Jiang Q, Dansey R, Jun S, Braun A (2010) Denosumab compared with zoledronic acid for the treatment of bone metastases in patients with advanced breast cancer: a randomized, double-blind study. J Clin Oncol 28(35):5132–5139. doi:10.1200/JCO.2010.29.7101
197. Amir E, Clemons M, Purdie CA, Miller N, Quinlan P, Geddie W, Coleman RE, Freedman OC, Jordan LB, Thompson AM (2012) Tissue confirmation of disease recurrence in breast cancer patients: pooled analysis of multi-centre, multi-disciplinary prospective studies. Cancer Treat Rev 38(6):708–714. doi:10.1016/j.ctrv.2011.11.006
198. Wu JM, Fackler MJ, Halushka MK, Molavi DW, Taylor ME, Teo WW, Griffin C, Fetting J, Davidson NE, De Marzo AM, Hicks JL, Chitale D, Ladanyi M, Sukumar S, Argani P (2008) Heterogeneity of breast cancer metastases: comparison of therapeutic target expression and promoter methylation between

primary tumors and their multifocal metastases. Clin Cancer Res 14(7):1938–1946. doi:10.1158/1078-0432.CCR-07-4082
199. Lo Nigro C, Vivenza D, Monteverde M, Lattanzio L, Gojis O, Garrone O, Comino A, Merlano M, Quinlan PR, Syed N, Purdie CA, Thompson A, Palmieri C, Crook T (2012) High frequency of complex TP53 mutations in CNS metastases from breast cancer. Br J Cancer 106(2):397–404. doi:10.1038/bjc.2011.464
200. Wikman H, Lamszus K, Detels N, Uslar L, Wrage M, Benner C, Hohensee I, Ylstra B, Eylmann K, Zapatka M, Sauter G, Kemming D, Glatzel M, Muller V, Westphal M, Pantel K (2012) Relevance of PTEN loss in brain metastasis formation in breast cancer patients. Breast Cancer Res 14(2):R49. doi:10.1186/bcr3150
201. Navin N, Krasnitz A, Rodgers L, Cook K, Meth J, Kendall J, Riggs M, Eberling Y, Troge J, Grubor V, Levy D, Lundin P, Maner S, Zetterberg A, Hicks J, Wigler M (2010) Inferring tumor progression from genomic heterogeneity. Genome Res 20(1):68–80. doi:10.1101/gr.099622.109
202. Stephens PJ, McBride DJ, Lin ML, Varela I, Pleasance ED, Simpson JT, Stebbings LA, Leroy C, Edkins S, Mudie LJ, Greenman CD, Jia M, Latimer C, Teague JW, Lau KW, Burton J, Quail MA, Swerdlow H, Churcher C, Natrajan R, Sieuwerts AM, Martens JW, Silver DP, Langerod A, Russnes HE, Foekens JA, Reis-Filho JS, van't Veer L, Richardson AL, Borresen-Dale AL, Campbell PJ, Futreal PA, Stratton MR (2009) Complex landscapes of somatic rearrangement in human breast cancer genomes. Nature 462 (7276):1005–1010. doi:10.1038/nature08645
203. Li X, Lewis MT, Huang J, Gutierrez C, Osborne CK, Wu MF, Hilsenbeck SG, Pavlick A, Zhang X, Chamness GC, Wong H, Rosen J, Chang JC (2008) Intrinsic resistance of tumorigenic breast cancer cells to chemotherapy. J Natl Cancer Inst 100(9):672–679. doi:10.1093/jnci/djn123
204. Naumov GN, Bender E, Zurakowski D, Kang SY, Sampson D, Flynn E, Watnick RS, Straume O, Akslen LA, Folkman J, Almog N (2006) A model of human tumor dormancy: an angiogenic switch from the nonangiogenic phenotype. J Natl Cancer Inst 98(5):316–325. doi:10.1093/jnci/djj068
205. Naumov GN, Townson JL, MacDonald IC, Wilson SM, Bramwell VH, Groom AC, Chambers AF (2003) Ineffectiveness of doxorubicin treatment on solitary dormant mammary carcinoma cells or late-developing metastases. Breast Cancer Res Treat 82(3):199–206. doi:10.1023/B:BREA.0000004377.12288.3c
206. Quail DF, Joyce JA (2013) Microenvironmental regulation of tumor progression and metastasis. Nat Med 19(11):1423–1437. doi:10.1038/nm.3394
207. Bissell MJ, Hines WC (2011) Why don't we get more cancer? A proposed role of the microenvironment in restraining cancer progression. Nat Med 17(3):320–329. doi:10.1038/nm.2328
208. Al-Ejeh F, Smart CE, Morrison BJ, Chenevix-Trench G, Lopez JA, Lakhani SR, Brown MP, Khanna KK (2011) Breast cancer stem cells: treatment resistance and therapeutic opportunities. Carcinogenesis 32(5):650–658. doi:10.1093/carcin/bgr028
209. Gerber B, Freund M, Reimer T (2010) Recurrent breast cancer: treatment strategies for maintaining and prolonging good quality of life. Deutsches Arzteblatt international 107(6):85–91. doi:10.3238/arztebl.2010.0085
210. Battula VL, Shi Y, Evans KW, Wang RY, Spaeth EL, Jacamo RO, Guerra R, Sahin AA, Marini FC, Hortobagyi G, Mani SA, Andreeff M (2012) Ganglioside GD2 identifies breast cancer stem cells and promotes tumorigenesis. J Clin Invest 122(6):2066–2078. doi:10.1172/JCI59735
211. Wei W, Lewis MT (2015) Identifying and targeting tumor-initiating cells in the treatment of breast cancer. Endocr Relat Cancer 22(3):R135–R155. doi:10.1530/ERC-14-0447
212. Lo PK, Kanojia D, Liu X, Singh UP, Berger FG, Wang Q, Chen H (2012) CD49f and CD61 identify Her2/neu-induced mammary tumor-initiating cells that are potentially derived from luminal progenitors and maintained by the integrin-TGFbeta signaling. Oncogene 31(21):2614–2626. doi:10.1038/onc.2011.439
213. Owens TW, Naylor MJ (2013) Breast cancer stem cells. Front Physiol 4:225. doi:10.3389/fphys.2013.00225
214. Eyler CE, Rich JN (2008) Survival of the fittest: cancer stem cells in therapeutic resistance and angiogenesis. J Clin Oncol 26(17):2839–2845. doi:10.1200/JCO.2007.15.1829
215. Jekimovs C, Bolderson E, Suraweera A, Adams M, O'Byrne KJ, Richard DJ (2014) Chemotherapeutic compounds targeting the DNA double-strand break repair pathways: the good, the bad, and the promising. Front Oncol 4:86. doi:10.3389/fonc.2014.00086
216. Chen K, Huang YH, Chen JL (2013) Understanding and targeting cancer stem cells: therapeutic implications and challenges. Acta Pharmacol Sin 34(6):732–740. doi:10.1038/aps.2013.27
217. Beelen K, Zwart W, Linn SC (2012) Can predictive biomarkers in breast cancer guide adjuvant endocrine therapy? Nat Rev Clin Oncol 9(9):529–541. doi:10.1038/nrclinonc.2012.121
218. Garcia-Becerra R, Santos N, Diaz L, Camacho J (2012) Mechanisms of resistance to endocrine therapy in breast cancer: focus on signaling pathways, miRNAs and genetically based resistance. Int J Mol Sci 14(1):108–145. doi:10.3390/ijms14010108
219. Symmans WF, Hatzis C, Sotiriou C, Andre F, Peintinger F, Regitnig P, Daxenbichler G, Desmedt C, Domont J, Marth C, Delaloge S, Bauernhofer T, Valero V, Booser DJ, Hortobagyi GN, Pusztai L (2010) Genomic index of sensitivity to endocrine therapy for breast cancer. J Clin Oncol 28(27):4111–4119. doi:10.1200/jco.2010.28.4273

220. Chen D, Washbrook E, Sarwar N, Bates GJ, Pace PE, Thirunuvakkarasu V, Taylor J, Epstein RJ, Fuller-Pace FV, Egly JM, Coombes RC, Ali S (2002) Phosphorylation of human estrogen receptor alpha at serine 118 by two distinct signal transduction pathways revealed by phosphorylation-specific antisera. Oncogene 21(32):4921–4931. doi:10.1038/sj.onc.1205420
221. Hartman J, Strom A, Gustafsson JA (2009) Estrogen receptor beta in breast cancer--diagnostic and therapeutic implications. Steroids 74(8):635–641. doi:10.1016/j.steroids.2009.02.005
222. Yang LH, Tseng HS, Lin C, Chen LS, Chen ST, Kuo SJ, Chen DR (2012) Survival benefit of tamoxifen in estrogen receptor-negative and progesterone receptor-positive low grade breast cancer patients. J Breast Cancer 15(3):288–295. doi:10.4048/jbc.2012.15.3.288
223. Dowsett M, Salter J, Zabaglo L, Mallon E, Howell A, Buzdar AU, Forbes J, Pineda S, Cuzick J (2011) Predictive algorithms for adjuvant therapy: TransATAC. Steroids 76(8):777–780. doi:10.1016/j.steroids.2011.02.032
224. Normanno N, Di Maio M, De Maio E, De Luca A, de Matteis A, Giordano A, Perrone F, Group NC-NBC (2005) Mechanisms of endocrine resistance and novel therapeutic strategies in breast cancer. Endocr Relat Cancer 12 (4):721–747. doi:10.1677/erc.1.00857
225. Gee JM, Robertson JF, Gutteridge E, Ellis IO, Pinder SE, Rubini M, Nicholson RI (2005) Epidermal growth factor receptor/HER2/insulin-like growth factor receptor signalling and oestrogen receptor activity in clinical breast cancer. Endocr Relat Cancer 12(Suppl 1):S99–S111. doi:10.1677/erc.1.01005
226. Ligresti G, Militello L, Steelman LS, Cavallaro A, Basile F, Nicoletti F, Stivala F, McCubrey JA, Libra M (2009) PIK3CA mutations in human solid tumors: role in sensitivity to various therapeutic approaches. Cell Cycle 8(9):1352–1358. doi:10.4161/cc.8.9.8255
227. Beelen K, Opdam M, Severson TM, Koornstra RH, Vincent AD, Wesseling J, Muris JJ, Berns EM, Vermorken JB, van Diest PJ, Linn SC (2014) PIK3CA mutations, phosphatase and tensin homolog, human epidermal growth factor receptor 2, and insulin-like growth factor 1 receptor and adjuvant tamoxifen resistance in postmenopausal breast cancer patients. Breast Cancer Res 16(1):R13. doi:10.1186/bcr3606
228. Laplante M, Sabatini DM (2012) mTOR signaling in growth control and disease. Cell 149(2):274–293. doi:10.1016/j.cell.2012.03.017
229. Higgins MJ, Baselga J (2011) Targeted therapies for breast cancer. J Clin Invest 121(10):3797–3803. doi:10.1172/JCI57152
230. Lee JY, Komatsu K, Lee BC, Miyata M, O'Neill Bohn A, Xu H, Yan C, Li JD (2015) Vinpocetine inhibits Streptococcus Pneumoniae-induced upregulation of mucin MUC5AC expression via induction of MKP-1 phosphatase in the pathogenesis of otitis media. J Immunol 194(12):5990–5998. doi:10.4049/jimmunol.1401489
231. Lundgren K, Brown M, Pineda S, Cuzick J, Salter J, Zabaglo L, Howell A, Dowsett M, Landberg G, Ai T (2012) Effects of cyclin D1 gene amplification and protein expression on time to recurrence in postmenopausal breast cancer patients treated with anastrozole or tamoxifen: a TransATAC study. Breast Cancer Res 14(2):R57. doi:10.1186/bcr3161
232. Jerevall PL, Jansson A, Fornander T, Skoog L, Nordenskjold B, Stal O (2010) Predictive relevance of HOXB13 protein expression for tamoxifen benefit in breast cancer. Breast Cancer Res 12(4):R53. doi:10.1186/bcr2612
233. Fox P, Balleine RL, Lee C, Gao B, Balakrishnar B, Menzies AM, Yeap SH, Ali SS, Gebski V, Provan P, Coulter S, Liddle C, Hui R, Kefford R, Lynch J, Wong M, Wilcken N, Gurney H (2016) Dose escalation of tamoxifen in patients with low Endoxifen level: evidence for therapeutic drug monitoring-the TADE study. Clin Cancer Res 22(13):3164–3171. doi:10.1158/1078-0432.CCR-15-1470
234. Nahta R, O'Regan RM (2012) Therapeutic implications of estrogen receptor signaling in HER2-positive breast cancers. Breast Cancer Res Treat 135(1):39–48. doi:10.1007/s10549-012-2067-8
235. Jansson A, Delander L, Gunnarsson C, Fornander T, Skoog L, Nordenskjold B, Stal O (2009) Ratio of 17HSD1 to 17HSD2 protein expression predicts the outcome of tamoxifen treatment in postmenopausal breast cancer patients. Clin Cancer Res 15(10):3610–3616. doi:10.1158/1078-0432.CCR-08-2599
236. Wilken JA, Maihle NJ (2010) Primary trastuzumab resistance: new tricks for an old drug. Ann N Y Acad Sci 1210:53–65. doi:10.1111/j.1749-6632.2010.05782.x
237. Wickenden JA, Watson CJ (2010) Key signalling nodes in mammary gland development and cancer. Signalling downstream of PI3 kinase in mammary epithelium: a play in 3 Akts. Breast Cancer Res 12(2):202. doi:10.1186/bcr2558
238. Kelloff GJ, Lippman SM, Dannenberg AJ, Sigman CC, Pearce HL, Reid BJ, Szabo E, Jordan VC, Spitz MR, Mills GB, Papadimitrakopoulou VA, Lotan R, Aggarwal BB, Bresalier RS, Kim J, Arun B, Lu KH, Thomas ME, Rhodes HE, Brewer MA, Follen M, Shin DM, Parnes HL, Siegfried JM, Evans AA, Blot WJ, Chow WH, Blount PL, Maley CC, Wang KK, Lam S, Lee JJ, Dubinett SM, Engstrom PF, Meyskens FL, Jr., O'Shaughnessy J, Hawk ET, Levin B, Nelson WG, Hong WK, Prevention ATFoC (2006) Progress in chemoprevention drug development: the promise of molecular biomarkers for prevention of intraepithelial neoplasia and cancer--a plan to move forward. Clin Cancer Res 12 (12):3661–3697. doi:10.1158/1078-0432.CCR-06-1104
239. Vinatzer U, Dampier B, Streubel B, Pacher M, Seewald MJ, Stratowa C, Kaserer K, Schreiber M

(2005) Expression of HER2 and the coamplified genes GRB7 and MLN64 in human breast cancer: quantitative real-time reverse transcription-PCR as a diagnostic alternative to immunohistochemistry and fluorescence in situ hybridization. Clin Cancer Res 11(23):8348–8357. doi:10.1158/1078-0432.CCR-05-0841

240. Yardley DA, Kaufman PA, Huang W, Krekow L, Savin M, Lawler WE, Zrada S, Starr A, Einhorn H, Schwartzberg LS, Adams JW, Lie Y, Paquet AC, Sperinde J, Haddad M, Anderson S, Brigino M, Pesano R, Bates MP, Weidler J, Bosserman L (2015) Quantitative measurement of HER2 expression in breast cancers: comparison with 'real-world' routine HER2 testing in a multicenter collaborative biomarker study and correlation with overall survival. Breast Cancer Res 17:41. doi:10.1186/s13058-015-0543-x
241. English DP, Roque DM, Santin AD (2013) HER2 expression beyond breast cancer: therapeutic implications for gynecologic malignancies. Mol Diagn Ther 17(2):85–99. doi:10.1007/s40291-013-0024-9
242. Green AR, Barros FF, Abdel-Fatah TM, Moseley P, Nolan CC, Durham AC, Rakha EA, Chan S, Ellis IO (2014) HER2/HER3 heterodimers and p21 expression are capable of predicting adjuvant trastuzumab response in HER2+ breast cancer. Breast Cancer Res Treat 145(1):33–44. doi:10.1007/s10549-014-2925-7
243. Lantz E, Cunningham I, Higa GM (2010) Targeting HER2 in breast cancer: overview of long-term experience. Int J Womens Health 1:155–171
244. Niikura N, Tomotaki A, Miyata H, Iwamoto T, Kawai M, Anan K, Hayashi N, Aogi K, Ishida T, Masuoka H, Iijima K, Masuda S, Tsugawa K, Kinoshita T, Nakamura S, Tokuda Y (2016) Changes in tumor expression of HER2 and hormone receptors status after neoadjuvant chemotherapy in 21,755 patients from the Japanese breast cancer registry. Ann Oncol 27(3):480–487. doi:10.1093/annonc/mdv611
245. Eiermann W, International Herceptin Study G (2001) Trastuzumab combined with chemotherapy for the treatment of HER2-positive metastatic breast cancer: pivotal trial data. Ann Oncol 12(Suppl 1):S57–S62
246. Carney WP, Bernhardt D, Jasani B (2013) Circulating HER2 extracellular domain: a specific and quantitative biomarker of prognostic value in all breast cancer patients? Biomark Cancer 5:31–39. doi:10.4137/BIC.S12389
247. Tchou J, Lam L, Li YR, Edwards C, Ky B, Zhang H (2015) Monitoring serum HER2 levels in breast cancer patients. Spring 4:237. doi:10.1186/s40064-015-1015-6
248. Christianson TA, Doherty JK, Lin YJ, Ramsey EE, Holmes R, Keenan EJ, Clinton GM (1998) NH2-terminally truncated HER-2/neu protein: relationship with shedding of the extracellular domain and with prognostic factors in breast cancer. Cancer Res 58(22):5123–5129
249. Thongsong N, Setthawatcharawanich S, Sathirapanya P, Limapichat K, Phabphal K (2012) An uncommon cause of compressive myelopathy misdiagnosed as transverse myelitis. J Med Assoc Thail 95(5):727–729
250. Scaltriti M, Rojo F, Ocana A, Anido J, Guzman M, Cortes J, Di Cosimo S, Matias-Guiu X, Ramon y Cajal S, Arribas J, Baselga J (2007) Expression of p95HER2, a truncated form of the HER2 receptor, and response to anti-HER2 therapies in breast cancer. J Natl Cancer Inst 99(8):628–638. doi:10.1093/jnci/djk134
251. Mitra D, Brumlik MJ, Okamgba SU, Zhu Y, Duplessis TT, Parvani JG, Lesko SM, Brogi E, Jones FE (2009) An oncogenic isoform of HER2 associated with locally disseminated breast cancer and trastuzumab resistance. Mol Cancer Ther 8(8):2152–2162. doi:10.1158/1535-7163.MCT-09-0295
252. Yonemori K, Tsuta K, Shimizu C, Hatanaka Y, Hirakawa A, Ono M, Kouno T, Katsumata N, Ando M, Tamura K, Hasegawa T, Kinoshita T, Fujiwara Y (2010) Immunohistochemical expression of HER1, HER3, and HER4 in HER2-positive breast cancer patients treated with trastuzumab-containing neoadjuvant chemotherapy. J Surg Oncol 101(3):222–227. doi:10.1002/jso.21486
253. Ritter CA, Perez-Torres M, Rinehart C, Guix M, Dugger T, Engelman JA, Arteaga CL (2007) Human breast cancer cells selected for resistance to trastuzumab in vivo overexpress epidermal growth factor receptor and ErbB ligands and remain dependent on the ErbB receptor network. Clin Cancer Res 13(16):4909–4919. doi:10.1158/1078-0432.CCR-07-0701
254. Moasser MM, Basso A, Averbuch SD, Rosen N (2001) The tyrosine kinase inhibitor ZD1839 ("Iressa") inhibits HER2-driven signaling and suppresses the growth of HER2-overexpressing tumor cells. Cancer Res 61(19):7184–7188
255. Wang SE, Xiang B, Guix M, Olivares MG, Parker J, Chung CH, Pandiella A, Arteaga CL (2008) Transforming growth factor beta engages TACE and ErbB3 to activate phosphatidylinositol-3 kinase/Akt in ErbB2-overexpressing breast cancer and desensitizes cells to trastuzumab. Mol Cell Biol 28(18):5605–5620. doi:10.1128/MCB.00787-08
256. Narayan M, Wilken JA, Harris LN, Baron AT, Kimbler KD, Maihle NJ (2009) Trastuzumab-induced HER reprogramming in "resistant" breast carcinoma cells. Cancer Res 69(6):2191–2194. doi:10.1158/0008-5472.CAN-08-1056
257. Park YH, Jung HA, Choi MK, Chang W, Choi YL, Do IG, Ahn JS, Im YH (2014) Role of HER3 expression and PTEN loss in patients with HER2-overexpressing metastatic breast cancer (MBC) who received taxane plus trastuzumab treatment. Br J Cancer 110(2):384–391. doi:10.1038/bjc.2013.757
258. Sartor CI, Zhou H, Kozlowska E, Guttridge K, Kawata E, Caskey L, Harrelson J, Hynes N, Ethier

S, Calvo B, Earp HS 3rd (2001) Her4 mediates ligand-dependent antiproliferative and differentiation responses in human breast cancer cells. Mol Cell Biol 21(13):4265–4275. doi:10.1128/MCB.21.13.4265-4275.2001
259. Shattuck DL, Miller JK, Carraway KL 3rd, Sweeney C (2008) Met receptor contributes to trastuzumab resistance of Her2-overexpressing breast cancer cells. Cancer Res 68(5):1471–1477. doi:10.1158/0008-5472.CAN-07-5962
260. Minuti G, Cappuzzo F, Duchnowska R, Jassem J, Fabi A, O'Brien T, Mendoza AD, Landi L, Biernat W, Czartoryska-Arlukowicz B, Jankowski T, Zuziak D, Zok J, Szostakiewicz B, Foszczynska-Kloda M, Tempinska-Szalach A, Rossi E, Varella-Garcia M (2012) Increased MET and HGF gene copy numbers are associated with trastuzumab failure in HER2-positive metastatic breast cancer. Br J Cancer 107(5):793–799. doi:10.1038/bjc.2012.335
261. Zhuang G, Brantley-Sieders DM, Vaught D, Yu J, Xie L, Wells S, Jackson D, Muraoka-Cook R, Arteaga C, Chen J (2010) Elevation of receptor tyrosine kinase EphA2 mediates resistance to trastuzumab therapy. Cancer Res 70(1):299–308. doi:10.1158/0008-5472.can-09-1845
262. Nahta R (2012) Deciphering the role of insulin-like growth factor-I receptor in trastuzumab resistance. Chemother Res Pract 2012:648965. doi:10.1155/2012/648965
263. Joshi JP, Brown NE, Griner SE, Nahta R (2011) Growth differentiation factor 15 (GDF15)-mediated HER2 phosphorylation reduces trastuzumab sensitivity of HER2-overexpressing breast cancer cells. Biochem Pharmacol 82(9):1090–1099. doi:10.1016/j.bcp.2011.07.082
264. Hikita ST, Kosik KS, Clegg DO, Bamdad C (2008) MUC1* mediates the growth of human pluripotent stem cells. PLoS One 3(10):e3312. doi:10.1371/journal.pone.0003312
265. Peiro G, Ortiz-Martinez F, Gallardo A, Perez-Balaguer A, Sanchez-Paya J, Ponce JJ, Tibau A, Lopez-Vilaro L, Escuin D, Adrover E, Barnadas A, Lerma E (2014) Src, a potential target for overcoming trastuzumab resistance in HER2-positive breast carcinoma. Br J Cancer 111(4):689–695. doi:10.1038/bjc.2014.327
266. Zhang S, Huang WC, Li P, Guo H, Poh SB, Brady SW, Xiong Y, Tseng LM, Li SH, Ding Z, Sahin AA, Esteva FJ, Hortobagyi GN, Yu D (2011) Combating trastuzumab resistance by targeting SRC, a common node downstream of multiple resistance pathways. Nat Med 17(4):461–469. doi:10.1038/nm.2309
267. Witkiewicz AK, Cox D, Knudsen ES (2014) CDK4/6 inhibition provides a potent adjunct to Her2-targeted therapies in preclinical breast cancer models. Genes Cancer 5 (7–8):261–272. doi:10.18632/genesandcancer.24
268. Nahta R, Esteva FJ (2006) HER2 therapy: molecular mechanisms of trastuzumab resistance. Breast Cancer Res 8(6):215. doi:10.1186/bcr1612
269. Lee-Hoeflich ST, Pham TQ, Dowbenko D, Munroe X, Lee J, Li L, Zhou W, Haverty PM, Pujara K, Stinson J, Chan SM, Eastham-Anderson J, Pandita A, Seshagiri S, Hoeflich KP, Turashvili G, Gelmon KA, Aparicio SA, Davis DP, Sliwkowski MX, Stern HM (2011) PPM1H is a p27 phosphatase implicated in trastuzumab resistance. Cancer Discov 1(4):326–337. doi:10.1158/2159-8290.CD-11-0062
270. Scaltriti M, Eichhorn PJ, Cortes J, Prudkin L, Aura C, Jimenez J, Chandarlapaty S, Serra V, Prat A, Ibrahim YH, Guzman M, Gili M, Rodriguez O, Rodriguez S, Perez J, Green SR, Mai S, Rosen N, Hudis C, Baselga J (2011) Cyclin E amplification/overexpression is a mechanism of trastuzumab resistance in HER2+ breast cancer patients. Proc Natl Acad Sci U S A 108(9):3761–3766. doi:10.1073/pnas.1014835108
271. Crawford A, Nahta R (2011) Targeting Bcl-2 in Herceptin-resistant breast cancer cell lines. Curr Pharmacogenomics Person Med 9(3):184–190
272. Gong C, Yao Y, Wang Y, Liu B, Wu W, Chen J, Su F, Yao H, Song E (2011) Up-regulation of miR-21 mediates resistance to trastuzumab therapy for breast cancer. J Biol Chem 286(21):19127–19137. doi:10.1074/jbc.M110.216887
273. Ichikawa T, Sato F, Terasawa K, Tsuchiya S, Toi M, Tsujimoto G, Shimizu K (2012) Trastuzumab produces therapeutic actions by upregulating miR-26a and miR-30b in breast cancer cells. PLoS One 7(2):e31422. doi:10.1371/journal.pone.0031422
274. Pernas Simon S (2014) Neoadjuvant therapy of early stage human epidermal growth factor receptor 2 positive breast cancer: latest evidence and clinical implications. Ther Adv Med Oncol 6(5):210–221. doi:10.1177/1758834014535650
275. de Albuquerque A, Kaul S, Breier G, Krabisch P, Fersis N (2012) Multimarker analysis of circulating tumor cells in peripheral blood of metastatic breast cancer patients: a step forward in personalized medicine. Breast Care 7(1):7–12. doi:10.1159/000336548
276. Kallergi G, Papadaki MA, Politaki E, Mavroudis D, Georgoulias V, Agelaki S (2011) Epithelial to mesenchymal transition markers expressed in circulating tumour cells of early and metastatic breast cancer patients. Breast Cancer Res 13(3):R59. doi:10.1186/bcr2896
277. Wallwiener M, Hartkopf AD, Baccelli I, Riethdorf S, Schott S, Pantel K, Marme F, Sohn C, Trumpp A, Rack B, Aktas B, Solomayer EF, Muller V, Janni W, Schneeweiss A, Fehm TN (2013) The prognostic impact of circulating tumor cells in subtypes of metastatic breast cancer. Breast Cancer Res Treat 137(2):503–510. doi:10.1007/s10549-012-2382-0
278. Shields BJ, Wiede F, Gurzov EN, Wee K, Hauser C, Zhu HJ, Molloy TJ, O'Toole SA, Daly RJ, Sutherland RL, Mitchell CA, McLean CA, Tiganis T (2013) TCPTP regulates SFK and STAT3 signaling and is lost in triple-negative breast cancers. Mol Cell Biol 33(3):557–570. doi:10.1128/MCB.01016-12

Studies on DNA Damage Repair and Precision Radiotherapy for Breast Cancer

5

Yanhui Jiang, Yimin Liu, and Hai Hu

Abstract

Radiotherapy acts as an important component of breast cancer management, which significantly decreases local recurrence in patients treated with conservative surgery or with radical mastectomy. On the foundation of technological innovation of radiotherapy setting, precision radiotherapy of cancer has been widely applied in recent years. DNA damage and its repair mechanism are the vital factors which lead to the formation of tumor. Moreover, the status of DNA damage repair in cancer cells has been shown to influence patient response to the therapy, including radiotherapy. Some genes can affect the radiosensitivity of tumor cell by regulating the DNA damage repair pathway. This chapter will describe the potential application of DNA damage repair in precision radiotherapy of breast cancer.

Keywords

Breast cancer • Precision radiotherapy • DNA damage repair

5.1 Introduction

Radiotherapy acts as an important component of breast cancer management, which significantly decreases a local recurrence in patients treated with conservative surgery or with radical mastectomy. Traditionally, the application of radiotherapy in breast cancer is based mainly on histopathological and clinical features. With the emerging of gene expression profiles which has created a new era of treating tumors at molecular level, breast cancer is divided into different subtypes according to tumor behavior and radiation

Y. Jiang • Y. Liu (✉)
Department of Radiotherapy, Sun Yat-sen Memorial Hospital, Sun Yat-sen University, Guangzhou, China
e-mail: liuyimincn@139.com

H. Hu
Sun Yat-sen Memorial Hospital, Sun Yat-sen University, Guangzhou, Guangdong 510120, China
e-mail: huhai@mail.sysu.edu.cn

E. Song, H. Hu (eds.), *Translational Research in Breast Cancer*, Advances in Experimental Medicine and Biology 1026, https://doi.org/10.1007/978-981-10-6020-5_5

response. Thus, personalized radiotherapy which embraces the biological behavior of tumors stands out with the aim to replace the strategy of "one size fits all." The concept of precision radiotherapy, which goes beyond the technical definition, is being accepted by radiation oncologists. The integrity of DNA is essential to maintain genome stability of human. However, diverse elements, including chemical carcinogens, oxidative free radicals, and ionizing radiation, can make damage to DNA, which plays a significant role in the tumorigenesis. In this chapter, we will describe the potential application of DNA damage repair in precision radiotherapy of breast cancer.

5.2 DNA Damage Repair

The integrity of DNA is essential to maintain genome stability, which is primarily important for the survival and proper functioning of all organisms [1]. In eukaryotic cells, the vast majority of DNA is associated with the presence of protein, most compounding the chromosome in the nucleus and few existing in other organelles, such as mitochondria [2]. Composed by guanine, adenine, thymine, cytosine, pentose, and phosphate, disparate base sequence of DNA encodes various different biological information [3–5].

Using DNA as a template for DNA replication based on base pairing is a rigorous and precise event [6]. However, spontaneous errors occasionally happen, whose probability could be quite high. The frequency of base pairing errors is about 10^{-1}–10^{-2}, which could be decreased to about 10^{-5}–10^{-6} with the presence of DNA replication enzyme. If there is an error in the replication process of nucleotide incorporation, DNA polymerase will suspend the catalytic effect with its 3′–5′ exonuclease activity followed by nucleotide excision error, ensuring the veracity during replication [7–10]. If malfunction occurs while DNA replicates, mutation will accumulate or lead to a potentially high risk of cancer or even other diseases [11].

In the meantime, DNA is vulnerable to damage resulting from diverse elements, such as chemical carcinogens from the environment or in the body, oxidative free radicals produced by normal metabolism, telomere shortening, telomerase activity change, proto-oncogene activation, tumor suppressor gene inactivation, ultraviolet (UV) radiation and ionizing radiation, and many pharmaceuticals, especially genotoxic anticancer drugs. Multiple factors attack DNA and induce different levels of lesions. Both spontaneous and environmental lesions can lead to the damage of DNA molecule, including DNA double-strand breaks (DSBs); DNA single-strand breaks (SSBs); mismatched, methylated, and oxidized bases; and intra- and interstrand DNA cross-links. Unfortunately, such lesions can interfere replication and transcription of genome, leading to mutations or wider-scale genome aberrations if they are not repaired accurately [13–15].

In normal cells, most DNA damage can be repaired by triggering the activation of numerous complex and precise regulatory mechanism, including the activation of cell cycle checkpoint, the initiation of repair system, and the induction of cell apoptosis [16]. Furthermore, to properly protect the genome, the cells must detect all types of DNA structural alterations and try their best to repair different forms of damage by diverse repair pathways, including nucleotide excision repair (NER), base excision repair (BER), double-strand breaks (DSBs), and mismatch repair (MMR) [12, 17–20]. DSBs are the most serious damages to cells, normal life activities, and survival of the threatened [21–23]. Activation of repair pathway is rather essential in repairing DNA lesions and maintaining the integrity of the genome of the cell. In addition, the loss of some important components in the repair pathway will increase the probability of mutation and cancer [24]. Previous studies showed that various kinds of DNA damage repairs were defected in some solid tumors [25]. Furthermore, the inaccurate rejoining of broken DNA ends at DSBs may affect the function or survival of the somatic cells, which may affect future generations. Therefore, the ability of biological cells to repair DNA damage is very important in the process of evolution [26, 27]. On the other hand, unrepaired DNA damage can be toxic, promoting pathways of cell elimination such as apoptotic and necrotic death, which is thought to suppress tumor pathways. Therefore, DNA damage can also be used as a means to cure tumor, which is advantageous in anticancer therapies.

Tumor is a kind of cell tissue with unlimited proliferation abilities. DNA damage and its repair mechanism are important factors which lead to the formation of tumor. In the precancerous lesions, DNA damage and DNA damage repair pathways activated by the injury are often detected, with the activation of checkpoint kinases like ATM, Chk1, Chk2, H2AX, p53, and p16 and the formation of DNA marker foci [28–30]. More than half of all human cancers are caused by various forms of p53 inactivation, such as mutation of p53 or its important regulatory factors (such as ATM), or dysfunction of DNA damage repair (DDR) pathway.

Among the many types of DNA damages that exist inside mammalian cells, DSBs may be the most dangerous one. DSBs can result from either exogenous factors such as certain chemotherapeutic drugs and irradiation (IR) or endogenous reactive oxygen species (ROS) [31]. DNA DSBs can be detected by the ATR-interacting protein (ATRIP) or Ku complexes, MRE11/RAD50/NBS1 (MRN), ataxia telangiectasia-mutated (ATM), RAD-3-related (ATR), and DNA-dependent protein kinase catalytic subunit (DNA-PKcs) and BRCA1, respectively [32]. Many downstream proteins, such as p53, are then targeted [34]. In addition, there are at least two major classical pathways for the repair of DSBs, which are homologous recombination (HR) and nonhomologous end joining (NHEJ) [20, 34]. NHEJ is involved in all the phases of the cell cycle. In contrast, the HR is involved in late S and G2 phases of the cell cycle [35–37].

SSBs are also one of the most dangerous lesions as a consequence of spontaneous DNA instability. Their repair is critical because they can terminate gene transcription and generate toxic DSBs during DNA replication [38, 39]. In order to prevent the formation of DSBs, efficient repair of SSBs must be completed before DNA replication, which means that cells need to detect unrepaired SSBs and delay cell cycle progression to allow more time for repair. A recent study has indicated that unrepaired SSBs induced by both spontaneous DNA damage and exogenous agents activate ATM, which then leads to a G1 cell cycle delay that provides additional time for DNA repair before replication and prevents formation of DSBs [40]. Other studies have confirmed that a poly (ADP) ribose polymerase (PARP), primarily PARP1, is involved in the SSBs repair pathway by binding to the broken DNA, which enhances its poly(ADP) ribose (PAR) polymerization activity and induces accumulated PARs' recruitment of XRCC1, a scaffold protein, to further recruit the downstream repair enzymes [41, 42].

5.3 The Precision Radiotherapy for Breast Cancer

Nowadays, with the technological innovation of radiotherapy setting, treatments from simulation to planning to delivery in every cancer phase have been remarkably improved, indicating that the era of the precision radiotherapy is coming. Precision identification is expressed not only in planning but also in delivering treatment, which will have a positive impact on toxicity and local control. The refinement of radiation techniques has allowed for advancement of local control, increase of overall survival, and decrease of radiation-related adverse reactions [43–45]. Continuous efforts have also been made to minimize normal organ toxicity and improve local control in radiotherapy.

The advent of high-precision radiotherapy for cancer is based on the computed tomography (CT) scan which emphasizes the passage from a two-dimensional to a three-dimensional (3D) perspective [46]. With implementation of new treatment planning systems with multi-leaf collimators (MLC) and beam-eye views, the high conformability of 3D radiotherapy not only improves dosimetric accuracy but also ensures precise reconstruction of the complex relationships of adjacent structures and organs. In the process of radiotherapy, hot spots were limited by optimization of dose distribution, which would result in severe late effects. Through precise 3D reconstruction, exposure of the organs at risk (OARs) is reduced, which will pave the way to safe dose escalation and decrease toxicity [47]. After careful analysis of dosimetric data and clinical toxicity data, the complex dose-response relationships can be charted, which will help doctors define the specific tolerance doses for OARs [48]. In the Quantitative Analysis of Normal Tissue Effect in the Clinic (QUANTEC) project, the specific tolerance and dose-effect correlations of most irradiated organs were clear defined [49]. Additionally, models linking tumor control and dose with toxicity were used to predict normal tissue toxicity and evaluate potential treatment outcome.

Accumulating evidence suggests that technological advancement contributes to treatment benefits in increasing overall survival, improving local control, and decreasing toxicity [45, 50]. Local failure rate in 5 years after conservative surgery was about 5% 20 years ago and has become 1% in recent years, which is attributed to the technological advances in aspects such as radiotherapy and surgery [51, 52]. Recent studies have shown that reduced dose to the coronary arteries and myocardium can cause the decrease of cardiac events even when internal mammary nodes have been included in radiotherapy [53].

There has been a continuous development for intensity-modulated radiotherapy (IMRT), stereotactic ablative radiotherapy (SART), and other advanced radiotherapy technology, which can further reduce exposure of OARs with a steep dose gradient. Many studies have confirmed that compared with a 3D technique, IMRT can reduce acute side effects, delay changes in appearance, and lower incidence of side effects in breast cancer patients [53]. With the help of IMRT technique, different dose levels are delivered simultaneously to different target volumes within a single treatment fraction. This method is also known as SIB (simultaneous integrated boost technique), which can be used to yield higher doses to the critical area, but not increase the overall treatment time [54].

In order to reflect individual recurrence risk in high-risk and low-risk breast cancer patients, researchers have investigated the dose modulation effect across the breast. In their studies, 3-level radiation dose was administered, which reflected dose escalation in high-risk breast cancer patients and dose/volume de-escalation in low-risk breast cancer patients [55, 56]. What makes people excited is the excellent treatment outcome of stereotactic ablative radiotherapy (SART) in breast cancer patients, especially in treatment for metastases [57–59]. SART has a lot of advantages, such as offering a high dose/fraction for good local control, exploiting steep dose gradient, and delivering treatment regimen in few fractions, which can provide an optimal quality of life for patients [60].

In high-precision radiotherapy, accurate identification of target and OARs is critical. Adequate contouring also plays a crucial role to ensure the safety and effectiveness of the treatment plan. In clinical practice, the tumor bed is topographically uncertain after operation, and the use of surgical clips to mark the lumpectomy site is recommended by many doctors [61]. However, radiological imaging, including MR, CT, and PET-CT, can provide more useful and accurate information for target volume delineation in high-precision radiotherapy [62, 63]. Nowadays, many consensus guidelines have been put forward to solve the inconsistency of target volume delineation for radiation oncologists. Furthermore, software and anatomic borders have been carefully considered by radiation oncologists in high-precision radiotherapy.

The advances in physics and technology have greatly improved every process of high-precision radiotherapy, especially in treatment execution. Changes including displacement of the target due to organ and anatomical motion can occur in the course of radiotherapy, which would affect dose distribution and lead to an inadequacy of target coverage and damage on the OARs [64, 65]. High-precision radiotherapy can provide high conformability, which means high sensitivity to any changes and will ensure treatment accuracy. The immobilization devices can help to maintain patients in a fixed position and decrease damage on the OARs. Additionally, with the help of various devices, the image-guided radiotherapy

(IGRT) allows treatment to be more accurate. In case of displacements caused by coughing, body relaxation, and other uncontrolled physiological behaviors, monitoring devices should be equipped in the treatment rooms to continuously monitor and compensate for target motion during radiotherapy. With the MR linear accelerator (MR-linac), the target and organs at risk can be monitored more precisely, which can narrow safety margins around the target and lower treated volumes [66].

In spite of the technological advances and various choice of treatment modalities and techniques, the tendency to treat all breast cancer with the same regime, irrespective of tumor biology, is still common [67, 68]. Although modern radiotherapy has achieved great precision in equipment, target, dose, and treatment execution, which brings better local control and decreases toxicity-related mortality, there is still an unmet need for incorporating genomic profiling in the decision-making process. Future radiotherapy should target not only the tumor volume but also the people to select patients who would benefit most from a specific treatment.

Speers et al. have contrived a breast cancer-specific molecular signature of radiation response from in vitro studies in 16 different breast cancer cell lines [69]. They have also developed a radiosensitivity signature (RSS), which comprises of 51 genes involved in cell cycle regulation, DNA damage, and DNA repair, by using a training dataset of 343 early breast cancer patients treated with breast-conserving surgery and adjuvant radiotherapy. This model was subsequently validated in an independent cohort of 228 breast cancer patients and was found to predict patients who would develop locoregional recurrence in 10 years with a sensitivity value of 84% and a negative predictive value of 89%. The RSS is a culmination of comprehensive preclinical work and rigorous statistical analysis, which results in the translation from bench to bedside. It has been found to be independent of biological subtype and outperform traditional clinic-pathological predictors such as grade, tumor size, and nodal status, in predicting local recurrence and overall survival. However, the study had some limitations. Firstly, the training and validation work were performed on nonrandomized datasets of early breast cancer patients who were mostly treated with breast conservation surgery and postoperative radiotherapy [70, 71]. This adds an additional layer of complexity in interpreting the results due to the presence of confounders. Even though multivariate analysis using Cox regression was performed, more robust techniques such as propensity scores could have been adopted to control potential confounders. Secondly, the patient cohorts were treated more than a decade ago, in a premodern era of radiotherapy, and before the common use of third-generation systemic therapy and trastuzumab, all of which have impact on risk of recurrence.

As mentioned above, the biological subtype did not exert impact on the risk of recurrence when RSS was included in the analysis, which contradicts present studies of similar field in developing radiation-specific assays. Recently, a radiotherapy-specific multigene expression model, also known as the radiation sensitivity index (RSI), was developed to predict radiation responsiveness [72]. When combined with intrinsic subtypes and age, RSI was able to identify patients who would least benefit from radiotherapy (ER negative/RSI resistant) or who need a dose escalation approach (luminal/RSI resistant). And for the triple-negative (TN) subtype, subpopulations with a different risk of local recurrence (LR) (RSI sensitive and RSI resistant) were detected, which indicates that personalized radiotherapy may be considered for this subgroup. Accumulating evidence suggests that the diversity of breast cancer calls for specific radiation strategies. Kyndi et al. retrospectively performed a study to examine the importance of estrogen receptor (ER), progesterone receptor (PgR), human epidermal growth factor receptor 2 (HER2), and constructed subtypes in a large clinical trial randomly assigning patients to receive or not receive postmastectomy radiotherapy (PMRT) [73]. The results showed that luminal tumors benefitted the most from radiotherapy, while the HER2-positive and TN subtypes were less likely to exhibit a reduction in LR due to the radiotherapy. In this trial population, a subsequent gene expression analysis on the frozen tumor samples identified a 7-gene predictive model for LR dubbed as the "DBCG-radiotherapy" profile, which could predict the benefit of PMRT more accurately than molecular subtypes [74]. Furthermore, the DBCG-RT profile has identified a group of patients with low risk of local recurrence rate (LRR) and no additional benefit from PMRT among all the subtypes, including a number of luminal A patients who would not experience any benefit from PMRT when compared to the no-PMRT randomized group, which is in contrast with the previous analysis.

There are some studies supporting the omission of radiotherapy in highly selected patients, who are women aged 50 or over, presenting T1 N0, low-intermediate grade, hormone-sensitive and HER2-negative symptoms, and planning to receive endocrine therapy. Favorable tumor profiles have been identified either by immunohistochemical, or Oncotype DX (Genomic Health, Redwood City, CA) recurrence score (RS), or Prosigna PAM50 assay [75].

By contrast, the luminal B subtypes is characterized by more aberrant genomes. It has presented a higher risk of LR and benefited more from radiotherapy. The addition of radiotherapy did not always bring better oncologic outcome, highlighting the need to investigate alternative regimens. It was found that the expression of Ki-67 was the only factor associated with the effect of axillary radiotherapy on the risk of recurrence [76, 77]. In patients not receiving axillary dissection, high Ki-67 acted as a successful indication for axillary radiotherapy, which benefited disease-free survival [78].

TN and HER2-positive subtypes were reported to have an increased risk of developing LR, irrespective of the type of surgery [79]. However, the outcome for HER2-positive subtypes is greatly improved with the use of trastuzumab. In preclinical studies, the combination of trastuzumab with radiation enhances the radiosensitivity of breast cancer cell lines by inhibiting DNA repair and stimulating the death of tumor cell [80]. In clinical practice, when adding trastuzumab into radiotherapy, improved outcome has been observed with no increased toxicity. In HER2-positive patients receiving trastuzumab, the LR was far lower compared to TN patients. In addition, whole breast radiotherapy resulted in optimal local control. Accelerated partial breast irradiation (APBI) or even mastectomy without postoperative radiotherapy worsened outcome, which indicated that extensive radiotherapy should be scheduled for this subtype [81].

Basal-like breast cancers tend to relapse early and present with poor disease-free and overall survival. There is now no robust evidence supporting that basal-like breast cancer patients should be administered with different strategy of radiotherapy [82, 83]. Currently, APBI is not suggested for treatment of basal-like/TN breast cancer according to ASTRO/ESTRO (American Society for Radiation Oncology/ European Society for Therapeutic Radiology and Oncology) due to its increased risk of LR [84, 86]. More aggressive treatments such as larger radiation fields or higher prescription dose are routine approach for this subtype by radiation oncologists. Thus, a basal-like/TN phenotype might be given a supraclavicular field if the axilla has not been dissected, and there are one or two positive sentinel nodes [87]. Even in early-stage breast cancer of this type [88], PMRT is always suggested.

In ductal carcinoma in situ (DCIS), biomarkers can be used to guide the selection of adjuvant treatment. High Ki-67 tumors benefit the most from radiotherapy, irrespective of nuclear grade and necrosis [89], whereas radiotherapy has been found to have no effect on luminal A DCIS with Ki-67 <14%. HER2-positive DCIS also benefits from radiotherapy, showing a significant decrease in all local recurrences [90]. Based on the results of the Eastern Cooperative Oncology Group E5194 trial, which enrolled low-/intermediate-grade DCIS or small high-grade DCIS, the 12-gene panel, known as the Oncotype DX DCIS Score, was able to discriminate between a low and a high risk of LR [91].

In clinical practice, to predict toxicity of patients is an important work for radiation oncologists, and predicting toxicity is affected by individual variability and many other factors. Meanwhile, alternative radiation schedules, target agents, and other innovative therapies increase the complexity of predicting toxicity of radiotherapy. Therefore, to further study predictive markers of radiation-induced toxicity can help identify patients who will suffer from severe normal tissue reactions to radiation, which will facilitate the design of personalized radiotherapy. Currently, the mechanism of radiation-induced toxicity is largely unknown. Previous studies have suggested that some late toxicities induced by radiation are associated with a low rate of radiation-induced CD8 T-lymphocyte apoptosis [92, 93], and the patients with more single-nucleotide polymorphisms in candidate genes will suffer more from radiation-induced late effects [92, 93]. However, more studies should be done to determine which patients are likely to suffer more from later normal tissue injury.

5.4 The Application of DNA Damage in Radiotherapy of Breast Cancer

The status of DNA damage repair in cancer cells has been shown to influence patient response to therapeutic schedules, including radiotherapy and chemotherapy. Moreover, specific DNA repair pathways are more active in certain cancers, and alterations of these pathways may characterize particular cancer types. In this part, we will discuss specific DNA damage repair machinery detected in breast cancer and present the current stage of development of various DNA damage response inhibitors for breast cancer mono- and combination therapy.

5.4.1 BRCA1 and BRCA2

Germline mutations in DNA repair genes confer an increased risk of familial breast cancer development. It has been shown that around 55–65% women inheriting a deleterious heterozygous BRCA1 mutation and approximately 45% inheriting a BRCA2 mutation will develop breast cancer by the age of 70 [94]. These mutations contribute to a relative risk of breast cancer 10–30 times that of women in the general population, resulting in a nearly 85% lifetime risk of breast cancer development.

BRCA1 and BRCA2 are critical in the process of HR repair of double-strand DNA breaks. BRCA1 binds to DSBs through its association with the abraxas–RAP80 macro-complex, which associates with ubiquitylated histones at DNA DSBs [95], and this process is dependent on phosphorylation of histone H2AX (γH2AX), mediator of DNA damage checkpoint protein 1 (MDC1) and RING finger protein 8 (RNF8). Next, BRCA1 is involved in processing DSBs through its interaction with CtIP and the MRN complex (which is comprised of MRE11, RAD50, and NBS1). The BRCA1–CtIP complex promotes CtIP-mediated 5′-end resection of DSBs [96], which is abrogated by three independent tumor-associated mutations in the BRCT domain of BRCA1 [97, 98]. BRCA1 is also required for RAD51 recruitment to the sites of DNA damage through its interactions with PALB2 and BRCA2. Additionally, BRCA1 has been shown to play a critical role in transcriptional regulation of NER and NHEJ [99]. Studies show that the deletion of BRCA1 and BRCA2 can significantly inhibit the efficiency of HR and increase the sensitivity of cells to DNA damage drugs, and BRCA1-deficient cells were shown to be sensitive to methyl methanesulfonate (MMS). To date, multiple DNA damage response markers have been examined in a series of BRCA-mutated tumors compared to sporadic breast cancer.

Irradiation can cause a large number of DNA DSBs in tumor. As a consequence of this defect in HR, tumors that arise in BRCA carriers are likely to be more sensitive to ionizing radiation [100]. Nowadays, a lot of evidence supporting the role of BRCA1 in radioresistance has been obtained from cell and animal models. Human cells with BRCA1 and BRCA2 mutations contribute to enhanced radiosensitivity by an impaired proliferative capacity after irradiation [101]. The transfection of wild-type BRCA1 in the BRCA1−/− human breast cancer line can decrease the irradiation sensitivity and increase the efficiency of DSBs repair [102]. The ovarian cancer cell line with defective BRCA1 has an increased sensitivity compared to parental cell lines [103]. However, inconsistent with the results of experimental models, current clinical studies have failed to demonstrate the association between BRCA mutations and the prognosis of breast cancers.

5.4.2 PARP

Poly (ADP-ribose) polymerases (PARPs) are nuclear proteins that function in single-strand DNA repair through base excision and represent a major alternative DNA repair pathway in the cell [104, 105]. Therefore, they are especially essential for SSB repair.

The PARP family of proteins is defined by the catalytic capacity to modify target proteins by the covalent addition of chains of poly(ADP-ribose) polymers. PARP-1 is by far the most abundant one among the PARP family. PARP-1 and PARP-2 possess DNA-binding domains, and their catalytic function is activated when bond to sites of DNA damage. By this mechanism, they function by detecting the presence of damaged DNA and activating signaling pathways that promote appropriate cellular response. PARP-1 is relatively abundant in nucleus. In response to DNA damage, it binds rapidly to DNA breaks, a process that activates its catalytic function resulting in the modification of histones and proteins associated with SSB repair, DSB repair, and DNA replication. Such modifications are transient because pADPr polymers are rapidly degraded by the action of poly glycohydrolase. PARP-1 function in both single- and double-strand DNA breaks. Binding of PARP-1 to SSBs, either being directly induced by BER or as intermediate products of BER, can protect the damaged site from improper recombination events. In addition, PARP-1 induces the recruitment and activation of various components of the BER repair complex, most notably the scaffold XRCC1, by directly interacting with poly (ADP-ribosylation).

PARP inhibitors (PARPi), which inhibit the PARP enzyme, are emerging as potentially valuable drugs in treatment of advanced breast cancer, particularly in hereditary breast cancer. PARP inhibitors are promising therapy for BRCA-associated breast cancer [106–108]. Ionizing radiation of cancer generates SSBs and DSBs in an approximately ratio of 25:1. The major influence of PARPi is to delay but not abolish the repair of SSBs. DSBs are the most critical cytotoxic lesions, while the effect of SSB repair has a small impact on the survival of non-replicating cells. PARPi induces the levels of unrepaired DSB in replicating cells by at least two mechanisms. Firstly, the delayed repair of radiation-induced SSB increases the probability of unrepaired lesions colliding with the DNA replication machinery, producing excess DSBs. Additionally, the inhibition of the catalytic activity of PARP prevents PARP from modifying itself by the addition of pADRr. Unmodified PARP remains bound to sites of DNA damage. Therefore, PARPi inhibits downstream processes, increasing the toxicity of DSB which is produced in their presence.

Hirai showed that treatment with a PARPi enhanced the cytotoxic effect of gamma-irradiation. PARP inhibitor treatment induced S phase arrest and enhanced subsequent G2/M arrest after irradiation, which caused sensitization to irradiation [109]. And the PARPi impairing the efficacy of DNA break repair has been exploited to enhance the cytotoxicity of anticancer drugs and radiotherapy. Multiple in vivo and in vitro studies using PARPi in conjunction with radiotherapy have been shown to decrease tumor proliferation or prolong animal survival in various cancers. Furthermore, PARP-1/2 inhibitors have been demonstrated to be effective in preclinical models in combination with platinum, alkylating and methylation agents, topoisomerase I inhibitors, and radiation therapy [110]. In breast cancer, MK-487, a PARP-1/2 inhibitor, was demonstrated to strongly enhance response of human cancer xenografts to radiation. Additionally, new drugs targeting PARPis have been developed. These compounds include olaparib, niraparib, rucaparib, veliparib, talazoparib, and Nu1025.

Olaparib (AZD 2281, Ku59436), an orally active PARP inhibitor, was licensed for treatment of ovarian cancer by both the European Medicines Agency and the US Food and Drug Administration in 2014. In the phase I trial, olaparib was given to ovarian, breast, and prostate cancer patients who were BRCA1 or BRCA2 mutation carriers. The antitumor effects were observed in 9 of 19 patients, while no response was seen in control non-BRCA mutation carriers (41 patients) [111]. A maximum tolerated dose of 400 mg olaparib twice daily was determined, and toxicities of less than grade 3 were observed in both mutation- and non-mutation-carrying patients [112]. A phase II trial of olaparib in BRCA-associated breast cancer then showed a response rate of 41% at 400 mg twice daily and 22% at 100 mg twice daily, with PFS of 5.7 months and 3.8 months at 400 mg and 100 mg, respectively [113]. In another phase II trial, 298 patients with various recurrent cancers (ovarian, breast, prostate, and pancreatic) and confirmed BRCA1/2 mutations were treated with olaparib. In the 62 breast cancer patients, tumor response rate was 12.9%, and 47% of patients had disease stabilization for >8 weeks [114]. However, the phase I trial studying the effect of olaparib on the prognosis of cancer patients, especially breast cancer patients, treated with radiotherapy, is still in progress.

Rucaparib is also a FDA-approved monotherapy for treating patients with deleterious BRCA mutation-associated advanced ovarian cancer who have been treated with two or more chemotherapies. A phase I study of intravenous and oral rucaparib in combination with chemotherapy in patients with advanced solid tumors (22 breast, 15 ovarian/peritoneal, and 48 other primary cancers) demonstrated that rucaparib/carboplatin combination had radiologic antitumor activity, primarily in BRCA1- or BRCA2-mutated breast and ovarian/peritoneal cancers [115]. These studies have demonstrated that PARPi will be an essential therapy for BRCA-associated breast cancer. Further comprehension of the mechanism of action of PARPi, and the relationship between changes in gene expression and homologous recombination function in a breast cancer background, is likely to shed light on developing effective precision radiotherapy to bring the most benefits to breast cancer populations.

5.4.3 Estrogen Receptor α

It is evident that high levels of estradiol in blood are associated with higher risk of development of breast cancer in postmenopausal women [115]. Estrogen receptor α (ERα) signaling induces genomic instability by interfering with DNA damage response and DNA repair effector kinases. On one hand, ERα signaling influences ATM/Chk2 and ATR/Chk1 pathways, which are principal regulators of cell cycle arrest, following DNA double-strand or single-strand breaks. ERα suppresses ATM kinase expression through upregulating miR-18a and miR-106a [116]. Overexpression of ATM correlates with local recurrence and radiotherapy resistance in ERα breast cancer. Similarly, ERα inhibits ATR activation and ATR-Chk1 signaling to G2/M phase of cell cycle progression. Furthermore, ERα interferes with ATR activation at the sites of DNA damage through inhibiting the interplay between ATR and TopBP1. On the other hand, ERα positively affects DNA-PKcs which is a core component of NHEJ. ERα can bind to DNA-PKcs and facilitate its stabilization and activation of DNA damage response via the NEHJ pathway. Overall, these results demonstrate that ERα expression is crucial for DNA damage response of breast cancer and suggests the impact of ERα in radiotherapy of breast cancer.

ERα has been recognized not only as a poor prognostic factor of breast cancer but also as an important therapeutic target for breast cancer, which is of great significance in the clinical diagnosis and treatment of breast cancer. Breast cancer with ER-driven tumor growth is responsive to endocrine therapy. The antiestrogen tamoxifen, which competes with estrogen for binding to ER, has been the backbone of adjuvant endocrine therapy for early breast cancer [117]. However, not all patients with ER-positive tumors benefit from endocrine therapy. Therefore, adjuvant chemotherapy followed by adjuvant endocrine therapy is recommended for most of ER-positive breast cancer patients. Chisamore [118] found that a synthetic estrogen receptor-related receptor α (ERRα)-specific antagonist can inhibit the signal transduction of breast cancer. And Liu [119] found that IC1182780, a ERRα antagonist, can inhibit breast cancer cell proliferation. By far, fighting against ERα has become one of the most important ways in the comprehensive treatment of breast cancer. Moreover, ERα antagonists have been widely used in treatment of breast cancer. However, besides the comprehensive studies of how ER statute affects endocrine therapy and chemotherapy, the role of ER in radiotherapy has not been fully understood. ER- negative tumor cells also need to be further studied.

5.4.4 DNA-PK

DNA protein kinase (DNA-PK), a member of the PIKK family of serine/threonine protein kinases, plays a significant role in NHEJ and IR-induced DSBs in human cells. DNA protein kinase C (DNA-PKcs) has relevance to multiple cancer types, most likely because of the role of NHEJ pathway in repair of damage [120, 121]. DNA-free ends are firstly bound by the Ku70/80 heterodimer, resulting in recruitment and activation of DNA-PKcs. DNA-PKcs stimulates end-processing enzymes such as the nuclease artemis, polynucleotide kinase phosphorylase, DNA polymerases, and MRE11, to form a complex with DNA that is bound and religated by the XRCC4/DNA ligase IV. PARP is suggested to have a role in inhibiting NHEJ pathway [122]. A substitutable end-joining pathway, which is slower and involves ligase III, XRCC1, PNK, and PARP1 [123], is also existent. Active DNA-PK, a target of ATM and ATR, is comprised of a catalytic subunit (DNA-PKcs) and a regulatory heterodimer (Ku70 and Ku80 subunits). Reduced DNA-PKcs expression in breast cancer was implicated in higher tumor grade, higher mitotic index, tumor de-differentiation, and poor patient survival.

DNA-PK has been studied as a target for cancer therapy, mainly in radiotherapy sensitization. DSB is the most important reason for tumor radiotherapy. And the maximum cytotoxicity of radiotherapy is caused by DSBs. Therefore, the expression or activity of DNA-PK may affect the radiosensitivity of tumor cell by reducing the repair of DSBs. Previous studies have shown that inhibiting the expression or activity of DNA-PK can increase the radiosensitivity of the cells [124–128].

DNA-PKcs is the catalytic site of DNA-PK, and it is also the main target of the research of radiosensitization. The DNA-PKcs inhibitors can inhibit expression of mRNA and protein, the kinase activity, as well as the binding site with Ku, hereby preventing the formation of the complex. The most classical inhibitors of DNA-PKcs are wortmannin and LY294002, both of which are noncompetitive small molecule inhibitors targeting the ATP binding site of DNA-PKcs. The mechanism of wortmannin is to promote the irreversible alkylation of 802 lysine, so as to prevent phosphorylation of the DNA-PKcs, which can make the cell repair ineffective. Wortmannin can be sensitive to radiation [129], but it is unstable and difficult to dissolve in the aqueous solution. Moreover, its toxic side effects are obvious, which seriously affects its clinical application. Meanwhile, LY294002 can inhibit Ras signal transduction pathway. It has been suggested that NU7441, a reforming complex based on LY294002, can induce the sensitivity of breast cancer cells to irradiation and doxorubicin [129].

5.4.5 APE1

Apurinic/apyrimidinic endonuclease1 (APE1) is a ubiquitous multifunctional DNA repair enzyme and a redox signaling protein. It is critical for BER pathway, and its deregulation has been demonstrated in multiple tumors. APE1 can act as an apurinic/apyrimidinic endonuclease, 3′-5′ exonuclease, 3′-phosphatase, and 3′-phosphodiesterase, as well as a redox activator of major transcription factors [130, 131]. R237A, L104R, and E126D were APE1 variants, which are involved in 40%–60% reduction in DNA repair activity in biochemical assays [132–135]. Abdel-Fatah TM et al. examined APE1 levels in 1285 breast cancer [136] and found that downregulated APE1 expression was correlated with aggressive histological features and triple-negative phenotype [121].

APE1 is highly expressed in a variety of malignant tumors, including breast cancer, and affects the occurrence, development, and prognosis of tumors. APE1 inhibitors are synthetically lethal in ATM- and BRCA2-deficient cell lines of breast cancer [137, 138]. And the inhibition of APE1 redox function might have therapeutic potential by modulating cell migration and invasion in metastatic breast cancer. Meanwhile, the expression of APE1 has been demonstrated to be important in regulating of the radiosensitivity of cervical carcinoma and osteosarcoma [139, 140]. However, there are still no published researches demonstrating the association between APE1 and breast cancer radiotherapy.

5.4.6 ATM and ATR

The ATM gene spans approximately 160 Kb of genomic DNA containing 66 exons and encodes ATM protein, a serine/threonine kinase mainly involved in DNA damage response pathways following DNA double-strand breaks. ATM and ATR play significant roles in DNA damage response. ATM is one of the central kinases involved in the cellular response to DNA DSB, which may arise through the collapse of stalled replication forks or through exposure to ionizing radiation [141]. ATM plays a significant role in the activation of the G1/S cell cycle checkpoint. It also phosphorylates p53 directly in response to the induction of DNA DSB. Checkpoint kinase 2 (CHK2), a key downstream target of ATM and mediator of ATM signaling, can also phosphorylate p53, which leads to its stabilization by preventing its Mdm2-mediated ubiquitinylation and degradation [142]. Furthermore, ATM contributes to the accumulation and stabilization of p53 by directly phosphorylating Mdm2 [143]. Once activated, ATR acts via its downstream targets to promote DNA repair, stabilization, and restart of stalled replication forks and transient cell cycle arrest [144]. Many of these functions are mediated through the ATR downstream target CHK1.

ATM is a known tumor suppressor which is frequently mutated in a broad range of human cancers including lung, colorectal, and breast cancer. Upregulated expression of ATR and cytoplasmic pChk1 was dramatically correlated with advanced tumor stage, higher mitotic index, pleomorphism, and lymphovascular and poor prognosis. Downregulated ATM levels were implicated in aggressive breast cancer including larger size tumors, higher tumor grade, higher mitotic index, pleomorphism, tumor type, lymphovascular invasion, ER–, PR–, AR–, triple negative, and basal-like phenotypes [145]. Further studies demonstrate that ATM or ATR inhibition is synthetically lethal in breast cancer cells with XRCC1 and ERCC1 deficiency [146]. Meanwhile, ATR inhibition is also effective in Ras, MRE11, and DNA-PK-overexpressing cells. The gene product of ATM plays a crucial role in biological response to ionizing radiation. It is involved in the detection of DNA double-strand breaks and initiation of pathways that lead to cycle arrest followed by DNA repair or apoptosis [141]. ATM-mediated apoptosis occurs primarily through controlling posttranslational modifications of p53 and phosphorylation of Mdm2, an ubiquitin ligase which binds to p53 in its native unphosphorylated state, thereby targeting the tumor suppressor protein for degradation [147]. Therefore, ATM and ATR protein kinases are critical intermediates in a number of cellular responses to ionizing radiation.

According to preclinical studies, inactivation of ATM-CHK2 and/or ATR-CHK2 pathways boosts the anticancer activity of multiple therapeutic agents. This sensitization was proven to be particularly successful in tumor cell lines defective for p53 or p53 signaling [148–151]. Up to now, inhibitors of ATR and CHK1 are the only two classes of compounds that have entered clinical trials. Preliminary phase I studies demonstrated that inhibitors of CHK1 (LY2603618, MK-8776, UCN-01, and CBP501) are well tolerated in individuals with advanced solid tumors or lymphomas [152–155]. Furthermore, in a phase I dose-escalation study, ATR inhibitor (AZD7762) demonstrated cardiac dose-limiting toxicities in patients with advanced solid tumors and observation that arrested the further development of this agent [156]. Additionally, a phase II trial showed that UCN-01 induced serious adverse effects, including anemia, neutropenia, vomiting, and fatigue in patients with hematologic neoplasms. In this trial, 27% of patients had a partial or complete response upon two cycles of intravenous infusion of UCN-01 (http://www.clinicaltrials.gov). Contrarily, UCN-01 didn't show marked antitumor activity as a stand-alone agent in two phase II trials performed in patients with renal cell carcinoma or metastatic melanoma [157, 158]. The results of several ongoing clinical trials testing the safety and antineoplastic activity of ATR or CHK inhibitors in cancer patients are likely to inform the future development of tumor treatment with ATR or CHK1 inhibitors. The clinical profile of LY2606368 is being investigated in patients with advanced solid tumors (NCT01115790) and breast or ovarian cancers (NCT02203513), whereas AZD6738 (ATR inhibitor) is being employed in patients with advanced solid tumors (NCT02223923), alone or together with radiotherapy. Moreover, the pharmacokinetics and pharmacodynamics of the CHK1 inhibitor GDC-0575 are being tested in patients with refractory solid tumors or lymphomas (NCT01564251) (http://www.clinicaltrials.gov).

Pharmacological inhibitors of the ATM protein kinase have been proposed as radiosensitizing agents since the inherited ataxia-telangiectasia syndrome results in a profound hypersensitivity to ionizing radiation [159–161]. KU-55933 is a novel, specific, and potent inhibitor of the ATM kinase, which significantly sensitizes cells to the cytotoxic effects of ionizing radiation [160]. The suitability of another nontoxic, specific, and rapidly reversible ATM inhibitor CP466722 for the purpose of radiosensitization has been reported recently [161]. CHK1 is another promising molecular target to enhance the cytotoxic effects of radiotherapy in treatment of certain cancers. CHK1 plays a major role in mediating S and G2 arrest in response to DNA damage. Inhibition of CHK1 enhances the cytotoxicity of DNA-damaging agents by ionizing through aggregation of these cell cycle checkpoints. Preclinical studies on radiosensitization have fulfilled the development of a range of pharmacological CHK1 inhibitors [162, 163]. Moreover, a clinical trial employing AZD6738 (ATR inhibitor) in patients with advanced solid tumors (NCT02223923), alone or together with radiotherapy, is underway (http://www.clinicaltrials.gov).

5.5 Conclusion

DNA damage is a universal characteristic of breast cancer cells from premalignant to invasive stages. Therefore, the use of DNA repair inhibitors, either single or in combinations, is promising. The optimal combination of each of these inhibitors with radiotherapy has yet to be determined for breast cancer due to limited clinical data. To achieve the most benefits from DNA damage response therapies, studies on functionally relevant biomarkers are crucial for developing truly personalized medicine.

Acknowledgments This work has been supported by grants from the National Key R&D Program (2016YFC1302301), Natural Science Foundation of China (81672738).

References

1. Tubbs A, Nussenzweig A (2017) Endogenous DNA damage as a source of genomic instability in cancer. Cell 168:644–656
2. Lukas J, Lukas C, Bartek J (2011) More than just a focus: the chromatin response to DNA damage and its role in genome integrity maintenance. Nat Cell Biol 13:1161–1169
3. Aboussekhra A, Biggerstaff M, Shivji MK, Vilpo JA, Moncollin V, Podust VN, Protic M, Hubscher U, Egly JM, Wood RD (1995) Mammalian DNA nucleotide excision repair reconstituted with purified protein components. Cell 80:859–868
4. Bartkova J, Horejsi Z, Koed K, Kramer A, Tort F, Zieger K, Guldberg P, Sehested M, Nesland JM, Lukas C, Orntoft T, Lukas J, Bartek J (2005) DNA damage response as a candidate anti-cancer barrier in early human tumorigenesis. Nature 434:864–870
5. Jackson SP, Bartek J (2009) The DNA-damage response in human biology and disease. Nature 461:1071–1078

6. Wojciechowski F, Leumann CJ (2011) Alternative DNA base-pairs: from efforts to expand the genetic code to potential material applications. Chem Soc Rev 40:5669–5679
7. Aparicio OM (2013) Location, location, location: It's all in the timing for replication origins. Genes Dev 27:117–128
8. Gilbert DM (2010) Evaluating genome-scale approaches to eukaryotic DNA replication. Nat Rev Genet 11:673–684
9. MacAlpine DM, Almouzni G (2013) Chromatin and DNA replication. Cold Spring Harb Perspect Biol 5:a10207
10. Mechali M, Yoshida K, Coulombe P, Pasero P (2013) Genetic and epigenetic determinants of DNA replication origins, position and activation. Curr Opin Genet Dev 23:124–131
11. Mechali M (2010) Eukaryotic DNA replication origins: many choices for appropriate answers. Nat Rev Mol Cell Biol 11:728–738
12. Lagerwerf S, Vrouwe MG, Overmeer RM, Fousteri MI, Mullenders LH (2011) DNA damage response and transcription. DNA Repair (Amst) 10:743–750
13. Hubscher U, Maga G (2011) DNA replication and repair bypass machines. Curr Opin Chem Biol 15:627–635
14. Larrea AA, Lujan SA, Nick MS, Mieczkowski PA, Resnick MA, Gordenin DA, Kunkel TA (2010) Genome-wide model for the normal eukaryotic DNA replication fork. Proc Natl Acad Sci U S A 107:17674–17679
15. Lehmann AR (2003) Replication of damaged DNA. Cell Cycle 2:300–302
16. Malumbres M (2011) Physiological relevance of cell cycle kinases. Physiol Rev 91:973–1007
17. Brady CA, Jiang D, Mello SS, Johnson TM, Jarvis LA, Kozak MM, Kenzelmann BD, Basak S, Park EJ, McLaughlin ME, Karnezis AN, Attardi LD (2011) Distinct p53 transcriptional programs dictate acute DNA-damage responses and tumor suppression. Cell 145:571–583
18. Fotouhi A, Cornella N, Ramezani M, Wojcik A, Haghdoost S (2015) Investigation of micronucleus induction in MTH1 knockdown cells exposed to UVA, UVB or UVC. Mutat Res Genet Toxicol Environ Mutagen 793:161–165
19. Joo W, Xu G, Persky NS, Smogorzewska A, Rudge DG, Buzovetsky O, Elledge SJ, Pavletich NP (2011) Structure of the FANCI-FANCD2 complex: insights into the Fanconi anemia DNA repair pathway. Science 333:312–316
20. Sobol RW, Horton JK, Kuhn R, Gu H, Singhal RK, Prasad R, Rajewsky K, Wilson SH (1996) Requirement of mammalian DNA polymerase-beta in base-excision repair. Nature 379:183–186
21. Chiolo I, Minoda A, Colmenares SU, Polyzos A, Costes SV, Karpen GH (2011) Double-strand breaks in heterochromatin move outside of a dynamic HP1a domain to complete recombinational repair. Cell 144:732–744
22. Staszewski O, Baker RE, Ucher AJ, Martier R, Stavnezer J, Guikema JE (2011) Activation-induced cytidine deaminase induces reproducible DNA breaks at many non-Ig loci in activated B cells. Mol Cell 41:232–242
23. Zhang J, Powell SN (2005) The role of the BRCA1 tumor suppressor in DNA double-strand break repair. Mol Cancer Res 3:531–539
24. Ali R, Rakha EA, Madhusudan S, Bryant HE (2017) DNA damage repair in breast cancer and its therapeutic implications. Pathology 49:156–165
25. Lord CJ, Ashworth A (2012) The DNA damage response and cancer therapy. Nature 481:287–294
26. Deem A, Keszthelyi A, Blackgrove T, Vayl A, Coffey B, Mathur R, Chabes A, Malkova A (2011) Break-induced replication is highly inaccurate. PLoS Biol 9:e1000594
27. Rai R, Peng G, Li K, Lin SY (2007) DNA damage response: the players, the network and the role in tumor suppression. Cancer Genomics Proteomics 4:99–106
28. Abraham RT (2001) Cell cycle checkpoint signaling through the ATM and ATR kinases. Genes Dev 15:2177–2196
29. Helmink BA, Tubbs AT, Dorsett Y, Bednarski JJ, Walker LM, Feng Z, Sharma GG, McKinnon PJ, Zhang J, Bassing CH, Sleckman BP (2011) H2AX prevents CtIP-mediated DNA end resection and aberrant repair in G1-phase lymphocytes. Nature 469:245–249
30. Lee JH, Choy ML, Ngo L, Venta-Perez G, Marks PA (2011) Role of checkpoint kinase 1 (Chk1) in the mechanisms of resistance to histone deacetylase inhibitors. Proc Natl Acad Sci U S A 108:19629–19634
31. Mills KD, Ferguson DO, Alt FW (2003) The role of DNA breaks in genomic instability and tumorigenesis. Immunol Rev 194:77–95
32. Hurley PJ, Bunz F (2007) ATM and ATR: components of an integrated circuit. Cell Cycle 6:414–417
33. Coutts AS, La Thangue N (2006) The p53 response during DNA damage: impact of transcriptional cofactors. Biochem Soc Symp 73:181–189
34. Takata M, Sasaki MS, Sonoda E, Morrison C, Hashimoto M, Utsumi H, Yamaguchi-Iwai Y, Shinohara A, Takeda S (1998) Homologous recombination and non-homologous end-joining pathways of DNA double-strand break repair have overlapping roles in the maintenance of chromosomal integrity in vertebrate cells. EMBO J 17:5497–5508
35. Rothkamm K, Krüger I, Thompson LH, Löbrich M (2003) Pathways of DNA double-strand break repair during the mammalian cell cycle. Mol Cell Biol 23:5706–5715
36. Sjögren C, Ström L (2010) S-phase and DNA damage activated establishment of sister chromatid cohesion—importance for DNA repair. Exp Cell Res 316:1445–1453

37. Bauerschmidt C, Arrichiello C, Burdak-Rothkamm S, Woodcock M, Hill MA, Stevens DL, Rothkamm K (2010) Cohesin promotes the repair of ionizing radiation-induced DNA double-strand breaks in replicated chromatin. Nucleic Acids Res 38:477–487
38. Kuzminov A (2001) Single-strand interruptions in replicating chromosomes cause double-strand breaks. Proc Natl Acad Sci U S A 98:8241–8246
39. Caldecott KW (2001) Mammalian DNA single-strand break repair: an X-ra(y)ted affair. Bioessays 23:447–455
40. Khoronenkova SV, Dianov GL (2015) ATM prevents DSB formation by coordinating SSB repair and cell cycle progression. Proc Natl Acad Sci U S A 112:3997–4002
41. Caldecott KW (2014) DNA single-strand break repair. Exp Cell Res 329:2–8
42. Caldecott KW (2014) Protein ADP-ribosylation and the cellular response to DNA strand breaks. DNA Repair 19:108–113
43. Darby S, McGale P, Correa C, Taylor C, Arriagada R, Clarke M, Cutter D, Davies C, Ewertz M, Godwin J, Gray R, Pierce L, Whelan T, Wang Y, Peto R (2011) Effect of radiotherapy after breast-conserving surgery on 10-year recurrence and 15-year breast cancer death: meta-analysis of individual patient data for 10,801 women in 17 randomised trials. Lancet 378:1707–1716
44. Clarke M, Collins R, Darby S, Davies C, Elphinstone P, Evans V, Godwin J, Gray R, Hicks C, James S, MacKinnon E, McGale P, McHugh T, Peto R, Taylor C, Wang Y (2005) Effects of radiotherapy and of differences in the extent of surgery for early breast cancer on local recurrence and 15-year survival: an overview of the randomised trials. Lancet 366:2087–2106
45. (2000) Favourable and unfavourable effects on long-term survival of radiotherapy for early breast cancer: an overview of the randomised trials. Early breast cancer Trialists' collaborative group. Lancet 355:1757–1770
46. Thariat J, Hannoun-Levi JM, Sun MA, Vuong T, Gerard JP (2013) Past, present, and future of radiotherapy for the benefit of patients. Nat Rev Clin Oncol 10:52–60
47. Graham P (2014) Cardiac dosimetry for adjuvant left-sided breast radiotherapy: patterns with 2D-versus 3D-era planning and correlates of coronary dose with maximum depth of myocardial exposure. J Med Imaging Radiat Oncol 58:517–522
48. Sugano Y, Mizuta M, Takao S, Shirato H, Sutherland KL, Date H (2015) Optimization of the fractionated irradiation scheme considering physical doses to tumor and organ at risk based on dose-volume histograms. Med Phys 42:6203–6210
49. Jackson A, Marks LB, Bentzen SM, Eisbruch A, Yorke ED, Ten HR, Constine LS, Deasy JO (2010) The lessons of QUANTEC: recommendations for reporting and gathering data on dose-volume dependencies of treatment outcome. Int J Radiat Oncol Biol Phys 76:S155–S160
50. Nagai A, Shibamoto Y, Yoshida M, Inoda K, Kikuchi Y (2017) Intensity-modulated radiotherapy using two static ports of tomotherapy for breast cancer after conservative surgery: Dosimetric comparison with other treatment methods and 3-year clinical results. J Radiat Res:1–8
51. Botteri E, Bagnardi V, Rotmensz N, Gentilini O, Disalvatore D, Bazolli B, Luini A, Veronesi U (2010) Analysis of local and regional recurrences in breast cancer after conservative surgery. Ann Oncol 21:723–728
52. Veronesi U, Cascinelli N, Mariani L, Greco M, Saccozzi R, Luini A, Aguilar M, Marubini E (2002) Twenty-year follow-up of a randomized study comparing breast-conserving surgery with radical mastectomy for early breast cancer. N Engl J Med 347:1227–1232
53. Whelan TJ, Olivotto IA, Parulekar WR, Ackerman I, Chua BH, Nabid A, Vallis KA, White JR, Rousseau P, Fortin A, Pierce LJ, Manchul L, Chafe S, Nolan MC, Craighead P, Bowen J, McCready DR, Pritchard KI, Gelmon K, Murray Y, Chapman JA, Chen BE, Levine MN (2015) Regional nodal irradiation in early-stage breast cancer. N Engl J Med 373:307–316
54. Orlandi E, Palazzi M, Pignoli E, Fallai C, Giostra A, Olmi P (2010) Radiobiological basis and clinical results of the simultaneous integrated boost (SIB) in intensity modulated radiotherapy (IMRT) for head and neck cancer: a review. Crit Rev Oncol Hematol 73:111–125
55. Coles C, Agrawal R, Ah-See ML, Algurafi H, Alhasso A, Brunt AM, Chan C, Griffin C, Harnett A, Hopwood P (2016) Partial breast radiotherapy for women with early breast cancer: first results of local recurrence data for IMPORT LOW (CRUK/06/003). Eur J Cancer 57:S4
56. Donovan EM, Ciurlionis L, Fairfoul J, James H, Mayles H, Manktelow S, Raj S, Tsang Y, Tywman N, Yarnold J, Coles C (2011) Planning with intensity-modulated radiotherapy and tomotherapy to modulate dose across breast to reflect recurrence risk (IMPORT high trial). Int J Radiat Oncol Biol Phys 79:1064–1072
57. Rusthoven KE, Kavanagh BD, Burri SH, Chen C, Cardenes H, Chidel MA, Pugh TJ, Kane M, Gaspar LE, Schefter TE (2009) Multi-institutional phase I/II trial of stereotactic body radiation therapy for lung metastases. J Clin Oncol 27:1579–1584
58. Rusthoven KE, Kavanagh BD, Cardenes H, Stieber VW, Burri SH, Feigenberg SJ, Chidel MA, Pugh TJ, Franklin W, Kane M, Gaspar LE, Schefter TE (2009) Multi-institutional phase I/II trial of stereotactic body radiation therapy for liver metastases. J Clin Oncol 27:1572–1578
59. Milano MT, Zhang H, Metcalfe SK, Muhs AG, Okunieff P (2009) Oligometastatic breast cancer treated with curative-intent stereotactic body radiation therapy. Breast Cancer Res Treat 115:601–608

60. Scorsetti M, Franceschini D, De Rose F, Comito T, Villa E, Iftode C, Navarria P, D'Agostino GR, Masci G, Torrisi R, Testori A, Tinterri C, Santoro A (2016) Stereotactic body radiation therapy: a promising chance for oligometastatic breast cancer. Breast 26:11–17
61. Sung S, Lee JH, Lee JH, Kim SH, Kwak YK, Lee SW, Jeon YW, Suh YJ (2016) Displacement of surgical clips during postoperative radiotherapy in breast cancer patients who received breast-conserving surgery. J Breast Cancer 19:417–422
62. Cho O, Chun M, Oh YT, Kim MH, Park HJ, Heo JS, Noh OK (2013) Can initial diagnostic PET-CT aid to localize tumor bed in breast cancer radiotherapy: feasibility study using deformable image registration. Radiat Oncol 8:163
63. Giezen M, Kouwenhoven E, Scholten AN, Coerkamp EG, Heijenbrok M, Jansen WP, Mast ME, Petoukhova AL, Struikmans H (2011) Magnetic resonance imaging- versus computed tomography-based target volume delineation of the glandular breast tissue (clinical target volume breast) in breast-conserving therapy: an exploratory study. Int J Radiat Oncol Biol Phys 81:804–811
64. Conroy L, Yeung R, Watt E, Quirk S, Long K, Hudson A, Phan T, Smith WL (2016) Evaluation of target and cardiac position during visually monitored deep inspiration breath-hold for breast radiotherapy. J Appl Clin Med Phys 17:25–36
65. van Heijst TC, Philippens ME, Charaghvandi RK, den Hartogh MD, Lagendijk JJ, van den Bongard HJ, van Asselen B (2016) Quantification of intra-fraction motion in breast radiotherapy using supine magnetic resonance imaging. Phys Med Biol 61:1352–1370
66. Lagendijk JJ, Raaymakers BW, Raaijmakers AJ, Overweg J, Brown KJ, Kerkhof EM, van der Put RW, Hardemark B, van Vulpen M, van der Heide UA (2008) MRI/linac integration. Radiother Oncol 86:25–29
67. Calvo FA, Sole CV, Rivera S, Meirino R, Lizarraga S, Infante MA, Boldo E, Ferrer C, Marsiglia H, Deutsch E (2014) The use of radiotherapy for early breast cancer in woman at different ages. Clin Transl Oncol 16:680–685
68. Marta GN, de Moraes FY (2015) Postoperative nodal irradiation in breast cancer patients with 1 to 3 axillary lymph nodes involved: The debate continues... Expert Rev Anticancer Ther 15:1257–1259
69. Speers C, Zhao S, Liu M, Bartelink H, Pierce LJ, Feng FY (2015) Development and validation of a novel radiosensitivity signature in human breast cancer. Clin Cancer Res 21:3667–3677
70. Servant N, Bollet MA, Halfwerk H, Bleakley K, Kreike B, Jacob L, Sie D, Kerkhoven RM, Hupe P, Hadhri R, Fourquet A, Bartelink H, Barillot E, Sigal-Zafrani B, van de Vijver MJ (2012) Search for a gene expression signature of breast cancer local recurrence in young women. Clin Cancer Res 18:1704–1715
71. van de Vijver MJ, He YD, Van'T VL, Dai H, Hart AA, Voskuil DW, Schreiber GJ, Peterse JL, Roberts C, Marton MJ, Parrish M, Atsma D, Witteveen A, Glas A, Delahaye L, van der Velde T, Bartelink H, Rodenhuis S, Rutgers ET, Friend SH, Bernards R (2002) A gene-expression signature as a predictor of survival in breast cancer. N Engl J Med 347:1999–2009
72. Torres-Roca JF, Fulp WJ, Caudell JJ, Servant N, Bollet MA, van de Vijver M, Naghavi AO, Harris EE, Eschrich SA (2015) Integration of a radiosensitivity molecular signature into the assessment of local recurrence risk in breast cancer. Int J Radiat Oncol Biol Phys 93:631–638
73. Kyndi M, Sorensen FB, Knudsen H, Overgaard M, Nielsen HM, Overgaard J (2008) Estrogen receptor, progesterone receptor, HER-2, and response to post-mastectomy radiotherapy in high-risk breast cancer: the Danish breast cancer cooperative group. J Clin Oncol 26:1419–1426
74. Tramm T, Kyndi M, Myhre S, Nord S, Alsner J, Sorensen FB, Sorlie T, Overgaard J (2014) Relationship between the prognostic and predictive value of the intrinsic subtypes and a validated gene profile predictive of loco-regional control and benefit from post-mastectomy radiotherapy in patients with high-risk breast cancer. Acta Oncol 53:1337–1346
75. Bellon JR (2015) Personalized radiation oncology for breast cancer: the new frontier. J Clin Oncol 33:1998–2000
76. Colleoni M, Rotmensz N, Peruzzotti G, Maisonneuve P, Viale G, Renne G, Casadio C, Veronesi P, Intra M, Torrisi R, Goldhirsch A (2004) Minimal and small size invasive breast cancer with no axillary lymph node involvement: the need for tailored adjuvant therapies. Ann Oncol 15:1633–1639
77. Selz J, Stevens D, Jouanneau L, Labib A, Le Scodan R (2012) Prognostic value of molecular subtypes, ki67 expression and impact of postmastectomy radiation therapy in breast cancer patients with negative lymph nodes after mastectomy. Int J Radiat Oncol Biol Phys 84:1123–1132
78. Zurrida S, Bagnardi V, Curigliano G, Mastropasqua MG, Orecchia R, Disalvatore D, Greco M, Cataliotti L, D'Aiuto G, Talakhadze N, Goldhirsch A, Viale G (2013) High Ki67 predicts unfavourable outcomes in early breast cancer patients with a clinically clear axilla who do not receive axillary dissection or axillary radiotherapy. Eur J Cancer 49:3083–3092
79. Lowery AJ, Kell MR, Glynn RW, Kerin MJ, Sweeney KJ (2012) Locoregional recurrence after breast cancer surgery: a systematic review by receptor phenotype. Breast Cancer Res Treat 133:831–841
80. Adamowicz K, Marczewska M, Jassem J (2009) Combining systemic therapies with radiation in breast cancer. Cancer Treat Rev 35:409–416
81. Brollo J, Kneubil MC, Botteri E, Rotmensz N, Duso BA, Fumagalli L, Locatelli MA, Criscitiello C, Lohsiriwat V, Goldhirsch A, Leonardi MC, Orecchia R, Curigliano G (2013) Locoregional recurrence in patients with HER2 positive breast cancer. Breast 22:856–862

82. Eiermann W, Bergh J, Cardoso F, Conte P, Crown J, Curtin NJ, Gligorov J, Gusterson B, Joensuu H, Linderholm BK, Martin M, Penault-Llorca F, Pestalozzi BC, Razis E, Sotiriou C, Tjulandin S, Viale G (2012) Triple negative breast cancer: proposals for a pragmatic definition and implications for patient management and trial design. Breast 21:20–26
83. Moran MS (2015) Radiation therapy in the locoregional treatment of triple-negative breast cancer. Lancet Oncol 16:e113–e122
84. Moser EC, Vrieling C (2012) Accelerated partial breast irradiation: the need for well-defined patient selection criteria, improved volume definitions, close follow-up and discussion of salvage treatment. Breast 21:707–715
85. Smith BD, Arthur DW, Buchholz TA, Haffty BG, Hahn CA, Hardenbergh PH, Julian TB, Marks LB, Todor DA, Vicini FA, Whelan TJ, White J, Wo JY, Harris JR (2009) Accelerated partial breast irradiation consensus statement from the American Society for Radiation Oncology (ASTRO). J Am Coll Surg 209:269–277
86. Polgar C, Van Limbergen E, Potter R, Kovacs G, Polo A, Lyczek J, Hildebrandt G, Niehoff P, Guinot JL, Guedea F, Johansson B, Ott OJ, Major T, Strnad V (2010) Patient selection for accelerated partial-breast irradiation (APBI) after breast-conserving surgery: recommendations of the Groupe Europeen de Curietherapie-European Society for Therapeutic Radiology and Oncology (GEC-ESTRO) breast cancer working group based on clinical evidence (2009). Radiother Oncol 94:264–273
87. Haffty BG, Hunt KK, Harris JR, Buchholz TA (2011) Positive sentinel nodes without axillary dissection: implications for the radiation oncologist. J Clin Oncol 29:4479–4481
88. Kunkler IH, Canney P, van Tienhoven G, Russell NS (2008) Elucidating the role of chest wall irradiation in 'intermediate-risk' breast cancer: the MRC/EORTC SUPREMO trial. Clin Oncol (R Coll Radiol) 20:31–34
89. Lazzeroni M, Guerrieri-Gonzaga A, Botteri E, Leonardi MC, Rotmensz N, Serrano D, Varricchio C, Disalvatore D, Del CA, Bassi F, Pagani G, DeCensi A, Viale G, Bonanni B, Pruneri G (2013) Tailoring treatment for ductal intraepithelial neoplasia of the breast according to Ki-67 and molecular phenotype. Br J Cancer 108:1593–1601
90. Curigliano G, Disalvatore D, Esposito A, Pruneri G, Lazzeroni M, Guerrieri-Gonzaga A, Luini A, Orecchia R, Goldhirsch A, Rotmensz N, Bonanni B, Viale G (2015) Risk of subsequent in situ and invasive breast cancer in human epidermal growth factor receptor 2-positive ductal carcinoma in situ. Ann Oncol 26:682–687
91. Solin LJ, Gray R, Baehner FL, Butler SM, Hughes LL, Yoshizawa C, Cherbavaz DB, Shak S, Page DL, Sledge GJ, Davidson NE, Ingle JN, Perez EA, Wood WC, Sparano JA, Badve S (2013) A multigene expression assay to predict local recurrence risk for ductal carcinoma in situ of the breast. J Natl Cancer Inst 105:701–710
92. Azria D, Gourgou S, Sozzi WJ, Zouhair A, Mirimanoff RO, Kramar A, Lemanski C, Dubois JB, Romieu G, Pelegrin A, Ozsahin M (2004) Concomitant use of tamoxifen with radiotherapy enhances subcutaneous breast fibrosis in hypersensitive patients. Br J Cancer 91:1251–1260
93. Azria D, Riou O, Castan F, Nguyen TD, Peignaux K, Lemanski C, Lagrange JL, Kirova Y, Lartigau E, Belkacemi Y, Bourgier C, Rivera S, Noel G, Clippe S, Mornex F, Hennequin C, Kramar A, Gourgou S, Pelegrin A, Fenoglietto P, Ozsahin EM (2015) Radiation-induced CD8 t-lymphocyte apoptosis as a predictor of breast fibrosis after radiotherapy: results of the prospective multicenter french trial. EBioMedicine 2:1965–1973
94. Chen S, Parmigiani G (2007) Meta-analysis of BRCA1 and BRCA2 penetrance. J Clin Oncol 25:1329–1333
95. Wang B, Matsuoka S, Ballif BA, Zhang D, Smogorzewska A, Gygi SP, Elledge SJ (2007) Abraxas and RAP80 form a BRCA1 protein complex required for the DNA damage response. Science 316:1194–1198
96. Yun MH, Hiom K (2009) CtIP-BRCA1 modulates the choice of DNA double-strand-break repair pathway throughout the cell cycle. Nature 459:460–463
97. Chen L, Nievera CJ, Lee AY, Wu X (2008) Cell cycle-dependent complex formation of BRCA1. CtIP.MRN is important for DNA double-strand break repair. J Biol Chem 283:7713–7720
98. Yu X, Wu LC, Bowcock AM, Aronheim A, Baer R (1998) The C-terminal (BRCT) domains of BRCA1 interact in vivo with CtIP, a protein implicated in the CtBP pathway of transcriptional repression. J Biol Chem 273:25388–25392
99. Bunting SF, Callen E, Wong N, Chen HT, Polato F, Gunn A, Bothmer A, Feldhahn N, Fernandez-Capetillo O, Cao L, Xu X, Deng CX, Finkel T, Nussenzweig M, Stark JM, Nussenzweig A (2010) 53BP1 inhibits homologous recombination in Brca1-deficient cells by blocking resection of DNA breaks. Cell 141:243–254
100. Sharan SK, Morimatsu M, Albrecht U, Lim DS, Regel E, Dinh C, Sands A, Eichele G, Hasty P, Bradley A (1997) Embryonic lethality and radiation hypersensitivity mediated by Rad51 in mice lacking Brca2. Nature 386:804–810
101. Foray N, Randrianarison V, Marot D, Perricaudet M, Lenoir G, Feunteun J (1999) Gamma-rays-induced death of human cells carrying mutations of BRCA1 or BRCA2. Oncogene 18:7334–7342
102. Scully R, Ganesan S, Vlasakova K, Chen J, Socolovsky M, Livingston DM (1999) Genetic analysis of BRCA1 function in a defined tumor cell line. Mol Cell 4:1093–1099

103. Abbott DW, Thompson ME, Robinson-Benion C, Tomlinson G, Jensen RA, Holt JT (1999) BRCA1 expression restores radiation resistance in BRCA1-defective cancer cells through enhancement of transcription-coupled DNA repair. J Biol Chem 274:18808–18812
104. Miki Y, Swensen J, Shattuck-Eidens D, Futreal PA, Harshman K, Tavtigian S, Liu Q, Cochran C, Bennett LM, Ding W (1994) A strong candidate for the breast and ovarian cancer susceptibility gene BRCA1. Science 266:66–71
105. Wooster R, Bignell G, Lancaster J, Swift S, Seal S, Mangion J, Collins N, Gregory S, Gumbs C, Micklem G (1995) Identification of the breast cancer susceptibility gene BRCA2. 378, pp 789–792
106. Rakha EA, Reis-Filho JS, Ellis IO (2008) Basal-like breast cancer: a critical review. J Clin Oncol 26:2568–2581
107. Phillips KA, Andrulis IL, Goodwin PJ (1999) Breast carcinomas arising in carriers of mutations in BRCA1 or BRCA2: are they prognostically different? J Clin Oncol 17:3653–3663
108. Boyd J, Sonoda Y, Federici MG, Bogomolniy F, Rhei E, Maresco DL, Saigo PE, Almadrones LA, Barakat RR, Brown CL, Chi DS, Curtin JP, Poynor EA, Hoskins WJ (2000) Clinicopathologic features of BRCA-linked and sporadic ovarian cancer. JAMA 283:2260–2265
109. Hirai T, Shirai H, Fujimori H, Okayasu R, Sasai K, Masutani M (2012) Radiosensitization effect of poly(ADP-ribose) polymerase inhibition in cells exposed to low and high liner energy transfer radiation. Cancer Sci 103:1045–1050
110. Zaremba T, Curtin NJ (2007) PARP inhibitor development for systemic cancer targeting. Anti Cancer Agents Med Chem 7:515–523
111. Fong PC, Boss DS, Yap TA, Tutt A, Wu P, Mergui-Roelvink M, Mortimer P, Swaisland H, Lau A, O'Connor MJ, Ashworth A, Carmichael J, Kaye SB, Schellens JH, de Bono JS (2009) Inhibition of poly(ADP-ribose) polymerase in tumors from BRCA mutation carriers. N Engl J Med 361:123–134
112. Fong PC, Yap TA, Boss DS, Carden CP, Mergui-Roelvink M, Gourley C, De Greve J, Lubinski J, Shanley S, Messiou C, A'Hern R, Tutt A, Ashworth A, Stone J, Carmichael J, Schellens JH, de Bono JS, Kaye SB (2010) Poly(ADP)-ribose polymerase inhibition: frequent durable responses in BRCA carrier ovarian cancer correlating with platinum-free interval. J Clin Oncol 28:2512–2519
113. Tutt A, Robson M, Garber JE, Domchek SM, Audeh MW, Weitzel JN, Friedlander M, Arun B, Loman N, Schmutzler RK, Wardley A, Mitchell G, Earl H, Wickens M, Carmichael J (2010) Oral poly(ADP-ribose) polymerase inhibitor olaparib in patients with BRCA1 or BRCA2 mutations and advanced breast cancer: a proof-of-concept trial. Lancet 376:235–244
114. Kaufman B (2015) SRSR, J B, G M, G F, SM S, a H, O R, M S, N L, K B, a F, SM D: Olaparib monotherapy in patients with advanced cancer and a germline BRCA1/2 mutation. J Clin Oncol 33:244–250
115. Wilson RH, Evans TJ, Middleton MR, Molife LR, Spicer J, Dieras V, Roxburgh P, Giordano H, Jaw-Tsai S, Goble S, Plummer R (2017) A phase I study of intravenous and oral rucaparib in combination with chemotherapy in patients with advanced solid tumours. Br J Cancer 116:884–892
116. Castellano L, Giamas G, Jacob J, Coombes RC, Lucchesi W, Thiruchelvam P, Barton G, Jiao LR, Wait R, Waxman J, Hannon GJ, Stebbing J (2009) The estrogen receptor-alpha-induced microRNA signature regulates itself and its transcriptional response. Proc Natl Acad Sci U S A 106:15732–15737
117. Davies C, Godwin J, Gray R, Clarke M, Cutter D, Darby S, McGale P, Pan HC, Taylor C, Wang YC, Dowsett M, Ingle J, Peto R (2011) Relevance of breast cancer hormone receptors and other factors to the efficacy of adjuvant tamoxifen: patient-level meta-analysis of randomised trials. Lancet 378:771–784
118. Chisamore MJ, Wilkinson HA, Flores O, Chen JD (2009) Estrogen-related receptor-alpha antagonist inhibits both estrogen receptor-positive and estrogen receptor-negative breast tumor growth in mouse xenografts. Mol Cancer Ther 8:672–681
119. Liu YH, He N, Jiang SQ, Li X, Zhang JS (2011) Regulation of orphan receptor ERRα by estrogen and estrogen-related receptor antagonist in MCF-7 cell line. J Environ Health 28:677–680
120. Ciszewski WM, Tavecchio M, Dastych J, Curtin NJ (2014) DNA-PK inhibition by NU7441 sensitizes breast cancer cells to ionizing radiation and doxorubicin. Breast Cancer Res Treat 143:47–55
121. Moll U, Lau R, Sypes MA, Gupta MM, Anderson CW (1999) DNA-PK, the DNA-activated protein kinase, is differentially expressed in normal and malignant human tissues. Oncogene 18:3114–3126
122. Patel AG, Sarkaria JN, Kaufmann SH (2011) Nonhomologous end joining drives poly(ADP-ribose) polymerase (PARP) inhibitor lethality in homologous recombination-deficient cells. Proc Natl Acad Sci U S A 108(8):3406–11
123. Mansour WY, Rhein T, Dahm-Daphi J (2010) The alternative end-joining pathway for repair of DNA double-strand breaks requires PARP1 but is not dependent upon microhomologies. Nucleic Acids Res 38:6065–6077
124. Kim CH, Park SJ, Lee SH (2002) A targeted inhibition of DNA-dependent protein kinase sensitizes breast cancer cells following ionizing radiation. J Pharmacol Exp Ther 303:753–759
125. Belenkov AI, Paiement JP, Panasci LC, Monia BP, Chow TY (2002) An antisense oligonucleotide targeted to human Ku86 messenger RNA sensitizes M059K malignant glioma cells to ionizing radiation, bleomycin, and etoposide but not DNA cross-linking agents. Cancer Res 62:5888–5896

126. Bladen CL, Lam WK, Dynan WS, Kozlowski DJ (2005) DNA damage response and Ku80 function in the vertebrate embryo. Nucleic Acids Res 33:3002–3010
127. Sturgeon CM, Knight ZA, Shokat KM, Roberge M (2006) Effect of combined DNA repair inhibition and G2 checkpoint inhibition on cell cycle progression after DNA damage. Mol Cancer Ther 5:885–892
128. Ayene IS, Ford LP, Koch CJ (2005) Ku protein targeting by Ku70 small interfering RNA enhances human cancer cell response to topoisomerase ii inhibitor and gamma radiation. Mol Cancer Ther 4:529–536
129. Losada R, Rivero MT, Slijepcevic P, Goyanes V, Fernández JL (2005) Effect of Wortmannin on the repair profiles of DNA double-strand breaks in the whole genome and in interstitial telomeric sequences of Chinese hamster cells. Mutat Res 570:119–128
130. Wilson DR, Simeonov A (2010) Small molecule inhibitors of DNA repair nuclease activities of APE1. Cell Mol Life Sci 67:3621–3631
131. Al-Safi RI, Odde S, Shabaik Y, Neamati N (2012) Small-molecule inhibitors of APE1 DNA repair function: an overview. Curr Mol Pharmacol 5:14–35
132. Moore DH, Michael H, Tritt R, Parsons SH, Kelley MR (2000) Alterations in the expression of the DNA repair/redox enzyme APE/ref-1 in epithelial ovarian cancers. Clin Cancer Res 6:602–609
133. Kelley MR, Cheng L, Foster R, Tritt R, Jiang J, Broshears J, Koch M (2001) Elevated and altered expression of the multifunctional DNA base excision repair and redox enzyme Ape1/ref-1 in prostate cancer. Clin Cancer Res 7:824–830
134. Hadi MZ, Coleman MA, Fidelis K, Mohrenweiser HW, Wilson DR (2000) Functional characterization of Ape1 variants identified in the human population. Nucleic Acids Res 28:3871–3879
135. Kakolyris S, Kaklamanis L, Engels K, Turley H, Hickson ID, Gatter KC, Harris AL (1997) Human apurinic endonuclease 1 expression in a colorectal adenoma-carcinoma sequence. Cancer Res 57:1794–1797
136. Abdel-Fatah TM, Perry C, Moseley P, Johnson K, Arora A, Chan S, Ellis IO, Madhusudan S (2014) Clinicopathological significance of human apurinic/apyrimidinic endonuclease 1 (APE1) expression in oestrogen-receptor-positive breast cancer. Breast Cancer Res Treat 143:411–421
137. Poletto M, Di Loreto C, Marasco D, Poletto E, Puglisi F, Damante G, Tell G (2012) Acetylation on critical lysine residues of Apurinic/apyrimidinic endonuclease 1 (APE1) in triple negative breast cancers. Biochem Biophys Res Commun 424:34–39
138. Sultana R, McNeill DR, Abbotts R, Mohammed MZ, Zdzienicka MZ, Qutob H, Seedhouse C, Laughton CA, Fischer PM, Patel PM, Wilson DM 3rd, Madhusudan S (2012) Synthetic lethal targeting of DNA double-strand break repair deficient cells by human apurinic/apyrimidinic endonuclease inhibitors. Int J Cancer 131:2433–2444
139. Wang D, Zhong ZY, Li MX, Xiang DB, Li ZP (2007) Vector-based APE1 small interfering RNA enhances the sensitivity of human osteosarcoma cells to endostatin in vivo. Cancer Sci 98:1993–2001
140. Herring CJ, West CM, Wilks DP, Davidson SE, Hunter RD, Berry P, Forster G, MacKinnon J, Rafferty JA, Elder RH, Hendry JH, Margison GP (1998) Levels of the DNA repair enzyme human apurinic/apyrimidinic endonuclease(APE1, APEX, ref-1) are associated with the intrinsic radio-sensitivity of cervical cancers. Br J Cancer 78:1128–1133
141. Van Gent DC, Hoeijmakers JHJ, Kanaar R (2001) Chromosomal stability and the DNA double-strand break connection. Nat Rev Genet 2:196–206
142. Marine JC, Lozano G (2010) Mdm2-mediated ubiquitylation: P53 and beyond. Cell Death Differ 17:93–102
143. Khosravi R, Maya R, Gottlieb T, Oren M, Shiloh Y, Shkedy D (1999) Rapid ATM-dependent phosphorylation of MDM2 precedes p53 accumulation in response to DNA damage. Proc Natl Acad Sci U S A 96:14973–14977
144. Errico A, Costanzo V (2012) Mechanisms of replication fork protection: a safeguard for genome stability. Crit Rev Biochem Mol Biol 47:222–235
145. Mohni KN, Kavanaugh GM, Cortez D (2014) ATR pathway inhibition is synthetically lethal in cancer cells with ERCC1 deficiency. Cancer Res 74:2835–2845
146. Sultana R, Abdel-Fatah T, Abbotts R, Hawkes C, Albarakati N, Seedhouse C, Ball G, Chan S, Rakha EA, Ellis IO, Madhusudan S (2013) Targeting XRCC1 deficiency in breast cancer for personalized therapy. Cancer Res 73:1621–1634
147. Shiloh Y (2003) ATM and related protein kinases: safeguarding genome integrity. Nat Rev Cancer 3:155–168
148. Biddlestone-Thorpe L, Sajjad M, Rosenberg E, Beckta JM, Valerie NC, Tokarz M, Adams BR, Wagner AF, Khalil A, Gilfor D, Golding SE, Deb S, Temesi DG, Lau A, O'Connor MJ, Choe KS, Parada LF, Lim SK, Mukhopadhyay ND, Valerie K (2013) ATM kinase inhibition preferentially sensitizes p53-mutant glioma to ionizing radiation. Clin Cancer Res 19:3189–3200
149. Yang H, Yoon SJ, Jin J, Choi SH, Seol HJ, Lee JI, Nam DH, Yoo HY (2011) Inhibition of checkpoint kinase 1 sensitizes lung cancer brain metastases to radiotherapy. Biochem Biophys Res Commun 406:53–58
150. Borst GR, McLaughlin M, Kyula JN, Neijenhuis S, Khan A, Good J, Zaidi S, Powell NG, Meier P, Collins I, Garrett MD, Verheij M, Harrington KJ (2013) Targeted radiosensitization by the Chk1 inhibitor SAR-020106. Int J Radiat Oncol Biol Phys 85:1110–1118
151. Peasland A, Wang LZ, Rowling E, Kyle S, Chen T, Hopkins A, Cliby WA, Sarkaria J, Beale G, Edmondson RJ, Curtin NJ (2011) Identification and evaluation of a potent novel ATR inhibitor, NU6027, in breast and ovarian cancer cell lines. Br J Cancer 105:372–381

152. Wickremsinhe ER, Hynes SM, Palmieri MD, Mitchell MI, Abraham TL, Rehmel JF, Chana E, Jost LM, Cassidy KC (2014) Disposition and metabolism of LY2603618, a Chk-1 inhibitor following intravenous administration in patients with advanced and/or metastatic solid tumors. Xenobiotica 44:827–841
153. Shapiro GI, Tibes R, Gordon MS, Wong BY, Eder JP, Borad MJ, Mendelson DS, Vogelzang NJ, Bastos BR, Weiss GJ, Fernandez C, Sutherland W, Sato H, Pierceall WE, Weaver D, Slough S, Wasserman E, Kufe DW, Von Hoff D, Kawabe T, Sharma S (2011) Phase I studies of CBP501, a G2 checkpoint abrogator, as monotherapy and in combination with cisplatin in patients with advanced solid tumors. Clin Cancer Res 17:3431–3442
154. Dees EC, Baker SD, O'Reilly S, Rudek MA, Davidson SB, Aylesworth C, Elza-Brown K, Carducci MA, Donehower RC (2005) A phase I and pharmacokinetic study of short infusions of UCN-01 in patients with refractory solid tumors. Clin Cancer Res 11:664–671
155. Sausville EA, Arbuck SG, Messmann R, Headlee D, Bauer KS, Lush RM, Murgo A, Figg WD, Lahusen T, Jaken S, Jing X, Roberge M, Fuse E, Kuwabara T, Senderowicz AM (2001) Phase I trial of 72-hour continuous infusion UCN-01 in patients with refractory neoplasms. J Clin Oncol 19:2319–2333
156. Sausville E, Lorusso P, Carducci M, Carter J, Quinn MF, Malburg L, Azad N, Cosgrove D, Knight R, Barker P, Zabludoff S, Agbo F, Oakes P, Senderowicz A (2014) Phase I dose-escalation study of AZD7762, a checkpoint kinase inhibitor, in combination with gemcitabine in US patients with advanced solid tumors. Cancer Chemother Pharmacol 73:539–549
157. Li T, Christensen SD, Frankel PH, Margolin KA, Agarwala SS, Luu T, Mack PC, Lara PJ, Gandara DR (2012) A phase II study of cell cycle inhibitor UCN-01 in patients with metastatic melanoma: a California cancer consortium trial. Investig New Drugs 30:741–748
158. Rini BI, Weinberg V, Shaw V, Scott J, Bok R, Park JW, Small EJ (2004) Time to disease progression to evaluate a novel protein kinase C inhibitor, UCN-01, in renal cell carcinoma. CANCER-AM CANCER SOC 101:90–95
159. Cowell IG, Durkacz BW, Tilby MJ (2005) Sensitization of breast carcinoma cells to ionizing radiation by small molecule inhibitors of DNA-dependent protein kinase and ataxia telangiectsia mutated. Biochem Pharmacol 71:13–20
160. Hickson I, Zhao Y, Richardson CJ, Green SJ, Martin NM, Orr AI, Reaper PM, Jackson SP, Curtin NJ, Smith GC (2004) Identification and characterization of a novel and specific inhibitor of the ataxia-telangiectasia mutated kinase ATM. Cancer Res 64:9152–9159
161. Rainey MD, Charlton ME, Stanton RV, Kastan MB (2008) Transient inhibition of ATM kinase is sufficient to enhance cellular sensitivity to ionizing radiation. Cancer Res 68:7466–7474
162. Mack PC, Jones AA, Gustafsson MH, Gandara DR, Gumerlock PH, Goldberg Z (2004) Enhancement of radiation cytotoxicity by UCN-01 in non-small cell lung carcinoma cells. Radiat Res 162:623–634
163. Ree AH, Bratland A, Nome RV, Stokke T, Fodstad O, Andersson Y (2004) Inhibitory targeting of checkpoint kinase signaling overrides radiation-induced cell cycle gene regulation: a therapeutic strategy in tumor cell radiosensitization? Radiother Oncol 72:305–310

6 Targeted Therapies Against Growth Factor Signaling in Breast Cancer

Juan Du, Yu Yu, Jun Zhan, and Hongquan Zhang

Abstract

Breast cancer is the most prevalent female malignancy throughout the world. Conventional treatment strategies for breast cancer consist of chemotherapy, radiation, surgery, chemoradiation, hormone therapy, and targeted therapies. Among them, targeted therapies show advantages to reduce cost and toxicity for being possible for individualized treatments based on the intrinsic subtypes of breast cancer. With deeper understanding of key signaling pathways concerning tumor growth and survival, growth factor-controlled signaling pathways are frequently dysregulated in the development and progression of breast cancer. Thus, targeted therapies against growth factor-mediated signaling pathways have been shown to have promising efficacy in both preclinical animal models and human clinical trials. In this chapter, we will briefly introduce inhibitors and monoclonal antibodies that target the main growth factor-modulated scenarios including epidermal growth factor receptor (EGFR), transforming growth factor beta (TGF-β), insulin-like growth factor 1 receptor (IGF1R), and fibroblast growth factor receptor (FGFR) signaling pathways in breast cancer therapy.

Keywords

Breast cancer • Target therapy • Growth factor • Monoclonal antibody • Inhibitor • EGFR • TGF-β • FGFR • IGF1R

J. Du • Y. Yu • J. Zhan • H. Zhang (✉)
Department of Anatomy, Histology and Embryology, Laboratory of Molecular Cell Biology and Tumor Biology, School of Basic Medical Sciences, Peking University Health Science Center, Beijing, China
e-mail: Hongquan.Zhang@bjmu.edu.cn

6.1 EGFR Signaling

6.1.1 Introduction of EGFR Signaling

The human epidermal growth factor receptor (EGFR) family, also called HER family or ErbB

E. Song, H. Hu (eds.), *Translational Research in Breast Cancer*, Advances in Experimental Medicine and Biology 1026, https://doi.org/10.1007/978-981-10-6020-5_6

family, consists of four structurally related receptors: EGFR/HER1/ErbB1, HER2/ErbB2, HER3/ErbB3, and HER4/ErbB4 [1, 2]. These HER family members belong to transmembrane receptor tyrosine kinases (RTKs), comprising a glycosylated extracellular domain, a single hydrophobic transmembrane segment, and an intracellular portion with a juxtamembrane segment, a protein kinase domain, and a carboxy-terminal tail [3]. When the ligands bind to the extracellular domain of HER receptors, with the exception of HER2, this HER signaling becomes activated by forming homo- and heterodimers and subsequent tyrosine autophosphorylation [4, 5]. Activated EGFR phosphorylates a number of important signaling molecules, such as phosphatidylinositol 3-kinase (PI3K), Ras, phospholipase C (PLCγ), and signal transducers and activators of transcription 3 (STAT3) [6–8].

Upon EGF-induced EGFR dimerization and receptor phosphorylation, the signaling is transmitted to growth factor receptor binding protein-2 (GRB2), son of sevenless (Sos), and Ras. Activated Ras in turn binds to and stimulates Raf which phosphorylates and activates MEK (MEK1 and MEK2). Activated MEK family phosphorylates and activates mitogen-activated protein kinase (MAPK) [9–13].

In addition to inducing the Raf/MEK/ERK pathway, activated Ras stimulates phosphatidylinositol 3-kinase (PI3K)/PDK1/Akt pathway [14–16]. Activated PI3K induces the production of phosphatidylinositol 3,4-bisphosphate [PI(3,4)P2] and phosphatidylinositol 3,4,5-trisphosphate [PI(3,4,5)P3], which recruit phosphoinositide-dependent kinase-1 (PDK1). Subsequently, PDK1 phosphorylates Akt family and induces the signal cascade [8].

STAT3 is a member of STAT family of transcription factors, and it can be activated by cytokines and growth factors [17, 18]. Activated STAT3 directly induces the transcriptional activation of downstream molecules, such as c-fos, c-myc, p21, cyclin D1, bcl-xl, and Fas [19, 20]. Thus, STAT3 plays a critical role in tumorigenesis [21], tumor metastasis [22], and angiogenesis [23].

Together, EGFR functions by activating the downstream signal transduction including PI3K/Akt (PKB) pathway, Ras/Raf/MEK/ERK1/2 pathway, PLCγ pathway, and STAT pathway. These pathways play important roles in mediating cell survival, cell proliferation, cell adhesion, cell motility, and angiogenesis [24].

6.1.2 Targeting Therapy of EGFR Signaling

The EGFR family has a crucial role in tumorigenesis [25], and dysregulation of EGFR family members is prevalent in human neoplasia [26]. The EGFR and its ligands have been found to result in the progression of breast cancer [27–29]. Various therapeutic strategies targeting these epidermal growth factor receptors have been studied for the treatment of breast cancer. The strategy of treating breast cancer greatly depends on its subtype. Luminal breast cancer is a subtype susceptible to endocrine therapy. However, the resistance of endocrine therapy is frequently produced in most luminal breast cancers. To overcome the resistance, treatment directed against EGFR is developed [30]. *HER2* overexpression happens in up to 30% of breast cancer, and thus HER2 is regarded as an important therapeutic target (Table 6.1). The humanized monoclonal antibody trastuzumab (Herceptin), the first anti-HER2 drug produced for treating HER2-positive breast cancer, can significantly improve the outcome [25, 31]. The other two HER2 antibodies, pertuzumab and trastuzumab-DM1, have also been assessed clinically with promising results in the treatment of HER2-positive metastatic breast cancer [31]. In addition to these humanized monoclonal antibodies, effective small-molecule inhibitors of EGFR/ErbB tyrosine kinases have been developed for treating breast cancer, including the EGFR inhibitors gefitinib and erlotinib and the dual EGFR/HER2 inhibitor lapatinib [30]. Lapatinib is highly active and approved in combination with chemotherapy for treating HER2-positive metastatic breast cancer [3]. The efficacy of gefitinib and erlotinib has been tested clinically and is showed to be limited.

Table 6.1 EGFR-targeted mAbs and inhibitors in breast cancer

Compound	Clinical development	Subtypes
mAbs		
Trastuzumab	Phase II	HER2-positive metastatic breast cancer
	Phase III	Plus adjuvant chemotherapy for early-stage HER2-positive metastatic breast cancer (NCT00004067; NCT00005970)
	Phase II	Plus BKM120 for HER2-positive primary breast cancer prior to definitive surgery (NCT01816594)
Pertuzumab	Phase II	Metastatic breast cancer (NCT02491892)
	Phase II	Plus trastuzumab for metastatic breast cancer (NCT01674062)
	Phase II	Plus trastuzumab and plus trastuzumab for metastatic or locally advanced breast cancer (NCT01565083)
Trastuzumab-DM1	Phase III	HER2-positive metastatic breast cancer (NCT01419197)
Tyrosine kinase inhibitors		
Lapatinib	Phase III	HER2-positive metastatic breast cancer (NCT00078572)
Erlotinib	Phase II	Metastatic cancer

6.1.2.1 Monoclonal Antibodies

Trastuzumab The humanized monoclonal antibody trastuzumab is the first anti-HER2 drug for the treatment of breast cancer overexpressing HER2, as it is directed against domain IV in the extracellular segment of HER2 receptor [32], which leads to the inhibition of downstream signaling of HER2 pathway. The mechanism of trastuzumab treating metastatic breast cancer is complex. Some researchers have reported that expression of HER2 is downregulated upon the treatment of trastuzumab. Furthermore, trastuzumab obviously inhibits the proliferation of HER2-overexpressing breast cancer cells [33]. In addition, trastuzumab blocks the cleavage of extracellular domain of HER2 and inhibits the activation of HER2 protein kinase domain [33]. Trastuzumab can also activate the immune natural killer (NK) cells to directly kill cancer cells [34].

The clinical effectiveness of trastuzumab has been demonstrated in metastatic HER2-positive breast cancer [35] and early-stage HER2- positive breast cancer [36]. In 2001 and 2005, two phase III adjuvant trials of trastuzumab confirmed that trastuzumab combined with cytotoxic chemotherapy resulted in a better outcome than chemotherapy alone in HER2-positive metastatic breast cancer. This combination leads to a significant prolongation of overall survival, converting the outcome of HER2-positive metastatic patients [37–39].

However, similar to other cases receiving targeted anticancer therapies, about 50–66% of HER2-positive breast cancer patients produce resistances to trastuzumab [40]. Thus, how to overcome the intrinsic and acquired resistance is the major problem [41]. Some molecular mechanisms have been provided to underlay such resistance. For example, PI3K/Akt signal transduction is inhibited because of loss of PTEN activity [42], and other alternative pathways are activated, such as insulin-like growth factor receptor and hepatic growth factor receptor (c-Met) [43]. For the relapsed patients who have previously been treated with trastuzumab, the therapeutic approach of lapatinib in combination with capecitabine is alternative and effective [44].

Pertuzumab Pertuzumab is also an anti-HER2 humanized monoclonal antibody for the treatment of breast cancer overexpressing HER2. This antibody targets the domain II of HER2 and functions by inhibiting the heterodimerization of HER2 with HER3 [45]. In 2012, a phase II study showed that the combination of pertuzumab plus trastuzumab plus docetaxel significantly improved the outcome in patients with HER2-positive metastatic breast cancer, compared with placebo plus trastuzumab plus docetaxel without additional cardiac toxicity and skin rash [46]. Upon this clinical trial, the combined regimen of pertuzumab with trastuzumab and docetaxel is approved for treating patients with ErbB2-positive metastatic breast cancer [47].

Trastuzumab Emtansine (T-DM1) Trastuzumab-DM1 is a HER2 antibody drug conjugate in which trastuzumab is fused to the microtubule-inhibitory agent DM1 (derivative of maytansine) [48, 49]. Traztuzumab-DM1 is used to treat HER2-positive metastatic breast cancer that is resistant to trastuzumab or lapatinib [50]. Such antibody drug conjugates (ADCs) can merge the antibody and the cytotoxic agent via chemical linkers, allowing the cytotoxicity delivering specifically to breast cancer cells overexpressing HER2 [51, 52]. Thus, ADCs can improve the therapeutic effectiveness of this drug and at the same time reduce its adverse effects.

6.1.2.2 Tyrosine Kinase Inhibitors

Lapatinib Lapatinib is a reversible and selective tyrosine kinase inhibitor of the intracellular domains of EGFR and HER2. It competitively binds to ATP-binding site of the receptor, resulting in the blocking of PI3K/AKT/mTOR pathway [53]. Lapatinib inhibits the phosphorylation of HER2 and downstream Erk1/Erk2 on breast cancer cell line in vitro and in animal xenografts [3, 54]. Lapatinib is a potent US Food and Drug Administration (FDA)-approved drug for breast cancer treatment of patients. It is used in combination with capecitabine in HER2-positive breast cancer patients who have received prior therapies including anthracycline, taxane, and trastuzumab [55, 56]. Lapatinib is also used in combination with letrozole in postmenopausal patients with hormone receptor-positive breast cancer [57]. Some clinical trials have identified the efficacy of lapatinib-induced inhibition of EGFR to be modest compared with its effect on HER2 [58]. HER2 overexpression is a key determinant of sensitivity to lapatinib [59–61], whereas EGFR expression does not appear to be predictive [62].

Metastatic breast cancer patients become resistant to the combination treatment of lapatinib with capecitabine or letrozole in tumor growth or invasion. To explore the resistance mechanisms, mutation of HER2 and activation of compensatory survival pathways have been studied and found to confer the resistance [63, 64]. Several strategies have been developed to solve the problem, such as using pertuzumab, ado-trastuzumab-DM1, and mammalian target of rapamycin (mTOR) inhibitors [3]. The mTOR inhibitors have been reported to have a modest activity in treatment of breast cancer. The combination of EGFR inhibition (lapatinib) and mTOR inhibition (rapamycin) results in a significant inhibition of tumor growth compared with either agent alone [65].

Gefitinib Gefitinib is a reversible and specific tyrosine kinase inhibitor of EGFR. Preclinical studies have indicated that gefitinib is effective for inhibiting the EGFR pathway and enhancing response to chemotherapy in TNBC and HER2-positive cell lines [66]. However, most clinical studies have revealed that gefitinib has a limited activity in monotherapy or combined with either anti-HER2 treatment or chemotherapy in metastatic breast cancer [67].

Erlotinib Erlotinib is an oral tyrosine kinase inhibitor of EGFR used in the treatment of non-small cell lung cancer and pancreatic cancer [30]. Erlotinib has a very limited activity in monotherapy of metastatic breast cancer. Preclinical studies have showed that the EGFR signaling may participate in the regulation of angiogenesis [68–71]. Thus, a phase II study was carried out to assess the efficacy of combination of anti-EGFR and anti-VEGF therapies in metastatic breast cancer. However, the results still showed a limited activity of the combination [72]. Subsequently, another phase II study was performed, and authors elucidated that the therapy of double blockade of EGFR/VEGF may be an active regimen [73].

6.2 TGF-β Signaling

6.2.1 TGF-β Signaling and Breast Cancer

Transforming growth factor beta (TGF-β) is a ubiquitous and essential regulator of cellular function. TGF-β binds to type I and type II receptors on the cell surface in dimer [74], leading to the activation of various downstream substrates and inducing transcription of different target genes [75]. Thereby, activation of TGF-β signaling controls developmental programs and cell behaviors including cell proliferation, differentiation, migration, survival, angiogenesis, and immune surveillance [75].

TGF-β regulates cell cycle progression by promoting synthesis of p15 and p21 proteins, which block the formation of cyclin-CDK complex. Thus, TGF-β blocks cell cycle forward at the G1 phase [76], with concomitant stopping of cell proliferation, inducing cell differentiation or apoptosis [77–79]. Therefore, TGF-β inhibits tumor growth at the early stages of cancer. However, at the later stages tumor cells lose sensitivity to the inhibitory effect of TGF-β, and TGF-β promotes metastasis and growth of the tumor [80–82]. Thus, preservation of the TGF-β effect on cell migration, differentiation, and extracellular matrix formation may promote tumorigenesis. TGF-β strongly accelerates formation of extracellular matrix and tumor stroma. Therefore, it has emerged as a major modulator of angiogenesis in cancer [83, 84], which promotes cancer metastasis and supports tumor growth in remote organs [85].

TGF-β signaling pathways are classified into canonical (SMAD-dependent) and noncanonical (SMAD-independent) pathways. In the canonical pathway, the TGF-β ligand binds to TGF-β receptor type II (TβRII), recruits receptor type I (TβRI), and leads to phosphorylation of SMAD2 and SMAD3. The phosphorylated SMAD2 and SMAD3 will bind with SMAD4 prior to nuclear translocation. TGF-β can activate the noncanonical pathways including mTOR, STAT3, AKT, WISP2, NF-κB, PTEN, Erk1/Erk2, and Src signaling pathways, affecting invasion and metastasis of breast cancer cells [86].

Clinically, overexpression of TGF-β is considered as a biomarker for triple-negative breast cancer (TNBC) patients [87]. In TNBC cells, expression of transcriptional factor GLI2 is increased by TGF-β/SMAD signaling pathway, and its target gene PTH-rPin participates in the pathogenesis of osteolytic metastases in TNBC. TGF-β/GLI2 pathway also promotes hedgehog (Hh) pathway and potentiates tumor growth and osteolysis in TNBC [87].

Now, TGF-β is a well-recognized key regulator of cancer progression. However, TGF-β is a double-edged sword. In the early stage, TGF-β suppresses tumor formation by inhibiting cell cycle progression and promoting apoptosis. TGF-β turns to be a stimulator of cancer cell invasion, promoting distant metastasis and formation of pre-metastatic niche. TGF-β and other members of the TGF-β signaling pathway are considered good candidates for prognostic or predictive markers of cancer patients. Targeting TGF-β signaling pathway provides new horizon for the chemoprevention and treatment of human cancers.

6.2.2 Targeting TGF-β Signaling for Cancer Therapy

Several pharmacological strategies have been developed to block TGF-β signaling, such as vaccines, monoclonal antibodies, antisense oligonucleotides, and small-molecule inhibitors. Galunisertib (LY2157299 monohydrate) is an oral small-molecule inhibitor of the TGF-β receptor I kinase that specifically downregulates the phosphorylation of SMAD2 and abrogates activation of the canonical pathway. Furthermore, galunisertib has antitumor activity in tumor-bearing animal models such as breast, colon, and lung cancers and hepatocellular carcinoma. Unfortunately, long-term application of galunisertib can cause cardiac toxicities in animals, and a thorough pharmacokinetic investigation is required before wide examination of

Table 6.2 TGF-β-targeted mAbs and inhibitors in breast cancer

Compound	Clinical development	Subtypes
mAbs		
Fresolimumab	Phase II	Metastatic breast cancer (NCT01401062)
Inhibitors		
TEW-7197	Phase I	Advanced or refractory breast cancer (NCT02160106)

galunisertib for therapeutic purposes. Details for experiences of galunisertib application in patients have been summarized [88] (Table 6.2).

6.2.2.1 Inhibitors of TGF-β Signaling

PF-03446962 Due to the structural and genetic similarities of the different TGF-β signaling pathways, inhibitors must be highly selected to block the activation of targeted pathway. For example, ALK5 (activin-like kinase 5) and ALK1 (activin-like kinase 1) pathways both increase tumor angiogenesis. PF-03446962, an ALK1-neutralizing antibody, does not bind other ALKs and shows specificity. A phase II clinical trial is now investigating PF-03446962 to determine its capacity to block angiogenesis in patients with solid tumors [89]. Dalantercept/ACE-041, a chimeric protein consisting of the ALK1 ligand binding extracellular domain and an Fc portion (ALK1-Fc), can efficiently block BMP-9- and BMP-10-induced SMAD1 phosphorylation and SMAD1-dependent transcription. Dalantercept is currently in phase II trials as monotherapy and in combination with vascular endothelial growth factor (VEGF) inhibitors [90]. In contrast to the ALK1 inhibitors, inhibition of the ALK5 pathway blocks activation of TGF-β signaling components (e.g., SMAD2/SMAD3) [91]. However, ALK5 inhibitors increase angiogenesis in cell cultures of normal endothelial cells [92].

Small-Molecule Inhibitors Among the TGF-β inhibitors, small-molecule inhibitors (SMIs) are chemicals designed to block the activation of the signaling components downstream of the TGF-β receptor type I kinase (TβRI or ALK5) or type II (TβRII) by inhibiting the serine/threonine kinase activity. There is a growing list of SMIs blocking the TGF-βRI activation [93, 94]. A large library of SMIs from Eli Lilly and Company was screened in vitro using a TGF-β-dependent cell-based assay. Selected compounds were further assessed for their ability to inhibit autophosphorylation of the human TβRI kinase domain, which is the constitutively active (T204D mutation) TβRI kinase domain and is expressed by Sf9 insect cells [95, 96]. For example, LY364947 (diheteroaryl-substituted pyrazole 1) was identified as a potent TβRI inhibitor (IC_{50} = 51 nM). Compounds were further evaluated by measuring their inhibitory effect in a TGF-β-dependent luciferase assay using mink lung cells and mouse fibroblasts (NIH3T3) [97]. Compounds LY580276 [98], LY364947 [99], and LY2109761 [100] resemble LY2157299 monohydrate for the selectivity profile and inhibition of the ALK5 kinase activity [101, 102]. Therefore, galunisertib may represent an inhibitor of TGF-β signaling pathway that meets the desired characteristics as a tumor inhibitor.

Monoclonal Antibodies Monoclonal antibodies were developed to selectively inhibit the TGF-β signaling pathway. Fresolimumab (formerly GC1008) represents the first pan-TGF-β ligand monoclonal antibody to be investigated in cancer patients [103]. Fresolimumab was first developed for patients with renal fibrosis and later for patients with metastatic cancer [103–106]. Further, a TGF-β1-specific monoclonal antibody (TβM1 or LY2382770) was evaluated in a phase I study in cancer patients. However, due to the toxicity and low therapeutic response, all these monoclonal antibodies were terminated [103–106]. Monoclonal antibodies have also been generated for blocking the TβRII receptor and are currently being investigated in a first-in-human dose (FHD) study [103–106]. Therefore, more works should be done before monoclonal antibodies can be used for blocking TGF-β signaling pathway in human trials.

6.3 FGF Signaling

6.3.1 Introduction of FGF Signaling

Fibroblast growth factors (FGFs) which bind and activate their receptors (FGFRs) regulate a wide range of biologic processes including the formation of new blood vessels, wound repair, and embryonic development by mainly signaling through the RAS/MEK/ERK and the PI3K/AKT pathways [107]. FGFRs comprise FGFR1, FGFR2, FGFR3, FGFR4, and FGFR5. FGFR1–4 belong to receptor tyrosine kinases (RTKs), whereas FGFR5, lacking the protein tyrosine kinase domain, inhibits cell proliferation and promotes cell differentiation [108, 109].

Aberrant activation of FGF signaling represents a key player in tumor cell proliferation, differentiation, mobility, and invasion and involves many cancers including breast cancer via overexpression or mutational activation [110, 111]. Gene amplification, chromosomal translocation, aberrant transcriptional regulation, or down-modulation of negative regulators induces FGFR mutations and FGFR overexpression, resulting in tumor growth and progression [111, 112].

6.3.2 Targeting Therapy of FGFR Signaling

FGFRs have been associated with breast cancer development. Amplification of FGFR1 was detected in 14.5% of breast cancers [113]. Recent studies have demonstrated that FGFR1 overexpression is robustly associated with FGFR1 amplification [114, 115]. FGFR1 was identified as a potential therapeutic target for classic lobular carcinoma [114]. FGFR1 amplification was a significant independent predictor of poor outcome, specifically in ER-positive breast cancer [116], and has been implicated in resistance to endocrine therapies [115]. Integration of aCGH (comparative genomic hybridization) and expression data revealed that FGFR2 was significantly overexpressed when amplified in 4% of triple-negative breast cancers [117]. FGFR2 was also highly expressed in BRCA2-associated cancers [118]. FGFR3 expression is significantly increased in tamoxifen-resistant breast tumors compared with sensitive tumors [107]. Activation of FGFR3 reduced sensitivity to tamoxifen and fulvestrant in MCF7 cells, stimulating activation of the mitogen-activated protein kinase (MAPK) and phosphatidylinositol 3-kinase (PI3K) signaling pathways, both of which have been implicated in tamoxifen resistance in breast cancer [119, 120]. FGFR3 may play a role in promoting resistance to endocrine therapy. FGFR4 mRNAs were highly expressed in 32% of breast tumors [118]. Overexpression of FGFR4 in breast cancer is frequent by immunohistochemistry and associated with poor overall survival [121]. Targeting FGFRs using TKI258 (FGFR TKI) induces apoptosis and inhibits proliferation and mammary tumor outgrowth and metastasis in a breast cancer model [122]. This growing body of evidence has indicated that FGFRs may be a valuable target for treatment of breast cancer patients and stimulated the development of FGFR-targeted agents that are currently being evaluated in clinical studies (Table 6.3).

Table 6.3 FGFR-targeted inhibitors in breast cancer

Compound	Clinical development	Subtypes
Inhibitors		
AZD4547	Phase II	FGFR1-amplified HER2-negative breast cancer (NCT01795768)
		ER-positive breast cancer (NCT01791985)
JNJ-42756493	Phase I	Advanced or refractory breast cancer (NCT01703481)
BGJ398	Phase I	Metastatic breast cancer (NCT01928459)

6.3.2.1 Tyrosine Kinase Inhibitors

AZD4547 AZD4547 is an orally bioavailable, potent, and selective inhibitor which competes with ATP for binding to FGFR tyrosine kinase 1, 2, and 3, thus inhibiting autophosphorylation and downstream signaling in tumor cell lines and xenografts with deregulated FGFR expression [123–126]. In vitro kinase assays have demonstrated that AZD4547 inhibits FGFR1, FGFR2, and FGFR3 with IC50 values of 0.2, 2.5, and 1.8 nM, respectively [124]. A phase I study of AZD4547 in patients with advanced solid tumors showed that administration of AZD4547 with 80 mg bid continuous dosing was generally tolerated (NCT00979134) [127]. A phase II multicenter proof of concept study of AZD4547 in FGFR1-/FGFR2-amplified tumors demonstrated high activity in FGFR2-amplified gastric cancer and lower activity in FGFR1-amplified HER2-negative breast cancer (NCT01795768) [128]. A phase II trial combining AZD4547 with either letrozole or anastrozole (NCT01791985) in ER-positive breast cancer patients who have progressed on treatment with letrozole or anastrozole is ongoing.

JNJ-42756493 JNJ-42756493 is a selective inhibitor of FGFR1, FGFR2, FGFR3, and FGFR4 with nanomolar affinity for targeted therapy [129]. JNJ-42756493 suppressed FGFR signaling in tumor cell lines dependent upon deregulated FGFR expression both in vitro and in vivo [130]. Two multipart phase I in-human studies of JNJ-42756493 were initiated in advanced solid tumor patients (NCT01962532) (NCT01703481) and showed no dose-limiting toxicities or drug-related severe adverse events [131, 132].

BGJ398 BGJ398, a potent, orally bioavailable, small-molecule pan-FGFR kinase inhibitor, was found to inhibit FGFR1, FGFR2, and FGFR3 with IC 50 = 1 nM and FGFR4 with IC 50 = 60 nM [133]. A phase I study (NCT01004224) revealed that BGJ398 had a tolerable safety profile and single-agent activity in patients with advanced solid tumors with genetic alterations of FGFR1, FGFR2, and/or FGFR3. A tumor reduction was observed in FGFR1-amplified breast cancer [134]. A phase Ib clinical trial (NCT01928459) combining BGJ398 with the PI3K inhibitor BYL719 in patients with advanced solid tumors, including metastatic breast cancer, which expressed mutations of PIK3CA with or without alterations of FGFR1–3, was completed on January 9, 2017 with no study results being posted. A phase II study in selective FGFR pathway-regulated solid tumors (NCT02160041) is under evaluation.

6.4 IGF Signaling

6.4.1 Introduction of IGF Signaling

The insulin-like growth factor (IGF) signaling plays an important role in normal developmental and physiology [135, 136]. The IGF system comprises two ligands, IGF-1 and IGF-2, which bind to IGF-1R and IGF-2R. Both IGF-1 and IGF-2 signal through IGF-1R which is the main receptor in the IGF system with tyrosine kinase activity. IGF-2R is only activated by IGF-2 and cannot transduce any signals due to lack of kinase domain [137]. The binding of the ligand to IGF-1R results in the phosphorylation of the tyrosine kinase domain, which triggers and activates various downstream oncogenic pathways such as the mitogen-activated protein kinase (MAPK) and phosphatidylinositol-3 kinase (PI3K) signaling cascades to control cell proliferation, growth, and motile behavior [138]. IGF-1R shares 70% homology with insulin receptor (IR) [139]. IGFs (IGF-1 and IGF-2) and insulin can cross-bind to each other's receptor with low affinity [140].

IGF signaling dysregulation has been shown to be associated to the development and progression of many cancers and dwindling response to current standard-of-care therapy for tumors, such as breast cancer [141, 142]. An elevated IGF-1R content has been detected in nearly 80% of breast cancers compared with normal breast tissue [143]. In HER2 receptor-positive breast cancer, IGF-1R high expression was associated with an inferior prognosis, where HR = 2.37 (95% CI 1.21. 4.64) and P = 0.012 [144]. Trastuzumab

(Herceptin) is a monoclonal antibody commonly used in HER2 receptor-positive breast cancer. In breast cancer cell models that overexpress HER2, trastuzumab activity is interfered by increased level of IGF-1R [145]. Overexpression of IGF-IR is also observed in tamoxifen-resistant cancer cells [146, 147]. Therefore, IGF-1R represents a potential therapeutic target in cancer therapy. Currently, the efficacy and tolerability of IGF-1R-targeting monoclonal antibodies (mAbs) and small-molecule tyrosine kinase inhibitors (TKIs) are being evaluated in different phases of clinical trials [147].

6.4.2 Targeting Therapy of IGF Signaling

6.4.2.1 Monoclonal Antibodies

Humanized or fully human IGF1R mAbs against IGF1R can compete with IGFs and block ligand-dependent receptor signaling by binding to IGF-1R. Several monoclonal antibodies have advanced to the stage of clinical trials, including figitumumab (CP-751,871), ganitumab (AMG 479), AVE1642 (EM164), R-1507, MM-141, cixutumumab (IMC-A12), and dalotuzumab (MK-0646) (Table 6.4).

Figitumumab Figitumumab (CP-751,871), a human monoclonal antibody, was proved to block IGF1R ligand binding, inhibit IGF-I-induced autophosphorylation of the IGF-IR, and induce the downregulation of IGF-1R in vitro and in tumor xenografts through internalization and degradation of receptor-antibody complex [148]. Figitumumab can inhibit tumor growth both as a single agent and in combination with Adriamycin, 5-fluorouracil, or tamoxifen in multiple tumor models by enhancing the efficacy of both cytotoxic and targeted agents. Both in vitro and in vivo experiments have found that the combination of figitumumab with the therapeutic anti-HER2 antibody trastuzumab and the pan-HER

Table 6.4 IGF1R-targeted mAbs and inhibitors in breast cancer

Compound	Clinical development	Subtypes
mAbs		
Figitumumab	Phase II	Metastatic estrogen receptor-positive breast cancer (NCT00372996)
Ganitumab	Phase II	Hormone receptor-positive, locally advanced, or metastatic breast cancer (NCT00626106)
		Locally advanced breast cancer (NCT01042379)
		PIK3CA-mutated advanced breast cancer (NCT01708161)
Cixutumumab	Phase II	Hormone receptor-positive advanced or metastatic breast cancer (NCT00728949)
		HER2-positive stages IIIB–IV breast cancer (NCT00684983)
		Locally recurrent or metastatic breast cancer (NCT00699491)
R-1507	Phase II	Advanced breast cancer (NCT00796107)
AVE1642	Phase II	Advanced hormone-dependent breast cancer (NCT00774878)
Dalotuzumab	Phase II	Advanced luminal B breast cancer (NCT01234857)
		High-proliferative advanced breast cancer (NCT01605396)
		Metastatic hormone receptor-positive HER2-negative breast cancer (NCT00903006)
Tyrosine kinase inhibitors		
BMS-754807	Phase II	ER-positive breast cancer patients (NCT01225172)
		Advanced or metastatic Her-2-positive breast cancer (NCT00788333)

family tyrosine kinase inhibitor neratinib displays a synergistic effect on promoting cell apoptosis and inhibiting cell proliferation and tumor growth in the HER2-overexpressing (BT474) and HER2-normal (MCF7) breast cancer cells lines [149]. A phase II trial (NCT00372996) was designed to evaluate and compare the activities of figitumumab combined with exemestane and exemestane alone for metastatic estrogen receptor-positive breast cancer in postmenopausal women. The two groups showed no significant difference in progression-free survival (PFS) (HR 0.912, 95% CI 0.744–1.118; P = 0.560).

Ganitumab Ganitumab (AMG 479) is a fully human monoclonal antibody directed against IGF-1R [150]. AMG 479 can inhibit IGF-1R signaling activity in vitro and in vivo both as a single agent and in combination with therapy agents in a broad spectrum of tumor cell lines [150–153]. In a phase I clinical trial (NCT00562380), AMG 479 can be administered safely at a dose of up to 20 mg/kg intravenously (IV) every 2 weeks (Q2W) in patients with advanced solid malignancies including breast cancer [154] or non-Hodgkin's lymphoma [155]. In a phase II trial (NCT00626106), additional administration of ganitumab in postmenopausal women with hormone receptor-positive, locally advanced, or metastatic breast cancer who were previously treated with endocrine treatment did not improve progression-free survival and overall survival compared with placebo (HR 1·78, 80% CI 1·27–2·50; p = 0·025) [156]. Results from the I-SPY phase II trial (NCT01042379) of AMG 479/metformin plus standard neoadjuvant therapy regimen with or without trastuzumab for locally advanced breast cancers showed that no breast cancer subtype came close to the efficacy threshold of 85% likelihood of success in phase III. Therefore, the therapy did not appear to impact upfront reduction of tumor burden, and the trial was closed for the neoadjuvant treatment of breast cancer [157]. A phase Ib/II study (NCT01708161) of the combination of BYL719 plus AMG 479 in adult patients with PIK3CA-mutated advanced solid tumors including breast carcinoma and ovarian carcinoma was started in 2012 and is still ongoing.

Cixutumumab Cixutumumab (IMC-A12) is a fully human monoclonal IgG1 antibody that binds to the IGF-1R and inhibits IGFs binding and downstream signaling mechanisms in MCF7 human breast cancer cells in vitro and in vivo [158]. In a phase II clinical trial (NCT00728949) completed in 2015, the efficacy and tolerability of cixutumumab as a single agent was assessed to test whether hormone receptor-positive breast cancer cells that developed resistance to antiestrogen therapy might benefit from IGF-1R blockade [159]. The antitumor effect of cixutumumab in combination with antiestrogen therapies was also evaluated in patients showing resistance to antiestrogen therapy. Cixutumumab administered at 10 mg/kg with or without antiestrogen every 2 weeks had an acceptable safety profile, but no significant clinical efficacy. A phase II trial, completed in 2017 (NCT00684983) in women with previously treated HER2-positive stages IIIB–IV breast cancer, has evaluated the effects of capecitabine and lapatinib ditosylate (cape/lap) with or without cixutumumab. Results first received in 2014 showed that progression-free survival between the two groups had no significant difference (HR 1.04, 95%CI 0.58–1.89; p = 0.89) [160].

PI3K-AKT-mTOR signaling pathway is frequently activated in breast cancer and plays a critical role in promoting tumor cell growth [161–163]. In preclinical and clinical studies, the antitumor activity of mTOR inhibitors is attenuated by feedback induction of AKT phosphorylation mediated in part by IGF-1R [164–166]. A phase I trial (NCT00699491) was initiated to determine the maximum tolerated dose (MTD) and pharmacodynamic effects of cixutumumab in combination with temsirolimus (mTOR inhibitor) in patients with metastatic breast cancer refractory to standard therapies [167]. A phase II study in women with metastatic breast cancer is ongoing.

R-1507 R-1507, a humanized IGF1R mAb, has been developed to inhibit IGF-1R autophosphorylation and subsequent signal transduction by binding to the extracellular domain of IGF-IR with high affinity and selectivity [168]. A phase I trial showed that weekly administration of R1507 was well tolerated at the maximal administered dose of 9 mg/kg with no significant drug-related toxicities and showed antitumor activity in patients with advanced solid neoplasms, in particular Ewing's sarcoma [169, 170]. R1507 also displayed antiproliferative and anti-invasive activity in both tamoxifen-responsive (wild-type MCF7) and tamoxifen-resistant (Tam-R MCF7) breast cancer cell lines [171]. Moreover, R1507 suppressed IGF1R expression and inhibited IGF-1-stimulated IGF1R and AKT phosphorylation in ER-positive breast cancer cells lines (MCF7, T47D, and HCC712) [172]. The results indicated that R1507 might have efficacy in patients with endocrine therapy-resistant tumors. A phase II trial (NCT00796107) completed in July 2016 combined R1507 with letrozole (nonsteroidal aromatase inhibitor) in postmenopausal women with advanced breast cancer. Mature data of the study was not yet available.

AVE1642 AVE1642 (EM164), an antagonistic monoclonal antibody, can bind specifically to the IGF-IR with high affinity and inhibit the proliferation and survival functions of the receptor in diverse human cancer cell lines in vitro, including breast, lung, colon, cervical, ovarian, pancreatic, melanoma, prostate, neuroblastoma, rhabdomyosarcoma, and osteosarcoma cancer lines [173–176]. Treatment with AVE1642, either alone or in combination with gemcitabine, inhibited the growth of BxPC-3 human pancreatic cancer xenografts in SCID mice [173]. A phase I dose-escalation study was undertaken in patients with refractory advanced solid tumors, which showed that AVE1642 was well tolerated both as a single agent and in combination with docetaxel [177, 178]. A phase II study (NCT00774878) initiated in 2008 was meant to evaluate the clinical activity of AVE1642 in combination with fulvestrant and of fulvestrant alone in postmenopausal patients with advanced hormone-dependent breast cancer. However, the study was terminated not due to any safety or efficacy concerns in 2011.

Dalotuzumab Dalotuzumab (MK-0646;h7C10), a recombinant humanized mAb targeted against IGFR1, inhibits IGF-mediated tumor cell proliferation, IGFR1 autophosphorylation, and Akt phosphorylation in multiple cancer cell lines [179–181]. In mouse xenograft models, dalotuzumab displayed significant antitumor activity in particular against NSCLC and breast cancer [179]. A phase I clinical trial (NCT00759785) in patients with stages I–IIIa breast cancer has suggested that dalotuzumab is safe and well tolerated. A phase I clinical trial (NCT00730379) evaluating dalotuzumab in combination with ridaforolimus (mTOR inhibitor) showed significant antitumor activity in heavily pretreated advanced cancer, particularly in ER^+/high-proliferative breast cancer [182, 183]. A phase II trial (NCT01234857) designed to compare the combination of ridaforolimus and dalotuzumab with endocrine therapy in patients with advanced luminal B breast cancer has been recently completed. It showed that the combination of ridaforolimus plus dalotuzumab was no more effective than exemestane but with higher incidence of adverse events [184–186]. A phase II trial (NCT01605396) comparing the combination of ridaforolimus, dalotuzumab, and exemestane with that of ridaforolimus and exemestane in patients with high-proliferative advanced breast cancer, which had progressed following treatment with a nonsteroidal aromatase inhibitor, showed no significant difference in median PFS (HR 1.18; 80% CI, 0.81–1.72; P = 0.565) [187]. A phase I–II trial (NCT00903006) combined dalotuzumab with Sprycel (dasatinib) and Faslodex (fulvestrant) to treat patients with metastatic hormone receptor-positive HER2-negative breast cancer.

6.4.2.2 Tyrosine Kinase Inhibitors

Tyrosine kinase inhibitors have been reported to target IGF-1R, among which BMS-754807 has entered into clinical evaluation for breast cancer treatment (Table 6.3).

BMS-754807 BMS-754807 is a potent and reversible inhibitor of the insulin-like growth factor 1 receptor/insulin receptor family kinases. It inhibits the phosphorylation of IGF-1R and the downstream targets in vitro and achieves tumor growth inhibition with strong efficacy in multiple (epithelial, mesenchymal, and hematopoietic) xenograft tumor models with minimal weight loss [188–190]. Compared with single-agent therapy, combined treatment of BMS-754807 with either tamoxifen or letrozole exhibited antiproliferative effects in vitro and tumor regressions in vivo without major side effects in a model of postmenopausal, estrogen-dependent breast cancer [190]. Two phase II studies, one in ER-positive breast cancer patients with locally advanced/metastatic and acquired resistance to nonsteroidal aromatase inhibitors in combination with the aromatase inhibitor, letrozole (NCT01225172), and the other in subjects with advanced or metastatic Her-2-positive breast cancer after trastuzumab failure in combination with trastuzumab (Herceptin) (NCT00788333), have been completed with no mature data available.

References

1. Lemmon MA, Schlessinger J (2010) Cell signaling by receptor tyrosine kinases. Cell 141(7):1117–1134. doi:10.1016/j.cell.2010.06.011
2. Cohen S (1983) The epidermal growth factor (EGF). Cancer 51(10):1787–1791
3. Roskoski R Jr (2014) The ErbB/HER family of protein-tyrosine kinases and cancer. Pharmacol Res 79:34–74. doi:10.1016/j.phrs.2013.11.002
4. Cohen S, Ushiro H, Stoscheck C, Chinkers M (1982) A native 170,000 epidermal growth factor receptor-kinase complex from shed plasma membrane vesicles. J Biol Chem 257(3):1523–1531
5. Cohen S, Fava RA, Sawyer ST (1982) Purification and characterization of epidermal growth factor receptor/protein kinase from normal mouse liver. Proc Natl Acad Sci U S A 79(20):6237–6241
6. Park OK, Schaefer TS, Nathans D (1996) In vitro activation of Stat3 by epidermal growth factor receptor kinase. Proc Natl Acad Sci U S A 93(24):13704–13708
7. Anderson D, Koch CA, Grey L, Ellis C, Moran MF, Pawson T (1990) Binding of SH2 domains of phospholipase C gamma 1, GAP, and Src to activated growth factor receptors. Science 250(4983):979–982
8. Navolanic PM, Steelman LS, McCubrey JA (2003) EGFR family signaling and its association with breast cancer development and resistance to chemotherapy (review). Int J Oncol 22(2):237–252
9. Yarden Y, Sliwkowski MX (2001) Untangling the ErbB signalling network. Nat Rev Mol Cell Biol 2(2):127–137. doi:10.1038/35052073
10. Vojtek AB, Hollenberg SM, Cooper JA (1993) Mammalian Ras interacts directly with the serine/threonine kinase Raf. Cell 74(1):205–214
11. Wu CJ, Qian X, O'Rourke DM (1999) Sustained mitogen-activated protein kinase activation is induced by transforming erbB receptor complexes. DNA Cell Biol 18(10):731–741. doi:10.1089/104454999314872
12. Hashimoto A, Kurosaki M, Gotoh N, Shibuya M, Kurosaki T (1999) Shc regulates epidermal growth factor-induced activation of the JNK signaling pathway. J Biol Chem 274(29):20139–20143
13. Voice JK, Klemke RL, Le A, Jackson JH (1999) Four human ras homologs differ in their abilities to activate Raf-1, induce transformation, and stimulate cell motility. J Biol Chem 274(24):17164–17170
14. Chan TO, Rodeck U, Chan AM, Kimmelman AC, Rittenhouse SE, Panayotou G, Tsichlis PN (2002) Small GTPases and tyrosine kinases coregulate a molecular switch in the phosphoinositide 3-kinase regulatory subunit. Cancer Cell 1(2):181–191
15. Pacold ME, Suire S, Perisic O, Lara-Gonzalez S, Davis CT, Walker EH, Hawkins PT, Stephens L, Eccleston JF, Williams RL (2000) Crystal structure and functional analysis of Ras binding to its effector phosphoinositide 3-kinase gamma. Cell 103(6):931–943
16. Walker EH, Perisic O, Ried C, Stephens L, Williams RL (1999) Structural insights into phosphoinositide 3-kinase catalysis and signalling. Nature 402(6759):313–320. doi:10.1038/46319
17. Darnell JE Jr, Kerr IM, Stark GR (1994) Jak-STAT pathways and transcriptional activation in response to IFNs and other extracellular signaling proteins. Science 264(5164):1415–1421
18. Fu XY (1999) From PTK-STAT signaling to caspase expression and apoptosis induction. Cell Death Differ 6(12):1201–1208. doi:10.1038/sj.cdd.4400613
19. Barre B, Avril S, Coqueret O (2003) Opposite regulation of myc and p21waf1 transcription by STAT3 proteins. J Biol Chem 278(5):2990–2996. doi:10.1074/jbc.M210422200
20. Lo HW, Hsu SC, Ali-Seyed M, Gunduz M, Xia W, Wei Y, Bartholomeusz G, Shih JY, Hung MC (2005) Nuclear interaction of EGFR and STAT3 in the activation of the iNOS/NO pathway. Cancer Cell 7(6):575–589. doi:10.1016/j.ccr.2005.05.007
21. Bromberg JF, Wrzeszczynska MH, Devgan G, Zhao Y, Pestell RG, Albanese C, Darnell JE Jr (1999) Stat3 as an oncogene. Cell 98(3):295–303
22. Wei D, Le X, Zheng L, Wang L, Frey JA, Gao AC, Peng Z, Huang S, Xiong HQ, Abbruzzese JL,

Xie K (2003) Stat3 activation regulates the expression of vascular endothelial growth factor and human pancreatic cancer angiogenesis and metastasis. Oncogene 22(3):319–329. doi:10.1038/sj.onc.1206122
23. Niu G, Wright KL, Huang M, Song L, Haura E, Turkson J, Zhang S, Wang T, Sinibaldi D, Coppola D, Heller R, Ellis LM, Karras J, Bromberg J, Pardoll D, Jove R, Yu H (2002) Constitutive Stat3 activity up-regulates VEGF expression and tumor angiogenesis. Oncogene 21(13):2000–2008. doi:10.1038/sj.onc.1205260
24. Yarden Y, Pines G (2012) The ERBB network: at last, cancer therapy meets systems biology. Nat Rev Cancer 12(8):553–563. doi:10.1038/nrc3309
25. Lianos GD, Vlachos K, Zoras O, Katsios C, Cho WC, Roukos DH (2014) Potential of antibody-drug conjugates and novel therapeutics in breast cancer management. Onco Targets Ther 7:491–500. doi:10.2147/OTT.S34235
26. Arteaga CL, Moulder SL, Yakes FM (2002) HER (erbB) tyrosine kinase inhibitors in the treatment of breast cancer. Semin Oncol 29(3 Suppl 11):4–10
27. Ciardiello F, Tortora G (2001) A novel approach in the treatment of cancer: targeting the epidermal growth factor receptor. Clin Cancer Res 7(10):2958–2970
28. Mendelsohn J, Baselga J (2003) Status of epidermal growth factor receptor antagonists in the biology and treatment of cancer. J Clin Oncol 21(14):2787–2799. doi:10.1200/JCO.2003.01.504
29. Normanno N, De Luca A, Bianco C, Strizzi L, Mancino M, Maiello MR, Carotenuto A, De Feo G, Caponigro F, Salomon DS (2006) Epidermal growth factor receptor (EGFR) signaling in cancer. Gene 366(1):2–16. doi:10.1016/j.gene.2005.10.018
30. Lluch A, Eroles P, Perez-Fidalgo JA (2014) Emerging EGFR antagonists for breast cancer. Expert Opin Emerg Drugs 19(2):165–181. doi:10.1517/14728214.2014.903919
31. Howe LR, Brown PH (2011) Targeting the HER/EGFR/ErbB family to prevent breast cancer. Cancer Prev Res (Phila) 4(8):1149–1157. doi:10.1158/1940-6207.CAPR-11-0334
32. Cho HS, Mason K, Ramyar KX, Stanley AM, Gabelli SB, Denney DW Jr, Leahy DJ (2003) Structure of the extracellular region of HER2 alone and in complex with the Herceptin fab. Nature 421(6924):756–760. doi:10.1038/nature01392
33. Baselga J, Albanell J, Molina MA, Arribas J (2001) Mechanism of action of trastuzumab and scientific update. Semin Oncol 28(5 Suppl 16):4–11
34. Varchetta S, Gibelli N, Oliviero B, Nardini E, Gennari R, Gatti G, Silva LS, Villani L, Tagliabue E, Menard S, Costa A, Fagnoni FF (2007) Elements related to heterogeneity of antibody-dependent cell cytotoxicity in patients under trastuzumab therapy for primary operable breast cancer overexpressing Her2. Cancer Res 67(24):11991–11999. doi:10.1158/0008-5472.CAN-07-2068
35. Vogel CL, Cobleigh MA, Tripathy D, Gutheil JC, Harris LN, Fehrenbacher L, Slamon DJ, Murphy M, Novotny WF, Burchmore M, Shak S, Stewart SJ, Press M (2002) Efficacy and safety of trastuzumab as a single agent in first-line treatment of HER2-overexpressing metastatic breast cancer. J Clin Oncol 20(3):719–726. doi:10.1200/JCO.2002.20.3.719
36. Smith I, Procter M, Gelber RD, Guillaume S, Feyereislova A, Dowsett M, Goldhirsch A, Untch M, Mariani G, Baselga J, Kaufmann M, Cameron D, Bell R, Bergh J, Coleman R, Wardley A, Harbeck N, Lopez RI, Mallmann P, Gelmon K, Wilcken N, Wist E, Sanchez Rovira P, Piccart-Gebhart MJ (2007) 2-year follow-up of trastuzumab after adjuvant chemotherapy in HER2-positive breast cancer: a randomised controlled trial. Lancet 369(9555):29–36. doi:10.1016/S0140-6736(07)60028-2
37. Slamon DJ, Leyland-Jones B, Shak S, Fuchs H, Paton V, Bajamonde A, Fleming T, Eiermann W, Wolter J, Pegram M, Baselga J, Norton L (2001) Use of chemotherapy plus a monoclonal antibody against HER2 for metastatic breast cancer that overexpresses HER2. N Engl J Med 344(11):783–792. doi:10.1056/NEJM200103153441101
38. Piccart-Gebhart MJ, Procter M, Leyland-Jones B, Goldhirsch A, Untch M, Smith I, Gianni L, Baselga J, Bell R, Jackisch C, Cameron D, Dowsett M, Barrios CH, Steger G, Huang CS, Andersson M, Inbar M, Lichinitser M, Lang I, Nitz U, Iwata H, Thomssen C, Lohrisch C, Suter TM, Ruschoff J, Suto T, Greatorex V, Ward C, Straehle C, McFadden E, Dolci MS, Gelber RD (2005) Trastuzumab after adjuvant chemotherapy in HER2-positive breast cancer. N Engl J Med 353(16):1659–1672. doi:10.1056/NEJMoa052306
39. Romond EH, Perez EA, Bryant J, Suman VJ, Geyer CE Jr, Davidson NE, Tan-Chiu E, Martino S, Paik S, Kaufman PA, Swain SM, Pisansky TM, Fehrenbacher L, Kutteh LA, Vogel VG, Visscher DW, Yothers G, Jenkins RB, Brown AM, Dakhil SR, Mamounas EP, Lingle WL, Klein PM, Ingle JN, Wolmark N (2005) Trastuzumab plus adjuvant chemotherapy for operable HER2-positive breast cancer. N Engl J Med 353(16):1673–1684. doi:10.1056/NEJMoa052122
40. Hudis CA (2007) Trastuzumab--mechanism of action and use in clinical practice. N Engl J Med 357(1):39–51. doi:10.1056/NEJMra043186
41. Roukos DH (2011) Trastuzumab and beyond: sequencing cancer genomes and predicting molecular networks. Pharmacogenomics J 11(2):81–92. doi:10.1038/tpj.2010.81
42. Shepard HM, Brdlik CM, Schreiber H (2008) Signal integration: a framework for understanding the efficacy of therapeutics targeting the human EGFR family. J Clin Invest 118(11):3574–3581. doi:10.1172/JCI36049

43. Rexer BN, Arteaga CL (2012) Intrinsic and acquired resistance to HER2-targeted therapies in HER2 gene-amplified breast cancer: mechanisms and clinical implications. Crit Rev Oncog 17(1):1–16
44. Geyer CE, Forster J, Lindquist D, Chan S, Romieu CG, Pienkowski T, Jagiello-Gruszfeld A, Crown J, Chan A, Kaufman B, Skarlos D, Campone M, Davidson N, Berger M, Oliva C, Rubin SD, Stein S, Cameron D (2006) Lapatinib plus capecitabine for HER2-positive advanced breast cancer. N Engl J Med 355(26):2733–2743. doi:10.1056/NEJMoa064320
45. Franklin MC, Carey KD, Vajdos FF, Leahy DJ, de Vos AM, Sliwkowski MX (2004) Insights into ErbB signaling from the structure of the ErbB2-pertuzumab complex. Cancer Cell 5(4):317–328
46. Baselga J, Cortes J, Kim SB, Im SA, Hegg R, Im YH, Roman L, Pedrini JL, Pienkowski T, Knott A, Clark E, Benyunes MC, Ross G, Swain SM (2012) Pertuzumab plus trastuzumab plus docetaxel for metastatic breast cancer. N Engl J Med 366(2):109–119. doi:10.1056/NEJMoa1113216
47. Blumenthal GM, Scher NS, Cortazar P, Chattopadhyay S, Tang S, Song P, Liu Q, Ringgold K, Pilaro AM, Tilley A, King KE, Graham L, Rellahan BL, Weinberg WC, Chi B, Thomas C, Hughes P, Ibrahim A, Justice R, Pazdur R (2013) First FDA approval of dual anti-HER2 regimen: pertuzumab in combination with trastuzumab and docetaxel for HER2-positive metastatic breast cancer. Clin Cancer Res 19(18):4911–4916. doi:10.1158/1078-0432.CCR-13-1212
48. Zheng Y, Zhang C, Croucher DR, Soliman MA, St-Denis N, Pasculescu A, Taylor L, Tate SA, Hardy WR, Colwill K, Dai AY, Bagshaw R, Dennis JW, Gingras AC, Daly RJ, Pawson T (2013) Temporal regulation of EGF signalling networks by the scaffold protein Shc1. Nature 499(7457):166–171. doi:10.1038/nature12308
49. Sapra P, Betts A, Boni J (2013) Preclinical and clinical pharmacokinetic/pharmacodynamic considerations for antibody-drug conjugates. Expert Rev Clin Pharmacol 6(5):541–555. doi:10.1586/17512433.2013.827405
50. Burris HA 3rd, Rugo HS, Vukelja SJ, Vogel CL, Borson RA, Limentani S, Tan-Chiu E, Krop IE, Michaelson RA, Girish S, Amler L, Zheng M, Chu YW, Klencke B, O'Shaughnessy JA (2011) Phase II study of the antibody drug conjugate trastuzumab-DM1 for the treatment of human epidermal growth factor receptor 2 (HER2)-positive breast cancer after prior HER2-directed therapy. J Clin Oncol 29(4):398–405. doi:10.1200/JCO.2010.29.5865
51. Barginear MF, John V, Budman DR (2013) Trastuzumab-DM1: a clinical update of the novel antibody-drug conjugate for HER2-overexpressing breast cancer. Mol Med 18:1473–1479. doi:10.2119/molmed.2012.00302
52. Senter PD (2009) Potent antibody drug conjugates for cancer therapy. Curr Opin Chem Biol 13(3):235–244. doi:10.1016/j.cbpa.2009.03.023
53. Opdam FL, Guchelaar HJ, Beijnen JH, Schellens JH (2012) Lapatinib for advanced or metastatic breast cancer. Oncologist 17(4):536–542. doi:10.1634/theoncologist.2011-0461
54. Xia W, Mullin RJ, Keith BR, Liu LH, Ma H, Rusnak DW, Owens G, Alligood KJ, Spector NL (2002) Anti-tumor activity of GW572016: a dual tyrosine kinase inhibitor blocks EGF activation of EGFR/erbB2 and downstream Erk1/2 and AKT pathways. Oncogene 21(41):6255–6263. doi:10.1038/sj.onc.1205794
55. Ryan Q, Ibrahim A, Cohen MH, Johnson J, Ko CW, Sridhara R, Justice R, Pazdur R (2008) FDA drug approval summary: lapatinib in combination with capecitabine for previously treated metastatic breast cancer that overexpresses HER-2. Oncologist 13(10):1114–1119. doi:10.1634/theoncologist.2008-0816
56. Cameron D (2007) Lapatinib plus capecitabine in patients with HER2-positive advanced breast cancer. Clin Adv Hematol Oncol 5(6):456–458
57. Riemsma R, Forbes CA, Amonkar MM, Lykopoulos K, Diaz JR, Kleijnen J, Rea DW (2012) Systematic review of lapatinib in combination with letrozole compared with other first-line treatments for hormone receptor positive(HR+) and HER2+ advanced or metastatic breast cancer(MBC). Curr Med Res Opin 28(8):1263–1279. doi:10.1185/03007995.2012.707643
58. Finn RS, Press MF, Dering J, Arbushites M, Koehler M, Oliva C, Williams LS, Di Leo A (2009) Estrogen receptor, progesterone receptor, human epidermal growth factor receptor 2 (HER2), and epidermal growth factor receptor expression and benefit from lapatinib in a randomized trial of paclitaxel with lapatinib or placebo as first-line treatment in HER2-negative or unknown metastatic breast cancer. J Clin Oncol 27(24):3908–3915. doi:10.1200/JCO.2008.18.1925
59. Burstein HJ, Storniolo AM, Franco S, Forster J, Stein S, Rubin S, Salazar VM, Blackwell KL (2008) A phase II study of lapatinib monotherapy in chemotherapy-refractory HER2-positive and HER2-negative advanced or metastatic breast cancer. Ann Oncol 19(6):1068–1074. doi:10.1093/annonc/mdm601
60. Di Leo A, Gomez HL, Aziz Z, Zvirbule Z, Bines J, Arbushites MC, Guerrera SF, Koehler M, Oliva C, Stein SH, Williams LS, Dering J, Finn RS, Press MF (2008) Phase III, double-blind, randomized study comparing lapatinib plus paclitaxel with placebo plus paclitaxel as first-line treatment for metastatic breast cancer. J Clin Oncol 26(34):5544–5552. doi:10.1200/JCO.2008.16.2578
61. Johnston S, Pippen J Jr, Pivot X, Lichinitser M, Sadeghi S, Dieras V, Gomez HL, Romieu G,

Manikhas A, Kennedy MJ, Press MF, Maltzman J, Florance A, O'Rourke L, Oliva C, Stein S, Pegram M (2009) Lapatinib combined with letrozole versus letrozole and placebo as first-line therapy for postmenopausal hormone receptor-positive metastatic breast cancer. J Clin Oncol 27(33):5538–5546. doi:10.1200/JCO.2009.23.3734

62. Press MF, Finn RS, Cameron D, Di Leo A, Geyer CE, Villalobos IE, Santiago A, Guzman R, Gasparyan A, Ma Y, Danenberg K, Martin AM, Williams L, Oliva C, Stein S, Gagnon R, Arbushites M, Koehler MT (2008) HER-2 gene amplification, HER-2 and epidermal growth factor receptor mRNA and protein expression, and lapatinib efficacy in women with metastatic breast cancer. Clin Cancer Res 14(23):7861–7870. doi:10.1158/1078-0432.CCR-08-1056
63. Rexer BN, Ghosh R, Narasanna A, Estrada MV, Chakrabarty A, Song Y, Engelman JA, Arteaga CL (2013) Human breast cancer cells harboring a gatekeeper T798M mutation in HER2 overexpress EGFR ligands and are sensitive to dual inhibition of EGFR and HER2. Clin Cancer Res 19(19):5390–5401. doi:10.1158/1078-0432.CCR-13-1038
64. Puglisi F, Minisini AM, De Angelis C, Arpino G (2012) Overcoming treatment resistance in HER2-positive breast cancer: potential strategies. Drugs 72(9):1175–1193. doi:10.2165/11634000-000000000-00000
65. Liu T, Yacoub R, Taliaferro-Smith LD, Sun SY, Graham TR, Dolan R, Lobo C, Tighiouart M, Yang L, Adams A, O'Regan RM (2011) Combinatorial effects of lapatinib and rapamycin in triple-negative breast cancer cells. Mol Cancer Ther 10(8):1460–1469. doi:10.1158/1535-7163.MCT-10-0925
66. Corkery B, Crown J, Clynes M, O'Donovan N (2009) Epidermal growth factor receptor as a potential therapeutic target in triple-negative breast cancer. Ann Oncol 20(5):862–867. doi:10.1093/annonc/mdn710
67. Baselga J, Albanell J, Ruiz A, Lluch A, Gascon P, Guillem V, Gonzalez S, Sauleda S, Marimon I, Tabernero JM, Koehler MT, Rojo F (2005) Phase II and tumor pharmacodynamic study of gefitinib in patients with advanced breast cancer. J Clin Oncol 23(23):5323–5333. doi:10.1200/JCO.2005.08.326
68. Bruns CJ, Solorzano CC, Harbison MT, Ozawa S, Tsan R, Fan D, Abbruzzese J, Traxler P, Buchdunger E, Radinsky R, Fidler IJ (2000) Blockade of the epidermal growth factor receptor signaling by a novel tyrosine kinase inhibitor leads to apoptosis of endothelial cells and therapy of human pancreatic carcinoma. Cancer Res 60(11):2926–2935
69. Ciardiello F, Caputo R, Bianco R, Damiano V, Fontanini G, Cuccato S, De Placido S, Bianco AR, Tortora G (2001) Inhibition of growth factor production and angiogenesis in human cancer cells by ZD1839 (Iressa), a selective epidermal growth factor receptor tyrosine kinase inhibitor. Clin Cancer Res 7(5):1459–1465
70. Hirata A, Ogawa S, Kometani T, Kuwano T, Naito S, Kuwano M, Ono M (2002) ZD1839 (Iressa) induces antiangiogenic effects through inhibition of epidermal growth factor receptor tyrosine kinase. Cancer Res 62(9):2554–2560
71. Perrotte P, Matsumoto T, Inoue K, Kuniyasu H, Eve BY, Hicklin DJ, Radinsky R, Dinney CP (1999) Anti-epidermal growth factor receptor antibody C225 inhibits angiogenesis in human transitional cell carcinoma growing orthotopically in nude mice. Clin Cancer Res 5(2):257–265
72. Dickler MN, Rugo HS, Eberle CA, Brogi E, Caravelli JF, Panageas KS, Boyd J, Yeh B, Lake DE, Dang CT, Gilewski TA, Bromberg JF, Seidman AD, D'Andrea GM, Moasser MM, Melisko M, Park JW, Dancey J, Norton L, Hudis CA (2008) A phase II trial of erlotinib in combination with bevacizumab in patients with metastatic breast cancer. Clin Cancer Res 14(23):7878–7883. doi:10.1158/1078-0432.CCR-08-0141
73. Montagna E, Cancello G, Bagnardi V, Pastrello D, Dellapasqua S, Perri G, Viale G, Veronesi P, Luini A, Intra M, Calleri A, Rampinelli C, Goldhirsch A, Bertolini F, Colleoni M (2012) Metronomic chemotherapy combined with bevacizumab and erlotinib in patients with metastatic HER2-negative breast cancer: clinical and biological activity. Clin Breast Cancer 12(3):207–214. doi:10.1016/j.clbc.2012.03.008
74. Hinck AP, Mueller TD, Springer TA (2016) Structural biology and evolution of the TGF-beta family. Cold Spring Harb Perspect Biol 8(12). doi:10.1101/cshperspect.a022103
75. Massague J (2012) TGFbeta signalling in context. Nat Rev Mol Cell Biol 13(10):616–630. doi:10.1038/nrm3434
76. Hanahan D, Weinberg RA (2000) The hallmarks of cancer. Cell 100(1):57–70
77. Blobe GC, Schiemann WP, Lodish HF (2000) Role of transforming growth factor beta in human disease. N Engl J Med 342(18):1350–1358. doi:10.1056/NEJM200005043421807
78. Cheung SY, Boey YJ, Koh VC, Thike AA, Lim JC, Iqbal J, Tan PH (2015) Role of epithelial-mesenchymal transition markers in triple-negative breast cancer. Breast Cancer Res Treat 152(3):489–498. doi:10.1007/s10549-015-3485-1
79. Akhurst RJ, Hata A (2012) Targeting the TGFbeta signalling pathway in disease. Nat Rev Drug Discov 11(10):790–811. doi:10.1038/nrd3810
80. Delolme F, Anastasi C, Alcaraz LB, Mendoza V, Vadon-Le Goff S, Talantikite M, Capomaccio R, Mevaere J, Fortin L, Mazzocut D, Damour O, Zanella-Cleon I, Hulmes DJ, Overall CM, Valcourt U, Lopez-Casillas F, Moali C (2015) Proteolytic control of TGF-beta co-receptor activity by BMP-1/tolloid-like proteases revealed by quantitative iTRAQ proteomics. Cellular and molecular life sciences : CMLS 72(5):1009–1027. doi:10.1007/s00018-014-1733-x

81. Derynck R, Akhurst RJ, Balmain A (2001) TGF-beta signaling in tumor suppression and cancer progression. Nat Genet 29(2):117–129. doi:10.1038/ng1001-117
82. Elliott RL, Blobe GC (2005) Role of transforming growth factor Beta in human cancer. Journal of clinical oncology : official journal of the American Society of Clinical Oncology 23(9):2078–2093. doi:10.1200/JCO.2005.02.047
83. Fenig E, Kanfi Y, Wang Q, Beery E, Livnat T, Wasserman L, Lilling G, Yahalom J, Wieder R, Nordenberg J (2001) Role of transforming growth factor beta in the growth inhibition of human breast cancer cells by basic fibroblast growth factor. Breast Cancer Res Treat 70(1):27–37
84. Li C, Guo B, Bernabeu C, Kumar S (2001) Angiogenesis in breast cancer: the role of transforming growth factor beta and CD105. Microsc Res Tech 52(4):437–449. doi:10.1002/1097-0029(20010215)52:4<437::AID-JEMT1029>3.0.CO;2-G
85. Wilson TJ, Nannuru KC, Futakuchi M, Singh RK (2010) Cathepsin G-mediated enhanced TGF-beta signaling promotes angiogenesis via upregulation of VEGF and MCP-1. Cancer Lett 288(2):162–169. doi:10.1016/j.canlet.2009.06.035
86. Chen W, Zhou S, Mao L, Zhang H, Sun D, Zhang J, Li J, Tang JH (2016) Crosstalk between TGF-beta signaling and miRNAs in breast cancer metastasis. Tumour biology : the journal of the International Society for Oncodevelopmental Biology and Medicine 37(8):10011–10019. doi:10.1007/s13277-016-5060-8
87. Amankulor NM, Hambardzumyan D, Pyonteck SM, Becher OJ, Joyce JA, Holland EC (2009) Sonic hedgehog pathway activation is induced by acute brain injury and regulated by injury-related inflammation. The Journal of neuroscience : the official journal of the Society for Neuroscience 29(33):10299–10308. doi:10.1523/JNEUROSCI.2500-09.2009
88. Herbertz S, Sawyer JS, Stauber AJ, Gueorguieva I, Driscoll KE, Estrem ST, Cleverly AL, Desaiah D, Guba SC, Benhadji KA, Slapak CA, Lahn MM (2015) Clinical development of galunisertib (LY2157299 monohydrate), a small molecule inhibitor of transforming growth factor-beta signaling pathway. Drug Des Devel Ther 9:4479–4499. doi:10.2147/DDDT.S86621
89. Tang Y, Yang X, Friesel RE, Vary CP, Liaw L (2011) Mechanisms of TGF-beta-induced differentiation in human vascular smooth muscle cells. J Vasc Res 48(6):485–494. doi:10.1159/000327776
90. Hawinkels LJ, Garcia de Vinuesa A, Ten Dijke P (2013) Activin receptor-like kinase 1 as a target for anti-angiogenesis therapy. Expert Opin Investig Drugs 22(11):1371–1383. doi:10.1517/13543784.2013.837884
91. Kano MR, Bae Y, Iwata C, Morishita Y, Yashiro M, Oka M, Fujii T, Komuro A, Kiyono K, Kaminishi M, Hirakawa K, Ouchi Y, Nishiyama N, Kataoka K, Miyazono K (2007) Improvement of cancer-targeting therapy, using nanocarriers for intractable solid tumors by inhibition of TGF-beta signaling. Proc Natl Acad Sci U S A 104(9):3460–3465. doi:10.1073/pnas.0611660104
92. Liu Z, Kobayashi K, van Dinther M, van Heiningen SH, Valdimarsdottir G, van Laar T, Scharpfenecker M, Lowik CW, Goumans MJ, Ten Dijke P, Pardali E (2009) VEGF and inhibitors of TGFbeta type-I receptor kinase synergistically promote blood-vessel formation by inducing alpha5-integrin expression. J Cell Sci 122(Pt 18):3294–3302. doi:10.1242/jcs.048942
93. Jin CH, Krishnaiah M, Sreenu D, Subrahmanyam VB, Rao KS, Lee HJ, Park SJ, Park HJ, Lee K, Sheen YY, Kim DK (2014) Discovery of N-((4-([1,2,4]triazolo[1,5-a]pyridin-6-yl)-5-(6-methylpyridin-2-yl)-1H-imidazol-2 -yl)methyl)-2-fluoroaniline (EW-7197): a highly potent, selective, and orally bioavailable inhibitor of TGF-beta type I receptor kinase as cancer immunotherapeutic/antifibrotic agent. J Med Chem 57(10):4213–4238. doi:10.1021/jm500115w
94. Son JY, Park SY, Kim SJ, Lee SJ, Park SA, Kim MJ, Kim SW, Kim DK, Nam JS, Sheen YY (2014) EW-7197, a novel ALK-5 kinase inhibitor, potently inhibits breast to lung metastasis. Mol Cancer Ther 13(7):1704–1716. doi:10.1158/1535-7163.MCT-13-0903
95. Wrana JL, Attisano L, Carcamo J, Zentella A, Doody J, Laiho M, Wang XF, Massague J (1992) TGF beta signals through a heteromeric protein kinase receptor complex. Cell 71(6):1003–1014
96. Wieser R, Wrana JL, Massague J (1995) GS domain mutations that constitutively activate T beta R-I, the downstream signaling component in the TGF-beta receptor complex. EMBO J 14(10):2199–2208
97. Leof EB, Proper JA, Goustin AS, Shipley GD, DiCorleto PE, Moses HL (1986) Induction of c-sis mRNA and activity similar to platelet-derived growth factor by transforming growth factor beta: a proposed model for indirect mitogenesis involving autocrine activity. Proc Natl Acad Sci U S A 83(8):2453–2457
98. Kim DK, Kim J, Park HJ (2004) Synthesis and biological evaluation of novel 2-pyridinyl-[1,2,3]triazoles as inhibitors of transforming growth factor beta 1 type 1 receptor. Bioorg Med Chem Lett 14(10):2401–2405. doi:10.1016/j.bmcl.2004.03.024
99. Sawyer JS, Anderson BD, Beight DW, Campbell RM, Jones ML, Herron DK, Lampe JW, McCowan JR, McMillen WT, Mort N, Parsons S, Smith EC, Vieth M, Weir LC, Yan L, Zhang F, Yingling JM (2003) Synthesis and activity of new aryl- and heteroaryl-substituted pyrazole inhibitors of the transforming growth factor-beta type I receptor kinase domain. J Med Chem 46(19):3953–3956. doi:10.1021/jm0205705

100. Li HY, McMillen WT, Heap CR, McCann DJ, Yan L, Campbell RM, Mundla SR, King CH, Dierks EA, Anderson BD, Britt KS, Huss KL, Voss MD, Wang Y, Clawson DK, Yingling JM, Sawyer JS (2008) Optimization of a dihydropyrrolopyrazole series of transforming growth factor-beta type I receptor kinase domain inhibitors: discovery of an orally bioavailable transforming growth factor-beta receptor type I inhibitor as antitumor agent. J Med Chem 51(7):2302–2306. doi:10.1021/jm701199p
101. Peng SB, Yan L, Xia X, Watkins SA, Brooks HB, Beight D, Herron DK, Jones ML, Lampe JW, McMillen WT, Mort N, Sawyer JS, Yingling JM (2005) Kinetic characterization of novel pyrazole TGF-beta receptor I kinase inhibitors and their blockade of the epithelial-mesenchymal transition. Biochemistry 44(7):2293–2304. doi:10.1021/bi048851x
102. Muraoka-Cook RS, Shin I, Yi JY, Easterly E, Barcellos-Hoff MH, Yingling JM, Zent R, Arteaga CL (2006) Activated type I TGFbeta receptor kinase enhances the survival of mammary epithelial cells and accelerates tumor progression. Oncogene 25(24):3408–3423. doi:10.1038/sj.onc.1208964
103. Morris JC, Tan AR, Olencki TE, Shapiro GI, Dezube BJ, Reiss M, Hsu FJ, Berzofsky JA, Lawrence DP (2014) Phase I study of GC1008 (fresolimumab): a human anti-transforming growth factor-beta (TGFbeta) monoclonal antibody in patients with advanced malignant melanoma or renal cell carcinoma. PLoS One 9(3):e90353. doi:10.1371/journal.pone.0090353
104. Lonning S, Mannick J, McPherson JM (2011) Antibody targeting of TGF-beta in cancer patients. Curr Pharm Biotechnol 12(12):2176–2189
105. Trachtman H, Fervenza FC, Gipson DS, Heering P, Jayne DR, Peters H, Rota S, Remuzzi G, Rump LC, Sellin LK, Heaton JP, Streisand JB, Hard ML, Ledbetter SR, Vincenti F (2011) A phase 1, single-dose study of fresolimumab, an anti-TGF-beta antibody, in treatment-resistant primary focal segmental glomerulosclerosis. Kidney Int 79(11):1236–1243. doi:10.1038/ki.2011.33
106. Stevenson JP, Kindler HL, Papasavvas E, Sun J, Jacobs-Small M, Hull J, Schwed D, Ranganathan A, Newick K, Heitjan DF, Langer CJ, McPherson JM, Montaner LJ, Albelda SM (2013) Immunological effects of the TGFbeta-blocking antibody GC1008 in malignant pleural mesothelioma patients. Oncoimmunology 2(8):e26218. doi:10.4161/onci.26218
107. Goetz R, Mohammadi M (2013) Exploring mechanisms of FGF signalling through the lens of structural biology. Nat Rev Mol Cell Biol 14(3):166–180. doi:10.1038/nrm3528
108. Beenken A, Mohammadi M (2009) The FGF family: biology, pathophysiology and therapy. Nat Rev Drug Discov 8(3):235–253. doi:10.1038/nrd2792
109. Trueb B (2011) Biology of FGFRL1, the fifth fibroblast growth factor receptor. Cell Mol Life Sci 68(6):951–964
110. Turner N, Grose R (2010) Fibroblast growth factor signalling: from development to cancer. Nat Rev Cancer 10(2):116–129. doi:10.1038/nrc2780
111. Helsten T, Elkin S, Arthur E, Tomson BN, Carter J, Kurzrock R (2016) The FGFR landscape in cancer: analysis of 4,853 tumors by next-generation sequencing. Clin Cancer Res 22(1):259–267
112. Giacomini A, Chiodelli P, Matarazzo S, Rusnati M, Presta M, Ronca R (2016) Blocking the FGF/FGFR system as a "two-compartment" antiangiogenic/antitumor approach in cancer therapy. Pharmacol Res 107:172–185. doi:10.1016/j.phrs.2016.03.024
113. Theillet C, Adelaide J, Louason G, Bonnet-Dorion F, Jacquemier J, Adnane J, Longy M, Katsaros D, Sismondi P, Gaudray P (1993) FGFRI and PLAT genes and DNA amplification at 8p 12 in breast and ovarian cancers. Genes Chromosom Cancer 7(4):219–226
114. Reis-Filho JS, Simpson PT, Turner NC, Lambros MB, Jones C, Mackay A, Grigoriadis A, Sarrio D, Savage K, Dexter T (2006) FGFR1 emerges as a potential therapeutic target for lobular breast carcinomas. Clin Cancer Res 12(22):6652–6662
115. Turner N, Pearson A, Sharpe R, Lambros M, Geyer F, Lopez-Garcia MA, Natrajan R, Marchio C, Iorns E, Mackay A, Gillett C, Grigoriadis A, Tutt A, Reis-Filho JS, Ashworth A (2010) FGFR1 amplification drives endocrine therapy resistance and is a therapeutic target in breast cancer. Cancer Res 70(5):2085–2094. doi:10.1158/0008-5472.CAN-09-3746
116. Elsheikh SE, Green AR, Lambros MB, Turner NC, Grainge MJ, Powe D, Ellis IO, Reis-Filho JS (2007) FGFR1 amplification in breast carcinomas: a chromogenic in situ hybridisation analysis. Breast Cancer Res 9(2):R23
117. Turner N, Lambros MB, Horlings HM, Pearson A, Sharpe R, Natrajan R, Geyer FC, van Kouwenhove M, Kreike B, Mackay A (2010) Integrative molecular profiling of triple negative breast cancers identifies amplicon drivers and potential therapeutic targets. Oncogene 29(14):2013–2023
118. Bane AL, Pinnaduwage D, Colby S, Bull SB, O'Malley FP, Andrulis IL (2009) Expression profiling of familial breast cancers demonstrates higher expression of FGFR2 in BRCA2-associated tumors. Breast Cancer Res Treat 117(1):183–191
119. Sugihara H, Ishimoto T, Yasuda T, Izumi D, Eto K, Sawayama H, Miyake K, Kurashige J, Imamura Y, Hiyoshi Y, Iwatsuki M, Iwagami S, Baba Y, Sakamoto Y, Miyamoto Y, Yoshida N, Watanabe M, Takamori H, Baba H (2015) Cancer-associated fibroblast-derived CXCL12 causes tumor progression in adenocarcinoma of the esophagogastric junction. Med Oncol 32(6):618. doi:10.1007/s12032-015-0618-7

120. Gee JM, Robertson JF, Ellis IO, Nicholson RI (2001) Phosphorylation of ERK1/2 mitogen-activated protein kinase is associated with poor response to antihormonal therapy and decreased patient survival in clinical breast cancer. Int J Cancer 95(4):247–254
121. Dallol A, Buhmeida A, Merdad A, Al-Maghrabi J, Gari MA, Abu-Elmagd MM, Elaimi A, Assidi M, Chaudhary AG, Abuzenadah AM, Nedjadi T, Ermiah E, Alkhayyat SS, Al-Qahtani MH (2015) Frequent methylation of the KLOTHO gene and overexpression of the FGFR4 receptor in invasive ductal carcinoma of the breast. Tumor Biol 36(12):9677–9683. doi:10.1007/s13277-015-3733-3
122. Dey JH, Bianchi F, Voshol J, Bonenfant D, Oakeley EJ, Hynes NE (2010) Targeting fibroblast growth factor receptors blocks PI3K/AKT signaling, induces apoptosis, and impairs mammary tumor outgrowth and metastasis. Cancer Res 70(10):4151–4162. doi:10.1158/0008-5472.CAN-09-4479
123. Thomas AP, Theoclitou ME, Buttar D, Ruston L, Wrigley G, Dennis M, Rudge DA, Coleman T, Smith R, Gavine PR, Klinowska T, Mooney L, Brooks N (2012) The discovery of AZD4547: an orally bioavailable, potent and selective N-(5-Pyrazolyl) benzamide FGFR1-3 inhibitor. Cancer Res 72. doi:10.1158/1538-7445.AM2012-3912
124. Gayine PR, Mooney L, Kilgour E, Thomas AP, Al-Kadhimi K, Beck S, Coleman T, Baker D, Mellor MJ, Brooks NAN, Klinowska T (2011) Characterization of AZD4547: an orally bioavailable, potent and selective inhibitor of FGFR tyrosine kinases 1, 2 and 3. Cancer Res 71. doi:10.1158/1538-7445.AM2011-3568
125. Gavine PR, Mooney L, Kilgour E, Thomas AP, Al-Kadhimi K, Beck S, Rooney C, Coleman T, Baker D, Mellor MJ, Brooks AN, Klinowska T (2012) AZD4547: an orally bioavailable, potent, and selective inhibitor of the fibroblast growth factor receptor tyrosine kinase family. Cancer Res 72(8):2045–2056. doi:10.1158/0008-5472.CAN-11-3034
126. Liu L, Ye TH, Han YP, Song H, Zhang YK, Xia Y, Wang NY, Xiong Y, Song XJ, Zhu YX, Li de L, Zeng J, Ran K, Peng CT, Wei YQ, Yu LT (2014) Reductions in myeloid-derived suppressor cells and lung metastases using AZD4547 treatment of a metastatic murine breast tumor model. Cellular physiology and biochemistry : international journal of experimental cellular physiology, biochemistry, and pharmacology 33(3):633–645. doi:10.1159/000358640
127. Andre F, Ranson M, Dean E, Varga A, van der Noll R, Stockman PK, Ghiorghiu D, Kilgour E, Smith PD, Macpherson M, Lawrence P, Hastie A, Schellens JHM (2013) Results of a phase I study of AZD4547, an inhibitor of fibroblast growth factor receptor (FGFR), in patients with advanced solid tumors. Cancer Res 73(8). doi:10.1158/1538-7445.AM2013-LB-145
128. Smyth EC, Turner NC, Peckitt C, Pearson A, Brown G, Chua S, Gillbanks A, Johnston SRD, Tarazona N, Cutts R, Kilgour E, Rooney C, Smith NR, Sumpter KA, Ajaz MA, Thomas AL, Watkins D, Chau I, Popat S, Cunningham D (2015) Phase II multicenter proof of concept study of AZD4547 in FGFR amplified tumours. J Clin Oncol 33(15)
129. Angibaud PR, Mevellec L, Saxty G, Adelinet C, Akkari R, Berdini V, Bonnet P, Bourgeois M, Bourdrez X, Cleasby A, Colombel H, Csoka I, Embrechts W, Freyne E, Gilissen R, Jovcheva E, King P, Lacrampe J, Lardeau D, Ligny Y, Mcclue S, Meerpoel L, Newell DR, Page M, Papanikos A, Pasquier E, Pilatte I, Poncelet V, Querolle O, Rees DC, Rich S, Roux B, Sement E, Simonnet Y, Squires M, Tronel V, Verhulst T, Vialard J, Willems M, Woodhead SJ, Wroblowski B, Murray CW, Perera T (2014) Discovery of JNJ-42756493, a potent fibroblast growth factor receptor (FGFR) inhibitor using a fragment based approach. Cancer Res 74(19). doi:10.1158/1538-7445.AM2014-4748
130. Perera T, Jovcheva E, Vialard J, Verhulst T, Esser N, Wroblowski B, Gilissen R, Freyne E, King P, Platero S, Querolle O, Mevellec L, Murray C, Fazal L, Saxty G, Ward G, Squires M, Thompson N, Newell D, Angibaud P (2014) JNJ-42756493 is an inhibitor of FGFR-1, 2, 3 and 4 with nanomolar affinity for targeted therapy. Cancer Res 74(19). doi:10.1158/1538-7445.AM2014-1738
131. Dienstmann R, Bahleda R, Adamo B, Rodon J, Varga A, Gazzah A, Platero S, Smit H, Perera T, Zhong B, Stuyckens K, Elsayed Y, Takimoto C, Peddareddigari V, Tabernero J, Luo FR, Soria JC (2014) First in human study of JNJ-42756493, a potent pan fibroblast growth factor receptor (FGFR) inhibitor in patients with advanced solid tumors. Cancer Res 74(19). doi:10.1158/1538-7445.AM2014-CT325
132. Bahleda R, Dienstmann R, Adamo B, Gazzah A, Infante JR, Zhong B, Platero SJ, Smit H, Perera T, Stuyckens K, Bussolari J, Poddareddigari V, Soria JC, Luo FR, Tabernero J (2014) Phase 1 study of JNJ-42756493, a pan-fibroblast growth factor receptor (FGFR) inhibitor, in patients with advanced solid tumors. J Clin Oncol 32(15)
133. Guagnano V, Furet P, Spanka C, Bordas V, Le Douget M, Stamm C, Brueggen J, Jensen MR, Schnell C, Schmid H, Wartmann M, Berghausen J, Drueckes P, Zimmerlin A, Bussiere D, Murray J, Graus Porta D (2011) Discovery of 3-(2,6-dichloro-3,5-dimethoxy-phenyl)-1-{6-[4-(4-ethyl-piperazin-1-yl)-phenylamin o]-pyrimidin-4-yl}-1-methyl-urea (NVP-BGJ398), a potent and selective inhibitor of the fibroblast growth factor receptor family of receptor tyrosine kinase. J Med Chem 54(20):7066–7083. doi:10.1021/jm2006222
134. Sequist LV, Cassier P, Varga A, Tabernero J, Schellens JH, Delord JP, LoRusso P, Camidge DR, Medina MH, Schuler M, Campone M, Tian GG, Wong S, Corral J, Isaacs R, Sen SK, Porta DG, Kulkarni SG, Lefebvre C, Wolf J (2014) Phase I study of BGJ398,

a selective pan-FGFR inhibitor in genetically preselected advanced solid tumors. Cancer Res 74(19). doi:10.1158/1538-7445.AM2014-CT326
135. Leroith D, Roberts CT (1993) Insulin-like growth-factors and their receptors in normal physiology and pathological states. J Pediatr Endocrinol 6(3–4):251–255
136. LeRoith D, King G, Flier JS (1997) Insulin-like growth factors. N Engl J Med 336(9):633–640
137. Oates AJ, Schumaker LM, Jenkins SB, Pearce AA, DaCosta SA, Arun B, Ellis MJ (1998) The mannose 6-phosphate/insulin-like growth factor 2 receptor (M6P/IGF2R), a putative breast tumor suppressor gene. Breast Cancer Res Treat 47(3):269–281
138. Tao Y, Pinzi V, Bourhis J, Deutsch E (2007) Mechanisms of disease: signaling of the insulin-like growth factor 1 receptor pathway--therapeutic perspectives in cancer. Nat Clin Pract Oncol 4(10):591–602. doi:10.1038/ncponc0934
139. Mauro L, Naimo GD, Ricchio E, Panno ML, Ando S (2015) Cross-talk between Adiponectin and IGF-IR in breast cancer. Front Oncol 5:157. doi:10.3389/fonc.2015.00157
140. Danielsen A, Larsen E, Gammeltoft S (1990) Chromaffin cells express two types of insulin-like growth factor receptors. Brain Res 518(1–2):95–100
141. Sachdev D, Yee D (2007) Disrupting insulin-like growth factor signaling as a potential cancer therapy. Mol Cancer Ther 6(1):1–12. doi:10.1158/1535-7163.MCT-06-0080
142. Denduluri SK, Idowu O, Wang Z, Liao Z, Yan Z, Mohammed MK, Ye J, Wei Q, Wang J, Zhao L, Luu HH (2015) Insulin-like growth factor (IGF) signaling in tumorigenesis and the development of cancer drug resistance. Genes & diseases 2(1):13–25. doi:10.1016/j.gendis.2014.10.004
143. Papa V, Gliozzo B, Clark GM, McGuire WL, Moore D, Fujita-Yamaguchi Y, Vigneri R, Goldfine ID, Pezzino V (1993) Insulin-like growth factor-I receptors are overexpressed and predict a low risk in human breast cancer. Cancer Res 53(16):3736–3740
144. Yerushalmi R, Gelmon KA, Leung S, Gao D, Cheang M, Pollak M, Turashvili G, Gilks BC, Kennecke H (2012) Insulin-like growth factor receptor (IGF-1R) in breast cancer subtypes. Breast Cancer Res Treat 132(1):131–142. doi:10.1007/s10549-011-1529-8
145. Lu YH, Zi XL, Zhao YH, Mascarenhas D, Pollak M (2001) Insulin-like growth factor-I receptor signaling and resistance to trastuzumab (Herceptin). J Nat Cancer Inst 93(24):1852–1857 10.1093/jnci/93.24.1852
146. Gallagher EJ, LeRoith D (2011) Minireview: IGF, insulin, and cancer. Endocrinology 152(7):2546–2551. doi:10.1210/en.2011-0231
147. Brahmkhatri VP, Prasanna C, Atreya HS (2015) Insulin-like growth factor system in cancer: novel targeted therapies. Biomed Res Int 2015:538019. doi:10.1155/2015/538019
148. Cohen BD, Baker DA, Soderstrom C, Tkalcevic G, Rossi AM, Miller PE, Tengowski MW, Wang F, Gualberto A, Beebe JS, Moyer JD (2005) Combination therapy enhances the inhibition of tumor growth with the fully human anti-type 1 insulin-like growth factor receptor monoclonal antibody CP-751,871. Clinical cancer research : an official journal of the American Association for Cancer Research 11(5):2063–2073. doi:10.1158/1078-0432.CCR-04-1070
149. Chakraborty AK, Zerillo C, DiGiovanna MP (2015) In vitro and in vivo studies of the combination of IGF1R inhibitor figitumumab (CP-751,871) with HER2 inhibitors trastuzumab and neratinib. Breast Cancer Res Treat 152(3):533–544. doi:10.1007/s10549-015-3504-2
150. Beltran PJ, Mitchell P, Chung YA, Cajulis E, Lu J, Belmontes B, Ho J, Tsai MM, Zhu M, Vonderfecht S, Baserga R, Kendall R, Radinsky R, Calzone FJ (2009) AMG 479, a fully human anti-insulin-like growth factor receptor type I monoclonal antibody, inhibits the growth and survival of pancreatic carcinoma cells. Mol Cancer Ther 8(5):1095–1105. doi:10.1158/1535-7163.MCT-08-1171
151. Beltran PJ, Chung YA, Moody G, Mitchell P, Cajulis E, Vonderfecht S, Kendall R, Radinsky R, Calzone FJ (2011) Efficacy of ganitumab (AMG 479), alone and in combination with rapamycin, in Ewing's and osteogenic sarcoma models. J Pharmacol Exp Ther 337(3):644–654. doi:10.1124/jpet.110.178400
152. Mendivil A, Zhou C, Cantrell LA, Gehrig PA, Malloy KM, Blok LJ, Burger CW, Bae-Jump VL (2011) AMG 479, a novel IGF-1-R antibody, inhibits endometrial cancer cell proliferation through disruption of the PI3K/Akt and MAPK pathways. Reprod Sci 18(9):832–841. doi:10.1177/1933719111398501
153. Cao ZA, Pinzon-Ortiz M, Chen Y, Li X, Beltran PJ, Gansert J, Peters M, Schlegel R, Schumacher KM, Huang A (2014) Targeting PIK3CA mutant breast cancer with the combination of PIK3CA-specific inhibitor, BYL719, and IGF1-R antibody, ganitumab. AACR
154. Murakami H, Doi T, Yamamoto N, Watanabe J, Boku N, Fuse N, Yoshino T, Ohtsu A, Otani S, Shibayama K, Takubo T, Loh E (2012) Phase 1 study of ganitumab (AMG 479), a fully human monoclonal antibody against the insulin-like growth factor receptor type I (IGF1R), in Japanese patients with advanced solid tumors. Cancer Chemother Pharmacol 70(3):407–414. doi:10.1007/s00280-012-1924-9
155. Tolcher AW, Sarantopoulos J, Patnaik A, Papadopoulos K, Lin CC, Rodon J, Murphy B, Roth B, McCaffery I, Gorski KS, Kaiser B, Zhu M, Deng H, Friberg G, Puzanov I (2009) Phase I, pharmacokinetic, and pharmacodynamic study of AMG 479, a fully human monoclonal antibody to insulin-like growth factor receptor 1. Journal of clinical oncology : official journal of the American Society of

Clinical Oncology 27(34):5800–5807. doi:10.1200/JCO.2009.23.6745
156. Robertson JFR, Ferrero J-M, Bourgeois H, Kennecke H, de Boer RH, Jacot W, McGreivy J, Suzuki S, Zhu M, McCaffery I, Loh E, Gansert JL, Kaufman PA (2013) Ganitumab with either exemestane or fulvestrant for postmenopausal women with advanced, hormone-receptor-positive breast cancer: a randomised, controlled, double-blind, phase 2 trial. Lancet Oncol 14(3):228–235. doi:10.1016/s1470-2045(13)70026-3
157. Yee D, Paoloni M, van't Veer L, Sanil A, Yau C, Forero A, Chien A, Wallace A, Moulder S, Albain K (2017) Abstract P6-11-04: the evaluation of ganitumab/metformin plus standard neoadjuvant therapy in high-risk breast cancer: results from the I-SPY 2 trial. AACR
158. Burtrum D, Zhu Z, Lu D, Anderson DM, Prewett M, Pereira DS, Bassi R, Abdullah R, Hooper AT, Koo H, Jimenez X, Johnson D, Apblett R, Kussie P, Bohlen P, Witte L, Hicklin DJ, Ludwig DL (2003) A fully human monoclonal antibody to the insulin-like growth factor I receptor blocks ligand-dependent signaling and inhibits human tumor growth in vivo. Cancer Res 63(24):8912–8921
159. Gradishar WJ, Yardley DA, Layman R, Sparano JA, Chuang E, Northfelt DW, Schwartz GN, Youssoufian H, Tang S, Novosiadly R, Forest A, Nguyen TS, Cosaert J, Grebennik D, Haluska P (2016) Clinical and translational results of a phase II, randomized trial of an anti-IGF-1R (Cixutumumab) in women with breast cancer that progressed on endocrine therapy. Clin Cancer Res 22(2):301–309. doi:10.1158/1078-0432.CCR-15-0588
160. Haluska P, Bernath AM, Ballman KV, Dueck AC, Linden HM, Goetz MP, Northfelt DW, Hou X, Tenner KS, Tienchaiananda P (2014) Randomized phase II trial of capecitabine and lapatinib with or without cixutumumab in patients with HER2+ breast cancer previously treated with trastuzumab and an anthracycline and/or a taxane: NCCTG N0733 (alliance). American Society of Clinical Oncology
161. Lee JW, Soung YH, Kim SY, Lee HW, Park WS, Nam SW, Kim SH, Lee JY, Yoo NJ, Lee SH (2005) PIK3CA gene is frequently mutated in breast carcinomas and hepatocellular carcinomas. Oncogene 24(8):1477–1480. doi:10.1038/sj.onc.1208304
162. Bachman KE, Argani P, Samuels Y, Silliman N, Ptak J, Szabo S, Konishi H, Karakas B, Blair BG, Lin C, Peters BA, Velculescu VE, Park BH (2004) The PIK3CA gene is mutated with high frequency in human breast cancers. Cancer Biol Ther 3(8):772–775
163. Bachman KE, Argani P, Samuels Y, Silliman N, Ptak J, Szabo S, Konishi H, Karakas B, Blair BG, Lin C, Peters BA, Velculescu VE, Park BH (2005) The PIK3CA gene is mutated with high frequency in human breast cancers (vol 3, pg 772, 2004). Cancer Biol Ther 4(2):133–133
164. Sun SY, Rosenberg LM, Wang XR, Zhou ZM, Yue P, Fu H, Khuri FR (2005) Activation of Akt and eIF4E survival pathways by rapamycin-mediated mammalian target of rapamycin inhibition. Cancer Res 65(16):7052–7058. doi:10.1158/0008-5472.CAN-05-0917
165. Shi YJ, Yan HJ, Frost P, Gera J, Lichtenstein A (2005) Mammalian target of rapamycin inhibitors activate the AKT kinase in multiple myeloma cells by up-regulating the insulin-like growth factor receptor/insulin receptor substrate-1/phosphatidylinositol 3-kinase cascade. Mol Cancer Ther 4(10):1533–1540. doi:10.1158/1535-7163.MCT-05-0068
166. Wan X, Harkavy B, Shen N, Grohar P, Helman LJ (2007) Rapamycin induces feedback activation of Akt signaling through an IGF-1R-dependent mechanism. Oncogene 26(13):1932–1940. doi:10.1038/sj.onc.1209990
167. Ma CX, Suman VJ, Goetz M, Haluska P, Moynihan T, Nanda R, Olopade O, Pluard T, Guo ZF, Chen HX, Erlichman C, Ellis MJ, Fleming GF (2013) A phase I trial of the IGF-1R antibody Cixutumumab in combination with temsirolimus in patients with metastatic breast cancer. Breast Cancer Res Treat 139(1):145–153. doi:10.1007/s10549-013-2528-8
168. Schnitzer T, Kuenkele K-P, Rebers F, Van Vugt M, Klein C, Lanzendoerfer M, Mundigl O, Parren P, van de Winkel J, Schumacher R (2006) 214 POSTER characterization of a recombinant, fully human monoclonal antibody directed against the human insulin-like growth factor-1 receptor. EJC Suppl 4(12):66–67
169. Kurzrock R, Patnaik A, Aisner J, Warren T, Leong S, Benjamin R, Eckhardt SG, Eid JE, Greig G, Habben K, McCarthy CD, Gore L (2010) A phase I study of weekly R1507, a human monoclonal antibody insulin-like growth factor-I receptor antagonist, in patients with advanced solid tumors. Clin Cancer Res 16(8):2458–2465. doi:10.1158/1078-0432.CCR-09-3220
170. Mahadevan D, Sutton GR, Arteta-Bulos R, Bowden CJ, Miller PJ, Swart RE, Walker MS, Haluska P, Munster PN, Marshall J, Hamid O, Kurzrock R (2014) Phase 1b study of safety, tolerability and efficacy of R1507, a monoclonal antibody to IGF-1R in combination with multiple standard oncology regimens in patients with advanced solid malignancies. Cancer Chemother Pharmacol 73(3):467–473. doi:10.1007/s00280-013-2372-x
171. Hutcheson IR, Pumford SL, Barrow D, Frankel SR, Nicholson RI (2009) Abstract B125: antiproliferative and anti-invasive effect of the novel anti-IGF-1R monoclonal antibody R1507 in endocrine-responsive and-resistant breast cancer cell lines. AACR

172. Sanchez C, Crowder R, Phommaly C, Ellis M (2009) The effect of the IGF1R antibody R1507 on ER positive breast cancer cell lines growth and survival. AACR
173. Maloney EK, McLaughlin JL, Dagdigian NE, Garrett LM, Connors KM, Zhou X-M, Blättler WA, Chittenden T, Singh R (2003) An anti-insulin-like growth factor I receptor antibody that is a potent inhibitor of cancer cell proliferation. Cancer Res 63(16):5073–5083
174. Descamps G, Wuilleme-Toumi S, Trichet V, Venot C, Debussche L, Hercend T, Collette M, Robillard N, Bataille R, Amiot M (2006) CD45neg but not CD45pos human myeloma cells are sensitive to the inhibition of IGF-1 signaling by a murine anti-IGF-1R monoclonal antibody, mAVE1642. J Immunol 177(6):4218–4223. doi:10.4049/jimmunol.177.6.4218
175. Geoerger B, Brasme JF, Daudigeos-Dubus E, Opolon P, Venot C, Debussche L, Vrignaud P, Vassal G (2010) Anti-insulin-like growth factor 1 receptor antibody EM164 (murine AVE1642) exhibits anti-tumour activity alone and in combination with temozolomide against neuroblastoma. Eur J Cancer 46(18):3251–3262. doi:10.1016/j.ejca.2010.06.005
176. Descamps G, Gomez-Bougie P, Venot C, Moreau P, Bataille R, Amiot M (2009) A humanised anti-IGF-1R monoclonal antibody (AVE1642) enhances Bortezomib-induced apoptosis in myeloma cells lacking CD45. Br J Cancer 100(2):366–369. doi:10.1038/sj.bjc.6604839
177. Soria JC, Massard C, Lazar V, Ozoux ML, Mery-Mignard D, Deslandes A, Tolcher AW (2013) A dose finding, safety and pharmacokinetic study of AVE1642, an anti-insulin-like growth factor-1 receptor (IGF-1R/CD221) monoclonal antibody, administered as a single agent and in combination with docetaxel in patients with advanced solid tumours. Eur J Cancer 49(8):1799–1807. doi:10.1016/j.ejca.2013.01.003
178. Macaulay VM, Middleton MR, Protheroe AS, Tolcher A, Dieras V, Sessa C, Bahleda R, Blay JY, LoRusso P, Mery-Mignard D, Soria JC (2013) Phase I study of humanized monoclonal antibody AVE1642 directed against the type 1 insulin-like growth factor receptor (IGF-1R), administered in combination with anticancer therapies to patients with advanced solid tumors. Annals of oncology : official journal of the European Society for Medical Oncology 24(3):784–791. doi:10.1093/annonc/mds511
179. Goetsch L, Gonzalez A, Leger O, Beck A, Pauwels PJ, Haeuw JF, Corvaia N (2005) A recombinant humanized anti-insulin-like growth factor receptor type I antibody (h7C10) enhances the antitumor activity of vinorelbine and anti-epidermal growth factor receptor therapy against human cancer xenografts. Int J Cancer 113(2):316–328. doi:10.1002/ijc.20543
180. Goetsch I, Gonzalez A, Beck A, Haeuw J, Corvaia N (2004) 286 generation of a recombinant humanized anti-insulin-like growth factor receptor type I antibody (h7C10) with an antitumor activity in a variety of human cancer xenograft models. EJC Suppl 2(8):87–88
181. Scartozzi M, Bianconi M, Maccaroni E, Giampieri R, Berardi R, Cascinu S (2010) Dalotuzumab, a recombinant humanized mAb targeted against IGFR1 for the treatment of cancer. Curr Opin Mol Ther 12(3):361–371
182. Di Cosimo S, Sathyanarayanan S, Bendell JC, Cervantes A, Stein MN, Brana I, Roda D, Haines BB, Zhang T, Winter CG, Jha S, Xu Y, Frazier J, Klinghoffer RA, Leighton-Swayze A, Song Y, Ebbinghaus S, Baselga J (2015) Combination of the mTOR inhibitor ridaforolimus and the anti-IGF1R monoclonal antibody dalotuzumab: preclinical characterization and phase I clinical trial. Clinical cancer research : an official journal of the American Association for Cancer Research 21(1):49–59. doi:10.1158/1078-0432.CCR-14-0940
183. Becker MA, Hou X, Tienchaianada P, Haines BB, Harrington SC, Weroha SJ, Sathyanarayanan S, Haluska P (2016) Ridaforolimus (MK-8669) synergizes with Dalotuzumab (MK-0646) in hormone-sensitive breast cancer. BMC Cancer 16(1):814. doi:10.1186/s12885-016-2847-3
184. Lu B, Blum J, Cortes J, Rugo H, Swanton C, Eaton L, Song Y, Zhang T, Ebbinghaus S, Baselga J (2011) OT3–01-16: a phase 2 study of Ridaforolimus (RIDA) and Dalotuzumab (DALO) in estrogen receptor positive (ER+) breast cancer. AACR
185. Baselga J, Morales S, Awada A, Blum J, Tan A, Ewertz M, Cortes J, Moy B, Ruddy K, Haddad T (2013) Abstract P2–16-04: a phase 2 study of ridaforolimus (RIDA) and dalotuzumab (DALO) in estrogen receptor positive (ER+) breast cancer. AACR
186. Baselga J, Morales SM, Awada A, Blum JL, Tan AR, Ewertz M, Cortes J, Moy B, Ruddy KJ, Haddad T, Ciruelos EM, Vuylsteke P, Ebbinghaus S, Im E, Eaton L, Pathiraja K, Gause C, Mauro D, Jones MB, Rugo HS (2017) A phase II study of combined ridaforolimus and dalotuzumab compared with exemestane in patients with estrogen receptor-positive breast cancer. Breast Cancer Res Treat:1–10. doi:10.1007/s10549-017-4199-3
187. Rugo HS, Tredan O, Ro J, Morales S, Musolino A, Afonso N, Ferreira M, Park KH, Cortes J, Tan AR (2015) Abstract PD5–1: results from the phase 2 trial of ridaforolimus, dalotuzumab, and exemestane compared to ridaforolimus and exemestane in advanced breast cancer. AACR
188. Wittman MD, Carboni JM, Yang Z, Lee FY, Antman M, Attar R, Balimane P, Chang C, Chen C, Discenza L, Frennesson D, Gottardis MM, Greer A, Hurlburt W, Johnson W, Langley DR, Li A, Li J, Liu P, Mastalerz H, Mathur A, Menard K, Patel

K, Sack J, Sang X, Saulnier M, Smith D, Stefanski K, Trainor G, Velaparthi U, Zhang G, Zimmermann K, Vyas DM (2009) Discovery of a 2,4-disubstituted pyrrolo[1,2-f][1,2,4]triazine inhibitor (BMS-754807) of insulin-like growth factor receptor (IGF-1R) kinase in clinical development. J Med Chem 52(23):7360–7363. doi:10.1021/jm900786r

189. Carboni JM, Wittman M, Yang Z, Lee F, Greer A, Hurlburt W, Hillerman S, Cao C, Cantor GH, Dell-John J, Chen C, Discenza L, Menard K, Li A, Trainor G, Vyas D, Kramer R, Attar RM, Gottardis MM (2009) BMS-754807, a small molecule inhibitor of insulin-like growth factor-1R/IR. Mol Cancer Ther 8(12):3341–3349. doi:10.1158/1535-7163.MCT-09-0499

190. Hou X, Huang F, Macedo LF, Harrington SC, Reeves KA, Greer A, Finckenstein FG, Brodie A, Gottardis MM, Carboni JM, Haluska P (2011) Dual IGF-1R/InsR inhibitor BMS-754807 synergizes with hormonal agents in treatment of estrogen-dependent breast cancer. Cancer Res 71(24):7597–7607. doi:10.1158/0008-5472.CAN-11-1080

Targeting Stemness: Implications for Precision Medicine in Breast Cancer

7

Zhi-Mei Liang, Yang Chen, and Man-Li Luo

Abstract

The genomic landscape of breast cancer has been delineated in recent years. Advances in molecular characterization and targeting strategies are making it feasible to integrate clinical, genome-based and phenotype-based diagnostic and therapeutic methods and apply them to individual patient in the era of precision medicine. Cancer stem cells (CSCs) are a subpopulation in the tumor which have the capability of self-renewal and differentiation. Breast CSCs have important clinical implications as they account for tumor initiation, maintenance, metastasis, therapy resistance, and relapse. In this chapter, we will introduce approaches used to characterize breast CSCs, crucial pathways involved in regulating cancer stemness, and implications of breast CSCs in the precision diagnosis and treatment of breast cancer. We will also discuss novel compounds and therapeutic strategies that selectively target breast CSCs. Integration of breast CSC-related molecular diagnosis and targeted therapy into the clinical workflow of precision medicine has the potential to deliver more effective treatment to breast cancer patients.

Keywords

Breast cancer • Cancer stem cell • Precision medicine

Zhi-Mei Liang and Yang Chen contributed equally to this work.

Z.-M. Liang • M.-L. Luo (✉)
Medical Research Center, Guangdong Provincial Key Laboratory of Malignant Tumor Epigenetics and Gene Regulation, Sun Yat-sen Memorial Hospital, Sun Yat-Sen University, Guangzhou, China
e-mail: luomli@mail.sysu.edu.cn

Y. Chen (✉)
Department of Laboratory, Sun Yat-Sen Memorial Hospital, Sun Yat-Sen University, 107 Yanjiang West Road, Guangzhou 510120, China

E. Song, H. Hu (eds.), *Translational Research in Breast Cancer*, Advances in Experimental Medicine and Biology 1026, https://doi.org/10.1007/978-981-10-6020-5_7

7.1 Introduction

In recent years, there have been significant advances in breast cancer diagnosis and treatment. However, resistance to therapy, metastasis, and relapse are still primary causes of cancer-related deaths. Breast cancer heterogeneity has added complexity to this problem [1]. Heterogeneity within tumors causes problems to therapies and compromises treatment outcomes. Currently, the origins of tumor heterogeneity, including intra-tumor heterogeneity and inter-tumor heterogeneity, are still not fully understood. The first theory proposed to explain these phenomena is the clonal evolution model, which depicts cancer as an evolutionary disease, driven by gene mutations and clonal selections. Recently, a large body of evidence suggests that breast cancers follow the CSC model [2], in which cancer is hierarchically organized, and only the population of CSCs is tumorigenic and can give rise to more CSCs through self-renewal and nontumorigenic cancer cells through differentiation [2]. These two models, both of which are supported by preclinical evidence, are not mutually exclusive. From a therapeutic perspective, the CSC model emphasizes the preexistence of a hierarchy, in which the CSCs are resistant to radiation and chemotherapy, and thought to be responsible for disease relapse. In the evolution model, although resistant cells might originally be present in the tumor at low frequency, these cells expand under the selective pressure imposed by therapies, resulting in the rapid outgrowth of drug-resistant clones. The CSC concept has important clinical implications because current therapies have been developed to target the bulk tumor. Albeit that these therapies may dramatically reduce tumor size, they are not likely to result in stable, long-lasting remission if the CSCs are not eradicated.

The consensus definition of a CSC is "a cell within a tumor that possesses the capacity to self-renew and to cause the heterogeneous lineages of cancer cells that comprises the tumor" [3]. The most important properties of CSCs are self-renewal and differentiation. Self-renewal allows CSCs to generate new CSCs and maintain the stem cell pool through symmetric division to produce two stem cells or through asymmetric division to produce a stem cell and a daughter progenitor cell [4, 5]. Differentiation allows CSCs to generate a heterogeneous progeny of neoplastic cells which constitute the bulk tumor [6]. Breast cancer is the first solid tumor that has been identified to have CSCs. These breast CSCs were isolated using flow cytometry by CD44, CD24, or ALDH and then characterized using mammosphere assay and transplantation assay into immunodeficient mice [7, 8]. Both clinical and basic studies have shown the role of breast CSCs in therapy resistance and tumor metastasis, which are characteristics tightly correlated with poor patient prognosis. Thus, targeting breast CSCs is a promising strategy to improve therapeutic outcome.

Over the past few years, the low sequencing price and the available targeting drugs have provided more options for cancer patients to receive therapies that target key molecular and genetic alterations in tumors. Remarkable advances in sequencing the genomes have made it possible to identify clinically relevant mutations in tumors. Growing understanding of inter-tumor heterogeneity in breast cancer has helped to direct suitable therapies to the individual patient. Many advances in personalized medicine occur in the field of breast cancer, such as immunohistochemistry for gene expression analysis, in situ hybridization (ISH) for gene amplification analysis, and estrogen receptor (ER) and human epidermal growth factor receptor2 (HER2)-blocking therapies. In recent years, several pathways of breast CSC maintenance have been discovered, and multiple druggable targets have been identified. Novel treatments targeting breast CSC are under development with a few therapeutic strategies being tested preclinically or in clinical trials. Meanwhile, CSC-related gene signatures have been proposed to predict therapy response, recurrence, and metastasis for patient stratification. Advances in the molecular characterization and therapeutic targeting of breast CSC, and incorporating them into clinical workflow, will deliver more effective precision medicine to breast cancer patients.

7.2 Methods to Identify and Enrich Breast CSCs

Precision medicine requires an understanding of molecular alterations that drive each cancer. The identification and characterization of CSCs in breast cancer provide insights to ways of enriching these cells for mechanism investigation and selective inhibition of CSCs as a treatment method. Breast cancer was the first solid tumor identified to have CSCs. In an elegant study, Al-Hajj et al. demonstrated that $Lin^{-}CD44^{+}CD24^{-/low}EpCAM^{+}$ human breast cancer cells were much more tumorigenic compared to other tumor cells in the xenograft transplantation to NOD/SCID mice [7]. Interestingly, tumors formed by these $Lin^{-}CD44^{+}CD24^{-/low}EpCAM^{+}$ cells exhibited heterogeneity of the primary tumor, and $Lin^{-}CD44^{+}CD24^{-/low}EpCAM^{+}$ cells could maintain their tumorigenic potential even after several generations. Thus, the $Lin^{-}CD44^{+}CD24^{-/low}EpCAM^{+}$ cells were identified as CSC-enriched population, because they have the properties of self-renewal, differentiation, and high tumorigenicity. Later on, several markers and methodologies were used to define or enrich the breast CSC subpopulation. However, the specific breast CSCs cannot be isolated by current methods, and all the abovementioned markers can only identify the CSC-enriched population. Therefore, besides sorting cells by breast CSC markers, subsequent in vitro and in vivo analysis is necessary to assess the function of the CSC-enriched population.

7.2.1 Markers for the Isolation of Breast CSCs

CSCs currently don't have universal markers because the maintenance of CSCs largely depends on the niche and organ-specific microenvironment. For breast cancer, some proteins have been identified as CSC markers, including CD44, CD24, EpCAM, and aldehyde dehydrogenase (ALDH). CD44 and CD24 have been used extensively with other surface markers to isolate CSCs from solid tumors. $CD44^{+}CD24^{-/low}EpCAM^{+}$ cells from breast cancer tissues exhibit CSC properties as they are capable of initiating tumors in NOD/SCID mice that recapitulate the primary tumor and can be serially passaged in vivo [7]. In human breast cancer, $ALDH^{+}$ cells also displayed tumor-initiating capacity and generated tumors that recapitulate the heterogeneity of the parental tumor in mice [8]. However, these markers can be only used to enrich for breast CSCs and do not result in the isolation of pure populations of breast CSCs.

CD44 is a specific receptor for hyaluronic acid expressed on cell surface. It belongs to the class I transmembrane glycoprotein family and interacts with the extracellular matrix (ECM), regulating cell adhesion, proliferation, survival, migration, angiogenesis, and differentiation [9]. Collagen, laminin, fibronectin, etc. can also be CD44 ligands [10]. CD44 protects CSCs from apoptosis [11], but its role in breast CSCs is still under investigation. The function of CD44 in cell-cell and cell-ECM adherence confers an advantage for CSCs when they travel in the blood vessel and arrive at distant metastasis sites [12]. The interaction between CD44 and hyaluronan activates Nanog transcription and proceeds to increase Rex1, SOX2, and MDR1 expression, which are all stem cell- or drug-resistant-related factors [11]. Knockdown of CD44 in breast CSC-enriched population sensitizes CSCs to the antitumor drug doxorubicin [13], alters the expression of stem cell-related genes, and induces CSC to differentiate into non-CSCs with lower tumorigenic potential [14].

CD24 is a heavily glycosylated GPI-anchored small protein. It was discovered as a B-cell surface protein and later found to be highly expressed in breast, ovarian, prostate, bladder, renal, and other human cancers [15]. It regulates the tumor cell adhesion with fibronectin, collagen, and lamin at cell-cell and cell-matrix interactions [16]. CD24 has been identified as an alternate ligand for P-selectin, an adhesion receptor on platelets and endothelial cells [17]. CD24 and P-selectin interaction facilitates tumor metastasis by promoting cells to pass through the blood vessel. Whether CD24 is simply a marker or plays a functional role in CSC has yet to be elucidated. $CD44^{+}CD24^{+}$ human breast cancer cells are found to be more differentiated and lack stem cell traits [18]. However, many contradicting studies show that CD24 expression enhances and inhibits breast cancer cell proliferation and metastasis [19]. Moreover, Stuelten et al. demonstrated that the levels of CD24 expression showed great variation between cell lines even in cells of the same cancer subtype [20]. These results suggest that CD24 might play distinct roles at different tumor stages. Few in vivo studies directly investigate the sorted cells with negative/lower positive/high CD24 expression in different subtypes of breast cancer. Thus, more validations are required.

Epithelial cell adhesion molecule (EpCAM)/epithelial-specific antigen (ESA) is a homophilic, calcium-independent cell adhesion molecule, broadly expressed on the basolateral surface of epithelial cells [21]. EpCAM belongs to the type I transmembrane glycoprotein family, which is also found to be expressed in cancer cells, cancer stem cells, embryonic stem cells, and germ cells [22]. In cancer development and progression, EpCAM shows controversial biological functions. In different cell types, both overexpression and knockdown of EpCAM decrease the oncogenic potential of cancer cells. Similarly, EpCAM acts as an adhesion molecule to mediate homophilic cell-cell adhesion and thus prevents metastasis. Meanwhile, EpCAM abrogates E-cadherin-mediated cell-cell adhesion and promotes metastasis. Its overexpression also correlates with both high and low survival rates of cancer patients [22]. Therefore, whether EpCAM is a tumor suppressor or an oncogene depends on the cancer cell type. In breast carcinoma, EpCAM overexpression associates with less differentiated tumors [23], larger sizes, lymph node metastasis, and poor survival [24]. Knockdown of EpCAM expression decreases cell proliferation, migration, and invasion in breast cancer cell lines [25]. In the seminal paper on breast CSCs, Al-Hajj et al. showed that the EpCAM^{+}CD44^{+}CD24^{-} fraction had a > tenfold higher tumor-forming capability than the EpCAM^{-}CD44^{+}CD24^{-} fraction [7]. Besides breast CSCs, EpCAM was reported to be expressed in CSCs from colon, pancreas, and prostate tumors. The potential role of EpCAM in CSCs is largely unknown. The mechanisms possibly involve the association of EpCAM overexpression with activation of the Wnt pathway, or induction of MYC, which are well-known factors in stem cell function [22].

Aldehyde dehydrogenases (ALDHs) are robust markers of CSCs for many cancers, including leukemia as well as brain, head and neck, breast, lung, liver, pancreatic, colon, bone, bladder, prostate, cervical cancer, and melanoma [26]. ALDH activity can be evaluated by the ALDEFLUOR assay. Based on its ability to oxidize intracellular aldehydes, this biochemical assay measures the enzymatic activity of aldehyde dehydrogenases [27]. The groundbreaking work by Ginestier et al. demonstrated that human breast cancer cells with high ALDH activity displayed tumor-initiating capacity and generated tumors that recapitulated the heterogeneity of the parental tumor in mice [8]. ALDH1 is a retinaldehyde dehydrogenase that oxidizes retinal to retinoic acid. ALDH1 overexpression in breast cancer correlates with worse clinical outcome in patients [28, 29]. Previously published immunohistological evidence showed that ALDH1A1 was responsible for ALDH activity in breast CSCs [8]. However, ALDH1A1 is not uniformly highly expressed in ALDH+ tumors. As the ALDH family includes 19 isoforms, Marcato et al. examined the expression of all 19 ALDH isoforms in breast cancer cell lines and tumor samples from patients and revealed that ALDH activity correlated best with expression of ALDH1A3. Knockdown of ALDH1A3 in breast cancer cells decreased ALDH activity by the ALDEFLUOR assay. Immunohistochemical staining also showed that ALDH1A3 expression was significantly associated with tumor grade, stage, and metastasis in breast cancer patients [30]. Thus, ALDHs' role as a breast CSC marker comes down to the specific isoform ALDH1A3.

Based on the surface markers, CSCs were first found to be around 2% in breast cancer [7]. Studies later demonstrated that breast CSC percentage could be as high as 25% among tumor cells, depending on the cancer subtype [31, 32]. A mathematical model showed that CSC population could be in any proportion of the tumor, and tumorigenic potential was directly related to the number of CSCs [33]. Moreover, a large body of literature demonstrated that $CD24^-CD44^+$ cells could not be detected in all breast cancer cell lines or patient tumors. $CD24^-CD44^+$ and $ALDH^+$ were identified to be overlapping but in different populations in breast cancer. And they were both capable of initiating tumors in NOD/SCID mice. As few as 20 $CD24^-CD44^+ALDH^+$ cells were shown to form tumors in mice. Thus, this population had the highest tumorigenic activity [8].

7.2.2 Enrichment and Characterization of Breast CSCs

In vitro and in vivo methods have been developed to enrich and examine breast CSCs. The sphere formation is a surrogate assay to test the self-renewal and differentiation capacity of stemlike cells at the single-cell level in vitro. The golden standard to determine CSC function is the serial transplantation assay which evaluates the ability of initiating new tumors that recapitulate the original tumor and can be passaged serially for generations in immunocompromised mice.

In the sphere-forming assay, both CSCs and non-CSCs are cultured in single-cell suspension, which only stem cells or progenitor cells survive to proliferate and differentiate to form a sphere structure, whereas non-stem or non-progenitor cells undergo anoikis within days of culture [34]. In the culture system of sphere-forming assay, cells are grown on low-attachment dishes in serum-free medium supplemented with epidermal growth factor (EGF), basic fibroblast growth factor (bFGF), and B27 supplement [35]. This method has been used to identify normal stem cells in fully differentiated tissues such as the breast, prostate, and brain, as well as CSCs [36]. Typically, cancer cells are serially diluted into single-cell suspension, and heparin is added to prevent aggregation. Therefore, the sphere structure could only be formed based on innate self-renewal and differentiation ability. Compared to adherent cells, the sphere-forming cells are of higher tumorigenic potential when transplanted into immunodeficient mice. Thus, the breast CSCs can be enriched by this assay [37]. Although the sphere-forming assays are of great value, certain critical considerations need to be taken into account [35]. Firstly, quiescent CSCs, which reside in a G0 state, may not be capable to form spheres, because quiescent stem cells do not respond to mitogens. Therefore, this assay only detects CSCs actively undergoing proliferation. Secondly, this assay is not a readout of CSC frequency, because both CSCs and progenitor cells can give rise to spheres [38]. The third caveat is that the size of a sphere is not a readout of self-renewal activity of stem cells. The size only reflects responsiveness to mitogens and the proliferation/differentiation capability of the parental clone-forming cells [35].

To circumvent the limitations described above and test the stem cell property of a given population of breast cancer cells, the in vivo limiting dilution assay and serial transplantation assay are developed to measure their ability to form new tumors in secondary and subsequent mice. In the limiting dilution assay of breast CSCs, a wide range of cancer cell dilutions are injected into immunocompromised mice as xenografts. Tumor-forming frequencies and CSC frequencies within serially diluted cells can be calculated by the single-hit Poisson hypothesis using a complementary log-log generalized linear model [39]. To determine the frequency of stem cells, this assay requires a serial dose of cell numbers for xenografts and a large number of replicates per dose. Dosage giving both positive and negative results should be included [40]. To further determine the self-renewal activity of breast CSC, the serial transplantation assay, adapted from adult stem cell research, is used in the xenograft model. The newly formed xenografts are digested into single-cell suspension and injected again into mice to form tumors for the second or more generations. Xenografts formed by CSCs are expected to replicate the phenotypic heterogeneity of the original tumor and contain CSC with self-renewal capability to initiate tumors in subsequent serial transplantations. The limitation of these approaches is that the immunodeficient mouse models do not have immune response, so the tumor microenvironment may lack factors critical for tumorigenesis.

7.3 Regulation Mechanisms of Breast CSCs

CSCs rely on critical signaling pathways for self-renewal and differentiation. Recently, intrinsic and extrinsic pathways for breast CSC maintenance have been exploited. Signals from the tumor microenvironment also regulate CSCs, including cytokines secreted from cancer cells and tumor-associated cells. Here we focus on the mechanisms of self-renewal pathways and cytokine pathways on the regulation of breast CSCs.

7.3.1 Self-renewal Signaling Pathways

Breast CSC needs self-renewal pathways for maintaining its population. Here we review the well-documented pathways used by both normal stem cells and breast CSCs, such as Wnt, Notch, and Hedgehog. In breast CSCs, accumulated mutations and epigenetic changes often cause aberrant activation of self-renewal pathways.

The Wnt pathway regulates embryonic development, tissue homeostasis, and stem cell maintenance [41]. Abnormal Wnt activation is critical for the initiation and progression of a variety of cancer, including mammary gland tumorigenesis [42, 43]. The Wnts are a family of secreted glycoproteins. In order to initiate signaling cascades, Wnt family ligands interact with a Frizzled (Fz) receptor, which belongs to the G-protein-coupled receptor family. The signaling transduction also requires lipoprotein receptor-related protein 5/6 (LRP-5/6), receptor tyrosine kinases (RTKs), and receptor tyrosine kinase-like or phan receptor 2 as co-receptors [42, 43]. In the canonical Wnt pathway, Wnt signal leads to GSK3β phosphorylation. As a result, β-catenin is stabilized and then translocates into the nucleus and activates the transcription of downstream oncogenes such as MYC and MMP7. In the noncanonical Wnt pathway, Wnt signal transduces through the Rho family GTPase or protein kinase A [41].

A large body of evidence suggests that Wnt signaling is involved in the breast CSC regulation. Multiple Wnt pathway components are expressed in the epithelium or stroma of mammary glands [44]. Wnt pathway genes are increasingly expressed in breast CSC-enriched populations compared with normal mammary stem cells (MsSCs) [42, 43]. In the MMTV-Wnt1 transgenic mice, Wnt signal was reported to promote the expansion of MaSC population in early tumorigenic lesions [45]. In the breast cancer, Wnt signaling hyperactivation often increases ligand production in an autocrine manner, which results in the enhancement of β-catenin stability [46]. Furthermore, both of the canonical and noncanonical Wnt pathways can cooperate with TGF-β pathway to expand the CSC population in breast cancer cells. Inhibition of these pathways results in decreased tumorigenicity and metastasis [47].

Notch signaling is a key regulator of stem cell maintenance and differentiation [48]. The Notch proteins include four single-pass transmembrane receptors and are expressed in various stem cells or progenitor cells. Notch is activated by serial cleavage upon interacting with the ligands Jagged1, Jagged2, Delta, and Delta-like [49]. Different tumors and subtypes can express different Notch receptors and ligands. In the canonical pathway, cleavage by a-disintegrin and metalloproteinase (ADAM) proteases and by enzymatic complex γ-secretase releases the Notch intracellular domain from cell membrane, which allows it to translocate to the nucleus and interact with the co-activator mastermind (MAM) and p300 to regulate target genes [50]. The noncanonical Notch signal is ligand or transcription independent. The most well-studied noncanonical Notch function is the regulation of Wnt/β-catenin signaling, and active β-catenin activity may serve as a readout for noncanonical Notch signals [51].

In recent years, dysregulated Notch activity has been implicated in a number of human malignancies, as well as in the CSC maintenance in various cancers [52]. Abnormal Notch signaling is involved in breast tumorigenesis through the deregulated self-renewal of normal MaSCs. In breast CSC-enriched $CD44^{+}CD24^{-}EpCAM^{+}$ cells, Notch-1 and Notch-4 activities were fourfold and eightfold higher, respectively, in comparison with bulk tumor cells. Either Notch1 or Notch4 inhibition decreased in vitro stem cell activity and reduced tumor formation in mice [53]. In the MaSC-enriched population, the Notch effector Cbf-1 knockdown increases stem cell activity, and constitutive Notch signaling specifically promotes luminal progenitor cell expansion, resulting in hyperplasia and tumor formation [54]. Moreover, combined treatment of trastuzumab with Notch pathway inhibition in ErbB2-positive xenograft of breast cancer is effective in preventing tumor relapse [55].

The Hedgehog (Hh) is implicated in regulating embryonic development, repair of normal tissues, and stem/progenitor cell maintenance [56]. Sonic hedgehog (Shh), desert hedgehog (Dhh), and Indian hedgehog (Ihh) ligands have been identified in mammals. Canonical Hh signaling follows the PTCH1-SMO-GLI axis, in which binding of one of the three ligands to the patched receptor (PTCH1) initiates the pathway activation, disabling the constitutive repression of smoothened (SMO) and leading to the translocation of GLI transcription factors (Gli1, Gli2, and Gli3) into the nucleus and expression of target genes, including cyclin D, cyclin E, c-Myc, EGF, and VEGF [57]. The noncanonical signaling mechanisms involve cellular responses to Hh ligand, which are mediated by either PTCH or SMO independent of GLI, or mechanisms leading to GLI activation independent of Hh ligand-mediated signaling. However, the potential role of these noncanonical pathways in breast carcinogenesis has not been investigated.

Emerging data indicate that Hh is required for CSC maintenance and tumorigenicity in various human cancers [58]. Increased expression of PTCH1, GLI1, and GLI2, which is the evidence of Hh signal activation, has been found in breast CSCs isolated from cancer patients [52]. Hh pathway activation by the overexpression of Hh ligands or GLI1/GLI2 increases sphere-forming activities and expansion of multi-lineage progenitors. Hh pathway inhibition by cyclopamine decreases the tumorigenic potential. GLI2 overexpression in human mammosphere-derived cells induces ductal hyperplasia when implanted into the humanized fat pads of NOD-SCID mice [59]. Goel et al. revealed a cascade downstream of the VEGF receptor Neuropilin-2 (NRP2) that regulated the breast CSC. GLI1 overexpression in the mammary gland in transgenic mice was also found to have an increased number of tumor initiation cells [60]. Moreover, Valenti et al. demonstrated an intracellular signaling module that synergistically regulated CAFs and breast CSCs by the Hedgehog signaling. CSCs secrete Shh and regulate cancer-associated fibroblasts (CAF) via paracrine activation, while CAFs subsequently secrete factors that promote expansion and self-renewal of breast CSCs [61].

The Hh, Notch, and Wnt pathways interact with each other through cross talk to regulate stemness [62]. Cross talk can lead to both enhancing and inhibitory interactions between pathways. Expression of the Notch target gene *Hes1* can be induced by the Hh signaling pathway, which requires the activation of both GLI1 and GLI2. On the other hand, activation of the Notch signaling can induce the expression of Shh [63]. Both Notch and Hh signaling activation occurs in chemotherapy-resistant CSCs. Notch and Hh inhibition depletes this subpopulation [64]. Moreover, the direct interaction of Notch-1 and β-catenin decreases Notch-1 ubiquitination, leading to increased expression of the Notch target gene Hes1 [63]. Importantly, the blockade of Notch ligand expression abrogates transformation of human mammary epithelial cells by Wnt1, demonstrating that there is a need for Notch-Wnt cross talk during mammary tumorigenesis [65]. Other cross talk between stemness-related pathways and important oncogenic pathways involves ErbB2, NF-κB, TGF-β, Jak/Stat, EGFR, etc. The interaction can happen by regulation of pathway components, usage of the same cofactors, or co-regulation of shared target genes. Better understanding of these cross talk mechanisms may allow for more effective design and development of combination antitumor therapies.

7.3.2 Signaling from the Tumor Microenvironment

Normal adult stem cells are regulated by local microenvironment or "niches." Similarly, multiple factors in the tumor microenvironment also regulate breast CSCs [66]. A variety of chemokines and cytokines released during inflammation by immune and tumor cells play a vital role in cancer progression, affecting key CSC properties [67].

Within the tumor microenvironment, IL-6 and IL-8 secreted by tumor-associated macrophages (TAMs), cancer-associated fibroblasts (CAFs), as well as other mesenchymal and immune cells regulate the inducible formation and maintenance of breast CSCs [68]. IL-6 has been shown to regulate the self-renewal of breast CSC directly, through the IL-6 receptor GP130 and STAT3 activation [69]. IL-6 is also a critical component of the positive feedback loop regulating breast CSCs and mesenchymal stem cells (MSCs). In the mouse xenografts, bone marrow-derived MSCs are recruited by gradients of IL-6 to sites of growing breast cancers, where they accelerate tumor growth by increasing the breast CSC population [66]. IL-8/CXCL8 is a pro-inflammatory chemokine. Its autocrine or paracrine signaling is transduced via two cell surface receptors CXCR1 and CXCR2 expressed on cancer cells, including breast CSCs [70]. Increased serum levels of IL-8 have been reported to increase the mammosphere formation of breast cancer cells [71]. IL-8 also enhances the cells' self-renewal capability, and its receptor CXCR1 is highly expressed on breast CSCs. Blocking CXCR1 significantly reduces the breast CSC proportion in mouse xenografts, resulting in decreased tumorigenicity and metastasis [72].

Nuclear factor-kappa B (NF-κB) signaling pathways provide one of the major links between CSCs and inflammation [73]. The canonical NF-κB pathway is activated by IκB kinase (IKK) α, β, and γ complex, which then activates RelA/p50. The noncanonical pathway is activated by NF-κB-inducing kinase (NIK) and IKKα, which then activates RelB/p52. NF-κB is predominantly localized in the cytoplasm with the IκB family in resting cells. Signals from the tumor microenvironment, including HIF-1α, TNF-α, etc., lead to IκB degradation mediated by ubiquitin ligase. Then NF-κB is activated and translocates to the nucleus to induce the expression of a number of cytokines, including IL-6 and IL-8, and EMT regulators, including Twist, Snail, and Slug [74]. This epithelial-mesenchymal transition (EMT) induced by NF-κB is accompanied by an acquisition of CSC properties [75].

While NF-κB regulates the transcription of a variety of cytokines, several cytokines drive NF-κB signaling, particularly IL-6 and IL-8, which are involved in CSC regulation, forming a positive feedback loop to promote CSC expansion and transformation. IL-6 maintains this feedback loop through STAT3, which in turn activates NF-κB and downstream microRNA lin28 and let7 [69]. NF-κB also plays an important role in oncoprotein-driven carcinogenesis by delaying the onset of HER2/neu-induced tumor in a mammary carcinogenesis model. NF-κB inhibition in mammary epithelium decreases angiogenesis and infiltration by macrophages, as well as the abundance of MaSCs [76]. IKKα phosphorylates p27 and promotes p27 nuclear export, leading to CSC expansion in a HER2/neu breast cancer model [77]. In vitro assay shows that knock-in of a kinase dead IKKα reduces the self-renewal of breast CSCs [78]. Notably, NF-κB constitutive activation is much higher in the basal-like breast cancer than that in the luminal subtype [79]. In non-CSC cancer cells, NF-κB induces the expression of Jagged 1 (JAG1), the Notch pathway ligand, in the basal-like, but not any other subtype of breast cancer. JAG1 is then secreted to the microenvironment and increases the abundance of breast CSCs in a Notch-dependent manner [79].

The C-X-C chemokine receptor type 4 (CXCR4) is a seven transmembrane (TM) GPCR involved in cancer development and immunodeficiency disorders. Besides $CD4^+$ T-helper cells, CXCR4 is expressed in various tissues. The expression of CXCR4 is higher in breast and lung tumor specimens compared to normal tissues and correlates with a poor prognosis of patients [80]. Recent studies have demonstrated an important role of CXCR4 and its ligand CXCL12 in the self-renewal of CSCs in vitro and in vivo. Activation of CXCR4 signaling by CXCL12 increases the abundance of CSCs, whereas blocking this pathway reduces the metastatic CSC activity [81].

Within the tumor microenvironment, the CXCR4/CXCL12 axis is critical for MSC recruitment to the tumors. CXCL12 is secreted by endothelium and stroma cells in the niche [82]. High intratumoral CXCL12 levels have been shown to attract $CXCR4^{+}$ stromal cells, inflammatory cells, and vascular cells into the tumors, where they secret growth factors, cytokines, and chemokines to support tumor cell growth. CXCR4 and HIFs expression is in turn upregulated in tumor cell to induce EMT as well as the expansion of CSCs [83, 84]. Moreover, due to the chemoattractive effect of CXCL12, $CXCR4^{+}$ circulating breast cancer cells can specifically home to certain metastatic sites. CXCR4 expression in CSCs confers increased invasiveness and metastatic potential as well as improved self-renewal and survival capacity [85].

Understanding the signaling pathways that regulate breast CSCs will help to develop the therapeutic strategy for abrogating CSCs. Conventional target therapy usually inhibits abnormal genes in the rapid proliferating tumor cells but misses the CSCs which are the source of cancer metastasis and relapse. Hence effective new therapeutic strategy for breast cancer must take the CSC pathway into account and target both bulk tumor cells and CSCs. We will discuss targeting the self-renewal pathways and blocking inflammatory pathways to eradicate the CSC population in the later section of this chapter.

7.4 Clinical Implications of Breast CSCs

In cancers that follow the stem cell model, the functional differences between tumorigenic and nontumorigenic cells have important clinical implications. Specific signature of breast CSC may have the potential for early detection and molecular characterization of breast cancer in patients. Moreover, a large body of evidence suggests that breast CSCs are highly resistant to radiation and chemotherapy. Mechanism of the treatment resistance includes abnormal expression of drug transport (efflux pumps), aberrations in drug pharmacokinetics and metabolism, modification of drug, changes in apoptotic signaling pathways, and altered DNA damage response.

7.4.1 Breast Cancer Subtypes and CSCs

Breast cancer has been classified into distinct subtypes by the gene expression profiling, namely, luminal A, luminal B, HER2+, basal-like, claudin-low, and normal-like [86]. The classification suggests that these subtypes originate from different cell types of the mammary gland. The issues relevant to CSCs include whether each subtype has its own CSCs, whether prognosis of patients relates to the stemness extent of the cell type, and whether different CSC type links to specific clinical phenotypes. These questions have not been resolved fully, but there is evidence showing that aggressive subtypes are associated with higher CSC frequencies. Hence breast CSCs may be responsible for aggressive behavior of breast cancer. Moreover, poorly differentiated tumors have a higher percentage of breast CSCs than better differentiated tumors [1]. Triple-negative breast cancers contain a higher proportion of the $CD24^{-/low}$ $CD44^{+}$ fraction compared to other subtypes [87].

CSCs specific for intrinsic subtypes of breast cancers are yet to be determined. The luminal progenitors are found to be the cells of origin for basal-like breast cancers, indicating that luminal and basal-like subtypes may harbor common CSCs. Through comparing gene signatures, Liu et al. demonstrated that breast CSCs of the luminal A subtype were most relevant to luminal progenitors or mature luminal cell, breast CSCs of the luminal B and HER2 subtype were most relevant to bipotent progenitors, and breast CSCs of the basal-like subtype were relevant to bipotent progenitors or mature luminal cell [88]. HER2+ CSCs display a distinct genotype from non-

HER2+ CSCs through altered epigenetic regulation. HER2 strongly regulates genes related to stem cell and progenitor cell control, while HER2+ CSCs increase expression of genes involved in S/G2/M transition and decrease expression of genes involved in differentiation and immune response [89].

Triple-negative breast cancers contain a higher proportion of the ALDH1+ compared to other subtypes [90]. ALDH1+ phenotype is correlated with basal and HER2+ breast cancers [91]. Independent of ER status, HER2 overexpression in cell lines causes an increase of the of ALDH1+ cell percentage. Liu et al. suggested that ALDH was a better marker for epithelial-like breast CSCs and $CD44^{+}CD24^{-/low}$ was a better indicator of mesenchymal-like CSCs. Epithelial-like CSCs proliferate rapidly and are more localized, whereas mesenchymal-like CSCs are more invasive and quiescent [92].

7.4.2 Breast CSCs and Treatment Resistance

Overcoming therapy resistance is a major challenge in treating cancers. Studies utilizing cell lines, animal models, and primary tumors have demonstrated that breast CSCs are more resistant to chemotherapy and radiation therapy than non-CSC cancer cells. Resistance to therapies can be either an innate characteristic of cancer cells or acquired later during treatments.

After neoadjuvant chemotherapy, the $CD24^{-}CD44^{+}$ population was increased in breast cancer patients [93]. The gene expression profile of post-chemotherapy tumors is similar to that of untreated $CD44^{+}CD24^{-}$ and mammosphere cells [93]. The multiple lines of defense that breast CSCs use to resist chemotherapy include: (1) being slow cycling and quiescent, which confers resistance to a variety of chemotherapy drugs that target the rapidly proliferating cells; (2) increased expression of ATP-binding cassette (ABC) transporters, including MDR1/ABCB1 and BCRP/ABCG2, which pump out chemotherapeutic drugs; (3) increased expression of ALDH enzymes which detoxify and inactivate chemo drugs; (4) increased DNA repair response because of the upregulation of DNA repair protein; (5) reduced apoptosis because of anti-apoptotic protein upregulation; and (6) increased activation of stemness pathways, such as Wnt, Notch, and Hedgehog signaling [94].

Based on these chemoresistance mechanisms, potential strategies can be developed to overcome CSC chemoresistance, by inhibiting ALDH, ABC transport proteins, and Wnt, Notch, and Hedgehog pathways, which will be discussed in the following section.

CSCs are also found to be more resistant to ionizing radiation (IR) than non-CSC cancer cells. IR therapy causes free radical reactive oxygen species (ROS) production in cells and induces DNA damage. Both stem cells and progenitor cells were intrinsically radioresistant. Normal breast progenitor cells have greater resistance than non-progenitor cells to radiation, and overexpression of the Wnt pathway components increase the radioresistance [95]. Breast CSCs in the MMTV-Wnt1 mouse model display lower levels of DNA damage than non-CSCs after radiation [96]. Enhanced DNA repair response, activated checkpoint signaling, overexpression of ROS scavengers and pro-survival molecules, and hypoxia in the tumor microenvironment contribute to IR resistance in CSCs [97]. To overcome the radioresistance of CSCs, in vitro and in vivo data provide promising results in targeting pro-survival and checkpoint signaling proteins. To improve the effect of radiation therapy, clinical trials aimed at CSCs are needed to bring current therapeutic strategies into the clinic.

CSC-associated treatment resistance is still one of the leading causes for treatment failure and tumor recurrence. A comprehensive understanding of the resistance mechanisms will help to develop more effective cancer therapeutics targeting CSCs. Evaluation of CSC-related factors before and after treatment has been incorporated into clinical trials to overcome treatment resistance and improve long-term survival of cancer patients.

7.5 Targeting Breast CSC for Cancer Therapy

Conventional target therapy usually inhibits abnormal genes in the rapid proliferating tumor cells but spares CSCs which are the source of cancer metastasis and recurrence. Targeting and elimination of breast CSCs may be a promising approach to treat breast cancer. Thus developing anti-CSC drugs is critical for decreasing cancer-related mortality. The strategies considered to eradicate breast CSCs include (1) inhibiting self-renewal signaling pathways which the breast CSCs depend on; (2) targeting proteins which are expressed in breast CSCs specifically, including surface markers, transporters, or enzymes; and (3) blocking the interaction of CSCs with supporting cells in the tumor microenvironment (see Fig. 7.1).

7.5.1 Targeting Stemness-Related Pathways

Breast CSCs rely on self-renewal pathways for maintaining their population. Targeting the key elements in self-renewal pathways offers attractive options, with a solid rationale. Currently there has been a surge in developing agents targeting self-renewal pathways and evaluating them in clinical trials. Herein, we discuss selected druggable targets and the clinical findings of agents targeting Notch, Wnt, and Hh pathway.

In the Notch signaling pathway, binding of Notch receptors (Notch 1–4) with ligands (Jagged1, Jagged2, Delta, and Delta-like) initiates a cascade of proteolytic cleavages mediated by the ADAM metalloprotease family members and internal cleavage by the γ-secretase. The strongest evidence to date for a role of Notch in CSCs is in glioma, breast cancer, and embryonal brain tumors [52]. Blocking the Notch pathway therapeutically can be achieved by Notch-targeted antibodies, DLL4 antibodies, and γ-secretase inhibitors (GSI), among which GSIs are currently the most advanced in clinical development.

OMP-59R5 (tarextumab) is an antibody targeting Notch2/3. The phase I clinical trials have shown preliminary evidence of efficacy and dosages well tolerated [98]. Notch1-targeted OMP-52 M51 is a humanized monoclonal antibody. Preclinical data have demonstrated that OMP-52 M51 reduces the percentage of breast CSCs in

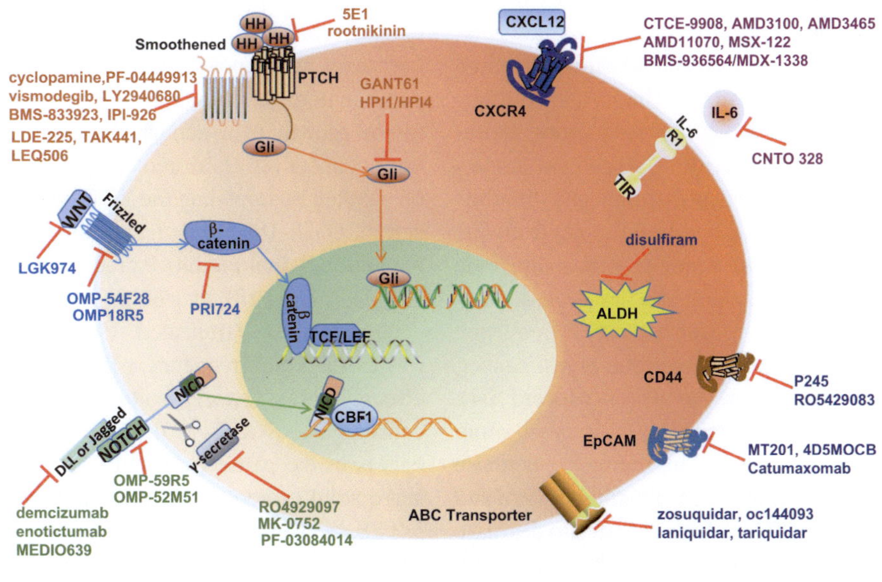

Fig. 7.1 Therapeutic targets and drugs developed against breast CSC-related signaling pathway, markers, and the tumor microenvironment

a xenograft mouse model [99]. The phase I clinical trials are evaluating this single agent in hematologic malignancy and solid tumors with aberrant activation of Notch1.

DLL4 is one of the Notch ligands, inhibition of which decreases the frequency of CSCs [100]. DLL4 monoclonal antibodies include demcizumab (OMP-21M18), enoticumab (REGN421), and MEDI0639. Early preclinical studies displayed suppression of Notch target gene expression by demcizumab. Combination treatment of demcizumab with paclitaxel reduced the breast CSC abundance and tumor growth in a mouse xenograft model [100]. In the phase I clinical trial, demcizumab caused grade III asymptomatic hypertension in 28% of patients, which was the most dangerous side effect. Enoticumab is being tested in the phase I clinical trial for advanced solid tumors, with several patients showing partial response or prolonged stable disease [101]. MEDI0639 is also being tested in the phase I clinical trial, with serious side effects occurring, including heart failure and gastrointestinal bleeding [102].

γ-secretase cleaves the Notch receptors and releases the active intracellular fragment. Among all Notch pathway inhibitors, γ-secretase inhibitors (GSI) are the most advanced agents under development at present. Several classes of GSI are currently in clinical trials for Alzheimer's disease, T-cell acute lymphoblastic leukemia (T-ALL), and breast cancer (http://clinicaltrials.gov/). These agents have displayed strong antitumor and anti-CSC activity in preclinical models [103]. DAPT targets presenilin and inhibits γ-secretase activity at a different site from the catalytic and substrate binding. It is a dipeptide non-transition state analog and has been demonstrated to deplete or decrease CSC frequency in vitro in lung cancer, metastatic breast cancer, nasopharyngeal cancer, and ovarian cancer [104]. RO4929097, derived from DAPT, is much more selective and potent in inhibiting γ-secretase activity and blocking Notch signaling in vitro and in vivo. RO4929097 has been reported to decrease tumor growth in xenograft mouse models [105]. MK0752 is a non-transition state analog containing sulfonamide. In combination with docetaxel, MK0752 treatment reduced the frequency of CSCs in tumor samples from cancer patients [106]. The selective tetralin amino imidazole GSI PF-03084014 inhibited the Notch pathway activated by docetaxel in a preclinical breast cancer study. Combination of PF-03084014 and docetaxel suppressed tumor growth synergistically in the breast cancer xenografts [107]. In clinical trials, the side effects of GSIs are secretory diarrhea and cutaneous rash [103]. Potential adverse events include off-target effects and systemic toxicity. Besides the Notch receptors, γ-secretase acts on more than 90 substrates, and GSIs block the cleavage of all of them. So far, clinical experience indicates that GSI should be administered in intermittent dosing regimens to prevent dose-limiting intestinal toxicity. However, the effectiveness of intermittent dosage for suppressing CSCs in patients is still unknown.

Wnt/β-catenin pathway can be dysregulated in many types of cancers and hence provides an excellent therapeutic target. Higher nuclear β-catenin level is observed in breast CSCs, indicating activation of Wnt signaling in CSCs. Drugs targeting Wnt signaling have been shown to decrease the tumorigenic potential of CSCs [108]. Preventing the secretion of Wnt ligands can block the Wnt signaling pathway. LGK974 is an inhibitor of porcupine (PORCN), which is an O-acetyltransferase on the cell membrane and necessary for proper secretion of Wnt ligands. In combination with paclitaxel, LGK974 reduces tumor growth in a xenograft model of human breast cancer [109]. Moreover, Wnt pathway can be blocked by impeding the WNT and FZD interaction. OMP-18R5 (vantictumab) is a monoclonal antibody that inhibits Fz receptors (Fz1, Fz2, Fz5, Fz7, and Fz8), as well as the fusion protein decoy receptor (truncated Fz8). Vantictumab lowers the CSC frequency and shows antitumor effects in breast cancer, colon cancer, pancreatic cancer, and NSCLC [110]. OMP-54F28 (ipafricept) is in the phase I clinical trial for advanced stage solid tumors. Preclinical testing of OMP-54F28 showed significant suppression of WNT-target gene expressions and antitumor effects in an MMTV-Wnt1 mouse model and teratoma cell lines [111]. Furthermore, transcription of β-catenin downstream genes can be inhibited by interfering the binding of β-catenin to the transcription factor TCF/LEF or the co-activator CREB-binding protein (CBP). PRI-724 suppresses the interaction of β-catenin with CBP and decreases CBP-dependent gene expression [112]. Its antitumor efficiency is being investigated in ongoing clinical studies. Meanwhile, two nonsteroidal anti-inflammatory drugs (NSAIDs) have been discovered to antagonize the Wnt pathway, sulindac targeting Dvl and celecoxib targeting β-catenin [103]. In summary, investigational drugs inhibiting the Wnt pathway are currently in early clinical studies. Some of the best preclinical results have been obtained from natural molecules for Wnt pathway inhibition.

Alterations of the Hedgehog (Hh) pathway contribute to CSC activation and tumorigenesis. Important druggable interactions in the Hh pathway include the binding of Shh, Dhh, or Ihh to PTCH, the interaction between PTCH and SMO, and the GLI-mediated transcription. GDC-0449 (vismodegib) directly antagonizes SMO and is the most advanced drug inhibiting the Hh pathway under clinical investigation [113]. Vismodegib was approved by the FDA for treatment of metastatic or locally advanced basal cell carcinoma (BCC) in 2012. Currently, various SMO inhibitors, such as PF-04449913, BMS-833923 (XL-139), LEQ506, TAK-441, LY 2940680 (Taladegib), SANT1-4, IPI-926, and LDE225, are under investigation in different tumor types in combination with a variety of chemotherapy agents. It appears that resistance to GDC-0449 does not confer to other SMO inhibitors.

Other Hedgehog pathway inhibitors include PTCH inhibitors and GLI-mediated transcription inhibitors. The monoclonal antibody 5E1 blocks the Hh interaction with PTCH. The small molecule robotnikinin binds to Shh and inhibits Shh interaction with PTCH [114]. HPI1 and HPI4 affect the stability and processing of GLI1 and GLI2 and thus prevent the GLI-dependent transcription [115]. GANT61 is an inhibitor of Hh signaling derived from hexahydropyrimidine. It alters the conformation of GLI1 and interferes the binding of DNA to GLI1 [116]. However, these inhibitors have not progressed into clinical trials for breast cancer treatment.

Cross talk among the stemness-related pathways has been reported in various cancer cells, which become a hurdle for drug development. Cross talk between these pathways can occur by several mechanisms. Firstly, physical interactions have been revealed between components of two pathways. Secondly, the component in one pathway can be a transcriptional target or enzymatic substrate of another pathway. Thirdly, one pathway may modulate a component of the other pathway or compete with other pathways [117]. Thus inhibition of one CSC pathway may cause compensatory activation of an interconnected pathway. To reduce this possibility, approaches to target multiple pathways through combination therapy are underway.

7.5.2 Targeting Breast CSC Markers

Considering that breast CSCs can be identified by the expression of markers, a potential strategy to target these cells is to block these specific proteins. Efforts to develop therapies against CD44, EpCAM, ALDH, and ATP-binding cassette (ABC) transporters are currently under investigation. In vitro and in vivo data appear to be promising in preclinical studies.

CD44 is expressed on the cell surface of a variety of CSCs. H90 is the first targeting monoclonal antibody of CD44, which displays CSC inhibition effects. P245 is also an anti-CD44 mAb, which prevents tumor growth in human breast cancer xenografts [118]. RO5429083, a humanized mAb directly targeting the extracellular epitope of CD44, has been evaluated on $CD44^+$ locally advanced and/or metastatic solid tumors in a phase I clinical trial. In recent studies, short DNA or RNA single strand with a high binding efficiency against CD44 and CD44 targeting aptamers was under exploration. However, such molecules have been tested only for their binding specificity and stability but not inhibition efficiency. They may conjugate anticancer agents and be utilized as delivery system specifically against breast CSCs [119].

EpCAM is significantly overexpressed in CSCs in breast, pancreatic, and colorectal cancer. Therapeutic antibodies of EpCAM are under development, with a few being tested as single agent or in combination in clinical trials. The clinical study of human monoclonal antibody MT201 (adecatumumab) is ongoing in patients with breast and prostate cancer [120]. The anti-EpCAM scFv MOC31 is produced by phage display cloning from the hybridoma. A second generation of humanized scFv 4D5MOCB displays appreciable targeting efficiency, good folding, and high thermal stability in vivo [121]. The anti-EpCAM × anti-CD3 bispecific mAb catumaxomab (Removab®) is generated from mouse and rat hybrid. It has been administered as an intraperitoneal infusion to treat patients with malignant ascites in EpCAM^{+} cancer [121]. The adverse effects are more apparent with high-affinity antibodies, because besides cancer cells, EpCAM is expressed on some normal epithelia, which causes systemic damage. Therefore, low-affinity antibodies are better tolerated in the tolerance experiments [120]. Improved therapeutic EpCAM antibodies are in need for further clinical studies.

ALDH enzyme activity is higher in breast CSCs than non-CSC cancer cells. Known inhibitors of ALDHs include DEAB, ampal, benzimidazole-based analogues, benomyl, CVT-10216, citral, chloral hydrate, chlorpropamide analogues, coprine, cyanamide, daidzin, disulfiram, gossypol, kynurenine tryptophan metabolites, molinate, nitroglycerin, and pargyline [122, 123]. Among these agents, disulfiram is the most extensively investigated for targeting CSCs and cancer treatment. In breast cancer, disulfiram inhibits ALDH activity and downregulates the expression of CSC markers. Disulfiram decreases tumor growth in MDA-MB-231 xenografts in mice and increases the sensitivity of xenografts to radiation therapy [123]. Ongoing clinical studies are testing whether disulfiram inhibits ALDH1 and increases the sensitivity of CSCs to chemotherapies and radiotherapies.

ATP-binding cassette (ABC) transporters are frequently highly expressed on the cell surface of both normal stem cells and CSCs. ABC transporters contribute to multidrug resistance due to the efflux of xenobiotic toxins. The first and second generations of inhibitors, such as R-verapamil, GF120918 (elacridar), MS-209 (dofequidar), PSC833 (valspodar), biricodar (INCEL, VX-710), and VX-853 (ortimcodar), block multidrug resistance protein 1 (MDR1, ABCBC1) [124]. The third generations of inhibitors, such as LY335979 (zosuquidar), oc144093 (ONT-093), R101933 (laniquidar), and XR-9576 (tariquidar), more specifically target ABCB1, ABCC1, and ABCG2 and are being tested to sensitize CSCs in different stages of clinical trials [125]. Dofequidar treatment in combination with doxorubicin, cyclophosphamide, and fluorouracil has shown some promising results in breast cancer patients who are premenopausal, not receiving any prior therapy, or at stage IV at diagnosis with an intact primary tumor [126].

Although therapies targeting surface proteins of CSC are very promising, some of them might affect normal stem cells. Future efforts should focus on increasing the efficiency and specificity that target CSCs, as well as diminishing damages to normal tissues. Improvements can also be achieved by developing new delivery methods and by enhancing the retention of drugs inside the CSCs.

7.5.3 Targeting the Tumor Microenvironment

Solid tumors are composed of cancer cells, immune cells, endothelial cells, fibroblasts, etc. In the tumor microenvironment, together with cytokines and growth factors, these cells communicate with each other and form a dynamic network. Interfering this network is a promising direction in drug discovery. Indeed, therapeutic strategies targeting proteins necessary for their interaction within microenvironment can be very effective at eradicating breast CSC in preclinical studies.

CNTO 328 is a humanized monoclonal antibody targeting IL-6. CNTO 328 has been tested as single agent or in combination in renal cell cancer and ovarian cancer patients in clinical trials, with some efficacy being observed [127]. Reparixin is an inhibitor of IL-8/CXCL8. Inhibitory effects of reparixin on tumor growth, angiogenesis, and tumor dissemination have been observed in vitro and in vivo. The safety profile of reparixin is being tested in early breast cancer patients. Ongoing clinical trials are investigating the combination therapy of reparixin with weekly paclitaxel in $HER2^-$ metastatic breast cancer patients. NF-κB is an intriguing transcription factor. Some NF-κBs target genes can be suppressed by aspirin and salicylates [128]. Given the role of NF-κB in inflammation, therapeutic inhibition of NF-κB needs to be transient to avoid immunosuppression [129]. Production of IL-1β and related cytokines during bacterial infections causes adverse effect of NF-κB targeting [129]. Although NF-κB inhibition still faces challenges, a few targeting agents will be tested in clinical trials in the next few years.

Strategies targeting the CXCL12/CXCR4 pathway may inhibit CSCs in cancer patients. In vitro and in vivo studies displayed that CXCR4 inhibition by peptides or small molecule inhibitors reduced the tumor growth of breast cancer. CXCR4 inhibitor CTCE-9908 treatment alleviated primary tumor burden and reduced the development of metastasis in mice-bearing human breast [130, 131]. The CXCR4 receptor inhibitor AMD3100 (plerixafor) was approved for clinical treatment in non-Hodgkin's lymphoma and multiple myelomas by the FDA in 2008. AMD3100 suppressed lung metastases in an orthotopically transplanted breast cancer model [132]. Overexpression of CXCR4 is correlated with resistance to chemotherapy, which can induce CXCR4 upregulation. Ongoing clinical trials are investigating whether AMD3100 in combination with conventional chemotherapy improves clinical outcomes by increasing the sensitivity of cancer cells to chemo drugs [80]. Moreover, based on the chemical scaffold of AMD3100, more potent CXCR4 inhibitors are under development, including MSX-122, AMD3465, and AMD11070, which block the cell surface binding of CXCL12. These inhibitors have been shown to reduce CXCR4/CXCL12-dependent cancer cell proliferation, migration, chemoresistance, and tumorigenicity in a dose-dependent manner [80]. In parallel, promising effects were observed in the fully humanized antibody MDX-1338 (BMS-936564, ulocuplumab), which targeted CXCR4 receptor selectively and prevented the binding of CXCL12 to $CXCR4^+$ cells [133].

Collectively, tumor microenvironment has great impact on cancer progression, with its involvement in supporting cells as well as autocrine and paracrine signaling. The review here only shows selective current treatments targeting the cytokines and factors in tumor microenvironment. More studies are required to further explore safer strategies targeting cytokines that are critical for immune response and effective eradication of breast CSCs.

7.6 Future Perspective

The incorporation of CSC concept in precision medicine is appealing in breast cancer clinical care. However, several challenges need to be addressed to improve patient outcome. The first is lack of well-characterized drivers. Successful targeting of CSCs requires a thorough understanding of CSC regulation and characterization of driver genes. The heterogeneity among tumors from different patients adds more complexity to this issue, which creates a new problem as to how to choose the most appropriate inhibitor for each patient. A comprehensive elucidation of regulatory mechanism and interaction networks of breast CSC will help to design safer and more effective combination regimens. The second challenge is how to select patients who will benefit more from targeting treatments. Reliable biomarkers are required to determine the activity of CSC pathway from specific tumors. The third but not the least challenge concerns the highly bioactive and specific drugs. Failure to deliver precision medicine is usually caused by the lack of proper drugs. Natural products with novel mechanisms are considered to be more effective than small molecules in targeting the same interaction [134]. However, research on natural product lags far behind targeted therapies. Screening natural molecules may help identify anti-CSC agents. Moreover, based on the natural molecules, semisynthetic natural compounds for targeted therapy can be generated to increase drug activity or avoid side effects.

Regarding the CSC regulatory mechanism and druggable targets, here we only focus on protein-coding genes. However, protein-coding DNA sequences account for only about 2% of human genome. The recent explosion in studies of non-coding RNAs (ncRNAs) has fostered a new view of the RNA function. Gene regulation networks are much more complex than expected. The fact that organismal complexity across species links well with the percentage of genomes being transcribed to ncRNAs suggests that RNA-based regulatory mechanisms are involved in the evolution of developmental complexity in eukaryotes. Great advance has been made in understanding human ncRNA in the last decade. Indeed, ncRNAs, such as microRNAs (miRNAs) and long noncoding RNAs (lncRNAs), are key players in cellular, physiological, and developmental processes, as well as in various diseases including cancers. In the field of CSC, a large body of evidence demonstrates that aberrant expression of miRNAs or lncRNAs contributes to cancer stemness. Moreover, miRNA or lncRNA profiles or signatures have the potential to be applied as effective biomarkers in clinical practice. Many problems may already have been solved during the development of technologies to target RNA. Thus, ncRNA is quite ready to be utilized as biomarkers and therapeutic targets. Currently, it is very desirable to investigate the mechanism of ncRNA regulating CSCs and translate the knowledge into clinical practice. Since the human genome expresses an enormous number of ncRNAs, precision medicine should take all the genomic information into account.

The characterization of breast CSC driver gene and therapy targets is still in its infancy, and a huge repertoire of ncRNAs is still unexplored. Comprehensive understanding of the breast CSC regulation, improved technologies in identification of molecular alterations, and high-throughput drug screenings may provide strong impetus for developing and applying precision medicine in breast cancer.

Acknowledgments We would like to thank the support from the Natural Science Foundation of China (81572890 to Luo ML), Guangdong Science and Technology Department (2015B050501004), Grant [2013] 163 from Key Laboratory of Malignant Tumor Molecular Mechanism and Translational Medicine of Guangzhou Bureau of Science and Information Technology, and Grant KLB09001 from the Key Laboratory of Malignant Tumor Gene Regulation and Target Therapy of Guangdong Higher Education Institutes.

References

1. Pece S, Tosoni D, Confalonieri S, Mazzarol G, Vecchi M, Ronzoni S, Bernard L, Viale G, Pelicci PG, Di Fiore PP (2010) Biological and molecular heterogeneity of breast cancers correlates with their cancer stem cell content. Cell 140(1):62–73. doi:10.1016/j.cell.2009.12.007
2. Rosen JM, Jordan CT (2009) The increasing complexity of the cancer stem cell paradigm. Science 324(5935):1670–1673. doi:10.1126/science.1171837
3. Clarke MF, Dick JE, Dirks PB, Eaves CJ, Jamieson CHM, Jones DL, Visvader J, Weissman IL, Wahl GM (2006) Cancer stem cells--perspectives on current status and future directions: AACR workshop on cancer stem cells. Cancer Res 66(19):9339–9344. doi:10.1158/0008-5472.can-06-3126
4. He SH, Nakada D, Morrison SJ (2009) Mechanisms of stem cell self-renewal. In: Annual review of cell and developmental biology, vol 25. Annual Review of Cell and Developmental Biology. Annual Reviews, Palo Alto, pp 377–406. doi:10.1146/annurev.cellbio.042308.113248
5. Morrison SJ, Kimble J (2006) Asymmetric and symmetric stem-cell divisions in development and cancer. Nature 441(7097):1068–1074. doi:10.1038/nature04956
6. Reya T, Morrison SJ, Clarke MF, Weissman IL (2001) Stem cells, cancer, and cancer stem cells. Nature 414(6859):105–111. doi:10.1038/35102167
7. Al-Hajj M, Wicha MS, Benito-Hernandez A, Morrison SJ, Clarke MF (2003) Prospective identification of tumorigenic breast cancer cells (vol 100, pg 3983, 2003). Proc Natl Acad Sci U S A 100(11):6890–6890. doi:10.1073/pnas.1131491100
8. Ginestier C, Hur MH, Charafe-Jauffret E, Monville F, Dutcher J, Brown M, Jacquemier J, Viens P, Kleer CG, Liu SL, Schott A, Hayes D, Birnbaum D, Wicha MS, Dontu G (2007) ALDH1 is a marker of normal and malignant human mammary stem cells and a predictor of poor clinical outcome. Cell Stem Cell 1(5):555–567. doi:10.1016/j.stem.2007.08.014
9. Naor D, Wallach-Dayan SB, Zahalka MA, Sionov RV (2008) Involvement of CD44, a molecule with a thousand faces, in cancer dissemination.

Semin Cancer Biol 18(4):260–267. doi:10.1016/j.semcancer.2008.03.015

10. Du L, Wang HY, He LY, Zhang JY, Ni BY, Wang XH, Jin HJ, Cahuzac N, Mehrpour M, Lu YY, Chen Q (2008) CD44 is of functional importance for colorectal cancer stem cells. Clin Cancer Res 14(21):6751–6760. doi:10.1158/1078-0432.ccr-08-1034
11. Bourguignon LYW, Spevak CC, Wong G, Xia WL, Gilad E (2009) Hyaluronan-CD44 interaction with protein kinase C epsilon promotes oncogenic signaling by the stem cell marker Nanog and the production of MicroRNA-21, leading to down-regulation of the tumor suppressor protein PDCD4, anti-apoptosis, and chemotherapy resistance in breast tumor cells. J Biol Chem 284(39):26533–26546. doi:10.1074/jbc.M109.027466
12. Jothy S (2003) CD44 and its partners in metastasis. Clin Exp Metastasis 20(3):195–201. doi:10.1023/a:1022931016285
13. Phuc PV, Phan LCN, Nhung TH, Tam NT, Hoang NM, Tue VG, Thuy DT, Ngoc PK (2011) Downregulation of CD44 reduces doxorubicin resistance of CD44(+)CD24(−) breast cancer cells. Onco Targets Ther 4:71–78. doi:10.2147/ott.s21431
14. Pham PV, Phan NLC, Nguyen NT, Truong NH, Duong TT, Le DV, Truong KD, Phan NK (2011) Differentiation of breast cancer stem cells by knockdown of CD44: promising differentiation therapy. J Transl Med 9:13. doi:10.1186/1479-5876-9-209
15. Kristiansen G, Winzer KJ, Mayordomo E, Bellach J, Schluns K, Denkert C, Dahl E, Pilarsky C, Altevogt P, Guski H, Dietel M (2003) CD24 expression is a new prognostic marker in breast cancer. Clin Cancer Res 9(13):4906–4913
16. Fang XF, Zheng P, Tang J, Liu Y (2010) CD24: from a to Z. Cell Mol Immunol 7(2):100–103. doi:10.1038/cmi.2009.119
17. Baumann P, Cremers N, Kroese FGM, Orend G, Chiquet-Ehrismann R, Uede T, Yagita H, Sleeman JP (2005) CD24 expression causes the acquisition of multiple cellular properties associated with tumor growth and metastasis. Cancer Res 65(23):10783–10793. doi:10.1158/0008-5472.can-05-0619
18. Ricardo S, Vieira AF, Gerhard R, Leitao D, Pinto R, Cameselle-Teijeiro JF, Milanezi F, Schmitt F, Paredes J (2011) Breast cancer stem cell markers CD44, CD24 and ALDH1: expression distribution within intrinsic molecular subtype. J Clin Pathol 64(11):937–946. doi:10.1136/jcp.2011.090456
19. Giatromanolaki A, Sivridis E, Fiska A, Koukourakis MI (2011) The CD44+/CD24-phenotype relates to 'triple-negative' state and unfavorable prognosis in breast cancer patients. Med Oncol 28(3):745–752. doi:10.1007/s12032-010-9530-3
20. Stuelten CH, Mertins SD, Busch JI, Gowens M, Scudiero DA, Burkett MW, Hite KM, Alley M, Hollingshead M, Shoemaker RH, Niederhuber JE (2010) Complex display of putative tumor stem cell markers in the NCI60 tumor cell line panel. Stem Cells 28(4):649–660. doi:10.1002/stem.324
21. Balzar M, Winter MJ, de Boer CJ, Litvinov SV (1999) The biology of the 17-1A antigen (ep-CAM). J Mol Med 77(10):699–712. doi:10.1007/s001099900038
22. van der Gun BTF, Melchers LJ, Ruiters MHJ, de Leij L, McLaughlin PMJ, Rots MG (2010) EpCAM in carcinogenesis: the good, the bad or the ugly. Carcinogenesis 31(11):1913–1921. doi:10.1093/carcin/bgq187
23. Schmidt M, Hasenclever D, Schaeffer M, Boehm D, Cotarelo C, Steiner E, Lebrecht A, Siggelkow W, Weikel W, Schiffer-Petry I, Gebhard S, Pilch H, Gehrmann M, Lehr HA, Koelbl H, Hengstler JG, Schuler M (2008) Prognostic effect of epithelial cell adhesion molecule overexpression in untreated node-negative breast cancer. Clin Cancer Res 14(18):5849–5855. doi:10.1158/1078-0432.ccr-08-0669
24. Gastl G, Spizzo G, Obrist P, Dunser M, Mikuz G (2000) Ep-CAM overexpression in breast cancer as a predictor of survival. Lancet 356(9246):1981–1982. doi:10.1016/s0140-6736(00)03312-2
25. Osta WA, Chen Y, Mikhitarian K, Mitas M, Salem M, Hannun YA, Cole DJ, Gillanders WK (2004) EpCAM is overexpressed in breast cancer and is a potential target for breast cancer gene therapy. Cancer Res 64(16):5818–5824. doi:10.1158/0008-5472.can-04-0754
26. Marcato P, Dean CA, Giacomantonio CA, Lee PWK (2011) Aldehyde dehydrogenase its role as a cancer stem cell marker comes down to the specific isoform. Cell Cycle 10(9):1378–1384. doi:10.4161/cc.10.9.15486
27. Cheung AMS, Wan TSK, Leung JCK, Chan LYY, Huang H, Kwong YL, Liang R, Leung AYH (2007) Aldehyde dehydrogenase activity in leukemic blasts defines a subgroup of acute myeloid leukemia with adverse prognosis and superior NOD/SCID engrafting potential. Leukemia 21(7):1423–1430. doi:10.1038/sj.leu.2404721
28. Bane A, Viloria-Petit A, Pinnaduwage D, Mulligan AM, O'Malley FP, Andrulis IL (2013) Clinical-pathologic significance of cancer stem cell marker expression in familial breast cancers. Breast Cancer Res Treat 140(1):195–205. doi:10.1007/s10549-013-2591-1
29. Kang EJ, Jung H, Woo OH, Park KH, Woo SU, Yang DS, Kim AR, Lee JB, Kim YH, Kim JS, Seo JH (2014) Association of aldehyde dehydrogenase 1 expression and biologically aggressive features in breast cancer. Neoplasma 61(3):352–362. doi:10.4149/neo_2014_045
30. Marcato P, Dean CA, Pan D, Araslanova R, Gillis M, Joshi M, Helyer L, Pan L, Leidal A, Gujar S, Giacomantonio CA, Lee PWK (2011) Aldehyde dehydrogenase activity of breast cancer stem cells is primarily due to isoform ALDH1A3 and its

expression is predictive of metastasis. Stem Cells 29(1):32–45. doi:10.1002/stem.563
31. Krivtsov AV, Twomey D, Feng ZH, Stubbs MC, Wang YZ, Faber J, Levine JE, Wang J, Hahn WC, Gilliland DG, Golub TR, Armstrong SA (2006) Transformation from committed progenitor to leukaemia stem cell initiated by MLL-AF9. Nature 442(7104):818–822. doi:10.1038/nature04980
32. Quintana E, Shackleton M, Sabel MS, Fullen DR, Johnson TM, Morrison SJ (2008) Efficient tumour formation by single human melanoma cells. Nature 456(7222):593–U533. doi:10.1038/nature07567
33. Johnston MD, Maini PK, Chapman SJ, Edwards CM, Bodmer WF (2010) On the proportion of cancer stem cells in a tumour. J Theor Biol 266(4):708–711. doi:10.1016/j.jtbi.2010.07.031
34. Dontu G, Abdallah WM, Foley JM, Jackson KW, Clarke MF, Kawamura MJ, Wicha MS (2003) In vitro propagation and transcriptional profiling of human mammary stem/progenitor cells. Genes Dev 17(10):1253–1270. doi:10.1101/gad.1061803
35. Pastrana E, Silva-Vargas V, Doetsch F (2011) Eyes wide open: a critical review of sphere-formation as an assay for stem cells. Cell Stem Cell 8(5):486–498. doi:10.1016/j.stem.2011.04.007
36. Dontu G, Wicha MS (2005) Survival of mammary stem cells in suspension culture: implications for stem cell biology and neoplasia. J Mammary Gland Biol Neoplasia 10(1):75–86. doi:10.1007/s10911-005-2542-5
37. Grimshaw MJ, Cooper L, Papazisis K, Coleman JA, Bohnenkamp HR, Chiapero-Stanke L, Taylor-Papadimitriou J, Burchell JM (2008) Mammosphere culture of metastatic breast cancer cells enriches for tumorigenic breast cancer cells. Breast Cancer Res 10(3):10. doi:10.1186/bcr2106
38. Pastrana E, Cheng LC, Doetsch F (2009) Simultaneous prospective purification of adult subventricular zone neural stem cells and their progeny. Proc Natl Acad Sci U S A 106(15):6387–6392. doi:10.1073/pnas.0810407106
39. Bonnefoix T, Bonnefoix P, Callanan M, Verdiel P, Sotto JJ (2001) Graphical representation of a generalized linear model-based statistical test estimating the fit of the single-hit Poisson model to limiting dilution assays. J Immunol 167(10):5725–5730
40. O'Brien CA, Kreso A, Jamieson CHM (2010) Cancer stem cells and self-renewal. Clin Cancer Res 16(12):3113–3120. doi:10.1158/1078-0432.ccr-09-2824
41. Clevers H, Loh KM, Nusse R (2014) An integral program for tissue renewal and regeneration: Wnt signaling and stem cell control. Science 346(6205):54–+. doi:10.1126/science.1248012
42. Polakis P (2012) Wnt signaling in cancer. Cold Spring Harb Perspect Biol 4(5):13. doi:10.1101/cshperspect.a008052
43. Reya T, Clevers H (2005) Wnt signalling in stem cells and cancer. Nature 434(7035):843–850. doi:10.1038/nature03319
44. Wend P, Holland JD, Ziebold U, Birchmeier W (2010) Wnt signaling in stem and cancer stem cells. Semin Cell Dev Biol 21(8):855–863. doi:10.1016/j.semcdb.2010.09.004
45. Shackleton M, Vaillant F, Simpson KJ, Stingl J, Smyth GK, Asselin-Labat ML, Wu L, Lindeman GJ, Visvader JE (2006) Generation of a functional mammary gland from a single stem cell. Nature 439(7072):84–88. doi:10.1038/nature04372
46. Bafico A, Liu GZ, Goldin L, Harris V, Aaronson SA (2004) An autocrine mechanism for constitutive Wnt pathway activation in human cancer cells. Cancer Cell 6(5):497–506. doi:10.1016/j.ccr.2004.09.032
47. Scheel C, Eaton EN, Li SHJ, Chaffer CL, Reinhardt F, Kah KJ, Bell G, Guo W, Rubin J, Richardson AL, Weinberg RA (2011) Paracrine and autocrine signals induce and maintain mesenchymal and stem cell states in the breast. Cell 145(6):926–940. doi:10.1016/j.cell.2011.04.029
48. Chiba S (2006) Notch signaling in stem cell systems. Stem Cells 24(11):2437–2447. doi:10.1634/stemcells.2005-0661
49. Roy M, Pear WS, Aster JC (2007) The multifaceted role of notch in cancer. Curr Opin Genet Dev 17(1):52–59. doi:10.1016/j.gde.2006.12.001
50. Ilagan MXG, Kopan R (2007) SnapShot: notch signaling pathway. Cell 128(6):1246
51. Andersen P, Uosaki H, Shenje LT, Kwon C (2012) Non-canonical notch signaling: emerging role and mechanism. Trends Cell Biol 22(5):257–265. doi:10.1016/j.tcb.2012.02.003
52. Pannuti A, Foreman K, Rizzo P, Osipo C, Golde T, Osborne B, Miele L (2010) Targeting notch to target cancer stem cells. Clin Cancer Res 16(12):3141–3152. doi:10.1158/1078-0432.ccr-09-2823
53. Harrison H, Farnie G, Howell SJ, Rock RE, Stylianou S, Brennan KR, Bundred NJ, Clarke RB (2010) Regulation of breast cancer stem cell activity by signaling through the Notch4 receptor. Cancer Res 70(2):709–718. doi:10.1158/0008-5472.can-09-1681
54. Bouras T, Pal B, Vaillant F, Harburg G, Asselin-Labat ML, Oakes SR, Lindeman GJ, Visvader JE (2008) Notch signaling regulates mammary stem cell function and luminal cell-fate commitment. Cell Stem Cell 3(4):429–441. doi:10.1016/j.stem.2008.08.001
55. Pandya K, Meeke K, Clementz AG, Rogowski A, Roberts J, Miele L, Albain KS, Osipo C (2011) Targeting both notch and ErbB-2 signalling pathways is required for prevention of ErbB-2-positive breast tumour recurrence. Br J Cancer 105(6):796–806. doi:10.1038/bjc.2011.321
56. Varjosalo M, Taipale J (2008) Hedgehog: functions and mechanisms. Genes Dev 22(18):2454–2472. doi:10.1101/gad.1693608
57. Amakye D, Jagani Z, Dorsch M (2013) Unraveling the therapeutic potential of the hedgehog pathway in cancer. Nat Med 19(11):1410–1422. doi:10.1038/nm.3389

58. Angeloni V, Tiberio P, Appierto V, Daidone MG (2015) Implications of stemness-related signaling pathways in breast cancer response to therapy. Semin Cancer Biol 31:43–51. doi:10.1016/j.semcancer.2014.08.004
59. Liu SL, Dontu G, Mantle ID, Patel S, Ahn NS, Jackson KW, Suri P, Wicha MS (2006) Hedgehog signaling and Bmi-1 regulate self-renewal of normal and malignant human mammary stem cells. Cancer Res 66(12):6063–6071. doi:10.1158/0008-5472.can-06-0054
60. Goel HL, Pursell B, Chang C, Shaw LM, Mao JH, Simin K, Kumar P, Vander Kooi CW, Shultz LD, Greiner DL, Norum JH, Toftgard R, Kuperwasser C, Mercurio AM (2013) GLI1 regulates a novel neuropilin-2/61 integrin based autocrine pathway that contributes to breast cancer initiation. EMBO Mol Med 5(4):488–508. doi:10.1002/emmm.201202078
61. Valenti G, Quinn HM, Heynen G, Lan LX, Holland JD, Vogel R, Wulf-Goldenberg A, Birchmeier W (2017) Cancer stem cells regulate cancer-associated fibroblasts via activation of hedgehog signaling in mammary gland tumors. Cancer Res 77(8):2134–2147. doi:10.1158/0008-5472.can-15-3490
62. Takebe N, Harris PJ, Warren RQ, Ivy SP (2011) Targeting cancer stem cells by inhibiting Wnt, notch, and hedgehog pathways. Nat Rev Clin Oncol 8(2):97–106. doi:10.1038/nrclinonc.2010.196
63. Borggrefe T, Lauth M, Zwijsen A, Huylebroeck D, Oswald F, Giaimo BD (2016) The notch intracellular domain integrates signals from Wnt, hedgehog, TGF beta/BMP and hypoxia pathways. Biochim Biophys Acta-Mol Cell Res 1863(2):303–313. doi:10.1016/j.bbamcr.2015.11.020
64. Domingo-Domenech J, Vidal SJ, Rodriguez-Bravo V, Castillo-Martin M, Quinn SA, Rodriguez-Barrueco R, Bonal DM, Charytonowicz E, Gladoun N, de la Iglesia-Vicente J, Petrylak DP, Benson MC, Silva JM, Cordon-Cardo C (2012) Suppression of acquired docetaxel resistance in prostate cancer through depletion of notch- and hedgehog-dependent tumor-initiating cells. Cancer Cell 22(3):373–388. doi:10.1016/j.ccr.2012.07.016
65. Ayyanan A, Civenni G, Ciarloni L, Morel C, Mueller N, Lefort K, Mandinova A, Raffoul W, Fiche M, Dotto GP, Brisken C (2006) Increased Wnt signaling triggers oncogenic conversion of human breast epithelial cells by a notch-dependent mechanism. Proc Natl Acad Sci U S A 103(10):3799–3804. doi:10.1073/pnas.0600065103
66. Liu SL, Ginestier C, Ou SJ, Clouthier SG, Patel SH, Monville F, Korkaya H, Heath A, Dutcher J, Kleer CG, Jung YH, Dontu G, Taichman R, Wicha MS (2011) Breast cancer stem cells are regulated by mesenchymal stem cells through cytokine networks. Cancer Res 71(2):614–624. doi:10.1158/0008-5472.can-10-0538
67. Lu HT, Ouyang WM, Huang CS (2006) Inflammation, a key event in cancer development. Mol Cancer Res 4(4):221–233. doi:10.1158/1541-7786.mcr-05-0261
68. Iliopoulos D, Hirsch HA, Wang GN, Struhl K (2011) Inducible formation of breast cancer stem cells and their dynamic equilibrium with non-stem cancer cells via IL6 secretion. Proc Natl Acad Sci U S A 108(4):1397–1402. doi:10.1073/pnas.1018898108
69. Iliopoulos D, Hirsch HA, Struhl K (2009) An epigenetic switch involving NF-kappa B, Lin28, let-7 MicroRNA, and IL6 links inflammation to cell transformation. Cell 139(4):693–706. doi:10.1016/j.cell.2009.10.014
70. Waugh DJJ, Wilson C (2008) The interleukin-8 pathway in cancer. Clin Cancer Res 14(21):6735–6741. doi:10.1158/1078-0432.ccr-07-4843
71. Singh JK, Farnie G, Bundred NJ, Simoes BM, Shergill A, Landberg G, Howell SJ, Clarke RB (2013) Targeting CXCR1/2 significantly reduces breast cancer stem cell activity and increases the efficacy of inhibiting HER2 via HER2-dependent and -independent mechanisms. Clin Cancer Res 19(3):643–656. doi:10.1158/1078-0432.ccr-12-1063
72. Ginestier C, Liu SL, Diebel ME, Korkaya H, Luo M, Brown M, Wicinski J, Cabaud O, Charafe-Jauffret E, Birnbaum D, Guan JL, Dontu G, Wicha MS (2010) CXCR1 blockade selectively targets human breast cancer stem cells in vitro and in xenografts. J Clin Invest 120(2):485–497. doi:10.1172/jc139397
73. Grivennikov SI, Greten FR, Karin M (2010) Immunity, inflammation, and cancer. Cell 140(6):883–899. doi:10.1016/j.cell.2010.01.025
74. Li CW, Xia WY, Huo LF, Lim SO, Wu Y, Hsu JL, Chao CH, Yamaguchi H, Yang NK, Ding QQ, Wang Y, Lai YJ, LaBaff AM, Wu TJ, Lin BR, Yang MH, Hortobagyi GN, Hung MC (2012) Epithelial-mesenchymal transition induced by TNF-alpha requires NF-kappa B-mediated transcriptional upregulation of Twist1. Cancer Res 72(5):1290–1300. doi:10.1158/0008-5472.can-11-3123
75. Korkaya H, Kim GI, Davis A, Malik F, Henry NL, Ithimakin S, Quraishi AA, Tawakkol N, D'Angelo R, Paulson AK, Chung S, Luther T, Paholak HJ, Liu SL, Hassan KA, Zen Q, Clouthier SG, Wicha MS (2012) Activation of an IL6 inflammatory loop mediates Trastuzumab resistance in HER2+breast cancer by expanding the cancer stem cell population. Mol Cell 47(4):570–584. doi:10.1016/j.molcel.2012.06.014
76. Liu M, Sakamaki T, Casimiro MC, Willmarth NE, Quong AA, Ju XM, Ojeifo J, Jiao XM, Yeow WS, Katiyar S, Shirley LA, Joyce D, Lisanti MP, Albanese C, Pestell RG (2010) The canonical NF-kappa B pathway governs mammary tumorigenesis in transgenic mice and tumor stem cell expansion. Cancer Res 70(24):10464–10473. doi:10.1158/0008-5472.can-10-0732
77. Zhang WZ, Tan W, Wu XF, Poustovoitov M, Strasner A, Li W, Borcherding N, Ghassemian M, Karin M (2013) A NIK-IKK alpha module expands

ErbB2-induced tumor-initiating cells by stimulating nuclear export of p27/Kip1. Cancer Cell 23(5):647–659. doi:10.1016/j.ccr.2013.03.012
78. Cao YX, Luo JL, Karin M (2007) I kappa B kinase a kinase activity is required for self-renewal of ErbB2/Her2-transformed mammary tumor-initiating cells. Proc Natl Acad Sci U S A 104(40):15852–15857. doi:10.1073/pnas.0706728104
79. Yamamoto M, Taguchi Y, Ito-Kureha T, Semba K, Yamaguchi N, Inoue J (2013) NF-kappa B non-cell-autonomously regulates cancer stem cell populations in the basal-like breast cancer subtype. Nat Commun 4:13. doi:10.1038/ncomms3299
80. Scala S (2015) Molecular pathways: targeting the CXCR4-CXCL12 Axis-untapped potential in the tumor microenvironment. Clin Cancer Res 21(19):4278–4285. doi:10.1158/1078-0432.ccr-14-0914
81. Ablett MP, O'Brien CS, Sims AH, Farnie G, Clarke RB (2014) A differential role for CXCR4 in the regulation of normal versus malignant breast stem cell activity. Oncotarget 5(3):599–612
82. Dar A, Goichberg P, Shinder V, Kalinkovich A, Kollet O, Netzer N, Margalit R, Zsak M, Nagler A, Hardan I, Resnick I, Rot A, Lapidot T (2005) Chemokine receptor CXCR4-dependent internalization and resecretion of functional chemokine SDF-1 by bone marrow endothelial and stromal cells. Nat Immunol 6(10):1038–1046. doi:10.1038/ni1251
83. Duda DG, Kozin SV, Kirkpatrick ND, Xu L, Fukumura D, Jain RK (2011) CXCL12 (SDF1 alpha)-CXCR4/CXCR7 pathway inhibition: an emerging sensitizer for anticancer therapies? Clin Cancer Res 17(8):2074–2080. doi:10.1158/1078-0432.ccr-10-2636
84. Domanska UM, Kruizinga RC, Nagengast WB, Timmer-Bosscha H, Huls G, de Vries EGE, Walenkamp AME (2013) A review on CXCR4/CXCL12 axis in oncology: no place to hide. Eur J Cancer 49(1):219–230. doi:10.1016/j.ejca.2012.05.005
85. Charles N, Holland EC (2010) The perivascular niche microenvironment in brain tumor progression. Cell Cycle 9(15):3012–3021. doi:10.4161/cc.9.15.12710
86. Koboldt DC, Fulton RS, McLellan MD, Schmidt H, Kalicki-Veizer J, McMichael JF, Fulton LL, Dooling DJ, Ding L, Mardis ER, Wilson RK, Ally A, Balasundaram M, Butterfield YSN, Carlsen R, Carter C, Chu A, Chuah E, Chun HJE, Coope RJN, Dhalla N, Guin R, Hirst C, Hirst M, Holt RA, Lee D, Li HYI, Mayo M, Moore RA, Mungall AJ, Pleasance E, Robertson AG, Schein JE, Shafiei A, Sipahimalani P, Slobodan JR, Stoll D, Tam A, Thiessen N, Varhol RJ, Wye N, Zeng T, Zhao YJ, Birol I, Jones SJM, Marra MA, Cherniack AD, Saksena G, Onofrio RC, Pho NH, Carter SL, Schumacher SE, Tabak B, Hernandez B, Gentry J, Nguyen H, Crenshaw A, Ardlie K, Beroukhim R, Winckler W, Getz G, Gabriel SB, Meyerson M, Chin L, Park PJ, Kucherlapati R, Hoadley KA, Auman JT, Fan C, Turman YJ, Shi Y, Li L, Topal MD, He XP, Chao HH, Prat A, Silva GO, Iglesia MD, Zhao W, Usary J, Berg JS, Adams M, Booker J, Wu JY, Gulabani A, Bodenheimer T, Hoyle AP, Simons JV, Soloway MG, Mose LE, Jefferys SR, Balu S, Parker JS, Hayes DN, Perou CM, Malik S, Mahurkar S, Shen H, Weisenberger DJ, Triche T, Lai PH, Bootwalla MS, Maglinte DT, Berman BP, Van den Berg DJ, Baylin SB, Laird PW, Creighton CJ, Donehower LA, Getz G, Noble M, Voet D, Saksena G, Gehlenborg N, DiCara D, Zhang JH, Zhang HL, Wu CJ, Liu SY, Lawrence MS, Zou LH, Sivachenko A, Lin P, Stojanov P, Jing R, Cho J, Sinha R, Park RW, Nazaire MD, Robinson J, Thorvaldsdottir H, Mesirov J, Park PJ, Chin L, Reynolds S, Kreisberg RB, Bernard B, Bressler R, Erkkila T, Lin J, Thorsson V, Zhang W, Shmulevich I, Ciriello G, Weinhold N, Schultz N, Gao JJ, Cerami E, Gross B, Jacobsen A, Sinha R, Aksoy BA, Antipin Y, Reva B, Shen RL, Taylor BS, Ladanyi M, Sander C, Anur P, Spellman PT, Lu YL, Liu WB, Verhaak RRG, Mills GB, Akbani R, Zhang NX, Broom BM, Casasent TD, Wakefield C, Unruh AK, Baggerly K, Coombes K, Weinstein JN, Haussler D, Benz CC, Stuart JM, Benz SC, Zhu JC, Szeto CC, Scott GK, Yau C, Paul EO, Carlin D, Wong C, Sokolov A, Thusberg J, Mooney S, Ng S, Goldstein TC, Ellrott K, Grifford M, Wilks C, Ma S, Craft B, Yan CH, Hu Y, Meerzaman D, Gastier-Foster JM, Bowen J, Ramirez NC, Black AD, Pyatt RE, White P, Zmuda EJ, Frick J, Lichtenberg T, Brookens R, George MM, Gerken MA, Harper HA, Leraas KM, Wise LJ, Tabler TR, McAllister C, Barr T, Hart-Kothari M, Tarvin K, Saller C, Sandusky G, Mitchell C, Iacocca MV, Brown J, Rabeno B, Czerwinski C, Petrelli N, Dolzhansky O, Abramov M, Voronina O, Potapova O, Marks JR, Suchorska WM, Murawa D, Kycler W, Ibbs M, Korski K, Spychala A, Murawa P, Brzezinski JJ, Perz H, Lazniak R, Teresiak M, Tatka H, Leporowska E, Bogusz-Czerniewicz M, Malicki J, Mackiewicz A, Wiznerowicz M, Le XV, Kohl B, Tien NV, Thorp R, Bang NV, Sussman H, Phu BD, Hajek R, Hung NP, Tran VTP, Thang HQ, Khan KZ, Penny R, Mallery D, Curley E, Shelton C, Yena P, Ingle JN, Couch FJ, Lingle WL, King TA, Gonzalez-Angulo AM, Mills GB, Dyer MD, Liu SY, Meng XL, Patangan M, Waldman F, Stoppler H, Rathmell WK, Thorne L, Huang M, Boice L, Hill A, Morrison C, Gaudioso C, Bshara W, Daily K, Egea SC, Pegram MD, Gomez-Fernandez C, Dhir R, Bhargava R, Brufsky A, Shriver CD, Hooke JA, Campbell JL, Mural RJ, Hu H, Somiari S, Larson C, Deyarmin B, Kvecher L, Kovatich AJ, Ellis MJ, King TA, Hu H, Couch FJ, Mural RJ, Stricker T, White K, Olopade O, Ingle JN, Luo CQ, Chen YQ, Marks JR, Waldman F, Wiznerowicz M, Bose R, Chang LW, Beck AH, Gonzalez-Angulo AM, Pihl T, Jensen M, Sfeir R, Kahn A, Chu A, Kothiyal P, Wang ZN, Snyder E, Pontius J, Ayala B, Backus M, Walton J, Baboud J, Berton D, Nicholls M, Srinivasan D,

Raman R, Girshik S, Kigonya P, Alonso S, Sanbhadti R, Barletta S, Pot D, Sheth M, Demchok JA, Shaw KRM, Yang LM, Eley G, Ferguson ML, Tarnuzzer RW, Zhang JS, Dillon LAL, Buetow K, Fielding P, Ozenberger BA, Guyer MS, Sofia HJ, Palchik JD, Canc Genome Atlas N (2012) Comprehensive molecular portraits of human breast tumours. Nature 490(7418):61–70. doi:10.1038/nature11412
87. Idowu MO, Kmieciak M, Dumur C, Burton RS, Grimes MM, Powers CN, Manjili MH (2012) CD44(+)/CD24(−/low) cancer stem/progenitor cells are more abundant in triple-negative invasive breast carcinoma phenotype and are associated with poor outcome. Hum Pathol 43(3):364–373. doi:10.1016/j.humpath.2011.05.005
88. Liu XY, Feng DF, Liu DM, Wang SY, Yu XX, Dai EY, Wang J, Wang LH, Jiang W (2016) Dissecting the origin of breast cancer subtype stem cell and the potential mechanism of malignant transformation. PLoS One 11(10):16. doi:10.1371/journal.pone.0165001
89. Liu JC, Voisin V, Bader GD, Deng T, Pusztai L, Symmans WF, Esteva FJ, Egan SE, Zacksenhaus E (2012) Seventeen-gene signature from enriched Her2/Neu mammary tumor-initiating cells predicts clinical outcome for human HER2(+):ER alpha(−) breast cancer. Proc Natl Acad Sci U S A 109(15):5832–5837. doi:10.1073/pnas.1201105109
90. Khoury T, Ademuyiwa FO, Chandraseekhar R, Jabbour M, DeLeo A, Ferrone S, Wang YY, Wang XH (2012) Aldehyde dehydrogenase 1A1 expression in breast cancer is associated with stage, triple negativity, and outcome to neoadjuvant chemotherapy. Mod Pathol 25(3):388–397. doi:10.1038/modpathol.2011.172
91. Nakshatri H, Srour EF, Badve S (2009) Breast cancer stem cells and intrinsic subtypes: controversies rage on. Curr Stem Cell Res Ther 4(1):50–60. doi:10.2174/157488809787169110
92. Liu SL, Cong Y, Wang D, Sun Y, Deng L, Liu YJ, Martin-Trevino R, Shang L, McDermott SP, Landis MD, Hong S, Adams A, D'Angelo R, Ginestier C, Charafe-Jauffret E, Clouthier SG, Birnbaum D, Wong ST, Zhan M, Chang JC, Wicha MS (2014) Breast cancer stem cells transition between epithelial and mesenchymal states reflective of their normal counterparts. Stem Cell Rep 2(1):78–91. doi:10.1016/j.stemcr.2013.11.009
93. Liu SL, Wicha MS (2010) Targeting breast cancer stem cells. J Clin Oncol 28(25):4006–4012. doi:10.1200/jco.2009.27.5388
94. Zhao JH (2016) Cancer stem cells and chemoresistance: the smartest survives the raid. Pharmacol Ther 160:145–158. doi:10.1016/j.pharmthera.2016.02.008
95. Morrison R, Schleicher SM, Sun Y, Niermann KJ, Kim S, Spratt DE, Chung CH, Lu B (2011) Targeting the mechanisms of resistance to chemotherapy and radiotherapy with the cancer stem cell hypothesis. J Oncol 2011:941876. doi:10.1155/2011/941876
96. Diehn M, Cho RW, Lobo NA, Kalisky T, Dorie MJ, Kulp AN, Qian DL, Lam JS, Ailles LE, Wong MZ, Joshua B, Kaplan MJ, Wapnir I, Dirbas FM, Somlo G, Garberoglio C, Paz B, Shen J, Lau SK, Quake SR, Brown JM, Weissman IL, Clarke MF (2009) Association of reactive oxygen species levels and radioresistance in cancer stem cells. Nature 458(7239):780–U123. doi:10.1038/nature07733
97. Kobayashi CI, Suda T (2012) Regulation of reactive oxygen species in stem cells and cancer stem cells. J Cell Physiol 227(2):421–430. doi:10.1002/jcp.22764
98. Spigel DR, Spira AI, Jotte RM, Gadgeel SM, Mita AC, Hart LL, Kapoun A, Xu L, Hill D, Zhou L, Dupont J, Pietanza MC (2014) Phase 1b of anticancer stem cell antibody OMP-59R5 (anti-Notch2/3) in combination with etoposide and cisplatin (EP) in patients (pts) with untreated extensive-stage small-cell lung cancer (ED-SCLC). J Clin Oncol 32(15):1
99. Cancilla B, Cain J, Wang M, Beviglia L, Shah J, Gurney A, Lewicki J, Esserman L, Hoey T, Kapoun AM (2013) Anti-Notch1 antibody (OMP-52M51) impedes tumor growth and cancer stem cell frequency (CSC) in a chemo-refractory breast cancer xenograft model with an activating Notch1 mutation and screening for activated Notch1 across multiple solid tumor types. Cancer Res 73(8):1. doi:10.1158/1538-7445.am2013-3728
100. Hoey T, Yen WC, Axelrod F, Basi J, Donigian L, Dylla S, Fitch-Bruhns M, Lazetic S, Park IK, Sato A, Satyal S, Wang XH, Clarke MF, Lewicki J, Gurney A (2009) DLL4 blockade inhibits tumor growth and reduces tumor-initiating cell frequency. Cell Stem Cell 5(2):168–177. doi:10.1016/j.stem.2009.05.019
101. Jimeno A, LoRusso P, Strother RM, Diamond JR, Plato L, Younger A, Messersmith WA, Kittaneh M, Sawyer D, Adriaens L, Liu LM, Kao RJ, DiCioccio AT, Brownstein CM, Lowy I, Trail P, Chiorean EG (2013) Phase I study of REGN421 (R)/SAR153192, a fully-human delta-like ligand 4 (Dll4) monoclonal antibody (mAb), in patients with advanced solid tumors. J Clin Oncol 31(15):1
102. Ryan PC, Huang JQ, Bao HF, Cho S, Brohawn P, Burke P, Lehmann K, Pilataxi F, Yao YH, McKeever K, Dixit R (2013) Nonclinical safety evaluation of MEDI0639 (anti-DLLA Mab) to support first time in human: linking DLL4-notch signaling blockade to exaggerated pharmacology effects in cynomolgus monkeys. Cancer Res 73(8):1. doi:10.1158/1538-7445.am2013-4424
103. Takebe N, Miele L, Harris PJ, Jeong W, Bando H, Kahn M, Yang S, Ivy SP (2015) Targeting notch, hedgehog, and Wnt pathways in cancer stem cells: clinical update. Nat Rev Clin Oncol 12(8):445–464. doi:10.1038/nrclinonc.2015.61
104. Morohashi Y, Kan T, Tominari Y, Fuwa H, Okamura Y, Watanabe N, Sato C, Natsugari H, Fukuyama T, Iwatsubo T, Tomita T (2006) C-terminal fragment of presenilin is the molecular target of a dipeptidic gamma-secretase-specific inhibitor

DAPT (N- N-(3,5-difluorophenacetyl)-L-alanyl -S-phenylglycine t-butyl ester). J Biol Chem 281(21):14670–14676. doi:10.1074/jbc.M513012200
105. Luistro L, He W, Smith M, Packman K, Vilenchik M, Carvajal D, Roberts J, Cai J, Berkofsky-Fessler W, Hilton H, Linn M, Flohr A, Jakob-Rotne R, Jacobsen H, Glenn K, Heimbrook D, Boylan JF (2009) Preclinical profile of a potent gamma-secretase inhibitor targeting notch signaling with in vivo efficacy and Pharmacodynamic properties. Cancer Res 69(19):7672–7680. doi:10.1158/0008-5472.can-09-1843
106. Schott AF, Landis MD, Dontu G, Griffith KA, Layman RM, Krop I, Paskett LA, Wong H, Dobrolecki LE, Lewis MT, Froehlich AM, Paranilam J, Hayes DF, Wicha MS, Chang JC (2013) Preclinical and clinical studies of gamma secretase inhibitors with docetaxel on human breast tumors. Clin Cancer Res 19(6):1512–1524. doi:10.1158/1078-0432.ccr-11-3326
107. Zhang CC, Yan ZM, Zong Q, Fang DD, Painter C, Zhang Q, Chen EH, Lira ME, John-Baptiste A, Christensen JG (2013) Synergistic effect of the gamma-secretase inhibitor PF-03084014 and docetaxel in breast cancer models. Stem Cells Transl Med 2(3):233–242. doi:10.5966/sctm.2012-0096
108. Khramtsov AI, Khramtsova GF, Tretiakova M, Huo DZ, Olopade OI, Goss KH (2010) Wnt/beta-catenin pathway activation is enriched in basal-like breast cancers and predicts poor outcome. Am J Pathol 176(6):2911–2920. doi:10.2353/ajpath.2010.091125
109. Liu J, Pan SF, Hsieh MH, Ng N, Sun FX, Wang T, Kasibhatla S, Schuller AG, Li AG, Cheng D, Li J, Tompkins C, Pferdekamper A, Steffy A, Cheng J, Kowal C, Phung V, Guo GR, Wang Y, Graham MP, Flynn S, Brenner JC, Li C, Villarroel MC, Schultz PG, Wu X, McNamara P, Sellers WR, Petruzzelli L, Boral AL, Seidel HM, McLaughlin ME, Che JW, Carey TE, Vanasse G, Harris JL (2013) Targeting Wnt-driven cancer through the inhibition of porcupine by LGK974. Proc Natl Acad Sci U S A 110(50):20224–20229. doi:10.1073/pnas.1314239110
110. Gurney A, Axelrod F, Bond CJ, Cain J, Chartier C, Donigan L, Fischer M, Chaudhari A, Ji M, Kapoun AM, Lam A, Lazetic S, Ma S, Mitra S, Park IK, Pickell K, Sato A, Satyal S, Stroud M, Tran H, Yen WC, Lewicki J, Hoey T (2012) Wnt pathway inhibition via the targeting of frizzled receptors results in decreased growth and tumorigenicity of human tumors. Proc Natl Acad Sci U S A 109(29):11717–11722. doi:10.1073/pnas.1120068109
111. DeAlmeida VI, Miao L, Ernst JA, Koeppen H, Polakis P, Rubinfeld B (2007) The soluble Wnt receptor Frizzled8CRD-hFc inhibits the growth of teratocarcinomas in vivo. Cancer Res 67(11):5371–5379. doi:10.1158/0008-5472.can-07-0266
112. Kahn M (2014) Can we safely target the WNT pathway? Nat Rev Drug Discov 13(7):513–532. doi:10.1038/nrd4233
113. Dirix L (2014) Discovery and exploitation of novel targets by approved drugs. J Clin Oncol 32(8):720–721. doi:10.1200/jco.2013.53.7118
114. Dockendorff C, Nagiec MM, Weiwer M, Buhrlage S, Ting A, Nag PP, Germain A, Kim HJ, Youngsaye W, Scherer C, Bennion M, Xue LL, Stanton BZ, Lewis TA, MacPherson L, Palmer M, Foley MA, Perez JR, Schreiber SL (2012) Macrocyclic hedgehog pathway inhibitors: optimization of cellular activity and mode of action studies. ACS Med Chem Lett 3(10):808–813. doi:10.1021/ml300172p
115. Hyman JM, Firestone AJ, Heine VM, Zhao Y, Ocasio CA, Han K, Sun M, Rack PG, Sinha S, Wu JJ, Solow-Cordero DE, Jiang J, Rowitch DH, Chen JK (2009) Small-molecule inhibitors reveal multiple strategies for hedgehog pathway blockade. Proc Natl Acad Sci U S A 106(33):14132–14137. doi:10.1073/pnas.0907134106
116. Lauth M, Bergstrom A, Shimokawa T, Toftgard R (2007) Inhibition of GLI-mediated transcription and tumor cell growth by small-molecule antagonists. Proc Natl Acad Sci U S A 104(20):8455–8460. doi:10.1073/pnas.0609699104
117. Katoh M (2007) Networking of WNT, FGF, notch, BMP, and hedgehog signaling pathways during carcinogenesis. Stem Cell Rev 3(1):30–38. doi:10.1007/s12015-007-0006-6
118. Marangoni E, Lecomte N, Durand L, de Pinieux G, Decaudin D, Chomienne C, Smadja-Joffe F, Poupon MF (2009) CD44 targeting reduces tumour growth and prevents post-chemotherapy relapse of human breast cancers xenografts. Br J Cancer 100(6):918–922. doi:10.1038/sj.bjc.6604953
119. Ababneh N, Alshaer W, Allozi O, Mahafzah A, El-Khateeb M, Hillaireau H, Noiray M, Fattal E, Ismail S (2013) In vitro selection of modified RNA aptamers against CD44 cancer stem cell marker. Nucl Acid Ther 23(6):401–407. doi:10.1089/nat.2013.0423
120. Munz M, Murr A, Kvesic M, Rau D, Mangold S, Pflanz S, Lumsden J, Volkland J, Fagerberg J, Riethmuller G, Ruttinger D, Kufer P, Baeuerle PA, Raum T (2010) Side-by-side analysis of five clinically tested anti-EpCAM monoclonal antibodies. Cancer Cell Int 10:12. doi:10.1186/1475-2867-10-44
121. Simon M, Stefan N, Pluckthun A, Zangemeister-Wittke U (2013) Epithelial cell adhesion molecule-targeted drug delivery for cancer therapy. Expert Opin Drug Deliv 10(4):451–468. doi:10.1517/17425247.2013.759938
122. Koppaka V, Thompson DC, Chen Y, Ellermann M, Nicolaou KC, Juvonen RO, Petersen D, Deitrich RA, Hurley TD, Vasiliou V (2012) Aldehyde dehydrogenase inhibitors: a comprehensive review of the pharmacology, mechanism of action, substrate

specificity, and clinical application. Pharmacol Rev 64(3):520–539. doi:10.1124/pr.111.005538
123. Pors K, Moreb JS (2014) Aldehyde dehydrogenases in cancer: an opporunity for biomaker and drug development? Drug Discov Today 19(12):1953–1963. doi:10.1016/j.drudis.2014.09.009
124. Karthikeyan S, Hoti SL (2015) Development of fourth generation ABC inhibitors from natural products: a novel approach to overcome cancer multidrug resistance. Anti Cancer Agents Med Chem 15(5):605–615
125. Dragu DL, Necula LG, Bleotu C, Diaconu CC, Chivu-Economescu M (2015) Therapies targeting cancer stem cells: current trends and future challenges. World J Stem Cells 7(9):1185–1201. doi:10.4252/wjsc.v7.i9.1185
126. Saeki T, Nomizu T, Toi M, Ito Y, Noguchi S, Kobayashi T, Asaga T, Minami H, Yamamoto N, Aogi K, Ikeda T, Ohashi Y, Sato W, Tsuruo T (2007) Dofequidar fumarate (MS-209) in combination with cyclophosphamide, doxorubicin, and fluorouracil for patients with advanced or recurrent breast cancer. J Clin Oncol 25(4):411–417. doi:10.1200/jco.2006.08.1646
127. Guo YQ, Xu F, Lu TJ, Duan ZF, Zhang Z (2012) Interleukin-6 signaling pathway in targeted therapy for cancer. Cancer Treat Rev 38(7):904–910. doi:10.1016/j.ctrv.2012.04.007
128. Karin M, Yamamoto Y, Wang QM (2004) The IKKNF-kappa B system: a treasure trove for drug development. Nat Rev Drug Discov 3(1):17–26. doi:10.1038/nrd1279
129. Baud V, Karin M (2009) OPINION is NF-kappa B a good target for cancer therapy? Hopes and pitfalls. Nat Rev Drug Discov 8(1):33–40. doi:10.1038/nrd2781
130. Hassan S, Buchanan M, Jahan K, Aguilar-Mahecha A, Gaboury L, Muller WJ, Alsawafi Y, Mourskaia AA, Siegel PM, Salvucci O, Basik M (2011) CXCR4 peptide antagonist inhibits primary breast tumor growth, metastasis and enhances the efficacy of anti-VEGF treatment or docetaxel in a transgenic mouse model. Int J Cancer 129(1):225–232. doi:10.1002/ijc.25665
131. Hotte SJ, Hirte HW, Iacobucci A, Wong D, Cantin L, Korz W, Miller WH (2007) Phase I/II study of CTCE-9908, a novel anticancer agent that inhibits CXCR4, in patients with advanced solid cancers. Mol Cancer Ther 6(12):3385S–3386S
132. Xu C, Zhao H, Chen HT, Yao QH (2015) CXCR4 in breast cancer: oncogenic role and therapeutic targeting. Drug Des Dev Ther 9:4953–4964. doi:10.2147/dddt.s84932
133. Kuhne MR, Mulvey T, Belanger B, Chen S, Pan C, Chong CL, Cao F, Niekro W, Kempe T, Henning KA, Cohen LJ, Korman AJ, Cardarelli PM (2013) BMS-936564/MDX-1338: a fully human anti-CXCR4 antibody induces apoptosis in vitro and shows antitumor activity in vivo in hematologic malignancies. Clin Cancer Res 19(2):357–366. doi:10.1158/1078-0432.ccr-12-2333
134. Ganesan A (2008) The impact of natural products upon modern drug discovery. Curr Opin Chem Biol 12(3):306–317. doi:10.1016/j.cbpa.2008.03.016

8 Disrupting Tumor Angiogenesis and "the Hunger Games" for Breast Cancer

Ziwei Zhou, Herui Yao, and Hai Hu

Abstract

Angiogenesis, one of the hallmarks of cancers, has become an attractive target for cancer therapy since decades ago. It is broadly thought that upregulation of angiogenesis is involved in tumor progression and metastasis. Though tumor vessels are tortuous, disorganized, and leaky, they deliver oxygen and nutrients for tumor development. Based on this knowledge, many kinds of drugs targeting angiogenesis pathways have been developed, such as bevacizumab. However, the clinical outcomes of anti-angiogenesis therapies are moderate in metastatic breast cancer as well as in metastatic colorectal cancer and non-small cell lung cancer, even combined with traditional chemotherapy. In this chapter, the morphologic angiogenesis patterns and the key molecular pathways regulating angiogenesis are elaborated. The FDA-approved anti-angiogenesis drugs and current challenges of anti-angiogenesis therapy are described. The strategies to overcome the barriers will also be elucidated.

Keywords

Angiogenesis • Breast cancer • VEGF • Anti-angiogenesis therapy

8.1 Introduction

Seventy years ago, tumor development relying on neovascularization was postulated for the first time. In the 1960s, Dr. Judah Folkman and his colleagues began to search for factors promoting angiogenesis. A decade later, Folkman published the classic paper formulating three momentous hypotheses: (1) angiogenesis is a critical factor for tumor progression; (2) tumors secrete "tumor angiogenesis factor" (TAF) to

Z. Zhou
Breast Tumor Center, Sun Yat-Sen Memorial Hospital, Sun Yat-Sen University, 107 Yanjiang West Road, Guangzhou 510120, People's Republic of China

H. Yao (✉)
Department of Medical Oncology, Sun Yat-Sen Memorial Hospital, Sun Yat-Sen University, Guangzhou 510120, People's Republic of China
e-mail: yaoherui@163.com

H. Hu (✉)
Sun Yat-sen Memorial Hospital, Sun Yat-sen University, Guangzhou, Guangdong 510120, China
e-mail: huhai@mail.sysu.edu.cn

E. Song, H. Hu (eds.), *Translational Research in Breast Cancer*, Advances in Experimental Medicine and Biology 1026, https://doi.org/10.1007/978-981-10-6020-5_8

initiate angiogenesis; (3) anti-angiogenesis therapy would be a promising strategy to treat cancers. Thirty years later, Folkman proposed the concept of "angiogenic switch" (the conversion from a dormant to an active tumor state), which has been listed as one of the hallmarks of cancers nowadays [1]. The contribution of Folkman and his colleagues laid the groundwork for the further exploration of anti-angiogenesis therapies. Considerable studies proved that angiogenesis is an early event during tumor development and is thought to precede the progression and metastasis [2]. The tumor vessels are tortuous, disorganized, and leaky, which facilitates tumor dissemination [3]. Though the studies of angiogenic regulators and anti-angiogenic molecules have been lasted for many years, the in-depth mechanism underlying tumor angiogenesis is still unclear. Hypoxia and cytokines may play a great role in tumor angiogenesis and may be a reasonable explanation for drug resistance and treatment failure. In this chapter, we briefly review main morphogenesis of vacuolization and significant signaling pathways, followed by the success and challenges of anti-angiogenic treatment.

8.2 The Cancerous Angiogenesis and Vascular Morphogenesis

Angiogenesis plays a critical role in both physiological and malignant events, such as the formation of granulation tissue and tumorigenesis. The typical angiogenesis is a process that sprouts new blood vessels from pre-existing vessels [1–4], the de novo formation of endothelial cells from mesoderm cell precursors. As one of the hallmarks of cancer, angiogenesis has several morphologic features, which will be elaborated as follows.

8.2.1 Sprouting Angiogenesis

The typical angiogenesis is a process that sprouts new blood vessels from pre-existing vessels [1–4], which is the de novo formation of endothelial cells from mesoderm cell precursors. The sprouting angiogenesis was the first main form of angiogenesis discovered in tumor. Generally, sprouting angiogenesis is initiated from the activation of endothelial cells by the angiogenic growth factors. Then, the activated endothelial cells begin to release enzymes called proteases that degrade the basement membrane to allow endothelial cells to escape from the original (parent) vessel walls, following which endothelial cells proliferate into the surrounding extracellular matrix (ECM) and extrude form solid sprouts connecting neighboring vessels, forming lumen within the vascular sprouts and creating vascular tubes. Finally, the nascent vascular tubes are covered with mature basement membrane related to supporting pericyte [5, 6]. During sprouting angiogenesis, the endothelial cell-ECM interaction plays a critical role in regulating endothelial cell moving, proliferation, lumen formation, and vessel formation.

In addition to the sprouting angiogenesis, tumor angiogenesis may occur through other mechanisms: (1) recruitment of bone marrow-derived endothelial progenitor cells (EPCs) to form new vessels (postnatal vasculogenesis) [7], (2) vasculogenic mimicry (the transdifferentiation of cancer cells allowing them to form tubular structures themselves) [8], (3) mosaic vessel formation (the incorporation of cancer cells into the vessel wall or vascular cooption) [9], and (4) intussusceptive microvascular growth (IMG) [10]. These mechanisms add to the complex of the tumor angiogenesis and may lead to the resistance of targeting tumor angiogenesis.

8.2.2 Endothelial Progenitor Cells

The endothelial progenitor cells (EPCs) promote tumor neovascularization by vasculogenesis, the embryonic process of de novo formation of blood vessels, which is distinguished from angiogenesis, the sprouting process of formation of new blood vessels from pre-existing ones. EPCs are derived from stem cells in the bone marrow and migrate to the tumor niche under chemotaxis, accelerating tumor growth, metastasis, and drug resistance by vascular development. The co-expression of CD34, CD309 (VEGFR2/KDR), and CD133 on the membrane defines the EPCs [11]. Recently, a considerable amount of evidence suggests that EPCs would play a significant role in tumor occurrence and progression, as higher levels of EPCs in circulation have been detected in both breast cancer xenograft model in mice [12, 13] and breast cancer patients [14, 15]. An increased number of EPCs was observed in advanced breast cancer patients in comparison with early-stage breast cancer patients [16]. Kuo et al. also found that breast cancer patients responding well to chemotherapy had more dramatic EPC reduction in circulation [17]. Besides, Dome B and his coworkers have published that higher levels of circulating EPCs before the treatment indicate higher death incidence [18]. Because EPCs paralleled with clinical prognosis and therapeutic effects, EPC might be a useful and noninvasive marker for antitumor therapy. Interestingly, a strong relationship between the quantity of circulating EPCs and the efficacy of various metronomic chemotherapy regimens has been observed according to several preclinical breast cancer models and recent studies [19–22]. This correlation demonstrates that EPCs might become a potential marker reflecting the effect of metronomic therapies and thus determine the optimal dose of metronomic therapies [23]. Furthermore, EPCs might help tumor cells escape from chemotherapy, the mechanism of which lies in that these chemotherapeutic agents may induce EPCs from the bone marrow to the tumor site, forming new vessels to support tumor survival and spread [24]. EPCs could be mobilized not only by chemotherapy but also by anti-angiogenesis agents. In breast cancer patient, the mobilization of EPCs from the bone marrow might be a reasonable explanation for the controversial effect of anti-angiogenic drugs, especially for drug resistance [25].

The reason why tumor could mobilize EPCs has been highlighted these days. Among considerable hypotheses of EPCs homing to the tumor sites, hypoxia is one of the most pivotal explanations. Hypoxia regulates HIF1-α translocating into nucleus and activates the transcription of VEGF-A, PDGF, stromal cell-derived factor (SDF1-α), and C-X-C chemokine receptor type 4 (CXCR4, the receptor of SDF-1) [26, 27]. The released cytokines in hypoxia and chronic inflammatory tumor sites attract EPCs, pericyte progenitors, and CD45+ vascular modulatory cells from the bone marrow to the tumor site [28]. Vascular endothelial growth factor (VEGF), one of the most strong factors promoting angiogenesis, can also stimulate EPC proliferation via activating the phosphatidyl-inositol-3-kinase (PI3K)-AKT/eNOS signaling pathway to increase NO level [29]. At the production of endothelial nitric oxide synthase (eNOS), NO stimulates matrix metalloproteinase 9 (MMP-9), which then cleaves the membrane Kit ligand (mKitL) into the soluble form (sKitL). The sKitL binds on the EPC membrane and contributes to its release and migration. Moreover, VEGF along with Ang-1 facilitates EPC mobilization by remodeling vascular architecture in the bone marrow [30]. Additionally, tumor-associated macrophages (TAMs), which have a negative correlation with patient prognosis, have an impact on EPC function and behavior in hypoxia microenvironment [31, 32]. It is thought that TAM could secrete transforming growth factor β (TGF-β), tumor necrosis factor α (TNF-α), interleukins, and other cytokines to influence the migration, proliferation, and survival of EPCs [33–37]. Particularly, TNF is one of the activators of Notch signaling pathway in the stem cells including EPCs [38–40]. The Notch receptors on EPCs linking with Notch ligands on the osteoblasts permit the revival of EPCs. The Notch pathway activation mainly leads to Jagged1 (JAG1, CD339) activation. The interaction between Jagged1 and Notch triggers a proteolytic cascade, resulting in the transcription of endothelial lineage genes and leading to EPC colony formation, migration, and the capability of vasculogenesis [40]. After the EPCs home to the chronic hypoxia and inflammatory state, EPCs not only form new vessel but also release proangiogenesis factors and inflammatory cytokines to maintain the angiogenesis and inflammation state in tumor site [1, 41]. Furthermore, the interaction between EPCs and its microenvironment provides a pre-metastatic niche before the arrival of the metastatic tumor cells [42, 43].

As EPCs play a critical role in tumor progression and metastasis, the possibility of targeting EPCs as therapeutic strategies would be raised. Additional administration of anti-CXCR4 antibody to weaken EPC mobilization capability enhances the efficacy of docetaxel, which has been proved in murine breast cancer model [44]. Another strategy aiming at EPCs is to reduce EPC quantity and alter phenotype via delivering therapeutic compounds using nanoparticles. Transplantation of modified bone marrow progenitors transduced by lentiviral vectors expressing genes from transcription-regulatory elements of Tie-2/Tek gene decreases the formation of new vessels in tumor sites [45]. Though the studies of using nanoparticle to target stem cells, including BM-derived EPCs, have been proved successful in vivo, they are still at preclinical stage and need more effective evidence and improvement to be translated into clinical practice [46, 47].

8.2.3 Mosaic Blood Vessels

Many studies have found that cancer cells participate in the walls of tumor blood vessels over the past five decades. This phenomenon was first reported in 1948, and the ultrastructural evidence of tumor-lined vessels was presented by Warren and Shubik and others in the 1960s. Since then, mosaic vessels have been highlighted for its great impact on tumor biological features. For example, the presence of cancer cells in the blood vessel wall could be an indication of tumor aggressiveness or metastasis. The absence of detectable CD31 and CD105 might be accompanied by other alterations in protein expression in tumor endothelial cells, possibly induced by the hostile tumor microenvironment. All these pathologic changes could influence barrier functions of endothelial growth factor and angiogenesis [48]. However, the occurrence and outcomes of mosaic blood vessels remain unclear for lack of further studies. There are several potential explanations of mosaic vessels. First, the rapid vessel growth would cause insufficient endothelial cells to cover the vessel lining, leaving tumor cells exposed to the lumen. Second, endothelial cells could drop from the vessel lining, leaving underlying cancer cells being explosive. Third, migrating malignant tumor cells would invade into the vessel walls, replacing endothelial cells from the lining [48]. Nevertheless, some studies doubt the claim that tumor cells mix into endothelial cells to form mosaic blood vessels, because the heterogeneity of blood vessel cells is detected by expression of endothelial markers (such as CD31 and CD105). It is possible that these regions could be lined by endothelial cells that do not express common endothelial markers, which have been consequently mistaken as tumor cells. Chang YS and his coworkers found that the majority (92%) of the mosaicism was explained by undetectable CD31 and CD105 expression rather than by missing endothelial cells in colon carcinoma xenografts implanted orthotopically. Though there is a forceful evidence supporting the non-tumor cell mosaicism hypothesis, different tumor cells are likely to present different mosaic vessel pattern [47–50]. Perhaps the development of new markers to distinct endothelial cells from tumor cells could figure out this problem.

Although many issues remain unclear in the basic research, some experimental results may shed light on preclinical studies. For instance, Yong S. Chang et al. proposed that the antivascular effects of some anticancer therapies could be explained by mosaic vessels, for killing exposed cancer cell could impair blood flow in 14% of the vessels, which caused significant antivascular effect [49]. This antitumor drug effect on antivascularity might also explain the anti-angiogenesis effects of metronomic chemotherapy, which is based on more frequent and low-dose drug administrations. Since tumor cells on mosaic vessels are directly exposed to chemotherapeutic agents, lower doses of administration may kill these tumor cells. The cancer cells in mosaic vessels might be shed much more quickly with more frequent administration, which would break the function of vessels. Additionally, holes of mosaic blood vessels due to missing endothelial cells might be responsible for the leakiness of tumor vessels, which can hinder uniform drug delivery [50, 51]. Since mosaic blood vessel is a special character of malignant cancer, mosaicism could be developed into a potentially useful biomarkers for early diagnosis before being detected by conventional imaging techniques [52], which might help earlier intervention and better prognosis.

Stephenson J. published a paper explaining the role played by mosaic vessels during transplantation of autologous hematopoietic progenitor cells (HPC) in breast cancer patient [53]. In the autologous HPC transplantation therapy, HPCs are mobilized by G-CSF into peripheral blood. Surprisingly, two studies evaluating breast cancer cell contamination in HPC collections found that patients receiving high-dose chemotherapy (cyclophosphamide) plus G-CSF tended to have higher contamination by neoplastic cells compared with patients mobilized by G-CSF alone [54, 55]. The explanation for this interesting finding is that tumor cell would be shed from mosaic vessels and thus contaminate the HPC collections. Previous data also offered strong support for this explanation. Chang [49] and his colleagues showed that nearly 15% of leaky vessels in a xenograft carcinoma were mosaic vessels and that replaced tumor cells occupied about 4% of the total vascular surface area, which used antibodies and green fluorescent protein to identify endothelial cells and xenograft carcinoma cells, respectively. These data conformed to the fact that about 10×10^6 cancer cells shed from neoplasia regions contained mosaic tumor vessels per gram every day, which indicates the antivascular prosperity of chemotherapeutic drugs, especially cyclophosphamide, and provides a possible calculational method for the mobilization of neoplastic cells in patients administrated with high-dose chemotherapy supported by autologous HPC transplantation [56].

8.2.4 Vasculogenic Mimicry

Vasculogenic mimicry (VM) is the formation of microvascular channels by aggressive, metastatic, and genetically deregulated tumor cells rather than endothelial cells [57, 58]. Maniotis et al. first reported this patterned network containing red blood cells without identifying endothelial cells from aggressive human intraocular and metastatic cutaneous melanomas in 1999 [8]. Later on, VM has been discovered in many other cancers, such as malignant mesothelioma [59], prostate cancer [60], and breast cancer [61]. With further studies on VM, three main characters of VM have been put forward: (1) VM specifically exists in aggressive and metastatic solid tumors, not poorly invasive ones, and these mimicry vessels are surrounded by highly invasive tumor cells [62]. (2) Endothelial cells cannot be detected in the mimicry lining using light microscope, electron microscope, and immunochemical technique, but basement membrane-like structure could be found. (3) VM is a tubular and network structure with periodic acid-Schiff staining, and the lumen of VM network could be linked to the vein cavity at tumor margin [63].

The finding of VM was a great contribution to understanding the biologic behavior of cancer cells and identifying diagnostic indicators and potential therapeutic targets. VM is considered to perfuse the rapidly growing tumors with remarkable efficiency. The formation of VM was found to be linked with some of the characteristics of highly invasive tumors, such as tumor cell plasticity. Tumor cells capable of VM indicate a multipotent phenotype with an extreme degree of plasticity. These tumor cells exhibit upregulated expression of genes related to embryonic progenitors, endothelial cells, vessel formation, matrix remodeling, and coagulation inhibitors, as well as downregulated gene expression associated with lineage-specific phenotype markers [64]. However, the difference between VM tumor cells and embryonic progenitors is that the former lack critical regulatory checkpoints, express multipotent stem cell markers, and thus characterize unregulated growth and aggressive features of malignancies [65]. Recent studies have found that the hypoxic microenvironment, which is a common and significant feature of malignant tumors, facilitates the phenotype switch from tumor mass to VM in human melanoma models [66]. The important initiator of hypoxia-related signaling pathway is hypoxia-inducible factor

(HIF) complex [67, 68]. HIF-1α is separated from von Hippel-Lindau protein (pVHL, a ubiquitin ligase) in the cytoplasm and binds to hypoxia response elements on gene regulatory regions [69]. Then HIF-1α activates transcription of hypoxia-target genes—VEGF-A, VEGFR1, EPHA2, Twist, Nodal, osteopontin, and COX-2 gene expression. Hypoxia can also influence Notch signaling pathway, the activation of which is thought to promote tumor cell plasticity and then underlie VM [70]. Besides, hypoxia can influence Notch signaling pathway, the cross talk between what are thought to promote tumor cell plasticity and then underlie VM [71]. Another mechanism of hypoxia-promoting VM is the increased mitochondria ROS in hypoxia stabilizing HIF-1α and activating *Met* oncogene, which has been demonstrated to induce VM in melanoma [72, 73]. Since hypoxia plays a critical role in VM, the resistance of anti-angiogenesis therapy and invalidation might be explained by VM. Therefore, anti-VM would become a promising target for killing tumor cells. Moreover, bevacizumab, sorafenib, and sunitinib have begun to reveal little benefit to the overall survival in breast cancer patients. According to preclinical and clinical results, the development of a hypoxic microenvironment within the tumor due to administration of the above medications, which contributes to the thriving of cancer stem cell, might account for the failure of anti-angiogenesis therapy [74]. The good news is that compounds affecting several vascular perfusion pathways could inhibit tumor growth. In recent studies, the nanostructured targeting epirubicin plus celecoxib liposomes could eliminate invasive human breast cancer (MDA-MB-435S in vitro cells and MDA-MB-435S xenografts in nude mice) along with the VM channels [75]. Isoxanthohumol, which inhibits TGF-β-inducible genes associated with angiogenesis and metastasis, has been proved to degenerate breast cancer VM in vitro model [76]. VM has been reported to exist across a wide range of malignant and aggressive tumors, specifically in those associated with metastasis. In this sense, detection of VM in tumor samples may assist in precise diagnosis and prognosis. Indicators related to VM, such as MMP-9, VE-Cad, FAK, EphA2, and HIF-1α, may reflect the activity of VM formation in tumors [75]. ABCB5, a chemoresistance gene, was observed to be expressed on tumor cells engaged in VM. Therefore, it is noteworthy to mark VM in cancer samples [77]. Nevertheless, more forceful and specific VM markers in different types of cancers should be further studied before VM analysis is accepted as a routine method for histopathology reports [71].

8.3 The Angiogenetic Pathways in Cancer

Folkman first raised the concept of "angiogenic switch" in 1971, proposing that once the balance between proangiogenesis and anti-angiogenesis breaks, the tumoral angiogenesis occurs and supports tumor growth and metastasis. The common proangiogenesis factors are VEGF family, Ang-1/Tie-2, platelet-derived growth factors, fibroblast growth factors (FGFs), neuropilin, transforming growth factor, insulin-like growth factor, chemokines, semaphorins/plexins/neuropilins, and slits/robo hedgehog. The anti-angiogenesis factors are endostatins, thrombospondin-1, angiostatin, and interferon-α. In breast cancers, VEGF, Ang-1/Tie-2, PDGF, and bFGF pathways are most actively involved [78].

8.3.1 VEGF Pathway

Vascular endothelial growth factor (VEGF) is the most common and potent proangiogenic protein in tumors. VEGF is usually secreted by tumor cells and binds to its receptor (VEGFR) on the endothelial cells. Senger and Dvorak intended to find out why ascites in pleural, pericardial, and peritoneal cavities is characteristic of most malignant solid tumors and worked to purify factors contributing to vascular hyperpermeability. In 1983, they reported vascular permeability factor (VPF) [79], which was found out to be identical with vascular endothelial growth factor (VEGF) later [80].

VEGFs are among the most important players in the regulation of blood and lymphatic vessel formation during physiological and pathological processes, such as embryonic development, wound healing, cancer, and macular degeneration [81]. VEGF family has five members: VEGF-A, VEGF-B, VEGF-C, VEGF-D (also known as FIGF), and placental growth factor (PIGF). VEGF-A, being the mostly studied molecule, plays the most important role in angiogenesis among these five factors and is thus referred to as VEGF commonly [82].

There are three receptors of VEGF family: VEGFR1, VEGFR2, and VEGFR3. VEGFRs are dimerized tyrosine kinases signaling through mitogen-activated protein kinases (MAPKs) and AKT. The kinase domains, splitting into two functional domains, give rise to the original name kinase domain insert receptor (KDR). Each kind of the receptors distributes over different tissues and binds several VEGF ligands with different specificities. VEGFR2, also known as kinase domain insert receptor (KDR), is the main receptor stimulated by VEGF-A and expressed on almost all endothelial cells. VEGFR2 mediates microvascular permeability and endothelial cell proliferation, invasion, migration, and survival [83, 84]. The critical role of VEGFR2 not only lies in malignancies but also in normal embryonic development. Studies found that heterozygous and homozygous VEGFR2 knockout mice would die in utero because of failure of vasculogenesis and blood island formation [85]. Phosphorylation of tyrosine residues is a key mediator in the signaling pathway of VEGFR2. Several phosphorylation sites have been identified to possess important functions. Phosphorylated Tyr1175 has been shown to bind to PLC-gamma directly, mediating activation of MAPK/ERK2 (extracellular signal-regulated kinase2) and endothelial cell migration [86]. Tyr951, a binding site for T-cell-specific adaptor (TSAd), has been shown to play an essential role in endothelial migration [87, 88] . Other phosphorylation sites including Tyr1054 and Tyr1059 are required for maximal kinase activity [89]. VEGFR1, expressed on endothelial cells as well as on several other cell types, is a receptor for VEGF-A, VEGF-B, and PIGF. In spite of its important role in developmental angiogenesis, the exact function of VEGFR1 on the endothelium of tumor remains unclear. In contrast, sflt1, a secreted soluble extracellular domain of VEGFR1, has been showed to negatively regulate angiogenesis by acting as a decoy receptor for VEGF or by downregulating VEGFR2-mediated signaling [82, 90, 91]. VEGFR3 (flt4) binds to VEGF-C and VEGF-D preferentially. It is expressed throughout the vasculature of embryonic development. In adults, it is thought to mainly regulate lymphangiogenesis. Activated VEGFR3 and upregulated VEGF-C and VEGF-D have been observed in many tumors on both lymphatic endothelium and blood vessels [92–95]. They have been verified to be associated with tumor lymphangiogenesis and lymph node metastasis [96]. Besides, VEGFR3 can promote endothelial cell migration and survival in the lymphatic system via protein kinase (PKC)-dependent activation of MAPK [97, 98].

8.3.2 Ang/Tie-2 Pathway

Ang/Tie-2 signaling pathway also plays an important role in angiogenesis, which provides a new targeting strategy to move beyond anti-VEGF therapies [99]. There are several signaling pathways that Ang-1/Tie-2 can affect, such as NF-κB [100, 101], ERK1/ERK2 [101, 102], focal adhesion kinase (FAK), [103] and PI3K-AKT pathways [104, 105]. Activation of Ang-1/Tie-2 prolongs endothelial cell survival, encourages endothelial cell migration, and increases interaction between perivascular cells and endothelial cells. However, the potential of Ang/Tie-2 in angiogenesis may not be very clear because studies have found that it may be pro- or anti-angiogenesis depending on the context [99]. Generally, in the presence of other proangiogenic factors, such as VEGF [106], Ang-1/Tie-2 contributes to endothelium proliferation and migration, forming new sprouting, distorted vessels. On the contrary, in the absence of proangiogenic factors, Ang-2 signaling leads to endothelial cell apoptosis and regression [107]. Tie-1, the other

Ang receptor, also presents contrary effect. Tie-1 can be activated by autophosphorylated Tie-2. The extracellular domain of Tie-1 in turn interferes with Ang-1/Tie-2 and then contributes to antagonism of Ang-2 signaling [108, 109]. The Ang-2 expression has been revealed to correlate with VEGF expression and microvessel density (MVD) among a series of 198 breast cancer samples. A high MVD induced by high VEGF and Ang-2 expression has a strong prognostic significance in breast cancer [110]. Higher levels of Ang-1, Ang-2, and VEGF mRNA suggested better responses to anti-angiogenesis therapy, particularly in familiar breast cancer patients carrying BRCA mutations and triple-negative breast cancer (TNBC) patients [111].

Strong evidences have suggested a key role of Ang in tumor angiogenesis, especially the role of Ang-2 in the control of tumor angiogenesis [99, 104, 105, 112–114]. Overexpression of Ang-2 correlates with cancer progression and poor outcome [112]. Ang-2 has been shown to stimulate breast cancer metastasis. The Ang-Tie pathway is crucial for the angiogenic switch in tumors and participates in the breast cancer metastasis through the avβ1 integrin-mediated pathway [113]. The Ang-Tie system together with VEGF-A (VEGF) promotes the initiation of angiogenesis and enhances the maturation of new vessels. The Ang-2/Tie-2 system also works beyond the tumoral angiogenesis. It is involved in inflammation and lymphangiogenesis which promote metastasis [112, 113]. In this regard, targeting Ang-2/Tie-2 pathway may represent a valuable therapeutic approach, which inhibits both angiogenic switch and inflammatory pathways [99].

Nevertheless, targeting this pathway is challenging for the following reasons: (i) The efficacy of tyrosine kinase inhibitor targeting both ligands or the receptor itself is likely to depend on the balance between Ang-1 and Ang-2, given the presence of the receptor agonist and its natural antagonist. (ii) Medications targeting the pathway may act as a double-edged sword with the possibility of either pro- or anti-angiogenesis effect depending on the context. In fact, preclinical models have demonstrated conflicting results from what has been anticipated [115–121]. In order to deal with this dilemma, several solutions have been put forward: (i) One is to use combination of inhibitors targeting the Ang-Tie-2 and inhibitors targeting VEGF receptor pathways, for VEGF plays a major role on the angiogenesis and VEGF also greatly affects the proangiogenic effect of Ang-2 [99]. (ii) Another is to develop Ang-2-specific agents rather than anti-Ang-1 drug, because recent studies have suggested that Ang-2 might be a more potential target for angiogenesis inhibition. Further on, dual Ang-1/Ang-2 blocker may bring more benefits to patients in some cases [122, 123].

8.3.3 PDGFR Pathway

Platelet-derived growth factors (PDGF) belong to a four-member family (PGDF A, B, C, and D). PDGF was first identified as a constituent of blood serum. It is produced by different types of cells, including fibroblasts. PDGF activates two tyrosine kinase receptors—PDGFR-α and PDFGR-β. PDGF signaling pathway is overactive in various malignancies, such as soft tissue sarcoma (except well-differentiated/dedifferentiated liposarcomas), gastrointestinal stromal tumor (GIST), and high-grade glioma [124]. In most common solid tumors, PDGF signaling appears to be most important for the pericytes of the tumor vessels and for the fibroblasts of the tumor stroma [125]. Some studies in breast cancer highlight the prognostic value of PDGF and its receptor. High PDGFR expression in stroma associates with remarkably shorter recurrence-free time and breast cancer-specific survival, especially in premenopausal women [126]. Patients with recurrence also have higher PDGF level in blood serum, suggesting that it may be a candidate of recurrent marker [127]. Besides, stromal PDGFR expression is correlated with other less favorable parameters, such as high histopathological grade, estrogen receptor negativity, and high HER2 expression [126].

In breast cancer, PDGF plays a critical role in the interaction between tumor cells and stroma. PDGF/PDGFR signaling pathway initiated by

TGF-β is significant for epithelial-mesenchymal transition and tumor metastasis [128]. Breast cancer desmoplasia, a myofibroblast-mediated fibrosis response exhibiting progression potential, is initiated mainly by breast cancer-secreted PDGF [129]. Luminal, a subtype of estrogen receptor-positive breast cancer, has been studied to uncover the function of PDGF during tumor progression. Tumor proliferation, hormone independence, and angiogenesis are thought to interact with each other. PDGF is suggested to be involved in paracrine fashion by stromal cells and promote tumor recurrence, estrogen-independent proliferation, and tumor angiogenesis. This effect can be suppressed by anti-PDGF agent like imatinib [127]. In addition, PDGF secreted by breast cancer cells can mobilize vascular smooth muscle cells (VSMCs) through NRP-1 signaling pathway. NRP-1, a coreceptor of VEGF and a transmembrane protein, is essential for normal angiogenesis and involved in tumor angiogenesis. The migration of VSMCs under the recruitment of PDGF participates in the process of angiogenesis and vessel remodeling, which in turn nourishes tumor growth [130].These findings promote ongoing clinical development of PDGF pathway inhibitors. Additionally, anti-PDGFR agents combined with anti-VEGFR agents have been demonstrated to force 40% of tumor vessels into regression [131]. There are several agents proved by FDA showing anti-PDGF capacity, most of which are tyrosine kinase inhibitors. Sorafenib, a broad-spectrum TKI, was approved for the treatment of advanced renal cell carcinoma, unresectable hepatocellular carcinoma, and progressive differentiated thyroid carcinoma refractory to radioactive iodine treatment [132 133]. Sunitinib is a multitargeted and potent TKI approved for the treatment of imatinib-resistant or imatinib-intolerant GIST, advanced renal cell carcinoma, and advanced pancreatic neuroendocrine tumors [134, 135]. Pazopanib as a second-generation, multitargeted TKI has received approval for the treatment of advanced renal cell carcinoma and advanced soft tissue sarcoma [136, 137]. Regorafenib is a small-molecule TKI and approved for the treatment of previously treated metastatic colorectal carcinoma and GIST [138]. Some drugs are still under clinical trials. Olaratumab, the monoclonal antibody selectively blocking PDGFRα, showed well toleration and preliminary antitumor efficacy. Twelve among nineteen patients had a best response of stable disease. Phase II studies of olaratumab as monotherapy or in combination with other agents are ongoing in some solid tumors [139]. PDGF has been reported as the biomarker indicating the efficacy of anti-PDGFR drugs. PDGFR A and B indicate the efficacy of regorafenib and sunitinib in cancer treatment [124].

8.3.4 FGF Pathway

Fibroblast growth factors (FGFs) are a family of growth factors with at least 20 members identified in human. There are four members of FGF receptors (FGFRs), including FGFR1, FGFR2, FGFR3, and FGFR4. The FGF-FGFR pathway is involved in angiogenesis, wound healing, embryonic development, and inflammation [140–142]. FGF1 and FGF2 are the classical FGFs, which lack cytoplasmic motif for extracellular export from their producer cells. However, there are several working models for FGF to be transported out of the cell rather than via the classical secretory apparatus [143]. In fact, basic fibroblast growth factor (bFGF, also known as FGF2) was the first identified proangiogenic molecule [144]. FGF1 and FGF2 can promote endothelial cell proliferation and the tube-like structure formation of endothelial cells, thus promoting angiogenesis [145]. FGF1 has been shown in clinical experimental studies to induce angiogenesis in the heart [146]. Several intracellular signaling pathways are activated by FGF-FGFR system, including the Ras pathway, Src family tyrosine kinases, phosphoinositide 3-kinase (PI3K), and the PLC pathway [147]. A recent study has also shown that FGF can enhance glycolysis to promote endothelial cell proliferation and migration via c-MYC glycolytic enzyme hexokinase 2 (HK2) pathway [148]. Besides directly promoting angiogenesis, FGF has cross talk with VEGF family during angiogenesis, lymphangiogenesis, and vasculogenesis. Evidence refers to the possibility that FGF2 induces neovascularization indirectly by activation of the VEGF/VEGFR system. FGF2 can modulate VEGF expression in endothelial cells [149]. On the other hand, anti-VEGF-A antibody dramatically reduces FGF2-induced vascularization [149].

FGF2 expression has been found in various tumor cell lines [150]. In a transgenic mouse model, the export of FGF2 from fibrosarcoma correlated with the appearance of an angiogenic phenotype [151]. Several studies have established a correlation between tumor FGF2 levels and intratumoral microvessel density (MVD) in cancer patients [147]. A positive correlation between MVD and cerebrospinal fluid FGF2 was also observed in children with brain tumors [152]. Anti-FGF2-neutralizing antibodies and the soluble FGFRs have been showed to suppress tumor growth under experimental conditions [153–155]. Targeting FGF-binding protein (FGF-BP) inhibits the growth and vascularization of xenografted tumors in mice in spite of the high levels of VEGF production in the tumors [156, 157]. Targeting FGF-FGFR system also inhibits tumor growth in an angiogenesis-independent manner. The inhibition of FGF-FGFR system by dominant negative FGFR transfection or by *fgf2* gene knockout results in a decrease of tumor growth by both angiogenesis-dependent and angiogenesis-independent mechanisms in glioma cells or prostate cancer [158, 159]. Genome-wide studies have revealed that *FGFR2* gene is a breast cancer-susceptibility gene [160], while variants in the other FGF receptors are not associated with risk of breast cancer [161]. The dual inhibitor lenvatinib, which targets VEGFR/FGFR, has shown broad antitumor activity in human tumor xenograft models [162]. There are also FGFR-specific inhibitors in clinical trials of breast, lung, and gastric cancers.

8.3.5 Angiogenesis in Breast Cancer

Tumor angiogenesis plays an essential role in breast cancer development, invasion, and metastasis [163–165]. The transplantation of breast cancer cells with angiogenic stimulatory factors increases tumor growth, invasion, and metastasis [166]. Conversely, transplantation of tumor cancer cells with anti-angiogenic stimulatory factors decreases tumor growth and metastasis [167]. The angiogenic switch is one of the key components that contributes to breast cancer progression. A number of factors in breast cancers have been shown to alter the angiogenic balance to promote cancer progression [168, 169]. The angiogenic growth factors include the activation of VEGF, Ang/Tie-2, PDGF, FGF, and MMPs or the inactivation of TSP-1, sVEGF receptors, and TIMPs. The angiogenic switch can be driven by the expression of oncogene or the loss of function of tumor suppressors [169, 170]. The anti-angiogenesis therapies have been widely tried in breast cancer in clinical trials.

The majority of breast cancers are endocrine-dependent diseases. Sex steroids have been proved to play a critical role in angiogenesis switch and the process of angiogenesis [171]. Clinical studies have revealed that the hormone replacement therapy (HRT) in postmenopausal women brings significant benefits on their cardiovascular system [172]. Expression of VEGF by the vascular epithelium can be induced by estradiol [173]. Furthermore, estradiol can induce endothelial proliferation and migration [174] through activation of estrogen receptor expressed in endothelial cells [174, 175]. Evidence also shows that progesterone plays a role in breast angiogenesis [173, 176]. Thus, both estrogen and progestin have positive effects on angiogenesis in breast cancer.

8.4 Discovery of Anti-angiogenetic Drugs and Challenge of Anti-angiogenetic Therapy

8.4.1 Discovery of Anti-angiogenetic Drugs

Rapid tumor growth requires a steady supply of nutrients, and blood vessel formation provides the growing tissue with nutrients to facilitate tumor development. Since the realization that tumor growth depends on angiogenesis, there has been a consistent interest in developing angiogenic drugs for cancer treatment. Judah Folkman raised the concept that anti-angiogenesis was a potential therapeutic strategy for treating cancer decades ago. Since then, studies have further proved anti-angiogenesis to be an effective strategy to target cancers. In this regard, the VEGF pathway and Ang/Tie-2 pathway are the most frequently targeted pathways.

VEGF is the most studied angiogenic factor and plays a significant role in the development of breast cancer. The anti-angiogenetic potential of several targets on VEGF/VEGFR axis has been examined. The ligand blockade has been studied most extensively. Bevacizumab is a recombinant humanized monoclonal antibody that targets all known isoforms of VEGF-A. It was the first and so far the most widely studied anti-angiogenic drug in breast cancer clinical trials [177]. Ramucirumab, the monoclonal antibody against the VEGFR2 external domain, has been investigated as well [178]. VEGFRs belong to a family that is closely related to receptor tyrosine kinase (TK) followed by a signal transduction cascade. Small-molecule TK inhibitors (TKIs) that block VEGFR intercellular catalytic activity are also developed in clinical trials. TKIs, such as axitinib, pazopanib, sorafenib, and sunitinib, have been shown to have anti-angiogenic effects. Some of the angiogenesis inhibitors have dual roles of anticancer effects. Cetuximab, vandetanib, and erlotinib target both VEGF and EGF pathways. More than ten angiogenesis inhibitors have been approved and widely used in clinical treatment (Table 8.1 shows the FDA-proved angiogenesis inhibitors).

Besides the targeting therapies, metronomic chemotherapy has shown antitumor effects through interfering with neoangiogenesis [179]. Metronomic chemotherapy is a low-dose chemotherapy, which is administered frequently for a long time. Low-dose administration of oral cyclophosphamide and methotrexate daily previously showed anti-angiogenic and antitumor effect in advanced breast cancer [180].

Bevacizumab has been approved by FDA for treating patients with advanced-stage colon cancer, non-small cell lung cancer, and breast cancer since 2008.This is the first anti-angiogenesis drug designed to "starve tumors." The effective proof of breast cancer was based on three randomized trials in the first-line HER2-negative metastatic breast cancer settings: E2100, AVADO, and RIBBON-1 [181]. All the settings met their primary endpoint of prolonging progression-free survival (PFS). However, the meta-analysis only confirmed a PFS of 2.5 months, which did not result into an overall survival (OS) benefit. FDA revoked the accelerated approval in 2011 because of the lack of an OS benefit and increased toxicity [182]. Then, there were several trails tying to apply bevacizumab into breast cancer treatment, including hormone receptor-positive subset (NSABP B-40) [182] and triple-negative breast cancer subset (GeparQuinto, CALGB 40603) [183, 184], which did not lead to an improvement in outcomes. The BEATRICE [185] and BETH studies in TNBC- and HER2-positive breast cancer using the adjuvant setting also did not improve the disease-free survival (DFS) or OS with the addition of bevacizumab. These trials of bevacizumab raise the critical question whether anti-angiogenesis therapy can benefit treatment for breast cancer. If so, how do we select patients from those who cannot take advantage from this treatment?

An encouraging progress is that the next-generation anti-angiogenesis inhibitor, ramucirumab, has been approved by the FDA for use in advanced gastric cancer on the basis of the REGARD [186] and RAINBOW [187] trials. Meanwhile, bevacizumab in the phase III AVAGAST trial for advanced gastric cancer did not improve overall survival and was not approved by FDA. Other next-generation anti-angiogenesis inhibitors are now in other trials for breast cancer. Interestingly, HER2 signaling is associated with induction of angiogenesis in breast cancers [188]. Trebananib (AMG 386), the recombinant peptide-Fc fusion protein that acts on the angiopoietin axis in angiogenesis, has shown primary effect on Her2-positive advanced breast cancer

Table 8.1 The FDA-approved anti-angiogenesis drugs

Drug	Class	Mechanism (cellular targets)	Year of approval	Indications
Bevacizumab (Avastin)	Anti-VEGF mAB	VEGF	2004	First- and second-line metastatic CRC
			2006	First-line NSCLC
			2009	Second-line GBM
			2009	Metastatic RCC
			2013	Second-line metastatic CRC
Ziv-aflibercept (Zaltrap, VEGF trap)	Anti-VEGF mAB	VEGF-A,VEGF-B, PlGF1, PlGF2	2012	Metastatic CRC (after prior oxaliplatin-containing regimen)
Sorafenib (Nexavar, BAY439006)	Small-molecule TKI	VEGFR2, VEGFR3, PDGFR, FLT3, c-kit	2005	Advanced RCC
			2007	Unresectable HCC
			2013	RAI-refractory DTC
Sunitinib (Sutent, SU11248)	Small-molecule TKI	VEGFR1, VEGFR2,VEGFR3, PDGFR,FLT3, c-kit, RET	2006	Imatinib-resistant or imatinib-intolerant GIST
			2006	Advanced RCC
			2011	Advanced pNET
Pazopanib (Votrient)	Small-molecule TKI	VEGFR1, VEGFR2, VEGFR3, PDGFR, Itk, Lck, c-Fms	2009	Advanced RCC
			2012	Advanced soft tissue sarcoma
Vandetanib (Caprelsa)	Small-molecule TKI	Ret, VEGFR, EGFR, BRK, TIE2	2011	Advanced MTC
Axitinib (Inlyta)	Small-molecule TKI	VEGFR1, VEGFR2, VEGFR3	2012	Advanced RCC (after failure of prior therapy)
Cabozantinib (XL184, Cometriq)	Small-molecule TKI	Met, VEGFR2, ret, kit, AXL, FLT3	2012	Progressive, metastatic MTC
Regorafenib (Stivarga)	Small-molecule TKI	Ret, VEGFR1, VEGFR2, VEGFR3, TIE2, kit, PDGFR	2012	Previously treated metastatic CRC
			2013	GIST
Temsirolimus (Torisel)	mTOR inhibitor	mTOR	2007	Advanced RCC
Everolimus (Afinitor, RAD001)	mTOR inhibitor	mTOR	2009	Second-line advanced RCC (after VEGFR TKI failure)
			2010	SEGA associated with TSC
			2011	pNET
			2012	Advanced HR+, HER2-breast cancer
			2012	AML associated with TSC
Endostatin (Endostar)	Endogenous anti-angiogenic factors	Endogenous anti-angiogenic factors	2005	First-line advanced NSCLC

[189]. Perhaps the next-generation anti-angiogenesis inhibitors represent a new opportunity of anti-angiogenesis therapies in breast cancer.

8.4.2 Defining the Challenges of Anti-angiogenetic Therapy

The anti-angiogenesis drugs have provided beneficial effects of cancer treatment. However, the striking benefits of anti-angiogenic treatment observed in mouse tumors have not been translated to clinical benefits. Only modest effects on human cancers have been observed by applying anti-angiogenic drugs. FDA revoked the approval for the clinical use of bevacizumab in metastatic breast cancer, because of the low gain and relative high risk in several randomized phase III trials [190]. Although combination of anti-angiogenic drugs and chemotherapy has shown improvement of clinical outcome, survival benefits of anti-angiogenic drugs in combination settings remain modest in most cancer types [191]. This situation suggests that there are still several challenges that need to be overcome. To enhance the efficacy of anti-angiogenesis therapy for cancer treatment, more efforts must be made to address various complex issues for designing anti-angiogenic strategies. Insights into the molecular mechanisms of angiogenesis in different contexts of cancers provide new opportunities for drug discovery. Moreover, anti-angiogenic therapies in combination with existing drugs open windows for cancer treatment. Therefore, understanding the combination mechanistic effects of anti-angiogenic and cytotoxic drugs will optimize the efficacy of treatment strategy.

8.4.2.1 Mechanistic Insights of Tumoral Angiogenesis

The limited effects of anti-angiogenesis therapy in clinic indicate the limited knowledge of the mechanism of angiogenesis. Unlike tumor cells, ECs are expected to be genetically stable. Thus, the resistance to anti-angiogenesis therapy would not be expected to occur. However, the observation that the majority of cancer patients display intrinsic resistance to VEGF inhibitors has challenged this hypothesis. Even though a proportion of cancer patients initially responded to the VEGF inhibitors, they subsequently developed apparent resistance. Both preclinical and clinical studies have shown the significant remodeling of tumor blood vessels in the presence of anti-VEGF drugs. Tumor becomes resistant to anti-angiogenic drugs, especially when only one angiogenic pathway is targeted (e.g.,, VEGF) [192]. Although VEGF is the mainly expressed angiogenesis factor, there are several other angiogenic factors being expressed. For example, up to six angiogenic proteins can be expressed in human breast cancers. The high-grade giant-cell tumors and angioblastomas produce bFGF instead of VEGF as the predominant angiogenic factors and do not benefit from the anti-VEGF drugs. Currently, most FDA-approved angiogenesis inhibitors target VEGF pathway. Developing angiogenesis inhibitors that target beyond VEGF pathway will provide solutions to anti-VEGF drug resistance. It also needs to be clarified whether inhibitors targeting broad-spectrum angiogenesis factors would develop less drug resistance than targeting a single angiogenic factor. For example, TNP-470, a synthetic analogue of fumagillin and caplostatin [193], did not induce drug resistance when administered to mice for prolonged periods of time [194].

Angiogenesis is now recognized as the product of evolving cross talk between different cell types within the tumor and its stroma [195]. Understanding the underlying mechanism of the interaction among tumor, stroma, and endothelium will provide new opportunity for anti-angiogenesis drug design. On the other hand, drugs that remodel tumor microenvironment would reduce the benefit of using anti-angiogenesis drug. Treatment of tumor with anti-G-CSF antibody recruits bone marrow-derived $CD11b^{+}Gr1^{+}$ myeloid cells and alters tumor environment, thus decreasing the benefit of anti-angiogenesis drugs.

8.4.2.2 Predictive Biomarkers for Anti-angiogenesis Therapy

Biomarkers for predicting the efficacy of targeted therapy have been wildly used in clinic. However, there are few reliable markers to direct the anti-angiogenesis treatment. Many studies have been performed to find predictive markers indicating clinical benefit, which allows more accurate patient selection for anti-angiogenic therapies. The biomarkers include those which are able to distinguish patients who are likely to benefit from anti-angiogenic therapy from nonresponders, those which accurately monitor the therapeutic efficacy and adverse effects and those which provide information for anti-angiogenetic drug selection. The levels of tumor-associated VEGF or circulating plasma VEGF have been used to predict patient outcome. In the trial of evaluating vinorelbine with bevacizumab, plasma VEGF levels were measured [196]. The patients with higher VEGF level had significantly shorter time to progression (TTP) than those with lower level of VEGF (3.7 months vs 9.3 months). The VEGFR level is also used as a biomarker, as plasma VEGF and VEGFR2 analysis in BEATRICE trial has shown that high baseline plasma VEGFR2 has a potentially predictive value for bevacizumab efficacy [185]. The genetic polymorphisms of the VEGF and VEGFR2 genes have been associated with the outcome of advanced breast cancer in the trial of paclitaxel compared with paclitaxel plus bevacizumab [197]. The gene polymorphism of VEGF-A and VEGFR2 also correlated with the treatment outcome following imatinib therapy, which is a tyrosine kinase inhibitor of BCR-ABL, c-Kit, and PDGFR [198]. Candidate markers may also include urinary metalloproteinases (MMPs), which are correlated with the brain tumor presence and response to therapy when combined with VEGF [199, 200]. Soluble KIT (sKIT) was found to have the potential to predict clinical outcome of sunitinib treatment. In a phase II trial evaluating the efficacy of sunitinib with an anthracycline and a taxane, decreases of sKIT levels by more than 50% from the start of the treatment to the end of the last treatment cycle showed significant longer TTP [201].

8.4.2.3 Combination of Anti-angiogenic Therapy with Cytotoxic Chemotherapy

Since 2008, bevacizumab has been approved by FDA for treating patients with advanced-stage colon cancer or non-small cell lung cancer, all in combination with chemotherapy. Angiogenesis inhibitors sensitize tumors, which brings an unexpected benefit to conventional chemotherapy. Studies have shown that anti-VEGF drugs induce the remodeling of tumor blood vessels, leading to a more normalized vasculature [51]. Bevacizumab has been shown to decrease tumor vascular leakage, lower the intratumoral-tissue pressure, and increase the delivery of chemotherapy to tumors [202]. Teicher et al. showed that anti-angiogenic treatment decreased the intratumoral pressure and resulted in increased oxygenation delivering to the tumor tissues, thus increasing the sensitivity of cancer tissue to ionizing radiation [203]. Therefore, a potential way to improve the efficacy of anti-angiogenic therapy for cancer patients is to combine various anti-angiogenic inhibitors with different chemotherapies.

Since the development of angiogenesis inhibitors has met bottleneck, considerable work is needed for future breakthrough. Drug resistance is one of the main barriers limiting the use of anti-angiogenesis agents. Activation of cancer signaling pathways, autoactivation of proangiogenesis pathways, tissue hypoxia, and vascular mimicry may explain the dilemma of drug resistance [204]. Administration of anti-angiogenesis agents is also a challenging issue. Problems such as how often should the drug be given, or how to combine the administration of drug with chemotherapies, should be settled via further studies. Besides, there is an urgent demand for biomarkers evaluating the efficacy of angiogenic inhibitors. VEGF and VEGFR levels could correlate with the outcome of bevacizumab treatment. VEGF and VEGFR2 genetic polymorphisms also correlate with the clinical outcomes in patients with metastatic breast cancer [205, 206]. Other conundrums needed to be overcome are adverse effects, pharmacokinetic changes of the drugs, the expense of life-long therapy, and the underlying mechanism of tumor angiogenesis. In all, collaboration between clinical oncologists and translational scientists is essentially important for the improvement of anti-angiogenic therapy [204].

Acknowledgments This work has been supported by grants from the National Key R&D Program (2016YFC1302301) by National Natural Science Foundation of China (81672738, U1601223).

References

1. Hanahan D, Weinberg RA (2011) Hallmarks of cancer: the next generation. Cell 144(5):646–674. doi:10.1016/j.cell.2011.02.013
2. Folkman J (1971) Tumor angiogenesis: therapeutic implications. N Engl J Med 285(21):1182–1186. doi:10.1056/NEJM197111182852108
3. Sundberg C, Nagy JA, Brown LF, Feng D, Eckelhoefer IA, Manseau EJ, Dvorak AM, Dvorak HF (2001) Glomeruloid microvascular proliferation follows adenoviral vascular permeability factor/vascular endothelial growth factor-164 gene delivery. Am J Pathol 158(3):1145–1160. doi:10.1016/S0002-9440(10)64062-X
4. Birbrair A, Zhang T, Wang ZM, Messi ML, Mintz A, Delbono O (2015) Pericytes at the intersection between tissue regeneration and pathology. Clin Sci (Lond) 128(2):81–93. doi:10.1042/CS20140278
5. Senger DR, Davis GE Angiogenesis. (1943–0264 (Electronic)). doi:D - NLM: PMC3140681 EDAT- 2011/08/03 06:00 MHDA- 2011/12/13 00:00 CRDT- 2011/08/03 06:00 AID - cshperspect.a005090 [pii] AID - 10.1101/cshperspect.a005090 [doi] PST - epublish
6. Ribatti D, Crivellato E "Sprouting angiogenesis", a reappraisal. (1095-564X (Electronic))
7. Guidolin D Fau - Nico B, Nico B Fau - Belloni AS, Belloni As Fau - Nussdorfer GG, Nussdorfer Gg Fau - Vacca A, Vacca A Fau - Ribatti D, Ribatti D Morphometry and mathematical modelling of the capillary-like patterns formed in vitro by bone marrow macrophages of patients with multiple myeloma. (0887–6924 (Print))
8. Maniotis AJ, Folberg R Fau - Hess A, Hess A Fau - Seftor EA, Seftor Ea Fau - Gardner LM, Gardner Lm Fau – Pe'er J, Pe'er J Fau - Trent JM, Trent Jm Fau - Meltzer PS, Meltzer Ps Fau - Hendrix MJ, Hendrix MJ Vascular channel formation by human melanoma cells in vivo and in vitro: vasculogenic mimicry. (0002–9440 (Print)). doi:D - NLM: PMC1866899 EDAT- 1999/09/17 09:00 MHDA- 2000/06/10 09:00 CRDT- 1999/09/17 09:00 AID - S0002–9440(10)65173–5 [pii] AID - 10.1016/S0002-9440(10)65173-5 [doi] PST - ppublish
9. Holash J, Maisonpierre Pc Fau - Compton D, Compton D Fau - Boland P, Boland P Fau - Alexander CR, Alexander Cr Fau - Zagzag D, Zagzag D Fau - Yancopoulos GD, Yancopoulos Gd Fau - Wiegand SJ, Wiegand SJ Vessel cooption, regression, and growth in tumors mediated by angiopoietins and VEGF. (0036–8075 (Print))
10. Ribatti D, Djonov V Intussusceptive microvascular growth in tumors. (1872–7980 (Electronic))
11. Hristov M, Weber C Endothelial progenitor cells: characterization, pathophysiology, and possible clinical relevance. (1582–1838 (Print))
12. Shirakawa K, Furuhata S Fau - Watanabe I, Watanabe I Fau - Hayase H, Hayase H Fau - Shimizu A, Shimizu A Fau - Ikarashi Y, Ikarashi Y Fau - Yoshida T, Yoshida T Fau - Terada M, Terada M Fau - Hashimoto D, Hashimoto D Fau - Wakasugi H, Wakasugi H Induction of vasculogenesis in breast cancer models. (0007–0920 (Print)). doi:D - NLM: PMC2376301 EDAT- 2002/11/28 04:00 MHDA- 2003/01/08 04:00 CRDT- 2002/11/28 04:00 PHST- 2002/04/04 [received] PHST- 2002/08/22 [revised] PHST- 2002/08/29 [accepted] AID - 10.1038/sj.bjc.6600610 [doi] AID - 6600610 [pii] PST - ppublish
13. Shaked Y, Bertolini F Fau - Man S, Man S Fau - Rogers MS, Rogers Ms Fau - Cervi D, Cervi D Fau - Foutz T, Foutz T Fau - Rawn K, Rawn K Fau - Voskas D, Voskas D Fau - Dumont DJ, Dumont Dj Fau - Ben-David Y, Ben-David Y Fau - Lawler J, Lawler J Fau - Henkin J, Henkin J Fau - Huber J, Huber J Fau - Hicklin DJ, Hicklin Dj Fau – D'Amato RJ, D'Amato Rj Fau - Kerbel RS, Kerbel RS Genetic heterogeneity of the vasculogenic phenotype parallels angiogenesis; Implications for cellular surrogate marker analysis of antiangiogenesis. (1535–6108 (Print))
14. Mancuso P, Antoniotti P Fau - Quarna J, Quarna J Fau - Calleri A, Calleri A Fau - Rabascio C, Rabascio C Fau - Tacchetti C, Tacchetti C Fau - Braidotti P, Braidotti P Fau - Wu H-K, Wu Hk Fau - Zurita AJ, Zurita Aj Fau - Saronni L, Saronni L Fau - Cheng JB, Cheng Jb Fau - Shalinsky DR, Shalinsky Dr Fau - Heymach JV, Heymach Jv Fau - Bertolini F, Bertolini F Validation of a standardized method for enumerating circulating endothelial cells and progenitors: flow cytometry and molecular and ultrastructural analyses. (1078–0432 (Print))
15. Kim HK, Song Ks Fau - Kim HO, Kim Ho Fau - Chung J-H, Chung Jh Fau - Lee KR, Lee Kr Fau - Lee Y-J, Lee Yj Fau - Lee DH, Lee Dh Fau - Lee ES, Lee Es Fau - Kim HK, Kim Hk Fau - Ryu KW, Ryu Kw Fau - Bae J-M, Bae JM Circulating numbers of endothelial progenitor cells in patients with gastric and breast cancer. (0304–3835 (Print))
16. Naik RP, Jin D Fau - Chuang E, Chuang E Fau - Gold EG, Gold Eg Fau - Tousimis EA, Tousimis Ea Fau - Moore AL, Moore Al Fau - Christos PJ, Christos Pj Fau - de Dalmas T, de Dalmas T Fau - Donovan D,

Donovan D Fau - Rafii S, Rafii S Fau - Vahdat LT, Vahdat LT Circulating endothelial progenitor cells correlate to stage in patients with invasive breast cancer. (1573–7217 (Electronic))
17. Kuo YH, Lin Ch Fau - Shau W-Y, Shau Wy Fau - Chen T-J, Chen Tj Fau - Yang S-H, Yang Sh Fau - Huang S-M, Huang Sm Fau - Hsu C, Hsu C Fau - Lu Y-S, Lu Ys Fau - Cheng A-L, Cheng AL Dynamics of circulating endothelial cells and endothelial progenitor cells in breast cancer patients receiving cytotoxic chemotherapy. (1471–2407 (Electronic)). doi:D - NLM: PMC3561193 EDAT- 2012/12/28 06:00 MHDA- 2013/07/03 06:00 CRDT- 2012/12/28 06:00 PHST- 2012/03/20 [received] PHST- 2012/12/18 [accepted] AID - 1471-2407-12-620 [pii] AID - 10.1186/1471–2407–12-620 [doi] PST - epublish
18. Dome B, Timar J Fau - Dobos J, Dobos J Fau - Meszaros L, Meszaros L Fau - Raso E, Raso E Fau - Paku S, Paku S Fau - Kenessey I, Kenessey I Fau - Ostoros G, Ostoros G Fau - Magyar M, Magyar M Fau - Ladanyi A, Ladanyi A Fau - Bogos K, Bogos K Fau - Tovari J, Tovari J Identification and clinical significance of circulating endothelial progenitor cells in human non-small cell lung cancer. (0008–5472 (Print))
19. Shaked Y, Emmenegger U Fau - Man S, Man S Fau - Cervi D, Cervi D Fau - Bertolini F, Bertolini F Fau - Ben-David Y, Ben-David Y Fau - Kerbel RS, Kerbel RS Optimal biologic dose of metronomic chemotherapy regimens is associated with maximum antiangiogenic activity. (0006–4971 (Print)). doi:D - NLM: PMC1895327 EDAT- 2005/07/07 09:00 MHDA- 2005/12/13 09:00 CRDT- 2005/07/07 09:00 AID - 2005-04-1422 [pii] AID - 10.1182/blood-2005-04-1422 [doi] PST - ppublish
20. Shaked Y, Emmenegger U Fau - Francia G, Francia G Fau - Chen L, Chen L Fau - Lee CR, Lee Cr Fau - Man S, Man S Fau - Paraghamian A, Paraghamian A Fau - Ben-David Y, Ben-David Y Fau - Kerbel RS, Kerbel RS Low-dose metronomic combined with intermittent bolus-dose cyclophosphamide is an effective long-term chemotherapy treatment strategy. (0008–5472 (Print))
21. Munoz R, Man S Fau - Shaked Y, Shaked Y Fau - Lee CR, Lee Cr Fau - Wong J, Wong J Fau - Francia G, Francia G Fau - Kerbel RS, Kerbel RS Highly efficacious nontoxic preclinical treatment for advanced metastatic breast cancer using combination oral UFT-cyclophosphamide metronomic chemotherapy. (0008–5472 (Print))
22. Ng SSW, Sparreboom A, Shaked Y, Lee C, Man S, Desai N, Soon-Shiong P, Figg WD, Kerbel RS (2006) Influence of formulation vehicle on metronomic Taxane chemotherapy: albumin-bound versus Cremophor EL–based paclitaxel. Clin Cancer Res 12(14):4331
23. Le Bourhis X, Romon R, Hondermarck H (2010) Role of endothelial progenitor cells in breast cancer angiogenesis: from fundamental research to clinical ramifications. Breast Cancer Res Treat 120(1):17–24. doi:10.1007/s10549-009-0686-5
24. Shaked Y, Henke E, Roodhart JM, Mancuso P, Langenberg MH, Colleoni M, Daenen LG, Man S, Xu P, Emmenegger U, Tang T, Zhu Z, Witte L, Strieter RM, Bertolini F, Voest EE, Benezra R, Kerbel RS (2008) Rapid chemotherapy-induced acute endothelial progenitor cell mobilization: implications for antiangiogenic drugs as chemosensitizing agents. Cancer Cell 14(3):263–273. doi:10.1016/j.ccr.2008.08.001
25. Furstenberger G, von Moos R, Lucas R, Thurlimann B, Senn HJ, Hamacher J, Boneberg EM (2006) Circulating endothelial cells and angiogenic serum factors during neoadjuvant chemotherapy of primary breast cancer. Br J Cancer 94(4):524–531. doi:10.1038/sj.bjc.6602952
26. Paliege A, Rosenberger C Fau - Bondke A, Bondke A Fau - Sciesielski L, Sciesielski L Fau - Shina A, Shina A Fau - Heyman SN, Heyman Sn Fau - Flippin LA, Flippin La Fau - Arend M, Arend M Fau - Klaus SJ, Klaus Sj Fau - Bachmann S, Bachmann S Hypoxia-inducible factor-2alpha-expressing interstitial fibroblasts are the only renal cells that express erythropoietin under hypoxia-inducible factor stabilization. (1523–1755 (Electronic))
27. Schioppa T, Uranchimeg B, Saccani A, Biswas SK, Doni A, Rapisarda A, Bernasconi S, Saccani S, Nebuloni M, Vago L, Mantovani A, Melillo G, Sica A (2003) Regulation of the chemokine receptor CXCR4 by hypoxia. J Exp Med 198(9):1391–1402. doi:10.1084/jem.20030267
28. Ahn GO, Brown JM (2009) Role of endothelial progenitors and other bone marrow-derived cells in the development of the tumor vasculature. Angiogenesis 12(2):159–164. doi:10.1007/s10456-009-9135-7
29. Chen TG, Zhong ZY, Sun GF, Zhou YX, Zhao Y (2011) Effects of tumour necrosis factor-alpha on activity and nitric oxide synthase of endothelial progenitor cells from peripheral blood. Cell Prolif 44(4):352–359. doi:10.1111/j.1365-2184.2011.00764.x
30. Hattori K, Dias S, Heissig B, Hackett NR, Lyden D, Tateno M, Hicklin DJ, Zhu Z, Witte L, Crystal RG, Moore MA, Rafii S (2001) Vascular endothelial growth factor and angiopoietin-1 stimulate postnatal hematopoiesis by recruitment of vasculogenic and hematopoietic stem cells. J Exp Med 193(9):1005–1014
31. Lewis JS, Lee Ja Fau - Underwood JC, Underwood Jc Fau - Harris AL, Harris Al Fau - Lewis CE, Lewis CE Macrophage responses to hypoxia: relevance to disease mechanisms. (0741–5400 (Print))
32. Bingle L, Brown Nj Fau - Lewis CE, Lewis CE The role of tumour-associated macrophages in tumour progression: implications for new anticancer therapies. (0022–3417 (Print))
33. Imamura H, Ohta T, Tsunetoshi K, Doi K, Nozaki K, Takagi Y, Kikuta K (2010) Transdifferentiation of bone marrow-derived endothelial progenitor

cells into the smooth muscle cell lineage mediated by tansforming growth factor-beta1. Atherosclerosis 211(1):114–121. doi:10.1016/j.atherosclerosis.2010.02.040
34. Sales VL, Engelmayr GC Jr, Mettler BA, Johnson JA Jr, Sacks MS, Mayer JE Jr (2006) Transforming growth factor-beta1 modulates extracellular matrix production, proliferation, and apoptosis of endothelial progenitor cells in tissue-engineering scaffolds. Circulation 114(1 Suppl):I193–I199. doi:10.1161/circulationaha.105.001628
35. Sasi SP, Yan X, Enderling H, Park D, Gilbert HY, Curry C, Coleman C, Hlatky L, Qin G, Kishore R, Goukassian DA (2012) Breaking the 'harmony' of TNF-alpha signaling for cancer treatment. Oncogene 31(37):4117–4127. doi:10.1038/onc.2011.567
36. Zeoli A, Dentelli P, Rosso A, Togliatto G, Trombetta A, Damiano L, di Celle PF, Pegoraro L, Altruda F, Brizzi MF (2008) Interleukin-3 promotes expansion of hemopoietic-derived CD45+ angiogenic cells and their arterial commitment via STAT5 activation. Blood 112(2):350–361. doi:10.1182/blood-2007-12-128215
37. Fan Y, Ye J, Shen F, Zhu Y, Yeghiazarians Y, Zhu W, Chen Y, Lawton MT, Young WL, Yang GY (2008) Interleukin-6 stimulates circulating blood-derived endothelial progenitor cell angiogenesis in vitro. J Cereb Blood Flow Metab 28(1):90–98. doi:10.1038/sj.jcbfm.9600509
38. Varnum-Finney B, Brashem-Stein C, Bernstein ID (2003) Combined effects of notch signaling and cytokines induce a multiple log increase in precursors with lymphoid and myeloid reconstituting ability. Blood 101(5):1784–1789. doi:10.1182/blood-2002-06-1862
39. Fernandez L, Rodriguez S, Huang H, Chora A, Fernandes J, Mumaw C, Cruz E, Pollok K, Cristina F, Price JE, Ferkowicz MJ, Scadden DT, Clauss M, Cardoso AA, Carlesso N (2008) Tumor necrosis factor-alpha and endothelial cells modulate notch signaling in the bone marrow microenvironment during inflammation. Exp Hematol 36(5):545–558. doi:10.1016/j.exphem.2007.12.012
40. Kwon SM, Eguchi M, Wada M, Iwami Y, Hozumi K, Iwaguro H, Masuda H, Kawamoto A, Asahara T (2008) Specific jagged-1 signal from bone marrow microenvironment is required for endothelial progenitor cell development for neovascularization. Circulation 118(2):157–165. doi:10.1161/circulationaha.107.754978
41. Yang DG, Liu L, Zheng XY (2008) Cyclin-dependent kinase inhibitor p16(INK4a) and telomerase may co-modulate endothelial progenitor cells senescence. Ageing Res Rev 7(2):137–146. doi:10.1016/j.arr.2008.02.001
42. Kaplan RN, Riba Rd Fau - Zacharoulis S, Zacharoulis S Fau - Bramley AH, Bramley Ah Fau - Vincent L, Vincent L Fau - Costa C, Costa C Fau - MacDonald DD, MacDonald Dd Fau - Jin DK, Jin Dk Fau - Shido K, Shido K Fau - Kerns SA, Kerns Sa Fau - Zhu Z, Zhu Z Fau - Hicklin D, Hicklin D Fau - Wu Y, Wu Y Fau - Port JL, Port Jl Fau - Altorki N, Altorki N Fau - Port ER, Port Er Fau - Ruggero D, Ruggero D Fau - Shmelkov SV, Shmelkov Sv Fau - Jensen KK, Jensen Kk Fau - Rafii S, Rafii S Fau - Lyden D, Lyden D VEGFR1-positive haematopoietic bone marrow progenitors initiate the pre-metastatic niche. (1476–4687 (Electronic)). doi: D - NLM: NIHMS236203 D - NLM: PMC2945882 EDAT- 2005/12/13 09:00 MHDA- 2005/12/31 09:00 CRDT- 2005/12/13 09:00 PHST- 2005/05/13 [received] PHST- 2005/08/19 [accepted] AID - nature04186 [pii] AID - 10.1038/nature04186 [doi] PST - ppublish
43. Jin F, Brockmeier U, Otterbach F, Metzen E (2012) New insight into the SDF-1/CXCR4 axis in a breast carcinoma model: hypoxia-induced endothelial SDF-1 and tumor cell CXCR4 are required for tumor cell intravasation. Mol Cancer Res 10(8):1021–1031. doi:10.1158/1541-7786.mcr-11-0498
44. Kim HK, Song KS, Kim HO, Chung JH, Lee KR, Lee YJ, Lee DH, Lee ES, Kim HK, Ryu KW, Bae JM (2003) Circulating numbers of endothelial progenitor cells in patients with gastric and breast cancer. Cancer Lett 198(1):83–88
45. De Palma M, Venneri MA, Roca C, Naldini L (2003) Targeting exogenous genes to tumor angiogenesis by transplantation of genetically modified hematopoietic stem cells. Nat Med 9(6):789–795. doi:10.1038/nm871
46. Kim JA, Lee HJ, Kang HJ, Park TH (2009) The targeting of endothelial progenitor cells to a specific location within a microfluidic channel using magnetic nanoparticles. Biomed Microdevices 11(1):287–296. doi:10.1007/s10544-008-9235-y
47. van Noort D, Ong SM, Zhang C, Zhang S, Arooz T, Yu H (2009) Stem cells in microfluidics. Biotechnol Prog 25(1):52–60. doi:10.1002/btpr.171
48. di Tomaso E, Capen D, Haskell A, Hart J, Logie JJ, Jain RK, McDonald DM, Jones R, Munn LL (2005) Mosaic tumor vessels: cellular basis and ultrastructure of focal regions lacking endothelial cell markers. Cancer Res 65(13):5740–5749. doi:10.1158/0008-5472.CAN-04-4552
49. Chang YS, di Tomaso E, McDonald DM, Jones R, Jain RK, Munn LL (2000) Mosaic blood vessels in tumors: frequency of cancer cells in contact with flowing blood. Proc Natl Acad Sci U S A 97(26):14608–14613. doi:10.1073/pnas.97.26.14608
50. Jain RK (2001) Normalizing tumor vasculature with anti-angiogenic therapy: a new paradigm for combination therapy. Nat Med 7(9):987–989. doi:10.1038/nm0901-987
51. Jain RK (2005) Normalization of tumor vasculature: an emerging concept in antiangiogenic therapy. Science (New York, NY) 307(5706):58–62. doi:10.1126/science.1104819

52. Folkman J (2001) Can mosaic tumor vessels facilitate molecular diagnosis of cancer? Proc Natl Acad Sci U S A 98(2):398–400. doi:10.1073/pnas.98.2.398
53. Stephenson J (2001) Mosaic vessels shed cancer clues by the million. Lancet Oncol 2(3):130. doi:10.1016/s1470-2045(00)00249-7
54. Bertolini F, Lanza A, Peccatori F, Zibera C, Gibelli N, Perotti C, Da Prada GA, Torretta L, Cocorocchio E, Martinelli G, Robustelli della Cuna G (1998) Hematopoietic progenitor cell collection and neoplastic cell contamination in breast cancer patients receiving chemotherapy plus granulocyte-colony stimulating factor (G-CSF) or G-CSF alone for mobilization. Ann Oncology 9 (8):913–916
55. Kleinman MB, Wiley El Fau - Guo M, Guo M Fau - Rademaker AW, Rademaker Aw Fau - Villa M, Villa M Fau - Tallman MS, Tallman Ms Fau - Newman SB, Newman Sb Fau - Gordon LI, Gordon Li Fau - Winter JN, Winter JN Immunohistochemical detection of breast cancer cells in paired peripheral blood progenitor cell specimens collected after cytokine or cytokine and myelosuppressive chemotherapy. (0268–3369 (Print))
56. Bertolini F, Martinelli G, Goldhirsch A (2001) Mosaic tumour blood vessels and high-dose chemotherapy for breast cancer. Lancet Oncol 2(10):595. doi:10.1016/s1470-2045(01)00514-9
57. Folberg R, Hendrix MJ, Maniotis AJ (2000) Vasculogenic mimicry and tumor angiogenesis. Am J Pathol 156(2):361–381. doi:10.1016/s0002-9440(10)64739-6
58. Folberg R, Maniotis AJ (2004) Vasculogenic mimicry. APMIS : acta pathologica, microbiologica, et immunologica Scandinavica 112(7–8):508–525. doi:10.1111/j.1600-0463.2004.apm11207-0810.x
59. Pulford E, Hocking A, Griggs K, McEvoy J, Bonder C, Henderson DW, Klebe S (2016) Vasculogenic mimicry in malignant mesothelioma: an experimental and immunohistochemical analysis. Pathology 48(7):650–659. doi:10.1016/j.pathol.2016.07.009
60. Wang H, Lin H, Pan J, Mo C, Zhang F, Huang B, Wang Z, Chen X, Zhuang J, Wang D, Qiu S (2016) Vasculogenic mimicry in prostate cancer: the roles of EphA2 and PI3K. J Cancer 7(9):1114–1124. doi:10.7150/jca.14120
61. Shirakawa K, Kobayashi H, Heike Y, Kawamoto S, Brechbiel MW, Kasumi F, Iwanaga T, Konishi F, Terada M, Wakasugi H (2002) Hemodynamics in vasculogenic mimicry and angiogenesis of inflammatory breast cancer xenograft. Cancer Res 62(2):560–566
62. Paulis YW, Soetekouw PM, Verheul HM, Tjan-Heijnen VC, Griffioen AW (2010) Signalling pathways in vasculogenic mimicry. Biochim Biophys Acta 1806(1):18–28. doi:10.1016/j.bbcan.2010.01.001
63. Maniotis AJ, Folberg R, Hess A, Seftor EA, Gardner LM, Pe'er J, Trent JM, Meltzer PS, Hendrix MJ (1999) Vascular channel formation by human melanoma cells in vivo and in vitro: vasculogenic mimicry. Am J Pathol 155(3):739–752. doi:10.1016/s0002-9440(10)65173-5
64. Seftor RE, Hess AR, Seftor EA, Kirschmann DA, Hardy KM, Margaryan NV, Hendrix MJ (2012) Tumor cell vasculogenic mimicry: from controversy to therapeutic promise. Am J Pathol 181(4):1115–1125. doi:10.1016/j.ajpath.2012.07.013
65. Postovit LM, Margaryan NV, Seftor EA, Kirschmann DA, Lipavsky A, Wheaton WW, Abbott DE, Seftor RE, Hendrix MJ (2008) Human embryonic stem cell microenvironment suppresses the tumorigenic phenotype of aggressive cancer cells. Proc Natl Acad Sci U S A 105(11):4329–4334. doi:10.1073/pnas.0800467105
66. Mihic-Probst D, Ikenberg K, Tinguely M, Schraml P, Behnke S, Seifert B, Civenni G, Sommer L, Moch H, Dummer R (2012) Tumor cell plasticity and angiogenesis in human melanomas. PLoS One 7(3):e33571. doi:10.1371/journal.pone.0033571
67. De Bock K, Mazzone M, Carmeliet P (2011) Antiangiogenic therapy, hypoxia, and metastasis: risky liaisons, or not? Nat Rev Clin Oncol 8(7):393–404. doi:10.1038/nrclinonc.2011.83
68. Benizri E, Ginouves A, Berra E (2008) The magic of the hypoxia-signaling cascade. Cell Mol Life Sci 65(7–8):1133–1149. doi:10.1007/s00018-008-7472-0
69. Ivan M, Kondo K, Yang H, Kim W, Valiando J, Ohh M, Salic A, Asara JM, Lane WS, Kaelin WG Jr (2001) HIFalpha targeted for VHL-mediated destruction by proline hydroxylation: implications for O_2 sensing. Science (New York, NY) 292(5516):464–468. doi:10.1126/science.1059817
70. Fernandez-Barral A, Orgaz JL, Gomez V, del Peso L, Calzada MJ, Jimenez B (2012) Hypoxia negatively regulates antimetastatic PEDF in melanoma cells by a hypoxia inducible factor-independent, autophagy dependent mechanism. PLoS One 7(3):e32989. doi:10.1371/journal.pone.0032989
71. Andrews S, Ford D, Martin P (2012) Knockdown of osteopontin reduces the inflammatory response and subsequent size of postsurgical adhesions in a murine model. Am J Pathol 181(4):1165–1172. doi:10.1016/j.ajpath.2012.06.027
72. Vartanian A, Stepanova E, Grigorieva I, Solomko E, Baryshnikov A, Lichinitser M (2011) VEGFR1 and PKCalpha signaling control melanoma vasculogenic mimicry in a VEGFR2 kinase-independent manner. Melanoma Res 21(2):91–98. doi:10.1097/CMR.0b013e328343a237
73. Comito G, Calvani M, Giannoni E, Bianchini F, Calorini L, Torre E, Migliore C, Giordano S, Chiarugi P (2011) HIF-1alpha stabilization by mitochondrial ROS promotes met-dependent invasive growth and vasculogenic mimicry in melanoma cells. Free Radic Biol Med 51(4):893–904. doi:10.1016/j.freeradbiomed.2011.05.042

74. Conley SJ, Gheordunescu E, Kakarala P, Newman B, Korkaya H, Heath AN, Clouthier SG, Wicha MS (2012) Antiangiogenic agents increase breast cancer stem cells via the generation of tumor hypoxia. Proc Natl Acad Sci U S A 109(8):2784–2789. doi:10.1073/pnas.1018866109
75. Ju RJ, Li XT, Shi JF, Li XY, Sun MG, Zeng F, Zhou J, Liu L, Zhang CX, Zhao WY, Lu WL (2014) Liposomes, modified with PTD(HIV-1) peptide, containing epirubicin and celecoxib, to target vasculogenic mimicry channels in invasive breast cancer. Biomaterials 35(26):7610–7621. doi:10.1016/j.biomaterials.2014.05.040
76. Serwe A, Rudolph K, Anke T, Erkel G (2012) Inhibition of TGF-beta signaling, vasculogenic mimicry and proinflammatory gene expression by isoxanthohumol. Investig New Drugs 30(3):898–915. doi:10.1007/s10637-011-9643-3
77. Frank NY, Schatton T, Kim S, Zhan Q, Wilson BJ, Ma J, Saab KR, Osherov V, Widlund HR, Gasser M, Waaga-Gasser AM, Kupper TS, Murphy GF, Frank MH (2011) VEGFR-1 expressed by malignant melanoma-initiating cells is required for tumor growth. Cancer Res 71(4):1474–1485. doi:10.1158/0008-5472.can-10-1660
78. Folkman J (2007) Angiogenesis: an organizing principle for drug discovery? Nat Rev Drug Discov 6(4):273–286. doi:10.1038/nrd2115
79. Senger DR, Galli SJ, Dvorak AM, Perruzzi CA, Harvey VS, Dvorak HF (1983) Tumor cells secrete a vascular permeability factor that promotes accumulation of ascites fluid. Science (New York, NY) 219(4587):983–985
80. Ferrara N, Henzel WJ (1989) Pituitary follicular cells secrete a novel heparin-binding growth factor specific for vascular endothelial cells. Biochem Biophys Res Commun 161(2):851–858
81. Olsson AK, Dimberg A, Kreuger J, Claesson-Welsh L (2006) VEGF receptor signalling - in control of vascular function. Nat Rev Mol Cell Biol 7(5):359–371. doi:10.1038/nrm1911
82. Tie J, Desai J (2012) Antiangiogenic therapies targeting the vascular endothelia growth factor signaling system. Crit Rev Oncog 17(1):51–67
83. Dvorak HF (2002) Vascular permeability factor/vascular endothelial growth factor: a critical cytokine in tumor angiogenesis and a potential target for diagnosis and therapy. J Clin Oncol 20(21):4368–4380. doi:10.1200/jco.2002.10.088
84. Hicklin DJ, Ellis LM (2005) Role of the vascular endothelial growth factor pathway in tumor growth and angiogenesis. J Clin Oncol 23(5):1011–1027. doi:10.1200/jco.2005.06.081
85. Shalaby F, Rossant J, Yamaguchi TP, Gertsenstein M, Wu XF, Breitman ML, Schuh AC (1995) Failure of blood-island formation and vasculogenesis in Flk-1-deficient mice. Nature 376(6535):62–66. doi:10.1038/376062a0
86. Takahashi T, Yamaguchi S, Chida K, Shibuya M (2001) A single autophosphorylation site on KDR/Flk-1 is essential for VEGF-A-dependent activation of PLC-gamma and DNA synthesis in vascular endothelial cells. EMBO J 20(11):2768–2778. doi:10.1093/emboj/20.11.2768
87. Matsumoto T, Bohman S, Dixelius J, Berge T, Dimberg A, Magnusson P, Wang L, Wikner C, Qi JH, Wernstedt C, Wu J, Bruheim S, Mugishima H, Mukhopadhyay D, Spurkland A, Claesson-Welsh L (2005) VEGF receptor-2 Y951 signaling and a role for the adapter molecule TSAd in tumor angiogenesis. EMBO J 24(13):2342–2353. doi:10.1038/sj.emboj.7600709
88. Zeng H, Sanyal S, Mukhopadhyay D (2001) Tyrosine residues 951 and 1059 of vascular endothelial growth factor receptor-2 (KDR) are essential for vascular permeability factor/vascular endothelial growth factor-induced endothelium migration and proliferation, respectively. J Biol Chem 276(35):32714–32719. doi:10.1074/jbc.M103130200
89. Dougher M, Terman BI (1999) Autophosphorylation of KDR in the kinase domain is required for maximal VEGF-stimulated kinase activity and receptor internalization. Oncogene 18(8):1619–1627. doi:10.1038/sj.onc.1202478
90. Dunk C, Ahmed A (2001) Vascular endothelial growth factor receptor-2-mediated mitogenesis is negatively regulated by vascular endothelial growth factor receptor-1 in tumor epithelial cells. Am J Pathol 158(1):265–273. doi:10.1016/s0002-9440(10)63965-x
91. Hiratsuka S, Minowa O, Kuno J, Noda T, Shibuya M (1998) Flt-1 lacking the tyrosine kinase domain is sufficient for normal development and angiogenesis in mice. Proc Natl Acad Sci U S A 95(16):9349–9354
92. Kaipainen A, Korhonen J, Mustonen T, van Hinsbergh VW, Fang GH, Dumont D, Breitman M, Alitalo K (1995) Expression of the fms-like tyrosine kinase 4 gene becomes restricted to lymphatic endothelium during development. Proc Natl Acad Sci U S A 92(8):3566–3570
93. Jussila L, Alitalo K (2002) Vascular growth factors and lymphangiogenesis. Physiol Rev 82(3):673–700. doi:10.1152/physrev.00005.2002
94. Valtola R, Salven P, Heikkila P, Taipale J, Joensuu H, Rehn M, Pihlajaniemi T, Weich H, deWaal R, Alitalo K (1999) VEGFR-3 and its ligand VEGF-C are associated with angiogenesis in breast cancer. Am J Pathol 154(5):1381–1390. doi:10.1016/s0002-9440(10)65392-8
95. Partanen TA, Alitalo K, Miettinen M (1999) Lack of lymphatic vascular specificity of vascular endothelial growth factor receptor 3 in 185 vascular tumors. Cancer 86(11):2406–2412
96. He Y, Kozaki K, Karpanen T, Koshikawa K, Yla-Herttuala S, Takahashi T, Alitalo K (2002) Suppression of tumor lymphangiogenesis and lymph node metastasis by blocking vascular endothelial growth factor receptor 3 signaling. J Natl Cancer Inst 94(11):819–825

97. Wang JF, Zhang X, Groopman JE (2004) Activation of vascular endothelial growth factor receptor-3 and its downstream signaling promote cell survival under oxidative stress. J Biol Chem 279(26):27088–27097. doi:10.1074/jbc.M314015200
98. Makinen T, Veikkola T, Mustjoki S, Karpanen T, Catimel B, Nice EC, Wise L, Mercer A, Kowalski H, Kerjaschki D, Stacker SA, Achen MG, Alitalo K (2001) Isolated lymphatic endothelial cells transduce growth, survival and migratory signals via the VEGF-C/D receptor VEGFR-3. EMBO J 20(17):4762–4773. doi:10.1093/emboj/20.17.4762
99. Cascone T, Heymach JV (2012) Targeting the angiopoietin/Tie2 pathway: cutting tumor vessels with a double-edged sword? J Clin Oncol 30(4):441–444. doi:10.1200/jco.2011.38.7621
100. Tadros A, Hughes DP, Dunmore BJ, Brindle NP (2003) ABIN-2 protects endothelial cells from death and has a role in the antiapoptotic effect of angiopoietin-1. Blood 102(13):4407–4409. doi:10.1182/blood-2003-05-1602
101. Kim I, Kim HG, So JN, Kim JH, Kwak HJ, Koh GY (2000) Angiopoietin-1 regulates endothelial cell survival through the phosphatidylinositol 3′-kinase/Akt signal transduction pathway. Circ Res 86(1):24–29
102. Papapetropoulos A, Fulton D, Mahboubi K, Kalb RG, O'Connor DS, Li F, Altieri DC, Sessa WC (2000) Angiopoietin-1 inhibits endothelial cell apoptosis via the Akt/survivin pathway. J Biol Chem 275(13):9102–9105
103. Wu F, Yang LY, Li YF, Ou DP, Chen DP, Fan C (2009) Novel role for epidermal growth factor-like domain 7 in metastasis of human hepatocellular carcinoma. Hepatology 50(6):1839–1850. doi:10.1002/hep.23197
104. Saharinen P, Eklund L, Miettinen J, Wirkkala R, Anisimov A, Winderlich M, Nottebaum A, Vestweber D, Deutsch U, Koh GY, Olsen BR, Alitalo K (2008) Angiopoietins assemble distinct Tie2 signalling complexes in endothelial cell-cell and cell-matrix contacts. Nat Cell Biol 10(5):527–537. doi:10.1038/ncb1715
105. Fukuhara S, Sako K, Minami T, Noda K, Kim HZ, Kodama T, Shibuya M, Takakura N, Koh GY, Mochizuki N (2008) Differential function of Tie2 at cell-cell contacts and cell-substratum contacts regulated by angiopoietin-1. Nat Cell Biol 10(5):513–526. doi:10.1038/ncb1714
106. Gale NW, Thurston G, Hackett SF, Renard R, Wang Q, McClain J, Martin C, Witte C, Witte MH, Jackson D, Suri C, Campochiaro PA, Wiegand SJ, Yancopoulos GD (2002) Angiopoietin-2 is required for postnatal angiogenesis and lymphatic patterning, and only the latter role is rescued by angiopoietin-1. Dev Cell 3(3):411–423
107. Fagiani E, Christofori G (2013) Angiopoietins in angiogenesis. Cancer Lett 328(1):18–26. doi:10.1016/j.canlet.2012.08.018
108. Hansen TM, Singh H, Tahir TA, Brindle NP (2010) Effects of angiopoietins-1 and -2 on the receptor tyrosine kinase Tie2 are differentially regulated at the endothelial cell surface. Cell Signal 22(3):527–532. doi:10.1016/j.cellsig.2009.11.007
109. Seegar TC, Eller B, Tzvetkova-Robev D, Kolev MV, Henderson SC, Nikolov DB, Barton WA (2010) Tie1-Tie2 interactions mediate functional differences between angiopoietin ligands. Mol Cell 37(5):643–655. doi:10.1016/j.molcel.2010.02.007
110. Tsutsui S, Inoue H, Yasuda K, Suzuki K, Takeuchi H, Nishizaki T, Higashi H, Era S, Mori M (2006) Angiopoietin 2 expression in invasive ductal carcinoma of the breast: its relationship to the VEGF expression and microvessel density. Breast Cancer Res Treat 98(3):261–266. doi:10.1007/s10549-005-9157-9
111. Danza K, Pilato B, Lacalamita R, Addati T, Giotta F, Bruno A, Paradiso A, Tommasi S (2013) Angiogenetic axis angiopoietins/Tie2 and VEGF in familial breast cancer. European journal of human genetics : EJHG 21(8):824–830. doi:10.1038/ejhg.2012.273
112. Lorger M, Felding-Habermann B (2010) Capturing changes in the brain microenvironment during initial steps of breast cancer brain metastasis. Am J Pathol 176(6):2958–2971. doi:10.2353/ajpath.2010.090838
113. Hashizume H, Falcon BL, Kuroda T, Baluk P, Coxon A, Yu D, Bready JV, Oliner JD, McDonald DM (2010) Complementary actions of inhibitors of angiopoietin-2 and VEGF on tumor angiogenesis and growth. Cancer Res 70(6):2213–2223. doi:10.1158/0008-5472.can-09-1977
114. Schulz P, Fischer C, Detjen KM, Rieke S, Hilfenhaus G, von Marschall Z, Bohmig M, Koch I, Kehrberger J, Hauff P, Thierauch KH, Alves F, Wiedenmann B, Scholz A (2011) Angiopoietin-2 drives lymphatic metastasis of pancreatic cancer. FASEB J 25(10):3325–3335. doi:10.1096/fj.11-182287
115. Tian S, Hayes AJ, Metheny-Barlow LJ, Li LY (2002) Stabilization of breast cancer xenograft tumour neovasculature by angiopoietin-1. Br J Cancer 86(4):645–651. doi:10.1038/sj.bjc.6600082
116. Stoeltzing O, Ahmad SA, Liu W, McCarty MF, Wey JS, Parikh AA, Fan F, Reinmuth N, Kawaguchi M, Bucana CD, Ellis LM (2003) Angiopoietin-1 inhibits vascular permeability, angiogenesis, and growth of hepatic colon cancer tumors. Cancer Res 63(12):3370–3377
117. Hawighorst T, Skobe M, Streit M, Hong YK, Velasco P, Brown LF, Riccardi L, Lange-Asschenfeldt B, Detmar M (2002) Activation of the tie2 receptor by angiopoietin-1 enhances tumor vessel maturation and impairs squamous cell carcinoma growth. Am J Pathol 160(4):1381–1392. doi:10.1016/s0002-9440(10)62565-5
118. Holopainen T, Huang H, Chen C, Kim KE, Zhang L, Zhou F, Han W, Li C, Yu J, Wu J, Koh GY, Alitalo K, He Y (2009) Angiopoietin-1 overexpression modulates vascular endothelium to facilitate tumor cell dissemination and metastasis establishment. Cancer Res 69(11):4656–4664. doi:10.1158/0008-5472.can-08-4654

119. Cao Y, Sonveaux P, Liu S, Zhao Y, Mi J, Clary BM, Li CY, Kontos CD, Dewhirst MW (2007) Systemic overexpression of angiopoietin-2 promotes tumor microvessel regression and inhibits angiogenesis and tumor growth. Cancer Res 67(8):3835–3844. doi:10.1158/0008-5472.can-06-4056
120. Kunz P, Hoffend J, Altmann A, Dimitrakopoulou-Strauss A, Koczan D, Eisenhut M, Bonaterra GA, Dengler TJ, Mier W, Haberkorn U, Kinscherf R (2006) Angiopoietin-2 overexpression in morris hepatoma results in increased tumor perfusion and induction of critical angiogenesis-promoting genes. J Nuclear Med 47(9):1515–1524
121. Ahmad SA, Liu W, Jung YD, Fan F, Wilson M, Reinmuth N, Shaheen RM, Bucana CD, Ellis LM (2001) The effects of angiopoietin-1 and -2 on tumor growth and angiogenesis in human colon cancer. Cancer Res 61(4):1255–1259
122. Oliner J, Min H, Leal J, Yu D, Rao S, You E, Tang X, Kim H, Meyer S, Han SJ, Hawkins N, Rosenfeld R, Davy E, Graham K, Jacobsen F, Stevenson S, Ho J, Chen Q, Hartmann T, Michaels M, Kelley M, Li L, Sitney K, Martin F, Sun JR, Zhang N, Lu J, Estrada J, Kumar R, Coxon A, Kaufman S, Pretorius J, Scully S, Cattley R, Payton M, Coats S, Nguyen L, Desilva B, Ndifor A, Hayward I, Radinsky R, Boone T, Kendall R (2004) Suppression of angiogenesis and tumor growth by selective inhibition of angiopoietin-2. Cancer Cell 6(5):507–516. doi:10.1016/j.ccr.2004.09.030
123. Coxon A, Bready J, Min H, Kaufman S, Leal J, Yu D, Lee TA, Sun JR, Estrada J, Bolon B, McCabe J, Wang L, Rex K, Caenepeel S, Hughes P, Cordover D, Kim H, Han SJ, Michaels ML, Hsu E, Shimamoto G, Cattley R, Hurh E, Nguyen L, Wang SX, Ndifor A, Hayward IJ, Falcon BL, McDonald DM, Li L, Boone T, Kendall R, Radinsky R, Oliner JD (2010) Context-dependent role of angiopoietin-1 inhibition in the suppression of angiogenesis and tumor growth: implications for AMG 386, an angiopoietin-1/2-neutralizing peptibody. Mol Cancer Ther 9(10):2641–2651. doi:10.1158/1535-7163.mct-10-0213
124. O'Sullivan B, Brierley J International union against cancer UICC manual of clinical oncology. Ninth edition edn
125. Lindahl P, Johansson BR, Leveen P, Betsholtz C (1997) Pericyte loss and microaneurysm formation in PDGF-B-deficient mice. Science 277(5323):242–245
126. Paulsson J, Sjoblom T, Micke P, Ponten F, Landberg G, Heldin CH, Bergh J, Brennan DJ, Jirstrom K, Ostman A (2009) Prognostic significance of stromal platelet-derived growth factor beta-receptor expression in human breast cancer. Am J Pathol 175(1):334–341. doi:10.2353/ajpath.2009.081030
127. Pinto MP, Dye WW, Jacobsen BM, Horwitz KB (2014) Malignant stroma increases luminal breast cancer cell proliferation and angiogenesis through platelet-derived growth factor signaling. BMC Cancer 14:735. doi:10.1186/1471-2407-14-735
128. Stuelten CH, DaCosta BS, Arany PR, Karpova TS, Stetler-Stevenson WG, Roberts AB (2005) Breast cancer cells induce stromal fibroblasts to express MMP-9 via secretion of TNF-alpha and TGF-beta. J Cell Sci 118(Pt 10):2143–2153. doi:10.1242/jcs.02334
129. Shao ZM, Nguyen M, Barsky SH (2000) Human breast carcinoma desmoplasia is PDGF initiated. Oncogene 19(38):4337–4345. doi:10.1038/sj.onc.1203785
130. Banerjee S, Sengupta K, Dhar K, Mehta S, D'Amore PA, Dhar G, Banerjee SK (2006) Breast cancer cells secreted platelet-derived growth factor-induced motility of vascular smooth muscle cells is mediated through neuropilin-1. Mol Carcinog 45(11):871–880. doi:10.1002/mc.20248
131. Erber R, Thurnher A, Katsen AD, Groth G, Kerger H, Hammes HP, Menger MD, Ullrich A, Vajkoczy P (2004) Combined inhibition of VEGF and PDGF signaling enforces tumor vessel regression by interfering with pericyte-mediated endothelial cell survival mechanisms. FASEB J 18(2):338–340. doi:10.1096/fj.03-0271fje
132. Escudier B, Eisen T, Stadler WM, Szczylik C, Oudard S, Siebels M, Negrier S, Chevreau C, Solska E, Desai AA, Rolland F, Demkow T, Hutson TE, Gore M, Freeman S, Schwartz B, Shan M, Simantov R, Bukowski RM (2007) Sorafenib in advanced clear-cell renal-cell carcinoma. N Engl J Med 356 (2):125–134. doi:10.1056/NEJMoa060655
133. Llovet JM, Ricci S, Mazzaferro V, Hilgard P, Gane E, Blanc JF, de Oliveira AC, Santoro A, Raoul JL, Forner A, Schwartz M, Porta C, Zeuzem S, Bolondi L, Greten TF, Galle PR, Seitz JF, Borbath I, Haussinger D, Giannaris T, Shan M, Moscovici M, Voliotis D, Bruix J (2008) Sorafenib in advanced hepatocellular carcinoma. N Engl J Med 359(4):378–390. doi:10.1056/NEJMoa0708857
134. Motzer RJ, Hutson TE, Tomczak P, Michaelson MD, Bukowski RM, Rixe O, Oudard S, Negrier S, Szczylik C, Kim ST, Chen I, Bycott PW, Baum CM, Figlin RA (2007) Sunitinib versus interferon alfa in metastatic renal-cell carcinoma. N Engl J Med 356 (2):115–124. doi:10.1056/NEJMoa065044
135. Raymond E, Dahan L, Raoul JL, Bang YJ, Borbath I, Lombard-Bohas C, Valle J, Metrakos P, Smith D, Vinik A, Chen JS, Horsch D, Hammel P, Wiedenmann B, Van Cutsem E, Patyna S, Lu DR, Blanckmeister C, Chao R, Ruszniewski P (2011) Sunitinib malate for the treatment of pancreatic neuroendocrine tumors. N Engl J Med 364(6):501–513. doi:10.1056/NEJMoa1003825
136. Motzer RJ, Hutson TE, Cella D, Reeves J, Hawkins R, Guo J, Nathan P, Staehler M, de Souza P, Merchan JR, Boleti E, Fife K, Jin J, Jones R, Uemura H, De Giorgi U, Harmenberg U, Wang J, Sternberg CN, Deen K, McCann L, Hackshaw MD, Crescenzo R, Pandite LN, Choueiri TK (2013) Pazopanib versus sunitinib in metastatic renal-cell carcinoma. N Engl J Med 369(8):722–731. doi:10.1056/NEJMoa1303989

137. Schutz FA, Choueiri TK, Sternberg CN (2011) Pazopanib: clinical development of a potent anti-angiogenic drug. Crit Rev Oncol Hematol 77(3):163–171. doi:10.1016/j.critrevonc.2010.02.012
138. Strumberg D, Schultheis B (2012) Regorafenib for cancer. Expert Opin Investig Drugs 21(6):879–889. doi:10.1517/13543784.2012.684752
139. Chiorean EG, Sweeney C, Youssoufian H, Qin A, Dontabhaktuni A, Loizos N, Nippgen J, Amato R (2014) A phase I study of olaratumab, an anti-platelet-derived growth factor receptor alpha (PDGFRalpha) monoclonal antibody, in patients with advanced solid tumors. Cancer Chemother Pharmacol 73(3):595–604. doi:10.1007/s00280-014-2389-9
140. Presta M, Dell'Era P, Mitola S, Moroni E, Ronca R, Rusnati M (2005) Fibroblast growth factor/fibroblast growth factor receptor system in angiogenesis. Cytokine Growth Factor Rev 16(2):159–178. doi:10.1016/j.cytogfr.2005.01.004
141. Bottcher RT, Niehrs C (2005) Fibroblast growth factor signaling during early vertebrate development. Endocr Rev 26 (1):63–77. doi:10.1210/er.2003-0040
142. Carmeliet P, Jain RK (2000) Angiogenesis in cancer and other diseases. Nature 407(6801):249–257. doi:10.1038/35025220
143. Dow JK, deVere WRW (2000) Fibroblast growth factor 2: its structure and property, paracrine function, tumor angiogenesis, and prostate-related mitogenic and oncogenic functions. Urology 55(6):800–806
144. Shing Y, Folkman J, Sullivan R, Butterfield C, Murray J, Klagsbrun M (1984) Heparin affinity: purification of a tumor-derived capillary endothelial cell growth factor. Science 223(4642):1296–1299
145. Desbaillets I, Ziegler U, Groscurth P, Gassmann M (2000) Embryoid bodies: an in vitro model of mouse embryogenesis. Exp Physiol 85(6):645–651
146. Stegmann TJ (1999) New approaches to coronary heart disease: induction of neovascularisation by growth factors. BioDrugs 11(5):301–308
147. Cross MJ, Claesson-Welsh L (2001) FGF and VEGF function in angiogenesis: signalling pathways, biological responses and therapeutic inhibition. Trends Pharmacol Sci 22(4):201–207
148. Yu P, Wilhelm K, Dubrac A, Tung JK, Alves TC, Fang JS, Xie Y, Zhu J, Chen Z, De Smet F, Zhang J, Jin SW, Sun L, Sun H, Kibbey RG, Hirschi KK, Hay N, Carmeliet P, Chittenden TW, Eichmann A, Potente M, Simons M (2017) FGF-dependent metabolic control of vascular development. Nature 545(7653):224–228. doi:10.1038/nature22322
149. Seghezzi G, Patel S, Ren CJ, Gualandris A, Pintucci G, Robbins ES, Shapiro RL, Galloway AC, Rifkin DB, Mignatti P (1998) Fibroblast growth factor-2 (FGF-2) induces vascular endothelial growth factor (VEGF) expression in the endothelial cells of forming capillaries: an autocrine mechanism contributing to angiogenesis. J Cell Biol 141(7):1659–1673
150. Moscatelli D, Presta M, Joseph-Silverstein J, Rifkin DB (1986) Both normal and tumor cells produce basic fibroblast growth factor. J Cell Physiol 129(2):273–276. doi:10.1002/jcp.1041290220
151. Kandel J, Bossy-Wetzel E, Radvanyi F, Klagsbrun M, Folkman J, Hanahan D (1991) Neovascularization is associated with a switch to the export of bFGF in the multistep development of fibrosarcoma. Cell 66(6):1095–1104. doi:0092-8674(91)90033-U [pii]
152. Li VW, Folkerth RD, Watanabe H, Yu C, Rupnick M, Barnes P, Scott RM, Black PM, Sallan SE, Folkman J (1994) Microvessel count and cerebrospinal fluid basic fibroblast growth factor in children with brain tumours. Lancet 344(8915):82–86
153. Baird A, Mormede P, Bohlen P (1986) Immunoreactive fibroblast growth factor (FGF) in a transplantable chondrosarcoma: inhibition of tumor growth by antibodies to FGF. J Cell Biochem 30(1):79–85. doi:10.1002/jcb.240300109
154. Gross JL, Herblin WF, Dusak BA, Czerniak P, Diamond MD, Sun T, Eidsvoog K, Dexter DL, Yayon A (1993) Effects of modulation of basic fibroblast growth factor on tumor growth in vivo. J Natl Cancer Inst 85(2):121–131
155. Hori A, Sasada R, Matsutani E, Naito K, Sakura Y, Fujita T, Kozai Y (1991) Suppression of solid tumor growth by immunoneutralizing monoclonal antibody against human basic fibroblast growth factor. Cancer Res 51(22):6180–6184
156. Czubayko F, Liaudet-Coopman ED, Aigner A, Tuveson AT, Berchem GJ, Wellstein A (1997) A secreted FGF-binding protein can serve as the angiogenic switch in human cancer. Nat Med 3(10):1137–1140
157. Rak J, Kerbel RS (1997) bFGF and tumor angiogenesis--back in the limelight? Nat Med 3(10):1083–1084
158. Auguste P, Gursel DB, Lemiere S, Reimers D, Cuevas P, Carceller F, Di Santo JP, Bikfalvi A (2001) Inhibition of fibroblast growth factor/fibroblast growth factor receptor activity in glioma cells impedes tumor growth by both angiogenesis-dependent and -independent mechanisms. Cancer Res 61(4):1717–1726
159. Polnaszek N, Kwabi-Addo B, Peterson LE, Ozen M, Greenberg NM, Ortega S, Basilico C, Ittmann M (2003) Fibroblast growth factor 2 promotes tumor progression in an autochthonous mouse model of prostate cancer. Cancer Res 63(18):5754–5760
160. Liang J, Chen P, Hu Z, Zhou X, Chen L, Li M, Wang Y, Tang J, Wang H, Shen H (2008) Genetic variants in fibroblast growth factor receptor 2 (FGFR2) contribute to susceptibility of breast cancer in Chinese women. Carcinogenesis 29(12):2341–2346. doi:10.1093/carcin/bgn235
161. Agarwal D, Pineda S, Michailidou K, Herranz J, Pita G, Moreno LT, Alonso MR, Dennis J, Wang Q, Bolla MK, Meyer KB, Menendez-Rodriguez P, Hardisson D, Mendiola M, Gonzalez-Neira A, Lindblom A, Margolin S, Swerdlow A, Ashworth A, Orr N, Jones

M, Matsuo K, Ito H, Iwata H, Kondo N, Hartman M, Hui M, Lim WY, Iau PT, Sawyer E, Tomlinson I, Kerin M, Miller N, Kang D, Choi J, Park SK, Noh D, Hopper JL, Schmidt DF, Makalic E, Southey MC, Teo SH, Yip CH, Sivanandan K, Tay W, Brauch H, Bruning T, Hamann U, Dunning AM, Shah M, Andrulis IL, Knight JA, Glendon G, Tchatchou S, Schmidt MK, Broeks A, Rosenberg EH, van't Veer LJ, Fasching PA, Renner SP, Ekici AB, Beckmann MW, Shen C, Hsiung C, Yu J, Hou M, Blot W, Cai Q, Wu AH, Tseng C, Van Den Berg D, Stram DO, Cox A, Brock IW, Reed MW, Muir K, Lophatananon A, Stewart-Brown S, Siriwanarangsan P, Zheng W, Deming-Halverson S, Shrubsole MJ, Long J, Shu X, Lu W, Gao Y, Zhang B, Radice P, Peterlongo P, Manoukian S, Mariette F, Sangrajrang S, McKay J, Couch FJ, Toland AE, Yannoukakos D, Fletcher O, Johnson N, dos Santos SI, Peto J, Marme F, Burwinkel B, Guenel P, Truong T, Sanchez M, Mulot C, Bojesen SE, Nordestgaard BG, Flyer H, Brenner H, Dieffenbach AK, Arndt V, Stegmaier C, Mannermaa A, Kataja V, Kosma V, Hartikainen JM, Lambrechts D, Yesilyurt BT, Floris G, Leunen K, Chang-Claude J, Rudolph A, Seibold P, Flesch-Janys D, Wang X, Olson JE, Vachon C, Purrington K, Giles GG, Severi G, Baglietto L, Haiman CA, Henderson BE, Schumacher F, Marchand LL, Simard J, Dumont M, Goldberg MS, Labreche F, Winqvist R, Pylkas K, Jukkola-Vuorinen A, Grip M, Devilee P, Tollenaar RA, Seynaeve C, Garcia-Closas M, Chanock SJ, Lissowska J, Figueroa JD, Czene K, Eriksson M, Humphreys K, Darabi H, Hooning MJ, Kriege M, Collee JM, Tilanus-Linthorst M, Li J, Jakubowska A, Lubinski J, Jaworska-Bieniek K, Durda K, Nevanlinna H, Muranen TA, Aittomaki K, Blomqvist C, Bogdanova N, Dork T, Hall P, Chenevix-Trench G, Easton DF, Pharroah PD, Arias-Perez JI, Zamora P, Benitez J, Milne RL (2014) FGF receptor genes and breast cancer susceptibility: results from the Breast Cancer Association Consortium. Br J Cancer 110(4):1088–1100. doi:10.1038/bjc.2013.769

162. Yamamoto Y, Matsui J, Matsushima T, Obaishi H, Miyazaki K, Nakamura K, Tohyama O, Semba T, Yamaguchi A, Hoshi SS, Mimura F, Haneda T, Fukuda Y, Kamata J, Takahashi K, Matsukura M, Wakabayashi T, Asada M, Nomoto K, Watanabe T, Dezso Z, Yoshimatsu K, Funahashi Y, Tsuruoka A (2014) Lenvatinib, an angiogenesis inhibitor targeting VEGFR/FGFR, shows broad antitumor activity in human tumor xenograft models associated with microvessel density and pericyte coverage. Vasc Cell 6:18. doi:10.1186/2045-824X-6-18
163. Schneider BP, Miller KD (2005) Angiogenesis of breast cancer. J Clin Oncol 23(8):1782–1790. doi:10.1200/jco.2005.12.017
164. Bareschino MA, Schettino C, Colantuoni G, Rossi E, Rossi A, Maione P, Ciardiello F, Gridelli C (2011) The role of antiangiogenetic agents in the treatment of breast cancer. Curr Med Chem 18(33):5022–5032
165. Khosravi Shahi P, Soria Lovelle A, Perez Manga G (2009) Tumoral angiogenesis and breast cancer. Clin Transl Oncol 11(3):138–142
166. Zhang HT, Craft P, Scott PA, Ziche M, Weich HA, Harris AL, Bicknell R (1995) Enhancement of tumor growth and vascular density by transfection of vascular endothelial cell growth factor into MCF-7 human breast carcinoma cells. J Natl Cancer Inst 87(3):213–219
167. Chen W, Wang S, Tian T, Bai J, Hu Z, Xu Y, Dong J, Chen F, Wang X, Shen H (2009) Phenotypes and genotypes of insulin-like growth factor 1, IGF-binding protein-3 and cancer risk: evidence from 96 studies. Eur J Hum Genet 17(12):1668–1675. doi:10.1038/ejhg.2009.86
168. Fox SB, Generali DG, Harris AL (2007) Breast tumour angiogenesis. Breast Cancer Res 9(6):216. doi:10.1186/bcr1796
169. Toffoli S, Roegiers A, Feron O, Van Steenbrugge M, Ninane N, Raes M, Michiels C (2009) Intermittent hypoxia is an angiogenic inducer for endothelial cells: role of HIF-1. Angiogenesis 12(1):47–67. doi:10.1007/s10456-009-9131-y
170. Naumov GN, Bender E, Zurakowski D, Kang SY, Sampson D, Flynn E, Watnick RS, Straume O, Akslen LA, Folkman J, Almog N (2006) A model of human tumor dormancy: an angiogenic switch from the nonangiogenic phenotype. J Natl Cancer Inst 98(5):316–325. doi:10.1093/jnci/djj068
171. Hyder SM (2006) Sex-steroid regulation of vascular endothelial growth factor in breast cancer. Endocr Relat Cancer 13(3):667–687. doi:10.1677/erc.1.00931
172. Rubanyi GM, Johns A, Kauser K (2002) Effect of estrogen on endothelial function and angiogenesis. Vasc Pharmacol 38(2):89–98
173. Greb RR, Heikinheimo O, Williams RF, Hodgen GD, Goodman AL (1997) Vascular endothelial growth factor in primate endometrium is regulated by oestrogen-receptor and progesterone-receptor ligands in vivo. Hum Reprod 12(6):1280–1292
174. Morales DE, McGowan KA, Grant DS, Maheshwari S, Bhartiya D, Cid MC, Kleinman HK, Schnaper HW (1995) Estrogen promotes angiogenic activity in human umbilical vein endothelial cells in vitro and in a murine model. Circulation 91(3):755–763
175. Kim-Schulze S, McGowan KA, Hubchak SC, Cid MC, Martin MB, Kleinman HK, Greene GL, Schnaper HW (1996) Expression of an estrogen receptor by human coronary artery and umbilical vein endothelial cells. Circulation 94(6):1402–1407
176. Thomas T, Rhodin J, Clark L, Garces A (2003) Progestins initiate adverse events of menopausal estrogen therapy. Climacteric 6(4):293–301
177. Kerbel RS (2012) Strategies for improving the clinical benefit of antiangiogenic drug based therapies for breast cancer. J Mammary Gland Biol Neoplasia 17(3–4):229–239. doi:10.1007/s10911-012-9266-0

178. Mackey J, Gelmon K, Martin M, McCarthy N, Pinter T, Rupin M, Youssoufian H (2009) TRIO-012: a multicenter, multinational, randomized, double-blind phase III study of IMC-1121B plus docetaxel versus placebo plus docetaxel in previously untreated patients with HER2-negative, unresectable, locally recurrent or metastatic breast cancer. Clin Breast Cancer 9(4):258–261. doi:10.3816/CBC.2009.n.044
179. Munoz R, Shaked Y, Bertolini F, Emmenegger U, Man S, Kerbel RS (2005) Anti-angiogenic treatment of breast cancer using metronomic low-dose chemotherapy. Breast 14(6):466–479. doi:10.1016/j.breast.2005.08.026
180. Colleoni M, Rocca A, Sandri MT, Zorzino L, Masci G, Nole F, Peruzzotti G, Robertson C, Orlando L, Cinieri S, de BF, Viale G, Goldhirsch A (2002) Low-dose oral methotrexate and cyclophosphamide in metastatic breast cancer: antitumor activity and correlation with vascular endothelial growth factor levels. Ann Oncol 13 (1):73–80
181. Rossari JR, Metzger-Filho O, Paesmans M, Saini KS, Gennari A, de Azambuja E, Piccart-Gebhart M (2012) Bevacizumab and breast cancer: a meta-analysis of first-line phase III studies and a critical reappraisal of available evidence. J Oncol 2012:417673. doi:10.1155/2012/417673
182. Montero AJ, Vogel C (2012) Fighting fire with fire: rekindling the bevacizumab debate. N Engl J Med 366(4):374–375. doi:10.1056/NEJMe1113368
183. Sikov WM, Berry DA, Perou CM, Singh B, Cirrincione CT, Tolaney SM, Kuzma CS, Pluard TJ, Somlo G, Port ER, Golshan M, Bellon JR, Collyar D, Hahn OM, Carey LA, Hudis CA, Winer EP (2015) Impact of the addition of carboplatin and/or bevacizumab to neoadjuvant once-per-week paclitaxel followed by dose-dense doxorubicin and cyclophosphamide on pathologic complete response rates in stage II to III triple-negative breast cancer: CALGB 40603 (Alliance). J Clin Oncol 33(1):13–21. doi:10.1200/JCO.2014.57.0572
184. von Minckwitz G, Eidtmann H, Rezai M, Fasching PA, Tesch H, Eggemann H, Schrader I, Kittel K, Hanusch C, Kreienberg R, Solbach C, Gerber B, Jackisch C, Kunz G, Blohmer JU, Huober J, Hauschild M, Fehm T, Muller BM, Denkert C, Loibl S, Nekljudova V, Untch M (2012) Neoadjuvant chemotherapy and bevacizumab for HER2-negative breast cancer. N Engl J Med 366(4):299–309. doi:10.1056/NEJMoa1111065
185. Cameron D, Brown J, Dent R, Jackisch C, Mackey J, Pivot X, Steger GG, Suter TM, Toi M, Parmar M, Laeufle R, Im YH, Romieu G, Harvey V, Lipatov O, Pienkowski T, Cottu P, Chan A, Im SA, Hall PS, Bubuteishvili-Pacaud L, Henschel V, Deurloo RJ, Pallaud C, Bell R (2013) Adjuvant bevacizumab-containing therapy in triple-negative breast cancer (BEATRICE): primary results of a randomised, phase 3 trial. Lancet Oncol 14(10):933–942. doi:10.1016/s1470-2045(13)70335-8
186. Fuchs CS, Tomasek J, Yong CJ, Dumitru F, Passalacqua R, Goswami C, Safran H, dos Santos LV, Aprile G, Ferry DR, Melichar B, Tehfe M, Topuzov E, Zalcberg JR, Chau I, Campbell W, Sivanandan C, Pikiel J, Koshiji M, Hsu Y, Liepa AM, Gao L, Schwartz JD, Tabernero J (2014) Ramucirumab monotherapy for previously treated advanced gastric or gastro-oesophageal junction adenocarcinoma (REGARD): an international, randomised, multicentre, placebo-controlled, phase 3 trial. Lancet (London, England) 383(9911):31–39. doi:10.1016/s0140-6736(13)61719-5
187. Wilke H, Muro K, Van Cutsem E, Oh SC, Bodoky G, Shimada Y, Hironaka S, Sugimoto N, Lipatov O, Kim TY, Cunningham D, Rougier P, Komatsu Y, Ajani J, Emig M, Carlesi R, Ferry D, Chandrawansa K, Schwartz JD, Ohtsu A (2014) Ramucirumab plus paclitaxel versus placebo plus paclitaxel in patients with previously treated advanced gastric or gastro-oesophageal junction adenocarcinoma (RAINBOW): a double-blind, randomised phase 3 trial. Lancet Oncol 15(11):1224–1235. doi:10.1016/s1470-2045(14)70420-6
188. Hoar FJ, Chaudhri S, Wadley MS, Stonelake PS (2003) Co-expression of vascular endothelial growth factor C (VEGF-C) and c-erbB2 in human breast carcinoma. Eur J Cancer 39(12):1698–1703
189. Dieras V, Wildiers H, Jassem J, Dirix LY, Guastalla JP, Bono P, Hurvitz SA, Goncalves A, Romieu G, Limentani SA, Jerusalem G, Lakshmaiah KC, Roche H, Sanchez-Rovira P, Pienkowski T, Segui Palmer MA, Li A, Sun YN, Pickett CA, Slamon DJ (2015) Trebananib (AMG 386) plus weekly paclitaxel with or without bevacizumab as first-line therapy for HER2-negative locally recurrent or metastatic breast cancer: a phase 2 randomized study. Breast (Edinburgh, Scotland) 24(3):182–190. doi:10.1016/j.breast.2014.11.003
190. Robert NJ, Dieras V, Glaspy J, Brufsky AM, Bondarenko I, Lipatov ON, Perez EA, Yardley DA, Chan SY, Zhou X, Phan SC, O'Shaughnessy J (2011) RIBBON-1: randomized, double-blind, placebo-controlled, phase III trial of chemotherapy with or without bevacizumab for first-line treatment of human epidermal growth factor receptor 2-negative, locally recurrent or metastatic breast cancer. J Clin Oncol 29(10):1252–1260. doi:10.1200/jco.2010.28.0982
191. Kerbel RS, Benezra R, Lyden DC, Hattori K, Heissig B, Nolan DJ, Mittal V, Shaked Y, Dias S, Bertolini F, Rafii S (2008) Endothelial progenitor cells are cellular hubs essential for neoangiogenesis of certain aggressive adenocarcinomas and metastatic transition but not adenomas. Proceedings of the National Academy of Sciences of the United States of America 105 (34):E54; author reply E55. doi:10.1073/pnas.0804876105
192. Folkman J (2006) Antiangiogenesis in cancer therapy--endostatin and its mechanisms of action. Exp Cell Res 312(5):594–607. doi:10.1016/j.yexcr.2005.11.015

193. Fernandez PM, Rickles FR (2002) Tissue factor and angiogenesis in cancer. Curr Opin Hematol 9(5):401–406
194. Satchi-Fainaro R, Puder M, Davies JW, Tran HT, Sampson DA, Greene AK, Corfas G, Folkman J (2004) Targeting angiogenesis with a conjugate of HPMA copolymer and TNP-470. Nat Med 10(3):255–261. doi:10.1038/nm1002
195. Sakurai T, Kudo M (2011) Signaling pathways governing tumor angiogenesis. Oncology 81(Suppl 1):24–29. doi:10.1159/000333256
196. Burstein HJ, Chen YH, Parker LM, Savoie J, Younger J, Kuter I, Ryan PD, Garber JE, Chen H, Campos SM, Shulman LN, Harris LN, Gelman R, Winer EP (2008) VEGF as a marker for outcome among advanced breast cancer patients receiving anti-VEGF therapy with bevacizumab and vinorelbine chemotherapy. Clin Cancer Res 14(23):7871–7877. doi:10.1158/1078-0432.ccr-08-0593
197. Schneider BP, Wang M, Radovich M, Sledge GW, Badve S, Thor A, Flockhart DA, Hancock B, Davidson N, Gralow J, Dickler M, Perez EA, Cobleigh M, Shenkier T, Edgerton S, Miller KD (2008) Association of vascular endothelial growth factor and vascular endothelial growth factor receptor-2 genetic polymorphisms with outcome in a trial of paclitaxel compared with paclitaxel plus bevacizumab in advanced breast cancer: ECOG 2100. J Clin Oncol 26(28):4672–4678. doi:10.1200/jco.2008.16.1612
198. Kim DH, Xu W, Kamel-Reid S, Liu X, Jung CW, Kim S, Lipton JH (2010) Clinical relevance of vascular endothelial growth factor (VEGFA) and VEGF receptor (VEGFR2) gene polymorphism on the treatment outcome following imatinib therapy. Ann Oncol 21(6):1179–1188. doi:10.1093/annonc/mdp452
199. Smith ER, Zurakowski D, Saad A, Scott RM, Moses MA (2008) Urinary biomarkers predict brain tumor presence and response to therapy. Clinical cancer research : an official journal of the American Association for Cancer Research 14(8):2378–2386. doi:10.1158/1078-0432.ccr-07-1253
200. Moses MA, Wiederschain D, Loughlin KR, Zurakowski D, Lamb CC, Freeman MR (1998) Increased incidence of matrix metalloproteinases in urine of cancer patients. Cancer Res 58(7):1395–1399
201. Burstein HJ, Elias AD, Rugo HS, Cobleigh MA, Wolff AC, Eisenberg PD, Lehman M, Adams BJ, Bello CL, DePrimo SE, Baum CM, Miller KD (2008) Phase II study of sunitinib malate, an oral multitargeted tyrosine kinase inhibitor, in patients with metastatic breast cancer previously treated with an anthracycline and a taxane. Journal of clinical oncology : official journal of the American Society of Clinical Oncology 26(11):1810–1816. doi:10.1200/jco.2007.14.5375
202. Jain RK (2005) Antiangiogenic therapy for cancer: current and emerging concepts. Oncology (Williston Park) 19(4 Suppl 3):7–16
203. Kallman RF, Dorie MJ (1986) Tumor oxygenation and reoxygenation during radiation therapy: their importance in predicting tumor response. Int J Radiat Oncol Biol Phys 12(4):681–685. doi:0360-3016(86)90080-5 [pii]
204. Cao Y, Arbiser J, D'Amato RJ, D'Amore PA, Ingber DE, Kerbel R, Klagsbrun M, Lim S, Moses MA, Zetter B, Dvorak H, Langer R (2011) Forty-year journey of angiogenesis translational research. Sci Transl Med 3(114):114rv113. doi:10.1126/scitranslmed.3003149
205. Lu H, Shu XO, Cui Y, Kataoka N, Wen W, Cai Q, Ruan ZX, Gao YT, Zheng W (2005) Association of genetic polymorphisms in the VEGF gene with breast cancer survival. Cancer Res 65(12):5015–5019. doi:10.1158/0008-5472.CAN-04-2786
206. Jin Q, Hemminki K, Enquist K, Lenner P, Grzybowska E, Klaes R, Henriksson R, Chen B, Pamula J, Pekala W, Zientek H, Rogozinska-Szczepka J, Utracka-Hutka B, Hallmans G, Forsti A (2005) Vascular endothelial growth factor polymorphisms in relation to breast cancer development and prognosis. Clin Cancer Res 11(10):3647–3653. doi:10.1158/1078-0432.CCR-04-1803

9 Key Factors in Breast Cancer Dissemination and Establishment at the Bone: Past, Present and Future Perspectives

Sioned Owen, Catherine Zabkiewicz, Lin Ye, Andrew J. Sanders, Chang Gong, and Wen G. Jiang

Abstract

Bone metastases associated with breast cancer remain a clinical challenge due to their associated morbidity, limited therapeutic intervention and lack of prognostic markers. With a continually evolving understanding of bone biology and its dynamic microenvironment, many potential new targets have been proposed. In this chapter, we discuss the roles of well-established bone markers and how their targeting, in addition to tumour-targeted therapies, might help in the prevention and treatment of bone metastases. There are a vast number of bone markers, of which one of the best-known families is the bone morphogenetic proteins (BMPs). This chapter focuses on their role in breast cancer-associated bone metastases, associated signalling pathways and the possibilities for potential therapeutic intervention. In addition, this chapter provides an update on the role receptor activator of nuclear factor-κB (RANK), RANK ligand (RANKL) and osteoprotegerin (OPG) play on breast cancer development and their subsequent influence during the homing and establishment of breast cancer-associated bone metastases. Beyond the well-established bone molecules, this chapter also

S. Owen • C. Zabkiewicz • L. Ye • A.J. Sanders • W.G. Jiang (✉)
Cardiff University School of Medicine, CCMRC, Cardiff University,
Henry Wellcome Building, Heath Park, Cardiff CF14 4XN, UK
e-mail: JiangW@cardiff.ac.uk

C. Gong
Cardiff University School of Medicine, CCMRC, Cardiff University,
Henry Wellcome Building, Heath Park, Cardiff CF14 4XN, UK

Sun Yat-sen Memorial Hospital, Sun Yat-sen University, Guangzhou 510120, China

E. Song, H. Hu (eds.), *Translational Research in Breast Cancer*, Advances in Experimental Medicine and Biology 1026, https://doi.org/10.1007/978-981-10-6020-5_9

explores the role of other potential factors such as activated leukocyte cell adhesion molecule (ALCAM) and its potential impact on breast cancer cells' affinity for the bone environment, which implies that ALCAM could be a promising therapeutic target.

Keywords
Breast cancer • Bone metastasis • BMPs • OPG • ALCAM

9.1 Introduction

The propensity of breast cancer to metastasise to bone is a well-noted phenomenon. In 1889, through post-mortem study, Paget [1] identified that breast cancer cases were associated with bone metastasis. In spite of this observed occurrence, clinical intervention remains limited and palliative. Breast cancer is the leading cancer in females in the UK and the USA and is a common cause of cancer-related deaths. The most common metastatic site is the bone, and 50–70% of patients develop bone metastases [2]. Though the survival rate of breast cancer patients diagnosed with bone metastases varies greatly in the literature, depending on if it is bone metastases alone or in combination with visceral metastasis (up to 72 months [3]), only 30% of these women are expected to achieve 5-year survival after their bone metastasis diagnosis [4].

The metastatic cascade is not a new concept, and the process is highly inefficient, with only 0.001–0.02% of cancer cells forming metastatic foci [5] of which, our understanding of the biological drivers remains poor. With the success of first-line therapies, breast cancer patients are surviving longer. However, the bone marrow provides a niche for metastasising breast cancer cells, which can be activated many years later. Evidence has shown that tumour cells are detectable in patient bone marrow, but these do not always result in metastatic foci due to a variety of factors including tumour cell dormancy or host response [6, 7]. A better understanding of what reactivates these cancer cells and their interplay, both mechanically and biologically, which occurs between the bone environment and these cells, is fundamental to future identification of patients most at risk of developing bone metastases and development of novel therapeutic interventions.

The bone is a dynamic tissue which not only provides structural support and protection but also acts as a reservoir for haematopoietic cells and an elaborate blood supply. As metastatic breast cancer cells colonise the bone environment utilising growth factors, such as transforming growth factor β (TGF-β) and insulin-like growth factors, to stimulate tumour growth and they also feedback into the bone environment through direct cell-to-cell contact and paracrine influence. Tumour cells secrete a range of factors, including interleukins, tumour necrosis factor (TNF)-alpha and parathyroid hormone-related protein (PTHrP), both osteo-inductive factors, which influence the physiological bone environment, enable tumour growth and stimulate osteoclastogenesis [8]. This bidirectional signalling and co-operation are referred to as the 'vicious cycle' and involve bone remodelling cells such as osteoclasts and osteoblasts, as well as the recruitment and modulation of other cell types including platelets, immune and nerve cells, which can further facilitate pro-tumorigenic processes, including angiogenesis [9–11].

The nature of bone metastases are heterogeneous ranging from bone destructing (osteolytic) to bone forming (osteoblastic), with potential for both cases to also occur at the same time (mixed lesions). Within breast cancer, it is the osteolytic phenotype which is frequently observed. Bisphosphonates, which bind to bone mineral and result in osteoclast apoptosis, and receptor activator of NF-κB ligand (RANKL)-targeted antibodies remain the main standard of care for skeletal-related events (SREs) occurring in breast cancer patients with bone metastases [12].

These interventions currently reduce the morbidities which are associated with bone metastases, including debilitating pain, fractures and hypercalcaemia. However, they do not target the tumorigenic process and only inhibit osteoclast function and osteoclastogenesis. Approximately half of patients continue to develop new bone metastases, and breast cancer patients who develop pathological fracture have a 32% increased risk of death compared to those who do not [13, 14]. Therefore, novel therapies aimed at recently identified targets are required to influence both the metastatic tumour cells and the bone environment.

It is vital to understand the molecular and cellular events involved in bone modelling and remodelling and the effects tumour cells have on the skeleton and vice versa. The key cells, osteoclasts and osteoblasts, from their two distinct lineages, haematopoietic and mesenchymal respectively, were heavily investigated in the 1980s and 1990s to elucidate their roles in bone remodelling, after Epker and Frost (1965) [15] demonstrated an interaction between these crucial cells in the remodelling process. This resulted in the subsequent identification of a trio of key molecules in the 1990s, receptor of activator of NF-κB (RANK), RANKL and osteoprotegerin (OPG), whose interplay is fundamental to the regulation of bone homeostasis [16–18]. Since then, these molecules have been under intense investigation in bone-related conditions, including bone metastases.

9.2 OPG, RANK and RANKL in Bone Metastasis Associated with Breast Cancer

RANKL, its receptor RANK and its naturally secreted decoy OPG are all members of the TNF receptor superfamily. Originally linked to bone remodelling and immunity, they have since been extensively studied in a wide range of solid cancers, including breast cancer, particularly focusing on their effects on the bone microenvironment. Furthermore, recent studies in breast cancer have shown that these bone-related molecules could be potential prognostic and therapeutic targets beyond the bone.

9.2.1 RANK/RANKL Signalling in Mammary Gland Development and Carcinogenesis

In the last decade, the RANK/RANKL pathway has come to prominence in breast cancer research beyond the bone environment due to several key observations. RANK and RANKL along with several hormones, including sex hormones, prolactin and PTHrP, have been linked with both normal mammary gland development and lactation, including ductal side branching, alveolar differentiation and lumen formation during pregnancy and carcinogenesis [19–22]. The interaction of progesterone and RANKL signalling has been particularly relevant given that this can occur in both progesterone-positive and progesterone-negative cells. In progesterone-responsive luminal cells, RANKL is upregulated, which helps stabilise RNA. Furthermore, evidence has shown that through paracrine RANKL signalling on oestrogen and/or progesterone receptor-negative breast cancer cells, proliferation can also be induced [23–27].

Further evidence has shown that RANK loss or overexpression contributes to disrupted mammary gland development and impaired lactation during pregnancy [28]. Furthermore, RANK has been shown to promote proliferation and survival of mammary epithelial cells as well as expansion of mammary stem and luminal progenitor cells [29]. Thus, RANK signalling, through enhanced activation of protein kinase B (PKB/AKT) and extracellular signal-related kinase (ERK) 1/2, causes mammary progenitor populations, which are potentially supported by the paracrine signalling of RANKL, in either its membrane bound or soluble forms, to promote breast tumour formation [30].

Beyond the role that RANK/RANKL signalling plays in mammary gland development, Blake et al. [31] demonstrated that MDA-MB-231 breast cancer cells overexpressing RANK resulted in greater bone colonisation and growth. Furthermore, Casimiro et al. [32] identified that RANK expressing bone-seeking subclones of MDA-MB-231 cells exhibited increased cell migration and invasion through RANKL-mediated c-jun N-terminal kinase (JNK) and ERK 1/2 signalling. We have previously demonstrated that the targeting of RANK expression in breast cancer cells in vitro reduced cell-matrix adhesion, migration and invasion [33]. Whilst Jones et al. [34] reported that mice with RANK deletion, specifically in mammary gland epithelial cells, exhibited decreased cell proliferation under progesterone stimulation compared to the wild-type mice. However, in spite of these observations, no RANK targeting agents currently appear to be in clinical trials for the treatment of primary breast tumours.

Denosumab, which targets osteoclastogenesis by blocking the actions of RANKL, is effective in the treatment of SREs. Literature shows that in vitro cocultured breast cancer cell lines devoid of RANKL could still stimulate RANKL expression in stromal osteoblasts, thus driving osteoclastogenesis and highlighting the relevance in targeting changes which occur in the microenvironment as well as the tumour cells and their related factors [19, 35].

9.2.2 The Pros and Cons of OPG in Breast Cancer

Given that OPG is the natural decoy of RANKL and a negative regulator of bone metabolism, the reversal of its downregulation has been considered for the treatment of breast cancer-associated bone metastasis. Studies have found that basal levels of OPG are expressed in breast cancer cells and tissues. Van Poznak et al. [36] has further demonstrated that in 55% of breast cancer cases studied, OPG expression was detected and correlated with oestrogen and progesterone receptors. However, subsequent functionality tests have suggested a negative correlation between OPG and the oestrogen receptor (ER), whereby activation of the ER results in a decrease in OPG expression. This was demonstrated in the ER-positive cell line MCF-7 when it was treated with 17β-oestradiol, which inhibited OPG at both mRNA and protein levels. The effect was reversed with the addition of the selective ER downregulator Fulvestrant (ICI-182,780) [37]. OPG also has a weak affinity for TNF-related apoptosis inducing ligand (TRAIL), which is believed to aid breast cancer cell survival by evasion of death receptor-induced apoptosis, as has been demonstrated in vitro but not in vivo [38]. It may be less effective in vivo due to evidence suggesting that excessive RANKL can reverse the effect of OPG on TRAIL-induced apoptosis [39, 40]. The identification of OPG in breast cancer and its associated metastasis in vivo appears mixed, potentially due to differing effects between the whole and truncated protein versions of OPG and its response to other factors such as hepatocyte growth factor (HGF) [33, 41–44]. It seems that outside of the bone microenvironment, OPG might have a tumour-proliferating effect and thus be a driver for metastasis to occur in other parts of the body [45].

Implications of OPG and breast cancer have also been conflicting, potentially due to different breast cancer subtypes [44, 46–49]. Therefore, OPG as a direct target may not provide the best solution for the treatment of breast cancer-associated bone metastasis. As Croft et al. highlighted in 2013, all clinical trials targeting OPG had been discontinued [50]. Perhaps modulation of OPG as an indirect consequence of other therapeutic intervention could be beneficial. If it is contained to the bone environment, its utilisation for breast cancer-associated bone metastasis treatment may be more advantageous.

9.3 Targeting Aberrant BMP Signalling in Bone Metastasis of Breast Cancer

Bone morphogenetic proteins (BMPs) belong to the TGF-β superfamily which play pivotal roles in embryonic and postnatal development as well as the homeostasis of tissues and organs by coordinating differentiation, proliferation, apoptosis and motility of cells in tissue and organ-specific structures. BMP signalling is relayed through a heteromeric receptor complex comprising two types of serine-threonine kinase transmembrane receptors. Type I receptors mediating BMP signalling include activin receptor-like kinase-1 (ALK-1), BMP receptor type IA (BMPR-IA, also known as ALK-3), BMP receptor type IB (BMPR-IB, or ALK-6), ALK-4, ALK-5 and activin A receptor type I (ActRI). The type II receptors include BMP receptor type II (BMPR-II), activin A receptor type IIA (ActRIIA) and activin A receptor type IIB (ActRIIB). Upon binding of BMP ligands, the type II receptors phosphorylate the glycine-serine (GS) domain of type I receptors, leading to the recruitment of the pathway-restricted Smads (R-Smads, Smads1, 5 and 8) to the complex. With assistance from Smad 4, the R-Smads intracellular signalling complex is translocated into the nucleus, leading to the induction of BMP-responsive genes. Smads 6 and 7 negatively regulate this Smad-dependent signalling. On the other hand, the Smad-independent pathways, such as mitogen-activated protein kinase (MAPK) pathway and the RAS pathway, also relay signals into cells. Thus, these diverse pathways orchestrate cellular responses to BMP ligands [51].

As a group of important regulators for bone formation and turnover, the possible implication of these proteins in bone metastasis of certain solid tumours has been investigated [51]. The attention to BMPs and their role in breast cancer arose nearly a decade ago, with a number of key findings being made in this area.

9.3.1 Aberrant Expression of BMPs in Breast Cancer

Our previous studies have shown that expression of BMPs, including BMP-2, BMP-4, BMP-6, BMP-7, BMP-12, BMP-15 and GDF9a, is reduced in breast cancer. The decreased expression of BMP-2, BMP-7, GDF9a and BMP-15 is associated with poor prognosis [52–54], which has also been found in other studies [55, 56]. BMP-7 is reduced in primary tumours with bone metastasis [57]. However, other studies have linked overexpression of BMP-2, BMP-4, BMP-5 and BMP-7 with breast cancer [52, 58–61]. These contrasting findings suggest that BMPs may play different roles during the development and progression of breast cancer.

Certain BMP receptors have been assessed for their involvement in breast cancer. An increased expression of BMPR-IB has been observed in poorly differentiated tumours, with a higher rate of proliferation and cytogenetic instability. The overexpression of BMPR-IB was also associated with poor prognosis in ER-positive carcinomas [62]. This suggests that the expression of this type I receptor may be associated with the ER status and is regulated by oestrogen. Moreover, our previous study in a breast cancer cohort from the University Hospital of Wales showed that a decreased expression of BMPR-IB was associated with poor prognosis [63]. Differences in the ER status may be a reason for these conflicting findings. Activated Smad1/5/8 and Smad 2 were observed in nuclei of breast cancer cells from both primary tumours and bone metastases. This is supported by findings from an in vivo murine model [64]. TGF-β3 and BMP-2 promoted invasion of MDA-231-D cells, where a blockage of the TGF-β and/or BMP signalling by expression of domain-negative receptors eliminated the TGF-β3- and BMP-2-induced invasion and TGF-β3- and BMP-2-associated bone metastasis. It suggests that BMPs and TGF-ß may synergistically work together to promote the invasion and bone metastasis of breast cancer [64].

9.3.2 Regulatory Aspects of BMP Signalling

The diversity of BMP expression and signalling occurs in malignancies throughout their development and progression, reflecting the temporal and contextual nature of BMP influence. The additional complexity is both the regulatory machinery for BMPs and their interactions with other factors. A number of hormones and growth factors have been indicated within the BMP signalling networks.

9.3.2.1 Crosstalk with Oestrogen Receptor Signalling

Hormone receptor status may have great influence on aberrations in BMP phenotype and signalling with self-adjustment by tumour cells themselves, according to their needs for development and progression at different stages. Indeed, epigenetic regulation of BMPs and BMPRs in breast cancer is associated with ER status [65]. Oestrogen represses the expression of BMPR-IA, BMPR-IB, ActRIIA and ActRIIB but not ActRI and BMPR-II [66]. The expression of BMP-7 has been found to highly correlate with the expression levels of ER, although BMP-7 expression is reduced in response to oestrogen [67, 68] and BMP-2 expression is significantly higher in ER-negative tumours [69].

Oestrogen and BMPs can influence each other's function through interactions between receptors and downstream signalling [70, 71]. For example, oestrogen interferes with the biological function of BMP-2 by inhibiting the activation of Smad, as a result of biochemical interaction between Smad and ER [70]. Conversely, BMP signalling can affect ER function, as Smad 4 prevents the transcriptional regulation mediated by cytoplasmic ER [71], and BMP-2 inhibits oestradiol-induced proliferation of ER-positive breast cancer cells via upregulation of cyclin kinase inhibitor p21, which in turn inhibits the oestradiol-induced cyclin D1-associated kinase activity [72].

Hypermethylation of BMP-6 and its reduced expression have been observed in ER-negative breast cancer tissues [65]. Methylation of the BMP-6 gene promoter has been detected in ER-negative cell lines, whilst in ER-positive cells, the BMP-6 gene promoter remains demethylated. Studies show overexpression of BMP-6 particularly in ER-positive cell lines and tumour samples [65, 73]. Further in vitro study has demonstrated the interaction of ER with sites on the BMP-6 promoter region [65]. This suggests that ER status is linked with BMP expression at epigenetic level.

9.3.2.2 Crosstalk with Androgen Receptor Signalling

In breast cancer, the androgen receptor (AR) has received increasing attention related to treatment resistance, and its expression has been linked to both good and poor prognosis [74]. In tumours responsive to neoadjuvant endocrine therapy, AR mRNA and protein expression is decreased, whilst this is not seen in treatment-resistant tumours. In a clinical cohort, a high AR/ER ratio was shown as an independent risk for failure of tamoxifen treatment and poor survival. The finding has been corroborated by both in vitro and in vivo studies on breast cancer models, whereby AR overexpression is shown to increase tamoxifen resistance [75]. The underlying mechanisms regarding interactions between BMP signalling and AR are not yet clear in breast cancer. However, it would be a novel area to explore for possible targeted therapy particularly for endocrine treatment-resistant breast cancers.

9.3.2.3 Crosstalk with Growth Factor Signalling

Several other factors and pathways have been indicated in the regulation of BMP expression and function. BMP-4 generally inhibits breast cancer cell growth but enhances proliferation of breast cancer cells induced by fibroblast growth factor (FGF), epidermal growth factor (EGF) and HGF [76]. EGF treatment of breast cancer cells in vitro upregulates BMP-4 signalling, leading to suppression of matrix metalloprotease (MMP) 9. This effect was reduced when treated with BMP-4 antagonists Gremlin or Smad 6 [77]. In addition, BMP-6 in breast cancer cells can be upregulated by EGF and other EGF receptor (EGFR) ligands [55]. Conversely, EGF-, FGF- and HGF-activated MAPK/ERKs phosphorylate a linking region of Smad1/5/8, resulting in reduced nuclear translocation and transcription of BMP target genes [78, 79]. BMPs also exert reciprocal effects, suppressing EGF-induced gene transcription through MAPK/ERK-1 signalling [80]. BMP-9 decreases human epidermal growth factor receptor 2 (HER2) expression, inactivating ERK1/2 and phosphoinositide 3-kinase (PI3K)/AKT signalling pathways and leading to reduced proliferation and metastasis of SK-BR-3 breast cancer cells [81].

There is also interaction between Wnt and BMP signalling. SOSTDC1, a secreted regulator of both pathways, is under-expressed in breast cancer tissue and breast cancer cells. SOSTDC1 increases Wnt3a signalling and decreases BMP-7 signalling whilst eliciting little effect on BMP-2-induced signalling [82].

Nacamuli et al. demonstrated that BMP-3 expression can be controlled by recombinant human FGF in calvarial osteoblasts [83]. Retinoid has been shown to induce expression of BMP-2 in the retinoid-sensitive cell lines [84]. Rapamycin induces BMP-4 and downregulates BMP antagonist Follistatin expression in a prostate cancer cell line (PC3) [85]. Our previous studies showed that HGF upregulated the expression of BMP-7 and BMP receptors in prostate cancer cells. These upregulations were blocked by NK4, an antagonist of HGF [86, 87]. HGF-regulated BMP and BMP signalling may form a part of its contribution to the disease progression and bone metastasis. These studies collectively indicate that BMPs, together with other growth factors, form a collaborative interacting network during the development and progression of cancer, which would be worthy of further study, particularly given how important receptor status has become in breast cancer prognosis and treatments.

9.3.3 BMP Signalling in the Predisposition of Metastasis to Bone and Formation of Osteolytic Lesions

BMPs are the most powerful bone inductive factors which are abundant in bone matrix. In a bone metastatic lesion, BMPs can be synthesised by osteoblasts and stored in bone matrix. In addition, cancer cells can release BMPs and their antagonists to coordinate their functions. Secreted from cancer cells, BMPs contribute to bone lesion by targeting bone cells and in turn enhance aggressiveness of cancer cells. BMPs can also indirectly support the colonisation and development of bone metastasis by promoting tumour-associated angiogenesis, which makes them key factors in the 'vicious cycle' of bone metastasis. Both clinical and experimental studies have suggested profound roles for BMPs in the bone metastasis of breast cancer.

9.3.3.1 Profile of BMPs in Bone Dissemination and the Metastatic Bone Microenvironment

Decreased expression of BMP-7 in primary tumours correlates with bone metastases, whilst BMP-7 is capable of inhibiting the growth of breast cancer tumours in bone in vivo [57]. Other studies have shown BMP-7 overexpression in primary tumours is associated with bone metastases [68]. In murine 4T1E/M3 mammary cells, which are highly metastatic to bone, expression of BMP-7, BMPR and phosphorylated Smad1/5/8 are upregulated. These highly invasive features are attenuated when BMP-7 is inhibited [57].

Other studies have found that BMP-induced transcriptional pathways are active in bone metastatic lesions in vivo, and xenograft tumours with dominant negative BMP receptors have fewer bone metastases [64].

BMP-9 suppresses the growth of breast tumour cells in bone, mediated by BMP-9-induced downregulation of connective tissue growth factor (CTGF) [88, 89]. Orthotropic implant of tumours with silk scaffolds coupled with BMP-2 and seeded with bone marrow stromal cells (BMSC), contributed to bone metastasis of breast cancer cells in vivo [90].

Breast cancer cells themselves can display an osteoblast-like phenotype by expressing bone matrix proteins such as bone sialoprotein (BSP), osteopontin (OPN), OPG and osteoblast-specific cadherins [91–93]. Breast cancer cells with induced epithelial-to-mesenchymal transition (EMT) exhibited an elevated level of bone-related genes (BRGs) and osteoblast-like features when exposed to BMP-2. Breast cancer cells expressing these BRGs favoured spread and survival in the bone. Interestingly, the cells were also more resistant to chemotherapy [93]. The BMP antagonist Noggin reversed these effects, as did knockdown of runt-related transcription factor 2 (RUNX2), which regulates bone remodelling and osteogenic differentiation [68, 93, 94]. This 'bone signature' induced by BMPs may be one of the reasons breast cancer cells home to bone tissue. Once the breast cancer cell is established in the bone, BMPs and their antagonists continue to influence survival of the tumour within the microenvironment [95].

9.3.3.2 Regulation of BMPs in Bone Microenvironment

Local factors such as sexual hormones may play a role in regulation and adaptable expression of BMPs in bone metastases. The selective oestrogen receptor modulator (SERM) raloxifene increases BMP-4 promoter activity in U-2 OS osteoblast-like cells. ER is thought to be indispensable for this effect [96]. In addition, oestradiol enhances BMP-4-induced expression of osteoblastic markers (RUNX2, osterix, osteocalcin) in osteoprogenitor cells [97].

In osteoblasts, BMP-6 reporter activity is increased with antioestrogen treatment and decreased with oestradiol treatment, which provides evidence that ER regulates BMP-6 differentially in the breast and bone. Patients with ER-positive breast tumours are more likely to develop skeletal metastases [73], and this interaction between ER and BMP signalling may be the key influence on skeletal secondary formation in breast cancer.

BMP antagonists also appear to have a significant role in bone metastasis. Conditioned medium (CM) from breast cancer cells (HT-39) resulted in upregulation of BSP mRNA expression in osteoprogenitor cells (MC3T3-E1 cells) and a promotion of their osteoblastic behaviour. This effect was blocked by the addition of BMP antagonist Noggin [98]. High expression levels of Noggin are associated with bone metastases in both cell line/murine models and clinical samples of breast cancer bone metastases [99]. Upregulation of Noggin and another antagonist Follistatin, by zinc finger E-box-binding homeobox 1 (ZEB1) in breast cancer cells, induces differentiation of osteoclasts in vitro, which suggests an osteolytic influence in the bone microenvironment [100].

Another recent study has also demonstrated that lack of Noggin expression in both breast and prostate cancer cells is associated with osteoblastic activities in bone metastases. Overexpression of Noggin in an osteo-inductive prostate cancer cell line (C4-2B) inhibited osteoblastic activities but had little effect on bone resorption and tumour growth [101]. BMPs and their antagonists evidently play a role in coordinating the osteoblastic and osteolytic activities in bone metastatic lesions and thus necessitate further study, particularly in regard to therapeutic potential.

9.3.4 Therapeutic Potential of Targeting BMPs

We require agents that act to prevent or resolve bone metastasis, and in this respect, BMPs/BMP antagonists are largely underexplored. Both clinical and experimental studies suggest profound potential for targeting BMPs in treating breast cancer and bone metastasis. BMPs not only directly affect cancer cells to coordinate their abilities during disease progression and bone metastasis but also indirectly contribute to bone metastasis through regulating tumour-related angiogenesis and the bone microenvironment.

In an in vivo bone tumour model, exposure of tumour-bearing subjects to Noggin, an antagonist of BMPs, reduces the size of bone lesions by a mechanism that involves both osteoblastic and osteolytic processes. The BMP antagonists, Noggin and Follistatin, are also determining factors of the cells response to BMPs. Expression of these antagonists can be regulated by BMPs themselves probably through an autocrine or paracrine feedback loop. A good example is BMP-7, whose endogenous expression is intimately linked to the levels of Noggin and Follistatin in the same cell [102]. These findings collectively indicate the value of BMPs and their antagonists in the management of tumour progression and bone metastasis.

9.3.4.1 BMP Receptor Inhibitors

The BMP type I receptor small molecule inhibitors dorsomorphin and LDN 193189 have been used in several breast cancer studies to abrogate BMP signalling, appearing to reverse stemlike features in breast cancer cells and reduce invasiveness. Their clinical testing is yet to be further developed, and targeting the pathway downstream of the receptors still needs to be explored.

However, clinical trials are already underway for blocking ALK-1. ALK-1 inhibitors block the interaction of BMP-9 and BMP-10 with ALK-1, interrupting the subsequent intracellular signalling pathway. PF-03446962 is an ALK-1-specific neutralising antibody currently being evaluated in Phase II trials for solid tumours as an anti-angiogenic treatment [103]. Dalantercept is a soluble chimeric ALK-1 receptor-like protein (ALK1-Fc), which displays high-affinity binding with BMP-9 and BMP-10, resulting in inhibition of angiogenesis and suppression of tumour growth [104]. Initial studies showed that ALK1-Fc decreased metastasis formation in a breast cancer model [105]. In mice, treatment with ALK1-Fc seemed to remodel tumour vasculature, with increased perfusion and reduced hypoxia. A temporary improvement of tumour perfusion could result in a better delivery and efficacy of chemotherapy. Indeed, pretreatment with ALK1-Fc allowed tumours to be more sensitive to cisplatin, which could repress disease progression [104].

9.3.4.2 BMP/DKK1 Inhibitors

Within the bone environment, Dickkopf 1 (DKK1) is a downstream molecule of BMP signalling that inhibits canonical Wnt signalling and therefore negatively regulates bone mass. Tumour production of DKK1 is thought to contribute to osteolytic bone lesions [14, 106]. A DKK1-neutralising antibody is in clinical trials for multiple myeloma. Bortezomib is a proteasome inhibitor which inhibits osteoclast formation and bone resorption whilst enhancing osteoblastic differentiation and mineralisation in vitro. The detailed mechanisms are unclear but may result from decreased DKK1. The fact that BMP signalling acts upstream makes BMP antagonism and interaction with Wnt signalling a future area of exploration for bone metastases therapeutics [14, 107].

9.3.4.3 mTOR Inhibitors

Another area of therapeutic interest more recently is the PI3K-Akt-mechanistic target of rapamycin (mTOR) pathway – a key mediator of cellular proliferation, apoptosis, migration and angiogenesis, which is commonly activated in breast cancer, conferring resistance to hormonal therapy and trastuzumab. In breast cancer models, BMP-2 induces PI3K in osteoblasts to regulate differentiation. Blocking the PI3K-Akt-mTOR pathway suppresses RANKL and increases OPG secretion by the bone marrow stroma, which reduces osteoclast activity. mTOR inhibitors are part of ongoing trials regarding hormone receptor-positive, treatment-resistant tumours. The relationship of BMPs with this pathway and the apparent involvement of PI3K/mTOR in the bone provide intriguing prospects for the treatment of bone metastasis [108, 109] .

9.4 Activated Leukocyte Cell Adhesion Molecule (ALCAM) in Bone Metastasis

Current projects within our laboratories have highlighted a number of proteins and pathways involved in regulating metastatic characteristics and their potential importance in the development of bone metastasis. One such candidate is activated leukocyte cell adhesion molecule (ALCAM).

9.4.1 Discovery and Characterisation of ALCAM

Bowen et al. first identified and characterised ALCAM in 1995 and subsequently mapped it to chromosome 3q13.1–q13.2 [110]. ALCAM has been reported to be identical to MEMD, a cell adhesion molecule found to be preferentially expressed in metastasising melanoma cell lines compared to non-metastasising lines [111]. ALCAM, also known as CD166, is a member of the immunoglobulin superfamily and is involved in mediating homophilic (ALCAM-ALCAM) and heterophilic (ALCAM-CD6) interactions [110, 111]. Members of this family are characterised by the presence of five NH_2 terminals, extracellular immunoglobulin domains comprising two membrane distal variable (V)-type folds and three membrane proximal constant (C2) folds, a transmembrane region and a short cytoplasmic region [112]. The membrane distal domain 1 appears to be important for homophilic binding, whilst the membrane proximal (C2 fold) domain 4 and 5 appear to be important for avidity and ALCAM clustering on the membrane [112, 113]. As with other cell adhesion molecule, ALCAM has been linked with a number of physiological functions but has also been implicated in cancer progression, attracting considerable scientific attention.

9.4.2 Metastatic Potential of ALCAM and Clinical Implications in Breast Cancer

The role played by ALCAM in cancer progression appears to be highly complex. Despite significant research into its expression profile in cancer progression, it still remains unclear as to the precise function or expressional alterations of ALCAM in cancer progression. One factor potentially influencing this complexity is the capacity for ALCAM to exist at a number of cellular and extracellular locations. None the less, there are many contrasting reports highlighting the prognostic potential of ALCAM expression. Such examples, focusing on cellular expression of ALCAM in breast cancer reports, have been summarised in Table 9.1, though similar observations are made within a number of other cancer types as well. Hence, it is apparent from such studies that ALCAM plays a significant, if somewhat unclear, role in breast cancer progression, and with further understanding, it could hold potential as a biomarker or therapeutic strategy. The potential of ALCAM as a prognostic factor is also strengthened due to the capacity of a shed/secreted form being detectable in patient serum. ALCAM can be proteolytically shed into the surrounding extracellular environment by proteases such as A disintegrin and A metalloproteinase 17/tumour necrosis factor-alpha-converting enzyme (ADAM17/TACE) [114]. Unlike cellular ALCAM, a clear trend has emerged within the literature, and elevated serum ALCAM has been detected in breast cancer patients. Serum ALCAM has also been shown to be enhanced in higher-grade breast cancers, and current data indicates it may be a more suitable serum marker than the current established markers, CA15-3 and CEA in breast cancer [115–117].

A number of cell-based studies have also explored the functional significance of ALCAM in breast cancer cell lines. ALCAM has been suggested as an important player in programmed cell death and apoptosis. A previous study has identified a protective effect of ALCAM against programmed cell death in breast cancer cells and demonstrated that the overexpression of BCL2 could enhance ALCAM expression and induce apoptosis/autophagy following the silencing of ALCAM. Furthermore, the study highlighted that ALCAM expression might be inhibited by tamoxifen and enhanced by 17-β oestradiol in MCF-7 cells [118]. A further study characterised ALCAM in MDA-MB-231 and MCF-7 breast cancer cell lines and generated knockdown and overexpression models, respectively. The study did not detect any differences in growth rates of such cells, though ALCAM did appear to influence apoptosis. The study also described an enhanced migratory potential of ALCAM-suppressed MDA-MB-231 cells and, in keeping, a reduced level of migration in MCF-7 cells overexpressing ALCAM. However, the invasive potential of MDA-MB-231 knockdown cells was reduced, and no significant impact on invasion was shown in the MCF-7 overexpression line [119]. Another study isolated a human monoclonal antibody, recognising ALCAM (scFv173),

Table 9.1 Potential prognostic implication of ALCAM expression in clinical breast cancer tissues

Key finding/prognostic value	References
IHC and qPCR analysis was used to examine ALCAM expression in breast cancer ($n = 120$) and normal breast ($n = 32$) tissues. Stronger membranous and cytoplasmic ALCAM staining intensity were seen in normal tissue compared to breast tumour tissue. Transcript analysis suggested slightly higher ALCAM levels in tumour vs. normal breast samples. Low ALCAM transcript analysis in primary tumour was associated with higher grade and NPI stage, nodal involvement, local recurrence, death due to breast cancer and poorer disease-free survival	[129]
IHC staining of breast carcinomas ($n = 162$) indicated higher ALCAM expression in invasive and intraductal cancers compared to normal breast tissues. High ALCAM cytoplasmic staining was associated with reduced patient disease-free survival times and earlier disease progression. High ALCAM membranous staining was associated with reduced overall patient survival	[130]
Laser scanning cytometry and confocal microscopy analysis of breast cancer samples ($n = 56$) indicating high ALCAM expression was significantly correlated with small tumour diameter, low grade and oestrogen and progesterone receptor positivity. Low ALCAM expression tended to associate with HER2 amplification. ALCAM/MMP-2 ratio may hold potential indicator of progression, with higher ratios seen in low-grade and small tumour diameter samples	[131]
ALCAM protein ($n = 160$) and mRNA ($n = 162$) expression in primary mammary carcinomas was analysed. ALCAM protein expression correlated with ER status. In patients treated with adjuvant chemotherapy, high ALCAM protein expression was generally associated with longer disease-free and overall survival, and this observation was more apparent at the mRNA level, where the association with overall survival was significant. Cox regression analysis indicated high ALCAM mRNA was a predictor of longer overall survival in chemotherapy-treated patients	[132]
IHC and qPCR analysis were used to assess ALCAM expression in primary breast cancer ($n = 243$) and non-neoplastic mammary tissues ($n = 34$). ALCAM staining was reduced in cancerous tissue and tissue from patients who developed skeletal metastasis compared to normal tissues. Lower ALCAM transcript expression was also associated with poorer patient prognosis, poor NPI, local recurrence, metastasis, skeletal metastasis and death	[124]
IHC analysis of FFPE tissues from 29 autopsy cases (primary $n = 25$ and related distant metastasis $n = 84$) demonstrated that ALCAM staining intensities in primary tumour and metastasis of the same patient were similar, and ALCAM expression in the primary tumour was prognostic for ALCAM staining in metastasis of the patient	[133]
mRNA analysis of ALCAM, osteopontin, HER2 and ER in breast cancer tissues ($n = 481$) identified that in low or negative HER2/ER samples, high osteopontin and low ALCAM transcript expression aided in identification of high-risk patients with shorter disease-free and overall survival rates	[134]
IHC staining on breast cancer patient samples ($n = 347$) demonstrated ALCAM was associated with oestrogen and progesterone receptor status. ALCAM overexpression significantly correlated with nodal involvement and tended to be associated with the presence of disseminated tumour cells in the bone marrow. Strong ALCAM expression also correlated with reduced recurrence and overall survivals in ductal carcinomas	[119]
Significant association was noted between ALCAM polymorphism (rs6437585T > C in ALCAM promoter) and risk of developing breast cancer in a Chinese population	[135]
IHC ALCAM staining intensity in tissue obtained from breast cancer metastasis ($n = 117$) was found to be higher in skin metastasis than in any other metastatic tissues examined (bone, liver, brain and lung), and ALCAM staining was also found to be higher in primary tumours which metastasised to the skin	[136]
Protein analysis of ALCAM expression in breast cancer patients ($n = 150$) indicated high expression of ALCAM at the membrane was associated with metastatic dissemination and lymph node metastasis, whereas increased ALCAM expression in the cytoplasm was associated with short-term local recurrence and shortened disease-free patient survival	[137]
Two SNPs within the ALCAM gene (re1044243 and rs1157) were associated with breast cancer-specific survival when analysed in a Swedish population-based series ($n = 783$) of breast cancer cases. However, no association was observed within a Polish population of familial/early-onset breast cancer cases ($n = 506$)	[138]

(continued)

Table 9.1 (continued)

Key finding/prognostic value	References
IHC analysis of ALCAM in 173 cases (African American $n = 78$ and Caucasian $n = 95$) of breast cancer indicated that, in both ethnic groups, lower ALCAM expression at intercellular junctions correlated with grade, ER, PR and triple negative status and contributed to a more aggressive phenotype. Ethnicity also significantly contributed to ALCAM expression after accounting for other factors, with the African American group more likely to have low ALCAM expression than the Caucasian ethnic group	[139]
IHC analysis of TMA ($n = 2197$) demonstrated membranous ALCAM expression in both normal and cancerous breast tissues. A decreased expression of ALCAM was found to be associated with negative ER and PR status, high Ki67 index, advanced tumour size and grade and cancers with loss of ALCAM expression resulted in significantly poorer disease-free and overall survival rates	[140]
IHC analysis of primary ER-positive breast cancer tissues from tamoxifen responders ($n = 16$) and nonresponders ($n = 20$) suggested higher ALCAM staining in the nonresponders	[141]

which could bind ALCAM on both cancer cell lines and in tumour tissues. It reported that the addition of scFv173 could inhibit the invasive potential of MDA-MB-231 in vitro and reduce tumour development of a colorectal carcinoma cell line (HCT116) in vivo [120].

To further understand the role of ALCAM in breast cancer, a number of studies have explored the potential mechanisms responsible for controlling ALCAM expression. King et al. have reported that DNA methylation of the ALCAM promoter is one such mechanism influencing ALCAM expression and that this may be significant factor in regulating ALCAM expression in tumour tissue. Furthermore, such a loss may inhibit adherence between circulating tumour cells, therefore supporting a role for ALCAM loss in enhancing metastatic potential [121]. Recently, the regulation of ALCAM expression by microRNAs has been reported in breast cancer. ALCAM was found to be one of a panel of genes whose expression was altered following expression of miR-125b. Expression of miR-125b enhanced both ALCAM mRNA and protein levels, which was found to influence MCF-7 growth using a further shRNA study [122]. It has also recently been reported that inhibition of miR-214, overexpression of miR-148b or a combination can inhibit tumour cell crossing of the vessel endothelium through a negative regulation of ALCAM and integrin α5 [123].

9.4.3 Potential Role for ALCAM in Bone Metastasis

Though complex, the literature supports a role for ALCAM in the progression and metastatic dissemination of cancer. To elucidate the potential of ALCAM in influencing the development of bone metastasis, our laboratories further explored ALCAM in a larger, combined breast cancer cohort and examined the association between ALCAM expression and the development of bone metastasis [124]. In keeping with our previous findings, through immunohistochemical analysis, lower ALCAM cytoplasmic expression was noted in breast cancer tissues and in tissue sections from patients who went on to develop skeletal metastasis compared to normal breast tissue. Furthermore, quantitative PCR analysis similarly indicated that significantly lower ALCAM transcript expression was associated with patients with poorer prognostic indicators and that low ALCAM expression was associated with those patients who went on to develop skeletal metastasis. This trend was similarly observed when focusing on ductal carcinoma cases alone [124]. To further explore this potential link, our laboratories examined ALCAM overexpression and knockdown models in MDA-MB-231 and ZR-751 cell lines, respectively, and explored the in vitro impact of culturing such cells in the presence of a bone matrix extract (BME) generated from ground and sonicated femoral heads. Such experiments highlighted a role for ALCAM expression in negatively regulating cell growth and matrix adhesion and migration and underscored a potential relationship between ALCAM expression and the responsiveness of MDA-MB-231 cells to the bone extract, particularly in terms of growth and migratory responses where ALCAM overexpression in the presence of bone matrix extracts brought about greater reductions in such traits [125]. Taken together, these two studies by Davies et al. suggest a potential inhibitory role for ALCAM in the development of bone metastasis. However, an additional study by Hein et al. examining ALCAM immunostaining in a tissue microarray suggested that high ALCAM staining correlated with ER positivity, nodal involvement and the presence of disseminated tumour cells within the bone marrow environment. Furthermore, alteration of ALCAM levels in MDA-MB-231 and MCF-7 cells caused differential expression of a number of molecules including cathepsin D and RUNX2 [119], both of which have implications in bone malignancies and metastasis [126, 127]. Further evidence supportive of a regulator role for ALCAM in bone dissemination was recently reported in a study by Hansen et al. in a prostate cancer model [128]. In their study, Hansen et al. reported that the shedding and detection of tumour-derived ALCAM was significantly elevated in tumour-bearing mice and that reduced ALCAM levels could significantly reduce the incident and metastatic burden of bone metastasis following intracardiac seeding of cells [128].

Given the significant impact of metastatic dissemination and establishment of tumour cells in the bone on patient well-being and ultimately survival, there is a great need to identify and utilise the responsible mechanisms for the development of new therapeutic strategies. ALCAM, a molecule linked to cancer progression and metastasis, though in a somewhat complex fashion, represents an interesting example of one such novel strategy. From early indications, ALCAM is likely to play a role in cancer cell metastasis and development in the bone environment, though additional work is required to further determine the effect of this molecule on this process. Furthermore, the detection of secreted or shed ALCAM in the serum of cancer patients may potentially present a relatively non-invasive biomarker to monitor patients. Therefore, further large-scale study is required to identify and utilise the full potential of ALCAM to monitor cancer dissemination to the bone.

9.5 Concluding Remarks

Targeting bone-associated molecules in the treatment of breast cancer-associated metastases is not a simple or quick fix. With such a rich and diverse environment for molecular targets, opportunities to target bone metastasis are vast. Consideration has been given to treat breast cancer patients in the first-line treatment with denosumab, in the hope of targeting any breast cancer cells which have already become resident in the bone environment. However, such a sledge hammer approach is not sustainable in the long term, as potentially identified targets such as OPG have demonstrated that the anti-tumorigenic benefits it exert on one area of the body may result in detrimental effects elsewhere. Therefore, ongoing efforts are essential for seeking factors which could identify patients at greatest risk of developing bone metastasis or a serum marker which could provide insight into the development of bone metastasis. The answers may not lie in classically identified molecules but in newly identified agents such as miRNAs or emerging regulators such as ALCAM. Complete elucidation of the molecules associated with the development of breast cancer-associated bone metastasis, their interactions and effects on the bone microenvironment is critical to achieve success of developing any future prognostic and therapeutic approach.

Acknowledgements Authors would like to thank Cancer Research Wales and the Life Sciences Research Network Wales for supporting their work.

References

1. Paget S (1989) The distribution of secondary growths in cancer of the breast. 1889. Cancer Metastasis Rev 8(2):98–101
2. Yardley DA (2016) Pharmacologic management of bone-related complications and bone metastases in postmenopausal women with hormone receptor-positive breast cancer. Breast Cancer (Dove Med Press) 8:73–82. doi:10.2147/BCTT.S97963
3. Briasoulis E, Karavasilis V, Kostadima L, Ignatiadis M, Fountzilas G, Pavlidis N (2004) Metastatic breast carcinoma confined to bone: portrait of a clinical entity. Cancer 101(7):1524–1528. doi:10.1002/cncr.20545
4. Harries M, Taylor A, Holmberg L, Agbaje O, Garmo H, Kabilan S, Purushotham A (2014) Incidence of bone metastases and survival after a diagnosis of bone metastases in breast cancer patients. Cancer Epidemiol 38(4):427–434. doi:10.1016/j.canep.2014.05.005
5. Fidler IJ (1970) Metastasis: quantitative analysis of distribution and fate of tumor emboli labeled with 125 I-5-iodo-2′-deoxyuridine. J Natl Cancer Inst 45(4):773–782
6. Kasimir-Bauer S (2009) Circulating tumor cells as markers for cancer risk assessment and treatment monitoring. Mol Diagn Ther 13(4):209–215. doi:10.2165/11315870-000000000-00000
7. Talmadge JE, Fidler IJ (2010) AACR centennial series: the biology of cancer metastasis: historical perspective. Cancer Res 70(14):5649–5669. doi:10.1158/0008-5472.CAN-10-1040
8. Weilbaecher KN, Guise TA, McCauley LK (2011) Cancer to bone: a fatal attraction. Nat Rev Cancer 11(6):411–425. doi:10.1038/nrc3055
9. Guise TA (2013) Breast cancer bone metastases: it's all about the neighborhood. Cell 154(5):957–959. doi:10.1016/j.cell.2013.08.020
10. Mundy GR (2002) Metastasis to bone: causes, consequences and therapeutic opportunities. Nat Rev Cancer 2(8):584–593. doi:10.1038/nrc867
11. Reddi AH, Roodman D, Freeman C, Mohla S (2003) Mechanisms of tumor metastasis to the bone: challenges and opportunities. J Bone Miner Res 18(2):190–194
12. Steger GG, Bartsch R (2011) Denosumab for the treatment of bone metastases in breast cancer: evidence and opinion. Ther Adv Med Oncol 3(5):233–243. doi:10.1177/1758834011412656
13. Roodman GD (2004) Mechanisms of bone metastasis. Discov Med 4(22):144–148
14. Lipton A, Uzzo R, Amato RJ, Ellis GK, Hakimian B, Roodman GD, Smith MR (2009) The science and practice of bone health in oncology: managing bone loss and metastasis in patients with solid tumors. J Natl Compr Canc Netw 7(Suppl 7):S1-29:quiz S30
15. Epker BN, Frost HM (1965) Correlation of bone resorption and formation with the physical behavior of loaded bone. J Dent Res 44:33–41
16. Simonet WS, Lacey DL, Dunstan CR, Kelley M, Chang MS, Luthy R, Nguyen HQ, Wooden S, Bennett L, Boone T, Shimamoto G, DeRose M, Elliott R, Colombero A, Tan HL, Trail G, Sullivan J, Davy E, Bucay N, Renshaw-Gegg L, Hughes TM, Hill D, Pattison W, Campbell P, Sander S, Van G, Tarpley J, Derby P, Lee R, Boyle WJ (1997) Osteoprotegerin: a novel secreted protein involved in the regulation of bone density. Cell 89(2):309–319
17. Tsuda E, Goto M, Mochizuki S, Yano K, Kobayashi F, Morinaga T, Higashio K (1997) Isolation of a novel cytokine from human fibroblasts that specifically inhibits osteoclastogenesis. Biochem Biophys Res Commun 234(1):137–142
18. Anderson DM, Maraskovsky E, Billingsley WL, Dougall WC, Tometsko ME, Roux ER, Teepe MC, DuBose RF, Cosman D, Galibert L (1997) A homologue of the TNF receptor and its ligand enhance T-cell growth and dendritic-cell function. Nature 390(6656):175–179. doi:10.1038/36593
19. Dougall WC (2012) Molecular pathways: osteoclast-dependent and osteoclast-independent roles of the RANKL/RANK/OPG pathway in tumorigenesis and metastasis. Clin Cancer Res 18(2):326–335. doi:10.1158/1078-0432.CCR-10-2507
20. Karsenty G (1999) The genetic transformation of bone biology. Genes Dev 13(23):3037–3051
21. Roodman GD (2001) Biology of osteoclast activation in cancer. J Clin Oncol 19(15):3562–3571. doi:10.1200/JCO.2001.19.15.3562
22. Ross FP (2000) RANKing the importance of measles virus in Paget's disease. J Clin Invest 105(5):555–558. doi:10.1172/JCI9557
23. Beleut M, Rajaram RD, Caikovski M, Ayyanan A, Germano D, Choi Y, Schneider P, Brisken C (2010) Two distinct mechanisms underlie progesterone-induced proliferation in the mammary gland. Proc Natl Acad Sci U S A 107(7):2989–2994. doi:10.1073/pnas.0915148107
24. Gonzalez-Suarez E, Jacob AP, Jones J, Miller R, Roudier-Meyer MP, Erwert R, Pinkas J, Branstetter D, Dougall WC (2010) RANK ligand mediates progestin-induced mammary epithelial proliferation and carcinogenesis. Nature 468(7320):103–107. doi:10.1038/nature09495
25. Hu H, Wang J, Gupta A, Shidfar A, Branstetter D, Lee O, Ivancic D, Sullivan M, Chatterton RT Jr, Dougall WC, Khan SA (2014) RANKL expression in normal and malignant breast tissue responds to progesterone and is up-regulated during the luteal phase. Breast Cancer Res Treat 146(3):515–523. doi:10.1007/s10549-014-3049-9
26. Kiesel L, Kohl A (2016) Role of the RANK/RANKL pathway in breast cancer. Maturitas 86:10–16. doi:10.1016/j.maturitas.2016.01.001

27. Mukherjee A, Soyal SM, Li J, Ying Y, He B, DeMayo FJ, Lydon JP (2010) Targeting RANKL to a specific subset of murine mammary epithelial cells induces ordered branching morphogenesis and alveologenesis in the absence of progesterone receptor expression. FASEB J 24(11):4408–4419. doi:10.1096/fj.10-157982
28. Fata JE, Kong YY, Li J, Sasaki T, Irie-Sasaki J, Moorehead RA, Elliott R, Scully S, Voura EB, Lacey DL, Boyle WJ, Khokha R, Penninger JM (2000) The osteoclast differentiation factor osteoprotegerin-ligand is essential for mammary gland development. Cell 103(1):41–50
29. Gonzalez-Suarez E, Branstetter D, Armstrong A, Dinh H, Blumberg H, Dougall WC (2007) RANK overexpression in transgenic mice with mouse mammary tumor virus promoter-controlled RANK increases proliferation and impairs alveolar differentiation in the mammary epithelia and disrupts lumen formation in cultured epithelial acini. Mol Cell Biol 27(4):1442–1454. doi:10.1128/MCB.01298-06
30. Pellegrini P, Cordero A, Gallego MI, Dougall WC, Munoz P, Pujana MA, Gonzalez-Suarez E (2013) Constitutive activation of RANK disrupts mammary cell fate leading to tumorigenesis. Stem Cells 31(9):1954–1965. doi:10.1002/stem.1454
31. Blake ML, Tometsko M, Miller R, Jones JC, Dougall WC (2014) RANK expression on breast cancer cells promotes skeletal metastasis. Clin Exp Metastasis 31(2):233–245. doi:10.1007/s10585-013-9624-3
32. Casimiro S, Mohammad KS, Pires R, Tato-Costa J, Alho I, Teixeira R, Carvalho A, Ribeiro S, Lipton A, Guise TA, Costa L (2013) RANKL/RANK/MMP-1 molecular triad contributes to the metastatic phenotype of breast and prostate cancer cells in vitro. PLoS One 8(5):e63153. doi:10.1371/journal.pone.0063153
33. Owen S, Sanders AJ, Mason MD, Jiang WG (2016) Importance of osteoprotegrin and receptor activator of nuclear factor kappaB in breast cancer response to hepatocyte growth factor and the bone microenvironment in vitro. Int J Oncol 48(3):919–928. doi:10.3892/ijo.2016.3339
34. Jones DH, Nakashima T, Sanchez OH, Kozieradzki I, Komarova SV, Sarosi I, Morony S, Rubin E, Sarao R, Hojilla CV, Komnenovic V, Kong YY, Schreiber M, Dixon SJ, Sims SM, Khokha R, Wada T, Penninger JM (2006) Regulation of cancer cell migration and bone metastasis by RANKL. Nature 440(7084):692–696. doi:10.1038/nature04524
35. Gonzalez-Suarez E, Sanz-Moreno A (2016) RANK as a therapeutic target in cancer. FEBS J 283(11):2018–2033. doi:10.1111/febs.13645
36. Van Poznak C, Cross SS, Saggese M, Hudis C, Panageas KS, Norton L, Coleman RE, Holen I (2006) Expression of osteoprotegerin (OPG), TNF related apoptosis inducing ligand (TRAIL), and receptor activator of nuclear factor kappaB ligand (RANKL) in human breast tumours. J Clin Pathol 59(1):56–63. doi:10.1136/jcp.2005.026534
37. Rachner TD, Schoppet M, Niebergall U, Hofbauer LC (2008) 17beta-estradiol inhibits osteoprotegerin production by the estrogen receptor-alpha-positive human breast cancer cell line MCF-7. Biochem Biophys Res Commun 368(3):736–741. doi:10.1016/j.bbrc.2008.01.118
38. Zinonos I, Labrinidis A, Lee M, Liapis V, Hay S, Ponomarev V, Diamond P, Findlay DM, Zannettino AC, Evdokiou A (2011) Anticancer efficacy of Apo2L/TRAIL is retained in the presence of high and biologically active concentrations of osteoprotegerin in vivo. J Bone Miner Res 26(3):630–643. doi:10.1002/jbmr.244
39. Holen I, Cross SS, Neville-Webbe HL, Cross NA, Balasubramanian SP, Croucher PI, Evans CA, Lippitt JM, Coleman RE, Eaton CL (2005) Osteoprotegerin (OPG) expression by breast cancer cells in vitro and breast tumours in vivo--a role in tumour cell survival? Breast Cancer Res Treat 92(3):207–215. doi:10.1007/s10549-005-2419-8
40. Rachner TD, Benad P, Rauner M, Goettsch C, Singh SK, Schoppet M, Hofbauer LC (2009) Osteoprotegerin production by breast cancer cells is suppressed by dexamethasone and confers resistance against TRAIL-induced apoptosis. J Cell Biochem 108(1):106–116. doi:10.1002/jcb.22232
41. Cody JJ, Rivera AA, Lyons GR, Yang SW, Wang M, Sarver DB, Wang D, Selander KS, Kuo HC, Meleth S, Feng X, Siegal GP, Douglas JT (2010) Arming a replicating adenovirus with osteoprotegerin reduces the tumor burden in a murine model of osteolytic bone metastases of breast cancer. Cancer Gene Ther 17(12):893–905. doi:10.1038/cgt.2010.47
42. Fisher JL, Thomas-Mudge RJ, Elliott J, Hards DK, Sims NA, Slavin J, Martin TJ, Gillespie MT (2006) Osteoprotegerin overexpression by breast cancer cells enhances orthotopic and osseous tumor growth and contrasts with that delivered therapeutically. Cancer Res 66(7):3620–3628. doi:10.1158/0008-5472.CAN-05-3119
43. Fradet A, Sorel H, Bouazza L, Goehrig D, Depalle B, Bellahcene A, Castronovo V, Follet H, Descotes F, Aubin JE, Clezardin P, Bonnelye E (2011) Dual function of ERRalpha in breast cancer and bone metastasis formation: implication of VEGF and osteoprotegerin. Cancer Res 71(17):5728–5738. doi:10.1158/0008-5472.CAN-11-1431
44. Weichhaus M, Segaran P, Renaud A, Geerts D, Connelly L (2014) Osteoprotegerin expression in triple-negative breast cancer cells promotes metastasis. Cancer Med 3(5):1112–1125. doi:10.1002/cam4.277
45. Zinonos I, Luo KW, Labrinidis A, Liapis V, Hay S, Panagopoulos V, Denichilo M, Ko CH, Yue GG, Lau CB, Ingman W, Ponomarev V, Atkins GJ, Findlay DM, Zannettino AC, Evdokiou A (2014) Pharmacologic inhibition of bone resorption prevents cancer-induced osteolysis but enhances soft tissue metastasis in a mouse model of osteolytic breast

cancer. Int J Oncol 45(2):532–540. doi:10.3892/ijo.2014.2468
46. Owen S, Ye L, Sanders AJ, Mason MD, Jiang WG (2013) Expression profile of receptor activator of nuclear-kappaB (RANK), RANK ligand (RANKL) and osteoprotegerin (OPG) in breast cancer. Anticancer Res 33(1):199–206
47. Park HS, Lee A, Chae BJ, Bae JS, Song BJ, Jung SS (2014) Expression of receptor activator of nuclear factor kappa-B as a poor prognostic marker in breast cancer. J Surg Oncol 110(7):807–812. doi:10.1002/jso.23737
48. Sanger N, Ruckhaberle E, Bianchini G, Heinrich T, Milde-Langosch K, Muller V, Rody A, Solomayer EF, Fehm T, Holtrich U, Becker S, Karn T (2014) OPG and PgR show similar cohort specific effects as prognostic factors in ER positive breast cancer. Mol Oncol 8(7):1196–1207. doi:10.1016/j.molonc.2014.04.003
49. Santini D, Schiavon G, Vincenzi B, Gaeta L, Pantano F, Russo A, Ortega C, Porta C, Galluzzo S, Armento G, La Verde N, Caroti C, Treilleux I, Ruggiero A, Perrone G, Addeo R, Clezardin P, Muda AO, Tonini G (2011) Receptor activator of NF-kB (RANK) expression in primary tumors associates with bone metastasis occurrence in breast cancer patients. PLoS One 6(4):e19234. doi:10.1371/journal.pone.0019234
50. Croft M, Benedict CA, Ware CF (2013) Clinical targeting of the TNF and TNFR superfamilies. Nat Rev Drug Discov 12(2):147–168. doi:10.1038/nrd3930
51. Ye L, Lewis-Russell JM, Kyanaston HG, Jiang WG (2007) Bone morphogenetic proteins and their receptor signaling in prostate cancer. Histol Histopathol 22(10):1129–1147
52. Davies SR, Watkins G, Douglas-Jones A, Mansel RE, Jiang WG (2008) Bone morphogenetic proteins 1 to 7 in human breast cancer, expression pattern and clinical/prognostic relevance. J Exp Ther Oncol 7(4):327–338
53. Hanavadi S, Martin TA, Watkins G, Mansel RE, Jiang WG (2007) The role of growth differentiation factor-9 (GDF-9) and its analog, GDF-9b/BMP-15, in human breast cancer. Ann Surg Oncol 14(7):2159–2166. doi:10.1245/s10434-007-9397-5
54. Li J, Ye L, Parr C, Douglas-Jones A, Kyanaston HG, Mansel RE, Jiang WG (2009) The aberrant expression of bone morphogenetic protein 12 (BMP-12) in human breast cancer and its potential prognostic value. Gene Ther Mol Biol 13:186–193
55. Clement JH, Sanger J, Hoffken K (1999) Expression of bone morphogenetic protein 6 in normal mammary tissue and breast cancer cell lines and its regulation by epidermal growth factor. Int J Cancer 80(2):250–256
56. Reinholz MM, Iturria SJ, Ingle JN, Roche PC (2002) Differential gene expression of TGF-beta family members and osteopontin in breast tumor tissue: analysis by real-time quantitative PCR. Breast Cancer Res Treat 74(3):255–269
57. Buijs JT, Henriquez NV, van Overveld PG, van der Horst G, Que I, Schwaninger R, Rentsch C, Ten Dijke P, Cleton-Jansen AM, Driouch K, Lidereau R, Bachelier R, Vukicevic S, Clezardin P, Papapoulos SE, Cecchini MG, Lowik CW, van der Pluijm G (2007) Bone morphogenetic protein 7 in the development and treatment of bone metastases from breast cancer. Cancer Res 67(18):8742–8751. doi:10.1158/0008-5472.CAN-06-2490
58. Bobinac D, Maric I, Zoricic S, Spanjol J, Dordevic G, Mustac E, Fuckar Z (2005) Expression of bone morphogenetic proteins in human metastatic prostate and breast cancer. Croat Med J 46(3):389–396
59. Raida M, Clement JH, Ameri K, Han C, Leek RD, Harris AL (2005) Expression of bone morphogenetic protein 2 in breast cancer cells inhibits hypoxic cell death. Int J Oncol 26(6):1465–1470
60. Alarmo EL, Rauta J, Kauraniemi P, Karhu R, Kuukasjarvi T, Kallioniemi A (2006) Bone morphogenetic protein 7 is widely overexpressed in primary breast cancer. Genes Chromosomes Cancer 45(4):411–419
61. Alarmo EL, Kuukasjarvi T, Karhu R, Kallioniemi A (2007) A comprehensive expression survey of bone morphogenetic proteins in breast cancer highlights the importance of BMP4 and BMP7. Breast Cancer Res Treat 103(2):239–246. doi:10.1007/s10549-006-9362-1
62. Helms MW, Packeisen J, August C, Schittek B, Boecker W, Brandt BH, Buerger H (2005) First evidence supporting a potential role for the BMP/SMAD pathway in the progression of oestrogen receptor-positive breast cancer. J Pathol 206(3):366–376
63. Bokobza S, Ye L, Kynaston H, Mansel RE, Jiang WG (2009) Reduced expression of BMPR-IB correlates with poor prognosis and increased proliferation of breast cancer cells. Cancer Genom Proteom 6(2):101–108
64. Katsuno Y, Hanyu A, Kanda H, Ishikawa Y, Akiyama F, Iwase T, Ogata E, Ehata S, Miyazono K, Imamura T (2008) Bone morphogenetic protein signaling enhances invasion and bone metastasis of breast cancer cells through Smad pathway. Oncogene 27(49):6322–6333. doi:10.1038/onc.2008.232
65. Zhang M, Wang Q, Yuan W, Yang S, Wang X, Yan JD, Du J, Yin J, Gao SY, Sun BC, Zhu TH (2007) Epigenetic regulation of bone morphogenetic protein-6 gene expression in breast cancer cells. J Steroid Biochem Mol Biol 105(1–5):91–97. doi:10.1016/j.jsbmb.2007.01.002
66. Takahashi M, Otsuka F, Miyoshi T, Otani H, Goto J, Yamashita M, Ogura T, Makino H, Doihara H (2008) Bone morphogenetic protein 6 (BMP6) and BMP7 inhibit estrogen-induced proliferation of breast cancer cells by suppressing p38 mitogen-activated protein kinase activation. J Endocrinol 199(3):445–455. doi:10.1677/JOE-08-0226
67. Schwalbe M, Sanger J, Eggers R, Naumann A, Schmidt A, Hoffken K, Clement JH (2003) Differential expression and regulation of bone mor-

phogenetic protein 7 in breast cancer. Int J Oncol 23(1):89–95
68. Alarmo EL, Kallioniemi A (2010) Bone morphogenetic proteins in breast cancer: dual role in tumourigenesis? Endocr Relat Cancer 17(2):R123–R139. doi:10.1677/ERC-09-0273
69. Julien S, Ivetic A, Grigoriadis A, QiZe D, Burford B, Sproviero D, Picco G, Gillett C, Papp SL, Schaffer L, Tutt A, Taylor-Papadimitriou J, Pinder SE, Burchell JM (2011) Selectin ligand sialyl-Lewis x antigen drives metastasis of hormone-dependent breast cancers. Cancer Res 71(24):7683–7693. doi:10.1158/0008-5472.CAN-11-1139
70. Yamamoto T, Saatcioglu F, Matsuda T (2002) Cross-talk between bone morphogenic proteins and estrogen receptor signaling. Endocrinology 143(7):2635–2642. doi:10.1210/endo.143.7.8877
71. Wu L, Wu Y, Gathings B, Wan M, Li X, Grizzle W, Liu Z, Lu C, Mao Z, Cao X (2003) Smad4 as a transcription corepressor for estrogen receptor alpha. J Biol Chem 278(17):15192–15200. doi:10.1074/jbc.M212332200
72. Ghosh-Choudhury N, Ghosh-Choudhury G, Celeste A, Ghosh PM, Moyer M, Abboud SL, Kreisberg J (2000) Bone morphogenetic protein-2 induces cyclin kinase inhibitor p21 and hypophosphorylation of retinoblastoma protein in estradiol-treated MCF-7 human breast cancer cells. Biochim Biophys Acta 1497(2):186–196
73. Ong DB, Colley SM, Norman MR, Kitazawa S, Tobias JH (2004) Transcriptional regulation of a BMP-6 promoter by estrogen receptor alpha. J Bone Miner Res 19(3):447–454. doi:10.1359/JBMR.0301249
74. Feng J, Li L, Zhang N, Liu J, Zhang L, Gao H, Wang G, Li Y, Zhang Y, Li X, Liu D, Lu J, Huang B (2016) Androgen and AR contribute to breast cancer development and metastasis: an insight of mechanisms. Oncogene. doi:10.1038/onc.2016.432
75. Cochrane DR, Bernales S, Jacobsen BM, Cittelly DM, Howe EN, D'Amato NC, Spoelstra NS, Edgerton SM, Jean A, Guerrero J, Gomez F, Medicherla S, Alfaro IE, McCullagh E, Jedlicka P, Torkko KC, Thor AD, Elias AD, Protter AA, Richer JK (2014) Role of the androgen receptor in breast cancer and preclinical analysis of enzalutamide. Breast Cancer Res 16(1):R7. doi:10.1186/bcr3599
76. Montesano R, Sarkozi R, Schramek H (2008) Bone morphogenetic protein-4 strongly potentiates growth factor-induced proliferation of mammary epithelial cells. Biochem Biophys Res Commun 374(1):164–168. doi:10.1016/j.bbrc.2008.07.007
77. Laulan NB, St-Pierre Y (2015) Bone morphogenetic protein 4 (BMP-4) and epidermal growth factor (EGF) inhibit metalloproteinase-9 (MMP-9) expression in cancer cells. Oncoscience 2(3):309–316. Doi:10.18632/oncoscience.144
78. Kretzschmar M, Doody J, Massague J (1997) Opposing BMP and EGF signalling pathways converge on the TGF-beta family mediator Smad1. Nature 389(6651):618–622. doi:10.1038/39348
79. Guo X, Wang XF (2009) Signaling cross-talk between TGF-beta/BMP and other pathways. Cell Res 19(1):71–88. doi:10.1038/cr.2008.302
80. Ghosh Choudhury G, Jin DC, Kim Y, Celeste A, Ghosh-Choudhury N, Abboud HE (1999) Bone morphogenetic protein-2 inhibits MAPK-dependent elk-1 transactivation and DNA synthesis induced by EGF in mesangial cells. Biochem Biophys Res Commun 258(2):490–496
81. Ren W, Liu Y, Wan S, Fei C, Wang W, Chen Y, Zhang Z, Wang T, Wang J, Zhou L, Weng Y, He T, Zhang Y (2014) BMP9 inhibits proliferation and metastasis of HER2-positive SK-BR-3 breast cancer cells through ERK1/2 and PI3K/AKT pathways. PLoS One 9(5):e96816. doi:10.1371/journal.pone.0096816
82. Clausen KA, Blish KR, Birse CE, Triplette MA, Kute TE, Russell GB, D'Agostino RB Jr, Miller LD, Torti FM, Torti SV (2011) SOSTDC1 differentially modulates Smad and beta-catenin activation and is downregulated in breast cancer. Breast Cancer Res Treat 129(3):737–746. doi:10.1007/s10549-010-1261-9
83. Nacamuli RP, Fong KD, Lenton KA, Song HM, Fang TD, Salim A, Longaker MT (2005) Expression and possible mechanisms of regulation of BMP3 in rat cranial sutures. Plast Reconstr Surg 116(5):1353–1362
84. Hallahan AR, Pritchard JI, Chandraratna RA, Ellenbogen RG, Geyer JR, Overland RP, Strand AD, Tapscott SJ, Olson JM (2003) BMP-2 mediates retinoid-induced apoptosis in medulloblastoma cells through a paracrine effect. Nat Med 9(8):1033–1038
85. van der Poel HG, Hanrahan C, Zhong H, Simons JW (2003) Rapamycin induces Smad activity in prostate cancer cell lines. Urol Res 30(6):380–386
86. Ye L, Lewis-Russell JM, Sanders AJ, Kynaston H, Jiang WG (2008) HGF/SF up-regulates the expression of bone morphogenetic protein 7 in prostate cancer cells. Urol Oncol 26(2):190–197. doi:10.1016/j.urolonc.2007.03.027
87. Ye L, Lewis-Russell JM, Davies G, Sanders AJ, Kynaston H, Jiang WG (2007) Hepatocyte growth factor up-regulates the expression of the bone morphogenetic protein (BMP) receptors, BMPR-IB and BMPR-II, in human prostate cancer cells. Int J Oncol 30(2):521–529
88. Ren W, Sun X, Wang K, Feng H, Liu Y, Fei C, Wan S, Wang W, Luo J, Shi Q, Tang M, Zuo G, Weng Y, He T, Zhang Y (2014) BMP9 inhibits the bone metastasis of breast cancer cells by downregulating CCN2 (connective tissue growth factor, CTGF) expression. Mol Biol Rep 41(3):1373–1383. doi:10.1007/s11033-013-2982-8
89. Wang K, Feng H, Ren W, Sun X, Luo J, Tang M, Zhou L, Weng Y, He TC, Zhang Y (2011) BMP9 inhibits the proliferation and invasiveness of breast cancer cells MDA-MB-231. J Cancer Res

Clin Oncol 137(11):1687–1696. doi:10.1007/s00432-011-1047-4
90. Moreau JE, Anderson K, Mauney JR, Nguyen T, Kaplan DL, Rosenblatt M (2007) Tissue-engineered bone serves as a target for metastasis of human breast cancer in a mouse model. Cancer Res 67 (21):10304–10308. doi:10.1158/0008-5472.CAN-07-2483
91. Kapoor P, Suva LJ, Welch DR, Donahue HJ (2008) Osteoprotegerin and the bone homing and colonization potential of breast cancer cells. J Cell Biochem 103(1):30–41. doi:10.1002/jcb.21382
92. Ibrahim T, Leong I, Sanchez-Sweatman O, Khokha R, Sodek J, Tenenbaum HC, Ganss B, Cheifetz S (2000) Expression of bone sialoprotein and osteopontin in breast cancer bone metastases. Clin Exp Metastasis 18(3):253–260
93. Tan CC, Li GX, Tan LD, Du X, Li XQ, He R, Wang QS, Feng YM (2016) Breast cancer cells obtain an osteomimetic feature via epithelial-mesenchymal transition that have undergone BMP2/RUNX2 signaling pathway induction. Oncotarget. doi:10.18632/oncotarget.12939
94. Carreira AC, Alves GG, Zambuzzi WF, Sogayar MC, Granjeiro JM (2014) Bone morphogenetic proteins: structure, biological function and therapeutic applications. Arch Biochem Biophys 561:64–73. doi:10.1016/j.abb.2014.07.011
95. Rucci N, Teti A (2010) Osteomimicry: how tumor cells try to deceive the bone. Front Biosci (Schol Ed) 2:907–915
96. van den Wijngaard A, Mulder WR, Dijkema R, Boersma CJ, Mosselman S, van Zoelen EJ, Olijve W (2000) Antiestrogens specifically up-regulate bone morphogenetic protein-4 promoter activity in human osteoblastic cells. Mol Endocrinol 14(5):623–633
97. Matsumoto Y, Otsuka F, Takano-Narazaki M, Katsuyama T, Nakamura E, Tsukamoto N, Inagaki K, Sada KE, Makino H (2013) Estrogen facilitates osteoblast differentiation by upregulating bone morphogenetic protein-4 signaling. Steroids 78(5):513–520. doi:10.1016/j.steroids.2013.02.011
98. Bunyaratavej P, Hullinger TG, Somerman MJ (2000) Bone morphogenetic proteins secreted by breast cancer cells upregulate bone sialoprotein expression in preosteoblast cells. Exp Cell Res 260(2):324–333. doi:10.1006/excr.2000.5019
99. Tarragona M, Pavlovic M, Arnal-Estape A, Urosevic J, Morales M, Guiu M, Planet E, Gonzalez-Suarez E, Gomis RR (2012) Identification of NOG as a specific breast cancer bone metastasis-supporting gene. J Biol Chem 287(25):21346–21355. doi:10.1074/jbc.M112.355834
100. Mock K, Preca BT, Brummer T, Brabletz S, Stemmler MP, Brabletz T (2015) The EMT-activator ZEB1 induces bone metastasis associated genes including BMP-inhibitors. Oncotarget
101. Schwaninger R, Rentsch CA, Wetterwald A, van der Horst G, van Bezooijen RL, van der Pluijm G, Lowik CW, Ackermann K, Pyerin W, Hamdy FC, Thalmann GN, Cecchini MG (2007) Lack of noggin expression by cancer cells is a determinant of the osteoblast response in bone metastases. Am J Pathol 170 (1):160–175. doi:10.2353/ajpath.2007.051276
102. Ye L, Lewis-Russell JM, Kynaston H, Jiang WG (2007) Endogenous BMP-7 controls the motility of prostate cancer cells through regulation of BMP antagonists. J Urol 178(3) Pt 1: 1086–1091
103. Herbertz S, Sawyer JS, Stauber AJ, Gueorguieva I, Driscoll KE, Estrem ST, Cleverly AL, Desaiah D, Guba SC, Benhadji KA, Slapak CA, Lahn MM (2015) Clinical development of galunisertib (LY2157299 monohydrate), a small molecule inhibitor of transforming growth factor-beta signaling pathway. Drug Des Devel Ther 9:4479–4499. doi:10.2147/DDDT.S86621
104. Hawinkels LJ, de Vinuesa AG, Paauwe M, Kruithof-de Julio M, Wiercinska E, Pardali E, Mezzanotte L, Keereweer S, Braumuller TM, Heijkants RC, Jonkers J, Lowik CW, Goumans MJ, ten Hagen TL, ten Dijke P (2016) Activin receptor-like kinase 1 ligand trap reduces microvascular density and improves chemotherapy efficiency to various solid tumors. Clin Cancer Res 22(1):96–106. doi:10.1158/1078-0432.CCR-15-0743
105. Cunha SI, Pietras K (2011) ALK1 as an emerging target for antiangiogenic therapy of cancer. Blood 117(26):6999–7006. doi:10.1182/blood-2011-01-330142
106. Chen G, Deng C, Li YP (2012) TGF-beta and BMP signaling in osteoblast differentiation and bone formation. Int J Biol Sci 8(2):272–288. doi:10.7150/ijbs.2929
107. Suvannasankha A, Chirgwin JM (2014) Role of bone-anabolic agents in the treatment of breast cancer bone metastases. Breast Cancer Res 16(6):484. doi:10.1186/s13058-014-0484-9
108. Zhang L, Ye Y, Long X, Xiao P, Ren X, Yu J (2016) BMP signaling and its paradoxical effects in tumorigenesis and dissemination. Oncotarget 7(47):78206–78218. doi:10.18632/oncotarget.12151
109. Royce ME, Osman D (2015) Everolimus in the treatment of metastatic breast cancer. Breast Cancer (Auckl) 9:73–79. doi:10.4137/BCBCR.S29268
110. Bowen MA, Patel DD, Li X, Modrell B, Malacko AR, Wang WC, Marquardt H, Neubauer M, Pesando JM, Francke U et al (1995) Cloning, mapping, and characterization of activated leukocyte-cell adhesion molecule (ALCAM), a CD6 ligand. J Exp Med 181(6):2213–2220
111. Degen WG, van Kempen LC, Gijzen EG, van Groningen JJ, van Kooyk Y, Bloemers HP, Swart GW (1998) MEMD, a new cell adhesion molecule in metastasizing human melanoma cell lines, is identical to ALCAM (activated leukocyte cell adhesion molecule). Am J Pathol 152(3):805–813
112. Swart GW (2002) Activated leukocyte cell adhesion molecule (CD166/ALCAM): developmen-

tal and mechanistic aspects of cell clustering and cell migration. Eur J Cell Biol 81(6):313–321. doi:10.1078/0171-9335-00256
113. van Kempen LC, Nelissen JM, Degen WG, Torensma R, Weidle UH, Bloemers HP, Figdor CG, Swart GW (2001) Molecular basis for the homophilic activated leukocyte cell adhesion molecule (ALCAM)-ALCAM interaction. J Biol Chem 276(28):25783–25790. doi:10.1074/jbc.M011272200
114. Micciche F, Da Riva L, Fabbi M, Pilotti S, Mondellini P, Ferrini S, Canevari S, Pierotti MA, Bongarzone I (2011) Activated leukocyte cell adhesion molecule expression and shedding in thyroid tumors. PLoS One 6(2):e17141. doi:10.1371/journal.pone.0017141
115. Kulasingam V, Zheng Y, Soosaipillai A, Leon AE, Gion M, Diamandis EP (2009) Activated leukocyte cell adhesion molecule: a novel biomarker for breast cancer. Int J Cancer 125(1):9–14. doi:10.1002/ijc.24292
116. Witzel I, Schroder C, Muller V, Zander H, Tachezy M, Ihnen M, Janicke F, Milde-Langosch K (2012) Detection of activated leukocyte cell adhesion molecule in the serum of breast cancer patients and implications for prognosis. Oncology 82(6):305–312. doi:10.1159/000337222
117. Al-Shehri FS, Abd El Azeem EM (2015) Activated leukocyte cell adhesion molecule (ALCAM) in Saudi breast cancer patients as prognostic and predictive indicator. Breast Cancer (Auckl) 9:81–86. doi:10.4137/BCBCR.S25563
118. Jezierska A, Matysiak W, Motyl T (2006) ALCAM/CD166 protects breast cancer cells against apoptosis and autophagy. Med Sci Monit 12(8):BR263–BR273
119. Hein S, Muller V, Kohler N, Wikman H, Krenkel S, Streichert T, Schweizer M, Riethdorf S, Assmann V, Ihnen M, Beck K, Issa R, Janicke F, Pantel K, Milde-Langosch K (2011) Biologic role of activated leukocyte cell adhesion molecule overexpression in breast cancer cell lines and clinical tumor tissue. Breast Cancer Res Treat 129(2):347–360. doi:10.1007/s10549-010-1219-y
120. Wiiger MT, Gehrken HB, Fodstad O, Maelandsmo GM, Andersson Y (2010) A novel human recombinant single-chain antibody targeting CD166/ALCAM inhibits cancer cell invasion in vitro and in vivo tumour growth. Cancer Immunol Immunother 59(11):1665–1674. doi:10.1007/s00262-010-0892-3
121. King JA, Tan F, Mbeunkui F, Chambers Z, Cantrell S, Chen H, Alvarez D, Shevde LA, Ofori-Acquah SF (2010) Mechanisms of transcriptional regulation and prognostic significance of activated leukocyte cell adhesion molecule in cancer. Mol Cancer 9:266. doi:10.1186/1476-4598-9-266
122. Akman HB, Selcuklu SD, Donoghue MT, Akhavantabasi S, Sapmaz A, Spillane C, Yakicier MC, Erson-Bensan AE (2015) ALCAM is indirectly modulated by miR-125b in MCF7 cells. Tumour Biol 36(5):3511–3520. doi:10.1007/s13277-014-2987-5
123. Orso F, Quirico L, Virga F, Penna E, Dettori D, Cimino D, Coppo R, Grassi E, Elia AR, Brusa D, Deaglio S, Brizzi MF, Stadler MB, Provero P, Caselle M, Taverna D (2016) miR-214 and miR-148b targeting inhibits dissemination of melanoma and breast cancer. Cancer Res 76(17):5151–5162. doi:10.1158/0008-5472.CAN-15-1322
124. Davies SR, Dent C, Watkins G, King JA, Mokbel K, Jiang WG (2008) Expression of the cell to cell adhesion molecule, ALCAM, in breast cancer patients and the potential link with skeletal metastasis. Oncol Rep 19(2):555–561
125. Davies S, Jiang WG (2010) ALCAM, activated leukocyte cell adhesion molecule, influences the aggressive nature of breast cancer cells, a potential connection to bone metastasis. Anticancer Res 30(4):1163–1168
126. Gemoll T, Epping F, Heinrich L, Fritzsche B, Roblick UJ, Szymczak S, Hartwig S, Depping R, Bruch HP, Thorns C, Lehr S, Paech A, Habermann JK (2015) Increased cathepsin D protein expression is a biomarker for osteosarcomas, pulmonary metastases and other bone malignancies. Oncotarget 6 (18):16517–16526. doi:10.18632/oncotarget.4140
127. Vishal M, Swetha R, Thejaswini G, Arumugam B, Selvamurugan N (2017) Role of Runx2 in breast cancer-mediated bone metastasis. Int J Biol Macromol 99:608–614. doi:10.1016/j.ijbiomac.2017.03.021
128. Hansen AG, Arnold SA, Jiang M, Palmer TD, Ketova T, Merkel A, Pickup M, Samaras S, Shyr Y, Moses HL, Hayward SW, Sterling JA, Zijlstra A (2014) ALCAM/CD166 is a TGF-beta-responsive marker and functional regulator of prostate cancer metastasis to bone. Cancer Res 74(5):1404–1415. doi:10.1158/0008-5472.CAN-13-1296
129. King JA, Ofori-Acquah SF, Stevens T, Al-Mehdi AB, Fodstad O, Jiang WG (2004) Activated leukocyte cell adhesion molecule in breast cancer: prognostic indicator. Breast Cancer Res 6(5):R478–R487. doi:10.1186/bcr815
130. Burkhardt M, Mayordomo E, Winzer KJ, Fritzsche F, Gansukh T, Pahl S, Weichert W, Denkert C, Guski H, Dietel M, Kristiansen G (2006) Cytoplasmic overexpression of ALCAM is prognostic of disease progression in breast cancer. J Clin Pathol 59(4):403–409. doi:10.1136/jcp.2005.028209
131. Jezierska A, Olszewski WP, Pietruszkiewicz J, Olszewski W, Matysiak W, Motyl T (2006) Activated leukocyte cell adhesion molecule (ALCAM) is associated with suppression of breast cancer cells invasion. Med Sci Monit 12(7):BR245–BR256
132. Ihnen M, Muller V, Wirtz RM, Schroder C, Krenkel S, Witzel I, Lisboa BW, Janicke F, Milde-Langosch K (2008) Predictive impact of activated leukocyte cell adhesion molecule (ALCAM/CD166) in breast cancer. Breast Cancer Res Treat 112(3):419–427. doi:10.1007/s10549-007-9879-y
133. Ihnen M, Kohler N, Kersten JF, Milde-Langosch K, Beck K, Holler S, Muller V, Witzel I, Janicke F,

Kilic E (2010) Expression levels of activated leukocyte cell adhesion molecule (ALCAM/CD166) in primary breast carcinoma and distant breast cancer metastases. Dis Markers 28(2):71–78. doi:10.3233/DMA-2010-0685
134. Ihnen M, Wirtz RM, Kalogeras KT, Milde-Langosch K, Schmidt M, Witzel I, Eleftheraki AG, Papadimitriou C, Janicke F, Briassoulis E, Pectasides D, Rody A, Fountzilas G, Muller V (2010) Combination of osteopontin and activated leukocyte cell adhesion molecule as potent prognostic discriminators in HER2- and ER-negative breast cancer. Br J Cancer 103(7):1048–1056. doi:10.1038/sj.bjc.6605840
135. Zhou P, Du LF, Lv GQ, Yu XM, Gu YL, Li JP, Zhang C (2011) Functional polymorphisms in CD166/ALCAM gene associated with increased risk for breast cancer in a Chinese population. Breast Cancer Res Treat 128(2):527–534. doi:10.1007/s10549-011-1365-x
136. Ihnen M, Kilic E, Kohler N, Loning T, Witzel I, Hagel C, Holler S, Kersten JF, Muller V, Janicke F, Milde-Langosch K (2011) Protein expression analysis of ALCAM and CEACAM6 in breast cancer metastases reveals significantly increased ALCAM expression in metastases of the skin. J Clin Pathol 64(2):146–152. doi:10.1136/jcp.2010.082602
137. Piao D, Jiang T, Liu G, Wang B, Xu J, Zhu A (2012) Clinical implications of activated leukocyte cell adhesion molecule expression in breast cancer. Mol Biol Rep 39(1):661–668. doi:10.1007/s11033-011-0783-5
138. Varadi V, Bevier M, Grzybowska E, Johansson R, Enquist-Olsson K, Henriksson R, Butkiewicz D, Pamula-Pilat J, Tecza K, Hemminki K, Lenner P, Forsti A (2012) Genetic variation in ALCAM and other chromosomal instability genes in breast cancer survival. Breast Cancer Res Treat 131(1):311–319. doi:10.1007/s10549-011-1765-y
139. Tan F, Mosunjac M, Adams AL, Adade B, Taye O, Hu Y, Rizzo M, Ofori-Acquah SF (2014) Enhanced down-regulation of ALCAM/CD166 in African-American breast cancer. BMC Cancer 14:715. doi:10.1186/1471-2407-14-715
140. Burandt E, Bari Noubar T, Lebeau A, Minner S, Burdelski C, Janicke F, Muller V, Terracciano L, Simon R, Sauter G, Wilczak W, Lebok P (2014) Loss of ALCAM expression is linked to adverse phenotype and poor prognosis in breast cancer: a TMA-based immunohistochemical study on 2,197 breast cancer patients. Oncol Rep 32(6):2628–2634. doi:10.3892/or.2014.3523
141. Chen MJ, Cheng YM, Chen CC, Chen YC, Shen CJ (2017) MiR-148a and miR-152 reduce tamoxifen resistance in ER+ breast cancer via downregulating ALCAM. Biochem Biophys Res Commun 483(2):840–846. doi:10.1016/j.bbrc.2017.01.012

Perspectives of Reprogramming Breast Cancer Metabolism

10

Yi-Ping Wang and Qun-Ying Lei

Abstract

Reprogramming of cellular metabolism is one of the hallmarks of breast cancer. Breast cancer cells remodel metabolic network to maintain their transformed state and survive in a harsh tumor microenvironment. Dysregulated metabolism further interacts with cellular signaling and epigenetics to promote breast cancer development. Meanwhile, breast cancer stem cells exhibit unique metabolic features, which are critical for therapeutic resistance and tumor recurrence. Besides, aberrant metabolism of breast cancer cells reshapes tumor microenvironment, such as promoting cancer vascularization and sabotaging tumor immunity, to accelerate tumor progression. These special metabolic traits not only open vulnerabilities of breast cancer by targeting essential metabolic pathways but also provide promising diagnostic and prognostic biomarkers to facilitate clinical investigations. Studies in the last few decades have significantly advanced our understanding of mechanisms underlying the reprogramming of breast cancer metabolism and metabolic regulation of breast cancer biology. Targeting tumor metabolism serves as a potentially effective therapeutic approach to suppress breast cancer.

Keywords

Metabolic reprogramming • Glycolysis • Amino acid metabolism • Fatty acid metabolism • Cell signaling • Cancer microenvironment • Breast cancer stem cell • Cancer immunity • Metabolic biomarker

10.1 Introduction

The last few decades have witnessed a marvelous prosperity in our understanding of cancer metabolism. Remodeling of cellular metabolism represents a fascinating hallmark of cancers, including

Y.-P. Wang (✉) • Q.-Y. Lei (✉)
Cancer Hospital, Institute of Biomedical Sciences, Fudan University, 131 Dong'an Road, Shanghai, China
e-mail: yiping_wang@fudan.edu.cn; qlei@fudan.edu.cn

E. Song, H. Hu (eds.), *Translational Research in Breast Cancer*, Advances in Experimental Medicine and Biology 1026, https://doi.org/10.1007/978-981-10-6020-5_10

breast tumor [1]. Living in a nutrient-deprived and hypoxic microenvironment, breast cancer cells vividly absorb and utilize nutrients with a remodeled metabolic network to maintain their transformed state and enhance cell proliferation [2]. Genetic alterations in oncogenes and tumor-suppressive genes deregulate metabolic pathways to support the initiation of cancer [3]. Dysregulated metabolism further intersects with signaling pathways and cellular epigenetics to promote cancer development [4]. A subset of cancer cells in breast tumor is cancer initiative and referred to as breast cancer stem cells [5]. In concordance with the remodeling of cancer metabolism, breast cancer stem cells also display unique metabolic features, which are critical for the contribution of cancer stems cells to therapy resistance and tumor recurrence [5]. Remodeled cellular metabolism not only provides sufficient building blocks for biosynthesis but also facilitates cancer cells to survive a harsh microenvironment by promoting tumor vascularization and sabotaging cancer immunity [6]. The unique metabolic phenotype opens vulnerabilities of breast cancer and allows us to selectively eliminate cancer cells by targeting essential metabolic pathways. Besides, during the reprogramming of cancer metabolism, the landscape of cancer metabolome is simultaneously reshaped. Breast cancer-specific metabolites serve as potential diagnostic and prognostic biomarkers.

Recent studies have provided tremendous insights in the reprogramming of metabolism and metabolic regulation of breast cancer biology. Targeting metabolism acts as a potentially effective therapeutic approach to suppress breast cancer.

10.2 Overview of Metabolic Alterations in Breast Cancer

In breast tumor, a reprogrammed metabolic network is essential for sustaining macromolecular biosynthesis and energy production. Aerobic glycolysis, designated as Warburg effect, is a principal metabolic characteristic shared by most types of cancers. In addition, breast cancer cells are addicted to glutamine, an anaplerotic precursor replenishing TCA cycle. Breast cancer cells also consume acetate and folate to support biosynthesis of lipids, nucleotides, etc. As shown in Fig. 10.1, dysregulation in metabolic processes of glucose, amino acids, lipids, and other carbon sources comprises the metabolic landscape of breast cancer.

10.2.1 Glucose Metabolism

Enhanced glucose uptake and aerobic glycolysis are shared features of almost all cancer cells. The primitive and committed reaction of glycolysis is catalyzed by hexokinase. Human hexokinases contain three members, HK1, HK2, and HK3 [7]. The expression hexokinase 2 (HK2) is limited to some specific normal adult tissues [8]. However, HK2 is markedly overexpressed in breast cancer cells. Conditional knockout of *HK2* in the animal model of ErbB2-driven mammary tumor disrupts the initiation and maintenance of cancer. Both in vitro assays and in vivo models demonstrate that depletion of *HK2* suppresses the neoplastic phenotype of breast cancer cells [9]. Collectively, glucose metabolism is necessary for the tumorigenesis and development of breast cancer.

10.2.2 Glutamine Metabolism

In addition to glucose, glutamine belongs to the most sufficient circulating amino acid and functions as another key carbon/energy source for cancer cells. It has been observed for a long time that the consumption of glutamine exceeds other amino acids by more than tenfold, indicating that glutamine is not solely a proteogenic amino acid or nitrogen donor [10]. Subsequent isotope-tracing studies suggest that glutamine-derived carbon is secreted as lactate, i.e., glutaminolysis. While the citrate produced in TCA cycle is transported into the cytosol, α-KG produced from glutaminolysis replenishes the truncated TCA cycle. Thus, glutamine acts as an anaplerotic precursor to support energy production and biosynthesis.

After entering into cells, glutamine is deaminated by glutaminases into glutamate. Glutaminases are overexpressed in advanced-

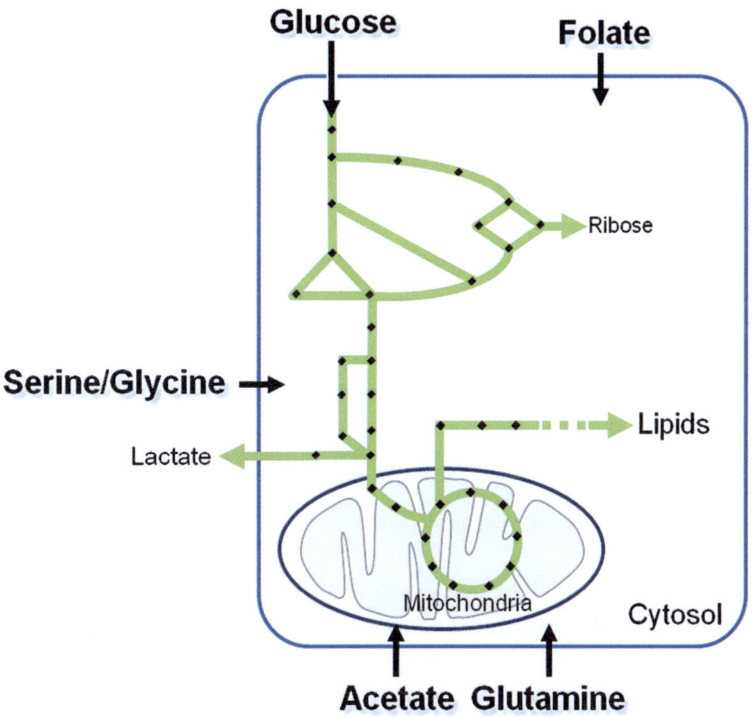

Fig. 10.1 Metabolic alterations in breast cancer

stage breast cancer. Screening for chemical inhibitors of Rho GTPases-induced transformation reveals that a small-molecule inhibitor against glutaminase efficiently suppresses the transformation and proliferation of breast cancer cells, suggesting that glutamine metabolism is necessary for the malignant transformation and growth of mammary tumor [11].

10.2.3 Acetate Metabolism

Besides the reliance on glycolysis and glutamine metabolism, cancer cells exhibit increased lipid demands to maintain membrane construction during rapid proliferation. However, cancer microenvironment is nutrient-lacking and poor-oxygenized. Certain types of cancer, including breast tumor and glioma, utilize acetate as compensatory source of carbon to support de novo lipogenesis [12]. Acetyl-CoA synthetase (ACSS) is responsible for the first step of acetate utilization. Through catalyzing the conversion of acetate to acetyl-CoA, ACSS provides the building block of fatty acids. In breast cancer, ACSS is overexpressed or even amplified in the genome. ACSS expression is correlated with breast cancer progression. Mechanistically, ACSS2 supports de novo lipogenesis by using acetate as a carbon source, especially in hypoxic and lipid-depleted conditions. The knockdown of *ACSS* delays the proliferation of multiple cancer cells [13]. Besides, acetyl-CoA generated by ACSS can be utilized as a donor of acetyl group to mediate histone acetylation and epigenetically regulate the expression of metabolic genes [14]. Thus, acetate functions as both a precursor and an epigenetic metabolite to support de novo lipogenesis.

10.2.4 Folate Metabolism

Folate belongs to water-soluble B vitamins and is a carbon donor for one-carbon metabolism. Folate supports NADPH production, nucleotide biosynthesis, and methylation reactions [15]. Cancer cells overexpress folate receptors, which are presumably linked to increased DNA synthesis and cell growth. In a cross-cancer profiling of metabolic gene expression, one carbon metabolism is identified as the highest-scoring pathway. Mitochondrial protein MTHFD2 (bifunctional methylenetetrahydrofolate dehydrogenase/cyclohydrolase) is the key enzyme for folate metabolism. Notably, both the mRNA and protein expressions of MTHFD2 are significantly enhanced in breast cancer. MTHFD2 expression links to poor survival of breast cancer patients, suggesting a fundamental role of folate catabolism in the development of breast cancer [16]. Of note, supplementation of folate may increase cancer risk and promote cancer growth in late stages of tumor [15].

10.2.5 Fatty Acid Metabolism

Cancer cells require large amounts of lipid and cholesterol, which are satisfied by either uptaking more exogenous lipids and lipoproteins or activating de novo lipogenesis and cholesterol biosynthesis. Lipid synthesis is critical for satisfying anabolic demands of cancer cells. Sufficient supply of lipids is particularly vital for the survival and proliferation of cancer cells in disadvantageous environment. An acyl-chain-targeted high-resolution profiling of specific lipid species in hypoxic cells reveals an increase in acyl chains with less carbon atoms and higher degree of saturation. The production of more saturated fatty acyl-CoA is a mark for increased de novo lipogenesis [17, 18]. Transcriptomic analysis of in vitro MCF-10A transformation model suggests that multiple genes related to lipid metabolism are upregulated. Depletion of these genes reduces morphological transition, anchorage-independent growth, and cell motility of transformed MCF-10A cells [19], but not untransformed cells [20].

In mammals, acetyl-CoA acts as a carbon precursor in the biosynthetic pathways producing lipid and cholesterol. Production of lipids depends on the cytosolic acetyl-CoA pool, which is supplied by two metabolic enzymes ACLY and ACSS. ACLY cleaves citrate into oxaloacetate and acetyl-coA, while ACSS ligates acetate and coenzyme A to produce acetyl-CoA. Both ACLY and ACSS are essential for lipogenesis, especially in nutrient-poor conditions. Inhibition of ACLY using chemical inhibitors or siRNAs markedly delays tumor growth. Interestingly, fatty acid synthase (FASN), the rate-limiting player in de novo lipogenic pathway, had been found to be strongly correlated with breast cancer recurrence, metastases, and survival in the early 1990s [21]. De novo lipogenesis is critical in maintaining the proliferative potential and the transformed state of breast cancer [22, 23].

10.2.6 Serine and Glycine Metabolism

Glucose-derived metabolic fluxes are not only directed toward lactate but also other metabolic processes. An isotope-labeling metabolomic study demonstrates that breast tumor cells divert a substantial amount of glycolysis-derived carbon to the pathways of glycine and serine metabolism. Phosphoglycerate dehydrogenase (PHGDH) functions in the branching point of glycolysis-dependent serine and glycine metabolism. Unlike normal mammary cell MCF-10A, breast tumor cells are dependent on PHGDH for rapid proliferation. Increased protein expression of PHGDH is linked with unfavorable outcome in clinical investigations of breast cancer [24]. Clinical breast cancer array data shows that PHGDH expression specifically associates with triple-negative and basal subtypes, but not other clinical parameters such as metastases and tumor size. Reconstituted basement membrane (Matrigel) model demonstrates that PHGDH expression promotes glucose-dependent glycine/serine metabolism, disrupts acinar morphogenesis, and supports anchorage-independent growth of MCF-10A cells, which potentiates cells to transformation [25].

10.3 Control of Metabolism Reprogramming by Oncogenic Signals

Gain of function of oncogenes and loss of function of tumor suppressor genes are key driving factors during tumorigenesis and development. Oncogenic events promote cancer development through not only activating cellular signaling pathways but also rewiring metabolic network. In particular, glucose metabolism and mitochondria function are modulated by multiple oncogenic signals to support malignant transformation, cell proliferation, and tumor progression.

10.3.1 SIRT3 Modulates Cellular Metabolism and the Expression of HIF1-α Target Genes

Sirtuins, which are evolutionarily homologous to Sir2 gene of yeast, mediate calorie-restriction-induced longevity. Sirtuins are potentially involved in tumorigenesis as cancer is a disease of aging. SIRT3, a mitochondrial member of human sirtuins, exerts tumor-suppressive functions in breast cancer. IHC staining demonstrates that SIRT3 mainly locates in normal mammary ductal cells. In contrast to normal mammary tissue, SIRT3 protein expression is statistically lower expressed in breast cancer specimen. Moreover, *SIRT3* gene is frequently deleted in breast cancers [26]. Clinically, the mRNA expression of *SIRT3* is significantly reduced in high-grade malignant mammary tumors [27]. Knockout mice model suggests that the loss of *SIRT3* results in the development of ER-/PR-positive breast cancer. Although *SIRT3* knockout is unable to induce spontaneous immortalization, the depletion of *SIRT3* in mouse embryonic fibroblasts (MEF) is linked to aberrant mitochondrial function and genome instability. Interestingly, the expression of either Myc or Ras oncogene promotes in vitro transformation of *SIRT3*-null MEF cells, indicating that deletion of *SIRT3* creates an immortalization-permissive phenotype. Transformed *SIRT3* knockout MEFs exhibit enhanced glycolysis and suppressed oxidative phosphorylation, i.e., Warburg effect, to support cell proliferation. In addition, SIRT3 modulates transcription regulatory program during tumorigenesis. Deletion of *SIRT3* exaggerates ROS production, which stabilizes HIF1-α and thereby activates the transcription of glycolytic genes. In clinical breast tumor samples, downregulation of SIRT3 correlates with overexpression of HIF1 target genes [28].

10.3.2 SKP2 Mediates AKT Activation and Promotes Aerobic Glycolysis

As a key player promoting tumorigenesis, in particular cancer metabolism, AKT kinase is a major driver for Warburg effect in various cancers [29]. Diverse growth factors signal through different E3 ligases to activate AKT. In response to epidermal growth factor (EGF), AKT is activated by ErbB receptors and Skp2 SCF ubiquitin ligase complex-induced ubiquitination-dependent membrane translocation. In HER2-positive breast cancers, SKP2 overexpression associates with the hyperactivation of AKT and tumor metastasis. Importantly, targeting SKP2 results in a more pronounced effect of Herceptin therapy in eliminating HER2-positive breast cancer [30]. Numerous downstream targets are regulated by AKT to promote aerobic glycolysis. For example, activated AKT upregulates the transcription of glucose transporters and enhances their membrane translocation to enhance glucose uptake. Moreover, AKT directly activates hexokinase and phosphofructokinase to boost glycolytic flux [31].

10.3.3 Pro-inflammatory MiRNA Upregulates Glycolytic Enzymes

In tumor microenvironment, pro-tumorigenic inflammation enhances glycolysis of cancer cells. microRNAs (miRNAs) serve as a new group of regulators that promotes inflammation. Pro-inflammatory cytokines, including IL-1β,

IL-6, IFN-γ, and TNFα, upregulate glycolytic activity of breast tumor cells. Activation of glycolysis by these cytokines is at least in part dependent on the posttranslationally regulatory role of miR-155, an inflammation-induced microRNA. miR-155 promotes STAT3-induced hexokinase 2 (HK2) transcription and suppresses C/EBPβ, a negative regulator of HK2. Of note, overexpression of miR-155 is frequently observed in multiple cancers, supporting the oncogenic role of mir-155 during malignant transformation and cancer metabolism [32].

10.3.4 TAp73 Enhances Pentose Phosphate Pathway

p73 belongs to p53-family proteins. *p73* gene is majorly expressed as two isoforms, TAp73 and ΔNp73. Unlike p53, which is frequently inactivated in cancers, TAp73 is overexpressed in various cancers. TAp73 promotes breast cancer proliferation by activating pentose phosphate pathway (PPP). Pentose shunt supports the production of ribose-5-phosphate, a precursor for de novo RNA and DNA biosynthesis [33], and NADPH, a reducing equivalent to support biosynthetic reactions and maintain redox homeostasis [34]. G6PD is the key enzyme catalyzing the first reaction of PPP and directs glucose to multiple biosynthesis pathways and the clearance of toxic reactive oxygen species (ROS). TAp73 enhances G6PD mRNA expression.G6PD overexpression has been shown to be a negative prognostic indicator in metastasis-free survival analysis of breast cancer patients [35].

10.3.5 Estrogen-Related Receptors Promote Warburg Effect

Orphan nuclear receptors belong to nuclear receptor superfamily. Estrogen-related receptors (ERRs), a subgroup of orphan receptors, modulate gene transcription in response to physiological signals [36]. Although ERRs are homologous to estrogen receptors, they do not bind to estrogens or other steroid hormones. In breast cancer, ERRs transcriptionally regulate the expression of metabolism-related genes that participate in glycolysis, glutaminolysis, pentose phosphate pathway, ROS detoxification, and de novo lipogenesis. Specifically, ERBB2 upregulates the expressions of glycolytic enzymes and glucose transporters to promote Warburg effect. Inhibition of ERBB2 sensitizes ERBB2-postive breast cancer cells to trastuzumab treatment. Besides, ERRα regulates the expressions of ERBB2 and Myc to induce metabolic reprogramming, which makes ERRα a potential marker for breast cancer prognosis.

10.3.6 P53 Suppresses Pyruvate Dehydrogenase Kinase 2 at Transcriptional Level

Pyruvate dehydrogenase (PDH) links glycolysis and TCA cycle through converting pyruvate to acetyl-CoA. However, cancer cells prefer to transform pyruvate into lactate, but not acetyl-CoA, to promote aerobic glycolysis. In cells, PDH is phosphorylated and inhibited by pyruvate dehydrogenase kinase. mRNA level of pyruvate dehydrogenase kinase 2 (PDK2) is downregulated by tumor suppressor p53, leading to inhibition of pyruvate-lactate conversion [37]. In breast cancer cells, p53 potentially inhibits PDK2 through both downregulating PDK2 transcription in an E2F-dependent manner and destabilizing PDK2 protein. Inhibition of PDK2 is necessary for the function of p53 on early apoptotic events.

10.3.7 Wnt Snail Signaling Modifies Mitochondria Respiration and Glucose Utilization

Wnt signaling pathway regulates multiple physiological events, such as epithelial-to-mesenchymal transition (EMT) and embryonic development. Aberrant Wnt signaling has been found in specific cancers, including colorectal cancer and breast tumor. In addition, cellular metabolic activity is regulated by Wnt pathway. Wnt signaling inhibits multiple key components of mitochondria respiration. On the other hand, Wnt signaling upregulates pyruvate carboxylase (PC), the rate-limiting enzyme in anaplerotic reactions, to promote aerobic glycolysis of breast cancer cells [38]. Reshaping of metabolism by Wnt pathway cooperates with snail-dependent EMT to promote tumor growth.

10.3.8 PGC-1α Mediates Mitochondria Biogenesis and Promotes Breast Cancer Metastasis

Bioenergetic profile of cancer cells is coupled with not only cell proliferation but also migration and invasion. While cancer cells divert metabolic fluxes into biosynthetic pathways to promote cell growth, enhanced mitochondrial respiration and ATP production support cell migration and invasion. PGC-1α is a transcriptional coactivator which upregulates oxidative phosphorylation and mitochondrial biogenesis through transcriptionally activating related genes [39]. In clinical breast cancer samples, PGC-1α expression strongly correlates with distal metastases. Depletion of PGC-1α suppresses cell migration, but not cell proliferation and tumor growth. Noteworthy, gene expression profiling reveals that transcripts related to oxidative phosphorylation are significantly increased in circulating breast cancer cells compared to the primary tumor. High expression of PGC-1α is indicative of EMT in breast cancer [40].

10.4 Metabolic Regulation of Cellular Signaling in Breast Cancer

Recent evidence suggests that reprogrammed metabolism serves as both the cause and the consequence of tumorigenesis. Dysregulation of metabolism intersects with cellular signaling to promote breast cancer. Aberrations in metabolites or metabolic enzymes modulate the activities of nutrient sensing, signal transduction, and gene transcription, to facilitate the survival and growth of breast tumor cells.

10.4.1 Nutrient Sensor OGT Coordinates Glucose Availability with Cell Cycle and Hypoxia Response

Hexosamine biosynthetic pathway (HBP) generates multiple nucleotide hexosamines, which are donors for protein O-GlcNAc modification. O-GlcNAc transferase (OGT) is responsible for O-GlcNAc modification reactions. In this regard, OGT functions as a nutrient sensor to regulate the activity of target proteins. In breast cancer cells, a substantial fraction of glucose is directed into HBP. Consequently, hexosamine synthesis is enhanced [41]. While the expression of OGT is upregulated in breast cancer, O-GlcNAcase (OGA), a counterpart of OGT mediating removal of O-GlcNAc modification, is decreased in breast cancer cells. Elevated OGT expression and decreased OGA expression result in hyper-O-GlcNAcation of many cytoplasmic and nuclear proteins. Specifically, OGT-mediated hyper-O-GlcNAcation suppresses cell cycle inhibitory FoxM1-SKP2-p21 pathway, thereby supporting the transformation and proliferation of breast cancer cells. These findings demonstrate that cell cycle progression is coupled with glucose availability through nutrient sensor OGT.

In contrast, hypo-O-GlcNAcation inhibits the initiation and progression of breast cancer through modulating HIF signaling. Reduction in O-GlcNAcation suppresses glycolysis of cancer cells, leading to an increase in α-ketoglutarate (α-KG). Accumulation of α-KG further causes hydroxylation and VHL-induced degradation of HIF-1. Decreased HIF signaling is incapable of sustaining the expression of GLUT1, thereby leading to ER stress and apoptosis of breast cancer cells. Collectively, O-GlcNAcation modulates HIF signaling to regulate hypoxia adaptation of cancer cells [42].

10.4.2 JMJD5 Modulates PKM2 Subcellular Localization and Remodels Glycolysis

Metabolic enzymes possibly involve in cellular signaling through moonlighting functions. Pyruvate kinase PKM2, a key glycolytic enzyme, is specifically enriched in cancer cells [43]. PKM2 remodels glycolysis to promote Warburg effect and moonlights as a protein kinase to modulate epigenetics. It also interacts with JMJD5 to enhance hypoxia response in breast cancer cells. JMJD5 is an epigenetic modifying enzyme that regulates histone methylation. Interestingly, JMJD5 exerts demethylase-independent effect on metabolism. JMJD5 associates with PKM2 and suppresses PKM2 enzymatic activity through disrupting PKM2 tetramerization. Furthermore, through promoting PKM2 nuclear localization and HIF1α-mediated transactivation, JMJD5 enhances glucose uptake and glycolytic activity of breast tumor cells [44].

10.4.3 Glycogen Accumulation Contributes to p53-Induced Senescence

Hypoxic microenvironment drives metabolic reprogramming of breast cancer cells to sustain energy production. Glycogen is a polysaccharide and serves as the principal energy storage in human cells. Strikingly, glycogen metabolism is enhanced in both breast tumor models and hypoxic cultured cancer cells [45]. Glycogen synthase (GS) and glycogen phosphorylase (GP) are responsible for the synthesis and breakdown of glycogen, respectively. Glycogen metabolism signals through AMPK pathway and p53 to regulate cell viability. Inactivation of PYGL, an isoform of GP, leads to deficient glycogen degradation and accumulation of glycogen. Furthermore, loss of function of PYGL stimulates cellular ROS generation, which potentially modulates AMPK signaling to promote p53-mediated senescence [45].

10.4.4 Mevalonate Pathway Modulates Hippo Signaling

Hippo pathway regulates cell growth and organ size. To coordinate the metabolic state and tissue proliferation, Hippo signaling potentially acts in concert with cellular metabolism. Interestingly, the activity of YAP and TAZ, the key components of Hippo pathway [46], is coordinated with mevalonate synthesis. Depleting HMG-CoA reductase, the key enzyme of mevalonate synthesis, suppresses the activity of YAP/TAZ. Mechanistically, the activation of YAP/TAZ by Rho GTPases is dependent on geranylgeranyl pyrophosphate, a specific metabolite from mevalonate biosynthesis [47, 48]. The intersection of mevalonate biosynthesis and Hippo pathway promotes cell proliferation and self-renewal of breast cancer. Elevation of mevalonate synthesis enzymes links to poor breast cancer prognosis, highlighting the therapeutic potential of mevalonate synthesis pathway in treating breast cancer.

10.5 Metabolism of Breast Cancer Stem Cell

In breast tumor, a subgroup of cells is termed as breast cancer stem cells, due to their stem cell-like properties, such as self-renewal, tumorigenicity, and potential of differentiation. Breast cancer stem cells are proposed to be involved in relapse, metastases, and chemoresistance/radioresistance of breast tumor [49]. Breast cancer stem cells establish their cell identity by expressing unique pattern of proteins and surface markers as fingerprints. Accordingly, breast cancer stem cells have distinctive metabolic properties to sustain their stemness and promote cancer progression.

10.5.1 Reduced ROS Level Promotes Radioresistance and EMT Phenotype of Cancer Stem Cells

Mammalian cells consume oxygen by mitochondrial respiratory chain to produce energy. Oxygen consumption or respiration is coupled with ROS production. In breast cancer stem cells ($CD44^{+}CD24^{-/low}Lin^{-}$), enhanced expression of ROS-clearing enzyme system leads to a significant reduction in cellular ROS level compared to adjacent non-tumorigenic cells. Enhanced ROS scavenging system protects breast cancer stem cells from DNA damage, which is critical in the radioresistance of breast cancer [50]. In addition, Snail-G9a-Dnmt1 complex regulates redox homeostasis and supports the EMT feature of breast cancer stem cells [51]. Snail-G9a-Dnmt1 complex suppresses E-cadherin expression and promotes DNA hypermethylation of *FBP1* (*fructose-1,6-bisphosphatase 1*), an important gluconeogenic gene. Specifically, FBP1 expression is significantly decreased in basal-like breast tumor, to promote glycolysis and suppress ROS production, contributing to breast cancer stem cell phenotype [52].

10.5.2 Notch Signaling Interacts with Cellular Metabolism to Promote Cancer Stem Cell

Hyperactive aldehyde dehydrogenase (ALDH) discriminates breast cancer stem cells from their non-tumorigenic counterparts. ALDH activity is closely related to the physiological properties of breast cancer stem cells. ALDH1A1, a key member of ALDH family, is the major enzyme responsible for converting retinaldehyde to retinoic acid. The catalytic activity of ALDH1A1 is controlled by lysine acetylation. PCAF-mediated acetylation of ALDH1A1 inhibits its activity, leading to decreased self-renewal ability and reduced population of breast cancer stem cells [53]. Notably, Notch pathway regulates the stem cell property of breast cancers by modulating ALDH1A1 acetylation.

Sphingolipid metabolism is involved in Notch signaling and cancer stem cell function. By using a cancer stem cell model with high-ALDH activity, metabolite profiling of these cells demonstrates that sphingosine-1-phosphate (S1P) is critical for cancer stem cell expansion [54]. Through S1P receptor 3 (S1PR3), S1P activates Notch signaling. Overexpression of sphingosine kinase 1 (SphK1), which generates S1P, remarkably enhances the tumorigenic ability of cancer stem cell in mice. $SphK1^{+}/ALDH1^{+}$ cells or $S1PR3^{+}/ALDH1^{+}$ cells also present in patient-derived mammospheres, supporting the notion that SphK1-SIPR3-Notch signaling regulates breast cancer stem cell [54].

10.6 Breast Cancer Metabolism and Microenvironment

Deregulated metabolism within cancer cells modulates the activities of multiple cells residing in malignant tissue, such as endothelial cells, inflammatory cells, and immune cells, to remodel tumor microenvironment. Metabolic remodeling of cancer microenvironment by cancer metabolism further regulates tumor angiogenesis, inflammation, and cancer immunity to promote cancer development.

10.6.1 Breast Cancer Cells Release Lactate to Promote Tumor Vascularization

Aerobic glycolysis is coupled with increased lactate production and secretion, which eventually leads to acidification of cancer microenvironment. The release of lactate into the tumor microenvironment further promotes cancer progression [55]. Studies using xenograft models suggest that MCT4, which transports monocarboxylate across cell membrane, mediates lactate secretion of breast tumor cells. The secreted lactate is further transported into endothelial cells that express monocarboxylate transporter MCT-1, triggering a NF-κB/IL-8 autocrine pathway. Afterward, lactate signaling in endothelial cell induces cell migration and tube formation, promoting tumor vascular morphogenesis and perfusion. Interestingly, lactate signaling can be blocked by 2-oxoglutarate and reactive oxygen species (ROS) inhibitors [56], suggesting that the proangiogenic effect of lactate connects with cellular redox homeostasis. In conclusion, metabolic remodeling of cancer cell reprograms the metabolism and signaling of endothelial cells in the tumor microenvironment to promote cancer development.

10.6.2 Breast Cancer Metabolism and Tumor Immunity

Cancer cells compete for nutrients with other cells in the microenvironment. Thus, dysregulated cancer metabolism potentially regulates the metabolism and function of inflammatory and immune cells [57, 58]. Several unfolded protein response (UPR) signaling components are activated in breast cancer, among which is glucose-regulated protein 78 (GRP78) [6]. Inhibition of GRP78 suppresses the mitochondrial transportation of fatty acids. Consequently, fatty acid oxidation is attenuated, leading to accumulation of essential polyunsaturated fatty acids in intracellular space. Alterations in intracellular fatty acids further increase serum level of MCP-1, a chemoattractant cytokine, and decrease expression of self-recognition identifier CD47 in tumor. Besides, suppression of GRP78 enhances macrophage infiltration, suggesting a potential link between fatty acid metabolism and cancer immunity [6].

10.7 Metabolism as a Target for Breast Cancer

As a hallmark of cancer, reprogrammed metabolic network exerts unique features which can be utilized as diagnostic targets. Deregulation of cancer metabolism also results in specific changes in local or systemic metabolites, serving as indicative biomarkers during cancer prognosis. More importantly, targeting metabolic reprogramming is evolving as a promising strategy for therapeutic intervention of breast cancer.

10.7.1 Therapeutically Targeting Breast Cancer Metabolism

10.7.1.1 Glucose Metabolism

Inhibiting glycolysis potentially leads to systematic toxicity, which makes anti-glycolytic drugs unfavorable for clinical cancer treatment. However, dual targeting of glycolysis, using 2-DG and mitochondria-targeted drugs, serves as a potential strategy to eliminate breast cancer cells [59]. Acquired resistance of Herceptin, an ErbB2-targeted antibody, hinders the clinical treatment of breast cancer. ErbB2 signaling upregulates LDHA in a HSF1-dependent manner, resulting in enhanced glycolysis. Targeting glucose metabolism can facilitate overcoming Herceptin resistance of breast cancer. Combined treatment with glycolytic inhibitors in ErbB2-positive breast cancer overcomes the resistance and leads to more potent inhibition of glycolysis [60].

10.7.1.2 Glutamine Metabolism

Rho GTPases promote malignant transformation and cell proliferation of human breast cancer. Interestingly, glutaminase, the enzyme responsible for glutamine hydrolysis, is remarkably upregulated in transformed breast cancers, compared with normal untransformed cells. Glutaminase inhibitor efficiently suppresses Rho GTPases-induced transformation of fibroblasts [11]. Glutamine metabolism is coupled with malate-aspartate shuttle. Aspartate aminotransferase (GOT) is a key component in malate-aspartate shuttle, which mediates the transportation of NADH between mitochondria and cytoplasm. GOT inhibitor oxamate and amino oxyacetate suppress TCA cycle and oxygen consumption without affecting lactate production. More importantly, aspartate aminotransferase inhibitors profoundly inhibit the proliferation of breast cancer cells [61].

10.7.1.3 Mevalonate Metabolism

Farnesyl pyrophosphate synthase (FPPS) locates at the first branching point of mevalonate pathway from lipid synthesis. FPPS is a well-established therapeutic target. Inhibitors of FPPS, i.e., bisphosphonates, have potential antiproliferative effects [62]. However, previous FPPS inhibitors, such as zoledronic acid, are highly affinitive to bone mineral, hindering its application in the treatment of tumors. Currently, allosteric non-bisphosphonate FPPS inhibitors have been developed to evaluate their antitumor effects [63].

10.8 Metabolites as Biomarkers: Opportunities and Challenges

10.8.1 Cancer Metabolism-Based Tumor Imaging

Fluorodeoxyglucose (FDG) positron emission tomography (PET) imaging has been serving as a powerful tool for monitoring cancer development (growth, staging, metastasis, therapy response) in clinical practice. Although hypoxia induces a higher rate of glycolysis, the overall metabolism is potentially decreased in a hypoxia microenvironment. The glucose uptake and oxygen availability can be monitored by ^{18}F-labeled fluorodeoxyglucose (FDG) and ^{18}F-labeled fluoromisonidazole (FMISO), respectively. Interestingly, positron emission tomography of cancer patients indicates that hypoxia broadly correlates with enhanced glucose metabolism. However, a proportion of hypoxic tumors display moderate glycolytic activity, while some highly glycolytic tumors are well oxygenated [64], suggesting a subpopulation-specific metabolic activity or a heterogeneous metabolic phenotype.

Of note, ^{18}F-fluorodeoxyglucose positron emission tomography significantly correlates with basal-like breast cancers that are overexpressing MYC [65]. A transcriptome enrichment analysis reveals that the uptake of ^{18}FDG radiotracer is linked to enhanced glycolysis and biosynthetic pathways, i.e., pentose shunt and one-carbon/folate metabolism. This FDG signature partially overlaps with the gene transcriptional signature of basal-like breast cancer and MYC-induced mouse tumor model. Moreover, human breast cancer with MYC overexpression or amplification shows strong ^{18}FDG uptake. Thus, subtypes of breast cancer should be taken into consideration during PET imaging.

In addition to PET imaging, optical metabolic imaging (OMI) serves as a noninvasive approach to monitor glycolytic levels of breast cancer cell lines and xenograft tumors. It can detect the fluorescence intensities of autofluorescent coenzymes within cells and rapidly assess the metabolic response at cellular cells, assisting studies on drug response and chemoresistance [66].

10.8.2 Metabolites as Diagnostic and Prognostic Markers

10.8.2.1 Urine Metabolome Facilitates Early Diagnosis of Breast Cancer

Nuclear magnetic resonance spectroscopy-based comparison of the urine metabolomes from early-/late-stage breast cancer patients and normal individuals indicates that breast cancer patients have unique pattern of metabolites. Interestingly, metabolites derived from TCA cycle, amino acid metabolism, and gut microbial metabolism are most significantly changed, which may aid in the clinical diagnosis of breast cancer [67].

10.8.2.2 Serum Metabolites Derived from Cholesterol and Vitamin D as Predictive Biomarkers

Some metabolites from the serum, including 25-hydroxyvitamin D and 27-hydroxycholesterol, correlate with tumor grade, endocrine therapy response, and overall survival of breast cancer patients [68]. Vitamin D metabolism is potentially related to cancer development. A perspective study of postmenopausal breast cancer patients suggests that reduced post-diagnostic serum 25-hydroxyvitamin D concentration is linked to both poor distant disease-free and overall survival.

Estrogen promotes ER-positive breast tumor growth. Hypercholesterolemia serves as a poor prognosis factor of ER-positive breast cancer patients. ER-positive tumors with hypercholesterolemia exhibit decreased response to endocrine therapies. Promotion of caner progression by cholesterol depends on its conversion to 27-hydroxycholesterol (27HC). 27HC abundance is upregulated in both normal mammary tissue and malignant tissue of ER-positive patients. Elevated 27HC is possibly caused by increased expression of CYP27A1, a 27HC-generating enzyme, and decreased expression of CYP781, a 27HC-metabolizing enzyme. CYP27A1 correlates with tumor grade and overall survival, while CYP781 links to reduced survival of breast cancer patients. 27HC promotes breast cancer growth through both ER-dependent and ER-independent mechanisms [69]. Inhibiting CYP27A1 or reducing serum cholesterol levels acts as a promising strategy to suppress breast cancer [70].

10.8.2.3 Breast Cancer-Specific Metabolites Link to Cancer Development

Metabolic profiling of breast tumors reveals an accumulation of oncometabolite 2HG. Interestingly, elevated 2HG level correlates with Myc activation and global DNA hypermethylation [71]. 2HG accumulation is frequently observed in African-American patients with poor prognosis. Specific accumulation of 2HG is also observed in ER-negative and basal-like tumors. Besides, breast tumor with higher 2HG has a stem cell-like transcriptional landscape, which overexpresses glutaminase and incorporates glutamine into 2HG. Importantly, 2HG accumulation significantly correlates with poor prognosis [72].

Additionally, global lipid profiling of breast cancer samples reveals that de novo lipid biosynthesis has a greater contribution to membrane phospholipids in tumors than in normal breast tissues. And this contribution correlates with cancer development and overall survival. Estrogen receptor-negative and grade 3 tumors have the highest concentration of lipids derived from de novo lipogenesis, suggesting the diagnostic values of clinical tumor sample-derived phospholipids [17].

10.8.3 Diet and Breast Cancer Development

Diet is a vital factor modulating cell metabolism and cancer development. In particular, obesity is a highly risky factor for cancer. In animal breast cancer model, diet-derived cholesterol accelerates tumorigenesis and promotes breast tumor growth. Besides, cholesterol-rich diet increases angiogenesis and promotes tumor progression [73]. Thus, cholesterol-reduced or cholesterol-depleted diet is possibly applicable to breast cancer patients.

10.9 Concluding Remarks and Future Perspectives

In conclusion, recent investigations have immensely expanded our understanding of the relation between metabolic remodeling and cancer development. Technical breakthroughs in high-throughput genomic, proteomic, and metabolomic approaches would further help us to discover the metabolic identity of breast cancer and develop new metabolic targets for precision cancer medicine. It is worthy to mention that a vast majority of metabolic characteristics of breast cancer remain unknown. Unveiling key regulatory events would further sharpen our knowledge during understanding breast cancer metabolism.

1. The spectrum of essential nutrients for breast cancer cells: acetate is employed by cancer cells to be an additional carbon source under hypoxic conditions. This metabolic plasticity possibly hinders therapeutic interventions against a specific metabolic pathway, due to the existence of compensatory mechanisms. Thus, identification of key nutrients for breast cancer would help us to develop more effective strategies for breast cancer therapy.
2. How breast cancer cells sense different nutrients and adapt to metabolic fluctuations of cancer microenvironment: to survive in a metabolically dynamic microenvironment, cancer cells need to efficiently sense nutrient status and correspondingly remodel its metabolic activities to sustain their proliferation. Thus, delineating the sensing/signaling pathway of different nutrients is a fundamental key question in this burgeoning field.
3. Understanding metabolic heterogeneity of breast cancer: the metabolic feature of different cells within a tumor is highly plastic, making it almost impossible to eliminate all cancer cells with a same metabolic target. The origin and regulatory mechanism of metabolic heterogeneity/plasticity remain open questions in cancer metabolism.

References

1. Benjamin DI, Cravatt BF, Nomura DK (2012) Global profiling strategies for mapping dysregulated metabolic pathways in cancer. Cell Metab 16(5):565–577. https://doi.org/10.1016/j.cmet.2012.09.013
2. Pavlova NN, Thompson CB (2016) The emerging hallmarks of cancer metabolism. Cell Metab 23(1):27–47. https://doi.org/10.1016/j.cmet.2015.12.006
3. Croce CM (2008) Oncogenes and cancer. N Engl J Med 358(5):502–511. https://doi.org/10.1056/NEJMra072367
4. Boroughs LK, DeBerardinis RJ (2015) Metabolic pathways promoting cancer cell survival and growth. Nat Cell Biol 17(4):351–359. https://doi.org/10.1038/ncb3124
5. Peiris-Pages M, Martinez-Outschoorn UE, Pestell RG, Sotgia F, Lisanti MP (2016) Cancer stem cell metabolism. Breast Cancer Res 18(1):55. https://doi.org/10.1186/s13058-016-0712-6
6. Cook KL, Soto-Pantoja DR, Clarke PA, Cruz MI, Zwart A, Warri A, Hilakivi-Clarke L, Roberts DD, Clarke R (2016) Endoplasmic reticulum stress protein GRP78 modulates lipid metabolism to control drug sensitivity and antitumor immunity in breast cancer. Cancer Res 76(19):5657–5670. https://doi.org/10.1158/0008-5472.CAN-15-2616
7. Patra KC, Hay N (2014) The pentose phosphate pathway and cancer. Trends Biochem Sci 39(8):347–354. https://doi.org/10.1016/j.tibs.2014.06.005
8. Wang L, Xiong H, Wu F, Zhang Y, Wang J, Zhao L, Guo X, Chang LJ, Zhang Y, You MJ, Koochekpour S, Saleem M, Huang H, Lu J, Deng Y (2014) Hexokinase 2-mediated Warburg effect is required for PTEN- and p53-deficiency-driven prostate cancer growth. Cell Rep 8(5):1461–1474. https://doi.org/10.1016/j.celrep.2014.07.053
9. Patra KC, Wang Q, Bhaskar PT, Miller L, Wang Z, Wheaton W, Chandel N, Laakso M, Muller WJ, Allen EL, Jha AK, Smolen GA, Clasquin MF, Robey RB, Hay N (2013) Hexokinase 2 is required for tumor initiation and maintenance and its systemic deletion is therapeutic in mouse models of cancer. Cancer Cell 24(2):213–228. https://doi.org/10.1016/j.ccr.2013.06.014
10. Wang YP, Zhou W, Wang J, Huang X, Zuo Y, Wang TS, Gao X, Xu YY, Zou SW, Liu YB, Cheng JK, Lei QY (2016) Arginine Methylation of MDH1 by CARM1 inhibits glutamine metabolism and suppresses pancreatic cancer. Mol Cell 64(4):673–687. https://doi.org/10.1016/j.molcel.2016.09.028
11. Wang JB, Erickson JW, Fuji R, Ramachandran S, Gao P, Dinavahi R, Wilson KF, Ambrosio AL, Dias SM, Dang CV, Cerione RA (2010) Targeting mitochondrial glutaminase activity inhibits oncogenic transformation. Cancer Cell 18(3):207–219. https://doi.org/10.1016/j.ccr.2010.08.009
12. Mashimo T, Pichumani K, Vemireddy V, Hatanpaa KJ, Singh DK, Sirasanagandla S, Nannepaga S, Piccirillo SG, Kovacs Z, Foong C, Huang Z, Barnett S, Mickey BE, DeBerardinis RJ, Tu BP, Maher EA, Bachoo RM (2014) Acetate is a bioenergetic substrate for human glioblastoma and brain metastases. Cell 159(7):1603–1614. https://doi.org/10.1016/j.cell.2014.11.025
13. Comerford SA, Huang Z, Du X, Wang Y, Cai L, Witkiewicz AK, Walters H, Tantawy MN, Fu A, Manning HC, Horton JD, Hammer RE, McKnight SL, Tu BP (2014) Acetate dependence of tumors. Cell 159(7):1591–1602. https://doi.org/10.1016/j.cell.2014.11.020
14. Gao X, Lin SH, Ren F, Li JT, Chen JJ, Yao CB, Yang HB, Jiang SX, Yan GQ, Wang D, Wang Y, Liu Y, Cai Z, Xu YY, Chen J, Yu W, Yang PY, Lei QY (2016) Acetate functions as an epigenetic metabolite to promote lipid synthesis under hypoxia. Nat Commun 7:11960. https://doi.org/10.1038/ncomms11960
15. Yang HB, Xu YY, Zhao XN, Zou SW, Zhang Y, Zhang M, Li JT, Ren F, Wang LY, Lei QY (2015) Acetylation of MAT II alpha represses tumour cell growth and is decreased in human hepatocellular cancer. Nat Commun 6:6973. https://doi.org/10.1038/ncomms7973
16. Nilsson R, Jain M, Madhusudhan N, Sheppard NG, Strittmatter L, Kampf C, Huang J, Asplund A, Mootha VK (2014) Metabolic enzyme expression highlights a key role for MTHFD2 and the mitochondrial folate pathway in cancer. Nat Commun 5:3128. https://doi.org/10.1038/ncomms4128
17. Hilvo M, Denkert C, Lehtinen L, Muller B, Brockmoller S, Seppanen-Laakso T, Budczies J, Bucher E, Yetukuri L, Castillo S, Berg E, Nygren H, Sysi-Aho M, Griffin JL, Fiehn O, Loibl S, Richter-Ehrenstein C, Radke C, Hyotylainen T, Kallioniemi O, Iljin K, Oresic M (2011) Novel theranostic opportunities offered by characterization of altered membrane lipid metabolism in breast cancer progression. Cancer Res 71(9):3236–3245. https://doi.org/10.1158/0008-5472.CAN-10-3894
18. Rysman E, Brusselmans K, Scheys K, Timmermans L, Derua R, Munck S, Van Veldhoven PP, Waltregny D, Daniels VW, Machiels J, Vanderhoydonc F, Smans K, Waelkens E, Verhoeven G, Swinnen JV (2010) De novo lipogenesis protects cancer cells from free radicals and chemotherapeutics by promoting membrane lipid saturation. Cancer Res 70(20):8117–8126. https://doi.org/10.1158/0008-5472.CAN-09-3871
19. Soule HD, Maloney TM, Wolman SR, Peterson WD Jr, Brenz R, McGrath CM, Russo J, Pauley RJ, Jones RF, Brooks SC (1990) Isolation and characterization of a spontaneously immortalized human breast epithelial cell line, MCF-10. Cancer Res 50(18):6075–6086
20. Hirsch HA, Iliopoulos D, Joshi A, Zhang Y, Jaeger SA, Bulyk M, Tsichlis PN, Shirley Liu X, Struhl K (2010) A transcriptional signature and common gene networks link cancer with lipid metabolism and

diverse human diseases. Cancer Cell 17(4):348–361. https://doi.org/10.1016/j.ccr.2010.01.022
21. Kuhajda FP, Jenner K, Wood FD, Hennigar RA, Jacobs LB, Dick JD, Pasternack GR (1994) Fatty acid synthesis: a potential selective target for antineoplastic therapy. Proc Natl Acad Sci U S A 91(14):6379–6383
22. Pizer ES, Jackisch C, Wood FD, Pasternack GR, Davidson NE, Kuhajda FP (1996) Inhibition of fatty acid synthesis induces programmed cell death in human breast cancer cells. Cancer Res 56(12):2745–2747
23. Alli PM, Pinn ML, Jaffee EM, McFadden JM, Kuhajda FP (2005) Fatty acid synthase inhibitors are chemopreventive for mammary cancer in neu-N transgenic mice. Oncogene 24(1):39–46. https://doi.org/10.1038/sj.onc.1208174
24. Pollari S, Kakonen SM, Edgren H, Wolf M, Kohonen P, Sara H, Guise T, Nees M, Kallioniemi O (2011) Enhanced serine production by bone metastatic breast cancer cells stimulates osteoclastogenesis. Breast Cancer Res Treat 125(2):421–430. https://doi.org/10.1007/s10549-010-0848-5
25. Locasale JW, Grassian AR, Melman T, Lyssiotis CA, Mattaini KR, Bass AJ, Heffron G, Metallo CM, Muranen T, Sharfi H, Sasaki AT, Anastasiou D, Mullarky E, Vokes NI, Sasaki M, Beroukhim R, Stephanopoulos G, Ligon AH, Meyerson M, Richardson AL, Chin L, Wagner G, Asara JM, Brugge JS, Cantley LC, Vander Heiden MG (2011) Phosphoglycerate dehydrogenase diverts glycolytic flux and contributes to oncogenesis. Nat Genet 43(9):869–874. https://doi.org/10.1038/ng.890
26. Beroukhim R, Mermel CH, Porter D, Wei G, Raychaudhuri S, Donovan J, Barretina J, Boehm JS, Dobson J, Urashima M, Mc Henry KT, Pinchback RM, Ligon AH, Cho YJ, Haery L, Greulich H, Reich M, Winckler W, Lawrence MS, Weir BA, Tanaka KE, Chiang DY, Bass AJ, Loo A, Hoffman C, Prensner J, Liefeld T, Gao Q, Yecies D, Signoretti S, Maher E, Kaye FJ, Sasaki H, Tepper JE, Fletcher JA, Tabernero J, Baselga J, Tsao MS, Demichelis F, Rubin MA, Janne PA, Daly MJ, Nucera C, Levine RL, Ebert BL, Gabriel S, Rustgi AK, Antonescu CR, Ladanyi M, Letai A, Garraway LA, Loda M, Beer DG, True LD, Okamoto A, Pomeroy SL, Singer S, Golub TR, Lander ES, Getz G, Sellers WR, Meyerson M (2010) The landscape of somatic copy-number alteration across human cancers. Nature 463(7283):899–905. https://doi.org/10.1038/nature08822
27. Kim HS, Patel K, Muldoon-Jacobs K, Bisht KS, Aykin-Burns N, Pennington JD, van der Meer R, Nguyen P, Savage J, Owens KM, Vassilopoulos A, Ozden O, Park SH, Singh KK, Abdulkadir SA, Spitz DR, Deng CX, Gius D (2010) SIRT3 is a mitochondria-localized tumor suppressor required for maintenance of mitochondrial integrity and metabolism during stress. Cancer Cell 17(1):41–52. https://doi.org/10.1016/j.ccr.2009.11.023
28. Finley LW, Carracedo A, Lee J, Souza A, Egia A, Zhang J, Teruya-Feldstein J, Moreira PI, Cardoso SM, Clish CB, Pandolfi PP, Haigis MC (2011) SIRT3 opposes reprogramming of cancer cell metabolism through HIF1alpha destabilization. Cancer Cell 19(3):416–428. https://doi.org/10.1016/j.ccr.2011.02.014
29. Elstrom RL, Bauer DE, Buzzai M, Karnauskas R, Harris MH, Plas DR, Zhuang H, Cinalli RM, Alavi A, Rudin CM, Thompson CB (2004) Akt stimulates aerobic glycolysis in cancer cells. Cancer Res 64(11):3892–3899. https://doi.org/10.1158/0008-5472.CAN-03-2904
30. Chan CH, Li CF, Yang WL, Gao Y, Lee SW, Feng Z, Huang HY, Tsai KK, Flores LG, Shao Y, Hazle JD, Yu D, Wei W, Sarbassov D, Hung MC, Nakayama KI, Lin HK (2012) The Skp2-SCF E3 ligase regulates Akt ubiquitination, glycolysis, Herceptin sensitivity, and tumorigenesis. Cell 149(5):1098–1111. https://doi.org/10.1016/j.cell.2012.02.065
31. Robey RB, Hay N (2009) Is Akt the "Warburg kinase"?-Akt-energy metabolism interactions and oncogenesis. Semin Cancer Biol 19(1):25–31. https://doi.org/10.1016/j.semcancer.2008.11.010
32. Jiang S, Zhang LF, Zhang HW, Hu S, Lu MH, Liang S, Li B, Li Y, Li D, Wang ED, Liu MF (2012) A novel miR-155/miR-143 cascade controls glycolysis by regulating hexokinase 2 in breast cancer cells. EMBO J 31(8):1985–1998. https://doi.org/10.1038/emboj.2012.45
33. Wang YP, Zhou LS, Zhao YZ, Wang SW, Chen LL, Liu LX, Ling ZQ, Hu FJ, Sun YP, Zhang JY, Yang C, Yang Y, Xiong Y, Guan KL, Ye D (2014) Regulation of G6PD acetylation by SIRT2 and KAT9 modulates NADPH homeostasis and cell survival during oxidative stress. EMBO J 33(12):1304–1320. https://doi.org/10.1002/embj.201387224
34. Xu SN, Wang TS, Li X, Wang YP (2016) SIRT2 activates G6PD to enhance NADPH production and promote leukaemia cell proliferation. Sci Rep 6:32734. https://doi.org/10.1038/srep32734
35. Du W, Jiang P, Mancuso A, Stonestrom A, Brewer MD, Minn AJ, Mak TW, Wu M, Yang X (2013) TAp73 enhances the pentose phosphate pathway and supports cell proliferation. Nat Cell Biol 15(8):991–1000. https://doi.org/10.1038/ncb2789
36. Deblois G, Giguere V (2013) Oestrogen-related receptors in breast cancer: control of cellular metabolism and beyond. Nat Rev Cancer 13(1):27–36. https://doi.org/10.1038/nrc3396
37. Contractor T, Harris CR (2012) p53 negatively regulates transcription of the pyruvate dehydrogenase kinase Pdk2. Cancer Res 72(2):560–567. https://doi.org/10.1158/0008-5472.CAN-11-1215
38. Lee SY, Jeon HM, Ju MK, Kim CH, Yoon G, Han SI, Park HG, Kang HS (2012) Wnt/snail signaling regulates cytochrome C oxidase and glucose metabolism. Cancer Res 72(14):3607–3617. https://doi.org/10.1158/0008-5472.CAN-12-0006

39. Cai R, Yu T, Huang C, Xia X, Liu X, Gu J, Xue S, Yeh ET, Cheng J (2012) SUMO-specific protease 1 regulates mitochondrial biogenesis through PGC-1alpha. J Biol Chem 287(53):44464–44470. https://doi.org/10.1074/jbc.M112.422626
40. LeBleu VS, O'Connell JT, Gonzalez Herrera KN, Wikman H, Pantel K, Haigis MC, de Carvalho FM, Damascena A, Domingos Chinen LT, Rocha RM, Asara JM, Kalluri R (2014) PGC-1alpha mediates mitochondrial biogenesis and oxidative phosphorylation in cancer cells to promote metastasis. Nat Cell Biol 16(10):992–1003., 1001-1015. https://doi.org/10.1038/ncb3039
41. Rao X, Duan X, Mao W, Li X, Li Z, Li Q, Zheng Z, Xu H, Chen M, Wang PG, Wang Y, Shen B, Yi W (2015) O-GlcNAcylation of G6PD promotes the pentose phosphate pathway and tumor growth. Nat Commun 6:8468. https://doi.org/10.1038/ncomms9468
42. Ferrer CM, Lynch TP, Sodi VL, Falcone JN, Schwab LP, Peacock DL, Vocadlo DJ, Seagroves TN, Reginato MJ (2014) O-GlcNAcylation regulates cancer metabolism and survival stress signaling via regulation of the HIF-1 pathway. Mol Cell 54(5):820–831. https://doi.org/10.1016/j.molcel.2014.04.026
43. Lv L, Li D, Zhao D, Lin R, Chu Y, Zhang H, Zha Z, Liu Y, Li Z, Xu Y, Wang G, Huang Y, Xiong Y, Guan KL, Lei QY (2011) Acetylation targets the M2 isoform of pyruvate kinase for degradation through chaperone-mediated autophagy and promotes tumor growth. Mol Cell 42(6):719–730. https://doi.org/10.1016/j.molcel.2011.04.025
44. Wang HJ, Hsieh YJ, Cheng WC, Lin CP, Lin YS, Yang SF, Chen CC, Izumiya Y, Yu JS, Kung HJ, Wang WC (2014) JMJD5 regulates PKM2 nuclear translocation and reprograms HIF-1alpha-mediated glucose metabolism. Proc Natl Acad Sci U S A 111(1):279–284. https://doi.org/10.1073/pnas.1311249111
45. Favaro E, Bensaad K, Chong MG, Tennant DA, Ferguson DJ, Snell C, Steers G, Turley H, Li JL, Gunther UL, Buffa FM, McIntyre A, Harris AL (2012) Glucose utilization via glycogen phosphorylase sustains proliferation and prevents premature senescence in cancer cells. Cell Metab 16(6):751–764. https://doi.org/10.1016/j.cmet.2012.10.017
46. Zhou X, Wang S, Wang Z, Feng X, Liu P, Lv XB, Li F, Yu FX, Sun Y, Yuan H, Zhu H, Xiong Y, Lei QY, Guan KL (2015) Estrogen regulates hippo signaling via GPER in breast cancer. J Clin Invest 125(5):2123–2135. https://doi.org/10.1172/JCI79573
47. Sorrentino G, Ruggeri N, Specchia V, Cordenonsi M, Mano M, Dupont S, Manfrin A, Ingallina E, Sommaggio R, Piazza S, Rosato A, Piccolo S, Del Sal G (2014) Metabolic control of YAP and TAZ by the mevalonate pathway. Nat Cell Biol 16(4):357–366. https://doi.org/10.1038/ncb2936
48. Wang Z, Wu Y, Wang H, Zhang Y, Mei L, Fang X, Zhang X, Zhang F, Chen H, Liu Y, Jiang Y, Sun S, Zheng Y, Li N, Huang L (2014) Interplay of mevalonate and hippo pathways regulates RHAMM transcription via YAP to modulate breast cancer cell motility. Proc Natl Acad Sci U S A 111(1):E89–E98. https://doi.org/10.1073/pnas.1319190110
49. Bierie B, Pierce SE, Kroeger C, Stover DG, Pattabiraman DR, Thiru P, Liu Donaher J, Reinhardt F, Chaffer CL, Keckesova Z, Weinberg RA (2017) Integrin-beta4 identifies cancer stem cell-enriched populations of partially mesenchymal carcinoma cells. Proc Natl Acad Sci U S A. https://doi.org/10.1073/pnas.1618298114
50. Diehn M, Cho RW, Lobo NA, Kalisky T, Dorie MJ, Kulp AN, Qian D, Lam JS, Ailles LE, Wong M, Joshua B, Kaplan MJ, Wapnir I, Dirbas FM, Somlo G, Garberoglio C, Paz B, Shen J, Lau SK, Quake SR, Brown JM, Weissman IL, Clarke MF (2009) Association of reactive oxygen species levels and radioresistance in cancer stem cells. Nature 458(7239):780–783. https://doi.org/10.1038/nature07733
51. Schieber MS, Chandel NS (2013) ROS links glucose metabolism to breast cancer stem cell and EMT phenotype. Cancer Cell 23(3):265–267. https://doi.org/10.1016/j.ccr.2013.02.021
52. Dong C, Yuan T, Wu Y, Wang Y, Fan TW, Miriyala S, Lin Y, Yao J, Shi J, Kang T, Lorkiewicz P, St Clair D, Hung MC, Evers BM, Zhou BP (2013) Loss of FBP1 by snail-mediated repression provides metabolic advantages in basal-like breast cancer. Cancer Cell 23(3):316–331. https://doi.org/10.1016/j.ccr.2013.01.022
53. Zhao D, Mo Y, Li MT, Zou SW, Cheng ZL, Sun YP, Xiong Y, Guan KL, Lei QY (2014) NOTCH-induced aldehyde dehydrogenase 1A1 deacetylation promotes breast cancer stem cells. J Clin Invest 124(12):5453–5465. https://doi.org/10.1172/JCI76611
54. Hirata N, Yamada S, Shoda T, Kurihara M, Sekino Y, Kanda Y (2014) Sphingosine-1-phosphate promotes expansion of cancer stem cells via S1PR3 by a ligand-independent Notch activation. Nat Commun 5:4806. https://doi.org/10.1038/ncomms5806
55. Doherty JR, Cleveland JL (2013) Targeting lactate metabolism for cancer therapeutics. J Clin Invest 123(9):3685–3692. https://doi.org/10.1172/JCI69741
56. Vegran F, Boidot R, Michiels C, Sonveaux P, Feron O (2011) Lactate influx through the endothelial cell monocarboxylate transporter MCT1 supports an NF-kappaB/IL-8 pathway that drives tumor angiogenesis. Cancer Res 71(7):2550–2560. https://doi.org/10.1158/0008-5472.CAN-10-2828
57. Ho PC, Bihuniak JD, Macintyre AN, Staron M, Liu X, Amezquita R, Tsui YC, Cui G, Micevic G, Perales JC, Kleinstein SH, Abel ED, Insogna KL, Feske S, Locasale JW, Bosenberg MW, Rathmell JC, Kaech SM (2015) Phosphoenolpyruvate is a metabolic checkpoint of anti-tumor T cell responses. Cell 162(6):1217–1228. https://doi.org/10.1016/j.cell.2015.08.012
58. Cui G, Staron MM, Gray SM, Ho PC, Amezquita RA, Wu J, Kaech SM (2015) IL-7-induced glycerol

transport and TAG synthesis promotes memory CD8+ T cell longevity. Cell 161(4):750–761. https://doi.org/10.1016/j.cell.2015.03.021
59. Cheng G, Zielonka J, Dranka BP, McAllister D, Mackinnon AC Jr, Joseph J, Kalyanaraman B (2012) Mitochondria-targeted drugs synergize with 2-deoxyglucose to trigger breast cancer cell death. Cancer Res 72(10):2634–2644. https://doi.org/10.1158/0008-5472.CAN-11-3928
60. Zhao Y, Liu H, Liu Z, Ding Y, Ledoux SP, Wilson GL, Voellmy R, Lin Y, Lin W, Nahta R, Liu B, Fodstad O, Chen J, Wu Y, Price JE, Tan M (2011) Overcoming trastuzumab resistance in breast cancer by targeting dysregulated glucose metabolism. Cancer Res 71(13):4585–4597. https://doi.org/10.1158/0008-5472.CAN-11-0127
61. Thornburg JM, Nelson KK, Clem BF, Lane AN, Arumugam S, Simmons A, Eaton JW, Telang S, Chesney J (2008) Targeting aspartate aminotransferase in breast cancer. Breast Cancer Res 10(5):R84. https://doi.org/10.1186/bcr2154
62. von Moos R, Costa L, Ripamonti CI, Niepel D, Santini D (2017) Improving quality of life in patients with advanced cancer: targeting metastatic bone pain. Eur J Cancer 71:80–94. https://doi.org/10.1016/j.ejca.2016.10.021
63. Jahnke W, Rondeau JM, Cotesta S, Marzinzik A, Pelle X, Geiser M, Strauss A, Gotte M, Bitsch F, Hemmig R, Henry C, Lehmann S, Glickman JF, Roddy TP, Stout SJ, Green JR (2010) Allosteric non-bisphosphonate FPPS inhibitors identified by fragment-based discovery. Nat Chem Biol 6(9):660–666. https://doi.org/10.1038/nchembio.421
64. Rajendran JG, Mankoff DA, O'Sullivan F, Peterson LM, Schwartz DL, Conrad EU, Spence AM, Muzi M, Farwell DG, Krohn KA (2004) Hypoxia and glucose metabolism in malignant tumors: evaluation by [18F] fluoromisonidazole and [18F]fluorodeoxyglucose positron emission tomography imaging. Clin Cancer Res 10(7):2245–2252
65. Palaskas N, Larson SM, Schultz N, Komisopoulou E, Wong J, Rohle D, Campos C, Yannuzzi N, Osborne JR, Linkov I, Kastenhuber ER, Taschereau R, Plaisier SB, Tran C, Heguy A, Wu H, Sander C, Phelps ME, Brennan C, Port E, Huse JT, Graeber TG, Mellinghoff IK (2011) 18F-fluorodeoxy-glucose positron emission tomography marks MYC-overexpressing human basal-like breast cancers. Cancer Res 71(15):5164–5174. https://doi.org/10.1158/0008-5472.CAN-10-4633
66. Walsh AJ, Cook RS, Manning HC, Hicks DJ, Lafontant A, Arteaga CL, Skala MC (2013) Optical metabolic imaging identifies glycolytic levels, subtypes, and early-treatment response in breast cancer. Cancer Res 73(20):6164–6174. https://doi.org/10.1158/0008-5472.CAN-13-0527
67. Slupsky CM, Steed H, Wells TH, Dabbs K, Schepansky A, Capstick V, Faught W, Sawyer MB (2010) Urine metabolite analysis offers potential early diagnosis of ovarian and breast cancers. Clin Cancer Res 16(23):5835–5841. https://doi.org/10.1158/1078-0432.CCR-10-1434
68. Vrieling A, Hein R, Abbas S, Schneeweiss A, Flesch-Janys D, Chang-Claude J (2011) Serum 25-hydroxyvitamin D and postmenopausal breast cancer survival: a prospective patient cohort study. Breast Cancer Res 13(4):R74. https://doi.org/10.1186/bcr2920
69. Wu Q, Ishikawa T, Sirianni R, Tang H, McDonald JG, Yuhanna IS, Thompson B, Girard L, Mineo C, Brekken RA, Umetani M, Euhus DM, Xie Y, Shaul PW (2013) 27-Hydroxycholesterol promotes cell-autonomous, ER-positive breast cancer growth. Cell Rep 5(3):637–645. https://doi.org/10.1016/j.celrep.2013.10.006
70. Nelson ER, Wardell SE, Jasper JS, Park S, Suchindran S, Howe MK, Carver NJ, Pillai RV, Sullivan PM, Sondhi V, Umetani M, Geradts J, McDonnell DP (2013) 27-Hydroxycholesterol links hypercholesterolemia and breast cancer pathophysiology. Science 342(6162):1094–1098. https://doi.org/10.1126/science.1241908
71. Wang Y, Xiao M, Chen X, Chen L, Xu Y, Lv L, Wang P, Yang H, Ma S, Lin H, Jiao B, Ren R, Ye D, Guan KL, Xiong Y (2015) WT1 recruits TET2 to regulate its target gene expression and suppress leukemia cell proliferation. Mol Cell 57(4):662–673. https://doi.org/10.1016/j.molcel.2014.12.023
72. Terunuma A, Putluri N, Mishra P, Mathe EA, Dorsey TH, Yi M, Wallace TA, Issaq HJ, Zhou M, Killian JK, Stevenson HS, Karoly ED, Chan K, Samanta S, Prieto D, Hsu TY, Kurley SJ, Putluri V, Sonavane R, Edelman DC, Wulff J, Starks AM, Yang Y, Kittles RA, Yfantis HG, Lee DH, Ioffe OB, Schiff R, Stephens RM, Meltzer PS, Veenstra TD, Westbrook TF, Sreekumar A, Ambs S (2014) MYC-driven accumulation of 2-hydroxyglutarate is associated with breast cancer prognosis. J Clin Invest 124(1):398–412. https://doi.org/10.1172/JCI71180
73. Llaverias G, Danilo C, Mercier I, Daumer K, Capozza F, Williams TM, Sotgia F, Lisanti MP, Frank PG (2011) Role of cholesterol in the development and progression of breast cancer. Am J Pathol 178(1):402–412. https://doi.org/10.1016/j.ajpath.2010.11.005

Metabolic Changes During Cancer Cachexia Pathogenesis

11

Ng Shyh-Chang

Abstract

Wasting of adipose tissue and skeletal muscle is a hallmark of metastatic cancer and a major cause of death. Like patients with cachexia caused by other chronic infections or inflammatory diseases, the cancer subject manifests both malnutrition and metabolic stress. Both carbohydrate utilization and amino acid incorporation are decreased in the muscles of cancer cachexia patients. Cancer cells affect host metabolism in two ways: (a) their own metabolism of nutrients into other metabolites and (b) circulating factors they secrete or induce the host to secrete. Accelerated glycolysis and lactate production, i.e., the Warburg effect and the resultant increase in Cori cycle activity, are the most widely discussed metabolic effects. Meanwhile, although a large number of pro-cachexia circulating factors have been found, such as TNFa, IL-6, myostatin, and PTHrp, none have been shown to be a dominant factor that can be targeted singly to treat cancer cachexia in humans. It is possible that given the complex multifactorial nature of the cachexia secretome, and the personalized differences between cancer patients, targeting any single circulating factor would always be insufficient to treat cachexia for all patients. Here we review the metabolic changes that occur in response to tumor growth and tumor-secreted factors during cachexia.

Keywords

Cachexia • Cancer • Metabolism • Circulating factors

11.1 Introduction to Cancer Cachexia

Involuntary weight loss, or cachexia, is a complication frequently observed in cancer. During cancer progression, the patient's muscle mass and

N. Shyh-Chang (✉)
Genome Institute of Singapore, Agency for Science Technology and Research, Singapore, Singapore
e-mail: ngsc1@gis.a-star.edu.sg

E. Song, H. Hu (eds.), *Translational Research in Breast Cancer*, Advances in Experimental Medicine and Biology 1026, https://doi.org/10.1007/978-981-10-6020-5_11

adipose mass are often depleted, while other organs tend to be spared. Concurrently, the patient develops severe muscle weakness and becomes bedridden due to loss of mobility and normal cardiopulmonary muscle function. In the late stage, when anorexia becomes severe, the patient's total body weight will decline rapidly with immune functions being severely compromised. Even though cachexia has been extensively studied in both humans and rodents for its impact on quality of life and survival of cancer patients, the topic remains poorly understood [1].

About 90% of cancer patient deaths are caused by metastasis [2, 3], and cachexia is found in 80% of late-stage cancer patients, frequently co-occurring with metastasis [4–6]. Cachexia weakens the body, which may worsen metastatic disease. Since cachectic patients cannot tolerate the side effects of chemotherapy, curative chemotherapies have to be terminated prematurely [7]. Conversely, metastatic tumors may promote cachexia by increasing the production of pro-cachexia factors. In oncology practice, cachexia is widely considered as a late-stage complication in lung or breast cancer (typically detected at early stages) and an early complication in gastric or pancreatic cancer (typically detected at late stages), which reflects the association between cachexia and metastasis in late-stage cancers. Generally, cachexia is correlated with late-stage cancer and thought to directly cause 20–40% of cancer deaths [8, 9].

At the point of diagnosis, 54% of cancer patients lost some weight, and 32% lost >5% weight within half a year. The incidence rate of cachectic weight loss ranges from 30% (non-Hodgkin's lymphoma) to 87% (gastric carcinoma) [10]. However, it is very likely that the true incidence rate of cancer cachexia is being underestimated in many types of cancer, including breast cancer [9]. In a study of over 8000 cancer patients, only 2.5% were documented with the ICD-9 code for cachexia, but a retrospective investigation revealed that 23% of the patients conformed to the clinical definition of cachexia, which indicates a tenfold underestimation [9].

Almost a century ago, Warren [11] attempted to determine the cause of mortality in half a thousand cancer patients that he autopsied. In 23% of these patients, the data demonstrated progressive wasting and weakness, and no other cause of death was presented upon autopsy. Warren concluded that these 23% patients had died directly from cachexia [11]. Cachexia was also diagnosed in many of the other cancer patients and thus could have compounded other causes of mortality. While some investigators failed to mention cachexia as a cause of mortality [12], others listed cachexia as the cause of death in about two thirds of cancer patients [13]. Regardless, cachexia is known to compromise the immune system, wound repair, heart, lung, liver, and kidney functions, all of which are likely to directly worsen the prognosis of cancer patients.

By relative proportions, cachexia is often documented in lung, head and neck, pancreatic, gastric, and colorectal cancers [14] but rarely in breast cancer. However, recent studies have shown that cachexia does develop in breast cancer patients [15], and it is closely correlated with bone metastases [7, 16]. By absolute numbers, breast cancer is actually the most common cancer (31%) among cancer cachexia patients, which is due to the high frequency of breast cancer patients [9].

It should be noted that cachexia is prevalent not only in cancer but also in many other chronic diseases. About 30–50% of hospital patients manifest some degree of undernutrition and weight loss [17], and most nonneoplastic chronic diseases terminate in cachexia, such as chronic disseminated infections and chronic cardiac, pulmonary, hepatic, or renal diseases.

11.2 Cancer Cachexia as a Degenerative Metabolic Syndrome

As mentioned above, skeletal muscle mass decreases progressively during the course of cancer growth and dissemination. Fast-twitch glycolytic myofibers usually undergo wasting more severely than slow-twitch oxidative myofibers, and myofibrillar proteins tend to be lost more rapidly than the sarcoplasmic proteins [18].

Incorporation of radioactively labeled amino acids [18, 19] into muscle proteins tends to be much lower in tumor-bearing animals than control animals. This occurs regardless of whether the animals were ad lib fed [19] or pair-fed [18], suggesting that the depressed anabolic flux of amino acids is independent of food intake. This depression is more marked in the fast-twitch muscles than in the slow-twitch muscles and more marked in myofibrillar than in sarcoplasmic proteins [18].

Glucose catabolism is also impeded in the skeletal muscles of cachectic cancer patients. The rates of uptake and subsequent conversion of glucose to glycogen, lactate, or CO_2 are all slowed down, and multiple enzymes involved in glycolysis and glucose oxidation are also suppressed [19, 20].

Another hallmark of cancer cachexia is the rapid loss of adipose mass, which even exceeds the rate observed during uncomplicated starvation [21]. Lipogenesis is lower in tumor-bearing animals with cachexia but normal in tumor-bearing animals without cachexia [22]. Lipolysis and fatty acid oxidation rates are similar in cancer patients and normal controls during fasting [23]. However, after a glucose infusion, cancer patients showed a much lower decrease in free fatty acid oxidation than in controls, suggesting a continuously high, fasting-like rate of fatty acid oxidation even in fed conditions. In fact, many clinical studies have suggested or indicated a higher basal rate of metabolism in cancer patients, than in undernourished patients, who show a lower basal rate of metabolism compared to normal subjects [24, 25] (Fig. 11.1).

Nevertheless, in most of these studies, it is impossible to rule out the contributions of recent surgery, infection, or fever to basal metabolism, all of which are also frequently observed in cancer patients. Thus the effect of cancer cachexia per se on whole-body basal metabolism has remained unclear.

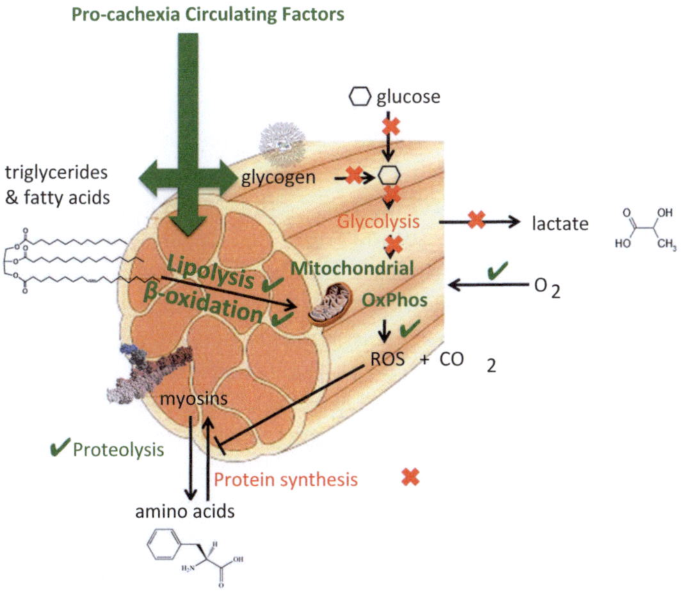

Fig. 11.1. Metabolic dysfunction in skeletal muscles during cancer cachexia. A complex mixture of pro-cachexia circulating factors converges to induce excessive lipolysis and fatty acid β-oxidation in myofibers. In contrast, glucose uptake and oxidation are suppressed. The net result is a higher basal rate of metabolism due to increased mitochondrial oxidative phosphorylation (OxPhos). Increased reactive oxygen species (ROS) and oxidative stress lead to decreased protein synthesis and increased proteolysis and specific loss of the myosin proteins and result in skeletal muscle wasting during cancer cachexia

11.3 Mechanisms for Cancer Cachexia

11.3.1 Protein Homeostasis

In the normal adult human, without environmental insults or stimuli such as exercise, skeletal muscle mass remains constant because protein degradation and synthesis are kept in balance. But during cancer cachexia, muscle wasting manifests rapidly. This must be due to increased protein degradation, decreased protein synthesis, or both. An early study suggested that the cachectic muscle wasting is primarily due to decreased protein synthesis [26]. Another study agrees with this conclusion, after quantifying myofibrillar protein degradation in cachectic and non-cachectic patients in vivo and finding no significant differences [27]. Yet other studies have ascribed the cachectic muscle wasting to an increased rate of protein degradation, by quantifying the amino acids released [28].

Findings from animal models of cachexia suggest that both protein synthesis and degradation are important, so any treatment of cachexia will need to address both processes. Fast-twitch muscles, such as gastrocnemius, undergo atrophy more rapidly than slow-twitch oxidative muscles in animal models of cachexia [29]. This is thought to be due to increased protein degradation in fast-twitch fibers [30]. The myosin isoform also undergoes a switch, as the slow-twitch isoform decreases and the fast-twitch isoform increases during cachexia in animal models [31]. However, human cancer cachexia patients manifest a loss of both slow- and fast-twitch myofibers [32–34], contradicting these findings in animal models.

Nonetheless, both animal models and human patient findings suggest that the dysfunctional homeostasis of specific muscle proteins plays an important role in cachectic muscle wasting. Indeed, myofibrillar proteins, which constitute nearly 50% of the total muscle protein, are lost most rapidly during muscle wasting. In cachectic muscle wasting, myosins are selectively lost, while other myofibrillar proteins such as sarcomeric actin A, troponins, and tropomyosins remain relatively constant [29].

11.3.1.1 Protein Synthesis

In muscles, protein synthesis is mainly controlled by the translation initiation phase. There are two primary control mechanisms. The first is via the 40S ribosomal subunit binding to the initiator methionyl-tRNA, a step that is controlled by eIF2. The eIF2B subunit can be inhibited by phosphorylation by four different kinases, including the double-stranded RNA-dependent protein kinase (PKR) which responds to extrinsic stimuli, leading to inhibition of protein translation initiation [35].

The second mechanism is controlled by eIF4F complex, which controls the binding of 40S ribosomal subunit to mRNA containing the 5′-cap and m7GpppX motif [36]. The eIF4E subunit binding the 5′-cap is one of the main regulatory factors. Its concentration is regulated via sequestration by its binding proteins (4EBP1/2; 163). Phosphorylation of 4EBP1/4EBP2 releases eIF4E for assembly of the eIF4F complex. 4EBP1 and 4EBP2's phosphorylation is mediated by PI3K and mTOR signaling [35, 36]. The mTOR kinase also activates the ribosomal protein S6 kinase (S6 K) to enhance the translation of 5′-poly-pyrimidine tract (5′-TOP) mRNAs. 5′-TOP mRNAs often encode proteins involved in protein translation [37]. Thus the PI3K-mTOR pathway regulates several mechanisms to control protein synthesis.

While some murine models of cachexia suggest that the decreased muscle protein synthesis is due to anorexia, it is also decreased in other cachexia models without anorexia [38]. This suggests a fundamental defect in protein synthesis during cachexia. When mice bearing the cachectic MAC16 tumor lose > 15% weight, their gastrocnemius muscles also show PKR phosphorylation, thus increasing phospho-eIF2B to decrease translation initiation [39]. In cachectic esophagogastric cancer patients, both PKR and eIF2B phosphorylations are also increased, compared with healthy controls [33]. The increased phospho-eIF2B could partially explain the loss of myofibrillar proteins, since the myosin abundance is linearly correlated withphospho-eIF2B. Weight loss in MAC16-bearing mice is also associated with decreased mTOR signaling, leading to a decrease in phospho-4EBP1, resulting in a decreased concentration of the active eIF4F complex for protein translation [40].

In addition to contributing directly toward protein synthesis, branched-chain amino acids (BCAAs) can also promote protein synthesis by inducing the mTOR pathway. Leucine is the most potent BCAA and could ameliorate the weight loss in MAC16-bearing mice, via increased mTOR signaling and muscle mass [40]. These results support the supplementation of leucine or other mTOR agonists in the diets of cancer cachexia patients, to boost protein synthesis and treat muscle atrophy, assuming the patients' cancers are not dependent on mTOR for growth.

However the counterargument to a simple protein synthesis theory is that, with a long half-life of 2–4 weeks in rodents, myofibrillar myosins are very stable [41, 42]. The half-life of human myosins is likely to be even longer, due to their longer lifespan and lower metabolic rate than rodents. Thus a decrease in protein synthesis is not consistent with the rapid muscle loss in cachexia. Furthermore, inhibition of protein synthesis via rapamycin, which inhibits mTOR-enhanced hypertrophy, did not cause muscle atrophy in mice [43, 44]. These findings suggest a complex interplay between protein synthesis and protein degradation exists to control muscle atrophy.

11.3.1.2 Proteasomes and Protein Degradation

The ubiquitin-proteasome system disassembles and degrades protein complexes in all cells [45, 46]. Results from murine models of cancer cachexia suggest that the ubiquitin-proteasome system plays an important role in myofibrillar protein degradation [47] and that the PI3K-AKT-FoxO3 pathway controls the ubiquitin-proteasome pathway in muscles [48].

In the ubiquitin-proteasome pathway, proteins are marked for degradation by the proteasome via a chain of ubiquitin. Ubiquitination occurs through the ubiquitin protein ligase (E3), which catalyzes the covalent attachment of ubiquitin to specific protein targets. Ubiquitinated proteins will then be transferred to the 26S proteasome, where they undergo degradation. Thus E3 ligases are the factors that recognize specific structural motifs and mediate target specificity. Two muscle-specific E3 ligases, muscle RING finger 1 (MuRF1) and atrogin-1/muscle atrophy F-box (MAFbx), are highly expressed in a wide range of muscle atrophy conditions [49, 50]. MAFbx overexpression causes myotube atrophy in vitro, while MAFbx and MuRF1 knockout mice are resistant to atrophy in vivo. These E3 ligases are also highly upregulated in murine models of cancer cachexia [51]. MAFbx targets sarcomeric proteins, including desmin, vimentin, and myosins, as well as protein translational machinery, glycolysis enzymes, mitochondrial proteins, and transcription factors such as MyoD [52]. MuRF1 ubiquitinates several myofibrillar proteins, including myosin heavy and light chains, sarcomeric actin A, and troponins [53–57].

However, the two E3 ligases' broad spectrum of targets is actually inconsistent with the specific loss of myosins, not other myofibrillar proteins, during cancer cachexia. Furthermore, no definitive data exists to prove their roles in causing cancer cachexia. In fact, although MuRF1 and MAFbx knockout mice are available, and even if they are protected from certain forms of atrophy, nobody has demonstrated that these knockout mice are protected from cancer cachexia [58]. Most importantly, several studies have shown that the ubiquitin-proteasome system is not activated in human cachexia patients [59, 60]. This would suggest that the ubiquitin-proteasome system is less important in human cachexia.

11.3.1.3 Lysosomes and Autophagy

The lysosomal system of protein degradation includes the aspartate protease cathepsin D and the cysteine proteases cathepsins B, H, and L. In some cancer cachexia patients, muscle biopsies showed increased lysosomal cathepsin B and no change in the ubiquitin-proteasome system [61]. Another early study showed that increased lysosomal cathepsin D in the muscle biopsies of cancer patients correlated with increased weight loss [62].

Although lysosomal protein degradation is well characterized in muscle wasting, the importance of autophagy was poorly understood until recently [63–66]. Autophagy is a conserved process whereby macromolecules and organelles are degraded with lysosomes and recycled for cellular survival. Autophagy proceeds through six specific steps: (1) induction, (2) commitment to phagophore formation, (3) elongation for autophagosome formation, (4) fusion of autophagosome with lysosomes, (5) lysosomal degradation and recycling, and (6) lysosomal regeneration.

As expected, autophagy is increased in the cachectic muscles of tumor-bearing rodents [67, 68] and in many other contexts [69–81]. In human patients, autophagy is activated in the muscles of esophageal cancer patients [82], whereas Atg5 and beclin1, both critical factors in autophagy activation, are upregulated in the cachectic muscles of gastrointestinal and pancreatic cancer patients [34]. Gabarap, which regulates autophagosome elongation and fusion [83], is also upregulated in the muscles of gastrointestinal cancer patients [84].

Intriguingly, aging-induced muscle wasting, or sarcopenia, is often characterized by a downregulation in autophagy instead [66, 70, 85, 86]. The extant literature suggests that autophagy is activated during cancer but has not addressed if autophagy is a cause or a consequence of cachexia. Indeed autophagy induction in cachectic muscles could be simply a metabolic response to the bioenergetic stresses induced in muscles during cancer. It was found that a downregulation of autophagy, by tamoxifen-inducible knockout of Atg7 in mouse models of lung cancer, greatly suppressed cancer cell proliferation. But it failed to block muscle and adipose wasting [87]. In fact, muscle-specific Atg7 deletion caused an accumulation of oxidative stress in muscle mitochondria, muscle wasting, and ultimately premature death [86–88]. Thus, autophagy is actually required for the survival and optimal metabolic function of myofibers. Seen in this light, the induction of lysosomal autophagy in the muscles of various cachexia models and human cancer patients is more likely to be an adaptive response, a consequence not a cause of cachexia, to preserve energy and ameliorate dysfunctional metabolism during cancer cachexia.

11.3.2 Circulating Factors

The myriad effects on muscle metabolism, protein synthesis, and degradation during cancer cachexia are likely due to extrinsic circulating factors. Some of these factors originate from the cancer cells themselves, while others originate from the host tissues' inflammatory response to cancer. The existence of circulating factors that influence host metabolism was first demonstrated by Lucke et al. in the 1950s through parabiosis experiments between tumor-bearing rats and tumor-free rats [89]. However, the list of new factors has been growing ever since and likely remains incomplete to this day.

11.3.2.1 TNFα

TNFα or tumor necrosis factor alpha is a proinflammatory cytokine that was originally named "cachectin" [90]. TNFα plays a critical role in regulating muscle regeneration and muscle protein degradation [91–97]. TNFα is required to activate muscle regeneration, but TNFα also appears to prevent normal myoblast differentiation [98, 99].

There is significant evidence that TNFα causes cachectic muscle wasting in animal models, even though its role in humans has remained controversial. Transplantation of TNFα-overexpressing cells produces a cachexia-like syndrome in animals [100]. Acute injection of recombinant TNFα into rats led to increased muscle protein degradation and decreased muscle protein synthesis [101]. TNFα receptor knockout mice showed reduced muscle wasting in response to Lewis lung carcinoma, compared to control mice, despite equivalent levels of plasma TNFα [93]. TNFα is known to activate NFKB, which also induces the ubiquitin-proteasome system [102]. Moreover, TNFα has been found to increase ROS and p38 MAPK activation, to induce expression of the E3 ligase MAFbx [102]. p38 MAPK has been found to regulate muscle catabolism and myogenesis [103–106].

However, the relevance of TNFα to human cachexia remains unclear, as questions remain over whether the levels of TNFα actually do increase in cachectic cancer patients [107].

Recent clinical trials of anti-TNFα antibodies for cancer cachexia have shown little to no efficacy [108]. It is possible that TNFα induces cachexia in only a specific subgroup of cancer patients and that effective treatment is only possible after personalized diagnosis of TNFα levels. An even more likely scenario is that cancer cachexia is due to multiple circulating factors, and inhibiting any particular factor singly will always be insufficient.

11.3.2.2 Interleukins

Many cancer types can produce pro-inflammatory interleukins (e.g., IL-1a/IL-1b, IL-6, IL-8), and their effects can synergize with the host immune system's interleukins as well. One prominent candidate that cooperates with TNFα to drive inflammation during cancer is IL-6. Serum levels of IL-6 are correlated with weight loss in cancer patients and, importantly, patient survival [109, 110]. Gain-of-function and loss-of-function studies also demonstrate that IL-6 is necessary to induce cachexia in murine models [111–113]. In rats, IL-6 injections induce both total and myofibrillar proteolyses in muscles [114].

However, IL-6 injection experiments suggest that only supraphysiological doses of IL-6 can induce muscle wasting [115]. Some other studies have failed to induce muscle wasting with IL-6 injections in mice, even after repeated administration [116]. Moreover, recent clinical trials of an anti-IL-6 antibody in cachectic lung cancer patients failed to protect against muscle wasting [117].

11.3.2.3 Myostatin, Activin, and GDF11

Myostatin, or GDF8, is a member of the TGFβ super family of ligands that causes muscle wasting [118–122]. Myostatin binds and activates the activin receptor IIb (ActRIIB) to drive its effects on muscle wasting [123]. Myostatin suppresses myocellular growth, as loss-of-function mutations lead to muscle hypertrophy whereas gain-of-function leads to muscle atrophy [124]. Myostatin's effects on muscular hypertrophy are mediated via its effects on synthesis and degradation of muscle protein [123]. Moreover, myostatin also inhibits muscle regeneration. Specifically, myostatin blocks MyoD expression in muscle stem cells and myoblasts [125]. Without MyoD activation, muscle regeneration is suppressed because myoblasts cannot commit to fuse and differentiate into myofibers to repair injured muscles. Few studies have found myostatin to increase during cancer cachexia, but one study has found that myostatin expression increases in pre-cachectic gastric cancer patients [126], and others have found that myostatin increases in other types of cachexia due to chronic heart failure, COPD, and HIV/AIDS [127, 128]. Interestingly, many homologues of myostatin and ligands of ActRIIB, such as activin A and GDF11, have been found to change during the course of cancer cachexia and aging, respectively [129–132].

Several drugs have been developed to inhibit myostatin/ActRIIB signaling, mostly via ActRIIB decoys or antibodies [133–137]. Specifically in cancer cachexia, multiple studies have shown that ActRIIB inhibition can prevent muscle atrophy [133–137]. However, a Phase IIb/PhaseIII clinical trial for a promising anti-ActRII antibody failed recently, unable to meet its primary end point for a rare but severe form of muscle atrophy [138]. Several clinical trials for myostatin inhibitors have all failed earlier as well [138]. Further work will be required to determine if the myostatin/ActRIIB pathway still represents a viable therapeutic target in cachexia.

11.3.2.4 Proteolysis-Inducing Factor

Proteolysis-inducing factor (PIF) is a small glycoprotein first derived from the cachectic mouse MAC16 cell line [139]. The mouse C26 adenocarcinoma was found to produce PIF, whereas a variant of C26 which had lost its PIF gene became non-cachectic [140]. PIF injection induces an extraordinarily rapid 10% weight loss within 24 h. Unlike TNFα, PIF-induced cachexia occurs without anorexia and directly affects muscle mass, via a 50% increase in protein degradation and a 50% decrease in protein synthesis [141]. In murine myotube cultures, PIF causes specific loss of myosin proteins, whereas sarcomeric actin A remains constant [142]. PIF also leads to an upregulation of ubiquitin-proteasome system via NFKB signaling [142].

All the findings above are based on studies of mouse PIF. Some labeling studies and enzymatic deglycosylation studies have suggested that the PIF produced by the cachectic human G361 cell line is identical to the mouse PIF in mass and contains similar N- and O-glycosylation chains [143]. Subsequent studies have uncovered the existence of human PIF in other metastatic human cell lines and the urine of cancer cachexia patients [144]. Nevertheless, the role and existence of human PIF in cancer cachexia patients remain controversial [145].

11.3.2.5 Parathyroid Hormone-Related Peptide

In the 1940s, Fuller Albright postulated the existence of a tumor factor that caused cancer-induced hypercalcemia [146]. Nearly 50 years later, several groups discovered a parathyroid hormone-like peptide (PTHrp) that was secreted from tumors and caused the hypercalcemia of malignancy [147–150]. Later studies showed that PTHrp correlated with cancer cachexia, and it was suggested that severe hypercalcemia caused muscle wasting [151].

A recent study suggested that during cancer cachexia, "browning" of white adipose tissue occurred to increase total energy expenditure and futile cycles of thermogenic oxidation [152]. In the Lewis lung carcinoma model of cachexia, PTHrp directly activated the thermogenesis genes in adipose tissues to cause "browning," while blocking of PTHrp prevented "browning" [153]. In lung and colorectal cancer patients, serum PTHrp correlated with muscle wasting and energy expenditure [153]. Although an anti-PTHrp antibody ameliorated cancer cachexia, it did not completely inhibit it, suggesting that other circulating factors synergize with PTHrp to cause browning and cachectic wasting [153].

11.3.2.6 Hypermetabolism

Total energy expenditure (TEE), estimated via oxygen consumption, can be subclassified into physically active energy expenditure (AEE), diet-induced energy expenditure (DEE), and resting energy expenditure (REE). In most sedentary people, ~70% of the TEE can be ascribed to REE. About half of cancer patients are considered hypermetabolic (REE > 110% of expected), and weight-stable patients tend to have a lower REE [154]. The REE of cancer patients is highly dependent on the cancer type. REE is significantly higher in both lung cancer and pancreatic cancer patients, but not as much in gastrointestinal cancer patients [155, 156]. Because malnourished patients near death show a higher REE [157], these REE correlations might be confounded by the patients' cancer stage and proximity to death at the time of diagnosis. Similarly, REE varies widely between different animal models of cachexia and different stages of observation, with several manifesting high REE during early stages of cancer cachexia, but which decreases with disease progression [158, 159].

The mechanisms causing hypermetabolism in cancer cachexia are very complex. Hypermetabolism has been linked to systemic inflammation [155] or an elevated adrenergic state [160]. Human tumors are known to manifest high glucose uptake, glycolysis, and lactate production even in the presence of oxygen, i.e., the Warburg effect [161]. Excess lactate produced by the tumor can be converted back to glucose via the Cori cycle in the liver, which consumes ATP. Cancer cachexia patients are known to manifest higher rates of ATP-consuming Cori cycle activity, although the effects are variable [162, 163].

Another major cause for a higher REE in cancer cachexia patients is the increased fatty acid oxidation in the brown adipose tissue (BAT) and skeletal muscles. Normally, BAT controls both body temperature and energy balance in many mammals, including humans, but generally there is little BAT remaining in humans by adulthood. However, an early study has shown that periadrenal BAT is present in 80% of cancer cachexia patients, compared with only 13% of age-matched controls [164]. And as mentioned above, mouse models of cancer cachexia also manifest adipose browning, in part due to increased PTHrp secretion by cachectic tumors [152, 153]. Nevertheless, it remained unclear how adipose browning could cause muscle wasting. In most studies, hypermetabolism and energy wasting were regarded as undesirable consequences of cancer cachexia.

Most recently, we have found that excessive fatty acid oxidation in skeletal muscles induces muscle atrophy during cancer cachexia [106]. Previous studies showed that fatty acid oxidation was increased during cachexia, but it was never considered a driving cause of cachexia [1, 4–6]. We found that complex pro-inflammatory factors converged to trigger excessive catabolism of fatty acids in muscles, causing excessive mitochondrial ROS (Fig. 11.1), and thus activation of the aging-associated p38 MAPK signaling pathway [104, 105], which led to progressive muscle wasting. In fact, many previous studies have already shown that multiple pro-inflammatory factors such as TNFα, IL-1, IL-6, IL-8, prostaglandins, and the zinc-alpha2-glycoprotein ZAG can induce lipolysis and fatty acid oxidation [165–168]. Moreover, it is well known that ROS can potently activate the stress-inducible p38 MAPK [169]. Pharmacological inhibition of fatty acid oxidation rescues p38 MAPK activation and prevents muscle atrophy during cancer cachexia [106].

ROS may be a common agent for inducing muscle atrophy under other conditions too (Fig. 11.1). Mice deficient in superoxide dismutase (SOD), the major Cu-/Zn-dependent antioxidant enzyme, show accelerated muscle wasting due to increased oxidative stress [170]. Hydrogen peroxide is also known to increase muscle protein degradation in murine myotubes [171].

11.4 Conclusions

Wasting of adipose tissue and skeletal muscle is a hallmark of metastatic cancer and a major cause of death. Like patients with cachexia caused by other chronic infections or inflammatory diseases, the cancer subject manifests both malnutrition and metabolic stress. Both carbohydrate utilization and amino acid incorporation are decreased in the muscles of cancer cachexia patients. However, the cancer cells' rapid proliferation alone is insufficient to cause cachexia, as fetal cells are also rapidly proliferating during uncomplicated pregnancy. The truly unique feature of cancer is the cancer cells themselves, which affect host metabolism in two ways: (a) their own metabolism of nutrients into other metabolites and (b) circulating factors they secrete or induce the host to secrete. Accelerated glycolysis and lactate production, i.e., the Warburg effect, and the resultant increase in Cori cycle activity are the most widely discussed metabolic effects. Meanwhile, although a large number of pro-cachexia circulating factors have been found, such as TNFα, IL-6, myostatin, and PTHrp, none have been shown to be a dominant factor that can be targeted singly to treat cancer cachexia in humans.

It is possible that given the complex multifactorial nature of the cachexia secretome, and the personalized differences between cancer patients, targeting any single circulating factor would always be insufficient to treat cachexia for all patients [1, 4–6]. Another approach would be to use broad-spectrum, anti-inflammatory drugs such as NSAIDs. However, owing to their chronic gastrointestinal and cardiovascular toxicity profiles, and weak efficacies, NSAIDs have been ineffective for treating cachexia [172]. One final approach would be to target the hypermetabolism and oxidative stress that these circulating factors convergently produce in skeletal muscles, to treat muscle wasting (Fig. 11.1). Our preclinical results [106] suggest that this approach might be feasible.

References

1. Argiles JM, Busquets S, Stemmler B, Lopez-Soriano FJ (2014) Cancer cachexia: understanding the molecular basis. Nat Rev Cancer 14(11):754–762. doi:10.1038/nrc3829
2. Chaffer CL, Weinberg RA (2011) A perspective on cancer cell metastasis. Science 331(6024):1559–1564. doi:10.1126/science.1203543
3. Massague J, Obenauf AC (2016) Metastatic colonization by circulating tumour cells. Nature 529(7586):298–306. doi:10.1038/nature17038

4. Tisdale MJ (2009) Mechanisms of cancer cachexia. Physiol Rev 89(2):381–410. doi:10.1152/physrev.00016.2008
5. Spano D, Heck C, De Antonellis P, Christofori G, Zollo M (2012) Molecular networks that regulate cancer metastasis. Semin Cancer Biol 22(3):234–249. doi:10.1016/j.semcancer.2012.03.006
6. Fearon KC, Glass DJ, Guttridge DC (2012) Cancer cachexia: mediators, signaling, and metabolic pathways. Cell Metab 16(2):153–166. doi:10.1016/j.cmet.2012.06.011
7. Waning DL, Guise TA (2014) Molecular mechanisms of bone metastasis and associated muscle weakness. Clin Cancer Res 20(12):3071–3077. doi:10.1158/1078-0432.CCR-13-1590
8. Guttridge DC (2015) A TGF-beta pathway associated with cancer cachexia. Nat Med 21(11):1248–1249. doi:10.1038/nm.3988
9. Fox KM, Brooks JM, Gandra SR, Markus R, Chiou CF (2009) Estimation of cachexia among cancer patients based on four definitions. J Oncol 2009:693458. doi:10.1155/2009/693458
10. Dewys WD, Begg C, Lavin PT, Band PR, Bennett JM, Bertino JR, Cohen MH, Douglass HO Jr, Engstrom PF, Ezdinli EZ, Horton J, Johnson GJ, Moertel CG, Oken MM, Perlia C, Rosenbaum C, Silverstein MN, Skeel RT, Sponzo RW, Tormey DC (1980) Prognostic effect of weight loss prior to chemotherapy in cancer patients. Eastern cooperative oncology group. Am J Med 69(4):491–497
11. Warren S (1932) The immediate causes of death in cancer. Am J Med Sci 184:610–615
12. Houten L, Reilley AA (1980) An investigation of the cause of death from cancer. J Surg Oncol 13(2):111–116
13. Harnett WL (1952) British empire cancer campaign: a survey of cancer in London. British Empire Cancer Campaign, London
14. Consul N, Guo X, Coker C, Lopez-Pintado S, Hibshoosh H, Zhao B, Kalinsky K, Acharyya S (2016) Monitoring metastasis and cachexia in a patient with breast cancer: a case study. Clin Med Insights Oncol 10:83–94. doi:10.4137/CMO.S40479
15. Kubo Y, Naito T, Mori K, Osawa G, Aruga E (2017) Skeletal muscle loss and prognosis of breast cancer patients. Support Care Cancer. doi:10.1007/s00520-017-3628-5
16. Waning DL, Guise TA (2015) Cancer-associated muscle weakness: What's bone got to do with it? Bonekey Rep 4:691. doi:10.1038/bonekey.2015.59
17. Heymsfield SB HJ, Lawson DH (1980) Enteral hyperalimentation. In: JE B (ed) Developments in digestive diseases. Lea and Febiger, Philadelphia, pp 59–83
18. Clark CM, Goodlad GA (1971) Depletion of proteins of phasic and tonic muscles in tumour-bearing rats. Eur J Cancer 7(1):3–9
19. Lundholm K, Edstrom S, Ekman L, Karlberg I, Bylund AC, Schersten T (1978) A comparative study of the influence of malignant tumor on host metabolism in mice and man: evaluation of an experimental model. Cancer 42(2):453–461
20. Lundholm K, Bylund AC, Holm J, Schersten T (1976) Skeletal muscle metabolism in patients with malignant tumor. Eur J Cancer 12(6):465–473
21. Begg RW (1958) Tumor-host relations. Adv Cancer Res 5:1–54
22. Ramaswamy KLI, Baker N (1980) Dietary control of lipogenesis in vivo in host tissues and tumours of mice bearing Ehrlich ascites carcinoma. Cancer Res 40:4606–4611
23. Waterhouse C, Kemperman JH (1971) Carbohydrate metabolism in subjects with cancer. Cancer Res 31(9):1273–1278
24. Warnold I, Lundholm K, Schersten T (1978) Energy balance and body composition in cancer patients. Cancer Res 38(6):1801–1807
25. Bozzetti F, Pagnoni AM, Del Vecchio M (1980) Excessive caloric expenditure as a cause of malnutrition in patients with cancer. Surg Gynecol Obstet 150(2):229–234
26. Emery PW, Edwards RH, Rennie MJ, Souhami RL, Halliday D (1984) Protein synthesis in muscle measured in vivo in cachectic patients with cancer. Br Med J (Clin Res Ed) 289(6445):584–586
27. Lundholm K, Bennegard K, Eden E, Svaninger G, Emery PW, Rennie MJ (1982) Efflux of 3-methylhistidine from the leg in cancer patients who experience weight loss. Cancer Res 42(11):4807–4811
28. O'Keefe SJ, Ogden J, Ramjee G, Rund J (1990) Contribution of elevated protein turnover and anorexia to cachexia in patients with hepatocellular carcinoma. Cancer Res 50(4):1226–1230
29. Acharyya S, Ladner KJ, Nelsen LL, Damrauer J, Reiser PJ, Swoap S, Guttridge DC (2004) Cancer cachexia is regulated by selective targeting of skeletal muscle gene products. J Clin Invest 114(3):370–378. doi:10.1172/JCI20174
30. Yu Z, Li P, Zhang M, Hannink M, Stamler JS, Yan Z (2008) Fiber type-specific nitric oxide protects oxidative myofibers against cachectic stimuli. PLoS One 3(5):e2086. doi:10.1371/journal.pone.0002086
31. Diffee GM, Kalfas K, Al-Majid S, McCarthy DO (2002) Altered expression of skeletal muscle myosin isoforms in cancer cachexia. Am J Physiol Cell Physiol 283(5):C1376–C1382. doi:10.1152/ajpcell.00154.2002
32. Schmitt TL, Martignoni ME, Bachmann J, Fechtner K, Friess H, Kinscherf R, Hildebrandt W (2007) Activity of the Akt-dependent anabolic and catabolic pathways in muscle and liver samples in cancer-related cachexia. J Mol Med (Berl) 85(6):647–654. doi:10.1007/s00109-007-0177-2
33. Eley HL, Skipworth RJ, Deans DA, Fearon KC, Tisdale MJ (2008) Increased expression of phosphorylated forms of RNA-dependent protein kinase and eukaryotic initiation factor 2alpha may signal

skeletal muscle atrophy in weight-losing cancer patients. Br J Cancer 98(2):443–449. doi:10.1038/sj.bjc.6604150
34. Johns N, Hatakeyama S, Stephens NA, Degen M, Degen S, Frieauff W, Lambert C, Ross JA, Roubenoff R, Glass DJ, Jacobi C, Fearon KC (2014) Clinical classification of cancer cachexia: phenotypic correlates in human skeletal muscle. PLoS One 9(1):e83618. doi:10.1371/journal.pone.0083618
35. Proud CG (2005) eIF2 and the control of cell physiology. Semin Cell Dev Biol 16(1):3–12. doi:10.1016/j.semcdb.2004.11.004
36. Proud CG (2007) Signalling to translation: how signal transduction pathways control the protein synthetic machinery. Biochem J 403(2):217–234. doi:10.1042/BJ20070024
37. Carlberg U, Nilsson A, Nygard O (1990) Functional properties of phosphorylated elongation factor 2. Eur J Biochem 191(3):639–645
38. Smith KL, Tisdale MJ (1993) Increased protein degradation and decreased protein synthesis in skeletal muscle during cancer cachexia. Br J Cancer 67(4):680–685
39. Eley HL, Tisdale MJ (2007) Skeletal muscle atrophy, a link between depression of protein synthesis and increase in degradation. J Biol Chem 282(10):7087–7097. doi:10.1074/jbc.M610378200
40. Eley HL, Russell ST, Tisdale MJ (2007) Effect of branched-chain amino acids on muscle atrophy in cancer cachexia. Biochem J 407(1):113–120. doi:10.1042/BJ20070651
41. Papageorgopoulos C, Caldwell K, Schweingrubber H, Neese RA, Shackleton CH, Hellerstein M (2002) Measuring synthesis rates of muscle creatine kinase and myosin with stable isotopes and mass spectrometry. Anal Biochem 309(1):1–10
42. Drexler HC, Ruhs A, Konzer A, Mendler L, Bruckskotten M, Looso M, Gunther S, Boettger T, Kruger M, Braun T (2012) On marathons and sprints: an integrated quantitative proteomics and transcriptomics analysis of differences between slow and fast muscle fibers. Mol Cell Proteomics 11(6):M111 010801. doi:10.1074/mcp.M111.010801
43. Pallafacchina G, Calabria E, Serrano AL, Kalhovde JM, Schiaffino S (2002) A protein kinase B-dependent and rapamycin-sensitive pathway controls skeletal muscle growth but not fiber type specification. Proc Natl Acad Sci U S A 99(14):9213–9218. doi:10.1073/pnas.142166599
44. Raffaello A, Milan G, Masiero E, Carnio S, Lee D, Lanfranchi G, Goldberg AL, Sandri M (2010) JunB transcription factor maintains skeletal muscle mass and promotes hypertrophy. J Cell Biol 191(1):101–113. doi:10.1083/jcb.201001136
45. Glickman MH, Ciechanover A (2002) The ubiquitin-proteasome proteolytic pathway: destruction for the sake of construction. Physiol Rev 82(2):373–428. doi:10.1152/physrev.00027.2001
46. Hasselgren PO, Wray C, Mammen J (2002) Molecular regulation of muscle cachexia: it may be more than the proteasome. Biochem Biophys Res Commun 290(1):1–10. doi:10.1006/bbrc.2001.5849
47. Khal J, Hine AV, Fearon KC, Dejong CH, Tisdale MJ (2005) Increased expression of proteasome subunits in skeletal muscle of cancer patients with weight loss. Int J Biochem Cell Biol 37(10):2196–2206. doi:10.1016/j.biocel.2004.10.017
48. Mammucari C, Milan G, Romanello V, Masiero E, Rudolf R, Del Piccolo P, Burden SJ, Di Lisi R, Sandri C, Zhao J, Goldberg AL, Schiaffino S, Sandri M (2007) FoxO3 controls autophagy in skeletal muscle in vivo. Cell Metab 6(6):458–471. doi:10.1016/j.cmet.2007.11.001
49. Bodine SC, Latres E, Baumhueter S, Lai VK, Nunez L, Clarke BA, Poueymirou WT, Panaro FJ, Na E, Dharmarajan K, Pan ZQ, Valenzuela DM, DeChiara TM, Stitt TN, Yancopoulos GD, Glass DJ (2001) Identification of ubiquitin ligases required for skeletal muscle atrophy. Science 294(5547):1704–1708. doi:10.1126/science.1065874
50. Gomes MD, Lecker SH, Jagoe RT, Navon A, Goldberg AL (2001) Atrogin-1, a muscle-specific F-box protein highly expressed during muscle atrophy. Proc Natl Acad Sci U S A 98(25):14440–14445. doi:10.1073/pnas.251541198
51. Lecker SH, Jagoe RT, Gilbert A, Gomes M, Baracos V, Bailey J, Price SR, Mitch WE, Goldberg AL (2004) Multiple types of skeletal muscle atrophy involve a common program of changes in gene expression. FASEB J 18(1):39–51. doi:10.1096/fj.03-0610com
52. Lokireddy S, Wijesoma IW, Sze SK, McFarlane C, Kambadur R, Sharma M (2012) Identification of atrogin-1-targeted proteins during the myostatin-induced skeletal muscle wasting. Am J Physiol Cell Physiol 303(5):C512–C529. doi:10.1152/ajpcell.00402.2011
53. Kedar V, McDonough H, Arya R, Li HH, Rockman HA, Patterson C (2004) Muscle-specific RING finger 1 is a bona fide ubiquitin ligase that degrades cardiac troponin I. Proc Natl Acad Sci U S A 101(52):18135–18140. doi:10.1073/pnas.0404341102
54. Fielitz J, Kim MS, Shelton JM, Latif S, Spencer JA, Glass DJ, Richardson JA, Bassel-Duby R, Olson EN (2007) Myosin accumulation and striated muscle myopathy result from the loss of muscle RING finger 1 and 3. J Clin Invest 117(9):2486–2495. doi:10.1172/JCI32827
55. Clarke BA, Drujan D, Willis MS, Murphy LO, Corpina RA, Burova E, Rakhilin SV, Stitt TN, Patterson C, Latres E, Glass DJ (2007) The E3 ligase MuRF1 degrades myosin heavy chain protein in dexamethasone-treated skeletal muscle. Cell Metab 6(5):376–385. doi:10.1016/j.cmet.2007.09.009
56. Polge C, Heng AE, Jarzaguet M, Ventadour S, Claustre A, Combaret L, Bechet D, Matondo M, Uttenweiler-Joseph S, Monsarrat B, Attaix D,

Taillandier D (2011) Muscle actin is polyubiquitinylated in vitro and in vivo and targeted for breakdown by the E3 ligase MuRF1. FASEB J 25(11):3790–3802. doi:10.1096/fj.11-180968

57. Cohen S, Brault JJ, Gygi SP, Glass DJ, Valenzuela DM, Gartner C, Latres E, Goldberg AL (2009) During muscle atrophy, thick, but not thin, filament components are degraded by MuRF1-dependent ubiquitylation. J Cell Biol 185(6):1083–1095. doi:10.1083/jcb.200901052
58. Sandri M (2016) Protein breakdown in cancer cachexia. Semin Cell Dev Biol 54:11–19. doi:10.1016/j.semcdb.2015.11.002
59. Stephens NA, Gallagher IJ, Rooyackers O, Skipworth RJ, Tan BH, Marstrand T, Ross JA, Guttridge DC, Lundell L, Fearon KC, Timmons JA (2010) Using transcriptomics to identify and validate novel biomarkers of human skeletal muscle cancer cachexia. Genome Med 2(1):1. doi:10.1186/gm122
60. D'Orlando C, Marzetti E, Francois S, Lorenzi M, Conti V, di Stasio E, Rosa F, Brunelli S, Doglietto GB, Pacelli F, Bossola M (2014) Gastric cancer does not affect the expression of atrophy-related genes in human skeletal muscle. Muscle Nerve 49(4):528–533. doi:10.1002/mus.23945
61. Jagoe RT, Redfern CP, Roberts RG, Gibson GJ, Goodship TH (2002) Skeletal muscle mRNA levels for cathepsin B, but not components of the ubiquitin-proteasome pathway, are increased in patients with lung cancer referred for thoracotomy. Clin Sci (Lond) 102(3):353–361
62. Schersten T, Lundholm K (1972) Lysosomal enzyme activity in muscle tissue from patients with malignant tumor. Cancer 30(5):1246–1251
63. Lapierre LR, Kumsta C, Sandri M, Ballabio A, Hansen M (2015) Transcriptional and epigenetic regulation of autophagy in aging. Autophagy 11(6):867–880. doi:10.1080/15548627.2015.1034410
64. Bechet D, Tassa A, Taillandier D, Combaret L, Attaix D (2005) Lysosomal proteolysis in skeletal muscle. Int J Biochem Cell Biol 37(10):2098–2114. doi:10.1016/j.biocel.2005.02.029
65. Deval C, Mordier S, Obled C, Bechet D, Combaret L, Attaix D, Ferrara M (2001) Identification of cathepsin L as a differentially expressed message associated with skeletal muscle wasting. Biochem J 360(Pt 1):143–150
66. Tassa A, Roux MP, Attaix D, Bechet DM (2003) Class III phosphoinositide 3-kinase--Beclin1 complex mediates the amino acid-dependent regulation of autophagy in C2C12 myotubes. Biochem J 376(Pt 3):577–586. doi:10.1042/BJ20030826
67. Penna F, Costamagna D, Pin F, Camperi A, Fanzani A, Chiarpotto EM, Cavallini G, Bonelli G, Baccino FM, Costelli P (2013) Autophagic degradation contributes to muscle wasting in cancer cachexia. Am J Pathol 182(4):1367–1378. doi:10.1016/j.ajpath.2012.12.023
68. Chacon-Cabrera A, Fermoselle C, Urtreger AJ, Mateu-Jimenez M, Diament MJ, de Kier Joffe ED, Sandri M, Barreiro E (2014) Pharmacological strategies in lung cancer-induced cachexia: effects on muscle proteolysis, autophagy, structure, and weakness. J Cell Physiol 229(11):1660–1672. doi:10.1002/jcp.24611
69. Mizushima N, Yamamoto A, Matsui M, Yoshimori T, Ohsumi Y (2004) In vivo analysis of autophagy in response to nutrient starvation using transgenic mice expressing a fluorescent autophagosome marker. Mol Biol Cell 15(3):1101–1111. doi:10.1091/mbc.E03-09-0704
70. Grumati P, Coletto L, Sabatelli P, Cescon M, Angelin A, Bertaggia E, Blaauw B, Urciuolo A, Tiepolo T, Merlini L, Maraldi NM, Bernardi P, Sandri M, Bonaldo P (2010) Autophagy is defective in collagen VI muscular dystrophies, and its reactivation rescues myofiber degeneration. Nat Med 16(11):1313–1320. doi:10.1038/nm.2247
71. Wohlgemuth SE, Seo AY, Marzetti E, Lees HA, Leeuwenburgh C (2010) Skeletal muscle autophagy and apoptosis during aging: effects of calorie restriction and life-long exercise. Exp Gerontol 45(2):138–148. doi:10.1016/j.exger.2009.11.002
72. Mofarrahi M, Sigala I, Guo Y, Godin R, Davis EC, Petrof B, Sandri M, Burelle Y, Hussain SN (2012) Autophagy and skeletal muscles in sepsis. PLoS One 7(10):e47265. doi:10.1371/journal.pone.0047265
73. Derde S, Vanhorebeek I, Guiza F, Derese I, Gunst J, Fahrenkrog B, Martinet W, Vervenne H, Ververs EJ, Larsson L, Van den Berghe G (2012) Early parenteral nutrition evokes a phenotype of autophagy deficiency in liver and skeletal muscle of critically ill rabbits. Endocrinology 153(5):2267–2276. doi:10.1210/en.2011-2068
74. Qiu J, Tsien C, Thapalaya S, Narayanan A, Weihl CC, Ching JK, Eghtesad B, Singh K, Fu X, Dubyak G, McDonald C, Almasan A, Hazen SL, Naga Prasad SV, Dasarathy S (2012) Hyperammonemia-mediated autophagy in skeletal muscle contributes to sarcopenia of cirrhosis. Am J Physiol Endocrinol Metab 303(8):E983–E993. doi:10.1152/ajpendo.00183.2012
75. Smuder AJ, Kavazis AN, Min K (1985) Powers SK (2011) exercise protects against doxorubicin-induced oxidative stress and proteolysis in skeletal muscle. J Appl Physiol 110(4):935–942. doi:10.1152/japplphysiol.00677.2010
76. Brocca L, Cannavino J, Coletto L, Biolo G, Sandri M, Bottinelli R, Pellegrino MA (2012) The time course of the adaptations of human muscle proteome to bed rest and the underlying mechanisms. J Physiol 590(20):5211–5230. doi:10.1113/jphysiol.2012.240267
77. O'Leary MF, Vainshtein A, Carter HN, Zhang Y, Hood DA (2012) Denervation-induced mitochondrial dysfunction and autophagy in skeletal muscle of apoptosis-deficient animals. Am J Physiol

Cell Physiol 303(4):C447–C454. doi:10.1152/ajpcell.00451.2011
78. Zhao J, Brault JJ, Schild A, Cao P, Sandri M, Schiaffino S, Lecker SH, Goldberg AL (2007) FoxO3 coordinately activates protein degradation by the autophagic/lysosomal and proteasomal pathways in atrophying muscle cells. Cell Metab 6(6):472–483. doi:10.1016/j.cmet.2007.11.004
79. Nascimbeni AC, Fanin M, Masiero E, Angelini C, Sandri M (2012) The role of autophagy in the pathogenesis of glycogen storage disease type II (GSDII). Cell Death Differ 19(10):1698–1708. doi:10.1038/cdd.2012.52
80. De Palma C, Morisi F, Cheli S, Pambianco S, Cappello V, Vezzoli M, Rovere-Querini P, Moggio M, Ripolone M, Francolini M, Sandri M, Clementi E (2012) Autophagy as a new therapeutic target in Duchenne muscular dystrophy. Cell Death Dis 3:e418. doi:10.1038/cddis.2012.159
81. Sandri M, Coletto L, Grumati P, Bonaldo P (2013) Misregulation of autophagy and protein degradation systems in myopathies and muscular dystrophies. J Cell Sci 126(Pt 23):5325–5333. doi:10.1242/jcs.114041
82. Tardif N, Klaude M, Lundell L, Thorell A, Rooyackers O (2013) Autophagic-lysosomal pathway is the main proteolytic system modified in the skeletal muscle of esophageal cancer patients. Am J Clin Nutr 98(6):1485–1492. doi:10.3945/ajcn.113.063859
83. Wang H, Sun HQ, Zhu X, Zhang L, Albanesi J, Levine B, Yin H (2015) GABARAPs regulate PI4P-dependent autophagosome:lysosome fusion. Proc Natl Acad Sci U S A 112(22):7015–7020. doi:10.1073/pnas.1507263112
84. Stephens NA, Skipworth RJ, Gallagher IJ, Greig CA, Guttridge DC, Ross JA, Fearon KC (2015) Evaluating potential biomarkers of cachexia and survival in skeletal muscle of upper gastrointestinal cancer patients. J Cachexia Sarcopenia Muscle 6(1):53–61. doi:10.1002/jcsm.12005
85. Wenz T, Rossi SG, Rotundo RL, Spiegelman BM, Moraes CT (2009) Increased muscle PGC-1alpha expression protects from sarcopenia and metabolic disease during aging. Proc Natl Acad Sci U S A 106(48):20405–20410. doi:10.1073/pnas.0911570106
86. Carnio S, LoVerso F, Baraibar MA, Longa E, Khan MM, Maffei M, Reischl M, Canepari M, Loefler S, Kern H, Blaauw B, Friguet B, Bottinelli R, Rudolf R, Sandri M (2014) Autophagy impairment in muscle induces neuromuscular junction degeneration and precocious aging. Cell Rep 8(5):1509–1521. doi:10.1016/j.celrep.2014.07.061
87. Karsli-Uzunbas G, Guo JY, Price S, Teng X, Laddha SV, Khor S, Kalaany NY, Jacks T, Chan CS, Rabinowitz JD, White E (2014) Autophagy is required for glucose homeostasis and lung tumor maintenance. Cancer Discov 4(8):914–927. doi:10.1158/2159-8290.CD-14-0363
88. Masiero E, Agatea L, Mammucari C, Blaauw B, Loro E, Komatsu M, Metzger D, Reggiani C, Schiaffino S, Sandri M (2009) Autophagy is required to maintain muscle mass. Cell Metab 10(6):507–515. doi:10.1016/j.cmet.2009.10.008
89. Lucke B, Berwick M, Zeckwer I (1952) Liver catalase activity in parabiotic rats with one partner tumor-bearing. J Natl Cancer Inst 13(3):681–686
90. Beutler B, Mahoney J, Le Trang N, Pekala P, Cerami A (1985) Purification of cachectin, a lipoprotein lipase-suppressing hormone secreted by endotoxin-induced RAW 264.7 cells. J Exp Med 161(5):984–995
91. Garcia-Martinez C, Agell N, Llovera M, Lopez-Soriano FJ, Argiles JM (1993) Tumour necrosis factor-alpha increases the ubiquitinization of rat skeletal muscle proteins. FEBS Lett 323(3):211–214
92. Garcia-Martinez C, Llovera M, Agell N, Lopez-Soriano FJ, Argiles JM (1994) Ubiquitin gene expression in skeletal muscle is increased by tumour necrosis factor-alpha. Biochem Biophys Res Commun 201(2):682–686
93. Llovera M, Carbo N, Lopez-Soriano J, Garcia-Martinez C, Busquets S, Alvarez B, Agell N, Costelli P, Lopez-Soriano FJ, Celada A, Argiles JM (1998) Different cytokines modulate ubiquitin gene expression in rat skeletal muscle. Cancer Lett 133(1):83–87
94. Llovera M, Garcia-Martinez C, Agell N, Lopez-Soriano FJ, Argiles JM (1997) TNF can directly induce the expression of ubiquitin-dependent proteolytic system in rat soleus muscles. Biochem Biophys Res Commun 230(2):238–241
95. van Hall G (2012) Cytokines: muscle protein and amino acid metabolism. Curr Opin Clin Nutr Metab Care 15(1):85–91. doi:10.1097/MCO.0b013e32834e6ea2
96. Tisdale MJ (1997) Biology of cachexia. J Natl Cancer Inst 89(23):1763–1773
97. Chen SE, Gerken E, Zhang Y, Zhan M, Mohan RK, Li AS, Reid MB, Li YP (2005) Role of TNF-{alpha} signaling in regeneration of cardiotoxin-injured muscle. Am J Physiol Cell Physiol 289(5):C1179–C1187. doi:10.1152/ajpcell.00062.2005
98. Chen SE, Jin B, Li YP (2007) TNF-alpha regulates myogenesis and muscle regeneration by activating p38 MAPK. Am J Physiol Cell Physiol 292(5):C1660–C1671. doi:10.1152/ajpcell.00486.2006
99. Miller SC, Ito H, Blau HM, Torti FM (1988) Tumor necrosis factor inhibits human myogenesis in vitro. Mol Cell Biol 8(6):2295–2301
100. Oliff A, Defeo-Jones D, Boyer M, Martinez D, Kiefer D, Vuocolo G, Wolfe A, Socher SH (1987) Tumors secreting human TNF/cachectin induce cachexia in mice. Cell 50(4):555–563

101. Garcia-Martinez C, Lopez-Soriano FJ, Argiles JM (1993) Acute treatment with tumour necrosis factor-alpha induces changes in protein metabolism in rat skeletal muscle. Mol Cell Biochem 125(1):11–18
102. Li YP, Reid MB (2000) NF-kappaB mediates the protein loss induced by TNF-alpha in differentiated skeletal muscle myotubes. Am J Physiol Regul Integr Comp Physiol 279(4):R1165–R1170
103. Keren A, Tamir Y, Bengal E (2006) The p38 MAPK signaling pathway: a major regulator of skeletal muscle development. Mol Cell Endocrinol 252(1–2):224–230. doi:10.1016/j.mce.2006.03.017
104. Bernet JD, Doles JD, Hall JK, Kelly Tanaka K, Carter TA, Olwin BB (2014) p38 MAPK signaling underlies a cell-autonomous loss of stem cell self-renewal in skeletal muscle of aged mice. Nat Med 20(3):265–271. doi:10.1038/nm.3465
105. Cosgrove BD, Gilbert PM, Porpiglia E, Mourkioti F, Lee SP, Corbel SY, Llewellyn ME, Delp SL, Blau HM (2014) Rejuvenation of the muscle stem cell population restores strength to injured aged muscles. Nat Med 20(3):255–264. doi:10.1038/nm.3464
106. Fukawa T, Yan-Jiang BC, Min-Wen JC, Jun-Hao ET, Huang D, Qian CN, Ong P, Li Z, Chen S, Mak SY, Lim WJ, Kanayama HO, Mohan RE, Wang RR, Lai JH, Chua C, Ong HS, Tan KK, Ho YS, Tan IB, Teh BT, Shyh-Chang N (2016) Excessive fatty acid oxidation induces muscle atrophy in cancer cachexia. Nat Med 22(6):666–671. doi:10.1038/nm.4093
107. Maltoni M, Fabbri L, Nanni O, Scarpi E, Pezzi L, Flamini E, Riccobon A, Derni S, Pallotti G, Amadori D (1997) Serum levels of tumour necrosis factor alpha and other cytokines do not correlate with weight loss and anorexia in cancer patients. Support Care Cancer 5(2):130–135
108. Jatoi A, Ritter HL, Dueck A, Nguyen PL, Nikcevich DA, Luyun RF, Mattar BI, Loprinzi CL (2010) A placebo-controlled, double-blind trial of infliximab for cancer-associated weight loss in elderly and/or poor performance non-small cell lung cancer patients (N01C9). Lung Cancer 68(2):234–239. doi:10.1016/j.lungcan.2009.06.020
109. Scott HR, McMillan DC, Crilly A, McArdle CS, Milroy R (1996) The relationship between weight loss and interleukin 6 in non-small-cell lung cancer. Br J Cancer 73(12):1560–1562
110. Moses AG, Maingay J, Sangster K, Fearon KC, Ross JA (2009) Pro-inflammatory cytokine release by peripheral blood mononuclear cells from patients with advanced pancreatic cancer: relationship to acute phase response and survival. Oncol Rep 21(4):1091–1095
111. Black K, Garrett IR, Mundy GR (1991) Chinese hamster ovarian cells transfected with the murine interleukin-6 gene cause hypercalcemia as well as cachexia, leukocytosis and thrombocytosis in tumor-bearing nude mice. Endocrinology 128(5):2657–2659. doi:10.1210/endo-128-5-2657
112. Strassmann G, Fong M, Kenney JS, Jacob CO (1992) Evidence for the involvement of interleukin 6 in experimental cancer cachexia. J Clin Invest 89(5):1681–1684. doi:10.1172/JCI115767
113. Strassmann G, Fong M, Freter CE, Windsor S, D'Alessandro F, Nordan RP (1993) Suramin interferes with interleukin-6 receptor binding in vitro and inhibits colon-26-mediated experimental cancer cachexia in vivo. J Clin Invest 92(5):2152–2159. doi:10.1172/JCI116816
114. Goodman MN (1994) Interleukin-6 induces skeletal muscle protein breakdown in rats. Proc Soc Exp Biol Med 205(2):182–185
115. Baltgalvis KA, Berger FG, Pena MM, Davis JM, Muga SJ, Carson JA (2008) Interleukin-6 and cachexia in ApcMin/+ mice. Am J Physiol Regul Integr Comp Physiol 294(2):R393–R401. doi:10.1152/ajpregu.00716.2007
116. Espat NJ, Auffenberg T, Rosenberg JJ, Rogy M, Martin D, Fang CH, Hasselgren PO, Copeland EM, Moldawer LL (1996) Ciliary neurotrophic factor is catabolic and shares with IL-6 the capacity to induce an acute phase response. Am J Phys 271(1 Pt 2):R185–R190
117. Bayliss TJ, Smith JT, Schuster M, Dragnev KH, Rigas JR (2011) A humanized anti-IL-6 antibody (ALD518) in non-small cell lung cancer. Expert Opin Biol Ther 11(12):1663–1668. doi:10.1517/14712598.2011.627850
118. Reardon KA, Davis J, Kapsa RM, Choong P, Byrne E (2001) Myostatin, insulin-like growth factor-1, and leukemia inhibitory factor mRNAs are upregulated in chronic human disuse muscle atrophy. Muscle Nerve 24(7):893–899
119. Carlson CJ, Booth FW, Gordon SE (1999) Skeletal muscle myostatin mRNA expression is fiber-type specific and increases during hindlimb unloading. Am J Phys 277(2 Pt 2):R601–R606
120. Zachwieja JJ, Smith SR, Sinha-Hikim I, Gonzalez-Cadavid N, Bhasin S (1999) Plasma myostatin-immunoreactive protein is increased after prolonged bed rest with low-dose T3 administration. J Gravit Physiol 6(2):11–15
121. Gustafsson T, Osterlund T, Flanagan JN, von Walden F, Trappe TA, Linnehan RM (1985) Tesch PA (2010) effects of 3 days unloading on molecular regulators of muscle size in humans. J Appl Physiol 109(3):721–727. doi:10.1152/japplphysiol.00110.2009
122. Shao C, Liu M, Wu X, Ding F (2007) Time-dependent expression of myostatin RNA transcript and protein in gastrocnemius muscle of mice after sciatic nerve resection. Microsurgery 27(5):487–493. doi:10.1002/micr.20392
123. Elkina Y, von Haehling S, Anker SD, Springer J (2011) The role of myostatin in muscle wasting: an overview. J Cachexia Sarcopenia Muscle 2(3):143–151. doi:10.1007/s13539-011-0035-5
124. Zimmers TA, Davies MV, Koniaris LG, Haynes P, Esquela AF, Tomkinson KN, McPherron AC,

Wolfman NM, Lee SJ (2002) Induction of cachexia in mice by systemically administered myostatin. Science 296(5572):1486–1488. doi:10.1126/science.1069525

125. Langley B, Thomas M, Bishop A, Sharma M, Gilmour S, Kambadur R (2002) Myostatin inhibits myoblast differentiation by down-regulating MyoD expression. J Biol Chem 277(51):49831–49840. doi:10.1074/jbc.M204291200

126. Aversa Z, Bonetto A, Penna F, Costelli P, Di Rienzo G, Lacitignola A, Baccino FM, Ziparo V, Mercantini P, Rossi Fanelli F, Muscaritoli M (2012) Changes in myostatin signaling in non-weight-losing cancer patients. Ann Surg Oncol 19(4):1350–1356. doi:10.1245/s10434-011-1720-5

127. George I, Bish LT, Kamalakkannan G, Petrilli CM, Oz MC, Naka Y, Sweeney HL, Maybaum S (2010) Myostatin activation in patients with advanced heart failure and after mechanical unloading. Eur J Heart Fail 12(5):444–453. doi:10.1093/eurjhf/hfq039

128. Elliott B, Renshaw D, Getting S, Mackenzie R (2012) The central role of myostatin in skeletal muscle and whole body homeostasis. Acta Physiol (Oxf) 205(3):324–340. doi:10.1111/j.1748-1716.2012.02423.x

129. Loumaye A, de Barsy M, Nachit M, Lause P, Frateur L, van Maanen A, Trefois P, Gruson D, Thissen JP (2015) Role of Activin a and myostatin in human cancer cachexia. J Clin Endocrinol Metab 100(5):2030–2038. doi:10.1210/jc.2014-4318

130. Sinha M, Jang YC, Oh J, Khong D, Wu EY, Manohar R, Miller C, Regalado SG, Loffredo FS, Pancoast JR, Hirshman MF, Lebowitz J, Shadrach JL, Cerletti M, Kim MJ, Serwold T, Goodyear LJ, Rosner B, Lee RT, Wagers AJ (2014) Restoring systemic GDF11 levels reverses age-related dysfunction in mouse skeletal muscle. Science 344(6184):649–652. doi:10.1126/science.1251152

131. Egerman MA, Cadena SM, Gilbert JA, Meyer A, Nelson HN, Swalley SE, Mallozzi C, Jacobi C, Jennings LL, Clay I, Laurent G, Ma S, Brachat S, Lach-Trifilieff E, Shavlakadze T, Trendelenburg AU, Brack AS, Glass DJ (2015) GDF11 increases with age and inhibits skeletal muscle regeneration. Cell Metab 22(1):164–174. doi:10.1016/j.cmet.2015.05.010

132. Zhou Y, Sharma N, Dukes D, Myzithras MB, Gupta P, Khalil A, Kahn J, Ahlberg JS, Hayes DB, Franti M, Criswell T (2017) GDF11 treatment attenuates the recovery of skeletal muscle function after injury in older rats. AAPS J 19(2):431–437. doi:10.1208/s12248-016-0024-x

133. Zhou X, Wang JL, Lu J, Song Y, Kwak KS, Jiao Q, Rosenfeld R, Chen Q, Boone T, Simonet WS, Lacey DL, Goldberg AL, Han HQ (2010) Reversal of cancer cachexia and muscle wasting by ActRIIB antagonism leads to prolonged survival. Cell 142(4):531–543. doi:10.1016/j.cell.2010.07.011

134. Busquets S, Toledo M, Orpi M, Massa D, Porta M, Capdevila E, Padilla N, Frailis V, Lopez-Soriano FJ, Han HQ, Argiles JM (2012) Myostatin blockage using actRIIB antagonism in mice bearing the Lewis lung carcinoma results in the improvement of muscle wasting and physical performance. J Cachexia Sarcopenia Muscle 3(1):37–43. doi:10.1007/s13539-011-0049-z

135. Murphy KT, Chee A, Gleeson BG, Naim T, Swiderski K, Koopman R, Lynch GS (2011) Antibody-directed myostatin inhibition enhances muscle mass and function in tumor-bearing mice. Am J Physiol Regul Integr Comp Physiol 301(3):R716–R726. doi:10.1152/ajpregu.00121.2011

136. Benny Klimek ME, Aydogdu T, Link MJ, Pons M, Koniaris LG, Zimmers TA (2010) Acute inhibition of myostatin-family proteins preserves skeletal muscle in mouse models of cancer cachexia. Biochem Biophys Res Commun 391(3):1548–1554. doi:10.1016/j.bbrc.2009.12.123

137. Gallot YS, Durieux AC, Castells J, Desgeorges MM, Vernus B, Plantureux L, Remond D, Jahnke VE, Lefai E, Dardevet D, Nemoz G, Schaeffer L, Bonnieu A, Freyssenet DG (2014) Myostatin gene inactivation prevents skeletal muscle wasting in cancer. Cancer Res 74(24):7344–7356. doi:10.1158/0008-5472.CAN-14-0057

138. Carroll J (2016) Novartis' 'breakthrough' muscle drug bimagrumab flunks a late-stage trial. http://www.fiercebiotech.com/biotech/novartis-breakthrough-muscle-drug-bimagrumab-flunks-a-late-stage-trial. Accessed 21 Apr 2016

139. Todorov P, Cariuk P, McDevitt T, Coles B, Fearon K, Tisdale M (1996) Characterization of a cancer cachectic factor. Nature 379(6567):739–742. doi:10.1038/379739a0

140. Hussey HJ, Todorov PT, Field WN, Inagaki N, Tanaka Y, Ishitsuka H, Tisdale MJ (2000) Effect of a fluorinated pyrimidine on cachexia and tumour growth in murine cachexia models: relationship with a proteolysis inducing factor. Br J Cancer 83(1):56–62. doi:10.1054/bjoc.2000.1278

141. Lorite MJ, Thompson MG, Drake JL, Carling G, Tisdale MJ (1998) Mechanism of muscle protein degradation induced by a cancer cachectic factor. Br J Cancer 78(7):850–856

142. Wyke SM, Tisdale MJ (2005) NF-kappaB mediates proteolysis-inducing factor induced protein degradation and expression of the ubiquitin-proteasome system in skeletal muscle. Br J Cancer 92(4):711–721. doi:10.1038/sj.bjc.6602402

143. Todorov PT, Field WN, Tisdale MJ (1999) Role of a proteolysis-inducing factor (PIF) in cachexia induced by a human melanoma (G361). Br J Cancer 80(11):1734–1737. doi:10.1038/sj.bjc.6690590

144. Tisdale MJ (2008) Re: Wieland BM et al. is there a human homologue to the murine proteolysis-inducing factor? Clin Cancer Res 14(7):2245;

author reply 2245–2246. doi:10.1158/1078-0432.CCR-07-4769
145. Wieland BM, Stewart GD, Skipworth RJ, Sangster K, Fearon KC, Ross JA, Reiman TJ, Easaw J, Mourtzakis M, Kumar V, Pak BJ, Calder K, Filippatos G, Kremastinos DT, Palcic M, Baracos VE (2007) Is there a human homologue to the murine proteolysis-inducing factor? Clin Cancer Res 13(17):4984–4992. doi:10.1158/1078-0432.CCR-07-0946
146. Case 27461 (1941) New Engl J Med 225:789–791
147. Moseley JM, Kubota M, Diefenbach-Jagger H, Wettenhall RE, Kemp BE, Suva LJ, Rodda CP, Ebeling PR, Hudson PJ, Zajac JD et al (1987) Parathyroid hormone-related protein purified from a human lung cancer cell line. Proc Natl Acad Sci U S A 84(14):5048–5052
148. Juppner H, Abou-Samra AB, Uneno S, Gu WX, Potts JT Jr, Segre GV (1988) The parathyroid hormone-like peptide associated with humoral hypercalcemia of malignancy and parathyroid hormone bind to the same receptor on the plasma membrane of ROS 17/2.8 cells. J Biol Chem 263(18):8557–8560
149. Strewler GJ, Stern PH, Jacobs JW, Eveloff J, Klein RF, Leung SC, Rosenblatt M, Nissenson RA (1987) Parathyroid hormone like protein from human renal carcinoma cells. Structural and functional homology with parathyroid hormone. J Clin Invest 80(6):1803–1807. doi:10.1172/JCI113275
150. Orland SM, Stewart AF, Livolsi VA, Wein AJ (1986) Detection of the hypercalcemic hormone of malignancy in an adrenal cortical carcinoma. J Urol 136(5):1000–1002
151. Iguchi H, Onuma E, Sato K, Sato K, Ogata E (2001) Involvement of parathyroid hormone-related protein in experimental cachexia induced by a human lung cancer-derived cell line established from a bone metastasis specimen. Int J Cancer 94(1):24–27. doi:10.1002/ijc.1425
152. Petruzzelli M, Schweiger M, Schreiber R, Campos-Olivas R, Tsoli M, Allen J, Swarbrick M, Rose-John S, Rincon M, Robertson G, Zechner R, Wagner EF (2014) A switch from white to brown fat increases energy expenditure in cancer-associated cachexia. Cell Metab 20(3):433–447. doi:10.1016/j.cmet.2014.06.011
153. Kir S, White JP, Kleiner S, Kazak L, Cohen P, Baracos VE, Spiegelman BM (2014) Tumour-derived PTH-related protein triggers adipose tissue browning and cancer cachexia. Nature 513(7516):100–104. doi:10.1038/nature13528
154. Bosaeus I, Daneryd P, Svanberg E, Lundholm K (2001) Dietary intake and resting energy expenditure in relation to weight loss in unselected cancer patients. Int J Cancer 93(3):380–383
155. Falconer JS, Fearon KC, Plester CE, Ross JA, Carter DC (1994) Cytokines, the acute-phase response, and resting energy expenditure in cachectic patients with pancreatic cancer. Ann Surg 219(4):325–331
156. Fredrix EW, Soeters PB, Wouters EF, Deerenberg IM, von Meyenfeldt MF, Saris WH (1991) Effect of different tumor types on resting energy expenditure. Cancer Res 51(22):6138–6141
157. Rigaud D, Hassid J, Meulemans A, Poupard AT, Boulier A (2000) A paradoxical increase in resting energy expenditure in malnourished patients near death: the king penguin syndrome. Am J Clin Nutr 72(2):355–360
158. Zylicz Z, Schwantje O, Wagener DJ, Folgering HT (1990) Metabolic response to enteral food in different phases of cancer cachexia in rats. Oncology 47(1):87–91
159. Bennani-Baiti N, Walsh D (2011) Animal models of the cancer anorexia-cachexia syndrome. Support Care Cancer 19(9):1451–1463. doi:10.1007/s00520-010-0972-0
160. Hyltander A, Drott C, Korner U, Sandstrom R, Lundholm K (1991) Elevated energy expenditure in cancer patients with solid tumours. Eur J Cancer 27(1):9–15
161. Vander Heiden MG, Cantley LC, Thompson CB (2009) Understanding the Warburg effect: the metabolic requirements of cell proliferation. Science 324(5930):1029–1033. doi:10.1126/science.1160809
162. Holroyde CP, Gabuzda TG, Putnam RC, Paul P, Reichard GA (1975) Altered glucose metabolism in metastatic carcinoma. Cancer Res 35(12):3710–3714
163. Holroyde CP, Axelrod RS, Skutches CL, Haff AC, Paul P, Reichard GA (1979) Lactate metabolism in patients with metastatic colorectal cancer. Cancer Res 39(12):4900–4904
164. Shellock FG, Riedinger MS, Fishbein MC (1986) Brown adipose tissue in cancer patients: possible cause of cancer-induced cachexia. J Cancer Res Clin Oncol 111(1):82–85
165. van Hall G, Steensberg A, Sacchetti M, Fischer C, Keller C, Schjerling P, Hiscock N, Moller K, Saltin B, Febbraio MA, Pedersen BK (2003) Interleukin-6 stimulates lipolysis and fat oxidation in humans. J Clin Endocrinol Metab 88(7):3005–3010. doi:10.1210/jc.2002-021687
166. Kawakami M, Murase T, Ogawa H, Ishibashi S, Mori N, Takaku F, Shibata S (1987) Human recombinant TNF suppresses lipoprotein lipase activity and stimulates lipolysis in 3T3-L1 cells. J Biochem 101(2):331–338
167. Green A, Dobias SB, Walters DJ, Brasier AR (1994) Tumor necrosis factor increases the rate of lipolysis in primary cultures of adipocytes without altering levels of hormone-sensitive lipase. Endocrinology 134(6):2581–2588. doi:10.1210/endo.134.6.8194485
168. Das SK, Eder S, Schauer S, Diwoky C, Temmel H, Guertl B, Gorkiewicz G, Tamilarasan KP, Kumari P, Trauner M, Zimmermann R, Vesely P, Haemmerle G, Zechner R, Hoefler G (2011) Adipose triglyceride lipase contributes to cancer-associated

cachexia. Science 333(6039):233–238. doi:10.1126/science.1198973
169. Son Y, Kim S, Chung HT, Pae HO (2013) Reactive oxygen species in the activation of MAP kinases. Methods Enzymol 528:27–48. doi:10.1016/B978-0-12-405881-1.00002-1
170. Muller FL, Song W, Liu Y, Chaudhuri A, Pieke-Dahl S, Strong R, Huang TT, Epstein CJ, Roberts LJ 2nd, Csete M, Faulkner JA, Van Remmen H (2006) Absence of CuZn superoxide dismutase leads to elevated oxidative stress and acceleration of age-dependent skeletal muscle atrophy. Free Radic Biol Med 40(11):1993–2004. doi:10.1016/j.freeradbiomed.2006.01.036
171. Gomes-Marcondes MC, Tisdale MJ (2002) Induction of protein catabolism and the ubiquitin-proteasome pathway by mild oxidative stress. Cancer Lett 180(1):69–74
172. Solheim TS, Fearon KC, Blum D, Kaasa S (2013) Non-steroidal anti-inflammatory treatment in cancer cachexia: a systematic literature review. Acta Oncol 52(1):6–17. doi:10.3109/0284186X.2012.724536

Cell Cycle Regulation in Treatment of Breast Cancer

12

Zijie Cai and Qiang Liu

Abstract

Cell cycle progression and cell proliferation are under precise and orchestrated control in normal cells. However, uncontrolled cell proliferation caused by aberrant cell cycle progression is a crucial characteristic of cancer. Understanding cell cycle progression and its regulation sheds light on cancer treatment. Agents targeting cell cycle regulators (such as CDKs) have been considered as promising candidates in cancer treatment. Although the first-generation pan-CDK inhibitors failed in clinical trials because of their adverse events and low efficacy, new selective CDK 4/6 inhibitors showed potent efficacy with tolerable safety in preclinical and clinical studies. Here we will review the mechanisms of cell cycle regulation and targeting key cell cycle regulators (such as CDKs) in breast cancer treatment. Particularly, we will discuss the mechanism of CDK inhibitors in disrupting cell cycle progression, the use of selective CDK4/6 inhibitors in treatment of advanced, hormone receptor (HR)-positive postmenopausal breast cancer patients, and other clinical trials that aim to extend the utilization of these agents.

Keywords

Cell cycle • CDK inhibitors • Breast cancer

12.1 Introduction

As the basic structural and functional unit of living organisms, cells reproduce themselves by means of cell cycle process, during which they duplicate their genetic materials and distribute their DNAs equally into two equal cells (also called daughter cells). In eukaryotic cells, cell cycle progression takes place in steps. The first

Z. Cai • Q. Liu (✉)
Breast Tumor Center, Sun Yat-sen Memorial Hospital, Sun Yat-sen University,
107 Yanjiang West Road, Guangzhou 510120, Guangdong, China
e-mail: victorlq@hotmail.com

E. Song, H. Hu (eds.), *Translational Research in Breast Cancer*, Advances in Experimental Medicine and Biology 1026, https://doi.org/10.1007/978-981-10-6020-5_12

step is called G1 phase, followed by the chromosomes replication in S phase. Then comes the G2 phase which is followed by chromosomes segregation in M phase [1]. Each step of cell cycle progresses in sequence, which is controlled by the actions of cyclins and their counterpart cyclin-dependent kinases (CDKs). Human cells contain a large family of CDKs and cyclins. However, only a few certain subsets of CDK-cyclin complexes are involved in cell cycle regulation [2]. The kinase activity of CDKs is controlled mainly by three different ways: the binding to their counterpart cyclins, the binding to negative regulators (CDK inhibitors, CKI), and phosphorylation/dephosphorylation of CDKs. The cell cycle is also supervised by checkpoints, which detect mistakes during DNA synthesis and chromosome segregation. CDKs activity interacts with checkpoints, which halts cell cycle progression and causes cell cycle arrest. This cell cycle progression brake enables cells to fix these mistakes, thus preventing defected DNA from transmitting to daughter cells [3]. Deregulation of CDKs leads to uncontrolled proliferation and increases genomic and chromosomal instability, which plays a significant role in carcinogenesis [4].

Deregulation of cell cycle, leading to aberrant cell proliferation, is a characteristic of cancer. Deranged CDK4/6 axis in the G1/S transition and perturbations in G2/M transition mediated by CDK1/2 are pivotal carcinogenesis events. Given their important role in cell cycle regulation, CDKs could be promising targets in cancer treatment. However, the first-generation pan-CDK inhibitors failed in preclinical/clinical trials because of the adverse events and low efficacy [3, 5]. In recent years, new selective CDK 4/6 inhibitors, including ribociclib, abemaciclib, and palbociclib, have been proved to be promising anticancer drugs with remarkable effects and manageable toxicity. Among these agents, palbociclib was the first CDK4/6 inhibitor that received FDA approval for treating postmenopausal women with estrogen receptor (ER)-positive, HER2-negative advanced breast cancer in combination with letrozole (February 2015) or with fulvestrant (February 2016) [6, 7].

In this review, we will introduce the mechanism of cell cycle progression, especially the aberrant cell cycle regulation in the development of breast cancer. We will also review the advantage and disadvantage of the first-generation pan-CDK inhibitors and the selective CDK4/6 inhibitors. Because of the high efficacy and tolerable adverse events of selective CDK4/6 inhibitors in treating advanced ER-positive breast cancer patients, we will also discuss the potential use of CDK4/6 inhibitors in treatment beyond current indication, with an aim to extend the utilization of these agents.

12.2 Cell Cycle and Its Regulation

Pioneer works by Lee Hartwell, Paul Nurse, and Tim Hunt demonstrated the mechanisms of mammalian cell division [1]. The well-established cell cycle regulation came from studies in yeast. Only one CDK (Cdc28 in *Saccharomyces cerevisiae* and Cdc2 in *Schizosaccharomyces pombe*) cooperated with its counterpart cyclins to regulate cell cycle progression in these simple cells. Although many new members of CDKs and cyclins have been identified in other species, only certain subsets of CDKs and cyclins are responsible for cell cycle regulation in human cells [2].

During cell cycle progression, each of the main events takes place sequentially. After cytokinesis is completed, daughter cells can either enter into the next stage of cell cycle or stay quiescence (also called G0). Cells initiate entry into cell cycle with the presence of extracellular signals such as growth factors. The cells that continue to divide need to go through the first stage (G1 phase) of the new cycle.

12.2.1 G1–S Phases

When cells enter into the cell cycle, mitogenic signals facilitate the synthesis of D-type cyclins (cyclin D1, D2, and D3) and relocation of CDK4/6 to nucleus, forming CDK4/6-cyclin D complexes. The interaction between CDK4/6 and cyclin D significantly enhances the kinase activity with a broader spectrum of substrate than other CDKs [8]. CDK4/6-cyclin D complexes phosphorylate retinoblastoma (Rb) protein family (including pRb, p107, and p130), which plays an important role in target gene suppression. Hypophosphorylated pRb prevents G1-S transition by blocking transcriptional activation of E2F and recruiting histone deacetylases to promoters of S-phase entry genes [9]. Once phosphorylated, inactivated pRb is released from E2F, which can then promote the transcription of E-type cyclins and other genes necessary for S-phase entry and DNA synthesis. Cyclin E binds to CDK2 and forms active CDK2-cyclin E complexes. At the end of G1, activated CDK2-cyclin E complexes facilitate Rb phosphorylation and cause the irreversible inactivation of Rb. This process, called the restriction point, is pivotal in carcinogenesis because alteration of the key regulators could lead to cell division without mitogenic stimuli [10]. In addition to Rb phosphorylation, CDK2-cyclin E complexes participate in phosphorylation of other substrates that involve DNA replication, histone modification, DNA repair, and centrosome duplication and maturation [11].

Inactivation of Rb also promotes expression of A-type and B-type cyclins. Once cells enter into S phase, cyclin E is rapidly degraded by SCF-Fbxw 7 ubiquitin ligase and then cleavage by proteasome, which leads to the inactivation of CDK2-cyclin E complexes [11]. With the accumulation of cyclin A during S phase, CDK2, detached from cyclin E, interacts with the newly synthesized cyclin A. CDK2-cyclin A complexes can phosphorylate numerous proteins necessary for finishing S phase, including transcription factors, proteins relevant to DNA synthesis, DNA repair, histone modification, and cell cycle checkpoints. After the completion of mitosis, CDK2 activity might still exist. Pre-mitotic levels of CDK2 and $p21^{CIP1}$ activity partially predict whether the postmitotic daughter cells continue to divide or become quiescent [12].

Another kinase, CDK3, which binds to cyclin C, may also be involved in Rb phosphorylation during G0-G1 transition. Considering that cyclin C expression is prior to cyclin D, Rb phosphorylation could be initiated by CDK3-cyclin C. CDK3 also interacts with cyclin E and cyclin A, but the role of CDK3 and its counterparts remains unclear [13] .

Cyclin-dependent kinase inhibitors (CKIs) also play a pivotal role in G1-S transition. The inhibitor of CDK4 (INK4) includes four structurally related proteins, $p16^{INK4A}$, $p15^{INK4B}$, $p18^{INK4C}$, and $p19^{INK4D}$, which consist of numerous ankyrin repeats. The INK4 proteins exclusively bind to CDK4/6 rather than other CDKs or cyclin D [14–18]. The cyclin-dependent kinase inhibitor 1/kinase inhibitory proteins (CIP/KIP), including, $p21^{CDKN1A/CIP1}$, $p27^{CDKN1B/KIP1}$, and $p57^{CDKN1C/KIP2}$, can bind to all CDKs in varying degree, which have an alternative positive or negative regulatory role.

The INK4 proteins disrupt the interaction between CDK4/6 and cyclin D, by binding to the catalytic domains of CDK4/6, which subsequently inhibits the kinase activity [19]. For example, diverse oncoproteins prevent neoplastic transformation by inducing $p16^{INK4A}$, which results in G1 arrest of the cell cycle and facilitates oncogene-induced senescence [20]. Similarly, $p15^{INK4B}$ suppresses epithelial cell proliferation with the presence of transforming growth factor-B [21]. Therefore, in the development of cancer, cells must evade the oncogene-induced senescence, which may occur through the loss of $p16^{INK4A}$ or loss of Rb [22, 23]. The loss of $p16^{INK4A}$ releases the CDK4/6 and subsequently activates Rb phosphorylation, leading to oncogenic proliferation, whereas the loss of Rb causes dysregulation of downstream signaling in the cell cycle. Therefore, Rb is necessary for the $p16^{INK4A}$-mediated cell cycle arrest, and Rb-negative cancer has intrinsic resistance to $p16^{INK4A}$ or the agents of CDK4/6 inhibitors [24].

In contrast to INK4 proteins in control of CDK4/6, the CIP/KIPs family binds to CDK2-cyclin E complexes, which potently inhibit kinase activity and thus stabilize cyclin E [11, 18]. $p21^{CIP1}$, one of the most important target genes of p53, serves as a DNA damage checkpoint which blocks DNA synthesis, whereas $p27^{KIP1}$ responses to mitogenic signaling and relates to deregulated proliferation [25, 26]. At the basal level, both $p21^{CIP1}$and $p27^{KIP1}$ can bind to and stabilize CDK4/6-cyclin D complexes without inhibiting their kinase activity. The sequestered $p21^{CIP1}$and $p27^{KIP1}$ released from CDK4/6-cylin D complexes indirectly inhibit

CDK2-cyclin E complexes, which form an interaction network between the cyclins and CDK inhibitors [18, 27]. In addition, the inhibitory function of p27 was confirmed to rescue cyclin D1-null mice that displayed defects without p27 ablation [18]. However, different studies showed that $p21^{CIP1}$and $p27^{KIP1}$ proteins had no direct inhibitory effects on CDK4/6-cyclin D complexes. Instead, they were found to promote the assembly and proper nuclear translocation of the complexes [28, 29]. The role of $p21^{CIP1}$and $p27^{KIP1}$ in carcinogenesis remains elusive.

12.2.2 G2–M Phases

In late S phase, cyclin A binds to CDK1, forming CDK1-cyclin A complexes. Sharing similar substrates with CDK2-cyclin B, CDK1-cylin A phosphorylates numerous proteins involved in DNA synthesis and cell cycle regulators [30, 31]. The precise roles of CDK1-cyclin A and CDK2-cyclin B in S to G2 transition and their difference still need further study. After the nuclear envelope breakdown, cyclin A is degraded via ubiquitin-mediated proteolysis, whereas cyclin B becomes evident. The newly synthesized cyclin B binds to CDK1, forming CDK1-cyclin B complexes that may control G2-M transition and trigger mitosis [32]. CDK1-cyclin B is presumed to phosphorylate abundant substrates including microtubule-binding proteins, proteins relevant to translation, ubiquitin-dependent proteolysis, replication, and other mitosis regulators. Cytoplasmic CDK1-cyclin B complexes also facilitate centrosome segregation through the phosphorylation of the centrosome-associated motor protein Eg5 during prophase. In order to exit from mitosis, CDK1-cylin B complexes are decomposed by the degradation of cyclin B regulated by the anaphase-promoting complex [33, 34].

12.2.3 Biological Function of Other CDKs

CDK5 Primarily active in postmitotic neurons, CDK5 interacts with p35 and p39, which are specific in brain tissue. CDK5-p35 and CDK5-p29 complexes can phosphorylate numerous substrates, which are relevant to neuronal cell cycle arrest and differentiation and apoptosis in neuronal diseases. These substrates are involved in transcription, neuronal function, migration, and synaptic transmission [35–38].

CDK7 As a component of the CDK-activating kinase (CAK), CDK7 interacts with cyclin H, forming CDK7-cyclin H complexes. The CDK7-cyclin H complexes are presumed to phosphorylate and facilitate all cell cycle CDKs. Given its interaction with TFIIH and RNA polymerase III, CDK7-cyclin H may also function in the regulation of transcription [39, 41].

CDK8 and CDK9 CDK8 and CDK9 cooperate with cyclin C and cyclin T, respectively. CDK8-cyclin C and CDK9-cyclin T complexes regulate transcription by phosphorylating the large subunit of RNA polymerase II. CDK8-cyclin C complexes can also inhibit CAK activity by phosphorylating cyclin H. Increased CDK8 kinase activity is relevant to expression of β-catenin transcriptional targets and the inhibition of E2F1 targets apoptotic genes [41]. On the other hand, CDK9 interacts with cyclin H and cyclin K, forming P-TEFb transcription factors that regulate transcriptional elongation [42].

CDK10 and CDK11 Although its cyclin partner has not been identified yet, CDK10 may function in the regulation of G2-M transition. CDK10 also modulates the trans-activation activity of Ets2 transcription factors, a regulator of CDK1 expression [43]. CDK11 binds to cyclin L and interacts with the general precursor mRNA splicing factors RNAPS1 and 9G8 and RNA polymerase II [44]. In addition to RNA process regulation, CDK11 is relevant to the duplication and maturation of centrosome, the assembly of spindle, the binding of chromatid, and the division of the cytoplasm at the end of mitosis [45–48].

CDK12 and CDK13 CDK12 (also called Crkrs) and CDK13 (also called CDC2L5) are involved in alternative splicing regulation by binding to cyclin L [45, 46].

12.3 Cell Cycle Dysregulation in Breast Cancer

Breast cancer is a heterogeneous disease generated from various genetic and epigenetic mutations of oncogenes and tumor suppressor genes that ruin homeostasis maintenance of proliferation, differentiation, and apoptosis in mammary epithelial cells. Under cell cycle dysregulation, decreased CDKs activities result in defective homeostasis, whereas hyperactivation of CDKs favors carcinogenesis by inducing uncontrolled cell division with subsequent development of malignant phenotypes. The mutations in CDKs and their regulators have been under extensive study. Dysregulation of the CDK4/6 axis and CDK2 has been emphasized in many human cancers including breast cancer due to its distinct mechanisms [49].

12.3.1 Cyclin D1 in ER-Positive Breast Cancer

Cyclin D1, encoded by CCND1 gene, was first described in carcinogenesis due to gene rearrangement—the chromosome 11p15:q13 inversion in parathyroid adenoma [50]. Overexpression of cyclin D1, with an incidence of 45–50% in primary ductal carcinomas, is one of the most common oncogenic events in breast cancer [51]. In patients with luminal estrogen receptor(ER)-positive breast cancer, activated ER signaling boosts the CCND1 transcription and leads to cyclin D1 overexpression [52]. In breast cancer cells, cyclin D1 is a direct target of estrogen signaling and enhances cell proliferation [53]. Cyclin D1 can also bind to ER and enhance transcriptional activity of ER through its CDK-independent function, which probably reinforces the interaction of cyclin D1 and ER signaling in ER-positive luminal breast cancer [54]. Additional dysregulation in ER-positive breast cancers includes cyclin D1 gene amplification and gene translocation [5, 55]. In patients with primary breast cancer, cyclin D1 overexpression is restricted to specific pathological subtypes. For example, cyclin D1 overexpression exists in almost exclusively estrogen receptor-positive ductal carcinoma and in vast majority of lobular carcinoma [56, 57].

In mouse mammary tumor virus (MMTV)-cyclin D1 transgenic mice model, overexpression of cyclin D1 results in mammary hyperplasia and development of mammary adenocarcinomas, implicating that cyclin D1 plays an important role in the development of breast cancer [58]. The distinction of cyclin D1 mRNA expression levels between benign and malignant lesion indicates that cyclin D1 overexpression is pivotal in the transition from ductal carcinoma in situ to invasive breast cancer [59]. Cyclin D1 protein overexpression in mammary hyperplasia and intraductal breast carcinoma suggests that cyclin D1 protein is important at the very early stage of breast carcinogenesis and continues to have a crucial role throughout the development of malignancy [60]. In human breast cancer cells, induction of cyclin D1 accelerates G1 phase, which makes it possible for the arrested cells to complete the cell cycle [61]. Cyclin D1 knockout mice are protected from breast cancer induced by Ras or Neu oncogenes, rather than c-myc or Wnt-1 oncogenes, revealing that cyclin D1 is a mediator in carcinogenesis [62]. The oncogenic action of Neu oncogenes seems to reflect a requirement for the cyclin D1-CDK4/6 interactions, since overexpression of p16 blocks carcinogenesis by Neu [63]. Taking together, cyclin D1 overexpression plays a critical role in evolution of breast cancer, and targeting cyclin D1 may be a feasible strategy in breast cancer treatment, specifically in patients with activated Neu-Ras pathways.

In addition to the CDK4/6-dependent activities, cyclin D1 has non-cell cycle-associated CDK-independent function, acting as transcriptional regulator in ER-positive breast cancer [64]. Cyclin D1 binds to the hormone-binding domain of ER and subsequently facilitates the interaction between ER and its coactivators, leading to upregulation of ER-mediated transcriptional activity through a CDK4/6-independent mechanism [65, 66].

12.3.2 Cyclin E in HER2-Positive Breast Cancer

In ER-positive breast cancer, cyclin E expression is at a low level. On the contrary, HER2-positive breast cancer is characterized by overexpression of cyclin E [67, 68]. Cyclin E overexpression also associates with poor differentiation [69], poor endocrine response [70], poor prognosis [71], and predicting sensitivity to cisplatin/Taxol chemotherapy and trastuzumab [72, 73]. In mouse model, cyclin E overexpression results in mammary hyperplasia and tumor formation at low incidence after long latency [74]. In breast cancer cell line, amplification of cyclin E results in a 64-fold increase of cyclin E mRNAs that express cyclin E throughout all stages of cell cycle [75]. In addition to the overexpression of full-length 50kD cyclin E, these cell lines overexpress other low molecular weight isoforms of cyclin E. These isoforms, lacking the amino terminus, are hyperactive in activating substrates and accelerating the cell cycle progression through G1/S phase. The level of cyclin E and the summation level of cyclin E isoforms are shown to be strongly associated with breast cancer patient survival [71]. Cyclin E overexpression coexists with HER2 gene amplification in some patients with HER2-positive breast cancers, which is generally associated with poor survival and probably trastuzumab resistance [76]. Previous studies showed contradictory prognostic effects of cyclin E in breast cancer patients, which was possibly due to the use of varying breast cancer phenotypes, different methods, and threshold values to evaluate the expression of cyclin E [77]. A recent study of 2494 patients with breast cancer shows that cytoplasmic cyclin E is a predictor of recurrence with the highest likelihood consistently across different patient cohort and subtypes, suggesting cyclin E as a critical target in breast cancer treatment [78].

Cyclin E and HER2 interact with each other by various mechanisms in patients with HER2-positive breast cancer. HER2 receptor-mediated carcinogenesis was shown to shorten G1 phase, resulting in aberrant cell cycle and subsequently uncontrolled proliferation, probably through upregulation of CDK2 activity [79]. Other studies demonstrated that HER2 straightly enhanced cyclin E activity since decreased HER2 signaling resulted in lower cyclin E expression, particularly the low molecular weight (LMW) isoforms, which in turn had prognostic and predictive roles in HER2-overexpressing breast cancer [80]. LMW-cyclin E binds to and activates CDK2 more strongly, leading to increased CDK2 activity and decreased sensitivity of the LMW-cyclinE-CDK2 complexes to inhibition by p21 and p27 [81]. The mammary tumorigenesis caused by LMW-cyclin E requires CDK2 activity, indicating that anti-CDK2 therapy may have potential role in LMW-overexpressing human breast tumors [82].

12.3.3 CKIs in Breast Cancer

Cyclin-dependent kinase inhibitors (CKIs) function as tumor suppressors predominately in the end of G1 phase, which trigger DNA damage checkpoint to block impaired cells and initiate repair progression or apoptosis. Despite distinct mechanisms of tumor suppressor genes, the interferences of these genes lead to accumulation of mutation and eventually cause carcinogenesis. The INK4 proteins play a pivotal role in carcinogenesis for the high incidence of $p16^{INK4A}$ and/or $p15^{INK4B}$ inactivation in various human cancers, including breast cancer [83].

$p16^{INK4A}$ In normal breast tissue, the absence of $p16^{INK4A}$ is associated with hyper-methylation of $p16^{INK4A}$ gene, whereas hypo-methylation of $p16^{INK4A}$ is associated with expression of $p16^{INK4A}$ mRNA in breast cancer [84]. Overexpression of $p16^{INK4A}$ occurs in both grade 1 and grade 2 breast carcinomas with a marked decline in grade 3 tumors [85]. A study in 14 breast cell lines showed that $p16^{INK4A}$ defect existed in 4 (29%) breast cell lines, 2 (14%) of which had $p16^{INK4A}$ gene methylation [86].These data suggest the role of $p16^{INK4A}$ is much more complex than previously hypothesized.

p15^{INK4B} In 14 breast cancer cell lines, 3 (21%) have p15^{INK4B} gene mutation, whereas no methylated one is found in primary breast carcinomas [86, 87]. Although the methylation of p15^{INK4B} gene is common in leukemia and glioma, this mutant was rare in breast cancer, which suggests that the mechanism of p15^{INK4B} gene inactivation may be more complicated in different organs [88].

p21^{CIP1} p21^{CIP1} has been long considered as a potential tumor suppressor gene, because p21^{CIP1}-null mice develop mammary tumor with the presence of Ras expression [89]. Nevertheless, the expression of p21^{CIP1} is suppressed in normal breast tissue, whereas the accumulation of p21^{CIP1} is observed in breast tumor tissues [85]. Clinical study implicates that the cytoplasmic localization of p21^{CIP1} is relevant to HER2-overexpression, both of which predict poor prognosis in breast cancer patients [90].

p27^{KIP1} The p27^{KIP1} acts as another important tumor suppressor gene for mice with deficient p27^{KIP1} generated pituitary adenomas and displays higher risk of carcinoma when exposed to carcinogens [91]. The expression of p27^{KIP1} is at relatively high levels in normal breast, whereas the expression of p27^{KIP1} is decreased in tumor tissues, particularly in high-grade tumors [85, 92].

12.4 Targeting CDKs in Breast Cancer Treatment

Breast cancer treatment is a combination of surgery, chemotherapy, endocrine therapy, radiotherapy, targeted therapy, and other therapies [93]. The critical role of CDKs and their counterparts in cell cycle regulation and carcinogenesis raises the possibility of targeting these molecules. The therapeutic value of targeting CDKs has been intensively investigated, especially the interphase CDKs (CDK1, CDK2, CDK4, CDK5, and CDK6). Nevertheless, their usages in breast cancer treatment as pharmaceutical targets still need further study [93]. The ideal CDK-targeted therapy requires interruption of specific CDKs signaling in malignant cells but spares other CDKs activities that are critical in normal cell cycle progress to achieve high efficacy and low toxicity. As mentioned above, dysregulation of the cyclin D-CDK4/6-Rb pathways may lead to acceleration of G1-S progression and uncontrolled proliferation. These observations enable the development of CDK4/6 inhibitors for specific transformed cells. Feasible CDK4/6 inhibitors are supposed to decrease Rb phosphorylation and block cell cycle progression in cells with Rb persistent activation. In cells that lose Rb function, these agents may be ineffective. Thus, selection of appropriate patients for specific anti-CDK4/6 therapy depends on whether the cancer mainly relies on CDK4/6 axis dysregulation to accelerate G1/S transition. Luminal ER-positive breast cancer, but not basal-like ER-negative breast cancer, is the subtype with amplication/overexpression of cyclin D1 and is suitable for anti-CDK4/6 therapy. Even for women with advanced ER-positive breast cancer who have developed resistance to endocrine therapy, most of them still rely on cyclin D1-CDK4/6 complexes to initiate the G1/S transition.

12.4.1 The First-Generation Pan-CDK Inhibitors

In the past two decades, numerous CDK inhibitors have been discovered as potential therapeutic agents and evaluated in preclinical/clinical trials in different tumor models. However, none of the first-generation pan-CDK inhibitors, including flavopiridol, olomoucine, and roscovitine, achieved permission in clinical application. These agents fail to meet the expectation in preclinical/clinical studies, exhibiting limited activity and severe toxicity.

Among these first-generation inhibitors, flavopiridol, also known as alvocidib, has been extensively investigated in more than 60 clinical trials up to now [49]. Flavopiridol derived from chromone alkaloid inhibits kinase activities of several CDKs (CDK1, CDK2, CDK4, CDK6, CDK7, and CDK9). Although flavopiridol has limited clinical effects in patients with hematological malignancies, including chronic lymphocytic leukemia, adverse events come out when the dose increases [94, 95]. Previous studies about flavopiridol showed disappointing results for the treatment of breast cancer. No evident antitumor response was observed in two patients (6%) with advance breast cancer in a phase I trial [96]. Another phase I trial showed that only one patient (5%) with breast cancer might benefit from the combination of flavopiridol with 5-fluorouracil, leucovorin, and oxaliplatin (FOLFOX) [97].The most common adverse events, including hypotension, neutropenia, fatigue, diarrhea, and nausea, often lead to discontinuation of the trials. Since flavopiridol did not achieve expected success as an ideal CDK inhibitor, no phase III trial was carried out, and the development of flavopiridol was given up.

In parallel with flavopiridol, a phase I trial was carried out to evaluate roscovitine (also called R-roscovitine, CY-202, or seliciclib), which had an inhibitory effect on CDK1, CDK2, CDK5, and CDK7. Of the 56 patients receiving roscovitine treatment, only one patient with hepatocellular carcinoma achieved a partial response and sustained tumor stabilization [98]. In breast cancer cell line MCF-7, roscovitine was shown to suppress cell proliferation and reduce cell survival of endocrine-resistant breast cancer cells [99]. In vivo model of MCF-7 cell line, roscovitine can synergize with the antitumor effect of doxorubicin without increasing toxicity [100]. These results reveal the potential therapeutic role of CDK2 inhibition in abrogating growth of endocrine-resistant breast cancer cells.

SNS032 (also called BMS-387032), with an inhibitory effect on CDK2, CDK7, and CDK9, has been shown to sensitize hypoxic and quiescent non-small cell lung cancer cells to radiation therapy. The inhibitory activity may rely on cell cycle independent of CDKs, including CDK7 and CDK9, which are presumed to modulate DNA double-strand break repair [101]. Other studies show that AML cells treated with SNS-032 are more susceptible to the cytotoxic effects of Ara-C, whereas SNS-032 fails to achieve expected clinical outcomes in patients with chronic lymphocytic leukemia and multiple myeloma [102, 103]. In a phase I trial of 21 patients with metastatic solid tumors, only 3 patients (15%) achieved a response of stable disease, while the results of 2 patients (10%) with advanced breast cancer were not published [104].

Dinaciclib (also called MK-7965 and SCH727965) is a potent pan-CDK inhibitor, with higher inhibitory effect on CDK1, CDK2, CDK5, and CDK9. It exhibits better inhibition of Rb phosphorylation, compared with flavopiridol [105]. Dinaciclib has been well tolerated in initial trials, and patients with advanced solid malignancies, myeloma, and chronic lymphocytic leukemia have received profitable clinical efficacy [105–107].However, a few studies on patients with advanced breast cancers showed disappointing outcomes. For example, a phase I trial in patients with metastatic triple-negative breast cancer revealed that the combination of dinaciclib and epirubicin might result in massive adverse events and failed to be an effective therapy for metastatic triple-negative breast cancer [108]. Randomized phase II trial also received disappointing results, which compared the therapeutic efficacy of dinaciclib with the chemotherapy drug capecitabine in patients with advanced breast cancer. Although dinaciclib monotherapy suppressed tumor progression with generally tolerated adverse events, its efficacy was inferior to capecitabine [109].

The reasons why the first-generation pan-CDK inhibitors fail in clinical trials may be explained by the following. Firstly, the first-generation pan-CDK inhibitors, with low specificity, may influence cell cycle progression in different aspects. It remains unknown what kind of CDKs are actually blocked in vivo and whether one may interfere with another. Secondly, the biomarkers for anti-CDKs therapy are unclear. Because of the inter- and intra-tumor heterogeneity of breast cancer, different subpopulations may respond totally differently to an identical agent. Therefore, the identification of sensitive subpopulations and the selection for appropriate agents need to be further optimized. Thirdly, some of these CDKs inhibitors can also target proteins (such CDK9) that are crucial in cellular transcription and thus influence cell proliferation and apoptosis in both cancer cell and normal cells. The inhibition of transcriptional CDKs may prevent carcinogenesis by inducing apoptosis of cancer cells. However, it limits the therapeutic dose of these nonselective agents because they fail to distinguish transformed cells from normal cells. As a result, severe adverse effects arise, such as hypotension, neutropenia, fatigue, diarrhea, and nausea [49, 93].

12.4.2 The Selective CDK4/6 Inhibitors

Since cyclin D-CDK4/6 pathways alteration provides a proliferative and survival advantage to various cancers, including breast cancer, targeting CDK4/6 may achieve more therapeutic benefits than targeting other CDKs. For example, CDK4/6 gene amplification and cyclin D1 gene amplification/translocation mainly exist in ER-positive breast cancer. Estrogen-mediated signaling can also lead to cyclin D1 overexpression. Preclinical studies in cell lines and xenografts have revealed that selective CDK4/6 inhibitors have potent inhibitory effects on malignancies with limited cytotoxicity [110].

Understanding the molecular structure of CDKs leads to the development of more selective CDK4/6 inhibitors [55].And up to now, three selective CDK4/6 inhibitors (palbociclib, ribociclib, and abemaciclib) have been widely studied in preclinical and clinical trials, with promising efficiency and manageable adverse events.

12.4.2.1 Palbociclib

Palbociclib (PD0332991) is one of the most well-known selective CDK4/6 inhibitors discovered from a subset of pyridopyrimidine compounds according to its unprecedented levels of selectivity for CDK4 as well as its superior physical and pharmaceutical properties. In vitro, it has a prior selectivity for CDK4 and CDK6 ($IC50 = 0.011$ μmol/L, 0.016 μmol/L, respectively) but has limited activity against other CDKs or tyrosine kinases. In preclinical studies, palbociclib was shown to arrest cells exclusively in G1, decrease phospho-Rb and Ki-67, and reduce expression of E2F target genes in Rb-positive tumors. Consistent with its mechanism of action, palbociclib failed to inhibit the growth of triple-negative breast cancer cell line with the feature of Rb deficiency [111]. Later study found that palbociclib might have inhibitory activity in Rb-deficient cells, probably because of the supplementary function of other phosphorylate retinoblastoma (Rb) proteins like p107 or p130 [112].

Two phase I studies investigating the dose-limiting toxicities (DLTs) and the maximum tolerated dose (MTD) of palbociclib were conducted in patients with relapsed or refractory cancer, including Rb-positive advanced solid tumors and non-Hodgkin's lymphoma [113, 114]. A phase I study of 33 patients, who received palbociclib in 2/1 schedule (palbociclib once daily for 2 weeks on treatment; 1 week rest), gained therapeutic benefits. A case of partial response was reported in the patient with testicular cancer, who received palbociclib 200 mg/d. Additional nine cases were reported to achieve stable disease for more than two cycles, and three cases maintained stable disease for more than ten cycles. Although treatment-related adverse events happened in 29 cases (88%), most of them were manageable [114]. Another study of 41 patients that administered palbociclib once daily for 21 of 28 days (3/1 schedule) revealed that 10 (27%) patients maintained stable disease for more than 4 cycles and 6 of them achieved prolonged benefit for more than 10 cycles with tolerated toxicities [113]. Similar dose-limiting adverse events were observed, and the most common adverse event was neutropenia. Based on these studies, the MTD, 200 mg/d and 125 mg/d, respectively, was recommended for phase II study.

A phase II study of palbociclib for monotherapy (125 mg/d; 3/1 schedule) was performed in 37 patients with Rb-positive advanced breast cancer [115]. Most of these patients had hormone receptor (HR)-positive disease and were pretreated with two or more prior hormonal therapy or chemotherapy. The overall median progression-free survival was 3.7 months, which was significantly related to HR level as well as progression on prior hormone treatment. Patients with HR-positive tumor had significantly longer progression-free survival than those with HR-negative tumors (4.5 months versus 1.5 months). The progression-free survival in those with progression disease was associated with the number of previous prior hormone treatment. As for the overall response rate, partial response (PR) was reported in two cases, and stable disease (SD) for more than 6 months existed in five cases. The clinical benefit rate (CBR = PR + 6 months SD) was 19% overall, 21% in HR positive, and 29% in patients with progress disease who previously received more than two prior hormone treatments. Notably, none of the markers (including Rb in nuclear, Ki67, p16 defect, and cyclin D1 overexpression) was relevant to either clinical benefit rate or progression-free survival. As for the safety and tolerability, 59 grade 3/4 adverse events were observed because of myelotoxicity. In addition, grade 3/4 neutropenia and leukopenia were observed in 19 cases (51%), grade 3/4 lymphopenia in 11 cases (30%), grade 3/4 thrombocytopenia in 7 cases (19%), and grade 3/4 anemia in 2 cases (5%). Nine patients (24%) suspended treatment, and 19 (51%) reduced drug dose due to cytopenias. Only one patient (3%) quitted the treatment after two cycles due to a moderate fatigue [115]. Taking together, this study has revealed that single-agent palbociclib is potent in patients with Rb-positive advanced breast cancer, particularly in HR-positive and endocrine-resistant patients, with manageable adverse events.

An open-label randomized phase I/II study (NCT00721409, also known as PALOMA-1/TRIO-18) aimed to verify the effect and safety of the combination of palbociclib and letrozole in women with estrogen receptor (ER)-positive and HER2-negative breast cancer [116]. Previous phase I study demonstrated tolerable treatment-related adverse events and no significant drug-drug interaction, suggesting a dose of oral palbociclib 125 mg/d on 3/1 schedule in combination with letrozole 2.5 mg/d orally [117]. The phase II study included 165 patients from 50 sites in 12 countries, who were divided into two separate cohorts. In cohort I, 66 women were recruited according to their estrogen receptor-positive and HER2-negative biomarker status alone. They received the combination of palbociclib (125 mg/d; 3/1 schedule) and letrozole (2.5 mg/d; continuously), compared with letrozole monotherapy. Meanwhile, in cohort II, 99 women with CCND1 amplification, loss of p16, or both were selected to receive the same treatment allocations. The primary intention of this study was to explore and analyze progression-free survival in cohort I. Unexpectedly, a remarkable improvement of progression-free survival was shown in cohort I with no evident association between prognosis and status of CCND1 or p16, leading to a combined analysis for both cohorts. Final analysis showed that median progression-free survival in combination therapy group versus monotherapy group was 20.2 months versus 10.2 months. In concert with previous studies, neutropenia, leucopenia, and fatigue had higher incidence in combination therapy group. Although slight increased incidence of adverse events was reported in combination therapy group, most of them were low grade [116]. The promising results from PLAOMA-1/TRIO 18 study allow FDA to speed up palbociclib approval. The combination therapy of palbociclib and letrozole is recommended as the prior endocrine-based therapy in postmenopausal women with estrogen receptor-positive, HER2-negative advanced breast cancer [115].

A randomized, multicenter phase III study PALOMA-2 (NTC01740427) was carried out to validate the results in a larger population. In this double-blinded study, 666 patients with estrogen receptor-positive, HER2-negative breast cancer, who had not received prior treatment, were recruited. These patients were randomly divided into two groups in a 2:1 ratio, with 444 patients to receive palbociclib plus letrozole and 222 patients to receive placebo plus letrozole for the same treatment allocations as PLAOMA-1/TRIO18. The progression-free survival was assessed, as well as other indexes such as overall survival and clinical benefit response. In combination therapy group (palbociclib plus letrozole), the median progression-free survival was 24.8 months comparing with 14.5 months in the monotherapy group (placebo plus letrozole). Neutropenia, leukopenia, and anemia mainly occur in palbociclib plus letrozole group with a higher incidence. Results from PALOMA-2 verified that in postmenopausal patients with ER-positive and HER2-negative advanced breast cancer, the combination therapy of palbociclib and letrozole significantly improved progression-free survival when compared with letrozole monotherapy. These findings indicate that selective CDK4/CDK6 inhibitor can be used as first-line treatment for the above patient group [118].

Another double-blinded, randomized phase III study PALOMA-3 (NCT01942135) confirmed the safety and efficacy of the combination of palbociclib and fulvestrant (a selective estrogen receptor degrader) in women with hormone receptor-positive, HER2-negative advanced breast cancer who were relapsed or refractory. Five hundred twenty-one patients were randomly divided in a 2:1 ratio into the combination therapy group of palbociclib plus fulvestrant and the fulvestrant monotherapy group. The median progression-free survival was 9.2 months in the combination therapy group versus 3.8 months in the fulvestrant monotherapy group. The most common adverse events in the combination therapy group were neutropenia, leukopenia, and anemia, with a much higher incidence than placebo plus fulvestrant group. PALOMA-3 was the first large trial to testify efficacy and safety of a selective CDK4/6 inhibitor in endocrine-resistant breast cancer [119, 120].

An increasing number of trials are assessing the safety and efficacy of palbociclib in different clinical conditions (including adjuvant therapy and neoadjuvant chemotherapy) and combining other drugs like trastuzumab with palbociclib in breast cancer treatment [121]. Given that palbociclib was synergistic with trastuzumab in HER2-positive breast cancer cell, addition of palbociclib to HER2 targeted therapy has raised great interest [122]. Preclinical breast cancer models revealed that CDK4/6 controlling downstream of HER2 served as a feasible therapeutic target in HER2-positive breast cancer. Selective CDK4/6 inhibitor palbociclib was synergistic with multiple HER2-targeted agents, which provided an additional mechanism to potently suppress the propagation of T-DM1-resistant HER2-positive cancer cells [123]. A phase 1b trial is ongoing to evaluate the combination therapy of palbociclib plus T-DM1.

12.4.2.2 Ribociclib

Ribociclib (LEE011) is another orally administered small molecular with high selectivity to inhibit CDK4/6 at nanomolar concentrations, which reduces Rb phosphorylation, blocks cell cycle progression, and induces G1 arrest. In preclinical studies, ribociclib was shown to have inhibitory activity in cancer cell lines and xenograft models of neuroblastoma, liposarcoma, and ER-positive breast cancer [124, 125].

In a phase I trial, ribociclib was tested as a monotherapy in 132 patients with Rb-positive malignancies, including 20 patients with breast cancer. This trial explored the maximum tolerated dose (MTD), recommended dose for expansion (RDE), and safety of ribociclib. The maximum tolerated dose (MTD) and the recommended phase 2 dose (RP2D) were 900 and 600 mg/day, respectively, on a 3/1 schedule. Among the 70 patients evaluated for MTD/RDE determination, 9 DLTs were observed during cycle 1. The most common DLTs were neutropenia and thrombocytopenia. The Ki67 levels of skin and tumor tissues were decreased due to the ribociclib-mediated antiproliferative activity. Stable disease was reported in 43 cases (including 8 progression-free cases for more than 6 months). Partial response was observed in three cases (including one patient with ER-positive, PIK3CA-mutant, and CCND1-amplified breast cancer) [126].

In a recent phase Ib/II study involving patients with ER-positive, HER2-negative advanced breast cancer, ribociclib (600 mg/d;3/1 schedule) in combination with letrozole showed an acceptable safety profile and exhibited promising clinical activity, particularly in patients who had never received previous systemic treatment for advanced disease. Between previously untreated patients and previously treated patients, the overall response rate (ORR) was 83% versus 5%, while the clinical benefit rate (CBR) was 73% versus 32% [127]. Since preclinical studies showed that ribociclib and the alpha-specific PI3K inhibitor alpelisib (BYL719) had synergistic activity in PIK3CA-mutant breast cancer, a phase 1b/2 study was carried out to access the safety and efficacy of the combination with ribociclib and PI3K inhibitors [128]. A triplet combination with ribociclib, letrozole, and alpelisib was administered in patients with HR-positive, HER2-negative advanced breast cancer (NCT01872260). There were 41 patients receiving ribociclib (300–500 mg QD in 3/1 schedule) plus letrozole (2.5 mg QD continuous), 21 patients receiving alpelisib (200–250 mg QD continuous) plus letrozole (2.5 mg QD continuous), and 36 patients receiving ribociclib (300–500 mg QD in 3/1 schedule) plus alpelisib (200–250 mg QD continuous) plus letrozole (2.5 mg QD continuous). Of the 27 patients evaluated for response, 2 (7%) patients had confirmed partial response, 4 (15%) patients had unconfirmed partial response, and 6 (22%) patients had stable disease with response of mild adverse events [129]. Another phase II study evaluated the biological activity of ribociclib plus letrozole compared with single-agent letrozole in the presurgical condition of breast cancer (NCT01919229). The combination of ribociclib and letrozole reduced Ki67-positive cell fraction potently with tolerate adverse event [130]. To explore potential inhibitory effect of CDK4/6 inhibitors in HER2-positive breast cancer, an ongoing open-label, phase 1b/2 clinical trial accesses the safety and efficacy of the combination of ribociclib and trastuzumab in comparison with T-DM1 monotherapy for patients with HER2-positive advanced or metastatic breast cancer (NCT02657343).

Three large, international, double-blinded, placebo-controlled phase III trials are evaluating the addition of ribociclib to endocrine therapy in patients with HR-positive, HER2-negative breast cancer. The MONALEESA-2 confirmed the combination therapy (ribociclib plus letrozole) as the prior treatment in patients with previously untreated HR-positive, HER2-negative advanced breast cancer (NCT01958021). A total of 668 patients were randomly divided in a 1:1 ratio into the combination therapy group (ribociclib plus letrozole) and the letrozole monotherapy group. The trial met its primary end point, with the median duration of progression-free survival not reached in the combination therapy group versus 14.7 months in the monotherapy group. The overall response rates were 40.7% in the combination therapy group and 27.5% in the monotherapy group, including 9 (2.7%) complete response versus 7 (2.1%) and 127 (38%) partial response versus 85 (25.4%). The clinical benefit rates were 79.6% in the combination therapy group versus 72.8% in the monotherapy group. As for safety, more grade 3/4 adverse events arise in the combination therapy group (81.2%) than in the monotherapy group (32.7%). The most common adverse events were neutropenia (74.3% versus 5.2%), nausea (51.5% versus 28.5%), infections (50.3% versus 42.4%), fatigue (36.5% versus 30.0%), and diarrhea (35.0% versus 22.1%). The MONALEESA-3 (NCT02422615) is another ongoing phase III trial, which accesses the efficacy and safety of the combination therapy of ribociclib plus fulvestrant for treatment of patients with untreated HR-positive, HER2-negative advanced breast cancer. The primary end point of the study is progression-free survival, and the secondary end points include overall survival, overall response rate, and safety. The MONALEESA-7 (NCT02278120) is another ongoing phase III trial, which aims to assess the safety and efficacy of ribociclib or placebo in combination with tamoxifen and goserelin or a nonsteroidal aromatase inhibitor (NSAI) and goserelin for the treatment of premenopausal women with HR-positive, HER2-negative advanced breast cancer.

12.4.2.3 Abemaciclib

Abemaciclib (LY283521), another oral selective CDK4/6 inhibitor characterized with its clinical safety profile, is currently in clinical development. At low nanomolar, abemaciclib strongly inhibits CDK4 and CDK6 and therefore reduces Rb phosphorylation, leading to cell cycle arrest in G1 and proliferation suppression, particularly in Rb-proficient breast cancer cell lines. Oral administration of abemaciclib suppressed tumor growth in human tumor xenografts including various tumor subtypes in tumor-bearing mice [131, 132].

The first-in-human phase I study evaluated the safety and efficacy of abemaciclib for the treatment of patients with solid tumors including breast cancer. In this trial, abemaciclib demonstrated promising single-agent activity, and limited toxicities occurred with the increase of drug dose. A total of 225 patients were recruited, including 33 patients in dose escalation and 192 patients in tumor-specific cohorts. The median progression-free survival was 8.8 months in HR-positive patients versus 1.1 months in HR-negative ones. Similarly, disease control rate could be associated with the HR status in patients who had been previously treated (HR positive, 29 in 36 cases (81%) versus HR negative, 3 in 9 cases (33%)). Based on the Rb inhibition and cell cycle arrest in normal cells and tumor cells, the maximum tolerated dose was 200 mg every 12 hours. Among the most common treatment-related toxicities, fatigue was manageable. Meanwhile, other toxicities occurred in gastrointestinal system, renal system, and hematopoietic system. A subgroup of 19 patients with HR-positive breast cancer received the combination therapy of abemaciclib plus fulvestrant. Partial responses were observed in four patients (21%) with no different adverse events compared to single-agent cohorts. The antitumor activity of abemaciclib in patients with HR-positive breast cancer was probably associated with TP53 rather than PIK3CA [133]. These results inspired the idea to test the combination of different therapies (letrozole, anastrozole, tamoxifen, exemestane, exemestane plus everolimus, trastuzumab) for patients with metastatic breast cancer in a phase 1b multiple cohorts study (NCT02057133). A total of 65 patients were assigned into 6 cohorts to receive the combination therapy of abemaciclib and other drugs (such as letrozole, anastrozole, tamoxifen, and trastuzumab). This study indicates that the combination of abemaciclib and different therapies is promising for patients with metastatic breast cancer [134].

The phase II study MONARCH-1 (NCT02102490) evaluated the safety and efficacy of abemaciclib as monotherapy for patients with previously treated, advanced, or metastatic HR-positive/HER2-negative breast cancer who had progressive disease on or after endocrine therapy and chemotherapy. In 132 eligible patients, the confirmed overall response rate was 19.7%, the clinical benefit rate was 42.4%, and the median PFS was 6.0 months, with a higher response rate than other CDK4/6 inhibitors [135]. Considering that abemaciclib can cross the blood-brain barrier, abemaciclib is supposed to have potential antitumor activity in patients with central nervous system metastases [136]. A currently ongoing phase II study (NCT02308020) is evaluating the efficacy of abemaciclib in patients with brain metastases from different solid primary tumors including HR-positive breast cancer. Another ongoing phase II study (NCT02675231) is exploring the efficacy of abemaciclib plus trastuzumab with or without fulvestrant or chemotherapy in patients with HR-positive, HER2-positive locally advanced or metastatic breast cancer. NeoMONARCH (NCT02441946) is a randomized, multicenter, open-label phase II neoadjuvant study comparing the biological effects of abemaciclib plus anastrozole, abemaciclib monotherapy, and anastrozole monotherapy in patients with early-stage HR-positive/HER2-negative breast cancer. Two hundred twenty-three patients were stratified by progesterone receptor status and tumor size and randomized into three groups at a ratio of 1:1:1 to receive abemaciclib (150 mg orally Q12H) plus anastrozole (1 mg orally QD), abemaciclib (150 mg orally Q12H), and anastrozole (1 mg orally QD) for 2 weeks followed by administration of abemaciclib (150 mg orally Q12H) plus anastrozole (1 mg QD) for the next 14 weeks. In a 9-month interim analysis, a single agent of abemaciclib or in combination with anastrozole exhibited significantly greater suppression of Ki67 after 2 weeks of dosing than anastrozole alone. Further results including safety, clinical efficacy, final Ki67, and RNA expression at surgery are not reported yet [137].

Two large, randomized, double-blinded, placebo-controlled, phase III studies are currently ongoing to confirm the effects of adding abemaciclib to fulvestrant and aromatase inhibitors, respectively. MONARCH-2 (NCT02107703) aims to compare progression-free survival for women with HR-positive (HR+)/HER2-negative advanced breast cancer who are randomized in a 2:1 ratio to receive either abemaciclib plus fulvestrant or fulvestrant alone. Another trial MONARCH-3 (NCT02246621) is to evaluate the effect of nonsteroidal aromatase inhibitors (anastrozole or letrozole) plus abemaciclib or placebo in postmenopausal women with breast cancer. Both trials use progression-free survival as primary end point and overall survival/ objective response rate as secondary end points.

12.5 Future Direction

Cell cycle dysregulation has been one of the most important therapeutic targets in cancer for many years. Selective CDK4/6 inhibitors have been recently approved by FDA to treat ER-positive advanced breast cancer, which takes more than two decades after the discovery of cyclin D1-CDK4/6 interaction. Many questions remain to be answered, including the biomarker, indication, and drug combination of anti-CDK4/6 therapy. Anti-CDK4/6 therapy could be a promising strategy for treating high-risk early breast cancer patients, HER2-positive patients, or even triple-negative breast cancer patients with functional Rb. More selective CDK2 inhibitors may also be useful in disrupting cyclin E-CDK2 function and treating a broader spectrum of cancer than CDK4/6 inhibitors.

References

1. Nurse PM (2002) Cyclin dependent kinases and cell cycle control. Bioscience Rep 22(5-6):487–499. doi: 10.1023/A:1022017701871
2. Malumbres M, Barbacid M (2005) Mammalian cyclin-dependent kinases. Trends Biochem Sci 30(11):630–641. doi:10.1016/j.tibs.2005.09.005
3. Malumbres M, Barbacid M (2009) Cell cycle, CDKs and cancer: a changing paradigm. Nat Rev Cancer 9(3):153–166. doi:10.1038/nrc2602
4. Kastan MB, Bartek J (2004) Cell-cycle checkpoints and cancer. Nature 432(7015):316–323. doi:10.1038/nature03097
5. Shapiro GI (2006) Cyclin-dependent kinase pathways as targets for cancer treatment. J Clin Oncol 24(11):1770–1783. doi:10.1200/Jco.2005.03.7689
6. Beaver JA, Amiri-Kordestani L, Charlab R, Chen W, Palmby T, Tilley A, Zirkelbach JF, Yu J, Liu Q, Zhao L, Crich J, Chen XH, Hughes M, Bloomquist E, Tang S, Sridhara R, Kluetz PG, Kim G, Ibrahim A, Pazdur R, Cortazar P (2015) FDA approval: Palbociclib for the treatment of postmenopausal patients with estrogen receptor-positive, HER2-negative metastatic breast cancer. Clin Cancer Res 21(21):4760–4766. doi:10.1158/1078-0432.CCR-15-1185
7. Walker AJ, Wedam S, Amiri-Kordestani L, Bloomquist E, Tang S, Sridhara R, Chen W, Palmby TR, Fourie Zirkelbach J, Fu W, Liu Q, Tilley A, Kim G, Kluetz PG, McKee AE, Pazdur R (2016) FDA approval of Palbociclib in combination with Fulvestrant for the treatment of hormone receptor-positive, HER2-negative metastatic breast cancer. Clin Cancer Res 22(20):4968–4972. doi:10.1158/1078-0432.CCR-16-0493
8. Anders L, Ke N, Hydbring P, Choi YJ, Widlund HR, Chick JM, Zhai H, Vidal M, Gygi SP, Braun P, Sicinski P (2011) A systematic screen for CDK4/6 substrates links FOXM1 phosphorylation to senescence suppression in cancer cells. Cancer Cell 20(5):620–634. doi:10.1016/j.ccr.2011.10.001
9. Harbour JW, Dean DC (2000) The Rb/E2F pathway: expanding roles and emerging paradigms. Genes Dev 14(19):2393–2409
10. Malumbres M, Barbacid M (2001) To cycle or not to cycle: a critical decision in cancer. Nat Rev Cancer 1(3):222–231. doi:10.1038/35106065
11. Hwang HC, Clurman BE (2005) Cyclin E in normal and neoplastic cell cycles. Oncogene 24(17):2776–2786. doi:10.1038/sj.onc.1208613
12. Spencer SL, Cappell SD, Tsai FC, Overton KW, Wang CL, Meyer T (2013) The proliferation-quiescence decision is controlled by a bifurcation in CDK2 activity at mitotic exit. Cell 155(2):369–383. doi:10.1016/j.cell.2013.08.062
13. Ren S, Rollins BJ (2004) Cyclin C/cdk3 promotes Rb-dependent G0 exit. Cell 117(2):239–251
14. Serrano M, Hannon GJ, Beach D (1993) A new regulatory motif in cell-cycle control causing specific inhibition of cyclin D/CDK4. Nature 366(6456):704–707. doi:10.1038/366704a0
15. Hannon GJ, Beach D (1994) p15INK4B is a potential effector of TGF-beta-induced cell cycle arrest. Nature 371(6494):257–261. doi:10.1038/371257a0
16. Chan FK, Zhang J, Cheng L, Shapiro DN, Winoto A (1995) Identification of human and mouse p19, a

novel CDK4 and CDK6 inhibitor with homology to p16ink4. Mol Cell Biol 15(5):2682–2688
17. Hirai H, Roussel MF, Kato JY, Ashmun RA, Sherr CJ (1995) Novel INK4 proteins, p19 and p18, are specific inhibitors of the cyclin D-dependent kinases CDK4 and CDK6. Mol Cell Biol 15(5):2672–2681
18. Sherr CJ, Roberts JM (1999) CDK inhibitors: positive and negative regulators of G1-phase progression. Genes Dev 13(12):1501–1512
19. Pavletich NP (1999) Mechanisms of cyclin-dependent kinase regulation: structures of Cdks, their cyclin activators, and Cip and INK4 inhibitors. J Mol Biol 287(5):821–828. doi:10.1006/jmbi.1999.2640
20. Serrano M, Blasco MA (2001) Putting the stress on senescence. Curr Opin Cell Biol 13(6):748–753
21. Reynisdottir I, Polyak K, Iavarone A, Massague J (1995) Kip/Cip and Ink4 Cdk inhibitors cooperate to induce cell cycle arrest in response to TGF-beta. Genes Dev 9(15):1831–1845
22. Witkiewicz AK, Knudsen KE, Dicker AP, Knudsen ES (2011) The meaning of p16(ink4a) expression in tumors: functional significance, clinical associations and future developments. Cell Cycle 10(15):2497–2503. doi:10.4161/cc.10.15.16776
23. LaPak KM, Burd CE (2014) The molecular balancing act of p16(INK4a) in cancer and aging. Mol Cancer Res 12(2):167–183. doi:10.1158/1541-7786.MCR-13-0350
24. Lukas J, Parry D, Aagaard L, Mann DJ, Bartkova J, Strauss M, Peters G, Bartek J (1995) Retinoblastoma-protein-dependent cell-cycle inhibition by the tumour suppressor p16. Nature 375(6531):503–506. doi:10.1038/375503a0
25. van den Heuvel S, Harlow E (1993) Distinct roles for cyclin-dependent kinases in cell cycle control. Science 262(5142):2050–2054
26. Polyak K, Lee MH, Erdjument-Bromage H, Koff A, Roberts JM, Tempst P, Massague J (1994) Cloning of p27Kip1, a cyclin-dependent kinase inhibitor and a potential mediator of extracellular antimitogenic signals. Cell 78(1):59–66
27. Coqueret O (2003) New roles for p21 and p27 cell-cycle inhibitors: a function for each cell compartment? Trends Cell Biol 13(2):65–70
28. LaBaer J, Garrett MD, Stevenson LF, Slingerland JM, Sandhu C, Chou HS, Fattaey A, Harlow E (1997) New functional activities for the p21 family of CDK inhibitors. Genes Dev 11(7):847–862
29. Cheng M, Olivier P, Diehl JA, Fero M, Roussel MF, Roberts JM, Sherr CJ (1999) The p21(Cip1) and p27(Kip1) CDK 'inhibitors' are essential activators of cyclin D-dependent kinases in murine fibroblasts. EMBO J 18(6):1571–1583. doi:10.1093/emboj/18.6.1571
30. Frank CJ, Hyde M, Greider CW (2006) Regulation of telomere elongation by the cyclin-dependent kinase CDK1. Mol Cell 24(3):423–432. doi:10.1016/j.molcel.2006.10.020
31. Pagano M (2004) Control of DNA synthesis and mitosis by the Skp2-p27-Cdk1/2 axis. Mol Cell 14(4):414–416
32. Nigg EA (2001) Mitotic kinases as regulators of cell division and its checkpoints. Nat Rev Mol Cell Biol 2(1):21–32. doi:10.1038/35048096
33. Harper JW, Burton JL, Solomon MJ (2002) The anaphase-promoting complex: it's not just for mitosis any more. Genes Dev 16(17):2179–2206. doi:10.1101/gad.1013102
34. Ubersax JA, Woodbury EL, Quang PN, Paraz M, Blethrow JD, Shah K, Shokat KM, Morgan DO (2003) Targets of the cyclin-dependent kinase Cdk1. Nature 425(6960):859–864. doi:10.1038/nature02062
35. Maestre C, Delgado-Esteban M, Gomez-Sanchez JC, Bolanos JP, Almeida A (2008) Cdk5 phosphorylates Cdh1 and modulates cyclin B1 stability in excitotoxicity. EMBO J 27(20):2736–2745. doi:10.1038/emboj.2008.195
36. Zhang J, Cicero SA, Wang L, Romito-Digiacomo RR, Yang Y, Herrup K (2008) Nuclear localization of Cdk5 is a key determinant in the postmitotic state of neurons. Proc Natl Acad Sci U S A 105(25):8772–8777. doi:10.1073/pnas.0711355105
37. Cruz JC, Tsai LH (2004) A Jekyll and Hyde kinase: roles for Cdk5 in brain development and disease. Curr Opin Neurobiol 14(3):390–394. doi:10.1016/j.conb.2004.05.002
38. Kesavapany S, Li BS, Amin N, Zheng YL, Grant P, Pant HC (2004) Neuronal cyclin-dependent kinase 5: role in nervous system function and its specific inhibition by the Cdk5 inhibitory peptide. Biochim Biophys Acta 1697(1–2):143–153. doi:10.1016/j.bbapap.2003.11.020
39. Fisher RP (2005) Secrets of a double agent: CDK7 in cell-cycle control and transcription. J Cell Sci 118(Pt 22):5171–5180. doi:10.1242/jcs.02718
40. Lolli G, Johnson LN (2005) CAK-Cyclin-dependent activating kinase: a key kinase in cell cycle control and a target for drugs? Cell Cycle 4(4):572–577
41. Morris EJ, Ji JY, Yang F, Di Stefano L, Herr A, Moon NS, Kwon EJ, Haigis KM, Naar AM, Dyson NJ (2008) E2F1 represses beta-catenin transcription and is antagonized by both pRB and CDK8. Nature 455(7212):552–556. doi:10.1038/nature07310
42. Garriga J, Grana X (2004) Cellular control of gene expression by T-type cyclin/CDK9 complexes. Gene 337:15–23. doi:10.1016/j.gene.2004.05.007
43. Kasten M, Giordano A (2001) Cdk10, a Cdc2-related kinase, associates with the Ets2 transcription factor and modulates its transactivation activity. Oncogene 20(15):1832–1838. doi:10.1038/sj.onc.1204295
44. Loyer P, Trembley JH, Katona R, Kidd VJ, Lahti JM (2005) Role of CDK/cyclin complexes in transcription and RNA splicing. Cell Signal 17(9):1033–1051. doi:10.1016/j.cellsig.2005.02.005
45. Yokoyama H, Gruss OJ, Rybina S, Caudron M, Schelder M, Wilm M, Mattaj IW, Karsenti E (2008)

Cdk11 is a RanGTP-dependent microtubule stabilization factor that regulates spindle assembly rate. J Cell Biol 180(5):867–875. doi:10.1083/jcb.200706189
46. Hu D, Valentine M, Kidd VJ, Lahti JM (2007) CDK11(p58) is required for the maintenance of sister chromatid cohesion. J Cell Sci 120(Pt 14):2424–2434. doi:10.1242/jcs.007963
47. Wilker EW, van Vugt MA, Artim SA, Huang PH, Petersen CP, Reinhardt HC, Feng Y, Sharp PA, Sonenberg N, White FM, Yaffe MB (2007) 14-3-3sigma controls mitotic translation to facilitate cytokinesis. Nature 446(7133):329–332. doi:10.1038/nature05584
48. Petretti C, Savoian M, Montembault E, Glover DM, Prigent C, Giet R (2006) The PITSLRE/CDK11p58 protein kinase promotes centrosome maturation and bipolar spindle formation. EMBO Rep 7(4):418–424. doi:10.1038/sj.embor.7400639
49. Asghar U, Witkiewicz AK, Turner NC, Knudsen ES (2015) The history and future of targeting cyclin-dependent kinases in cancer therapy. Nat Rev Drug Discov 14(2):130–146. doi:10.1038/nrd4504
50. Motokura T, Bloom T, Kim HG, Juppner H, Ruderman JV, Kronenberg HM, Arnold A (1991) A novel cyclin encoded by a bcl1-linked candidate oncogene. Nature 350(6318):512–515. doi:10.1038/350512a0
51. Sutherland RL, Musgrove EA (2002) Cyclin D1 and mammary carcinoma: new insights from transgenic mouse models. Breast Cancer Res 4(1):14–17
52. Gillett C, Fantl V, Smith R, Fisher C, Bartek J, Dickson C, Barnes D, Peters G (1994) Amplification and overexpression of cyclin D1 in breast cancer detected by immunohistochemical staining. Cancer Res 54(7):1812–1817
53. Foster JS, Henley DC, Ahamed S, Wimalasena J (2001) Estrogens and cell-cycle regulation in breast cancer. Trends Endocrinol Metab 12(7):320–327
54. Musgrove EA, Caldon CE, Barraclough J, Stone A, Sutherland RL (2011) Cyclin D as a therapeutic target in cancer. Nat Rev Cancer 11(8):558–572. doi:10.1038/nrc3090
55. Choi YJ, Anders L (2014) Signaling through cyclin D-dependent kinases. Oncogene 33(15):1890–1903. doi:10.1038/onc.2013.137
56. Herschkowitz JI, He X, Fan C, Perou CM (2008) The functional loss of the retinoblastoma tumour suppressor is a common event in basal-like and luminal B breast carcinomas. Breast Cancer Res 10(5):R75. doi:10.1186/bcr2142
57. Buckley MF, Sweeney KJ, Hamilton JA, Sini RL, Manning DL, Nicholson RI, deFazio A, Watts CK, Musgrove EA, Sutherland RL (1993) Expression and amplification of cyclin genes in human breast cancer. Oncogene 8(8):2127–2133
58. Wang TC, Cardiff RD, Zukerberg L, Lees E, Arnold A, Schmidt EV (1994) Mammary hyperplasia and carcinoma in MMTV-cyclin D1 transgenic mice. Nature 369(6482):669–671. doi:10.1038/369669a0
59. Weinstat-Saslow D, Merino MJ, Manrow RE, Lawrence JA, Bluth RF, Wittenbel KD, Simpson JF, Page DL, Steeg PS (1995) Overexpression of cyclin D mRNA distinguishes invasive and in situ breast carcinomas from non-malignant lesions. Nat Med 1(12):1257–1260
60. Alle KM, Henshall SM, Field AS, Sutherland RL (1998) Cyclin D1 protein is overexpressed in hyperplasia and intraductal carcinoma of the breast. Clin Cancer Res 4(4):847–854
61. Musgrove EA, Lee CS, Buckley MF, Sutherland RL (1994) Cyclin D1 induction in breast cancer cells shortens G1 and is sufficient for cells arrested in G1 to complete the cell cycle. Proc Natl Acad Sci U S A 91(17):8022–8026
62. Yu Q, Geng Y, Sicinski P (2001) Specific protection against breast cancers by cyclin D1 ablation. Nature 411(6841):1017–1021. doi:10.1038/35082500
63. Yang C, Ionescu-Tiba V, Burns K, Gadd M, Zukerberg L, Louis DN, Sgroi D, Schmidt EV (2004) The role of the cyclin D1-dependent kinases in ErbB2-mediated breast cancer. Am J Pathol 164(3):1031–1038. doi:10.1016/S0002-9440(10)63190-2
64. Roy PG, Thompson AM (2006) Cyclin D1 and breast cancer. Breast 15(6):718–727. doi:10.1016/j.breast.2006.02.005
65. Neuman E, Ladha MH, Lin N, Upton TM, Miller SJ, DiRenzo J, Pestell RG, Hinds PW, Dowdy SF, Brown M, Ewen ME (1997) Cyclin D1 stimulation of estrogen receptor transcriptional activity independent of cdk4. Mol Cell Biol 17(9):5338–5347
66. Zwijsen RM, Wientjens E, Klompmaker R, van der Sman J, Bernards R, Michalides RJ (1997) CDK-independent activation of estrogen receptor by cyclin D1. Cell 88(3):405–415
67. Sorlie T, Tibshirani R, Parker J, Hastie T, Marron JS, Nobel A, Deng S, Johnsen H, Pesich R, Geisler S, Demeter J, Perou CM, Lonning PE, Brown PO, Borresen-Dale AL, Botstein D (2003) Repeated observation of breast tumor subtypes in independent gene expression data sets. Proc Natl Acad Sci U S A 100(14):8418–8423. doi:10.1073/pnas.0932692100
68. Nielsen NH, Arnerlov C, Emdin SO, Landberg G (1996) Cyclin E overexpression, a negative prognostic factor in breast cancer with strong correlation to oestrogen receptor status. Br J Cancer 74(6):874–880
69. Donnellan R, Kleinschmidt I, Chetty R (2001) Cyclin E immunoexpression in breast ductal carcinoma: pathologic correlations and prognostic implications. Hum Pathol 32(1):89–94. doi:10.1053/hupa.2001.21141
70. Span PN, Tjan-Heijnen VC, Manders P, Beex LV, Sweep CG (2003) Cyclin-E is a strong predictor of endocrine therapy failure in human breast cancer. Oncogene 22(31):4898–4904. doi:10.1038/sj.onc.1206818

71. Keyomarsi K, Tucker SL, Buchholz TA, Callister M, Ding Y, Hortobagyi GN, Bedrosian I, Knickerbocker C, Toyofuku W, Lowe M, Herliczek TW, Bacus SS (2002) Cyclin E and survival in patients with breast cancer. N Engl J Med 347(20):1566–1575. doi:10.1056/NEJMoa021153
72. Scaltriti M, Eichhorn PJ, Cortes J, Prudkin L, Aura C, Jimenez J, Chandarlapaty S, Serra V, Prat A, Ibrahim YH, Guzman M, Gili M, Rodriguez O, Rodriguez S, Perez J, Green SR, Mai S, Rosen N, Hudis C, Baselga J (2011) Cyclin E amplification/overexpression is a mechanism of trastuzumab resistance in HER2+ breast cancer patients. Proc Natl Acad Sci U S A 108(9):3761–3766. doi:10.1073/pnas.1014835108
73. Smith ML, Seo YR (2000) Sensitivity of cyclin E-overexpressing cells to cisplatin/taxol combinations. Anticancer Res 20(4):2537–2539
74. Sutherland RL, Musgrove EA (2004) Cyclins and breast cancer. J Mammary Gland Biol Neoplasia 9(1):95–104. doi:10.1023/B:JOMG.0000023591.45568.77
75. Barton MC, Akli S, Keyomarsi K (2006) Deregulation of cyclin E meets dysfunction in p53: closing the escape hatch on breast cancer. J Cell Physiol 209(3):686–694. doi:10.1002/jcp.20818
76. Luhtala S, Staff S, Tanner M, Isola J (2016) Cyclin E amplification, over-expression, and relapse-free survival in HER-2-positive primary breast cancer. Tumour Biol 37(7):9813–9823. doi:10.1007/s13277-016-4870-z
77. Gao S, Ma JJ, Lu C (2013) Prognostic value of cyclin E expression in breast cancer: a meta-analysis. Tumour Biol 34(6):3423–3430. doi:10.1007/s13277-013-0915-8
78. Hunt KK, Karakas C, Ha MJ, Biernacka A, Yi M, Sahin A, Adjapong O, Hortobogyi GN, Bondy ML, Thompson PA, Cheung KL, Ellis IO, Bacus S, Symmans WF, Do KA, Keyomarsi K (2016) Cytoplasmic Cyclin E predicts recurrence in patients with breast cancer. Clin Cancer Res. doi:10.1158/1078-0432.CCR-16-2217
79. Timms JF, White SL, O'Hare MJ, Waterfield MD (2002) Effects of ErbB-2 overexpression on mitogenic signalling and cell cycle progression in human breast luminal epithelial cells. Oncogene 21(43):6573–6586. doi:10.1038/sj.onc.1205847
80. Mittendorf EA, Liu Y, Tucker SL, McKenzie T, Qiao N, Akli S, Biernacka A, Meijer L, Keyomarsi K, Hunt KK (2010) A novel interaction between HER2/neu and cyclin E in breast cancer. Oncogene 29(27):3896–3907. doi:10.1038/onc.2010.151
81. Akli S, Zheng PJ, Multani AS, Wingate HF, Pathak S, Zhang N, Tucker SL, Chang S, Keyomarsi K (2004) Tumor-specific low molecular weight forms of cyclin E induce genomic instability and resistance to p21, p27, and antiestrogens in breast cancer. Cancer Res 64(9):3198–3208
82. Akli S, Van Pelt CS, Bui T, Meijer L, Keyomarsi K (2011) Cdk2 is required for breast cancer mediated by the low-molecular-weight isoform of cyclin E. Cancer Res 71(9):3377–3386. doi:10.1158/0008-5472.CAN-10-4086
83. Robinson WA, Elefanty AG, Hersey P (1996) Expression of the tumour suppressor genes p15 and p16 in malignant melanoma. Melanoma Res 6(4):285–289
84. Van Zee KJ, Calvano JE, Bisogna M (1998) Hypomethylation and increased gene expression of p16INK4a in primary and metastatic breast carcinoma as compared to normal breast tissue. Oncogene 16(21):2723–2727. doi:10.1038/sj.onc.1201794
85. Wong SC, Chan JK, Lee KC, Hsiao WL (2001) Differential expression of p16/p21/p27 and cyclin D1/D3, and their relationships to cell proliferation, apoptosis, and tumour progression in invasive ductal carcinoma of the breast. J Pathol 194(1):35–42. doi:10.1002/path.838
86. Bisogna M, Calvano JE, Ho GH, Orlow I, Cordon-Cardo C, Borgen PI, Van Zee KJ (2001) Molecular analysis of the INK4A and INK4B gene loci in human breast cancer cell lines and primary carcinomas. Cancer Genet Cytogenet 125(2):131–138
87. Zariwala M, Liu E, Xiong Y (1996) Mutational analysis of the p16 family cyclin-dependent kinase inhibitors p15INK4b and p18INK4c in tumor-derived cell lines and primary tumors. Oncogene 12(2):451–455
88. Herman JG, Jen J, Merlo A, Baylin SB (1996) Hypermethylation-associated inactivation indicates a tumor suppressor role for p15INK4B. Cancer Res 56(4):722–727
89. Gartel AL, Radhakrishnan SK (2005) Lost in transcription: p21 repression, mechanisms, and consequences. Cancer Res 65(10):3980–3985. doi:10.1158/0008-5472.CAN-04-3995
90. Walsh A, Cook RS, Rexer B, Arteaga CL, Skala MC (2012) Optical imaging of metabolism in HER2 overexpressing breast cancer cells. Biomed Opt Express 3(1):75–85. doi:10.1364/BOE.3.000075
91. Musgrove EA, Davison EA, Ormandy CJ (2004) Role of the CDK inhibitor p27 (Kip1) in mammary development and carcinogenesis: insights from knockout mice. J Mammary Gland Biol Neoplasia 9(1):55–66. doi:10.1023/B:JOMG.0000023588.55733.84
92. Katayose Y, Kim M, Rakkar AN, Li Z, Cowan KH, Seth P (1997) Promoting apoptosis: a novel activity associated with the cyclin-dependent kinase inhibitor p27. Cancer Res 57(24):5441–5445
93. Lapenna S, Giordano A (2009) Cell cycle kinases as therapeutic targets for cancer. Nat Rev Drug Discov 8(7):547–566. doi:10.1038/nrd2907
94. Bose P, Simmons GL, Grant S (2013) Cyclin-dependent kinase inhibitor therapy for hematologic malignancies. Expert Opin Investig Drugs 22(6):723–738. doi:10.1517/13543784.2013.789859

95. Lin TS, Blum KA, Fischer DB, Mitchell SM, Ruppert AS, Porcu P, Kraut EH, Baiocchi RA, Moran ME, Johnson AJ, Schaaf LJ, Grever MR, Byrd JC (2010) Flavopiridol, fludarabine, and rituximab in mantle cell lymphoma and indolent B-cell lymphoproliferative disorders. J Clin Oncol 28(3):418–423. doi:10.1200/JCO.2009.24.1570
96. Ramaswamy B, Phelps MA, Baiocchi R, Bekaii-Saab T, Ni W, Lai JP, Wolfson A, Lustberg ME, Wei L, Wilkins D, Campbell A, Arbogast D, Doyle A, Byrd JC, Grever MR, Shah MH (2012) A dose-finding, pharmacokinetic and pharmacodynamic study of a novel schedule of flavopiridol in patients with advanced solid tumors. Investig New Drugs 30(2):629–638. doi:10.1007/s10637-010-9563-7
97. Hegeman RB, Mulkerin D, Thomas J, Alberti D, Binger K, Marnocha R, Kolesar J, Wilding G (2005) Phase I study of oxaliplatin in combination with 5-fluorouracil (5-FU), leucovorin (LV) and capecitabine (ORAL FOLFOX-6) in patients with advanced or metastatic solid tumors. J Clin Oncol 23(16):149s–149s
98. Le Tourneau C, Faivre S, Laurence V, Delbaldo C, Vera K, Girre V, Chiao J, Armour S, Frame S, Green SR, Gianella-Borradori A, Dieras V, Raymond E (2010) Phase I evaluation of seliciclib (R-roscovitine), a novel oral cyclin-dependent kinase inhibitor, in patients with advanced malignancies. Eur J Cancer 46(18):3243–3250. doi:10.1016/j.ejca.2010.08.001
99. Nair BC, Vallabhaneni S, Tekmal RR, Vadlamudi RK (2011) Roscovitine confers tumor suppressive effect on therapy-resistant breast tumor cells. Breast Cancer Res 13(3):R80. doi:10.1186/bcr2929
100. Appleyard MV, O'Neill MA, Murray KE, Paulin FE, Bray SE, Kernohan NM, Levison DA, Lane DP, Thompson AM (2009) Seliciclib (CYC202, R-roscovitine) enhances the antitumor effect of doxorubicin in vivo in a breast cancer xenograft model. Int J Cancer 124(2):465–472. doi:10.1002/ijc.23938
101. Kodym E, Kodym R, Reis AE, Habib AA, Story MD, Saha D (2009) The small-molecule CDK inhibitor, SNS-032, enhances cellular radiosensitivity in quiescent and hypoxic non-small cell lung cancer cells. Lung Cancer 66(1):37–47. doi:10.1016/j.lungcan.2008.12.026
102. Walsby E, Lazenby M, Pepper C, Burnett AK (2011) The cyclin-dependent kinase inhibitor SNS-032 has single agent activity in AML cells and is highly synergistic with cytarabine. Leukemia 25(3):411–419. doi:10.1038/leu.2010.290
103. Tong WG, Chen R, Plunkett W, Siegel D, Sinha R, Harvey RD, Badros AZ, Popplewell L, Coutre S, Fox JA, Mahadocon K, Chen T, Kegley P, Hoch U, Wierda WG (2010) Phase I and pharmacologic study of SNS-032, a potent and selective Cdk2, 7, and 9 inhibitor, in patients with advanced chronic lymphocytic leukemia and multiple myeloma. J Clin Oncol 28(18):3015–3022. doi:10.1200/JCO.2009.26.1347
104. Heath EI, Bible K, Martell RE, Adelman DC, Lorusso PM (2008) A phase 1 study of SNS-032 (formerly BMS-387032), a potent inhibitor of cyclin-dependent kinases 2, 7 and 9 administered as a single oral dose and weekly infusion in patients with metastatic refractory solid tumors. Investig New Drugs 26(1):59–65. doi:10.1007/s10637-007-9090-3
105. Kumar SK, LaPlant B, Chng WJ, Zonder J, Callander N, Fonseca R, Fruth B, Roy V, Erlichman C, Stewart AK (2015) Dinaciclib, a novel CDK inhibitor, demonstrates encouraging single-agent activity in patients with relapsed multiple myeloma. Blood 125(3):443–448. doi:10.1182/blood-2014-05-573741
106. Flynn J, Jones J, Johnson AJ, Andritsos L, Maddocks K, Jaglowski S, Hessler J, Grever MR, Im E, Zhou H, Zhu Y, Zhang D, Small K, Bannerji R, Byrd JC (2015) Dinaciclib is a novel cyclin-dependent kinase inhibitor with significant clinical activity in relapsed and refractory chronic lymphocytic leukemia. Leukemia 29(7):1524–1529. doi:10.1038/leu.2015.31
107. Nemunaitis JJ, Small KA, Kirschmeier P, Zhang D, Zhu Y, Jou YM, Statkevich P, Yao SL, Bannerji R (2013) A first-in-human, phase 1, dose-escalation study of dinaciclib, a novel cyclin-dependent kinase inhibitor, administered weekly in subjects with advanced malignancies. J Transl Med 11:259. doi:10.1186/1479-5876-11-259
108. Mitri Z, Karakas C, Wei C, Briones B, Simmons H, Ibrahim N, Alvarez R, Murray JL, Keyomarsi K, Moulder S (2015) A phase 1 study with dose expansion of the CDK inhibitor dinaciclib (SCH 727965) in combination with epirubicin in patients with metastatic triple negative breast cancer. Investig New Drugs 33(4):890–894. doi:10.1007/s10637-015-0244-4
109. Mita MM, Joy AA, Mita A, Sankhala K, Jou YM, Zhang D, Statkevich P, Zhu Y, Yao SL, Small K, Bannerji R, Shapiro CL (2014) Randomized phase II trial of the cyclin-dependent kinase inhibitor dinaciclib (MK-7965) versus capecitabine in patients with advanced breast cancer. Clin Breast Cancer 14(3):169–176. doi:10.1016/j.clbc.2013.10.016
110. O'Leary B, Finn RS, Turner NC (2016) Treating cancer with selective CDK4/6 inhibitors. Nat Rev Clin Oncol 13(7):417–430. doi:10.1038/nrclinonc.2016.26
111. Fry DW, Harvey PJ, Keller PR, Elliott WL, Meade M, Trachet E, Albassam M, Zheng X, Leopold WR, Pryer NK, Toogood PL (2004) Specific inhibition of cyclin-dependent kinase 4/6 by PD 0332991 and associated antitumor activity in human tumor xenografts. Mol Cancer Ther 3(11):1427–1438
112. Rivadeneira DB, Mayhew CN, Thangavel C, Sotillo E, Reed CA, Grana X, Knudsen ES (2010) Proliferative suppression by CDK4/6 inhibition: complex function of the retinoblastoma pathway in liver tissue and hepatoma cells. Gastroenterology 138(5):1920–1930. doi:10.1053/j.gastro.2010.01.007

113. Flaherty KT, Lorusso PM, Demichele A, Abramson VG, Courtney R, Randolph SS, Shaik MN, Wilner KD, O'Dwyer PJ, Schwartz GK (2012) Phase I, dose-escalation trial of the oral cyclin-dependent kinase 4/6 inhibitor PD 0332991, administered using a 21-day schedule in patients with advanced cancer. Clin Cancer Res 18(2):568–576. doi:10.1158/1078-0432.CCR-11-0509
114. Schwartz GK, LoRusso PM, Dickson MA, Randolph SS, Shaik MN, Wilner KD, Courtney R, O'Dwyer PJ (2011) Phase I study of PD 0332991, a cyclin-dependent kinase inhibitor, administered in 3-week cycles (schedule 2/1). Br J Cancer 104(12):1862–1868. doi:10.1038/bjc.2011.177
115. DeMichele A, Clark AS, Tan KS, Heitjan DF, Gramlich K, Gallagher M, Lal P, Feldman M, Zhang P, Colameco C, Lewis D, Langer M, Goodman N, Domchek S, Gogineni K, Rosen M, Fox K, O'Dwyer P (2015) CDK 4/6 inhibitor palbociclib (PD0332991) in Rb+ advanced breast cancer: phase II activity, safety, and predictive biomarker assessment. Clin Cancer Res 21(5):995–1001. doi:10.1158/1078-0432.CCR-14-2258
116. Finn RS, Crown JP, Lang I, Boer K, Bondarenko IM, Kulyk SO, Ettl J, Patel R, Pinter T, Schmidt M, Shparyk Y, Thummala AR, Voytko NL, Fowst C, Huang X, Kim ST, Randolph S, Slamon DJ (2015) The cyclin-dependent kinase 4/6 inhibitor palbociclib in combination with letrozole versus letrozole alone as first-line treatment of oestrogen receptor-positive, HER2-negative, advanced breast cancer (PALOMA-1/TRIO-18): a randomised phase 2 study. Lancet Oncol 16(1):25–35. doi:10.1016/S1470-2045(14)71159-3
117. Finn RS, Hurvitz SA, Allison MA, Applebaum S, Glaspy J, DiCarlo B, Courtney R, Shaik N, Kim ST, Fowst C, Slamon DJ (2009) Phase I study of PD 0332991, a novel, oral, Cyclin-D kinase (CDK) 4/6 inhibitor in combination with Letrozole, for first-line treatment of metastatic post-menopausal, estrogen receptor-positive (ER plus), human epidermal growth factor receptor 2 (HER2)-negative breast cancer. Cancer Res 69(24):788s–788s
118. Finn RS, Martin M, Rugo HS, Jones S, Im SA, Gelmon K, Harbeck N, Lipatov ON, Walshe JM, Moulder S, Gauthier E, Lu DR, Randolph S, Dieras V, Slamon DJ (2016) Palbociclib and Letrozole in advanced breast cancer. N Engl J Med 375(20):1925–1936. doi:10.1056/NEJMoa1607303
119. Cristofanilli M, Turner NC, Bondarenko I (2016) Fulvestrant plus palbociclib versus fulvestrant plus placebo for treatment of hormone-receptor-positive, HER2-negative metastatic breast cancer that progressed on previous endocrine therapy (PALOMA-3): final analysis of the multicentre, double-blind, phase 3 randomised controlled trial (vol 17, pg 431, 2016). Lancet Oncol 17(7):E270–E270
120. Turner NC, Huang Bartlett C, Cristofanilli M (2015) Palbociclib in hormone-receptor-positive advanced breast cancer. N Engl J Med 373(17):1672–1673. doi:10.1056/NEJMc1510345
121. Murphy CG, Dickler MN (2015) The role of CDK4/6 inhibition in breast cancer. Oncologist 20(5):483–490. doi:10.1634/theoncologist.2014-0443
122. Finn RS, Dering J, Conklin D, Kalous O, Cohen DJ, Desai AJ, Ginther C, Atefi M, Chen I, Fowst C, Los G, Slamon DJ (2009) PD 0332991, a selective cyclin D kinase 4/6 inhibitor, preferentially inhibits proliferation of luminal estrogen receptor-positive human breast cancer cell lines in vitro. Breast Cancer Res 11(5):R77. doi:10.1186/bcr2419
123. Witkiewicz AK, Cox D, Knudsen ES (2014) CDK4/6 inhibition provides a potent adjunct to Her2-targeted therapies in preclinical breast cancer models. Genes Cancer 5(7-8):261–272. doi:10.18632/genesandcancer.24
124. Rader J, Russell MR, Hart LS, Nakazawa MS, Belcastro LT, Martinez D, Li Y, Carpenter EL, Attiyeh EF, Diskin SJ, Kim S, Parasuraman S, Caponigro G, Schnepp RW, Wood AC, Pawel B, Cole KA, Maris JM (2013) Dual CDK4/CDK6 inhibition induces cell-cycle arrest and senescence in neuroblastoma. Clin Cancer Res 19(22):6173–6182. doi:10.1158/1078-0432.CCR-13-1675
125. O'Brien NA, Tomaso ED, Ayala R, Tong L, Issakhanian S, Linnartz R, Finn RS, Hirawat S, Slamon DJ (2014) In vivo efficacy of combined targeting of CDK4/6, ER and PI3K signaling in ER plus breast cancer. Cancer Res 74(19). doi:10.1158/1538-7445.AM2014-4756
126. Infante JR, Cassier PA, Gerecitano JF, Witteveen PO, Chugh R, Ribrag V, Chakraborty A, Matano A, Dobson JR, Crystal AS, Parasuraman S, Shapiro GI (2016) A phase I study of the Cyclin-dependent kinase 4/6 inhibitor Ribociclib (LEE011) in patients with advanced solid tumors and lymphomas. Clin Cancer Res 22(23):5696–5705. doi:10.1158/1078-0432.CCR-16-1248
127. Juric DMP, Campone M et al (2016) Ribociclib (LEE011) and letrozole in estrogen receptor-positive (ER+), HER2-negative (HER2–) advanced breast cancer (aBC): phase Ib safety, preliminary efficacy and molecular analysis. Presented at the 2016 annual meeting of the American Society of Clinical Oncology, Chicago
128. Vora SR, Juric D, Kim N, Mino-Kenudson M, Huynh T, Costa C, Lockerman EL, Pollack SF, Liu M, Li X, Lehar J, Wiesmann M, Wartmann M, Chen Y, Cao ZA, Pinzon-Ortiz M, Kim S, Schlegel R, Huang A, Engelman JA (2014) CDK 4/6 inhibitors sensitize PIK3CA mutant breast cancer to PI3K inhibitors. Cancer Cell 26(1):136–149. doi:10.1016/j.ccr.2014.05.020
129. Juric D, Ismail-Khan R, Campone M, Garcia-Estevez L, Becerra C, De Boer R, Hamilton E, Mayer IA, Hui R, Lathrop KI, Pagani O, Asano S, Bhansali SG, Zhang V, Hewes B, Munster P (2016) Phase Ib/II study of ribociclib and alpelisib and letrozole

in ER+, HER2-breast cancer: safety, preliminary efficacy and molecular analysis. Cancer Res 76. doi:10.1158/1538-7445.SABCS15-P3-14-01

130. Curigliano G, Gomez Pardo P, Meric-Bernstam F, Conte P, Lolkema MP, Beck JT, Bardia A, Martinez Garcia M, Penault-Llorca F, Dhuria S, Tang Z, Solovieff N, Miller M, Di Tomaso E, Hurvitz SA (2016) Ribociclib plus letrozole in early breast cancer: a presurgical, window-of-opportunity study. Breast 28:191–198. doi:10.1016/j.breast.2016.06.008
131. Gelbert LM, Cai S, Lin X, Sanchez-Martinez C, Del Prado M, Lallena MJ, Torres R, Ajamie RT, Wishart GN, Flack RS, Neubauer BL, Young J, Chan EM, Iversen P, Cronier D, Kreklau E, de Dios A (2014) Preclinical characterization of the CDK4/6 inhibitor LY2835219: in-vivo cell cycle-dependent/independent anti-tumor activities alone/in combination with gemcitabine. Investig New Drugs 32(5):825–837. doi:10.1007/s10637-014-0120-7
132. Tate SC, Cai S, Ajamie RT, Burke T, Beckmann RP, Chan EM, De Dios A, Wishart GN, Gelbert LM, Cronier DM (2014) Semi-mechanistic pharmacokinetic/pharmacodynamic modeling of the antitumor activity of LY2835219, a new cyclin-dependent kinase 4/6 inhibitor, in mice bearing human tumor xenografts. Clin Cancer Res 20(14):3763–3774. doi:10.1158/1078-0432.CCR-13-2846
133. Patnaik A, Rosen LS, Tolaney SM, Tolcher AW, Goldman JW, Gandhi L, Papadopoulos KP, Beeram M, Rasco DW, Hilton JF, Nasir A, Beckmann RP, Schade AE, Fulford AD, Nguyen TS, Martinez R, Kulanthaivel P, Li LQ, Frenzel M, Cronier DM, Chan EM, Flaherty KT, Wen PY, Shapiro GI (2016) Efficacy and safety of Abemaciclib, an inhibitor of CDK4 and CDK6, for patients with breast cancer, non-small cell lung cancer, and other solid tumors. Cancer Discov 6(7):740–753. doi:10.1158/2159-8290.CD-16-0095
134. Goetz MP, Beeram M, Beck T, Conlin AK, Dees EC, Dickler MN, Helsten TL, Conkling PR, Edenfield WJ, Richards DA, Turner PK, Cai N, Chan EM, Pant S, Becerra CH, Kalinsky K, Puhalla SL, Rexer BN, Burris HA, Tolaney SM (2016) Abemaciclib, an inhibitor of CDK4 and CDK6, combined with endocrine and HER2-targeted therapies for women with metastatic breast cancer. Cancer Res 76. doi:10.1158/1538-7445.SABCS15-P4-13-25
135. Dickler MN TS, Rugo HS et al (2016) MONARCH1: results from a phase II study of abemaciclib, a CDK4 and CDK6 inhibitor, as monotherapy, in patients with HR+/HER2- breast cancer, after chemotherapy for advanced disease. J Clin Oncol 34(Suppl, abstr 510)
136. Raub TJ, Wishart GN, Kulanthaivel P, Staton BA, Ajamie RT, Sawada GA, Gelbert LM, Shannon HE, Sanchez-Martinez C, De Dios A (2015) Brain exposure of two selective dual CDK4 and CDK6 inhibitors and the antitumor activity of CDK4 and CDK6 inhibition in combination with Temozolomide in an intracranial glioblastoma Xenograft. Drug Metab Dispos 43(9):1360–1371. doi:10.1124/dmd.114.062745
137. Hurvitz S MM, Fernández Abad M, Chan D, Rostorfer R, Petru E, Barriga S, Costigan TM, Caldwell CW, Nguyen T, Press M, Slamon D (2016) Biological effects of abemaciclib in a phase 2 neoadjuvant study for postmenopausal patients with HR+, HER2- breast cancer. Presented at the 2016 San Antonio breast cancer symposium

BRCA Gene Mutations and Poly(ADP-Ribose) Polymerase Inhibitors in Triple-Negative Breast Cancer

13

Hitomi Sumiyoshi Okuma and Kan Yonemori

Abstract

Breast cancer is the most common cancer in women worldwide. Treatment is chosen according to its hormone receptor status and human epidermal growth factor receptor 2 (HER2) status. Among the four main clinically set subtypes, hormone receptor-negative/HER2-negative subtype, also called triple-negative subtype (TNBC), is the most aggressive type with limited choices of therapy. However, recent research has provided important new insights into effective treatments for this subtype. One molecular target that has gained attention is the BRCA gene. BRCA proteins are involved in the maintenance of genomic integrity, therefore playing an important role as a "caretaker" DNA repair protein. Approximately 5% of all breast cancer patients are BRCA mutation carriers, and among the patients with BRCA mutations, 57.1% have the clinical TNBC subtype, showing a high association between BRCA mutations and TNBCs. When cells lack either BRCA1 or BRCA2, all types of homology-directed repairs are compromised, and poly(ADP-ribose) (PAR) polymerase (PARP) acts as a backup system to maintain the genome, consequently making the cells highly sensitive to PARP1 inhibitors. PARP inhibitors have shown promising activity in preclinical and early clinical trials, and today, phase III trials are ongoing. In this chapter, we discuss the mechanism and the role of PARP inhibitors in BRCA-mutated breast cancers and further elaborate the clinical potential of PARP inhibitors as well as their barriers.

H.S. Okuma • K. Yonemori (✉)
Department of Breast and Medical Oncology,
National Cancer Center Hospital,
5-1-1 Tsukiji, Chuo-ku, Tokyo, Japan
e-mail: kyonemor@ncc.go.jp

E. Song, H. Hu (eds.), *Translational Research in Breast Cancer*, Advances in Experimental Medicine and Biology 1026, https://doi.org/10.1007/978-981-10-6020-5_13

Keywords
Triple-negative breast cancer • BRCA mutation • PARP inhibitor • Synthetic lethality

13.1 Introduction

Recent researches have provided new insights into the effective treatments for breast cancer, which is the most common cancer in women worldwide. Clinically treated according to its subtype, breast cancer has four subtypes identified as follows: (1) hormone receptor positive/human epidermal growth factor receptor 2 (HER2) negative, (2) hormone receptor positive/HER2 positive, (3) hormone receptor negative/HER2 negative, and (4) hormone receptor negative/HER2 negative. The last subtype is also called triple-negative breast cancer (TNBC), one of the most aggressive types of breast cancer. Unlike hormone receptor-positive (luminal-like) subtypes, there are no targeted therapies available for patients with TNBC, which shows aggressive behaviors. Therefore, many researchers are investigating the molecular background of TNBCs, with a particular focus on *BRCA1/BRCA2* mutations.

In this chapter, we will discuss TNBC and the effects of *BRCA* mutations in this type of cancer. The roles of poly(ADP-ribose) polymerase (PARP) inhibitors in breast cancer treatment will also be elucidated.

13.2 TNBC

TNBC is defined based on immunohistochemical staining criteria. In the clinical setting, TNBC is defined to be estrogen receptor (ER) negative, progesterone receptor (PgR) negative, and human epidermal growth factor receptor 2 (HER2) negative. However, TNBC remains a heterogeneous disease that includes several intrinsic subtypes. Moreover, TNBC is known for its highly aggressive behavior and poor prognosis compared with other breast cancer subtypes [1], such as ER-positive, PgR-positive, and/or HER2-positive diseases.

13.2.1 Molecular Biological Features of TNBC

TNBC accounts for approximately 15% of all breast cancers. Compared with other subtypes, TNBC tends to occur in younger patients and exhibit large tumor burden, high nuclear grade, low BCL-2 expression, and high p53 and/or Ki-67 expression.

In 2000, Perou et al. performed a complementary DNA microarray gene profiling analysis in breast cancer and identified different molecular patterns, called "molecular portraits," among breast cancers [2]. In this analysis, they classified breast cancers into five different intrinsic subtypes: luminal A, luminal B, HER2-enriched, basal-like, and normal. Seventy-five percent of clinically proven TNBC can be classified into the basal-like subtype. In a later publication, researchers confirmed that among TNBCs, 80% were the basal-like subtype, 3% were the luminal subtype, and 9% were the HER2-enriched subtype [3].

Among basal-like subtypes, molecules such as cytokeratin 5/6, vimentin, and laminin have been shown to be highly expressed, whereas Bcl-2 has been shown to exhibit low expression [4]. Moreover, loss of phosphatase and tensin homolog (PTEN) and the disappearance of phosphatidylinositol-4,5-bisphosphate 3-kinase catalytic subunit alpha (PIK3CA) expression, retinoblastoma (RB) 1 mutations, or KRAS mutations are commonly observed in basal-like TNBC [5, 6].

Mutations or deletions in the *BRCA* gene (*BRCA1* and *BRCA2*) are also found in TNBCs.

Among basal-like subtypes, 75% have been reported to be BRCA1-associated breast cancers [7], whereas 19.5% of all TNBCs show BRCA germ line mutations [8].

13.2.2 Optimal Strategies for Treatment with Currently Approved Agents

Despite the findings of molecular subtypes among TNBC, no predictive values of the molecular subtypes have been established. Treatment is therefore selected from currently recommended agents that are approved in general breast cancer population.

Anthracyclines and taxanes remain the primary therapeutic approaches for TNBC, although there is limited evidence of success in patients treated with anthracycline- and/or taxane-containing regimens in the perioperative setting [9]. Patients who show primary or acquired resistance to key drugs may be given further chemotherapeutic agents that are not crossresistant, such as capecitabine, eribulin, gemcitabine, or vinorelbine [10–12]. The use of multidrug regimens in patients with metastatic cancer is controversial, and guidelines, such as those issued by European Breast Cancer Conference [13], recommend sequential monotherapy for advanced breast cancer. In cases where the aggressive nature of the disease calls for the need to stabilize the symptoms and reduce the risk of inner organ dysfunction, which is often noted in patients with TNBC, a multidrug regimen may be recommended rather than a single-drug regimen. Other agents that are sometimes used in TNBC therapy include platinum-based regimens [14–16] and PARP inhibitors (which are being investigated). The use of these agents has been supported by the strong association of TNBC with germ line *BRCA1* mutations.

Nonetheless, TNBC shows an aggressive behavior and very poor prognosis with limited treatment options. A biomarker-based understanding of molecular targets is required to facilitate further improvements in treatment strategies for TNBC.

13.3 BRCA Mutations

13.3.1 Functions and Mechanisms of BRCA

BRCA was first discovered in the 1990s and has been one of the most notorious and well-known cancer-related genes identified to date. It was originally considered as a tumor-suppressor gene [17]. However, further evidence shows that BRCA proteins are involved in the maintenance of genomic integrity. Therefore, instead of functioning as "gatekeeper" proteins of tumor suppressor, the BRCA family of proteins acts as "caretaker" proteins of DNA repair. Moreover, BRCA proteins are known to function in concert with other proteins, such as RAD50/Mre11 and RAD51, which play important roles in repairing DNA breaks caused by ionizing radiation [18].

During DNA replication, DNA molecules are particularly vulnerable to breakage in the single-stranded molecule portions that have not yet undergone replication near the replication fork. When an accidental breakage of the still unreplicated single-stranded DNA occurs at the replication forks, the resulting breaks are functionally equivalent to double-stranded breaks occurring in an already formed double helix. These double-stranded breaks are usually fixed by homology-directed repair (HDR). At sites of stalled replication forks where double-stranded breaks are observed, BRCA1 is located along with proliferating cell nuclear antigen (PCNA) and other DNA repair proteins, including RAD50 and RAD51 [19]. BRCA2 protein is also found at the same location, providing evidence of its collaboration in the DNA repair process [20]. When cells lack either BRCA1 or BRCA2, all types of HDR are compromised.

In mice, genetic disruption of BRCA1 function causes death during early embryogenesis, whereas mutant germ line alleles of *BRCA2* cause only partial loss of function, which results in susceptibility to lymphoid malignancies and unusual chromosomal aberrations [18]. In humans, mutant germ line alleles of either *BRCA1* or *BRCA2* lead to a natural susceptibility to breast and ovarian carcinomas [21]. In ovarian cancer, an estimated 70–80% of cases are caused by *BRCA* mutations. Some somatic mutations in *BRCA2* are associated with prostate and colon carcinomas. Additionally, female cells lacking BRCA1 function cannot properly inactivate one of the two X chromosomes. The mechanism of X-inactivation is essential in cells of early female embryos and must persist in all linear descendants. How this loss of BRCA function intersects with its DNA repair functions and how *BRCA1* mutation inclines to generate cancer primarily in women remain unknown.

13.3.2 BRCA Mutations in TNBCs

According to an analysis published by the International Breast Cancer Linkage Consortium, 0.12% of the general population carries *BRCA1* germ line mutations [22]. In patients with breast cancer, approximately 5% of patients are *BRCA* mutation carriers. According to a retrospective study, among patients with *BRCA* mutations, 57.1% have the clinical TNBC subtype [23]. Additionally, 19.5% of TNBCs have been shown to have germ line *BRCA* mutations [8]. When the population is narrowed down to those who have familial breast cancers, defined as breast cancer with a family history of one or more first- or second-degree relatives with breast cancer that does not fit the hereditary breast cancer definition, almost half of cancers are associated with germ line transmission of *BRCA1* or *BRCA2* mutations. In addition to germ line mutations, methylation of *BRCA1* is also known to be frequently found in TNBCs [24]. In all, the findings have shown that *BRCA* mutations are highly associated with TNBCs.

13.4 Function of PARP1

Among the many backup mechanisms required for proper repair or maintenance of the genome, poly(ADP-ribose) (PAR) synthesis is one of the earliest responses to DNA strand breakage. PARP1 is an abundant and stable component of chromatin and facilitates DNA repair by binding to DNA breaks and attracting other repairing proteins [25–29]. It is comprised of three functional domains: the amino-terminal DNA-binding domain which is important for binding of PARP1 to DNA breaks, the central automodification domain which allows the enzyme to PARate itself, and the C-terminal catalytic domain which transfers ADP-ribose subunits from NAD^+ to protein acceptors (Fig. 13.1) [30]. Among the seven main pathways used for DNA repair, PARP plays an important role in base excision repair (BER). At sites of single-stranded DNA breaks in which PARP binds to the DNA, PARP is activated and converts nicotinamide adenine dinucleotide (NAD) into ADP-ribose polymers (PAR) by attracting XRCC1, a scaffold protein that interacts with and recruits, stabilizes, or stimulates multiple enzymatic components involved in single-stranded breakage. For short patch repair and long patch repair at lesions that are more difficult to repair, the breakage goes through a single-stranded break intermediate and then arrives at a ligation stage to yield repaired DNA. PARP1 and PARP3 are among the 17 PARP isoforms that are also involved in double-stranded break repair [31].

For cells that lack BRCA1 or BRCA2 function, PARP acts as a backup system to maintain the genome and plays a critical role following accidental breaks that occur at replication forks during the S phase. Consequently, the cells become highly sensitive to killing by pharmacologic inhibitors of PARP1 [32]. However, *Parp−/−* mice are viable and fertile, which explains the redundant DNA repair systems. Therefore, PARP inhibition has little if any effect on normal tissues.

13.5 PARP Inhibitors and Their Effects on Cancer

PARP inhibitors exhibit competitive inhibition with NAD by blocking the catalytic PARP domain. PARP inhibitors show single-stranded DNA breakage repair activity, inducing apoptosis through accumulation of damaged DNA in the cells. By inhibiting PARP1, the repair

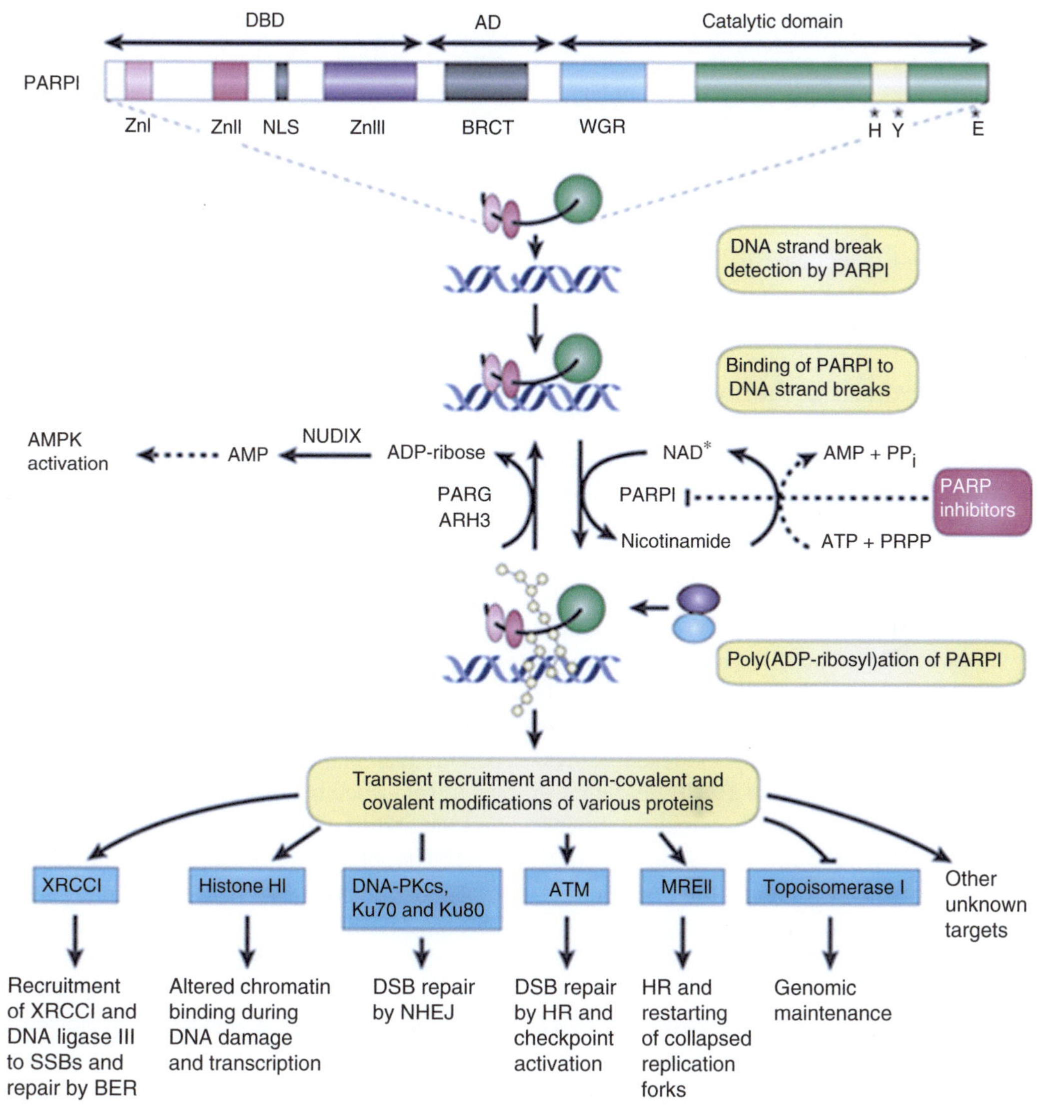

Abbreviations: *ATM*: ataxia telangiectasia-mutated, *BER*: base excision repair, *BRCT*: BRCA1 carboxy-terminal repeat motif, *DNA-PKcs*: DNA-protein kinase catalytic subunit, *DSB*: double-strand break, *HR*: homologous recombination, *NHEJ*: non-homologous end joining, *NLS*: nuclear localization signal, *PPi*: : inorganic pyrophosphate, *SSB*: single-strand break, *Zn*: zinc finger. [30]

Fig. 13.1 Function of PARP1 in DNA repair (Reprinted by permission from Macmillan Publishers Ltd: Nat Rev Cancer. (10(4): 293–301), copyright (2010))
Abbreviations: *ATM*, ataxia telangiectasia mutated; *BER*, base excision repair; *BRCT*, BRCA1 carboxy-terminal repeat motif; *DNA-PKcs*, DNA-protein kinase catalytic subunit; *DSB*, double-stranded break; *HR*, homologous recombination; *NHEJ*, nonhomologous end joining; *NLS*, nuclear localization signal; *PPi*, inorganic pyrophosphate; *SSB*, single-stranded break; *Zn*, zinc finger [30]

phenomenon can be trapped at the single-stranded intermediate state, thereby blocking ligation. PARP inhibitors bind to the catalytic site and prevent the release of PARP1 from DNA by "trapping" PARP1 at the site and removing PARP1 from the normal catalytic cycle [27, 28, 33]. When BER does not function properly, single-stranded breaks are left unrepaired, leading to the formation of double-stranded breaks due to stalling of the replication fork. Since double-stranded breaks are repaired by either nonhomologous end joining or homologous recombination, inhibition of PARP alone does not lead to efficient cell death. Therefore, for PARP inhibitors to exert beneficial effects on DNA repair, another repair pathway other than BER must be functionally damaged by PARP inhibition.

13.5.1 Synthetic Lethality

Synthetic lethality was introduced nearly a century ago by geneticists. It involves the combined knockout of two genes, which leads to a lethal form of genetic interactions that can selectively kill cancer cells while sparing normal cells [34]. The concept of synthetic lethality involving PARP and BRCA is related to the observation that both proteins are normally nonessential but critical for the survival of cancer cells. The most striking evidence of synthetic lethality is the use of PARP inhibitors in homologous recombination-defective tumors [32, 35]. As BRCA1 and BRCA2 are associated with homologous recombination, PARP inhibitors have been used for monotherapy in treating patients with *BRCA1*- or *BRCA2*-mutated cancers. Other genes associated with homologous recombination are *RAD51*, *RAD54*, *DSS1*, *PRA1*, *NBS*, *ATR*, *ATM*, *CHK1*, *CHK2*, *FANCD2*, *FANCA*, and *FANCC*. Cells with a deficiency in one of these genes show sensitivity to PARP inhibitors, confirming the concept of synthetic lethality [33].

13.6 Clinical Application of PARP Inhibitors in Cancer

In PARP1-knockout mice, deficiencies in PARP1 function result in impaired DNA repair, which consequently leads to a higher sensitivity to anticancer agents. It indicates that PARP1 inhibition may induce sensitivity to DNA damage by anticancer agents and therefore act as a radiosensitizer or chemosensitizer in the treatment of cancers. PARP1 is also known for its strong activation by radiotherapy or DNA methylating anticancer agents. Based on available evidence, along with the development of PARP inhibitors in patients with germ line BRCA mutations, new therapeutic approaches using PARP inhibitors combined with DNA-damaging anticancer agents have been evaluated. Approximately 30 years ago, small-molecule nicotinamide analogs were found to enhance the cytotoxicity of dimethyl sulfate, a DNA-damaging agent, by inhibiting PARylation [36–38]. Subsequently, clinical PARP inhibitors, including veliparib, rucaparib, olaparib, and niraparib, were developed. A more potent second-generation PARP inhibitor, talazoparib, has also been developed [39]. The difference among these agents is the ability to "trap" PARP1, an essential mechanism of PARP inhibitors. Talazoparib is approximately 100 times more potent than niraparib and is therefore more potent than olaparib and rucaparib [40]. The chemical structures of clinical PARP inhibitors and the ability of each PARP inhibitor to "trap" PARP1 is thought to broadly correlate with its cytotoxic potency [33]. Among currently available PARP inhibitors, olaparib (Lynparza) was the first to be approved by the US Food and Drug Administration (FDA) for treating patients with germ line *BRCA* mutations in advanced ovarian cancer in February 2014. The development of olaparib in breast cancer will be further discussed in this chapter.

13.7 PARP Inhibitors in the Field of Breast Cancer

During clinical development, PARP inhibitors have been investigated in combination with DNA-damaging anticancer agents or radiotherapy, or as monotherapy, in cancers that show decreased BRCA1 or BRCA2 functions, mainly TNBC. In the field of breast cancer, BSI-201 was the first PARP1 inhibitor to be reported [41]. In a phase I (and Ib) trial, this compound showed safety and effectiveness and was later tested in a randomized phase II trial, which compared combined treatment of gemcitabine plus carboplatin (GC) plus BSI-201 and GC alone in patients with metastatic TNBC with two or less prior regimens [42, 43]. The progression-free survival (PFS) was 6.9 months versus 3.3 months, and overall survival was 9.2 months versus 5.7 months, indicating a statistically longer survival for the GC plus BSI-201 arm. The overall response rate was also higher in the GC plus BSI-201 arm (48% versus 16%, $p = 0.002$). There were high expectations for the phase III trial, but the primary endpoint was not achieved.

Alternatively, olaparib has been developed as another promising PARP inhibitor for the treatment of breast cancer, which will be discussed below.

13.8 Development of Olaparib (Lynparza) in Breast Cancer

Olaparib is a PARP1 inhibitor first discovered during a screening test for agents that induce sensitivity of cells to cytotoxic agents, such as topoisomerase I inhibitors and alkylating agents. It has showed antitumor activity in cells with homologous recombination deficiency, which implies its role as a promising agent for the treatment of *BRCA*-mutated cancer. Moreover, olaparib was first approved by the FDA for treatment of *BRCA*-mutated ovarian cancers. In this section, we will discuss the development of olaparib studies in the field of breast cancer.

13.8.1 Preclinical Study

Through an in vitro study, olaparib monotherapy demonstrated strong antitumor activity in breast cancer cells with *BRCA1* mutations [44]. In an in vivo study of *BRCA1−/−* tumor-bearing mice, olaparib inhibited tumor growth without signs of toxicity, which significantly increased the survival rate. In a similar analysis with *BRCA2−/−* murine mammary epithelium, daily exposure to olaparib for 28 days caused significant regression or growth inhibition in 46 of 52 tumors [45]. The same analysis was conducted with olaparib in combination with carboplatin. Although no advantage over carboplatin monotherapy was observed, a significant increase in time to tumor relapse or death was observed if PARP inhibitors were continuously administered [46]. In combination therapy, temozolomide or dacarbazine plus olaparib was shown to have antitumor activity. Similarly, olaparib with topoisomerase I inhibitors or platinum agents also showed activity in vitro and in vivo.

13.8.2 Clinical Phase I Monotherapy Trials

Olaparib was first tested in early phase clinical trials for advanced solid tumors with no further standard therapy [47]. However, as the activity of this agent against *BRCA*-mutated cancers became clearer, the protocol was amended to include patients with *BRCA* mutations. Later, during the expansion phase, patients with *BRCA* mutations were specifically enrolled, and a total of 60 patients were eventually included. The dose of oral administration started at 10 mg either once daily or twice daily for 14 consecutive days in a 21-day cycle. During the higher dose phase, the drug was taken twice daily for 28 consecutive days. Dose-limiting toxicity was confirmed at doses of 400 and 600 mg twice daily. In the cohort receiving 400 mg, one patient experienced grade 3 agitation and grade 3 fatigue, and in the cohort receiving 600 mg, one patient experienced grade 4 thrombocytopenia and another patient experienced grade 3 somnolence. Overall, 21 patients with *BRCA* mutations were enrolled, and among the 19 patients with breast, ovarian, or prostate cancers, nine patients (47%) achieved a partial response, and 12 patients (63%) achieved a clinical benefit (partial response or stable disease). This was a surprisingly high response rate for a cohort that included patients with relapsed breast, ovarian, or prostate cancer. Furthermore, patients with *BRCA* mutations did not show higher incidences of adverse events than patients having wild-type *BRCA*.

13.8.3 Clinical Phase II Monotherapy Trials

To date, three phase II trials of olaparib monotherapy have been published in the field of breast cancer. The first trial was an international collaborative trial undertaken in six countries. This trial included patients with advanced breast cancer with *BRCA1* or *BRCA2* mutations who had been given at least one prior chemotherapy regimen [48]. The study was comprised of two different dosage cohorts: 400 mg twice daily (phase I maximum tolerated dose) and 100 mg twice daily (a dose that showed activity in the phase I trial). Objective responses were observed in 11 of 27 patients (41%; 95% confidence interval [CI]: 25–59) in the first cohort and 6 of 27 patients (22%; 95% CI: 11–41) in the latter cohort. The toxicities were mainly at low grade. Therefore, these results provided positive evidence for the concept of PARP inhibition in *BRCA*-deficient breast cancers. The second trial was a multicenter trial conducted in Canada and included patients with recurrent high-grade serous or poorly differentiated ovarian carcinoma or TNBC, regardless of *BRCA1* or *BRCA2* mutation status [49]. Patients received olaparib 400 mg twice daily. Ninety-one patients were enrolled (65 with ovarian cancer and 26 with breast cancer), and among the 63 evaluable patients, objective responses were observed in 7 of 17 patients (41%; 95% CI: 22–64) with *BRCA1* or *BRCA2* mutations and 11 of 46 patients (24%; 95% CI: 14–38) without mutations. Although no objective responses were reported in patients with breast cancer, 30% of patients achieved stable disease for at least 8 weeks, with a median PFS of 54 days. The third phase II trial was an international collaborative trial that enrolled patients with germ line *BRCA1* or *BRCA2* mutations with recurrent breast, ovarian, pancreatic, or prostate cancer [50]. Patients with breast cancer had to have at least three prior-chemotherapy regimens for metastatic disease. Olaparib was administered at 400 mg twice daily. Among the 298 patients treated and evaluated, an objective response was achieved in 78 of 298 patients (26.2%; 95% CI: 21.3–31.6) and in eight of 62 patients (12.9%; 95% CI: 5.7–23.9) with breast cancer. Stable disease was observed in 47% (95% CI: 34.0–59.9) of patients with breast cancer. Table 13.1 summarizes the phase II trials that included patients with breast cancer.

Table 13.1 Clinical phase II studies of olaparib monotherapy in breast cancer

Published year	Author	Eligibility	Olaparib dose (twice daily)	N	Response rate	PFS	Notes
2010	Tutt et al.	Advanced, BRCA mutation	400 mg	27	41%	5.7 months	
			100 mg	27	22%	3.8 months	
2011	Gelmon et al.	Advanced, BRCA mutation or TNBC	400 mg	23	0%	54 days	Stable disease of over 8 weeks: 30%
2015	Kaufman et al.	Advanced, BRCA mutation	400 mg	62	13%	3.7 months	Partial response + stable disease of over 8 weeks: 60%

PFS progression-free survival

13.8.4 Clinical Phase III Monotherapy Trials

Three phase III trials of olaparib monotherapy have been initiated in patients with germ line *BRCA* mutation-positive breast cancer. They are OlympiA (NCT02032823), Neo-Olympia (D081EC00005), and OlympiAD (NCT0000622). OlympiA is a randomized double-blind study which assesses the efficacy of olaparib at a dose of 300 mg twice daily. In this study, olaparib was administered with and without placebo as adjuvant treatment in patients with *BRCA1/BRCA2* mutations and high-risk HER2-negative breast cancer. The patients were divided into two groups, with one completing definitive local treatment and the other undergoing either neoadjuvant or adjuvant chemotherapy. Neo-Olympia is a randomized three-arm trial comparing olaparib monotherapy at a dose of 300 mg twice daily, placebo therapy plus weekly paclitaxel (80 mg/m2), and olaparib therapy at a dose of 100 mg twice daily plus weekly paclitaxel (80 mg/m2) in the neoadjuvant setting in patients with *BRCA1/BRCA2* mutations and operable, locally advanced, or inflammatory breast cancer. OlympiAD is a randomized open-label trial which assesses the efficacy of olaparib at a dose of 300 mg twice daily. It compares olaparib monotherapy with treatment of physician's choice (TPC) of capecitabine, vinorelbine, or eribulin in patients with *BRCA1/BRCA2* mutations and metastatic breast cancer. Two of the trials began enrolment in 2014, and findings from the OlympiAD trial were recently reported at the 2017 ASCO Annual Meeting [51]. At 77% data maturity, PFS was significantly longer in the olaparib arm [7.0 vs 4.2 months, hazard ratio (HR) 0.58; 95% CI: 0.43–0.80; $p = 0.0009$] with a higher objective response rate of 59.9% in the olaparib arm compared to 28.8% in the TPC arm (HR 0.57; 95CI: 0.40–0.83). The safety profile of olaparib was consistent with prior studies. These promising results were the first to demonstrate improved outcomes with a PARP inhibitor in breast cancer. Table 13.2 summarizes the phase III trials that included patients with breast cancer.

13.8.5 Combination Therapy

Olaparib has been tested with several other agents, such as paclitaxel, temozolomide, dacarbazine, topotecan, bevacizumab, paclitaxel plus carboplatin, and newer agents (e.g., phosphoinositol 3-kinase [PI3K] inhibitors).

13.8.5.1 Paclitaxel Plus Olaparib

In a phase I/II trial, patients with advanced TNBC were treated with olaparib at a dose of 200 mg twice daily in combination with paclitaxel (90 mg/m^2, days 1, 8, and 15) on a 28-day cycle [52]. Patients were treated with either first-line or second-line chemotherapy. The response rate was high, with seven (37%) out of 19 patients achieving an objective response. Although the toxicities were relatively well tolerated, severe neutropenia was observed at a greater frequency than expected. In the second cohort, the dose intensity of paclitaxel was not retained, even with the use of prophylactic granulocyte colony-stimulating factor.

13.8.5.2 Paclitaxel Plus Carboplatin Plus Olaparib

In a cohort of patients with advanced solid tumors including breast cancer, a phase I study was conducted to investigate the treatment of olaparib with either paclitaxel (80 mg/m^2, days 1, 8, and 15) or carboplatin (AUC 4–5, day 1) or both paclitaxel (90–175 mg/m^2, day 1) plus carboplatin (AUC 4–5, day 1; TC). Olaparib was given at a dose of 50–200 mg twice a day every day or 200–400 mg twice a day for 5 or 10 consecutive days [53]. The hematological toxicities were too strong to maintain the dose in the cohorts taking olaparib every day plus carboplatin or taking olaparib everyday plus TC. However, olaparib given at a dose of 100 mg twice a day every day in combination with PTX was well tolerated, as was olaparib given at 200 mg twice a day for 10 consecutive days plus TC. The overall objective response rate was 16.1% (14/87 patients), whereas the response rate in patients with *BRCA1/BRCA2* mutations was 50% (6/12 patients).

Table 13.2 Clinical phase III studies of olaparib monotherapy in breast cancer

Trial	Eligibility	Setting	Olaparib monotherapy arm	Comparator arm(s)	Primary endpoint
OlympiA	High-risk after definitive local treatment, BRCA mutation	Adjuvant	300 mg twice daily	Placebo	Invasive disease-free survival
Neo-Olympia	Operable, BRCA mutation	Neoadjuvant	300 mg twice daily (arm A)	Placebo + weekly PTX (arm B)	Pathological complete response
				Olaparib 100 mg twice daily + weekly PTX (arm C)	
OlympiAD	Advanced, BRCA mutation	Metastatic	300 mg twice daily	Capecitabine or vinorelbine or eribulin (physician's choice)	PFS

PTX paclitaxel, *PFS* progression-free survival

13.8.5.3 Eribulin Plus Olaparib

Eribulin mesylate is a nontaxane inhibitor of microtubule dynamics of the halichondrin class of antitumor agents. Eribulin is currently recognized as a global standard treatment for metastatic or recurrent breast cancer following the use of anthracyclines and taxanes. Pooled analyses of two phase III trials of eribulin monotherapy in patients with metastatic or recurrent breast cancer suggested favorable survival benefits, particularly in patients with TNBC [11, 12]. In a cohort of patients with TNBC, a phase I/II trial was conducted in Japan to investigate the safety profiles and efficacy of olaparib in combination with eribulin under the assumption that this combination may be a favorable regimen for patients with metastatic or recurrent TNBC [54]. Patients who had received both anthracycline- and taxane-containing regimens were enrolled to be treated with eribulin at a dose of 1.4 mg/m^2 (days 1 and 8) plus olaparib twice daily every day at a dose of 25–300 mg. The recommended phase II dose of olaparib was 300 mg twice daily. Pharmacokinetic (PK)/pharmacodynamic (PD) analysis also showed that the C_{max} and area under the curve (AUC) of olaparib were dose dependent and that both parameters of eribulin and olaparib were not influenced by each other. An objective response was observed in seven of the 18 evaluable patients, indicating a relatively high response rate of 38.9% (95% CI: 17.3–64.3). Six patients maintained their responses for over a year, and the median PFS was 4.22 months (95% CI: 2.99–7.36). The most frequent adverse events were the occurrences of neutropenia (grade 3 or more: 83.3%), but the drug was overall well tolerated.

13.9 Development of Other PARP Inhibitors: Talazoparib

Talazoparib has a much higher potency for "trapping" PARP inhibitors than olaparib. In a recent phase I study, talazoparib has shown some promise in treating 13 early-stage patients with germ line *BRCA1* or *BRCA2* mutations. The patients were treated for 2 months with talazoparib before neoadjuvant chemotherapy and surgery [55]. Decreased tumor volume was observed in all 13 patients following the 2-month treatment with talazoparib, and the average volume reduction was 78% (range: 30–98%). The toxicity of this drug also proved to be well tolerated, as no grade 4 toxicities were observed, and only one patient required dose reduction due to grade 3 neutropenia. The study is ongoing, and researchers will next

Table 13.3 Clinical studies of PARP inhibitors including breast cancer

Drug	Phase	Eligibility	Concomitant therapy	Notes
Olaparib	I	Breast cancer or ovarian cancer	Carboplatin	BRCA1 or BRCA2 mutation
	I	Breast cancer or women's cancer	Carboplatin	
	I	TNBC or ovarian cancer	BKM120	
	I	Solid tumors, including TNBC	Carboplatin and/or PTX	
	I/II	TNBC	PTX	
	I/II	TNBC or ovarian cancer	Cediranib	
Iniparib	II	TNBC with brain lesion	Irinotecan	
	II	TNBC	Gemcitabine and carboplatin	Iniparib twice weekly versus weekly
	II	TNBC	PTX	Neoadjuvant
Veliparib	I	Solid tumors	TMZ	BRCA1- or BRCA2-mutated breast cancer
	I	TNBC or gynecologic cancer	Pegylated liposomal doxorubicin	
	I	Breast cancer	Radiation therapy	Loco-regionally recurrent
	II	Breast cancer	TMZ	BRCA1- or BRCA2-mutated breast cancer
	II	TNBC or ovarian or non-Hodgkin's lymphoma	Cyclophosphamide	
Talazoparib	I	Solid tumors		
	III	Breast cancer		BRCA1 or BRCA2 mutation (versus physician's choice)
Rucaparib	II	TNBC	Cisplatin	BRCA1 or BRCA2 mutation (versus cisplatin)
E7449	I/II	Solid tumors, including TNBC	Alone or plus TMZ or plus carboplatin and PTX	

TNBC triple-negative breast cancer, *PTX* paclitaxel, *TMZ* temozolomide

investigate the pathological response to talazoparib alone for 4–6 months.

Although talazoparib can kill *BRCA*-mutated cells in vitro at a 200-fold lower dose than olaparib or rucaparib, the in vitro therapeutic ratio achieved in *BRCA1-/BRCA2*-defective cells is similar with that in wild-type cells for all three PARP inhibitors. Therefore, it is still too early to draw any conclusion regarding which PARP inhibitor is most effective. Table 13.3 shows the clinical trials conducted with PARP inhibitors in patients with breast cancer (excluding the clinical trial of olaparib monotherapy discussed above).

13.10 Acquired Resistance to PARP Inhibitors

Multiple potential mechanisms of resistance have been identified through in vitro experiments. Even though homologous recombination repair is defective, the restoration of homologous recombination repair in *BRCA1*-mutant tumor cells has been identified through loss of 53BP1 and REV7 proteins [56, 57]. Moreover, the loss of PARP1 [58] has been proposed to cause resistance, as with other proteins that are important for maintaining replication fork stability [59]. Secondary mutations in *BRCA1* or *BRCA2* can also occur, leading to restoration of sufficient homologous recombination repair function and resulting in PARP inhibitor resistance [60, 61]. Additionally, this secondary mutation is known to cause clinical resistance to platinum-based chemotherapy [62, 63].

13.10.1 Genetic Deficiencies Other Than BRCA1/BRCA2

Not long after the discovery that *BRCA1* and *BRCA2* mutant cells were highly susceptible to PARP inhibitors, deficiencies in a number of tumor-suppressor genes, such as *ATM*, *ATR*, *PALB2*, and *FANC*, which are all involved in homologous recombination repair, have been shown to confer sensitivity to PARP inhibitors [63, 64].

In an in vitro experiment, wild-type BRCA1/BRCA2 breast cancer cells (i.e., MCF-7 and ZR-75–1 cells) that were genetically manipulated to knockdown ATM expression were treated with olaparib [65]. ATM depletion sensitized both cell lines, as assessed by short- and long-term survival assays. These data indicated that ATM depletion could sensitize breast cancer cells to PARP inhibitors and that cancers, such as those arising in mutant ATM heterozygous carriers, may be potential targets for PARP inhibitors. A similar phenomenon has been discovered for other tumor cells, such as gastric cancer cell lines and colorectal cell lines, and studies have highlighted the clinical utility of ATM expression as a predictive marker for the sensitivity of gastric cancer cells to PARP inhibitors [66].

The Fanconi anemia (FA) repair pathway is also known to play a collaborative role with *BRCA* genes. Patients with FA have a high incidence of malignancies, and their cells show hypersensitivity to DNA cross-linking agents, such as mitomycin C (MMC) and cisplatin. Cancers with defective FA/BRCA pathways are likely to be more sensitive to these types of therapy or to treatments in which an additional repair mechanism is targeted, such as treatment with PARP inhibitors. In a recent study, researchers developed a new assay to identify patients with FA functional defects using FA triple-stain immunofluorescence (FATSI, FancD2/DAPI/Ki67) [67]. The study was also conducted to verify the safety and feasibility of veliparib as monotherapy and in combination with MMC. According to FATSI screening, 28.7% (185/643) of patients were FATSI-negative, demonstrating that a substantial number of tumors exhibited FA functional deficiency. Among the 61 FATSI-negative patients who received treatment, six antitumor responses were observed with five in the combination arm. However, some clinical benefits were observed, and a better understanding of this mechanism is needed.

13.11 Concluding Remarks and Future Perspectives

Many studies have investigated the use of PARP inhibitors in breast cancer, with a particular focus on TNBC with *BRCA* mutations. So far, one trial of olaparib monotherapy has shown promising results for breast cancer. However, given the relatively small size of the study, it is difficult to tell which subset of patients would benefit the most from olaparib. Determining the optimal use of PARP inhibitors within drug combinations has been challenging, and new biomarkers may be needed to identify appropriate populations who may benefit most from PARP inhibitors. In addition, resistance to PARP inhibitors can arise in advanced disease, and further studies are needed to elucidate the related mechanisms.

References

1. Ismail-Khan R, Bui MM (2010) A review of triple-negative breast cancer. Cancer Control 17(3):173–176
2. Perou CM, Sorlie T, Eisen MB, van de Rijn M, Jeffrey SS, Rees CA, Pollack JR, Ross DT, Johnsen H, Akslen LA, Fluge O, Pergamenschikov A, Williams C, Zhu SX, Lonning PE, Borresen-Dale AL, Brown PO, Botstein D (2000) Molecular portraits of human breast tumours. Nature 406(6797):747–752. https://doi.org/10.1038/35021093
3. Cancer Genome Atlas N (2012) Comprehensive molecular portraits of human breast tumours. Nature 490(7418):61–70. https://doi.org/10.1038/nature11412
4. Han W, Jung EM, Cho J, Lee JW, Hwang KT, Yang SJ, Kang JJ, Bae JY, Jeon YK, Park IA, Nicolau M, Jeffrey SS, Noh DY (2008) DNA copy number alterations and expression of relevant genes in triple-negative breast cancer. Genes Chromosomes Cancer 47(6):490–499. https://doi.org/10.1002/gcc.20550
5. Perren A, Weng LP, Boag AH, Ziebold U, Thakore K, Dahia PL, Komminoth P, Lees JA, Mulligan LM, Mutter GL, Eng C (1999) Immunohistochemical evidence of loss of PTEN expression in primary ductal adenocarcinomas of the breast. Am J Pathol 155(4):1253–1260. https://doi.org/10.1016/S0002-9440(10)65227-3
6. Kotsopoulos J, Lubinski J, Lynch HT, Neuhausen SL, Ghadirian P, Isaacs C, Weber B, Kim-Sing C, Foulkes WD, Gershoni-Baruch R, Ainsworth P, Friedman E, Daly M, Garber JE, Karlan B, Olopade OI, Tung N, Saal HM, Eisen A, Osborne M, Olsson H, Gilchrist D, Sun P, Narod SA (2005) Age at menarche and the risk of breast cancer in BRCA1 and BRCA2 mutation carriers. Cancer Causes Control 16(6):667–674. https://doi.org/10.1007/s10552-005-1724-1
7. Foulkes WD, Brunet JS, Stefansson IM, Straume O, Chappuis PO, Begin LR, Hamel N, Goffin JR, Wong N, Trudel M, Kapusta L, Porter P, Akslen LA (2004) The prognostic implication of the basal-like (cyclin E high/p27 low/p53+/glomeruloid-microvascular-proliferation+) phenotype of BRCA1-related breast cancer. Cancer Res 64(3):830–835
8. Gonzalez-Angulo AM, Timms KM, Liu S, Chen H, Litton JK, Potter J, Lanchbury JS, Stemke-Hale K, Hennessy BT, Arun BK, Hortobagyi GN, Do KA, Mills GB, Meric-Bernstam F (2011) Incidence and outcome of BRCA mutations in unselected patients with triple receptor-negative breast cancer. Clin Cancer Res 17(5):1082–1089. https://doi.org/10.1158/1078-0432.CCR-10-2560
9. Palmieri C, Krell J, James CR, Harper-Wynne C, Misra V, Cleator S, Miles D (2010) Rechallenging with anthracyclines and taxanes in metastatic breast cancer. Nat Rev Clin Oncol 7(10):561–574. https://doi.org/10.1038/nrclinonc.2010.122
10. Jassem J, Carroll C, Ward SE, Simpson E, Hind D (2009) The clinical efficacy of cytotoxic agents in locally advanced or metastatic breast cancer patients pretreated with an anthracycline and a taxane: a systematic review. Eur J Cancer 45(16):2749–2758. https://doi.org/10.1016/j.ejca.2009.05.035
11. Kaufman PA, Awada A, Twelves C, Yelle L, Perez EA, Velikova G, Olivo MS, He Y, Dutcus CE, Cortes J (2015) Phase III open-label randomized study of eribulin mesylate versus capecitabine in patients with locally advanced or metastatic breast cancer previously treated with an anthracycline and a taxane. J Clin Oncol 33(6):594–601. https://doi.org/10.1200/JCO.2013.52.4892
12. Cortes J, O'Shaughnessy J, Loesch D, Blum JL, Vahdat LT, Petrakova K, Chollet P, Manikas A, Dieras V, Delozier T, Vladimirov V, Cardoso F, Koh H, Bougnoux P, Dutcus CE, Seegobin S, Mir D, Meneses N, Wanders J, Twelves C, investigators E (2011) Eribulin monotherapy versus treatment of physician's choice in patients with metastatic breast cancer (EMBRACE): a phase 3 open-label randomised study. Lancet 377 (9769):914–923. doi:https://doi.org/10.1016/S0140-6736(11)60070-6
13. Cardoso F, Bedard PL, Winer EP, Pagani O, Senkus-Konefka E, Fallowfield LJ, Kyriakides S, Costa A, Cufer T, Albain KS, Force E-MT (2009) International guidelines for management of metastatic breast cancer: combination vs sequential single-agent chemotherapy. J Natl Cancer Inst 101(17):1174–1181. https://doi.org/10.1093/jnci/djp235
14. Foulkes WD, Smith IE, Reis-Filho JS (2010) Triple-negative breast cancer. N Engl J Med 363(20):1938–1948. https://doi.org/10.1056/NEJMra1001389
15. Uhm JE, Park YH, Yi SY, Cho EY, Choi YL, Lee SJ, Park MJ, Lee SH, Jun HJ, Ahn JS, Kang WK, Park K, Im YH (2009) Treatment outcomes and clinicopathologic characteristics of triple-negative breast cancer patients who received platinum-containing chemotherapy. Int J Cancer 124(6):1457–1462. https://doi.org/10.1002/ijc.24090
16. Tutt A, Ellis P, Kilburn L, Gilett C, Pinder S, Abraham J, Barrett S, Barrett-Lee P, Chan S, Cheang M, Dowsett M, Fox L, Gazinska P, Grigoriadis A, Gutin A, Harper-Wynne C, Hatton M, Kernaghan S, Lanchbury J, Morden J, Owen J, Parikh J, Parker P, Rahman N, Roylance R, Shaw A, Smith I, Thompson R, Timms K, Tovey H, Wardley A, Wilson G, Harries M, Bliss J (2015) Abstract S3-01: the TNT trial: a randomized phase III trial of carboplatin (C) compared with docetaxel (D) for patients with metastatic or recurrent locally advanced triple negative or <em>BRCA1/2</em> breast cancer (CRUK/07/012). Cancer Res 75(9 Suppl):S3-01-S03-01. doi:https://doi.org/10.1158/1538-7445.sabcs14-s3-01
17. Scully R, Livingston DM (2000) In search of the tumour-suppressor functions of BRCA1 and BRCA2. Nature 408(6811):429–432. https://doi.org/10.1038/35044000
18. Starita LM, Parvin JD (2003) The multiple nuclear functions of BRCA1: transcription, ubiquitination and DNA repair. Curr Opin Cell Biol 15(3):345–350

19. Gudmundsdottir K, Ashworth A (2006) The roles of BRCA1 and BRCA2 and associated proteins in the maintenance of genomic stability. Oncogene 25(43):5864–5874. https://doi.org/10.1038/sj.onc.1209874
20. Wilson JH, Elledge SJ (2002) Cancer. BRCA2 enters the fray. Science 297(5588):1822–1823. https://doi.org/10.1126/science.1077171
21. Chen S, Parmigiani G (2007) Meta-analysis of BRCA1 and BRCA2 penetrance. J Clin Oncol 25(11):1329–1333. https://doi.org/10.1200/JCO.2006.09.1066
22. Ford D, Easton DF, Peto J (1995) Estimates of the gene frequency of BRCA1 and its contribution to breast and ovarian cancer incidence. Am J Hum Genet 57(6):1457–1462
23. Atchley DP, Albarracin CT, Lopez A, Valero V, Amos CI, Gonzalez-Angulo AM, Hortobagyi GN, Arun BK (2008) Clinical and pathologic characteristics of patients with BRCA-positive and BRCA-negative breast cancer. J Clin Oncol 26(26):4282–4288. https://doi.org/10.1200/JCO.2008.16.6231
24. Birgisdottir V, Stefansson OA, Bodvarsdottir SK, Hilmarsdottir H, Jonasson JG, Eyfjord JE (2006) Epigenetic silencing and deletion of the BRCA1 gene in sporadic breast cancer. Breast Cancer Res 8(4):R38. https://doi.org/10.1186/bcr1522
25. De Vos M, Schreiber V, Dantzer F (2012) The diverse roles and clinical relevance of PARPs in DNA damage repair: current state of the art. Biochem Pharmacol 84(2):137–146. https://doi.org/10.1016/j.bcp.2012.03.018
26. Krishnakumar R, Kraus WL (2010) The PARP side of the nucleus: molecular actions, physiological outcomes, and clinical targets. Mol Cell 39(1):8–24. https://doi.org/10.1016/j.molcel.2010.06.017
27. Eustermann S, Wu WF, Langelier MF, Yang JC, Easton LE, Riccio AA, Pascal JM, Neuhaus D (2015) Structural basis of detection and signaling of DNA single-strand breaks by human PARP-1. Mol Cell 60(5):742–754. https://doi.org/10.1016/j.molcel.2015.10.032
28. Dawicki-McKenna JM, Langelier MF, DeNizio JE, Riccio AA, Cao CD, Karch KR, McCauley M, Steffen JD, Black BE, Pascal JM (2015) PARP-1 activation requires local unfolding of an autoinhibitory domain. Mol Cell 60(5):755–768. https://doi.org/10.1016/j.molcel.2015.10.013
29. Satoh MS, Lindahl T (1992) Role of poly(ADP-ribose) formation in DNA repair. Nature 356(6367):356–358. https://doi.org/10.1038/356356a0
30. Rouleau M, Patel A, Hendzel MJ, Kaufmann SH, Poirier GG (2010) PARP inhibition: PARP1 and beyond. Nat Rev Cancer 10(4):293–301. https://doi.org/10.1038/nrc2812
31. Ame JC, Spenlehauer C, de Murcia G (2004) The PARP superfamily. BioEssays 26(8):882–893. https://doi.org/10.1002/bies.20085
32. Farmer H, McCabe N, Lord CJ, Tutt AN, Johnson DA, Richardson TB, Santarosa M, Dillon KJ, Hickson I, Knights C, Martin NM, Jackson SP, Smith GC, Ashworth A (2005) Targeting the DNA repair defect in BRCA mutant cells as a therapeutic strategy. Nature 434(7035):917–921. https://doi.org/10.1038/nature03445
33. Lord CJ, Ashworth A (2017) PARP inhibitors: synthetic lethality in the clinic. Science 355(6330):1152–1158. https://doi.org/10.1126/science.aam7344
34. Helleday T, Petermann E, Lundin C, Hodgson B, Sharma RA (2008) DNA repair pathways as targets for cancer therapy. Nat Rev Cancer 8(3):193–204. https://doi.org/10.1038/nrc2342
35. Bryant HE, Schultz N, Thomas HD, Parker KM, Flower D, Lopez E, Kyle S, Meuth M, Curtin NJ, Helleday T (2005) Specific killing of BRCA2-deficient tumours with inhibitors of poly(ADP-ribose) polymerase. Nature 434(7035):913–917. https://doi.org/10.1038/nature03443
36. Shall S (1975) Seminar on poly(ADP-ribose) and ADP-ribosylation of protine. J Biochem 77(Suppl):2
37. M R Purnell WJW (1980) Novel inhibitors of poly(ADP-ribose) synthetase. Biochem J 185(3):775–777
38. Terada M, Fujiki H, Marks PA, Sugimura T (1979) Induction of erythroid differentiation of murine erythroleukemia cells by nicotinamide and related compounds. Proc Natl Acad Sci U S A 76(12):6411–6414
39. Shen Y, Rehman FL, Feng Y, Boshuizen J, Bajrami I, Elliott R, Wang B, Lord CJ, Post LE, Ashworth A (2013) BMN 673, a novel and highly potent PARP1/2 inhibitor for the treatment of human cancers with DNA repair deficiency. Clin Cancer Res 19(18):5003–5015. https://doi.org/10.1158/1078-0432.CCR-13-1391
40. Murai J, Huang SY, Renaud A, Zhang Y, Ji J, Takeda S, Morris J, Teicher B, Doroshow JH, Pommier Y (2014) Stereospecific PARP trapping by BMN 673 and comparison with olaparib and rucaparib. Mol Cancer Ther 13(2):433–443. https://doi.org/10.1158/1535-7163.MCT-13-0803
41. J.J. Mahany NL, E.I. Heath et al. (2008) A phase IB study evaluating BSI-201 in combination with chemotherapy in subjects with advanced solid tumors. J Clin Oncol 26(Suppl; abstr 3579)
42. O'Shaughnessy J, Osborne C, Pippen JE, Yoffe M, Patt D, Rocha C, Koo IC, Sherman BM, Bradley C (2011) Iniparib plus chemotherapy in metastatic triple-negative breast cancer. N Engl J Med 364(3):205–214. https://doi.org/10.1056/NEJMoa1011418
43. O'Shaughnessy J, Schwartzberg L, Danso MA, Miller KD, Rugo HS, Neubauer M, Robert N, Hellerstedt B, Saleh M, Richards P, Specht JM, Yardley DA, Carlson RW, Finn RS, Charpentier E, Garcia-Ribas I, Winer EP (2014) Phase III study of iniparib plus gemcitabine and carboplatin versus gemcitabine and carboplatin in patients with metastatic triple-negative breast cancer. J Clin Oncol 32(34):3840–3847. https://doi.org/10.1200/JCO.2014.55.2984
44. Menear KA, Adcock C, Boulter R, Cockcroft XL, Copsey L, Cranston A, Dillon KJ, Drzewiecki J, Garman S, Gomez S, Javaid H, Kerrigan F, Knights C, Lau A, Loh VM Jr, Matthews IT, Moore S, O'Connor MJ, Smith GC, Martin NM (2008) 4-[3-(4-cyclopropanecarbonylpiperazine-1-carbonyl)-4-fluorobenzyl]

-2H-phthalazin- 1-one: a novel bioavailable inhibitor of poly(ADP-ribose) polymerase-1. J Med Chem 51(20):6581–6591. https://doi.org/10.1021/jm8001263

45. Hay T, Matthews JR, Pietzka L, Lau A, Cranston A, Nygren AO, Douglas-Jones A, Smith GC, Martin NM, O'Connor M, Clarke AR (2009) Poly(ADP-ribose) polymerase-1 inhibitor treatment regresses autochthonous Brca2/p53-mutant mammary tumors in vivo and delays tumor relapse in combination with carboplatin. Cancer Res 69(9):3850–3855. https://doi.org/10.1158/0008-5472.CAN-08-2388
46. Rottenberg S, Jaspers JE, Kersbergen A, van der Burg E, Nygren AO, Zander SA, Derksen PW, de Bruin M, Zevenhoven J, Lau A, Boulter R, Cranston A, O'Connor MJ, Martin NM, Borst P, Jonkers J (2008) High sensitivity of BRCA1-deficient mammary tumors to the PARP inhibitor AZD2281 alone and in combination with platinum drugs. Proc Natl Acad Sci U S A 105(44):17079–17084. https://doi.org/10.1073/pnas.0806092105
47. Fong PC, Boss DS, Yap TA, Tutt A, Wu P, Mergui-Roelvink M, Mortimer P, Swaisland H, Lau A, O'Connor MJ, Ashworth A, Carmichael J, Kaye SB, Schellens JH, de Bono JS (2009) Inhibition of poly(ADP-ribose) polymerase in tumors from BRCA mutation carriers. N Engl J Med 361(2):123–134. https://doi.org/10.1056/NEJMoa0900212
48. Tutt A, Robson M, Garber JE, Domchek SM, Audeh MW, Weitzel JN, Friedlander M, Arun B, Loman N, Schmutzler RK, Wardley A, Mitchell G, Earl H, Wickens M, Carmichael J (2010) Oral poly(ADP-ribose) polymerase inhibitor olaparib in patients with BRCA1 or BRCA2 mutations and advanced breast cancer: a proof-of-concept trial. Lancet 376(9737):235–244. https://doi.org/10.1016/S0140-6736(10)60892-6
49. Gelmon KA, Tischkowitz M, Mackay H, Swenerton K, Robidoux A, Tonkin K, Hirte H, Huntsman D, Clemons M, Gilks B, Yerushalmi R, Macpherson E, Carmichael J, Oza A (2011) Olaparib in patients with recurrent high-grade serous or poorly differentiated ovarian carcinoma or triple-negative breast cancer: a phase 2, multicentre, open-label, non-randomised study. Lancet Oncol 12(9):852–861. https://doi.org/10.1016/S1470-2045(11)70214-5
50. Kaufman B, Shapira-Frommer R, Schmutzler RK, Audeh MW, Friedlander M, Balmana J, Mitchell G, Fried G, Stemmer SM, Hubert A, Rosengarten O, Steiner M, Loman N, Bowen K, Fielding A, Domchek SM (2015) Olaparib monotherapy in patients with advanced cancer and a germline BRCA1/2 mutation. J Clin Oncol 33(3):244–250. https://doi.org/10.1200/JCO.2014.56.2728
51. Robson ME, Im SA, Senkus E, Xu B, Domchek SM, Masuda N, Delaloge S, Li W, Tung NM, Armstrong A, Wu W, Goessl CD, Runswick S, Conte PF (2017) OlympiAD: Phase III trial of olaparib monotherapy versus chemotherapy for patients (pts) with HER2-negative metastatic breast cancer (mBC) and a germline BRCA mutation (gBRCAm). J Clin Oncol (Suppl; abstr LBA4):35
52. Dent RA, Lindeman GJ, Clemons M, Wildiers H, Chan A, McCarthy NJ, Singer CF, Lowe ES, Watkins CL, Carmichael J (2013) Phase I trial of the oral PARP inhibitor olaparib in combination with paclitaxel for first- or second-line treatment of patients with metastatic triple-negative breast cancer. Breast Cancer Res 15(5):R88. https://doi.org/10.1186/bcr3484
53. van der Noll R AJ, Jager A, et al. (2013) Phase I study of olaparib in combination with carboplatin and/or paclitaxel in patients with advanced solid tumors. J Clin Oncol 31(Suppl; abstr 2579)
54. Takahashi M YK, Yamamoto H, et al. (2016) A phase I/II trial of olaparib in combination with eribulin in patients with advanced or metastatic triple negative breast cancer (TNBC) previously treated with anthracyclines and taxanes: the analyses of efficacy and safety from phase II. J Clin Oncol 34(Suppl; abstr 1080)
55. Litton JK, Scoggins M, Ramirez DL, Murthy RK, Whitman GJ, Hess KR, Adrada BE, Moulder SL, Barcenas CH, Valero V, Booser D, Gomez JS, Mills GB, Piwnica-Worms H, Arun BK (2016) A pilot study of neoadjuvant talazoparib for early-stage breast cancer patients with a BRCA mutation. Ann Oncol 27(Suppl 6):vi43–vi67
56. Jaspers JE, Kersbergen A, Boon U, Sol W, van Deemter L, Zander SA, Drost R, Wientjens E, Ji J, Aly A, Doroshow JH, Cranston A, Martin NM, Lau A, O'Connor MJ, Ganesan S, Borst P, Jonkers J, Rottenberg S (2013) Loss of 53BP1 causes PARP inhibitor resistance in Brca1-mutated mouse mammary tumors. Cancer Discov 3(1):68–81. https://doi.org/10.1158/2159-8290.CD-12-0049
57. Xu G, Chapman JR, Brandsma I, Yuan J, Mistrik M, Bouwman P, Bartkova J, Gogola E, Warmerdam D, Barazas M, Jaspers JE, Watanabe K, Pieterse M, Kersbergen A, Sol W, Celie PH, Schouten PC, van den Broek B, Salman A, Nieuwland M, de Rink I, de Ronde J, Jalink K, Boulton SJ, Chen J, van Gent DC, Bartek J, Jonkers J, Borst P, Rottenberg S (2015) REV7 counteracts DNA double-strand break resection and affects PARP inhibition. Nature 521(7553):541–544. https://doi.org/10.1038/nature14328
58. Pettitt SJ, Rehman FL, Bajrami I, Brough R, Wallberg F, Kozarewa I, Fenwick K, Assiotis I, Chen L, Campbell J, Lord CJ, Ashworth A (2013) A genetic screen using the PiggyBac transposon in haploid cells identifies Parp1 as a mediator of olaparib toxicity. PLoS One 8(4):e61520. https://doi.org/10.1371/journal.pone.0061520
59. Chaudhuri AR, Callen E, Ding X, Gogola E, Duarte AA, Lee JE, Wong N, Lafarga V, Calvo JA, Panzarino NJ, John S, Day A, Crespo AV, Shen B, Starnes LM, de Ruiter JR, Daniel JA, Konstantinopoulos PA, Cortez D, Cantor SB, Fernandez-Capetillo O, Ge K, Jonkers J, Rottenberg S, Sharan SK, Nussenzweig A (2016) Erratum: replication fork stability confers chemoresistance in BRCA-deficient cells. Nature 539(7629):456. https://doi.org/10.1038/nature19826
60. Edwards SL, Brough R, Lord CJ, Natrajan R, Vatcheva R, Levine DA, Boyd J, Reis-Filho JS, Ashworth A

(2008) Resistance to therapy caused by intragenic deletion in BRCA2. Nature 451(7182):1111–1115. https://doi.org/10.1038/nature06548
61. Barber LJ, Sandhu S, Chen L, Campbell J, Kozarewa I, Fenwick K, Assiotis I, Rodrigues DN, Reis Filho JS, Moreno V, Mateo J, Molife LR, De Bono J, Kaye S, Lord CJ, Ashworth A (2013) Secondary mutations in BRCA2 associated with clinical resistance to a PARP inhibitor. J Pathol 229(3):422–429. https://doi.org/10.1002/path.4140
62. Sakai W, Swisher EM, Karlan BY, Agarwal MK, Higgins J, Friedman C, Villegas E, Jacquemont C, Farrugia DJ, Couch FJ, Urban N, Taniguchi T (2008) Secondary mutations as a mechanism of cisplatin resistance in BRCA2-mutated cancers. Nature 451(7182):1116–1120. https://doi.org/10.1038/nature06633
63. Lord CJ, Ashworth A (2016) BRCAness revisited. Nat Rev Cancer 16(2):110–120. https://doi.org/10.1038/nrc.2015.21
64. McCabe N, Turner NC, Lord CJ, Kluzek K, Bialkowska A, Swift S, Giavara S, O'Connor MJ, Tutt AN, Zdzienicka MZ, Smith GC, Ashworth A (2006) Deficiency in the repair of DNA damage by homologous recombination and sensitivity to poly(ADP-ribose) polymerase inhibition. Cancer Res 66(16):8109–8115. https://doi.org/10.1158/0008-5472.CAN-06-0140
65. Gilardini Montani MS, Prodosmo A, Stagni V, Merli D, Monteonofrio L, Gatti V, Gentileschi MP, Barila D, Soddu S (2013) ATM-depletion in breast cancer cells confers sensitivity to PARP inhibition. J Exp Clin Cancer Res 32:95. https://doi.org/10.1186/1756-9966-32-95
66. Subhash VV, Tan SH, Yeo MS, Yan FL, Peethala PC, Liem N, Krishnan V, Yong WP (2016) ATM expression predicts Veliparib and Irinotecan sensitivity in gastric cancer by mediating P53-independent regulation of cell cycle and apoptosis. Mol Cancer Ther 15(12):3087–3096. https://doi.org/10.1158/1535-7163.MCT-15-1002
67. Villalona-Calero MA, Duan W, Zhao W, Shilo K, Schaaf LJ, Thurmond J, Westman JA, Marshall J, Xiaobai L, Ji J, Rose J, Lustberg M, Bekaii-Saab T, Chen A, Timmers C (2016) Veliparib alone or in combination with Mitomycin C in patients with solid tumors with functional deficiency in homologous recombination repair. J Natl Cancer Inst 108(7). https://doi.org/10.1093/jnci/djv437

Targeting the Epigenome as a Novel Therapeutic Approach for Breast Cancer

14

Sumin Oh, Je Yeong Ko, Chaeun Oh, and Kyung Hyun Yoo

Abstract

Breast cancer is one of complex diseases that are influenced by environment. Various genetic and epigenetic alterations are provoking causes of breast carcinogenesis. Dynamic epigenetic regulation including DNA methylation and histone modification induces dysregulation of genes related to proliferation, apoptosis, and metastasis in breast cancer. DNA methylation is strongly associated with the repression of transcription through adding to the methyl group by DNA methyltransferases (DNMTs), and tumor suppressor genes such as CCND2 and RUNX3 have been investigated to undergo hypermethylation at promoter region in breast cancer. In addition, histone deacetylases (HDACs) contribute to transcriptional repression by removing acetyl group at lysine residues leading to tumorigenesis. Since epigenetic changes are reversible, therapeutic approaches have been applied with epigenetic modification drugs such as DNMT inhibitors and HDAC inhibitors. In this chapter, we will summarize the feature of epigenetic markers in breast cancer cells and the effect of single or combination of epigenetic reagents for breast cancer therapy.

Keywords

Epigenetic regulation • DNMT inhibitors • HDAC inhibitors • Therapeutic targets • Breast cancer

14.1 Introduction

Breast cancer is a heterogeneous disease characterized by levels of hormone receptors, including estrogen receptor (ER), progesterone receptor (PR), and human epidermal growth factor receptor 2 (HER2). Depending on the presence or absence of hormone receptors, breast cancer is

S. Oh • J.Y. Ko • C. Oh • K.H. Yoo (✉)
Department of Biological Sciences, Sookmyung Women's University, 100, Cheongpa-ro 47-gil, Yongsan-gu, Seoul 04310, South Korea
e-mail: khryu@sookmyung.ac.kr

E. Song, H. Hu (eds.), *Translational Research in Breast Cancer*, Advances in Experimental Medicine and Biology 1026, https://doi.org/10.1007/978-981-10-6020-5_14

subdivided into the luminal A, luminal B, HER2-enriched, and basal-like subtypes [1]. These different subtypes of breast cancer have distinct prognoses and responses to therapies [2]. However, various combinations of these molecular markers have been found in patients with breast cancer. This diversity makes it difficult to identify the individual progress of the disease and to select the appropriate treatment. Therefore, it is necessary to identify a clearer basis for precision therapies.

Breast cancer is known to have a genetic component, such as BRCA1 and BRCA2 [3–19], but epigenetic alterations frequently occur to change gene expression, which could be a cause of breast cancer symptoms [20–48]. Tumorigenesis is actually a multistep process involving both genetic and posttranslational changes, the latter of which corresponds to independent gene expression without mutation of the DNA sequence. Epigenetic alterations are commonly involved in DNA methylation and histone modification at promoter regions of target genes.

DNA methylation is an inherited epigenetic mark that is mainly associated with the repression of gene expression through transferring to the covalent bond of the methyl group on the C-5 of the cytosine ring of DNA by DNA methyltransferases (DNMTs) [49, 50]. DNA methylation is rare in the mammalian genome, but it has often been found in CpG islands typically located in the promoter region of genes. In mammals, there are three active DNMTs: DNMT1, DNMT3A, and DNMT3B [51–53]. DNMT1 methylates hemimethylated DNA during the S phase, while DNMT3A and DNMT3B are associated with de novo methylation during development. DNA methylation is important during development, such as X chromosome inactivation and imprinting. When DNA methylation is dysregulated, it results in the inappropriate silencing of genes and contributes to cancer development. The global distribution of methylation in mammals has posed a challenge to researchers in terms of determining whether methylation is a default state or is targeted at specific gene sequences. Hypermethylation of the promoters of genetic factors and tumor suppressor genes in breast cancer cells has been reported [54–56].

Histone modification plays another key role in epigenetics, and the status of histone is maintained through a balance between modifying enzymes, which can modulate specific modifications. In general, histone acetylation is associated with the open chromatin structure inducing transcriptional activation [57–60]. Histone deacetylases (HDACs) remove the acetyl group at lysine residues, whereas histone acetyltransferases (HATs) add the acetyl group to lysine residues. In mammals, 11 HDAC proteins have a highly conserved deacetylase domain, and they can be subdivided into four groups (Class I, Class IIa, Class IIb, and Class IV) depending on enzymatic function, structure, and pattern of localization and expression [61–65]. Class I consists of HDAC1, HDAC2, HDAC3, and HDAC8, and their localization is mainly detected at nuclei with high enzymatic activity. Class IIa includes HDAC4, HDAC5, HDAC7, and HDAC9, which have conserved binding sites for the transcription factor myocyte enhancer factor 2 (MEF2) and chaperone protein 14-3-3. Their expression is identified in specific tissues. For example, HDAC4 is highly expressed in the brain, whereas HDAC is significantly expressed in thymocytes. HDAC6 and HDAC10 belong to Class IIb. Cytoskeletal proteins and transmembrane proteins are direct targets of HDAC6, which is mainly located in the cytoplasm. HDAC11 belongs to Class IV and has been identified in the brain, heart, skeletal muscle, and kidney.

Epigenetic therapy is defined as the use of drugs or other epigenome-influencing techniques to treat medical conditions. Breast cancer is also influenced by epigenetic mechanisms, and epigenetic therapy offers a potential way to influence those pathways directly. Here, we will discuss the epigenetic aberrations and the dysregulation of genes that control the epigenome in breast cancers. DNA methylation and histone modification patterning and next-generation sequencing (NGS) have vastly increased our understanding of epigenetic deregulation. Epigenetic alterations, in contrast to genetic lesions, are themselves pharmacologically reversible and therefore represent strategies for novel therapeutic approaches to breast cancer.

14.2 DNA Hypermethylation in Breast Cancer

Since breast cancer has various causes, new signatures are required for therapy. The DNA methylation pattern could be used as a biomarker for cancer diagnosis and treatment selection. Whole genome approaches have been applied to discover the specific DNA methylation status, while hypermethylation and hypomethylation of CpG at the promoter have been investigated in genes related to proliferation, apoptosis, and metastasis.

The reduced expression of CyclinD2, which is important to cell cycle regulation, was detected in breast tumors. Many studies revealed that hypermethylated CpG islands at the promoter of the *CCND2* gene were reversely correlated with their gene expression [8–12, 14–16, 19]. For example, Truong et al. identified that a total of nine CpG islands of the *CCND2* promoter and four CpG islands were hypermethylated in breast cancer [66]. Approximately 62% of Vietnamese patients with breast tumors had hypermethylated CpG islands at these regions. In addition, it has been identified that changes of the CyclinD2 methylation pattern are more related to the late stage of breast cancer cell transformation, suggesting that it might be a biomarker for early detection [67].

Alterations of the estrogen receptor (ER) and progesterone receptor (PGR) are usually used as prognostic markers for breast cancer. DNA methylation of ER and PGR promoters was investigated in various breast cancer types [68–75]. Maekawa et al. have investigated whether or not the DNA methylation of the tissue-dependent and differentially methylated region (T-DMR) at the *ESR1* gene affects their gene expression in breast cancer [69]. They found two T-DMRs, T-DMR1 and T-DMR2, at the *ESR1* locus. The hypermethylation of T-DMR1 located far from TSS was highly related to downregulated gene expression, whereas that of T-DMR2 located close to TSS had only a modest effect on gene repression. However, the methylation status of *ESR1* and *PGR* promoters was not significantly associated with histological subtypes based on the presence of hormone receptors [68, 76, 77]. Ramezani et al. confirmed that there was no difference in the DNA methylation pattern between non-triple-negative cells and triple-negative cells [68]. Even though hypermethylations of ESR1 and PGR were detected in ERα- and PR-negative breast cancers, a strong correlation between low expression and hypermethylation at the *ESR1* and *PGR* locus could not be detected. It should be considered that the methylation status of these genes could be used for the diagnosis of breast cancer.

Runt-related transcription factor 3 (*RUNX3*), known as a tumor suppressor in various cancers including breast cancers, is involved in cell survival, proliferation, and differentiation. The contribution of hypermethylated *RUNX3* to breast cancer progression has been identified [78–85]. Stronger hypermethylation of RUNX3 was observed in ductal carcinoma in situ (DCIS) and invasive ductal carcinoma (IDC) compared to non-breast tumors and benign tumors. However, researchers confirmed that the DNA methylation statuses of DCIS and IDC were not significantly different, suggesting this hypermethylation might not be involved in the progression from DCIS to IDC [79]. In addition, the hypermethylation of the RUNX3 promoter was frequently detected in patients with ER-positive breast cancer, and *RUNX3* mRNA levels corresponded with their methylation status [80]. This suggests that RUNX3 methylation could be an early biomarker and therapeutic target for breast cancer.

Even though it has been identified that approximately 50% of cases of familial breast cancer are related to BRCA1 mutations, only a small percentage of sporadic breast cancer cases with BRCA1 mutations have been detected. However, expression of the BRCA1 gene was clearly decreased in sporadic breast cancer. Many studies have identified the relationship between gene expression level and methylation status at the promoter of BRCA1 [8–12, 14–16, 19, 86]. Hypermethylation of the BRCA1 promoter region was detected in over 50% of sporadic breast cancer cases, but it was not strongly correlated with clinicopathological characteristics. Rather, it was more associated with ER and PR status [19]. Interestingly, it has been reported that patients with hypermethylated BRCA1 have a good prognosis, suggesting it could be a potential therapeutic target for epigenetic treatment.

14.3 Global Analysis of Histone Modification in Breast Cancer

As mentioned above, histone modification is critical to the regulation of gene expression. It can be altered by DNA methylation patterns and is associated with the characterization of the chromatin structure [10, 19–21, 25, 35, 37, 47, 56, 81, 87–128]. Moreover, histone modification may cause the differentially regulated gene expression and different phenotypes between individuals with and without cancer and/or among cancer types [26, 129, 130]. Notably, research has revealed that the activity of enzymes known as histone modifiers is correlated with the regulation of gene expression [131, 132]. The most popular breast cancer drugs are also related with the alteration of the modification of histones, especially HDACs. The drugs and their effects are discussed in Sect. 4. Even in breast cancers, histone modifications differ depending on cancer subtypes, which are defined by ER, PR, and HER2 status [37, 93, 133–137].

Since the NGS technique was developed, it has become possible to discover the differences in histone modification not only in nearby regulated genes but also in unexpected regions in individuals with and without cancer. The method of chromatin immunoprecipitation sequencing (ChIP-Seq) has been used to show histone modification by detecting protein-DNA interactions globally. Published NGS experimental data, including ChIP-Seq and RNA-Seq, which will be referred to in the below sections, can be downloaded from the Gene Expression Omnibus (GEO). In addition, RNA-Seq data on cancer cells are deposited in The Cancer Genome Atlas (TCGA). Below, previous studies using global analysis, especially ChIP-Seq, for identifying the alteration of histone modification are presented.

14.3.1 Previous Studies

Cancers have both common and specific features. The first study introduced here focused on exploring the common features between two widespread types of cancers, breast cancer and ovarian cancer, which were associated with high mortality in women in the US. TCGA dataset were applied to compare and find common carcinogenesis factors between these two cancers. Gene expression level, miRNA expression level, and histone modification alteration were also investigated in the study. TP53, which was responsible for anti-proliferation or apoptosis, and BRCA1 were found to be highly mutated in both cells. Moreover, the study demonstrated similar patterns of abnormally expressed microRNAs and alteration of histone modification in two cancer types, implicating oncogene regulation in both cancer cells. Therefore, it was concluded that different tissue types can share a common epigenetic carcinogenesis mechanism [48].

Even though cancers have shared characteristics, the goal of most cancer-related studies is to discover cancer-specific features for a novel therapy target. In human mammary epithelial cells (HMECs) and MDA-MB-231 cells, a positive correlation between cell type-specific chromatin structures and gene expression was discovered. H3K4me1, which was considered as an enhancer marker, was utilized to identify enhancer regions by histone modification in HMECs and MDA-MB-231 cells. Nearby genes were then matched from the identified specific enhancers. Enhancers were classified as active and poised sites by the co-localization of H3K27ac, an activation marker. As a result, it has been confirmed that most highly expressed genes have cell type-specific and active enhancers and that they are involved in the gene functions of proteolysis, epidermis development, mitosis, and cell cycle.

Using specific enhancers, a motif search was performed to predict the transcription factors binding to the enhancers to determine cell type-specific characteristics. Consequentially, most of the revealed transcription factors were known as breast cancer target genes, such as TP63 [138]. Histone methyltransferase and demethylase, such as H3K9me2 and H3K9me3, are repressive markers in the genome, and KDM3A/JMJD1A is the enzyme for the demethylase of H3K9me1 and H3K9me2. These histone modification markers were shown to have a reversed correlation with KDM3A/JMJD1A in in vivo and in vitro experiments. In MCF7 and T47D, KDM3A/JMJD1A was progressively increased during cancer transformation compared to non-breast cancer cell lines. In KDM3A/JMJD1A knockdown cells, decreased regulated genes were known oncogenes, such as *MYC and PAX3,* in KDM3A/JMJD1A. Moreover, the deficiency of KDM3A/JMJD1A was implicated in tumor growth and increasing cancer cell migration [33]. The association of RACK7, which was considered as a pioneer binder for the activated protein kinase C-binding protein, and H3K4me3-specific demethylase KDM5C was recognized at active enhancers. The co-localization of two proteins was observed in active enhancer regions, including those of super-enhancers. Their general function was revealed to be negatively regulatory for enhancer. In ZR-75-30, an ER-positive breast cancer cell line, the relation between RACK7 and KDM5C was shown to be such that a deficiency of RACK7 led to a decrease of KDM5C at active enhancers. The loss of RACK7 and KDM5C was implicated in an increase of H3K4me3 and enhancer RNA (eRNA) transcription and a decrease of H3K4m, at active enhancers in ZR-75-30 and MCF7. Consequentially, the transcriptional activity of RACK7-bound genes was enhanced. Upregulated genes were classified as cell adhesion genes by gene ontology (GO) analysis in RACK7 and KDM5C knockout cells. Among differentially expressed genes, the S100A gene family, one of the well-known tumor suppressor families, was suggested as the indicator of RACK7-related tumor genesis. Interestingly, between the S100A genes in Clusters I and II, only members of Cluster I genes had RACK7 and KDM5C occupation at enhancers and reacted to ablated RACK7 and KDM5C in ZR-75-30 [27].

The above studies have indicated that the pathogenesis mechanism is caused by different types of histone modification, which would help predict fundamental targets for cancer therapies.

14.3.2 Public Database for Genomic Study

As mentioned above, several databases have released published experimental data for researchers. Published data can be downloaded with/without permission and used for reanalysis by comparing other samples. Of the many databases, we will introduce three well-known databases: GEO, TCGA, and ENCODE.

14.3.2.1 GEO

The Gene Expression Omnibus (GEO; https://www.ncbi.nlm.nih.gov/geo/) is the biggest database for sharing genomic data. This database is managed by the National Center for Biotechnology Information (NCBI), in which is affiliated to the National Institutes of Health (NIH). From the website, researchers can easily search for and download published data by keywords. In addition, all researchers are able to upload data. When high-throughput sequencing data, like RNA-Seq data, is uploaded, permission from GEO managers is needed to obtain GSE permission/access numbers of experimental data series. As with the group number, the GSE, each sample has its own GSM number. According to the policy of GEO, uploaded data should be released to all researchers. However, the uploaders can determine the data release date. Nowadays, a GEO accession number on the deposited data must be specified for paper submissions to most journals containing genomic studies.

14.3.2.2 TCGA

The Cancer Genome Atlas (TCGA; https://cancergenome.nih.gov) has been launched to share high-throughput data, especially on cancer disease tissues/cells. It is also funded by the NIH and the National Cancer Institute and National Human Genome Research Institute (NHGRI). The purpose of the database is to share cancer genomic data to investigate the genomic causes or differences in cancer diseases. Therefore, it contains many genomic sequencing samples, like Exome-seq and whole genome sequencing samples. To download raw data, rather than processed sequencing data, permission is required. Data types are divided into tiers related with permission level. Only tier 3, processed data and a few tier 1 samples are public in this database. Ways of obtaining permission are not described here.

14.3.2.3 ENCODE

The Encyclopedia of DNA Elements (ENCODE; https://www.encodeproject.org) Consortium has contributed to the generation of various types of NGS data for epigenetic study using comprehensive platforms, especially ChIP-Seq in humans. More than 7500 ChIP-Seq samples have been released. Samples include histone modification and transcription factor DNA-binding samples. Only a few groups have conducted experiments to generate ENCODE data based on their stringent experimental process. All ENCODE data is accessible to anyone, and it is deposited in GEO.

14.3.3 Example Analysis Using ENCODE Data

Histone modification can be examined globally by ChIP-Seq experiments. In ENCODE, 145 ChIP-Seq samples in MCF7 are now available, among which 14 are histone modification samples. In MCF7, RNA-Seq, ChIP-Seq, H3K27ac, H3K4me1, H3K4me3, and H3K27me3 were downloaded and analyzed for the global analysis example. Each mark is known as activator, enhancer, promoter, and deactivator, respectively. However, a detailed analysis process will not be described in this section.

Each histone tail status reflects different histone modification status, and the result of analysis is shown in Fig. 14.1. Figure 14.1a–e heatmap represents each marker's density from ±20 kb of the promoter or peak center. Figure 14.1a illustrates the gene promoter sites sorted by gene expression level. H3K27ac and H3K4me3 were clearly enriched in highly expressed genes. The occupancy of each histone modification was identified using the HOMER (http://homer.ucsd.edu/homer/) peak-calling method, one of the ChIP-Seq analysis programs. Among the total peaks identified, 90% were included in Fig. 14.1b–e. Of the peaks, even though H3K4me1 and H3K27me3 occupied many of the genome sites, H3K4me1 and H3K27me3 were shown to have different DNA-binding patterns compared with the others (Fig. 14.1b–e). While almost all of H3K27ac and H3K4me3 occupied each other and/or H3K4me1 (Fig. 14.1d, e), approximately half of H3K4me1 and more than 80% of H3K27me3 were localized alone (Fig. 14.1b, c). In only H3K4me1-occupied regions, 95% were intron or intergenic, which could be putative enhancer sites since H3K4me1 had been considered as enhancer marker (Fig. 14.1b). Figure 14.1c, d shows that H3K27ac and H3K4me1 co-occupied the sites suspected as active enhancer sites, and H3K27ac, H3K4me1, and H3K4me3 enriched regions supposed to be active promoter sites. Figure 14.1f shows one of the breast cancer target genes, BRCA1. By visualizing the experimental data, we confirmed that H3K27ac and H3K4me3 were extremely enriched, and H3K4me1 was occupied. Meanwhile, H3K27me3 did not exist in the promoter sites of highly expressed genes. This result indicates that the complexity of the histone modification mark signifies a different genomic structure. Hence, an understanding of the histone modification mechanism can promote the development of a compatible disease-targeted therapy.

14.4 Epigenetic Regulators as Therapeutic Targets

14.4.1 DNMT Inhibitors

DNMT inhibitors are widely used as epigenetic drugs for the treatment of tumorigenic diseases including breast cancer [139]. However, for precision therapy, specific inhibitors of DNMTs must be used due to their toxicity, side effects, and chemical instability [128]. The overexpression of DNMTs has been identified in breast cancers.

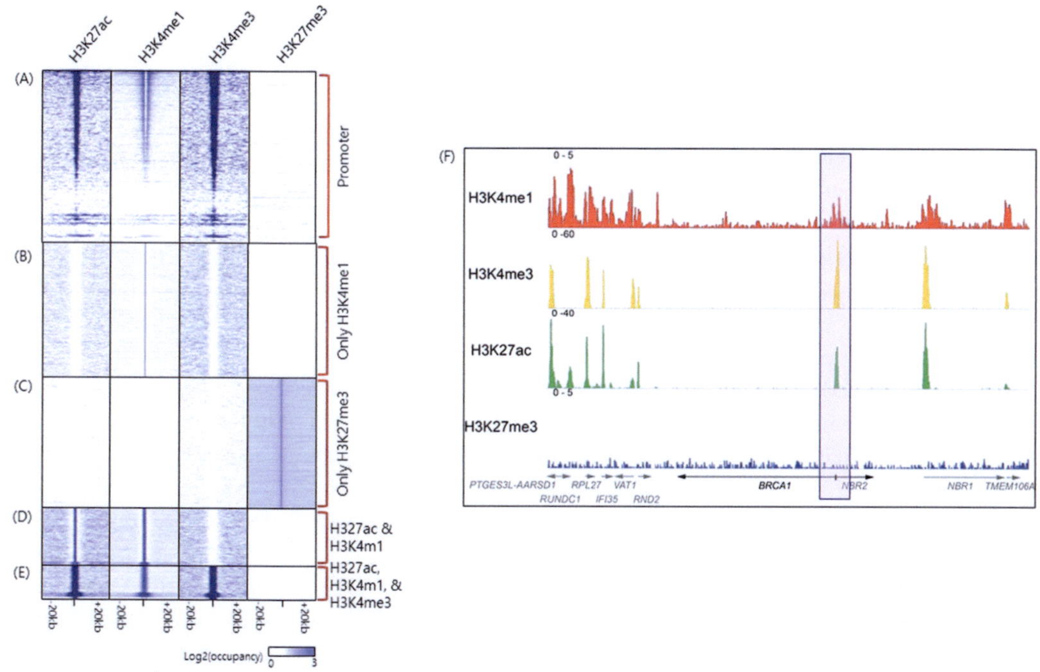

Fig. 14.1 Global analysis of H3K27ac, H3K4me1, H3K4me3, and H3K27me3 in MCF-7. (**a**–**e**) Heatmap representing DNA-binding density of histone modification ±20 kb around promoters or peak centers. (**a**) Enriched H3K27ac and H3K4me3 in all promoter sites of human genome sorted by gene expression level but not H3K4me1 and H3K27me3. (**b**) Most of peak sites were detected by only H3K4me1 in whole genome. (**c**) No detection of H3K27ac, H3K4me1, or H3K4me3 in almost of 80% of H3K27me3 peaks sites. (**d**) Considered as active enhancers by the occupancy sites of H3K27ac and H3K4me1. (**e**) Detected peaks as active promoters by H3K27ac, H3K4me1, and H3K4me3. (**f**) Genome browser snapshots on MCF-7. Enriched H3K4me1, H3K4me3, and H3K27ac marks in the promoter sites of BRCA1

In particular, DNMT3B is highly expressed in breast carcinoma, and more than 30% of patients with breast cancer show an increase in DNMT3B [56, 128, 140]. Here, we will summarize the effects of DNMT inhibitors in breast cancer cells.

14.4.1.1 5-Azacytidine

The effect of combination therapy with the DNMT inhibitor, 5-azacitidine, and the HDAC inhibitor, entinostat, has been investigated in patients with advanced hormone-resistant or triple-negative breast cancer (TNBC) [136]). Only a limited portion of the hormone-resistant group responded to the combination therapy, and there was no effect in the TNBC group. It suggests that combination epigenetic therapy could benefit patients with hormone-resistant diseases.

The effect of 5-azacitidine on the *GPER1* (seven-transmembrane G-protein-coupled estrogen receptor) locus was investigated in MCF7 and MDA-MB-231 cells [141]. They checked the methylation status of the 3′ region and coding region CpG islands at the GPER1 locus, and different methylation patterns were observed at the 3′ region of *GPER1* between MCF7 and MDA-MB-231 cells. Hypomethylation of this region was identified in MCF7 cells, and their expression was higher than that in MDA-MB-231 cells. Conversely, MDA-MB-231 cells showed hypermethylation of the 3′ region CpG islands of *GPER1,* and its expression remained at low levels. RNA-Seq data from TCGA also revealed that the *GPER1* mRNA level was downregulated in primary breast tumors compared to normal tissues. Thus, 5-azacitidine affects *GPER1* expression in MDA-MB-231 cells, suggesting a potential mechanism of GPER1 repression in TNBCs.

Sadikovic et al. identified genome-wide methylation patterns in breast cancer cell lines, MCF7, HCC1806, MDA-MB-231, and MDA-MB-468, with treatment of 5-azacytidine using amplification of intermethylated sites (AIMS) [142]. Approximately 50 different bands were identified in cells treated with 5-azacytidine compared to untreated cells. Among them, 12 sites appeared to have common susceptible regions that could be targeted to other carcinomas, such as choriocarcinoma and melanoma cells, and over 60% of them were associated with genomic instability in carcinomas. For example, hypomethylation events within *SANP190*, *BC13982* loci were observed in MCF7, MDA-MB-231, and MDA-MB-468 cells. It indicates that 5-azacytidine targets genes related with genomic instability and could affect therapy by changing the DNA methylation patterns of these target loci [142].

The in vivo and in vitro effects of 5-axacytidine in ERBB2-positive breast cancers have been investigated [143]. PTPRO (PTP receptor type O) has a role in tumor suppressor genes in several tumors, and the hypermethylation of this gene was observed in ERBB-positive breast cancers. Physical interaction between PTPRO and ERBB2 was investigated, and the activation of PTPRO reduced ERBB2 activity through increased endocytotoxic degradation. After 5-azacytidine treatment, the expression of PTPRO was increased by hypomethylation, and cell proliferation was reduced in ERBB2-overexpressed breast cancer cells. This suggests that PTPRO is a potential therapeutic target regulated by 5-azacytidine for the HER2-positive subtype of breast cancer.

The relationship between DNA methylation and RNA Pol II stalling in tumor suppressor genes was identified in TNBC cells, MDA-MB-231 [144]. The hypermethylation of *CLDN6*, *MAL*, *RIN1*, *PRA*, and *VGF* was found, and their expression was significantly recovered after 5-azacytidine treatment. Furthermore, it was discovered that RNA Pol II stalling was released in MDA-MB-231 cells treated with 5-azacytidine corresponding to their gene expression. It was identified that a change of RNA Pol II stalling could affect the modification of the histones H3K27me3 and H3K4me3. Occupancy of EZH2 and SUZ12, which are polycomb proteins, was critically decreased, whereas H3K4me3 levels were increased at tumor suppressor genes. Thus, 5-azacytidine treatment influences RNA Pol II stalling and histone modification as well as DNA methylation in breast cancer.

High doses of **5-azacytidine** lead to cytotoxicity in MDA-MB-435 cells, so proper doses of this epigenetic drug need to be obtained for therapy [145]. Hellreich et al. identified that a low concentration (1.95 ug/mL) did not have a cytotoxic effect and generated resistant populations. This study has demonstrated the potential use of 5-azacytidine in breast cancer therapy.

14.4.1.2 5-Aza-2′-Deoxycytidine

According to a genome-wide association studies (GWAS) analysis, single-nucleotide polymorphisms (SNPs) at 16q12 could contribute to breast cancer [39]. There are two genes, *TOX3* and *CASC16*, at this locus. *TOX3*, known as a nuclear protein, negatively controls *BRCA1* genes by binding to the promoter of *BRCA1*. Han et al. confirmed the expression of *TOX3* and *CASC16* genes in luminal, basal-like subtypes and normal breast tissues. *TOX3* was more highly expressed in luminal subtypes compared to normal tissues and basal-like subtypes. In addition, a TCGA analysis revealed that an inverse correlation between DNA methylation and gene expression was observed in basal-like breast tumors. 5-aza-2′-deoxycytidine treatment contributed to *TOX3* gene regulation through the demethylation of the CpG region of the *TOX3* gene. However, there is no strong correlation between SNPs and DNA methylation to regulate the *TOX3* gene, suggesting that genetic and epigenetic mechanisms have distinct influences on the expression of *TOX3*.

Xiang et al. investigated the effect of 5-aza-2′-deoxycytidine on *DACT2* genes, which are antagonists of Wnt signaling, and their loss was frequently related to carcinogenesis [146]. The expression of DACT2 was confirmed in breast cancer cell lines and tissues, and their expression was significantly decreased in breast cancers. Moreover, approximately 90% of cell lines and 70% of tissues showed hypermethylation of the promoter region of DACT2 in breast cancers. Upon 5-aza-2′-deoxycytidine treatment, DACT2 expression was recovered by demethylation of the promoter in BT549 and T47D cell lines. This resulted in induced apoptosis, decreased cell proliferation, migration, and EMT by blocking Wnt signaling. These epigenetic changes caused by 5-aza-2′-deoxycytidine could control breast carcinogenesis.

In addition, miRNA expression levels are regulated by 5-aza-2′-deoxycytidine [147]. The expression of miR-31 is significantly increased in MDA-MB-231 cells treated with 5-aza-CdR, and it has an effect on inhibition of invasion through decreased SATB2, which is a target of miR-31.

14.4.1.3 Zebularine

More effective and less toxic DNMT inhibitors and zebularine were tested in two breast cancer cell lines, MCF-7 and MDA-MB-231 [148], and the effect of zebularine on cell growth and apoptosis was confirmed. After zebularine treatment at different concentrations, the inhibition of proliferation was observed in both cell lines through upregulated P21, downregulated cyclin D, and induced S-phase arrest. However, the apoptosis pathway regulated by zebularine treatment differed in a cell line-specific manner. MDA-MB-231 cells showed activation of both the extrinsic and intrinsic pathways through decreased BCL2 and increased BAX and caspase-3, while decreased BCL2 and caspase3 were observed in MCF-7 cells. In addition, DNA demethylation and histone acetylation occurred after low levels of zebularine treatment, and it led to the depression of ER mRNA levels in ER-negative cells, MDA-MB-231.

The effect of zebularine in vivo has also been investigated [149], and mice treated with 5 mg/ml in drinking water showed significantly delayed tumor growth. In general, *DNMT1* and *DNMT3b* expressions were decreased after zebularine treatment, indicating their involvement in epigenetic regulation. A microarray analysis indicated that the expression of *Twist2*, *Becn1*, *Sfrp1*, *Cdkn1a*, and *H9* genes was increased in mice with zebularine treatment, and pyrosequencing results revealed demethylation of the Twist2 promoter region. In addition, the expression of genes, such as *CCND2*, *Sfrp1*, and *Gsn*, which has hypermethylated regions at the promoter, was upregulated in mice treated with zebularine, suggesting that zebularine could be a potential drug for effective breast cancer treatment.

14.4.2 HDAC Inhibitors

HDAC inhibitors affect transcriptional silencing through the removal of the acetyl groups at the lysine residues, which results in condensed chromatin. Based on their chemical structures, HDAC inhibitors are divided into four groups: hydroxamic acids, cyclic tetrapeptides, short-chain fatty acids, and benzamides [150–153]. Even though HDAC inhibitors are nonselective, some inhibitors are targeting the zinc cofactor at HDAC's active site and selectively regulate gene expression. For this reason, they have been applied to cancer therapy field and produced encouraging results in the clinic for breast cancer treatment.

14.4.2.1 Suberoylanilide Hydroxamic Acid (SAHA)

Suberoylanilide hydroxamic acid (SAHA) is clinically approved for the treatment of T-cell lymphoma, and there are continuing attempts to study the anticancer effect of SAHA on breast cancer. The responsiveness to SAHA was identified in various breast cancer cell lines [154]. BT474 cells were resistant to SAHA, while MCF7 cells were sensitive. After SAHA treatment, GSH levels were increased, and the expression of genes related with antioxidant enzymes, selenoprotein P plasma 1 (*SEPP1*), and nicotinamide nucleotide transhydrogenase (*NNT*) was upregulated in BT474 cells. However, all glutathione pathway genes were downregulated in MCF7 cells, indicating that the glutathione metabolic pathway is important for acquired resistance in SAHA treatment.

The effect of SAHA in TNBC cells has also been studied [155]. Interestingly, epithelial-mesenchymal transition (EMT) was induced in MDA-MB-231 and BT-549 cells with SAHA treatment supported by the aberrant expression of E-cadherin (*E-cad*), N-cadherin (*N-cad*), Vimentin (*Vim*), and fibronectin (*FN*). In addition, FOXA1, a critical transcription factor for hormone response, was downregulated, and nuclear transition was impaired in cells in which EMT was induced after SAHA treatment. These data suggest that side effects should be reviewed before SAHA is used as a therapeutic approach for breast cancer.

Feng et al. investigated the effect of SAHA in leptin-induced breast cancer cells, which led to enhanced proliferation and stimulated cells' entry into the S phase [156]. Obviously, inhibition of cell growth and increased apoptosis were observed in MCF-7 and MDA-MB 231 cells after SAHA treatment. In addition, SAHA influenced the transcription activity of $p21^{WAF1/CIP1}$, which is a negative regulator of the cell cycle, in both cell lines. However, SAHA affected CDK4 and Cyclin E expression in MCF-7 cells, whereas CDK2 and Cyclin E were affected in MDA-MB-231 cells. This indicates that distinct regulatory signaling is required for the G1-S transition in each type of cell. Histone acetylation patterns at H3 or H4 were checked upstream of $p21^{WAF1/CIP1}$, and only H3K27ac was affected by SAHA in MCF-7 cells, while MDA-MB-23a cells showed altered acetylation levels at H3 and H4 at the promoter region of $p21^{WAF1/CIP1}$. This suggests that SAHA regulates $p21^{WAF1/CIP1}$ through epigenetic modification, but the mechanism differs in a cell type-specific manner.

14.4.2.2 Trichostatin A

The effect of trichostatin A (TSA), known as an HDAC inhibitor treatment for several cancer cell lines, including breast cancer cells, has been elucidated [157]. TSA treatment induces mesenchymal-like morphological changes in both the human breast cancer cell line MCF-7 and the human gastric cancer cell line BGC-823. In addition, inhibition of cell migration and cancer cell colony formation were observed in TSA-treated MCF-7 and BGC-823 cells coupled with the alteration of β-catenin expression, suggesting that TSA has a dual function: a negative effect in cancer cells and EMT induction.

Chang et al. evaluated distinct sensitivities to HDAC inhibitors, pan-inhibitor TSA, and the Class I selective inhibitor depsipeptide, in various cancer cells [158]. A cell viability assay showed that breast, lung, and melanoma cell lines have different responses to TSA compared to depsipeptide. In addition, the authors showed that while purified recombinant HDAC 1, 2, and 5 were sensitive to TSA treatment, depsipeptide was inhibited in cellular extracts but not in purified HDACs. Despite the similar activities of these HDAC inhibitors, TSA and depsipeptide, these HDAC modulators exert distinct activity and selectivity on cancer cells.

In general, aberrantly expressed microRNAs (miRNAs) are involved in cancer development and drug resistance, so it is important to identify the function of dysregulated miRNAs in cancer. Liu et al. identified that TSA induced ERa expression in breast cancer cell lines, MCF-7 and MDA-MB-231, through the reduction of miR-204 [159]. In addition, decreased miR-204-enhanced sensitivity to tamoxifen (TAM) was observed in breast cancer cells. Furthermore, the combination treatment of TSA and TAM exerted a synergistic effect to inhibit tumor size in vivo compared to TSA treatment alone, indicating that miR-204 plays a key role in regulating drug resistance and may be an effective target for cancer therapy related to TSA.

The effect of TSA on the apoptosis of breast cancer cell lines, MCF-7 and MDA-MB-231, was studied [160]. It was found that TSA induced the apoptosis of MCF-7 and MDA-MB-231 cell lines via cell cycle arrest in the G2/M phase, and apoptosis suppressed by TSA treatment is dependent on mitochondrial reactive oxygen species (ROSs) derived from the reduced activity of the mitochondrial respiratory chain. Overall, these results suggest that TSA plays a role in inhibiting breast cancer cells via the induction of apoptosis.

ER is a critical factor for breast cancer development and endocrine therapy resistance. In addition to ER, hypoxia is a crucial physiological condition in tumorigenesis. Therefore, molecular mechanism studies in hypoxic microenvironments are essential for understanding breast cancer development and progression. Accumulating evidence has shown the effect of TSA on ERa repression in breast cancer cells under hypoxia conditions. Noh et al. have found that TSA affects ESR1 mRNA and ERa protein expression of ER-positive MCF-7 cells under hypoxia conditions and that the ubiquitin proteasome-mediated pathway is involved in ERa degradation induced by TSA treatment [161]. Their study also demonstrated that TSA treatment suppressed the cell proliferation of MCF-7 cells under normal and hypoxia conditions, indicating the effect of TSA on cell proliferation via the regulation of ERa.

14.4.2.3 Suberoyl Bis-Hydroxamic Acid (SBHA)

Suberoyl bis-hydroxamic acid (SBHA), one of the HDAC inhibitors, exerts an anticancer effect in several cancer types, including breast cancer. Zhuang et al. studied the effect of SBHA on the apoptosis of breast cancer cell line MCF-7 [162]. The results showed that SBHA treatment promoted apoptosis via the increased expression of p53, p24, PUMA, and Bax in MCF-7 cells. Interestingly, the authors demonstrated that the knockdown of p53 using siRNA attenuated SBHA-induced apoptosis and p53, p24, PUMA, and Bax, indicating that SBHA has an anticancer effect on breast cancer through the p53 pathway.

There is other evidence showing the effect of SBHA on the cell proliferation and apoptosis of breast cancer cells from a study conducted by Yang et al. [163]. The suppression of cell proliferation and cell cycle arrest in the G0/G1 phase were observed in SBHA-treated MCF-7 cells in an SBHA concentration-dependent manner. In addition to the inhibition of cell proliferation, SBHA treatment increased apoptotic cell death and the expression of Bax and decreased the expression of Bcl-2. Taken together, the above findings demonstrate that SBHA plays an anticancer role in breast cancer cells via the induction of cell cycle arrest and apoptosis.

14.4.2.4 Panobinostat

TNBC, a highly aggressive subtype of breast cancer, is correlated with decreased levels of E-cadherin due to the epigenetic inactivation of the CDH1 gene and the lack of ERa-regulated signaling. It has been reported that panobinostat, known as an antiproliferative agent, increases the membrane expression of E-cadherin without significantly affecting ERa and ERa-related signaling in TNBC cells and inhibits cell migration and invasion ability [164]. In addition, increased promoter activity of CDH1 was observed in panobinostat-treated TNBC cells. In conclusion, these results show the potential therapeutic role of panobinostat in aggressive breast cancer via repressing cell survival and invasiveness. There is other evidence showing the anticancer effect of panobinostat in TNBC. Tate et al. have found that panobinostat treatment inhibits cell proliferation via blockage of the G2/M phase and induces apoptosis [165]. In an in vivo study, panobinostat significantly reduced tumor volume and induced cell morphology change accompanied by the increased expression of CDH1, indicating the potential role of panobinostat in TNBC therapy.

Although aromatase inhibitors (AIs) are known as effective drugs for curing hormone receptor-positive breast cancer patients, acquired AI resistance has emerged as a serious problem in breast cancer therapy. According to a study conducted by Kubo et al., panobinostat treatment inhibits AI-resistant cell proliferation via inducing cell cycle arrest in the G2/M phase and apoptosis and decreases levels of NF-jB1 commonly overexpressed in AI-resistant cells [166]. In addition to these in vitro findings, the authors confirmed the anticancer effect of panobinostat treatment on AI-resistant tumors accompanied by the decreased expression of NF-jB1 in vivo, suggesting the use of panobinostat as a novel therapeutic agent for AI-resistant breast cancer patients.

One study focused on the effect of panobinostat on metastatic ability in TNBC cells [167]. Panobinostat treatment in TNBC cells induced changes in cell morphology and repressed the expression of several genes involved in EMT. Interestingly, panobinostat had a more inhibitory effect on EMT compared to other HDAC inhibitors, such as SAHA and TMP269. Panobinostat also reduced the metastatic ability of MDA-MB-231 cells in vivo, indicating its inhibitory effect on the metastasis of TNBC via regulating EMT.

14.4.2.5 Entinostat

The effect of the Class I selective HDAC inhibitor, entinostat, on TNBC cells was confirmed [168, 169]. Entinostat treatment reduced the population of $CD44^{high}/CD24^{low}$, which were considered as a tumor-initiating cell (TIC) marker, in TNBC cell lines, such as MDA-MB-231, BT-549, and MDA-MB-436, and led to the reduction of mammosphere formation. In addition to the decreased TIC population, the expression of TIC markers, including Bmi-1, Nanog, and Oct-4, was reduced in entinostat-treated TNBC cells. To determine the effect of entinostat on tumor formation and metastasis in vivo, entinostat was used to treat MDA-MB-231-inoculated NSG mice, and the treatment significantly inhibited tumor development and lung metastasis. Overall, these pieces of evidence demonstrate that the HDAC inhibitor entinostat may help inhibit breast cancer formation and metastasis.

14.4.2.6 Valproic Acid (VPA)

The effect of different concentrations of VPA on the cell viability of MCF-7 was examined, and a significant correlation between VPA treatment and MCF-7 viability was confirmed [170]. MCF-7 cell viability was found to be decreased in a VPA dose-dependent manner, suggesting that VPA has an anticancer effect in human breast cancer cell lines. Artacho-Cordón et al. evaluated whether ionizing radiation (IR) exposure would affect matrix metalloproteinase (MMP) activity and breast cancer cell invasion [171]. After IR exposure, compared to controls, the breast cancer cell line MDA-MB-231 significantly increased its invasion ability and mRNA levels of MMP-1, MMP-3, and MMP-13 along with their regulators, which resulted in the induction of the collagenolytic and gelatinolytic activity of MDA-MB-231. Interestingly, in this group, VPA treatment was found to inhibit IR-induced MMP expression and invasion ability. Taken together, the results of this study demonstrate that VPA has an anti-invasive effect on human breast cancer cell lines under IR exposure.

Accumulating evidence has showed the effect of VPA treatment on telomerase activity and Bax/Bcl-2 ratio in a human breast cancer cell line [172]. As telomerase activity is induced in most cancer types, researchers begin to examine whether VPA has an effect on not only cell viability and apoptosis but also telomerase activity. They found that VPA treatment reduced the cell viability and telomerase activity of MCF-7 and increased the Bax/Bcl-2 ratio, which indicates that VPA functions as a negative regulator of cell proliferation by reducing telomerase activity and increasing the Bax/Bcl-2 ratio. Therefore, this study suggests that the reduced activity of telomerase may be used as an indicator to predict VPA's anticancer effect on breast cancer cells.

Mawatari et al. investigated the antiproliferative effect of VPA on human breast cancer cell lines with different subtypes: SKBR3, BT474, MDA-MB-231, and MCF-7. VPA treatment inhibited the cell growth of four breast cancer cell lines, but the HER2-overexpressed, ER-negative breast cancer cell line, SKBR3, exhibited a dramatic antiproliferative effect after VPA treatment [173]. This antiproliferative effect of VPA on SKBR3 via cell cycle arrest and the induction of apoptosis was accompanied by the increased expression of $p21^{WAF}$ and cleaved caspase-3 and heat shock protein (Hsp) 70 acetylation, which indicated the anticancer effect of VPA on human breast cancer cells. The biological function of the chromatin remodeling action of VPA on HMECs and breast cancer cell lines, MCF7 (ERa-positive) and MDA-MB-231 (ERa-negative), was also studied [174]. VPA treatment induced cell differentiation and cell cycle arrest in the G0/G1 phase accompanied by decreased phosphorylated Rb and increased expression of p21. In addition, decreased expression of ERa and pS2, which is known as an invasive breast cancer prognostic marker, was observed in MCF-7. However, HMECs and MDA-MB-231 cells exhibited re-expression of ERa without an increase of pS2 under VPA treatment. These results illustrate that in both ERa-positive and

ERa-negative breast cancer cells, VPA induces cell cycle arrest and the physiological phenotype to enhance sensitivity to endocrine and chemotherapeutic agents.

14.4.2.7 Sodium Butyrate

Chopin et al. have reported that sodium butyrate (NaB) inhibits cell growth and induces differentiation via its activity of deacetylase inhibition [175]. After studying the effect of NaB on the growth of breast cancer cells, they found that the treatment of butyrate significantly reduced cell growth in various breast cancer cell lines cultured in monolayer, collagen gel and soft agar. In addition, butyrate induced G1 cell cycle arrest and apoptosis in MCF-7, MCF-7ras, T47-D, and BT-20 cells and G2/M phase arrest in MDA-MB-231 cells, indicating that butyrate acts as a negative regulator of cell growth in both hormone-dependent and hormone-independent breast cancer cells. Moreover, the study demonstrated that butyrate-induced growth inhibition occurred in a P53-independent manner and Fas/FasL signaling was involved in butyrate-induced apoptosis. In summary, the authors have identified the wide-spectrum anticancer effects of sodium butyrate in breast cancer cell lines.

Louis et al. have also examined whether sodium butyrate has an effect on apoptosis in the caspase-3-deficient breast cancer cell line MCF-7 [176]. It was found that sodium butyrate treatment suppressed the cell viability of MCF-7, but restoration of sodium butyrate in MCF-7 did not change sodium butyrate-induced apoptosis, indicating that sodium butyrate regulates cell proliferation in a dose-dependent fashion independent of caspase-3. This antiproliferative effect of sodium butyrate was caused by the arrest in the G2/M phase accompanied by the increased level of P21. Sodium butyrate treatment in MCF-7 increased pro-apoptotic Bax and reduced anti-apoptotic Bcl-2 expression. In addition to the changes in levels of apoptotic proteins, several antioxidant enzymes were shown to be involved in sodium butyrate-induced apoptosis, suggesting that the pro-apoptotic effect induced by sodium butyrate treatment is related to oxidative stress.

14.4.2.8 SK7041 and FTY720

The antitumor effect of SK-7041, known as a novel HDAC inhibitor in human lung and breast cancer cell lines and normal human bronchial epithelial (NHBE) cells, was compared with that of SAHA treatment [177]. SK-7041 treatment reduced more cell proliferation accompanied by histone hyperacetylation than SAHA treatment. In addition, the authors showed that SK-7041 had an antiproliferative effect via the induction of apoptotic cell death and selectively inhibited more cell proliferation of lung cancer cells than that of NHBE cells, suggesting that SK-7041 may be considered a potential anticancer drug.

Hait et al. have indicated that FTY720 (fingolimod, Gilenya), previously known as the Food and Drug Administration (FDA)-approved prodrug for the treatment of multiple sclerosis, has an anticancer effect in breast cancer cells [178]. The authors found that the nuclear accumulation of FTY720-P produced by sphingosine kinase 2 (SphK2) in breast cancer cells inhibited Class I HDACs and regulated gene expression independently of S1PRs. To validate the anticancer effect of FTY720 treatment, the authors used high-fat diet (HFD)-induced breast tumors in MMTV-PyMT transgenic mice that had increases of advanced lesions of breast tumors and triple-negative spontaneous breast tumors. As a result, FTY720 treatment reduced tumor volume and HDAC activity and induced estrogen receptor alpha (ERa) and PR expression. In addition, FTY720 treatment induced ERa expression and sensitivity to TAM in ERa-negative breast cancer in vitro and in vivo more efficiently than a known HDAC inhibitor. Taken together, these results suggest that FTY720 has an anticancer effect in ERa-negative breast cancer via the reactivation of epigenetic ERa.

14.4.2.9 N-(2-Hydroxyphenyl)-2-Propylpentanamide

N-(2-hydroxyphenyl)-2-propylpentanamide is a VPA aryl derivative. Prestegui-Martel and colleagues designed this drug in silico [179]. It was experimented in three cell lines, HeLa, rhabdomyosarcoma, and breast cancer, in vitro. Its inhibition level of the tumor cell proliferation was more efficient than that of reported drugs, such as VPA, a hepatotoxic drug, in vivo. In particular, N-(2-hydroxyphenyl)-2-propylpentanamide has showed a significant antiproliferative effect in TNBC.

14.4.2.10 Scriptaid

Previous studies have revealed that the expression of ERa is controlled by the DNMT inhibitor and the HDAC inhibitor to regulate its promoter using epigenetic mechanisms in breast cancer. Scriptaid, which is a novel HDAC inhibitor, has showed positive effects on the induction of cell growth and increased ER expression level in ER-negative breast cancer cells. In three TNBCs, MDA-MB-231, MDA-MB-435, and Hs578T, it was confirmed that Scriptaid induced cell growth, increased ER expression, and increased acetylated H3 and H4 proteins. In those cell types, **5-azacytidine** 2′-deoxycytidine (5-aza) contributed to the powerful effect on increased ER expression when used in combined treatment with Scriptaid compared to Scriptaid or **5-azacytidine** alone [180]. In another study, the effects of Scriptaid were shown to significantly induce not only growth inhibition (only anti-proliferation) but also apoptosis in breast cancer cells (MCF-7, SK-BR-3, and MDA-MB-231). Even though TAM was not effective in MDA-MB-231 and SK-BR-3, those cells could obtain the responsibility of TAM by Scriptaid treatment. Apoptosis was obviously increased in treated cells using the combination of Scriptaid and TAM compared to Scriptaid alone [181].

14.4.2.11 YCW1

Bcl-2/adenovirus E1B 19 kDa protein-interacting protein 3 (BNIP3) is a breast cancer-specific upregulated gene. A previous study demonstrated that YCM1, which is one type of HDACi developed for cancer therapy, is a potential target drug for TNBC by controlling BNIP3 regulation. The expression level of BNIP3 was downregulated in 4 T1 treated with YCM1, especially with the combination of YCW1 and IR. The additional effect of the combination treatment altered the growth inhibition of 4 T1 and MDA-MB-231 significantly. It could be the reason why the effect of downregulated BNIP3 is related with autophagy and cytotoxicity increases [182].

14.4.2.12 Santacruzamate A

Santacruzamate A (SCA) is a natural bioactive product that is extracted from Panamanian marine cyanobacterium. As with most HDAC inhibitors, SCA is also constructed with a zinc-binding group (ZBG), a cap terminus, and a linker region. It was found that two analogues were associated with anti-proliferation and immune reaction in MCF7. Their inhibition of cell growth and immune response level was correlated with the degranulation of cytotoxic T-cells (CTLs). However, the effective activity was not observed in MDA-MB-231. A compound with the reversed functional role of prior compounds was also detected as a suppressor of immune response [183].

14.4.2.13 Ferrocenyl

The capability of ferrocenyl to regulate tumor growth using synthesized formation with known effectible drugs, especially in TNBC, was investigated. A series of ferrocenyl catechols were synthesized, and its antiproliferative activity was inspected in MDA-MB-231. Novel compounds had a similar or greater ability to induce anti-proliferation by comparison with their phenolic analogues through chemical oxidization [184]. To develop an effective drug for regulating ER activity, TAM, ferrocenyl, and selective ER modulators (SERMs) were synthesized as novel compounds. In MCF7 and MDA-MB-231, cell proliferation was evaluated depending on their ERα- and ERβ-binding affinities, respectively. A cell proliferation assay was used to prove the antiproliferative effects of the compounds. As a result, a significantly induced antiproliferative ability was demonstrated in both cells through ligand activity caused by the cytotoxic process [185]. In TNBC, modified ferrocenyl, ferrocenyl tamoxifen derivative (FcOHTAM), which has a strong antiproliferative effect in MDA-MB-231, was investigated. Laine and colleagues developed stealth FcOHTAM loaded lipid nanocapsules (LNCs). To confirm the effects of these FcOHTAM-LNCs in vitro, TNBC cells were injected into mice for a xenograft mouse model. The result showed that tumor growth was significantly decreased in FcOHTAM-treated mice. In the treated group, the tumor volume was evaluated as 36%, which was less than that of the untreated group after 38 days [186].

Li et al. have found that SERMs are efficient for antitumor growth in ER-positive breast cancer but not ER-negative breast cancer. They also designed a novel hybrid ferrocenyl complex (FcOBHS-HDACi) by synthesizing compounds of dual-acting ER and HDAC inhibitors and incorporating the ferrocenyl unit for the antiproliferative influence in both ER-positive and ER-negative breast cancer cells. Cell proliferation was more significantly inhibited by the sulfonate unit of FcOBHS-HDACi in MDA-MB-231 than that in MCF7 and DU145 (metastatic prostate cancer cell line) [187]. Atmaca et al. investigated the effects of antitumor activity of two different synthesized ferrocenyl pyrazole (FP) derivatives, 5-ferrocenyl-1-phenyl-1Hpyrazole (FP-Ph) and 5-ferrocenyl-1H-pyrazole (FP-H), in MCF7 and MDA-MB-231. FP-Ph and FP-H treatment inhibited cell viability in both cell lines and increased apoptosis and necrotic cells in MCF7 and MDA-MB-231, respectively [76].

14.5 Synergistic Effect of Combination Treatment

14.5.1 Combination with DNMT Inhibitor

5-azacytidine and ING1 Satbir et al. studied the synergistic effect of a combination therapy, an HDAC inhibitor and a DNMT inhibitor, in breast cancer cells [188], by comparing the effect of two HDAC inhibitors (LBH589 and ING1) and 5-azacytidine. They showed that LBS589 and 5-azacytidine did not have a significant effect. However, ING1 and 5-azacytidine showed a better effect in terms of increasing apoptosis and decreasing DNA damage as well as inhibiting cell growth. This suggests that the combination effect of epigenetic regulators should be considered for therapeutic targets.

5-aza-2′-deoxycytidine and DZNep Yu et al. demonstrated the effect of an EZH2 inhibitor and DNA methylation inhibitor on RASSF2A in breast cancer cells [78]. Upon treatment with the EZH2 inhibitor, DZNep, RASSF2A expression was decreased, and EZH2 expression was downregulated. However, in cells treated with 5-aza-2′-deoxycytidine, *RASSF2A* was induced approximately by 20-fold, while EZH2 levels were not significantly affected. Hypomethylation of CpG at the RASSF2A locus was also detected after inhibitor treatment. They showed that DZNep and 5-aza-2′-deoxydytidine reduced cell proliferation, migration, and invasion in breast cancer cells. This suggests that these epigenetic regulators are involved in the robust induction of RASSF2A through a synergistic effect.

14.5.2 Combination with HDAC Inhibitors

SAHA and TRAIL The combination effect of SAHA and tumor necrosis factor-related apoptosis-inducing ligand (TRAIL) was evaluated in two different types of breast cancer cells, MCF-7 (ER-positive) and MDA-MB-231 (ER-negative) [189]. The researchers identified that ER-negative breast cancer cells were sensitive to TRAIL, whereas ER-positive cells were resistant. They then determined whether the sensitivity of TRAIL was recovered in resistant cells with SAHA treatment. Cells treated with SAHA or TRAIL alone showed only a minimal effect on the inhibition of cell proliferation and cell viability, but the combination of SAHA and TRAIL significantly decreased cell numbers in MCF-7 and MDA-MB-231 cell lines. This combination treatment can regulate the cell cycle through the inhibition of entry into the S phase. Cells in the G0/G1 phase were observed after treatment with SAHA and TRAIL. In addition, this combination treatment induced apoptosis through increased BAX protein in both cell lines. MDA-MB-231 cells treated with SAHA and TRAIL additionally induced caspase-3 and CDKN1A. Even though SAHA or TRAIL alone has a distinct effect on cell proliferation and apoptosis, the combinatorial treatment of SAHA and TRAIL has a synergistic effect, suggesting this is a suitable therapy for breast cancer.

SAHA and Olaparib The effect of a poly (ADP-ribose) polymerase (PARP) inhibitor (olaparib) has been studied in various cancers targeting the therapeutic effect on DNA repair-defective tumors. Min et al. determined whether SAHA could enhance the antitumor effects of olaparib in breast cancer cell lines using a cytotoxic assay and cell cycle analysis [190]. A combination treatment of SAHA and olaparib contributed to the inhibition of cell proliferation in TNBC cell lines. The PARP inhibitor was sensitive to the absence of PTEN, but a significant relationship between PARP inhibitor and PTEN deficiency was not confirmed in two TNBC cell lines, HCC70 and MDA-MB-468. However, after the combination treatment, PTEN was reactivated, and AKT and ERK signaling was decreased, suggesting that the proliferation pathway was modulated by this combination treatment. This suggests that SAHA and olaparib offer a therapeutic target for TNBC through recovered PTEN.

TSA and BEZ235 Chen et al. determined the anticancer effect of the combination treatment of TSA and PI3K/mTOR dual inhibitor BEZ235 on breast cancer cells [191]. They showed that this combination treatment exerted a synergistic effect on the growth inhibition of breast cancer cell lines via the induction of apoptosis in a caspase-dependent manner. Moreover, they observed that the combination treatment induced autophagic cell death through the increased expression of Beclin-1 and LC3B-II. An in vivo breast cancer xenograft model indicated that the combined treatment of TSA and BEZ235 completely inhibited the growth of tumors, which was unlike those treated with TSA or BEZ235 alone. Taken together, these data demonstrate that the combination treatment of TSA and BEZ235 has a synergistic effect on breast cancer inhibition, and it is suggested as a new selective strategy for breast cancer therapy.

SBHA and Bortezomib Yang et al. examined the effect of SBHA-only treatment and the combination treatment of SBHA and other agents to enhance the anticancer effect of SBHA on breast cancer cells [192]. They discovered the synergistic effect of the combination of SBHA and proteasome inhibitors, such as bortezomib and MG-132, in the breast cancer cell lines MCF-7 and MDA-MB-231. The co-treatment of this agent's potentiated suppression of cell proliferation and colony formation was accompanied by the increased expression of p53, Bax, Bcl-xS, and Bak and the decreased expression of Bcl-2 in breast cancer cells compared to single agent-treated cells, indicating the synergic antitumor effect of the combination treatment of SBHA and proteasome inhibitors.

Panobinostat and Mevastatin Lin et al. examined the synergic effect of the combination treatment of a 3-hydroxy-3-methylglutaryl-CoA reductase (HMGCR) inhibitor, mevastatin, and HDAC inhibitor, panobinostat (LBH-589), on **TNBC** cells [193]. Inhibited autophagic flux by preventing Vps34/Beclin1 complex formation was observed in mevastatin and panobinostat co-treated TNBC cells. In addition, they showed that this combination treatment inhibited mTOR via the activation of LKB1/AMPK signaling and resulted in cell cycle arrest in the G2/M phase. Furthermore, MDA-MB-231 xenograft mice treated by panobinostat plus mevastatin exhibited increased cell death, resulting in decreased tumor size in vivo. These findings strongly support a future therapeutic strategy using a combination treatment with an HDAC inhibitor to enhance the anticancer effect in breast cancer.

Panobinostat and Salinomycin Another study showing the effect of a combination treatment with panobinostat was conducted by the research group of Kai et al. [194]. The authors evaluated the efficacy of a combination of panobinostat and salinomycin, known as a breast cancer stem cells (BCSC)-targeting agent, on TNBC. In this study, TNBC cell proliferation and mammosphere formation were inhibited by the combination treatment in TNBC cells. In addition, this combination treatment induced cell cycle arrest, apoptosis, and the alteration of EMT in TNBC BCSCs, indicating the synergistic anticancer effect of panobinostat and salinomycin co-treatment.

Panobinostat and Letrozole Tan et al. gathered the clinical data of a phase I study showing the effect of the combination of panobinostat and letrozole on metastatic breast cancer patients [195]. They demonstrated that this combination therapy was clinically effective in endocrine- and chemotherapy-resistant metastatic breast cancer patients. Furthermore, they found a tolerable and safe dose of this combination treatment for patients and recommended panobinostat 20 mg orally three times weekly and oral letrozole 2.5 mg daily as a phase II starting dose.

Entinostat and Rapamycin Single-drug treatments have exhibited a limited ability to reduce tumor size, because various types of oncogenic signaling occur in many cancer types. To enhance the antitumor activity of drugs or agents, synergistic combination treatments have been developed. Ou et al. used the loss-of-function RNAi screening method of rapamycin, known as an mTOR inhibitor, to maximize the clinical potential of rapamycin and identified six candidate genes, AURKB, PLK1, PIK3R1, MAPK12, PRKD2, and PTK6, as sensitizers of rapamycin [196]. The authors used the HDAC inhibitor entinostat to reduce the expression of these rapamycin-sensitizing genes and examined the effect of the combination treatment of rapamycin and entinostat on in vivo tumor growth in an MDA-MB-231 xenograft model. This combination treatment dramatically inhibited the growth of MDA-MB-231 xenografts compared to the treatment of rapamycin or entinostat alone, thereby indicating that the combination treatment of rapamycin and the HDAC inhibitor enhances the anticancer effect of rapamycin compared to the single treatments in breast cancer.

Entinostat and All-Trans Retinoic Acid (ATRA) In addition to the combination treatment research reported by Ou et al., an additional study showed the anticancer effect of entinostat in combination with all-trans retinoic acid (ATRA) in breast cancer [196]. Schech et al. found that the combination of entinostat and ATRA inhibited the expression of HER2 involved in resistance to AIs and proliferation in AI-resistant cells [169]. In addition, this combination treatment inhibited the mammosphere formation of letrozole-resistant cells, LTLT-Ca, and reduced the expression of TIC markers. Letrozole-resistant MCF-7Ca cell-inoculated xenograft tumor size and TIC characteristics were more significantly inhibited by the treatment with entinostat, ATRA, and letrozole compared with those treated with a single agent or the combination of entinostat and ATRA, suggesting that the combination treatment of entinostat and ATRA synergized the anticancer effect through TIC inhibition in breast cancer.

VPA and Capecitabine Terranova-Barberio et al. demonstrated the effect of the combination treatment of an HDAC inhibitor and anticancer agent in breast cancer [197]. Among the various HDAC inhibitors and anticancer agents, they evaluated the synergistic antitumor effect of VPA and capecitabine, known as the oral prodrug 5-fluorourcacil (5-FU). They found that HDAC inhibitors including VPA increased the expression of thymidine phosphorylase (TP), which functions in the conversion of 5′-deoxy-5-fluorouridine (5′-DFUR) into active 5-FU in breast cancer cells but not in the non-tumorigenic cell line MCF10A. Furthermore, they confirmed that the combination treatment with VPA and capecitabine showed a powerful anticancer effect in breast cancer cells and that TP was a critical factor for the synergistic antitumor effect of the HDAC inhibitor and anticancer agent.

Sodium Butyrate and Etoposide Li et al. evaluated the effect of sodium butyrate alone and in combination with etoposide, known as a DNA-damaging agent, on breast cancer cells MCF-7 and normal human embryonic kidney 293 (HEK293) cells [198]. Sodium butyrate treatment suppressed the proliferation of both cell lines, but it had a more dramatic antiproliferative effect in MCF-7 cells than in HEK293 cells.

Sodium butyrate more effectively inhibited cell proliferation in combination with etoposide in both cell lines, but this synergic effect was more obvious in MCF-7. In addition, sodium butyrate induced the formation of γ-H2AX foci under the treatment of etoposide to a greater lesion in MCF-7 compared to those in HEK293 cells and differential patterns of nuclear expression of double strand break (DSB) repair-related protein. Taken together, these studies suggest that sodium butyrate increases sensitivity to the etoposide-induced cytotoxic effect and reduces DSB repair capacity in cancer cells but not in non-cancer cells.

Sodium Butyrate and MET siRNA Sun et al. identified a specific breast cancer cell population that showed resistance to sodium butyrate treatment [199]. These cells had cancer stem cell characteristics, such as self-renewal and high tumor initiation ability, and expressed CD133, known as a cancer stem cell marker. In addition, endogenous c-MET was found to be critical to the survival of breast cancer cells after sodium butyrate treatment and highly expressed in CD133-positive cells. Based on these findings, the effect of the combination treatment of MET siRNA and sodium butyrate was examined. As a result, this combination treatment efficiently inhibited the breast cancer incidence rate and progression. Taken together, the results of this study suggest the novel therapeutic strategy of a combination treatment with sodium butyrate and the regulation of cancer stemness in breast cancer.

14.6 Concluding Remarks and Future Perspectives

Breast cancer is a heterogeneous disease resulting from the ablation of various genetic and epigenetic factors. Epigenetic alteration has been investigated for therapeutic targets of prognostic and predictive factors in breast cancer [20, 22–24, 26, 34, 200]. Targeting epigenetic changes such as DNA methylation and histone modification focuses on current standard-of-care therapies for breast cancer. NGS techniques allow us to understand epigenetic deregulation in whole genome, and the effect of epigenetic drugs could be investigated easily and accurately. Here, we have summarized the landscapes of DNA methylation and histone modification and described the therapeutic effects of epigenetic drugs in breast cancer, especially DNMTs and HDACi.

Development of epigenetic therapies is expected to be of great help in the treatment of breast cancer. However, there are issues to be addressed in the future. Since, in part, aberrant epigenetic modification is not strongly correlated with gene expression, identification of influential epigenetic target as a biomarker for diagnosis and prognostication should be required for epi-drug therapies. In addition, in vivo studies are essential for checking the effect of epi-drugs with microenvironmental issue. Finally, epigenetic targets should be considered for precision care.

References

1. Dai X, Xiang L, Li T, Bai Z (2016) Cancer hallmarks, biomarkers and breast cancer molecular subtypes. J Cancer 7(10):1281–1294. doi:10.7150/jca.13141
2. Schnitt SJ (2010) Classification and prognosis of invasive breast cancer: from morphology to molecular taxonomy. Mod Pathol 23(Suppl 2):S60–S64. doi:10.1038/modpathol.2010.33
3. Weitzel JN, Lagos VI, Cullinane CA, Gambol PJ, Culver JO, Blazer KR, Palomares MR, Lowstuter KJ, MacDonald DJ (2007) Limited family structure and BRCA gene mutation status in single cases of breast cancer. JAMA 297(23):2587–2595. doi:10.1001/jama.297.23.2587
4. Ferrone CR, Levine DA, Tang LH, Allen PJ, Jarnagin W, Brennan MF, Offit K, Robson ME (2009) BRCA germline mutations in Jewish patients with pancreatic adenocarcinoma. J Clin Oncol 27(3):433–438. doi:10.1200/JCO.2008.18.5546
5. Chen S, Parmigiani G (2007) Meta-analysis of BRCA1 and BRCA2 penetrance. J Clin Oncol 25(11):1329–1333. doi:10.1200/JCO.2006.09.1066
6. Tai YC, Domchek S, Parmigiani G, Chen S (2007) Breast cancer risk among male BRCA1 and BRCA2 mutation carriers. J Natl Cancer Inst 99(23):1811–1814. doi:10.1093/jnci/djm203
7. Antoniou A, Pharoah PD, Narod S, Risch HA, Eyfjord JE, Hopper JL, Loman N, Olsson H, Johannsson O, Borg A, Pasini B, Radice P, Manoukian S, Eccles DM, Tang N, Olah E, Anton-Culver H, Warner E,

Lubinski J, Gronwald J, Gorski B, Tulinius H, Thorlacius S, Eerola H, Nevanlinna H, Syrjakoski K, Kallioniemi OP, Thompson D, Evans C, Peto J, Lalloo F, Evans DG, Easton DF (2003) Average risks of breast and ovarian cancer associated with BRCA1 or BRCA2 mutations detected in case series unselected for family history: a combined analysis of 22 studies. Am J Hum Genet 72(5):1117–1130. doi:10.1086/375033
8. Archey WB, Arrick BA (2017) Transactivation of the estrogen receptor promoter by BRCA1. Cancer Cell Int 17:33. doi:10.1186/s12935-017-0401-2
9. Li Q, Wei W, Jiang YI, Yang H, Liu J (2015) Promoter methylation and expression changes of BRCA1 in cancerous tissues of patients with sporadic breast cancer. Oncol Lett 9(4):1807–1813. doi:10.3892/ol.2015.2908
10. Al-Moghrabi N, Nofel A, Al-Yousef N, Madkhali S, Bin Amer SM, Alaiya A, Shinwari Z, Al-Tweigeri T, Karakas B, Tulbah A, Aboussekhra A (2014) The molecular significance of methylated BRCA1 promoter in white blood cells of cancer-free females. BMC Cancer 14:830. doi:10.1186/1471-2407-14-830
11. Rice JC, Ozcelik H, Maxeiner P, Andrulis I, Futscher BW (2000) Methylation of the BRCA1 promoter is associated with decreased BRCA1 mRNA levels in clinical breast cancer specimens. Carcinogenesis 21(9):1761–1765
12. Hsu NC, Huang YF, Yokoyama KK, Chu PY, Chen FM, Hou MF (2013) Methylation of BRCA1 promoter region is associated with unfavorable prognosis in women with early-stage breast cancer. PLoS One 8(2):e56256. doi:10.1371/journal.pone.0056256
13. Jing F, Zhang J, Tao J, Zhou Y, Jun L, Tang X, Wang Y, Hai H (2007) Hypermethylation of tumor suppressor genes BRCA1, p16 and 14-3-3sigma in serum of sporadic breast cancer patients. Onkologie 30(1–2):14–19. doi:10.1159/000096892
14. Birgisdottir V, Stefansson OA, Bodvarsdottir SK, Hilmarsdottir H, Jonasson JG, Eyfjord JE (2006) Epigenetic silencing and deletion of the BRCA1 gene in sporadic breast cancer. Breast Cancer Res 8(4):R38. doi:10.1186/bcr1522
15. Otani Y, Miyake T, Kagara N, Shimoda M, Naoi Y, Maruyama N, Shimomura A, Shimazu K, Kim SJ, Noguchi S (2014) BRCA1 promoter methylation of normal breast epithelial cells as a possible precursor for BRCA1-methylated breast cancer. Cancer Sci 105(10):1369–1376. doi:10.1111/cas.12506
16. Snell C, Krypuy M, Wong EM, kConFab investigators, Loughrey MB, Dobrovic A (2008) BRCA1 promoter methylation in peripheral blood DNA of mutation negative familial breast cancer patients with a BRCA1 tumour phenotype. Breast Cancer Res 10(1):R12. doi:10.1186/bcr1858
17. Niwa Y, Oyama T, Nakajima T (2000) BRCA1 expression status in relation to DNA methylation of the BRCA1 promoter region in sporadic breast cancers. Jpn J Cancer Res 91(5):519–526
18. Pirouzpanah S, Taleban FA, Mehdipour P, Atri M (2015) Association of folate and other one-carbon related nutrients with hypermethylation status and expression of RARB, BRCA1, and RASSF1A genes in breast cancer patients. J Mol Med (Berl) 93(8):917–934. doi:10.1007/s00109-015-1268-0
19. Zhang L, Long X (2015) Association of BRCA1 promoter methylation with sporadic breast cancers: evidence from 40 studies. Sci Rep 5:17869. doi:10.1038/srep17869
20. Fucito A, Lucchetti C, Giordano A, Romano G (2008) Genetic and epigenetic alterations in breast cancer: what are the perspectives for clinical practice? Int J Biochem Cell Biol 40(4):565–575. doi:10.1016/j.biocel.2007.10.018
21. Lo PK, Sukumar S (2008) Epigenomics and breast cancer. Pharmacogenomics 9(12):1879–1902. doi:10.2217/14622416.9.12.1879
22. Dworkin AM, Huang TH, Toland AE (2009) Epigenetic alterations in the breast: implications for breast cancer detection, prognosis and treatment. Semin Cancer Biol 19(3):165–171. doi:10.1016/j.semcancer.2009.02.007
23. Jovanovic J, Ronneberg JA, Tost J, Kristensen V (2010) The epigenetics of breast cancer. Mol Oncol 4(3):242–254. doi:10.1016/j.molonc.2010.04.002
24. Byler S, Goldgar S, Heerboth S, Leary M, Housman G, Moulton K, Sarkar S (2014) Genetic and epigenetic aspects of breast cancer progression and therapy. Anticancer Res 34(3):1071–1077
25. Karsli-Ceppioglu S, Dagdemir A, Judes G, Ngollo M, Penault-Llorca F, Pajon A, Bignon YJ, Bernard-Gallon D (2014) Epigenetic mechanisms of breast cancer: an update of the current knowledge. Epigenomics 6(6):651–664. doi:10.2217/epi.14.59
26. Basse C, Arock M (2015) The increasing roles of epigenetics in breast cancer: implications for pathogenicity, biomarkers, prevention and treatment. Int J Cancer 137(12):2785–2794. doi:10.1002/ijc.29347
27. Shen H, Xu W, Guo R, Rong B, Gu L, Wang Z, He C, Zheng L, Hu X, Hu Z, Shao ZM, Yang P, Wu F, Shi YG, Shi Y, Lan F (2016) Suppression of enhancer Overactivation by a RACK7-histone demethylase complex. Cell 165(2):331–342. doi:10.1016/j.cell.2016.02.064
28. Leroy G, Dimaggio PA, Chan EY, Zee BM, Blanco MA, Bryant B, Flaniken IZ, Liu S, Kang Y, Trojer P, Garcia BA (2013) A quantitative atlas of histone modification signatures from human cancer cells. Epigenetics Chromatin 6(1):20. doi:10.1186/1756-8935-6-20
29. McCullough LE, Chen J, Cho YH, Khankari NK, Bradshaw PT, White AJ, Teitelbaum SL, Terry MB, Neugut AI, Hibshoosh H, Santella RM, Gammon MD (2017) Modification of the association between recreational physical activity and survival after breast cancer by promoter methylation in breast

cancer-related genes. Breast Cancer Res 19(1):19. doi:10.1186/s13058-017-0811-z
30. Messier TL, Gordon JA, Boyd JR, Tye CE, Browne G, Stein JL, Lian JB, Stein GS (2016) Histone H3 lysine 4 acetylation and methylation dynamics define breast cancer subtypes. Oncotarget 7(5):5094–5109. doi:10.18632/oncotarget.6922
31. Monteiro FL, Vitorino R, Wang J, Cardoso H, Laranjeira H, Simoes J, Caldas M, Henrique R, Amado F, Williams C, Jeronimo C, Helguero LA (2017) The histone H2A isoform Hist2h2ac is a novel regulator of proliferation and epithelial-mesenchymal transition in mammary epithelial and in breast cancer cells. Cancer Lett 396:42–52. doi:10.1016/j.canlet.2017.03.007
32. Damaskos C, Valsami S, Kontos M, Spartalis E, Kalampokas T, Kalampokas E, Athanasiou A, Moris D, Daskalopoulou A, Davakis S, Tsourouflis G, Kontzoglou K, Perrea D, Nikiteas N, Dimitroulis D (2017) Histone deacetylase inhibitors: an attractive therapeutic strategy against breast cancer. Anticancer Res 37(1):35–46. doi:10.21873/anticanres.11286
33. Zhao QY, Lei PJ, Zhang X, Zheng JY, Wang HY, Zhao J, Li YM, Ye M, Li L, Wei G, Wu M (2016) Global histone modification profiling reveals the epigenomic dynamics during malignant transformation in a four-stage breast cancer model. Clin Epigenetics 8:34. doi:10.1186/s13148-016-0201-x
34. Huang Y, Nayak S, Jankowitz R, Davidson NE, Oesterreich S (2011) Epigenetics in breast cancer: what's new? Breast Cancer Res 13(6):225. doi:10.1186/bcr2925
35. Atalay C (2013) Epigenetics in breast cancer. Exp Oncol 35(4):246–249
36. Connolly R, Stearns V (2012) Epigenetics as a therapeutic target in breast cancer. J Mammary Gland Biol Neoplasia 17(3–4):191–204. doi:10.1007/s10911-012-9263-3
37. Vo AT, Millis RM (2012) Epigenetics and breast cancers. Obstet Gynecol Int 2012:602720. doi:10.1155/2012/602720
38. Lustberg MB, Ramaswamy B (2011) Epigenetic therapy in breast cancer. Curr Breast Cancer Rep 3(1):34–43. doi:10.1007/s12609-010-0034-0
39. Lustberg MB, Ramaswamy B (2009) Epigenetic targeting in breast cancer: therapeutic impact and future direction. Drug News Perspect 22(7):369–381. doi:10.1358/dnp.2009.22.7.1405072
40. Ai L, Kim WJ, Kim TY, Fields CR, Massoll NA, Robertson KD, Brown KD (2006) Epigenetic silencing of the tumor suppressor cystatin M occurs during breast cancer progression. Cancer Res 66(16):7899–7909. doi:10.1158/0008-5472.CAN-06-0576
41. Boyanapalli SS, Li W, Fuentes F, Guo Y, Ramirez CN, Gonzalez XP, Pung D, Kong AN (2016) Epigenetic reactivation of RASSF1A by phenethyl isothiocyanate (PEITC) and promotion of apoptosis in LNCaP cells. Pharmacol Res 114:175–184. doi:10.1016/j.phrs.2016.10.021
42. Sinha S, Shukla S, Khan S, Tollefsbol TO, Meeran SM (2015) Epigenetic reactivation of p21CIP1/WAF1 and KLOTHO by a combination of bioactive dietary supplements is partially ERalpha-dependent in ERalpha-negative human breast cancer cells. Mol Cell Endocrinol 406:102–114. doi:10.1016/j.mce.2015.02.020
43. Klarmann GJ, Decker A, Farrar WL (2008) Epigenetic gene silencing in the Wnt pathway in breast cancer. Epigenetics 3(2):59–63
44. Perri F, Longo F, Giuliano M, Sabbatino F, Favia G, Ionna F, Addeo R, Della Vittoria Scarpati G, Di Lorenzo G, Pisconti S (2017) Epigenetic control of gene expression: potential implications for cancer treatment. Crit Rev Oncol Hematol 111:166–172. doi:10.1016/j.critrevonc.2017.01.020
45. Connolly RM, Rudek MA, Piekarz R (2017) Entinostat: a promising treatment option for patients with advanced breast cancer. Future Oncol. doi:10.2217/fon-2016-0526
46. Deb M, Sengupta D, Kar S, Rath SK, Parbin S, Shilpi A, Roy S, Das G, Patra SK (2014) Elucidation of caveolin 1 both as a tumor suppressor and metastasis promoter in light of epigenetic modulators. Tumour Biol 35(12):12031–12047. doi:10.1007/s13277-014-2502-z
47. Ambatipudi S, Horvath S, Perrier F, Cuenin C, Hernandez-Vargas H, Le Calvez-Kelm F, Durand G, Byrnes G, Ferrari P, Bouaoun L, Sklias A, Chajes V, Overvad K, Severi G, Baglietto L, Clavel-Chapelon F, Kaaks R, Barrdahl M, Boeing H, Trichopoulou A, Lagiou P, Naska A, Masala G, Agnoli C, Polidoro S, Tumino R, Panico S, Dolle M, Peeters PH, Onland-Moret NC, Sandanger TM, Nost TH, Weiderpass E, Quiros JR, Agudo A, Rodriguez-Barranco M, Huerta Castano JM, Barricarte A, Fernandez AM, Travis RC, Vineis P, Muller DC, Riboli E, Gunter M, Romieu I, Herceg Z (2017) DNA methylome analysis identifies accelerated epigenetic ageing associated with postmenopausal breast cancer susceptibility. Eur J Cancer 75:299–307. doi:10.1016/j.ejca.2017.01.014
48. Longacre M, Snyder NA, Housman G, Leary M, Lapinska K, Heerboth S, Willbanks A, Sarkar S (2016) A comparative analysis of genetic and epigenetic events of breast and ovarian cancer related to tumorigenesis. Int J Mol Sci 17(5). doi:10.3390/ijms17050759
49. Schubeler D (2015) Function and information content of DNA methylation. Nature 517(7534):321–326. doi:10.1038/nature14192
50. Jin B, Li Y, Robertson KD (2011) DNA methylation: superior or subordinate in the epigenetic hierarchy? Genes Cancer 2(6):607–617. doi:10.1177/1947601910393957
51. Cheng X, Blumenthal RM (2008) Mammalian DNA methyltransferases: a structural perspective. Structure 16(3):341–350. doi:10.1016/j.str.2008.01.004

52. Okano M, Xie S, Li E (1998) Cloning and characterization of a family of novel mammalian DNA (cytosine-5) methyltransferases. Nat Genet 19(3):219–220. doi:10.1038/890
53. Bestor TH (2000) The DNA methyltransferases of mammals. Hum Mol Genet 9(16):2395–2402
54. Szyf M (2012) DNA methylation signatures for breast cancer classification and prognosis. Genome Med 4(3):26. doi:10.1186/gm325
55. Martens JW, Margossian AL, Schmitt M, Foekens J, Harbeck N (2009) DNA methylation as a biomarker in breast cancer. Future Oncol 5(8):1245–1256. doi:10.2217/fon.09.89
56. Brooks J, Cairns P, Zeleniuch-Jacquotte A (2009) Promoter methylation and the detection of breast cancer. Cancer Causes Control 20(9):1539–1550. doi:10.1007/s10552-009-9415-y
57. Sterner DE, Berger SL (2000) Acetylation of histones and transcription-related factors. Microbiol Mol Biol Rev 64(2):435–459
58. Bannister AJ, Kouzarides T (2011) Regulation of chromatin by histone modifications. Cell Res 21(3):381–395. doi:10.1038/cr.2011.22
59. Chervona Y, Arita A, Costa M (2012) Carcinogenic metals and the epigenome: understanding the effect of nickel, arsenic, and chromium. Metallomics 4(7):619–627. doi:10.1039/c2mt20033c
60. Lawrence M, Daujat S, Schneider R (2016) Lateral thinking: how histone modifications regulate gene expression. Trends Genet 32(1):42–56. doi:10.1016/j.tig.2015.10.007
61. Dokmanovic M, Clarke C, Marks PA (2007) Histone deacetylase inhibitors: overview and perspectives. Mol Cancer Res 5(10):981–989. doi:10.1158/1541-7786.MCR-07-0324
62. Marks PA, Xu WS (2009) Histone deacetylase inhibitors: potential in cancer therapy. J Cell Biochem 107(4):600–608. doi:10.1002/jcb.22185
63. Yang XJ, Seto E (2008) The Rpd3/Hda1 family of lysine deacetylases: from bacteria and yeast to mice and men. Nat Rev Mol Cell Biol 9(3):206–218. doi:10.1038/nrm2346
64. Longworth MS, Laimins LA (2006) Histone deacetylase 3 localizes to the plasma membrane and is a substrate of Src. Oncogene 25(32):4495–4500. doi:10.1038/sj.onc.1209473
65. Valenzuela-Fernandez A, Cabrero JR, Serrador JM, Sanchez-Madrid F (2008) HDAC6: a key regulator of cytoskeleton, cell migration and cell-cell interactions. Trends Cell Biol 18(6):291–297. doi:10.1016/j.tcb.2008.04.003
66. Truong PK, Lao TD, Doan TP, Le TA (2015) Loss of expression of cyclin d2 by aberrant DNA methylation: a potential biomarker in vietnamese breast cancer patients. Asian Pac J Cancer Prev 16(6):2209–2213
67. Lewis CM, Cler LR, Bu DW, Zochbauer-Muller S, Milchgrub S, Naftalis EZ, Leitch AM, Minna JD, Euhus DM (2005) Promoter hypermethylation in benign breast epithelium in relation to predicted breast cancer risk. Clin Cancer Res 11(1):166–172
68. Ramezani F, Salami S, Omrani MD, Maleki D (2012) CpG island methylation profile of estrogen receptor alpha in Iranian females with triple negative or non-triple negative breast cancer: new marker of poor prognosis. Asian Pac J Cancer Prev 13(2):451–457
69. Maekawa R, Sato S, Okada M, Lee L, Tamura I, Jozaki K, Kajimura T, Asada H, Yamagata Y, Tamura H, Yamamoto S, Sugino N (2016) Tissue-specific expression of estrogen receptor 1 is regulated by DNA methylation in a T-DMR. Mol Endocrinol 30(3):335–347. doi:10.1210/me.2015-1058
70. Gao L, Qi X, Hu K, Zhu R, Xu W, Sun S, Zhang L, Yang X, Hua B, Liu G (2016) Estrogen receptor beta promoter methylation: a potential indicator of malignant changes in breast cancer. Arch Med Sci 12(1):129–136. doi:10.5114/aoms.2016.57588
71. Zhang W, Chang Z, Shi KE, Song L, Cui LI, Ma Z, Li X, Ma W, Wang L (2016) The correlation between DNMT1 and ERalpha expression and the methylation status of ERalpha, and its clinical significance in breast cancer. Oncol Lett 11(3):1995–2000. doi:10.3892/ol.2016.4193
72. Dewi DL, Mohapatra SR, Blanco Cabanes S, Adam I, Somarribas Patterson LF, Berdel B, Kahloon M, Thurmann L, Loth S, Heilmann K, Weichenhan D, Mucke O, Heiland I, Wimberger P, Kuhlmann JD, Kellner KH, Schott S, Plass C, Platten M, Gerhauser C, Trump S, Opitz CA (2017) Suppression of indoleamine-2,3-dioxygenase 1 expression by promoter hypermethylation in ER-positive breast cancer. Oncoimmunology 6(2):e1274477. doi:10.1080/2162402X.2016.1274477
73. Mao X, Qiao Z, Fan C, Guo A, Yu X, Jin F (2016) Expression pattern and methylation of estrogen receptor alpha in breast intraductal proliferative lesions. Oncol Rep 36(4):1868–1874. doi:10.3892/or.2016.4988
74. Piva R, Rimondi AP, Hanau S, Maestri I, Alvisi A, Kumar VL, del Senno L (1990) Different methylation of oestrogen receptor DNA in human breast carcinomas with and without oestrogen receptor. Br J Cancer 61(2):270–275
75. Hori M, Iwasaki M, Yoshimi F, Asato Y, Itabashi M (1999) Hypermethylation of the estrogen receptor alpha gene is not related to lack of receptor protein in human breast cancer. Breast Cancer 6(2):79–86
76. Medina-Jaime AD, Reyes-Vargas F, Martinez-Gaytan V, Zambrano-Galvan G, Portillo-Delcampo E, Burciaga-Nava JA, Reyes-Romero M, Sifuentes-Alvarez A (2014) ESR1 and PGR gene promoter methylation and correlations with estrogen and progesterone receptors in ductal and lobular breast cancer. Asian Pac J Cancer Prev 15(7):3041–3044
77. Gaudet MM, Campan M, Figueroa JD, Yang XR, Lissowska J, Peplonska B, Brinton LA, Rimm DL, Laird PW, Garcia-Closas M, Sherman ME (2009) DNA hypermethylation of ESR1 and PGR in breast

cancer: pathologic and epidemiologic associations. Cancer Epidemiol Biomark Prev 18(11):3036–3043. doi:10.1158/1055-9965.EPI-09-0678
78. Jiang Y, Tong D, Lou G, Zhang Y, Geng J (2008) Expression of RUNX3 gene, methylation status and clinicopathological significance in breast cancer and breast cancer cell lines. Pathobiology 75(4):244–251. doi:10.1159/000132385
79. Lu DG, Ma YM, Zhu AJ, Han YW (2016) An early biomarker and potential therapeutic target of RUNX 3 hypermethylation in breast cancer, a system review and meta-analysis. Oncotarget. doi:10.18632/oncotarget.13125
80. Song XY, Li BY, Zhou EX, Wu FX (2016) The clinicopathological significance of RUNX3 hypermethylation and mRNA expression in human breast cancer, a meta-analysis. Onco Targets Ther 9:5339–5347. doi:10.2147/OTT.S77828
81. Yu YY, Chen C, Kong FF, Zhang W (2014) Clinicopathological significance and potential drug target of RUNX3 in breast cancer. Drug Des Devel Ther 8:2423–2430. doi:10.2147/DDDT.S71815
82. Kang HF, Dai ZJ, Bai HP, Lu WF, Ma XB, Bao X, Lin S, Wang XJ (2013) RUNX3 gene promoter demethylation by 5-Aza-CdR induces apoptosis in breast cancer MCF-7 cell line. Onco Targets Ther 6:411–417. doi:10.2147/OTT.S43744
83. Subramaniam MM, Chan JY, Omar MF, Ito K, Ito Y, Yeoh KG, Salto-Tellez M, Putti TC (2010) Lack of RUNX3 inactivation in columnar cell lesions of breast. Histopathology 57(4):555–563. doi:10.1111/j.1365-2559.2010.03675.x
84. Subramaniam MM, Chan JY, Soong R, Ito K, Ito Y, Yeoh KG, Salto-Tellez M, Putti TC (2009) RUNX3 inactivation by frequent promoter hypermethylation and protein mislocalization constitute an early event in breast cancer progression. Breast Cancer Res Treat 113(1):113–121. doi:10.1007/s10549-008-9917-4
85. Hwang KT, Han W, Bae JY, Hwang SE, Shin HJ, Lee JE, Kim SW, Min HJ, Noh DY (2007) Downregulation of the RUNX3 gene by promoter hypermethylation and hemizygous deletion in breast cancer. J Korean Med Sci 22(Suppl):S24–S31. doi:10.3346/jkms.2007.22.S.S24
86. Li Y, Melnikov AA, Levenson V, Guerra E, Simeone P, Alberti S, Deng Y (2015) A seven-gene CpG-island methylation panel predicts breast cancer progression. BMC Cancer 15:417. doi:10.1186/s12885-015-1412-9
87. Cho YH, McCullough LE, Gammon MD, Wu HC, Zhang YJ, Wang Q, Xu X, Teitelbaum SL, Neugut AI, Chen J, Santella RM (2015) Promoter Hypermethylation in white blood cell DNA and breast cancer risk. J Cancer 6(9):819–824. doi:10.7150/jca.12174
88. Chen ST, Lin SY, Yeh KT, Kuo SJ, Chan WL, Chu YP, Chang JG (2004) Mutational, epigenetic and expressional analyses of caveolin-1 gene in breast cancers. Int J Mol Med 14(4):577–582
89. Hong CP, Choe MK, Roh TY (2012) Characterization of chromatin structure-associated histone modifications in breast cancer cells. Genomics Inform 10(3):145–152. doi:10.5808/GI.2012.10.3.145
90. Costello JF, Fruhwald MC, Smiraglia DJ, Rush LJ, Robertson GP, Gao X, Wright FA, Feramisco JD, Peltomaki P, Lang JC, Schuller DE, Yu L, Bloomfield CD, Caligiuri MA, Yates A, Nishikawa R, Su Huang H, Petrelli NJ, Zhang X, O'Dorisio MS, Held WA, Cavenee WK, Plass C (2000) Aberrant CpG-island methylation has non-random and tumour-type-specific patterns. Nat Genet 24(2):132–138. doi:10.1038/72785
91. Esteller M (2007) Cancer epigenomics: DNA methylomes and histone-modification maps. Nat Rev Genet 8(4):286–298. doi:10.1038/nrg2005
92. Lo PK, Mehrotra J, D'Costa A, Fackler MJ, Garrett-Mayer E, Argani P, Sukumar S (2006) Epigenetic suppression of secreted frizzled related protein 1 (SFRP1) expression in human breast cancer. Cancer Biol Ther 5(3):281–286
93. Dagdemir A, Judes G, Lebert A, Echegut M, Karsli-Ceppioglu S, Rifai K, Daures M, Ngollo M, Dubois L, Penault-Llorca F, Bignon YJ, Bernard-Gallon D (2016) Epigenetic modifications with DZNep, NaBu and SAHA in luminal and mesenchymal-like breast cancer subtype cells. Cancer Genomics Proteomics 13(4):291–303
94. Agathanggelou A, Dallol A, Zochbauer-Muller S, Morrissey C, Honorio S, Hesson L, Martinsson T, Fong KM, Kuo MJ, Yuen PW, Maher ER, Minna JD, Latif F (2003) Epigenetic inactivation of the candidate 3p21.3 suppressor gene BLU in human cancers. Oncogene 22(10):1580–1588. doi:10.1038/sj.onc.1206243
95. Asiaf A, Ahmad ST, Aziz SA, Malik AA, Rasool Z, Masood A, Zargar MA (2014) Loss of expression and aberrant methylation of the CDH1 (E-cadherin) gene in breast cancer patients from Kashmir. Asian Pac J Cancer Prev 15(15):6397–6403
96. Alvarez C, Tapia T, Cornejo V, Fernandez W, Munoz A, Camus M, Alvarez M, Devoto L, Carvallo P (2013) Silencing of tumor suppressor genes RASSF1A, SLIT2, and WIF1 by promoter hypermethylation in hereditary breast cancer. Mol Carcinog 52(6):475–487. doi:10.1002/mc.21881
97. Askari M, Sobti RC, Nikbakht M, Sharma SC (2013) Promoter hypermethylation of tumour suppressor genes (p14/ARF and p16/INK4a): case-control study in north Indian population. Mol Biol Rep 40(8):4921–4928. doi:10.1007/s11033-013-2592-5
98. Bachman KE, Herman JG, Corn PG, Merlo A, Costello JF, Cavenee WK, Baylin SB, Graff JR (1999) Methylation-associated silencing of the tissue inhibitor of metalloproteinase-3 gene suggest a suppressor role in kidney, brain, and other human cancers. Cancer Res 59(4):798–802
99. Bae YK, Shim YR, Choi JH, Kim MJ, Gabrielson E, Lee SJ, Hwang TY, Shin SO (2005) Gene promoter

hypermethylation in tumors and plasma of breast cancer patients. Cancer Res Treat 37(4):233–240. doi:10.4143/crt.2005.37.4.233
100. Bagadi SA, Prasad CP, Kaur J, Srivastava A, Prashad R, Gupta SD, Ralhan R (2008) Clinical significance of promoter hypermethylation of RASSF1A, RARbeta2, BRCA1 and HOXA5 in breast cancers of Indian patients. Life Sci 82(25–26):1288–1292. doi:10.1016/j.lfs.2008.04.020
101. Ballestar E, Paz MF, Valle L, Wei S, Fraga MF, Espada J, Cigudosa JC, Huang TH, Esteller M (2003) Methyl-CpG binding proteins identify novel sites of epigenetic inactivation in human cancer. EMBO J 22(23):6335–6345. doi:10.1093/emboj/cdg604
102. Boily G, Saikali Z, Sinnett D (2004) Methylation analysis of the glypican 3 gene in embryonal tumours. Br J Cancer 90(8):1606–1611. doi:10.1038/sj.bjc.6601716
103. Celebiler Cavusoglu A, Sevinc AI, Saydam S, Canda T, Baskan Z, Kilic Y, Sakizli M (2010) Promoter methylation and expression changes of CDH1 and P16 genes in invasive breast cancer and adjacent normal breast tissue. Neoplasma 57(5):465–472
104. Chekhun VF, Kulik GI, Yurchenko OV, Tryndyak VP, Todor IN, Luniv LS, Tregubova NA, Pryzimirska TV, Montgomery B, Rusetskaya NV, Pogribny IP (2006) Role of DNA hypomethylation in the development of the resistance to doxorubicin in human MCF-7 breast adenocarcinoma cells. Cancer Lett 231(1):87–93. doi:10.1016/j.canlet.2005.01.038
105. Chen CM, Chen HL, Hsiau TH, Hsiau AH, Shi H, Brock GJ, Wei SH, Caldwell CW, Yan PS, Huang TH (2003) Methylation target array for rapid analysis of CpG island hypermethylation in multiple tissue genomes. Am J Pathol 163(1):37–45. doi:10.1016/S0002-9440(10)63628-0
106. Chimonidou M, Tzitzira A, Strati A, Sotiropoulou G, Sfikas C, Malamos N, Georgoulias V, Lianidou E (2013) CST6 promoter methylation in circulating cell-free DNA of breast cancer patients. Clin Biochem 46(3):235–240. doi:10.1016/j.clinbiochem.2012.09.015
107. Crucianelli F, Tricarico R, Turchetti D, Gorelli G, Gensini F, Sestini R, Giunti L, Pedroni M, Ponz de Leon M, Civitelli S, Genuardi M (2014) MLH1 constitutional and somatic methylation in patients with MLH1 negative tumors fulfilling the revised Bethesda criteria. Epigenetics 9(10):1431–1438. doi:10.4161/15592294.2014.970080
108. Dallol A, Forgacs E, Martinez A, Sekido Y, Walker R, Kishida T, Rabbitts P, Maher ER, Minna JD, Latif F (2002) Tumour specific promoter region methylation of the human homologue of the drosophila roundabout gene DUTT1 (ROBO1) in human cancers. Oncogene 21(19):3020–3028. doi:10.1038/sj.onc.1205421
109. Dimitrakopoulos L, Vorkas PA, Georgoulias V, Lianidou ES (2012) A closed-tube methylation-sensitive high resolution melting assay (MS-HRMA) for the semi-quantitative determination of CST6 promoter methylation in clinical samples. BMC Cancer 12:486. doi:10.1186/1471-2407-12-486
110. Virmani A, Rathi A, Heda S, Sugio K, Lewis C, Tonk V, Takahashi T, Roth JA, Minna JD, Euhus DM, Gazdar AF (2003) Aberrant methylation of the cyclin D2 promoter in primary small cell, non-small cell lung and breast cancers. Int J Cancer 107(3):341–345. doi:10.1002/ijc.11393
111. Virmani A, Rathi A, Sugio K, Sathyanarayana UG, Toyooka S, Kischel FC, Tonk V, Padar A, Takahashi T, Roth JA, Euhus DM, Minna JD, Gazdar AF (2003) Aberrant methylation of TMS1 in small cell, non small cell lung cancer and breast cancer. Int J Cancer 106(2):198–204. doi:10.1002/ijc.11206
112. Virmani AK, Rathi A, Sathyanarayana UG, Padar A, Huang CX, Cunnigham HT, Farinas AJ, Milchgrub S, Euhus DM, Gilcrease M, Herman J, Minna JD, Gazdar AF (2001) Aberrant methylation of the adenomatous polyposis coli (APC) gene promoter 1A in breast and lung carcinomas. Clin Cancer Res 7(7):1998–2004
113. Wang S, Ding YB, Chen GY, Xia JG, Wu ZY (2004) Hypermethylation of Syk gene in promoter region associated with oncogenesis and metastasis of gastric carcinoma. World J Gastroenterol 10(12):1815–1818
114. Wang S, Dorsey TH, Terunuma A, Kittles RA, Ambs S, Kwabi-Addo B (2012) Relationship between tumor DNA methylation status and patient characteristics in African-American and European-American women with breast cancer. PLoS One 7(5):e37928. doi:10.1371/journal.pone.0037928
115. Weissenborn C, Ignatov T, Nass N, Kalinski T, Dan Costa S, Zenclussen AC, Ignatov A (2017) GPER promoter methylation controls GPER expression in breast cancer patients. Cancer Investig 35(2):100–107. doi:10.1080/07357907.2016.1271886
116. Widschwendter M, Jones PA (2002) DNA methylation and breast carcinogenesis. Oncogene 21(35):5462–5482. doi:10.1038/sj.onc.1205606
117. Worm J, Kirkin AF, Dzhandzhugazyan KN, Guldberg P (2001) Methylation-dependent silencing of the reduced folate carrier gene in inherently methotrexate-resistant human breast cancer cells. J Biol Chem 276(43):39990–40000. doi:10.1074/jbc.M103181200
118. Wu HC, Southey MC, Hibshoosh H, Santella RM, Terry MB (2017) DNA methylation in breast tumor from high-risk women in the breast cancer family registry. Anticancer Res 37(2):659–664. doi:10.21873/anticanres.11361
119. Xiang YY, Ladeda V, Filmus J (2001) Glypican-3 expression is silenced in human breast cancer. Oncogene 20(50):7408–7412. doi:10.1038/sj.onc.1204925
120. Xu J, Shetty PB, Feng W, Chenault C, Bast RC Jr, Issa JP, Hilsenbeck SG, Yu Y (2012) Methylation of HIN-1, RASSF1A, RIL and CDH13 in breast cancer is associated with clinical characteristics, but only

RASSF1A methylation is associated with outcome. BMC Cancer 12:243. doi:10.1186/1471-2407-12-243
121. Yamamoto N, Nakayama T, Kajita M, Miyake T, Iwamoto T, Kim SJ, Sakai A, Ishihara H, Tamaki Y, Noguchi S (2012) Detection of aberrant promoter methylation of GSTP1, RASSF1A, and RARbeta2 in serum DNA of patients with breast cancer by a newly established one-step methylation-specific PCR assay. Breast Cancer Res Treat 132(1):165–173. doi:10.1007/s10549-011-1575-2
122. Yang J, Niu H, Huang Y, Yang K (2016) A systematic analysis of the relationship of CDH13 promoter methylation and breast cancer risk and prognosis. PLoS One 11(5):e0149185. doi:10.1371/journal.pone.0149185
123. Yang ZQ, Liu G, Bollig-Fischer A, Haddad R, Tarca AL, Ethier SP (2009) Methylation-associated silencing of SFRP1 with an 8p11-12 amplification inhibits canonical and non-canonical WNT pathways in breast cancers. Int J Cancer 125(7):1613–1621. doi:10.1002/ijc.24518
124. Yazici H, Terry MB, Cho YH, Senie RT, Liao Y, Andrulis I, Santella RM (2009) Aberrant methylation of RASSF1A in plasma DNA before breast cancer diagnosis in the breast cancer family registry. Cancer Epidemiol Biomark Prev 18(10):2723–2725. doi:10.1158/1055-9965.EPI-08-1237
125. Yeo W, Wong WL, Wong N, Law BK, Tse GM, Zhong S (2005) High frequency of promoter hypermethylation of RASSF1A in tumorous and non-tumourous tissue of breast cancer. Pathology 37(2):125–130
126. Yu P, Guo Y, Yusufu M, Liu Z, Wang S, Yin X, Peng G, Wang L, Zhao X, Guo H, Huang T, Liu C (2016) Decreased expression of EZH2 reactivates RASSF2A by reversal of promoter methylation in breast cancer cells. Cell Biol Int 40(10):1062–1070. doi:10.1002/cbin.10646
127. Zurita M, Lara PC, del Moral R, Torres B, Linares-Fernandez JL, Arrabal SR, Martinez-Galan J, Oliver FJ, Ruiz de Almodovar JM (2010) Hypermethylated 14-3-3-sigma and ESR1 gene promoters in serum as candidate biomarkers for the diagnosis and treatment efficacy of breast cancer metastasis. BMC Cancer 10:217. doi:10.1186/1471-2407-10-217
128. Zwergel C, Valente S, Mai A (2016) DNA methyltransferases inhibitors from natural sources. Curr Top Med Chem 16(7):680–696
129. Elsheikh SE, Green AR, Rakha EA, Powe DG, Ahmed RA, Collins HM, Soria D, Garibaldi JM, Paish CE, Ammar AA, Grainge MJ, Ball GR, Abdelghany MK, Martinez-Pomares L, Heery DM, Ellis IO (2009) Global histone modifications in breast cancer correlate with tumor phenotypes, prognostic factors, and patient outcome. Cancer Res 69(9):3802–3809. doi:10.1158/0008-5472.CAN-08-3907
130. Dalvai M, Bystricky K (2010) The role of histone modifications and variants in regulating gene expression in breast cancer. J Mammary Gland Biol Neoplasia 15(1):19–33. doi:10.1007/s10911-010-9167-z
131. Yoo KH, Hennighausen L (2012) EZH2 methyltransferase and H3K27 methylation in breast cancer. Int J Biol Sci 8(1):59–65
132. Tsai WW, Wang Z, Yiu TT, Akdemir KC, Xia W, Winter S, Tsai CY, Shi X, Schwarzer D, Plunkett W, Aronow B, Gozani O, Fischle W, Hung MC, Patel DJ, Barton MC (2010) TRIM24 links a non-canonical histone signature to breast cancer. Nature 468(7326):927–932. doi:10.1038/nature09542
133. Li Y, Li S, Chen J, Shao T, Jiang C, Wang Y, Chen H, Xu J, Li X (2014) Comparative epigenetic analyses reveal distinct patterns of oncogenic pathways activation in breast cancer subtypes. Hum Mol Genet 23(20):5378–5393. doi:10.1093/hmg/ddu256
134. Droog M, Nevedomskaya E, Dackus GM, Fles R, Kim Y, Hollema H, Mourits M, Nederlof PM, van Boven HH, Linn SC, van Leeuwen FE, Wessels LF, Zwart W (2017) Estrogen receptor alpha wields treatment-specific enhancers between morphologically similar endometrial tumors. Proc Natl Acad Sci U S A 114(8):E1316–E1325. doi:10.1073/pnas.1615233114
135. Bustos MA, Salomon MP, Nelson N, Hsu SC, DiNome ML, Hoon DS, Marzese DM (2017) Genome-wide chromatin accessibility, DNA methylation and gene expression analysis of histone deacetylase inhibition in triple-negative breast cancer. Genom Data 12:14–16. doi:10.1016/j.gdata.2017.01.002
136. Connolly RM, Li H, Jankowitz RC, Zhang Z, Rudek MA, Jeter SC, Slater SA, Powers P, Wolff AC, Fetting JH, Brufsky A, Piekarz R, Ahuja N, Laird PW, Shen H, Weisenberger DJ, Cope L, Herman JG, Somlo G, Garcia AA, Jones PA, Baylin SB, Davidson NE, Zahnow CA, Stearns V (2016) Combination epigenetic therapy in advanced breast cancer with 5-Azacitidine and Entinostat: a phase II National Cancer Institute/stand up to cancer study. Clin Cancer Res. doi:10.1158/1078-0432.CCR-16-1729
137. Oshiro MM, Futscher BW, Lisberg A, Wozniak RJ, Klimecki WT, Domann FE, Cress AE (2005) Epigenetic regulation of the cell type-specific gene 14-3-3sigma. Neoplasia 7(9):799–808
138. Rhie SK, Hazelett DJ, Coetzee SG, Yan C, Noushmehr H, Coetzee GA (2014) Nucleosome positioning and histone modifications define relationships between regulatory elements and nearby gene expression in breast epithelial cells. BMC Genomics 15:331. doi:10.1186/1471-2164-15-331
139. Gnyszka A, Jastrzebski Z, Flis S (2013) DNA methyltransferase inhibitors and their emerging role in epigenetic therapy of cancer. Anticancer Res 33(8):2989–2996
140. Girault I, Tozlu S, Lidereau R, Bieche I (2003) Expression analysis of DNA methyltransferases 1, 3A, and 3B in sporadic breast carcinomas. Clin Cancer Res 9(12):4415–4422

141. Manjegowda MC, Gupta PS, Limaye AM (2017) Hyper-methylation of the upstream CpG island shore is a likely mechanism of GPER1 silencing in breast cancer cells. Gene 614:65–73. doi:10.1016/j.gene.2017.03.006
142. Sadikovic B, Haines TR, Butcher DT, Rodenhiser DI (2004) Chemically induced DNA hypomethylation in breast carcinoma cells detected by the amplification of intermethylated sites. Breast Cancer Res 6(4):R329–R337. doi:10.1186/bcr799
143. Dong H, Ma L, Gan J, Lin W, Chen C, Yao Z, Du L, Zheng L, Ke C, Huang X, Song H, Kumar R, Yeung SC, Zhang H (2017) PTPRO represses ERBB2-driven breast oncogenesis by dephosphorylation and endosomal internalization of ERBB2. Oncogene 36(3):410–422. doi:10.1038/onc.2016.213
144. Tao Y, Liu S, Briones V, Geiman TM, Muegge K (2011) Treatment of breast cancer cells with DNA demethylating agents leads to a release of pol II stalling at genes with DNA-hypermethylated regions upstream of TSS. Nucleic Acids Res 39(22):9508–9520. doi:10.1093/nar/gkr611
145. Hellreich J, Gasparek J, O'Donnell R (2014) Effects of 5-azacytidine on the in vitro colony growth of the MDA-MB 435 cancer cell line (1047.11). The. FASEB J 28(1 Suppl)
146. Xiang T, Fan Y, Li C, Li L, Ying Y, Mu J, Peng W, Feng Y, Oberst M, Kelly K, Ren G, Tao Q (2016) DACT2 silencing by promoter CpG methylation disrupts its regulation of epithelial-to-mesenchymal transition and cytoskeleton reorganization in breast cancer cells. Oncotarget 7(43):70924–70935. doi:10.18632/oncotarget.12341
147. Luo LJ, Yang F, Ding JJ, Yan DL, Wang DD, Yang SJ, Ding L, Li J, Chen D, Ma R, Wu JZ, Tang JH (2016) MiR-31 inhibits migration and invasion by targeting SATB2 in triple negative breast cancer. Gene 594(1):47–58. doi:10.1016/j.gene.2016.08.057
148. Billam M, Sobolewski MD, Davidson NE (2010) Effects of a novel DNA methyltransferase inhibitor zebularine on human breast cancer cells. Breast Cancer Res Treat 120(3):581–592. doi:10.1007/s10549-009-0420-3
149. Chen M, Shabashvili D, Nawab A, Yang SX, Dyer LM, Brown KD, Hollingshead M, Hunter KW, Kaye FJ, Hochwald SN, Marquez VE, Steeg P, Zajac-Kaye M (2012) DNA methyltransferase inhibitor, zebularine, delays tumor growth and induces apoptosis in a genetically engineered mouse model of breast cancer. Mol Cancer Ther 11(2):370–382. doi:10.1158/1535-7163.MCT-11-0458
150. Ceccacci E, Minucci S (2016) Inhibition of histone deacetylases in cancer therapy: lessons from leukaemia. Br J Cancer 114(6):605–611. doi:10.1038/bjc.2016.36
151. Miller L, Abdalla A (2003) The role of endoscopy in the treatment of esophageal varices, 2002-2003. Curr Opin Gastroenterol 19(5):483–486
152. Yoshida M, Shimazu T, Matsuyama A (2003) Protein deacetylases: enzymes with functional diversity as novel therapeutic targets. Prog Cell Cycle Res 5:269–278
153. Marks PA, Breslow R (2007) Dimethyl sulfoxide to vorinostat: development of this histone deacetylase inhibitor as an anticancer drug. Nat Biotechnol 25(1):84–90. doi:10.1038/nbt1272
154. Chiaradonna F, Barozzi I, Miccolo C, Bucci G, Palorini R, Fornasari L, Botrugno OA, Pruneri G, Masullo M, Passafaro A, Galimberti VE, Fantin VR, Richon VM, Pece S, Viale G, Di Fiore PP, Draetta G, Pelicci PG, Minucci S, Chiocca S (2015) Redox-mediated Suberoylanilide Hydroxamic acid sensitivity in breast cancer. Antioxid Redox Signal 23(1):15–29. doi:10.1089/ars.2014.6189
155. Wu S, Luo Z, Yu PJ, Xie H, He YW (2016) Suberoylanilide hydroxamic acid (SAHA) promotes the epithelial mesenchymal transition of triple negative breast cancer cells via HDAC8/FOXA1 signals. Biol Chem 397(1):75–83. doi:10.1515/hsz-2015-0215
156. Feng X, Han H, Zou D, Zhou J, Zhou W (2017) Suberoylanilide hydroxamic acid-induced specific epigenetic regulation controls leptin-induced proliferation of breast cancer cell lines. Oncotarget 8(2):3364–3379. doi:10.18632/oncotarget.13764
157. Han RF, Li K, Yang ZS, Chen ZG, Yang WC (2014) Trichostatin a induces mesenchymal-like morphological change and gene expression but inhibits migration and colony formation in human cancer cells. Mol Med Rep 10(6):3211–3216. doi:10.3892/mmr.2014.2594
158. Chang J, Varghese DS, Gillam MC, Peyton M, Modi B, Schiltz RL, Girard L, Martinez ED (2012) Differential response of cancer cells to HDAC inhibitors trichostatin a and depsipeptide. Br J Cancer 106(1):116–125. doi:10.1038/bjc.2011.532
159. Liu J, Li Y (2015) Trichostatin a and tamoxifen inhibit breast cancer cell growth by miR-204 and ERalpha reducing AKT/mTOR pathway. Biochem Biophys Res Commun 467(2):242–247. doi:10.1016/j.bbrc.2015.09.182
160. Sun S, Han Y, Liu J, Fang Y, Tian Y, Zhou J, Ma D, Wu P (2014) Trichostatin a targets the mitochondrial respiratory chain, increasing mitochondrial reactive oxygen species production to trigger apoptosis in human breast cancer cells. PLoS One 9(3):e91610. doi:10.1371/journal.pone.0091610
161. Noh H, Park J, Shim M, Lee Y (2016) Trichostatin a enhances estrogen receptor-alpha repression in MCF-7 breast cancer cells under hypoxia. Biochem Biophys Res Commun 470(3):748–752. doi:10.1016/j.bbrc.2016.01.022
162. Zhuang ZG, Fei F, Chen Y, Jin W (2008) Suberoyl bis-hydroxamic acid induces p53-dependent apoptosis of MCF-7 breast cancer cells. Acta Pharmacol Sin 29(12):1459–1466. doi:10.1111/j.1745-7254.2008.00906.x

163. Yang X, Zhang N, Shi Z, Yang Z, Hu X (2015) Histone deacetylase inhibitor suberoyl bis-hydroxamic acid suppresses cell proliferation and induces apoptosis in breast cancer cells. Mol Med Rep 11(4):2908–2912. doi:10.3892/mmr.2014.3076
164. Fortunati N, Marano F, Bandino A, Frairia R, Catalano MG, Boccuzzi G (2014) The pan-histone deacetylase inhibitor LBH589 (panobinostat) alters the invasive breast cancer cell phenotype. Int J Oncol 44(3):700–708. doi:10.3892/ijo.2013.2218
165. Tate CR, Rhodes LV, Segar HC, Driver JL, Pounder FN, Burow ME, Collins-Burow BM (2012) Targeting triple-negative breast cancer cells with the histone deacetylase inhibitor panobinostat. Breast Cancer Res 14(3):R79. doi:10.1186/bcr3192
166. Kubo M, Kanaya N, Petrossian K, Ye J, Warden C, Liu Z, Nishimura R, Osako T, Okido M, Shimada K, Takahashi M, Chu P, Yuan YC, Chen S (2013) Inhibition of the proliferation of acquired aromatase inhibitor-resistant breast cancer cells by histone deacetylase inhibitor LBH589 (panobinostat). Breast Cancer Res Treat 137(1):93–107. doi:10.1007/s10549-012-2332-x
167. Rhodes LV, Tate CR, Segar HC, Burks HE, Phamduy TB, Hoang V, Elliott S, Gilliam D, Pounder FN, Anbalagan M, Chrisey DB, Rowan BG, Burow ME, Collins-Burow BM (2014) Suppression of triple-negative breast cancer metastasis by pan-DAC inhibitor panobinostat via inhibition of ZEB family of EMT master regulators. Breast Cancer Res Treat 145(3):593–604. doi:10.1007/s10549-014-2979-6
168. Schech A, Kazi A, Yu S, Shah P, Sabnis G (2015) Histone deacetylase inhibitor Entinostat inhibits tumor-initiating cells in triple-negative breast cancer cells. Mol Cancer Ther 14(8):1848–1857. doi:10.1158/1535-7163.MCT-14-0778
169. Schech AJ, Shah P, Yu S, Sabnis GJ, Goloubeva O, Rosenblatt P, Kazi A, Chumsri S, Brodie A (2015) Histone deacetylase inhibitor entinostat in combination with a retinoid downregulates HER2 and reduces the tumor initiating cell population in aromatase inhibitor-resistant breast cancer. Breast Cancer Res Treat 152(3):499–508. doi:10.1007/s10549-015-3442-z
170. Fortunati N, Bertino S, Costantino L, Bosco O, Vercellinatto I, Catalano MG, Boccuzzi G (2008) Valproic acid is a selective antiproliferative agent in estrogen-sensitive breast cancer cells. Cancer Lett 259(2):156–164. doi:10.1016/j.canlet.2007.10.006
171. Artacho-Cordon F, Rios-Arrabal S, Olivares-Urbano MA, Storch K, Dickreuter E, Munoz-Gamez JA, Leon J, Calvente I, Torne P, Salinas Mdel M, Cordes N, Nunez MI (2015) Valproic acid modulates radiation-enhanced matrix metalloproteinase activity and invasion of breast cancer cells. Int J Radiat Biol 91(12):946–956. doi:10.3109/09553002.2015.1087067
172. Vafaiyan Z, Gharaei R, Asadi J (2015) The correlation between telomerase activity and Bax/Bcl-2 ratio in valproic acid-treated MCF-7 breast cancer cell line. Iran J Basic Med Sci 18(7):700–704
173. Mawatari T, Ninomiya I, Inokuchi M, Harada S, Hayashi H, Oyama K, Makino I, Nakagawara H, Miyashita T, Tajima H, Takamura H, Fushida S, Ohta T (2015) Valproic acid inhibits proliferation of HER2-expressing breast cancer cells by inducing cell cycle arrest and apoptosis through Hsp70 acetylation. Int J Oncol 47(6):2073–2081. doi:10.3892/ijo.2015.3213
174. Travaglini L, Vian L, Billi M, Grignani F, Nervi C (2009) Epigenetic reprogramming of breast cancer cells by valproic acid occurs regardless of estrogen receptor status. Int J Biochem Cell Biol 41(1):225–234. doi:10.1016/j.biocel.2008.08.019
175. Chopin V, Toillon RA, Jouy N, Le Bourhis X (2002) Sodium butyrate induces P53-independent, Fas-mediated apoptosis in MCF-7 human breast cancer cells. Br J Pharmacol 135(1):79–86. doi:10.1038/sj.bjp.0704456
176. Louis M, Rosato RR, Brault L, Osbild S, Battaglia E, Yang XH, Grant S, Bagrel D (2004) The histone deacetylase inhibitor sodium butyrate induces breast cancer cell apoptosis through diverse cytotoxic actions including glutathione depletion and oxidative stress. Int J Oncol 25(6):1701–1711
177. Lee KW, Kim JH, Park JH, Kim HP, Song SH, Kim SG, Kim TY, Jong HS, Jung KH, Im SA, Kim TY, Kim NK, Bang YJ (2006) Antitumor activity of SK-7041, a novel histone deacetylase inhibitor, in human lung and breast cancer cells. Anticancer Res 26(5A):3429–3438
178. Hait NC, Avni D, Yamada A, Nagahashi M, Aoyagi T, Aoki H, Dumur CI, Zelenko Z, Gallagher EJ, Leroith D, Milstien S, Takabe K, Spiegel S (2015) The phosphorylated prodrug FTY720 is a histone deacetylase inhibitor that reactivates ERalpha expression and enhances hormonal therapy for breast cancer. Oncogene 4:e156. doi:10.1038/oncsis.2015.16
179. Prestegui-Martel B, Bermudez-Lugo JA, Chavez-Blanco A, Duenas-Gonzalez A, Garcia-Sanchez JR, Perez-Gonzalez OA, Padilla M, II, Fragoso-Vazquez MJ, Mendieta-Wejebe JE, Correa-Basurto AM, Mendez-Luna D, Trujillo-Ferrara J, Correa-Basurto J (2016) N-(2-hydroxyphenyl)-2-propylpentanamide, a valproic acid aryl derivative designed in silico with improved anti-proliferative activity in HeLa, rhabdomyosarcoma and breast cancer cells. J Enzyme Inhib Med Chem 31(Suppl 3):140–149. doi:10.1080/14756366.2016.1210138
180. Keen JC, Yan L, Mack KM, Pettit C, Smith D, Sharma D, Davidson NE (2003) A novel histone deacetylase inhibitor, scriptaid, enhances expression of functional estrogen receptor alpha (ER) in ER negative human breast cancer cells in combination with 5-aza 2′-deoxycytidine. Breast Cancer Res Treat 81(3):177–186. doi:10.1023/A:1026146524737
181. Giacinti L, Giacinti C, Gabellini C, Rizzuto E, Lopez M, Giordano A (2012) Scriptaid effects on breast

cancer cell lines. J Cell Physiol 227(10):3426–3433. doi:10.1002/jcp.24043

182. Chiu HW, Yeh YL, Wang YC, Huang WJ, Ho SY, Lin P, Wang YJ (2016) Combination of the novel histone deacetylase inhibitor YCW1 and radiation induces autophagic cell death through the downregulation of BNIP3 in triple-negative breast cancer cells in vitro and in an orthotopic mouse model. Mol Cancer 15(1):46. doi:10.1186/s12943-016-0531-5
183. Gromek SM, deMayo JA, Maxwell AT, West AM, Pavlik CM, Zhao Z, Li J, Wiemer AJ, Zweifach A, Balunas MJ (2016) Synthesis and biological evaluation of santacruzamate a analogues for anti-proliferative and immunomodulatory activity. Bioorg Med Chem 24(21):5183–5196. doi:10.1016/j.bmc.2016.08.040
184. Tan YL, Pigeon P, Top S, Labbe E, Buriez O, Hillard EA, Vessieres A, Amatore C, Leong WK, Jaouen G (2012) Ferrocenyl catechols: synthesis, oxidation chemistry and anti-proliferative effects on MDA-MB-231 breast cancer cells. Dalton Trans 41(25):7537–7549. doi:10.1039/c2dt30700f
185. Zheng Y, Wang C, Li C, Qiao J, Zhang F, Huang M, Ren W, Dong C, Huang J, Zhou HB (2012) Discovery of novel SERMs with a ferrocenyl entity based on the oxabicyclo[2.2.1]heptene scaffold and evaluation of their antiproliferative effects in breast cancer cells. Org Biomol Chem 10(48):9689–9699. doi:10.1039/c2ob26226f
186. Laine AL, Adriaenssens E, Vessieres A, Jaouen G, Corbet C, Desruelles E, Pigeon P, Toillon RA, Passirani C (2013) The in vivo performance of ferrocenyl tamoxifen lipid nanocapsules in xenografted triple negative breast cancer. Biomaterials 34(28):6949–6956. doi:10.1016/j.biomaterials.2013.05.065
187. Li C, Tang C, Hu Z, Zhao C, Li C, Zhang S, Dong C, Zhou HB, Huang J (2016) Synthesis and structure-activity relationships of novel hybrid ferrocenyl compounds based on a bicyclic core skeleton for breast cancer therapy. Bioorg Med Chem 24(13):3062–3074. doi:10.1016/j.bmc.2016.05.019
188. Thakur S, Feng X, Qiao Shi Z, Ganapathy A, Kumar Mishra M, Atadja P, Morris D, Riabowol K (2012) ING1 and 5-azacytidine act synergistically to block breast cancer cell growth. PLoS One 7(8):e43671. doi:10.1371/journal.pone.0043671
189. Zhou W, Feng X, Han H, Guo S, Wang G (2016) Synergistic effects of combined treatment with histone deacetylase inhibitor suberoylanilide hydroxamic acid and TRAIL on human breast cancer cells. Sci Rep 6:28004. doi:10.1038/srep28004
190. Min A, Im SA, Kim DK, Song SH, Kim HJ, Lee KH, Kim TY, Han SW, Oh DY, Kim TY, O'Connor MJ, Bang YJ (2015) Histone deacetylase inhibitor, suberoylanilide hydroxamic acid (SAHA), enhances anti-tumor effects of the poly (ADP-ribose) polymerase (PARP) inhibitor olaparib in triple-negative breast cancer cells. Breast Cancer Res 17:33. doi:10.1186/s13058-015-0534-y
191. Chen L, Jin T, Zhu K, Piao Y, Quan T, Quan C, Lin Z (2017) PI3K/mTOR dual inhibitor BEZ235 and histone deacetylase inhibitor Trichostatin A synergistically exert anti-tumor activity in breast cancer. Oncotarget. doi:10.18632/oncotarget.14442
192. Yang X, Shi Z, Zhang N, Ou Z, Fu S, Hu X, Shen Z (2014) Suberoyl bis-hydroxamic acid enhances cytotoxicity induced by proteasome inhibitors in breast cancer cells. Cancer Cell Int 14:107. doi:10.1186/s12935-014-0107-7
193. Lin Z, Zhang Z, Jiang X, Kou X, Bao Y, Liu H, Sun F, Ling S, Qin N, Jiang L, Yang Y (2017) Mevastatin blockade of autolysosome maturation stimulates LBH589-induced cell death in triple-negative breast cancer cells. Oncotarget. doi:10.18632/oncotarget.14868
194. Kai M, Kanaya N, Wu SV, Mendez C, Nguyen D, Luu T, Chen S (2015) Targeting breast cancer stem cells in triple-negative breast cancer using a combination of LBH589 and salinomycin. Breast Cancer Res Treat 151(2):281–294. doi:10.1007/s10549-015-3376-5
195. Tan WW, Allred JB, Moreno-Aspitia A, Northfelt DW, Ingle JN, Goetz MP, Perez EA (2016) Phase I study of Panobinostat (LBH589) and Letrozole in postmenopausal metastatic breast cancer patients. Clin Breast Cancer 16(2):82–86. doi:10.1016/j.clbc.2015.11.003
196. Ou O, Huppi K, Chakka S, Gehlhaus K, Dubois W, Patel J, Chen J, Mackiewicz M, Jones TL, Pitt JJ, Martin SE, Goldsmith P, Simmons JK, Mock BA, Caplen NJ (2014) Loss-of-function RNAi screens in breast cancer cells identify AURKB, PLK1, PIK3R1, MAPK12, PRKD2, and PTK6 as sensitizing targets of rapamycin activity. Cancer Lett 354(2):336–347. doi:10.1016/j.canlet.2014.08.043
197. Terranova-Barberio M, Roca MS, Zotti AI, Leone A, Bruzzese F, Vitagliano C, Scogliamiglio G, Russo D, D'Angelo G, Franco R, Budillon A, Di Gennaro E (2016) Valproic acid potentiates the anticancer activity of capecitabine in vitro and in vivo in breast cancer models via induction of thymidine phosphorylase expression. Oncotarget 7(7):7715–7731. doi:10.18632/oncotarget.6802
198. Li L, Sun Y, Liu J, Wu X, Chen L, Ma L, Wu P (2015) Histone deacetylase inhibitor sodium butyrate suppresses DNA double strand break repair induced by etoposide more effectively in MCF-7 cells than in HEK293 cells. BMC Biochem 16:2. doi:10.1186/s12858-014-0030-5
199. Sun B, Liu R, Xiao ZD, Zhu X (2012) C-MET protects breast cancer cells from apoptosis induced by sodium butyrate. PLoS one 7 (1):e30143. doi:10.1371/journal.pone.0030143
200. Cava C, Bertoli G, Castiglioni I (2015) Integrating genetics and epigenetics in breast cancer: biological insights, experimental, computational methods and therapeutic potential. BMC Syst Biol 9:62. doi:10.1186/s12918-015-0211-x

Progress in Vaccine Therapies for Breast Cancer

15

Xiaoyu Li and Xia Bu

Abstract

Therapeutic cancer vaccines aim to treat pre-existing cancer by boosting the patient's own immune system, which is an attractive strategy for cancer treatment. The cancer vaccines have mainly been designed to elicit antitumor T-cell immune responses that recognize and eradicate cancer. The advantages of cancer immunotherapy with cancer vaccines include a) high specificity of tumor antigen, b) minimal vaccine-related adverse events, and c) long-lasting immunity boosted by cancer vaccine which is important to control tumor relapse. In this chapter, we discuss identification of tumor antigens in breast cancer (e.g., cancer-testis antigens, neoantigens, HER2/neu, MUC1), the vaccine delivery systems utilized in breast cancer treatment (e.g., peptide vaccines, dendritic cell-based vaccines, and whole tumor cell-based vaccines), as well as clinical trials with therapeutic breast cancer vaccines. Moreover, new-generation clinical trials of breast cancer vaccines will aim at employing personalized vaccines designed to harness robust immune response to a custom-made neoantigen in the patient with breast cancer. Combination of vaccination and other forms of cancer therapy such as chemotherapy, radiotherapy, targeted therapy with monoclonal antibody, or immune checkpoint blockade will be required to achieve potent and durable antitumor clinical benefits.

X. Li (✉)
Department of Hematology, The First Affiliated Hospital, Henan University Cancer Center, School of Medicine, Henan University, Kaifeng, People's Republic of China
e-mail: xiaoyulimail@gmail.com

X. Bu
Department of Medical Oncology, The First Affiliated Hospital, Henan University Cancer Center, School of Medicine, Henan University, Kaifeng, People's Republic of China

E. Song, H. Hu (eds.), *Translational Research in Breast Cancer*, Advances in Experimental Medicine and Biology 1026, https://doi.org/10.1007/978-981-10-6020-5_15

Keywords
Breast cancer • Cancer vaccines • Cancer immunotherapy • Clinical trials

15.1 Introduction

Cancer immunotherapy aims to harness and enhance immune response in order to eradicate cancer. Recent dramatic clinical successes with agents targeting cytotoxic T-lymphocyte antigen-4 (CTLA-4) and programmed death-1 (PD-1) co-inhibitory molecules have provoked a significant increase in enthusiasm for cancer immunotherapeutic methods such as cancer vaccine therapy. There are many forms of cancer immunotherapy that are currently being used to treat cancer: immune checkpoint blockade, adoptive T-cell transfer, monoclonal antibody (mAb) immunotherapy, and cancer vaccines.

The objective of therapeutic cancer vaccines is to treat pre-existing cancer by enhancing a patient's own immune system against cancer, which is an attractive approach to treat cancer. Therapeutic cancer vaccines can be designed to specifically activate (a) T lymphocytes including CD8 + cytotoxic T lymphocytes (CTLs) as well as CD4+ T helper cells and direct them to recognize and attack cancer and (b) B cells to elicit humoral immune response and professional antigen-presenting cells (APCs) to induce both cellular and humoral antitumor immunity [1]. T-cell-mediated antitumor immune response can be elicited directly either by peptide-based vaccines, dendritic cell (DC)-based vaccines that are loaded with tumor-associated antigens (TAA)/tumor-specific antigens (TSA), or genetically modified whole tumor cells secreting GM-CSF or other cytokines. The US FDA has approved sipuleucel-T for prostate therapeutic cancer vaccine since 2011. Sipuleucel-T is an autologous active immunotherapy. The use of sipuleucel-T vaccine can elongate overall survival periods and activate T-cell immunity against tumor for the patients with metastatic castration-resistant prostate cancer.

The advantages of cancer immunotherapy with cancer vaccines are as follows: a) cancer vaccines have high tumor antigen specificity, thus inducing a robust immune response against antigens specifically expressed by tumor cells; b) cancer vaccines cause minimal immune-related adverse events; c) the elicited immunity by cancer vaccine is long-lasting due to the induction of immunological memory response, which is important to control tumor relapse. A large number of breast cancer vaccine strategies are under active clinical evaluation for their efficacy, safety, and toxicity, even though no vaccine for breast cancer therapy is currently approved by the US FDA.

15.2 Selection of Tumor Antigens in Breast Cancer

Identification of breast cancer antigens can enhance the development of either passive immunotherapy or active immunotherapy for the treatment of breast cancer. In general, tumor antigens used for tumor vaccines are derived from tumors. Therapeutic tumor vaccines have mainly been designed to provoke T-cell-mediated immune response that can recognize and eradicate existing tumors. The strategy is different from the prophylactic vaccines for infectious disease, which stimulate B-cell immunity. Thus, the objective of tumor vaccines is to elicit a cellular immune response mediated by T cells.

Tumor antigens can be broadly categorized into two classes: a) tumor-specific antigens (TSA), which are expressed exclusively in tumor cells, such as cancer-testis antigens and neoantigens that are mutated tumor antigens and tumor virus antigens, and b) tumor-associated antigens (TAA), which are overexpressed in tumors but can also exist in normal tissues, such as overexpressed antigens and differentiation antigens. The majority of breast tumor antigens are tumor-associated antigens such as HER2/neu, MUC1, p53, carbohydrate antigens, CEA, and hTERT antigens, some of which are commonly expressed in various types of tumors such as hTERT.

15.2.1 Cancer-Testis Antigens

Cancer-testis (CT) antigens are also known as cancer-germline antigens. These protein antigens are aberrantly expressed in a wide variety of human cancers, and they are normally silenced in normal tissues except in germ cells of testis. The CT antigens such as MAGE-1 and NY-ESO-1 have been demonstrated to be recognized by the immune system of cancer patients and to elicit a spontaneous cytotoxic T-cell response as well as humoral immune response in the patients, indicating that they are immunogenic and highly specific to tumors [2–4]. Thus, the identification of the CT antigens, which can be used as therapeutic cancer vaccine targets, has led to the development of antigen-specific cancer vaccines.

The first human tumor antigen recognized by antitumor cytotoxic T lymphocytes (CTL) was revealed in 1991, when Dr. Boon's group cloned the gene encoding melanoma-associated antigen, MAGE-1 [5] (later renamed MAGE-A1), which is a member of cancer-testis antigens family. One year later, the first epitope of tumor antigen recognized by tumor-specific CD8+ cytotoxic T lymphocytes (CTLs) was identified from the tumor antigen MAGE1 by the same research group [6]. The results from this study also showed that when the peptide was presented by mouse cells transfected with an HLA-A1 gene, it can be recognized by CTL as well, suggesting that tumor antigen MAGE-1 is associated with the HLA-A1 molecule. The sequence of the CTL-specific epitope was EADPTGHSY. Since then, a large amount of cancer-testis tumor antigens have been successfully identified in various cancers by T-cell epitope cloning, serological analysis of expression cDNA libraries (SEREX) technique, gene expression profiling, and comparative proteome analysis including NY-ESO-1, MAGE-A2, MAGE-A3, BAGE, and GAGE-1 [7–10].

A growing number of cancer-testis tumor antigens were also identified in human breast cancers, including MAGE tumor antigen [11], BAGE tumor antigen [12], GAGE tumor antigen [9], XAGE tumor antigens [13], etc. Several studies demonstrated that multiple cancer-testis antigens were preferentially expressed in triple-negative breast tumors (TNBC), and significantly higher expression of cancer-testis antigens such as NY-ESO-1 was detected in triple-negative breast cancers when compared with estrogen-sensitive tumors [14–16]. Thus, cancer-testis antigen vaccines can be a promising strategy for the treatment of TNBC.

15.2.2 Antigens Overexpressed in Breast Cancer

15.2.2.1 HER2/Neu Antigens

Human epidermal growth factor receptor-2/neu (HER2/neu, ERBB2), the most well-studied tumor-associated antigens for breast cancer vaccines in clinical trials, is a 185 kDa transmembrane tyrosine kinase molecule in the HER family. HER2/neu is overexpressed in one quarter of all primary breast cancers [17]. It is also overexpressed in many epithelial tumors including the stomach, ovarian, colorectal, and pancreatic carcinomas.

HER2/neu was originally thought to be less immunogenic in human breast cancers because this oncogene was only amplified but not mutated in breast cancers. Subsequently, the concept has been changed. Patients with HER-2/neu-positive tumors [18–20] have shown to have spontaneous cellular and humoral immunity, thus suggesting that HER2/neu is highly immunogenic. This finding further proves that HER2/neu is a potential tumor antigen for active vaccination. The intracellular domain (hICD) of human HER-2/neu oncogenic protein, homologous to rat neu ICD that is less immunogenic to rats, was used to immunize rats in a rat animal model by

Disis et al. [21]. The results from this study showed that the potent neu-specific humoral and cellular immune responses were elicited in the rats immunized with hICD, suggesting that vaccination with foreign antigens might be an effective therapeutic strategy for “self” tumor antigens.

Multiple HER-2 /neu peptides containing extracellular domain (ECD)/intracellular domain (ICD) were evaluated for antitumor immunity in a pilot clinical study. In this study, granulocyte macrophage colony-stimulating factor (GM-CSF) was used as an adjuvant. The results indicated that the mixture of peptides were capable of provoking HER2/neu antigen-specific CD4+ T-cell immunity, suggesting that the human body vaccinated with peptide vaccines could develop immunity to recognize and attack self-tumor antigens such as HER2/neu [22].

15.2.2.2 MUC-1 Antigens

Mucin 1 (MUC1), a member of mucin family, is a membrane-associated glycoprotein overexpressed in many types of epithelia carcinomas, including the pancreas, breast, lung, and gastrointestinal carcinomas [23–25]. It plays an important role involved in immune modulation and multiple biological processes including proliferation, differentiation, apoptosis, and metastasis. The biological function of MUC1 is defined by the posttranslational modification of the protein. In normal epithelial cells, MIC1 is hyperglycosylated, whereas in malignant cells lacking luminal polarity, MUC1 is underglycosylated, thus making the protein highly immunogenic.

More than 70% of tumors express MUC1, suggesting this tumor antigen an attractive target for vaccination [26, 27]. Several preclinical studies demonstrated that tumor cells overexpressing MUC-1 protein could elicit antitumor immunity [28–31]. Due to its highly repetitive and multivalent structure, the mucin epitope has been concluded to be an effective target antigen for cytotoxic T cells. This claim is founded on the ability of the mucin epitope to bind and stimulate the T-cell receptor without MHC presentation. The MUC-1 tumor vaccines have been tested in patients with breast cancer in several clinical trials and are discussed in the following section.

15.2.3 Neoantigens

Tumor-specific neoantigens are actively being explored as targets for personalized cancer vaccines. It has been reported that high mutational loads are strongly associated with increased tumor antigenicity (or immunogenic neoantigens) as well as high frequency of tumor-infiltrating lymphocytes such as CD8+ T cells [32–35]. Sjoblom T et al. [36] determined genetic alterations in more than 10 breast and colorectal cancers to identify somatic gene mutations including passenger mutations and driver mutations, and the data revealed that mutation of about 11 genes occurred in each tumor. Two years later, Allison’s group screened over one thousand peptides derived from these mutations to identify MHC class I T-cell epitopes. They found that an average of ten and seven unknown HLA-A0201 epitopes existed in breast cancer and colorectal cancer, respectively. This study provides an encouraging strategy for the individualized cancer vaccines [37].

An ongoing clinical feasibility study is underway to identify MHC class I and MHC class II tumor neoepitopes in breast cancer lymph nodes. Two different approaches have been employed. One is an elution HPLC method to be conducted in tumor cell lines. The second one is to use predictive algorithms on tumor sequencing data. This clinical trial will also investigate the frequency of the identified neoantigen-specific CD4+ and CD8+ T lymphocytes in the blood and tumor of breast cancer patients (ClinicalTrials.gov identifier: NCT02831634).

Two phase I clinical trials for personalized cancer vaccines in patients with triple-negative breast cancer following neoadjuvant chemotherapy are being performed by William E. Gillanders. One trial is to determine immune-related adverse events and antigenicity of individualized naked plasmid DNA vaccines (ClinicalTrials.gov identifier: NCT02348320 (see Table 15.1)). Another trial is to determine the safety and immunogenicity of a personalized synthetic long peptide breast cancer vaccine (Poly ICLC) strategy (ClinicalTrials.gov identifier: NCT02427581 (see Table 15.2)). The personalized cancer

Table 15.1 Ongoing clinical trials of DNA-/RNA-based vaccines in breast cancer

Phase	Study	BC subtype	Trial ID	PI
I	Personalized Polyepitope DNA vaccine +chemo	TNBC	NCT02348320	Gillanders
I	Mammaglobin-A DNA vaccine	BC	NCT02204098	Gillanders
I	CD105/Yb-1/SOX2/CDH3/ MDM2-polyepitope plasmid DNA vaccine	HER2-BC	NCT02157051	Disis
I	pUMVC3-IGFBP2-HER2-IGF1R plasmid DNA vaccine	HER2-BC	NCT02780401	Wisinski
I/II	PolyICLC in situ vaccine + PD-L1 + CTLA-4 blockade	BC	NCT02643303	Slingluff, Bhardwaj
I	IVAC_W_bre1_uID and IVAC_M_ uID RNA vaccine	TNBC	NCT02316457	Sahin

TNBC Triple-negative breast cancer, *BC* Breast cancer, *Chemo* Chemotherapy.

Table 15.2 Ongoing clinical trials of peptide-based vaccines in breast cancer

Phase	Study	BC subtype	Trial ID	PI
I	Personalized synthetic long peptide vaccine +chemo	TNBC	NCT02427581	Gillanders
I/II	HER2 peptide vaccine + GM-CSF + chemo +adoptive HER2-specific T cells	HER2 + BC	NCT00791037	Disis
I	Ad-sig-hMUC-1/ecdCD40L vaccine	BC	NCT02140996	Toh
II	Folate receptor alpha + GM-CSF	TNBC	NCT02593227	Kenney
II	E75 + GM-CSF	DCIS	NCT02636582	Mittendorf
I	hTERT/survivin multi-peptide vaccine + Basiliximab (anti-CD25)	BC	NCT01660529	Fox
I	PVX-410 + PD-L1 inhibitor	TNBC	NCT02826434	Multiple PI
II	E75 + GM-CSF + Herceptin	HER2 low BC	NCT01570036	Peoples
I	HER2 peptide vaccine	BC	NCT01376505	Kaumaya
I/II	HER2 ICD peptide vaccine + trastuzumab	HER2 + BC	NCT01922921	Salaar
I	HER2 peptide + GM-CSF + chemo + imiquimod	HER2 + BC	NCT02276300	Krackhardt
II	E75 + GM-CSF + trastuzumab	HER2 + BC	NCT02297698	Peoples
I	Sialyl Lewis^{a}-KLH + QS21	BC	NCT00470574	Gilewski + Dickler
I	MUC-1 peptide vaccine + PolyICLC	TNBC	NCT00986609	Baar
I/II	Folate-binding protein peptide vaccines (E39 and J65) + GM-CSF	BC	NCT02019524	Peoples
II	GP2Z + GM-CSF vs AE37 + GM-CSF vs GM-CSF	BC	NCT00524277	Peoples
I	HER2 ECD + TM protein vaccine	BC	NCT01526473	Lyerly

TNBC Triple-negative breast cancer, *BC* Breast cancer, *DCIS* Ductal carcinoma in situ.

vaccines are proposed to be feasible for human immunization and to have capacity to boost CD8 cytotoxic T-cell immunity against neoantigens. In addition, these studies hypothesize that clinical responses could be correlated with immune responses elicited by the personalized cancer vaccines.

15.2.4 Other Types of Tumor Antigens

Other types of cancer antigens include oncofetal antigens such as CEA and AFP and differentiation antigens such as NY-BR-1 and Wilms' tumor antigen (WT1). Carcinoembryonic antigen (CEA) is a membrane glycosylated protein that is expressed in 50% of patients with breast carcinomas [38]. Numerous clinical trials with CEA therapeutic cancer vaccines have been conducted for breast cancer patients. WT1 plays an important role in various cellular functions such as proliferation, differentiation, and apoptosis in tumors including breast cancer [39]. Several WT1 antigenic HLA class I-restricted peptides have been identified. In a phase I/II clinical trials enrolling 116 patients with various types of WT1+ tumors, the therapeutic efficacy of WT1 peptide vaccine was evaluated, and the enhanced immune response and favorable clinical outcome were observed in 50% of patients receiving immunization [40].

15.3 Vaccine Delivery Systems in Breast Cancer

15.3.1 Peptide-/Protein-Based Vaccines

Peptide-based vaccines utilize immunogenic epitopes derived from TAA/TSA to enhance activation of CD8 + cytotoxic T lymphocytes (CTLs) and CD4 + helper T cells and production of antibodies by B cells. The activated CTLs are able to recognize and fight tumor cells. Two major factors that control the elicited T-cell immune response by TAA/TSA peptides are 1) the capability of the peptide to bind to MHC complex and 2) the ability of the peptide-MHC class I/II molecule to bind to the T-cell receptor (TCR). While minor subsets of TAA epitopes are presented by MHC class II molecules to CD4+ T helper cells, more common TAA peptides are presented to tumor-specific CD8+ cytotoxic T cells by MHC class I molecules [41–45]. Many clinical trials using different peptide-/protein-based vaccines for breast cancer patients have been attempted. Some of these trials have been completed, while others are under investigation (see Table 15.2).

15.3.1.1 Multi-peptide Vaccines

A large number of clinical studies have investigated the therapeutic efficacies of tumor peptide vaccines derived from ECD and ICD of HER2/neu oncoprotein [46–49]. Disis et al. [46] conducted a phase I clinical trial enrolling 64 patients with HER2/neu + breast cancer, non-small cell lung cancer, or ovarian cancer, which detected tumor antigen-specific T-cell immune response using three distinct HER2/neu vaccines: the first vaccine consists of ECD peptides, p42 (aa 42–56), p98 (aa 98–114), and p328 (aa 328–345); the second vaccine is composed of ICD-peptides, p776 (aa 776–790), p927 (aa 927–941), and p1166 (aa 1166–1180); and the third vaccine consists of MHC class I HLA-A2 binding motifs, p369 (aa 369–386), p688 (aa 688–703), and p971 (aa 971–984), which is directed to provoke CD8+ cytotoxic T cells. GM-CSF as an immune adjuvant was used in combination with each vaccine. The results of vaccination in most immunized patients showed that the peptide vaccines described above were able to boost potent and durable antigen-specific T-cell responses [46].

The cytokine GM-CSF is widely used as a potent immune adjuvant in a variety of cancer vaccines. It is an important immune modulator which can promote differentiation, maturation, and recruitment of dendritic cells and enhance antigen presentation to CD4+ helper T cells and cross-presentation to CD8+ cytotoxic T cells [50–54].

15.3.1.2 NeuVax Peptide Vaccine

NeuVax, comprised of the E75 HER2/neu peptide and immune adjuvant GM-CSF, is engineered to lower the relapse of HER2-negative breast cancer patients not eligible for trastuzumab treatment. Peoples's group has performed two concurrent phase II trials [55] with 168 patients (HLA-A2 and HLA-A3) in an effort to reduce relapse in patients with breast cancer including both node + and node –. At a median of 20-month follow-up, research data illustrated that relapse rate of patients who received vaccination was much lower than that of control patients. NeuVax showed an acceptable safety profile and ability to stimulate E75-specific immune responses. The clinical trial with NeuVax alone has been recently completed [54]. Two clinical trials investigating NeuVax vaccine in combination with trastuzumab are currently underway. One is a phase IIb trial enrolling patients with triple-negative HER2 IHC1+/2+ and node + (ClinicalTrials.gov identifier: NCT01570036), and another one is a phase II clinical trial recruiting patients with high-risk, negative HER2 IHC 3+ or node + (ClinicalTrials.gov identifier: NCT02297698).

15.3.1.3 GP2 Peptide Vaccine

GP2 (aa 654–662, epitope: IISAVVGIL), the second HER2-derived peptide, is made up of 9 amino acids and HLA-A2/A3-restricted. Moreover, it is a MHC class I peptide derived from the transmembrane domain of HER2/neu. GP2 was initially believed to bind to human leukocyte antigen A2 with less affinity than E75 peptide [20]. However, years later, studies depicted that GP2 peptide was highly immunogenic to the immune system [56]. To elucidate whether HLA-A2-restricted GP2 peptide had high tumor immunogenicity, an in vitro study using prevaccination peripheral blood mononuclear cell (PBMC) was performed. The PBMC were isolated from breast cancer patients with HLA-A2 positive. CD14+ monocytes were purified for preparation of myeloid-derived dendritic cells. Meanwhile, highly pure CD8+ T cells were isolated from HLA-A2 healthy individuals. The PBMC were triggered with dendritic cells pulsed with GP2 peptide, and the Cr-51 cytotoxicity assays with HER2/neu-positive tumor cells were conducted. The results of this study demonstrated that the potent cytotoxic activity of GP2 peptide-boosted PBMC from breast cancer patients was detectable when compared to the control. Moreover, GP2-peptide-elicited antigen-specific CD8 T cells were cytotoxic to HER2/neu tumor targets. These findings clearly showed that GP2, similar to E75, is highly immunogenic. Strong evidence was also provided to support GP2 peptide as an effective vaccine strategy for HLA-A2-expressing breast cancer patients.

15.3.1.4 AE37 Peptide Vaccine

Clinical trials showed that MHC class II-restricted tumor antigen presentation to CD4+ T helper cells could induce the activation of both antigen-specific CD4+ T helper cells and CD8+ cytotoxic T cells, promoting long-lasting immune responses [57, 58].

The third HER2-derived peptide is the hybrid AE37 (aa776–790 + Ii-Key [LRMK] to enhance its immunogenicity). Derived from HER2 ICD, this peptide is MHC class II-restricted tumor antigen engineered to activate CD4 + Th cell immunity. Helmos et al. [59] published the results of the AE37 peptide vaccine in 15 patients with breast cancer, which was the first human phase I trial. They found that the hybrid AE37 vaccine along with GM-CSF was safe and tolerable with minimal toxicity. Moreover, the vaccine is capable of stimulating AE37-specific, long-lived CD4+ T-cell-mediated immunity, even in the absence of an immune adjuvant. Early clinical data illustrate that AE37 vaccination might prevent breast cancer relapse [52, 59, 60].

15.3.1.5 sTn-KLH Peptide Vaccine

Sialyl-Tn (sTn), a carbohydrate epitope identified on various glycoproteins, is expressed in a wide variety types of tumor cells including breast cancer. sTn plays an important role in tumor cell proliferation and metastasis [61]. The sTn antigen is conjugated to the keyhole limpet hemocyanin (KLH) to formulate a vaccine. Immunization of mice with either synthetic single sTn-KLH or clustered sTn-KLH [sTn(c)-KLH] conjugates coupled with immune adjuvant

QS-21 induced IgM and IgG antibodies reactive with OSM and the respective synthetic antigens [62]. sTn(c)-KLH plus QS-21was evaluated in a phase I clinical trial with 27 high-risk breast cancer patients. This trial aimed to determine (a) side effects of sTn antigen and (b) humoral immune response specific for sTn and sTn-overexpressing tumor cells such as IgM and IgG. According to the study results, all patients enrolled in the trial generated high titers of IgM and IgG antibodies specific to sTn(c). These findings demonstrated that sTn(c)-KLH peptide-based vaccine was highly immunogenic and had acceptable safety profile in patients with advanced breast cancer [63]. This clinical trial is currently under investigation (ClinicalTrial.gov identifier: NCT00470574).

15.3.1.6 MUC-1 Peptide Vaccine

MUC1, a cell wall-based mucin protein with high immunogenicity, is ubiquitously expressed across many types of tumors [64, 65]. Given this, MUC1 has been well studied as an attractive vaccine target. Overexpression of aberrantly glycosylated MUC1 on breast cancer cells can elicit a cytotoxic T-cell immune response and B cell-mediated antibody production in patients with early-stage breast cancer. MUC1-elicited immune responses are correlated with favorable clinical outcome [66]. And MUC1 could be used as an immunogenic tumor antigen in cancer patients with any HLA genotype [67, 68].

One study reported that majority of early-stage, high-grade triple-negative breast cancer (TNBC) expressed high levels of MUC1 [69], suggesting that these tumors might be sensitive to MUC-1 peptide vaccine-induced immune attack. A pilot phase III study in patients with early-stage ER+ breast cancer has demonstrated that administration of MUC-1 vaccine successfully extends patients' survival rate [70]. One active clinical trial is underway to determine the efficacy of MUC-1 peptide vaccine coupled with poly-IC in boosting the immune response to MUC1 in patients with triple-negative breast cancer (ClinicalTrial.gov identifier: NCT00986609).

15.3.2 Dendritic Cell-Based Vaccines

Dendritic cells (DCs), the professional antigen-presenting cells, have high capacity to capture, process, and present antigens to T cells and play critical roles in protective immunity [71, 72]. DCs express HLA class I and class II molecules, co-stimulatory proteins such as CD80 and CD86, and produce cytokines such as IL-12 that are necessary for T-cell activation [73, 74]. The strategy of DC-based vaccines involves the loading of whole tumor cells or defined antigen to autologous DCs ex vivo followed by inoculation into patients. Murine models have provided evidence that antitumor activity induced by DC-based vaccination is associated with the approach of DC administration [75–77]. DCs are involved in the stimulation of innate and adaptive immunity. The DC vaccine-mediated tumor rejection has been observed in many cancer patients receiving vaccination [78].

Numerous ongoing therapeutic DC-based cancer vaccines are being tested in clinical trials in breast cancer patients (see Table 15.3). A pilot clinical study in ten patients with advanced breast and ovarian cancer was performed [79]. Mature DC cells pulsed with HLA-A2-restricted MUC1 or HER2 peptides were administered subcutaneously to the patients for a total of three times. Fifty percent of patients who received the immunization developed HER2-specific CTLs, which efficiently inhibited HER2-expressing cancer cell line.

In a phase I clinical trial study, DCs pulsed with HER2 peptide vaccine were employed to immunize patients with ductal carcinoma in situ (DCIS), a preinvasive breast cancer. DC1 polarization culture techniques, including TLR agonist exposure, were also employed to study IL-12 secretion at the time of vaccination. Vaccination was administered by injecting the vaccines directly into lymph nodes of breast cancer patients. Vaccinated patients developed antigen-specific immunity and presented high levels of peptide-specific CD4 + Th cells and CD8+ cytotoxic T cells. This vaccination strategy has generated complement-fixing, tumor-lytic antibodies. Reductions in extent of DCIS and levels of

Table 15.3 Ongoing clinical trials of dendritic cell-based vaccines in breast cancer

Phase	Study	BC subtype	Trial ID	PI
I	Tumor blood vessel antigen peptide-pulsed DC	BC	NCT02479230	Baar
I	Ad HER2-pulsed DC	Her2/neu + BC	NCT01730118	Wood
III	DC + Chemo	TNBC	NCT02018458	Shaughnessy
I	HER2-pulsed DC	Her2/neu + BC	NCT02063724	Czerniecki
I	HER2-pulsed DC + Chemo	Her2/neu + BC	NCT02061423	Czerniecki
I/II	HER2-pulsed DC	DCIS	NCT02061332	Czerniecki

TNBC Triple-negative breast cancer, *BC* Breast cancer, *DCIS* Ductal carcinoma in situ, *Chemo* Chemotherapy, *DC* Dendritic cells

expression of HER2 after vaccination were observed. Furthermore, HER2/neu-pulsed DC vaccines led to accumulation of lymphocytes in the breast and changes in residual DCIS and induced complement-dependent antibody-mediated cell cytotoxicity [80]. A randomized selection clinical trial enrolling 54 patients with metastatic HER2-positive breast cancer or DCIS was recently conducted using HER2 peptide-pulsed dendritic cell vaccine [81]. Patients were randomized to intralesional (IL), intranodal (IN), or both intralesional and intranodal (ILN) injection. Vaccination by all injection routes was well tolerated. The result of the study suggests that anti-HER2 DC vaccination is a safe and immunogenic treatment to induce tumor-specific T-cell immune responses in HER2-positive patients. Similar clinical investigation with autologous DC-based vaccine administered to 27 breast cancer patients supports previous reports [82].

15.3.3 Whole Tumor Cell-Based Vaccines

The use of whole tumor cells-based vaccines is another strategy to induce immune responses via TAAs/TSAs in tumor cells. The tumor cells can be derived from either patient (autologous cells) or cell-line culture (allogeneic cells). To generate whole tumor cell-based cancer vaccines, the cells are usually genetically modified via virus-mediated transduction to express immune-activating cytokines or co-stimulatory molecules to enhance the effect of vaccination. GM-CSF has been the most potent protein in augmenting protective antitumor immunity [83–85]. Vaccination using irradiated, genetically modified GM-CSF-secreting tumor cells (GVAX) consistently enhanced antitumor immunity across a wide range of experimental tumor models [84]. Phase I/II clinical trials with GVAX have been conducted in various types of tumors [86–88]. The feature of whole tumor cell-based vaccines is to offer an entire pool of multiple tumor-associated antigens that can boost CD4+ and CD8+ T-cell immunity directed against these antigens. Emens et al. [89]conducted a phase II trial using an allogeneic breast tumor vaccine engineered to secrete GM-CSF in combination with chemotherapy for 28 patients with HER-2/neu-overexpressing metastatic breast cancer. The results have revealed two main findings: a) the administration of DC vaccine alone or in combination with chemotherapy is safe; and b) the vaccine provokes HER2-antigen-specific immune response in patients [89]. There are two clinical trials with whole tumor cell-based cancer vaccines currently being conducted in breast cancer patients (see Table 15.4).

15.3.4 DNA-/RNA-Based Vaccines

DNA-/RNA-based vaccine strategy has been increasingly recognized as an attractive cancer immunotherapy approach due to its cost-effectiveness, safety, and stability. DNA/RNA

Table 15.4 Ongoing clinical trials of whole tumor cell-based vaccines in breast cancer

Phase	Study	BC subtype	Trial ID	PI
II	Allogeneic whole tumor cell vaccine +GM-CSF + chemo	BC	NCT00971737	Emens
I/II	Whole tumor cell BriaVax vaccine +Chemo + IFNa	BC	NCT03066947	Peoples

BC Breast cancer., *Chemo* Chemotherapy

vaccines contain the genetic information for TAA or TSA, which can be injected alone into a patient as a naked nucleic acid vaccine. DNA-/RNA-based vaccines represent an inspiring strategy to harness specific and potent immune responses, including humoral and cellular immunity [90–94].

Several ongoing clinical trials with therapeutic DNA vaccines are being actively carried out in patients with breast cancer (see Table 15.1). Gillanders's group have recently completed a phase I clinical trial investigating immunogenicity and safety of a DNA vaccine directed to express the human mammaglobin-A antigen (SCGB2A2), which is used as a marker to detect metastatic breast cancer [95]. In this trial, the plasmid mammaglobin-A DNA vaccine is formulated as a naked plasmid DNA vaccine (WUSM-MGBA-01). Altogether 14 patients were vaccinated. Following the DNA vaccination, the frequency of mammaglobin-A-specific CD8+ T cells and the numbers of IFN-gamma-producing CD8+ T cells were significantly increased. Furthermore, the results of the clinical trial showed that the breast cancer patients who received mammaglobin-A DNA vaccination had improved progression-free survival rate when compared to the control patients. The mammaglobin-A DNA vaccine was safe due to its minimum toxicities [96].

An ongoing phase I clinical trial, the first clinical study with RNA-based vaccines (the combination of IVAC_W_bre1_uID and IVAC_M_uID) for personalized therapy in patients with TNBC (see Table 15.1), is currently being conducted. The Mutanome Engineered RNA Immunotherapy (MERIT) trial combines two RNA vaccine strategies: the IVAC® WAREHOUSE and the IVAC® MUTANOME. This combinatorial therapy produces two individualized IVAC® investigational medicinal products (IMPs) (IVAC_W_bre1_uID and IVAC_M_uID) for each patient.

The IVAC_W_bre1_uID (also known as IVAC Warehouse), an individualized, therapeutic cancer vaccine (IVAC), is composed of liposomes that contain RNA-encoding TAAs expressed in patient's cancer, which are selected from a warehouse and p53 RNA. These antigens have been demonstrated by immunogenicity testing. IVAC_M_uID (also known as IVAC MUTANOME) is founded on the characterization of mutations specific to tumors by next-generation sequencing (NGS). This mechanism is used to identify several neoantigens from mutant epitopes. Upon administration of RNA vaccines, the RNA is translated by APCs and the protein is presented by MHC class I/II molecules, leading to the activation of CTL and memory T-cell immune responses directed against tumor-specific antigens.

15.4 Conclusions and Future Perspectives

Vaccines have been traditionally employed as a preventive strategy in infectious diseases by producing neutralizing antibodies against foreign pathogens. More recently, therapeutic vaccines have been designed to induce the immune system to elicit tumor antigen-specific T-cell-mediated immune responses against infected cells and tumors. An optimal therapeutic cancer vaccine is intended to eradicate established tumors and generate immune memory responses to prevent future recurrence. Although various strategies of therapeutic breast cancer vaccines have been rapidly developed to mount effective antitumor immune responses, the issues of immune tolerance, immunosuppression (e.g., Tregs, MDSCs), and tumor escape need to be overcome for the development of more efficient cancer vaccines.

By far, several cancer vaccines have received US FDA approval, including two preventive cancer vaccines against hepatitis B and the human papilloma virus for prevention of hepatocellular carcinoma and cervical cancer, respectively, and one therapeutic vaccine sipuleucel-T for treatment of metastatic prostate cancer. Such successes will spark an interest in the generation of preventive vaccine for primary immunoprevention of breast cancer, because approved preventive cancer vaccines would induce antigen-specific immune response to prevent initiation and progression of cancer. The development of breast cancer preventive vaccine will require optimal factors including antigen selection, strategy for immune stimulation, surrogate for vaccine efficacy, and most importantly, informative biomarkers. Moreover, new generations of clinical trials of breast cancer vaccines will aim at designing and employing personalized vaccines to boost a robust immune response to a custom-made neoantigen in the patient with breast cancer.

Chemotherapy is believed to suppress the immune system in cancer patients. However, certain chemotherapy can trigger an immunologic response [97]. Accumulating evidence indicates that conventional cancer therapy not only exerts direct cytotoxic effects but also impacts on antitumor immune responses. Chemotherapy can facilitate immunity by increasing the immunogenicity of tumor cells, suggesting a potential combinatorial strategy with immunotherapeutic agents [97, 98]. Thus, combining vaccination and other forms of cancer therapy, such as chemotherapy, radiotherapy, targeted therapy with monoclonal antibody, or immune checkpoint blockade, is suggested as a promising approach to achieve potent and durable antitumor benefit.

Acknowledgments Xiaoyu Li is supported by a grant from National Natural Science Foundation of China (grant number: NSFC 81670163).

References

1. Greten TF, Jaffee EM (1999) Cancer vaccines. J Clin Oncol 17(3):1047–1060. doi:10.1200/JCO.1999.17.3.1047
2. Boon T, Coulie PG, Van den Eynde B (1997) Tumor antigens recognized by T cells. Immunol Today 18(6):267–268
3. Scanlan MJ, Gure AO, Jungbluth AA, Old LJ, Chen YT (2002) Cancer/testis antigens: an expanding family of targets for cancer immunotherapy. Immunol Rev 188:22–32
4. Simpson AJ, Caballero OL, Jungbluth A, Chen YT, Old LJ (2005) Cancer/testis antigens, gametogenesis and cancer. Nat Rev Cancer 5(8):615–625. doi:10.1038/nrc1669
5. van der Bruggen P, Traversari C, Chomez P, Lurquin C, De Plaen E, Van den Eynde B, Knuth A, Boon T (1991) A gene encoding an antigen recognized by cytolytic T lymphocytes on a human melanoma. Science 254(5038):1643–1647
6. Traversari C, van der Bruggen P, Luescher IF, Lurquin C, Chomez P, Van Pel A, De Plaen E, Amar-Costesec A, Boon T (1992) A nonapeptide encoded by human gene MAGE-1 is recognized on HLA-A1 by cytolytic T lymphocytes directed against tumor antigen MZ2-E. J Exp Med 176(5):1453–1457
7. Chen YT, Scanlan MJ, Sahin U, Tureci O, Gure AO, Tsang S, Williamson B, Stockert E, Pfreundschuh M, Old LJ (1997) A testicular antigen aberrantly expressed in human cancers detected by autologous antibody screening. Proc Natl Acad Sci U S A 94(5):1914–1918
8. Boel P, Wildmann C, Sensi ML, Brasseur R, Renauld JC, Coulie P, Boon T, van der Bruggen P (1995) BAGE: a new gene encoding an antigen recognized on human melanomas by cytolytic T lymphocytes. Immunity 2(2):167–175
9. De Backer O, Arden KC, Boretti M, Vantomme V, De Smet C, Czekay S, Viars CS, De Plaen E, Brasseur F, Chomez P, Van den Eynde B, Boon T, van der Bruggen P (1999) Characterization of the GAGE genes that are expressed in various human cancers and in normal testis. Cancer Res 59(13):3157–3165
10. Gaugler B, Van den Eynde B, van der Bruggen P, Romero P, Gaforio JJ, De Plaen E, Lethe B, Brasseur F, Boon T (1994) Human gene MAGE-3 codes for an antigen recognized on a melanoma by autologous cytolytic T lymphocytes. J Exp Med 179(3):921–930
11. Russo V, Traversari C, Verrecchia A, Mottolese M, Natali PG, Bordignon C (1995) Expression of the MAGE gene family in primary and metastatic human breast cancer: implications for tumor antigen-specific immunotherapy. Int J Cancer 64(3):216–221
12. Fujie T, Mori M, Ueo H, Sugimachi K, Akiyoshi T (1997) Expression of MAGE and BAGE genes in Japanese breast cancers. Ann Oncol 8(4):369–372

13. Egland KA, Kumar V, Duray P, Pastan I (2002) Characterization of overlapping XAGE-1 transcripts encoding a cancer testis antigen expressed in lung, breast, and other types of cancers. Mol Cancer Ther 1(7):441–450
14. Badovinac Crnjevic T, Spagnoli G, Juretic A, Jakic-Razumovic J, Podolski P, Saric N (2012) High expression of MAGE-A10 cancer-testis antigen in triple-negative breast cancer. Med Oncol 29(3):1586–1591. doi:10.1007/s12032-011-0120-9
15. Chen YT, Ross DS, Chiu R, Zhou XK, Chen YY, Lee P, Hoda SA, Simpson AJ, Old LJ, Caballero O, Neville AM (2011) Multiple cancer/testis antigens are preferentially expressed in hormone-receptor negative and high-grade breast cancers. PLoS One 6(3):e17876. doi:10.1371/journal.pone.0017876
16. Curigliano G, Viale G, Ghioni M, Jungbluth AA, Bagnardi V, Spagnoli GC, Neville AM, Nole F, Rotmensz N, Goldhirsch A (2011) Cancer-testis antigen expression in triple-negative breast cancer. Ann Oncol 22(1):98–103. doi:10.1093/annonc/mdq325
17. Slamon DJ, Godolphin W, Jones LA, Holt JA, Wong SG, Keith DE, Levin WJ, Stuart SG, Udove J, Ullrich A et al (1989) Studies of the HER-2/neu proto-oncogene in human breast and ovarian cancer. Science 244(4905):707–712
18. Baxevanis CN, Sotiropoulou PA, Sotiriadou NN, Papamichail M (2004) Immunobiology of HER-2/neu oncoprotein and its potential application in cancer immunotherapy. Cancer Immunol Immunother 53(3):166–175. doi:10.1007/s00262-003-0475-7
19. Disis ML, Calenoff E, McLaughlin G, Murphy AE, Chen W, Groner B, Jeschke M, Lydon N, McGlynn E, Livingston RB et al (1994) Existent T-cell and antibody immunity to HER-2/neu protein in patients with breast cancer. Cancer Res 54(1):16–20
20. Peoples GE, Goedegebuure PS, Smith R, Linehan DC, Yoshino I, Eberlein TJ (1995) Breast and ovarian cancer-specific cytotoxic T lymphocytes recognize the same HER2/neu-derived peptide. Proc Natl Acad Sci U S A 92(2):432–436
21. Disis ML, Shiota FM, Cheever MA (1998) Human HER-2/neu protein immunization circumvents tolerance to rat neu: a vaccine strategy for 'self' tumour antigens. Immunology 93(2):192–199
22. Disis ML, Grabstein KH, Sleath PR, Cheever MA (1999) Generation of immunity to the HER-2/neu oncogenic protein in patients with breast and ovarian cancer using a peptide-based vaccine. Clin Cancer Res 5(6):1289–1297
23. Kimura T, McKolanis JR, Dzubinski LA, Islam K, Potter DM, Salazar AM, Schoen RE, Finn OJ (2013) MUC1 vaccine for individuals with advanced adenoma of the colon: a cancer immunoprevention feasibility study. Cancer Prev Res (Phila) 6(1):18–26. doi:10.1158/1940-6207.CAPR-12-0275
24. Weiner LM, Surana R, Murray J (2010) Vaccine prevention of cancer: can endogenous antigens be targeted? Cancer Prev Res (Phila) 3(4):410–415. doi:10.1158/1940-6207.CAPR-10-0040
25. Kovjazin R, Horn G, Smorodinsky NI, Shapira MY, Carmon L (2014) Cell surface-associated anti-MUC1-derived signal peptide antibodies: implications for cancer diagnostics and therapy. PLoS One 9(1):e85400. doi:10.1371/journal.pone.0085400
26. Kohlgraf KG, Gawron AJ, Higashi M, VanLith ML, Shen X, Caffrey TC, Anderson JM, Hollingsworth MA (2004) Tumor-specific immunity in MUC1.Tg mice induced by immunization with peptide vaccines from the cytoplasmic tail of CD227 (MUC1). Cancer Immunol Immunother 53(12):1068–1084
27. Chen D, Xia J, Tanaka Y, Chen H, Koido S, Wernet O, Mukherjee P, Gendler SJ, Kufe D, Gong J (2003) Immunotherapy of spontaneous mammary carcinoma with fusions of dendritic cells and mucin 1-positive carcinoma cells. Immunology 109(2):300–307
28. Ding L, Lalani EN, Reddish M, Koganty R, Wong T, Samuel J, Yacyshyn MB, Meikle A, Fung PY, Taylor-Papadimitriou J et al (1993) Immunogenicity of synthetic peptides related to the core peptide sequence encoded by the human MUC1 mucin gene: effect of immunization on the growth of murine mammary adenocarcinoma cells transfected with the human MUC1 gene. Cancer Immunol Immunother 36(1):9–17
29. Apostolopoulos V, Xing PX, McKenzie IF (1994) Murine immune response to cells transfected with human MUC1: immunization with cellular and synthetic antigens. Cancer Res 54(19):5186–5193
30. Zhang S, Graeber LA, Helling F, Ragupathi G, Adluri S, Lloyd KO, Livingston PO (1996) Augmenting the immunogenicity of synthetic MUC1 peptide vaccines in mice. Cancer Res 56(14):3315–3319
31. Acres RB, Hareuveni M, Balloul JM, Kieny MP (1993) Vaccinia virus MUC1 immunization of mice: immune response and protection against the growth of murine tumors bearing the MUC1 antigen. J Immunother Emphasis Tumor Immunol 14(2):136–143
32. Joffroy CM, Buck MB, Stope MB, Popp SL, Pfizenmaier K, Knabbe C (2010) Antiestrogens induce transforming growth factor beta-mediated immunosuppression in breast cancer. Cancer Res 70(4):1314–1322. doi:10.1158/0008-5472.CAN-09-3292
33. Castle JC, Kreiter S, Diekmann J, Lower M, van de Roemer N, de Graaf J, Selmi A, Diken M, Boegel S, Paret C, Koslowski M, Kuhn AN, Britten CM, Huber C, Tureci O, Sahin U (2012) Exploiting the mutanome for tumor vaccination. Cancer Res 72(5):1081–1091. doi:10.1158/0008-5472.CAN-11-3722
34. Robbins PF, Lu YC, El-Gamil M, Li YF, Gross C, Gartner J, Lin JC, Teer JK, Cliften P, Tycksen E, Samuels Y, Rosenberg SA (2013) Mining exomic sequencing data to identify mutated antigens recognized by adoptively transferred tumor-reactive T cells. Nat Med 19(6):747–752. doi:10.1038/nm.3161
35. Wick DA, Webb JR, Nielsen JS, Martin SD, Kroeger DR, Milne K, Castellarin M, Twumasi-Boateng K, Watson PH, Holt RA, Nelson BH (2014) Surveillance

of the tumor mutanome by T cells during progression from primary to recurrent ovarian cancer. Clin Cancer Res 20(5):1125–1134. doi:10.1158/1078-0432.CCR-13-2147
36. Sjoblom T, Jones S, Wood LD, Parsons DW, Lin J, Barber TD, Mandelker D, Leary RJ, Ptak J, Silliman N, Szabo S, Buckhaults P, Farrell C, Meeh P, Markowitz SD, Willis J, Dawson D, Willson JK, Gazdar AF, Hartigan J, Wu L, Liu C, Parmigiani G, Park BH, Bachman KE, Papadopoulos N, Vogelstein B, Kinzler KW, Velculescu VE (2006) The consensus coding sequences of human breast and colorectal cancers. Science 314(5797):268–274. doi:10.1126/science.1133427
37. Segal NH, Parsons DW, Peggs KS, Velculescu V, Kinzler KW, Vogelstein B, Allison JP (2008) Epitope landscape in breast and colorectal cancer. Cancer Res 68(3):889–892. doi:10.1158/0008-5472.CAN-07-3095
38. Thompson JA, Grunert F, Zimmermann W (1991) Carcinoembryonic antigen gene family: molecular biology and clinical perspectives. J Clin Lab Anal 5(5):344–366
39. Cheever MA, Allison JP, Ferris AS, Finn OJ, Hastings BM, Hecht TT, Mellman I, Prindiville SA, Viner JL, Weiner LM, Matrisian LM (2009) The prioritization of cancer antigens: a national cancer institute pilot project for the acceleration of translational research. Clin Cancer Res 15(17):5323–5337. doi:10.1158/1078-0432.CCR-09-0737
40. Keilholz U, Letsch A, Busse A, Asemissen AM, Bauer S, Blau IW, Hofmann WK, Uharek L, Thiel E, Scheibenbogen C (2009) A clinical and immunologic phase 2 trial of Wilms tumor gene product 1 (WT1) peptide vaccination in patients with AML and MDS. Blood 113(26):6541–6548. doi:10.1182/blood-2009-02-202598
41. Topalian SL, Gonzales MI, Parkhurst M, Li YF, Southwood S, Sette A, Rosenberg SA, Robbins PF (1996) Melanoma-specific CD4+ T cells recognize nonmutated HLA-DR-restricted tyrosinase epitopes. J Exp Med 183(5):1965–1971
42. Manici S, Sturniolo T, Imro MA, Hammer J, Sinigaglia F, Noppen C, Spagnoli G, Mazzi B, Bellone M, Dellabona P, Protti MP (1999) Melanoma cells present a MAGE-3 epitope to CD4(+) cytotoxic T cells in association with histocompatibility leukocyte antigen DR11. J Exp Med 189(5):871–876
43. Pieper R, Christian RE, Gonzales MI, Nishimura MI, Gupta G, Settlage RE, Shabanowitz J, Rosenberg SA, Hunt DF, Topalian SL (1999) Biochemical identification of a mutated human melanoma antigen recognized by CD4(+) T cells. J Exp Med 189(5):757–766
44. Wang RF, Wang X, Rosenberg SA (1999) Identification of a novel major histocompatibility complex class II-restricted tumor antigen resulting from a chromosomal rearrangement recognized by CD4(+) T cells. J Exp Med 189(10):1659–1668
45. Chiari R, Hames G, Stroobant V, Texier C, Maillere B, Boon T, Coulie PG (2000) Identification of a tumor-specific shared antigen derived from an Eph receptor and presented to CD4 T cells on HLA class II molecules. Cancer Res 60(17):4855–4863
46. Disis ML, Gooley TA, Rinn K, Davis D, Piepkorn M, Cheever MA, Knutson KL, Schiffman K (2002) Generation of T-cell immunity to the HER-2/neu protein after active immunization with HER-2/neu peptide-based vaccines. J Clin Oncol 20(11):2624–2632. doi:10.1200/JCO.2002.06.171
47. Disis ML, Schiffman K, Guthrie K, Salazar LG, Knutson KL, Goodell V, dela Rosa C, Cheever MA (2004) Effect of dose on immune response in patients vaccinated with an her-2/neu intracellular domain protein--based vaccine. J Clin Oncol 22 (10):1916–1925. doi:10.1200/JCO.2004.09.005
48. Disis ML, Goodell V, Schiffman K, Knutson KL (2004) Humoral epitope-spreading following immunization with a HER-2/neu peptide based vaccine in cancer patients. J Clin Immunol 24(5):571–578. doi:10.1023/B:JOCI.0000040928.67495.52
49. Knutson KL, Schiffman K, Disis ML (2001) Immunization with a HER-2/neu helper peptide vaccine generates HER-2/neu CD8 T-cell immunity in cancer patients. J Clin Invest 107(4):477–484. doi:10.1172/JCI11752
50. Bowne WB, Wolchok JD, Hawkins WG, Srinivasan R, Gregor P, Blachere NE, Moroi Y, Engelhorn ME, Houghton AN, Lewis JJ (1999) Injection of DNA encoding granulocyte-macrophage colony-stimulating factor recruits dendritic cells for immune adjuvant effects. Cytokines Cell Mol Ther 5(4):217–225
51. Disis ML, Bernhard H, Shiota FM, Hand SL, Gralow JR, Huseby ES, Gillis S, Cheever MA (1996) Granulocyte-macrophage colony-stimulating factor: an effective adjuvant for protein and peptide-based vaccines. Blood 88(1):202–210
52. Benavides LC, Sears AK, Gates JD, Clifton GT, Clive KS, Carmichael MG, Holmes JP, Mittendorf EA, Ponniah S, Peoples GE (2011) Comparison of different HER2/neu vaccines in adjuvant breast cancer trials: implications for dosing of peptide vaccines. Expert Rev Vaccines 10(2):201–210. doi:10.1586/erv.10.167
53. Tagliamonte M, Petrizzo A, Tornesello ML, Buonaguro FM, Buonaguro L (2014) Antigen-specific vaccines for cancer treatment. Hum Vaccin Immunother 10(11):3332–3346. doi:10.4161/21645515.2014.973317
54. Mittendorf EA, Clifton GT, Holmes JP, Schneble E, van Echo D, Ponniah S, Peoples GE (2014) Final report of the phase I/II clinical trial of the E75 (nelipepimut-S) vaccine with booster inoculations to prevent disease recurrence in high-risk breast cancer patients. Ann Oncol 25(9):1735–1742. doi:10.1093/annonc/mdu211

55. Peoples GE, Holmes JP, Hueman MT, Mittendorf EA, Amin A, Khoo S, Dehqanzada ZA, Gurney JM, Woll MM, Ryan GB, Storrer CE, Craig D, Ioannides CG, Ponniah S (2008) Combined clinical trial results of a HER2/neu (E75) vaccine for the prevention of recurrence in high-risk breast cancer patients: U.S. military cancer institute clinical trials group study I-01 and I-02. Clin Cancer Res 14(3):797–803. doi:10.1158/1078-0432.CCR-07-1448
56. Mittendorf EA, Storrer CE, Foley RJ, Harris K, Jama Y, Shriver CD, Ponniah S, Peoples GE (2006) Evaluation of the HER2/neu-derived peptide GP2 for use in a peptide-based breast cancer vaccine trial. Cancer 106(11):2309–2317. doi:10.1002/cncr.21849
57. Knutson KL, Disis ML (2005) Augmenting T helper cell immunity in cancer. Curr Drug Targets Immune Endocr Metabol Disord 5(4):365–371
58. Knutson KL, Disis ML (2005) Tumor antigen-specific T helper cells in cancer immunity and immunotherapy. Cancer Immunol Immunother 54(8):721–728. doi:10.1007/s00262-004-0653-2
59. Holmes JP, Benavides LC, Gates JD, Carmichael MG, Hueman MT, Mittendorf EA, Murray JL, Amin A, Craig D, von Hofe E, Ponniah S, Peoples GE (2008) Results of the first phase I clinical trial of the novel II-key hybrid preventive HER-2/neu peptide (AE37) vaccine. J Clin Oncol 26(20):3426–3433. doi:10.1200/JCO.2007.15.7842
60. Mittendorf EA, Ardavanis A, Litton JK, Shumway NM, Hale DF, Murray JL, Perez SA, Ponniah S, Baxevanis CN, Papamichail M, Peoples GE (2016) Primary analysis of a prospective, randomized, single-blinded phase II trial evaluating the HER2 peptide GP2 vaccine in breast cancer patients to prevent recurrence. Oncotarget 7 (40):66192–66201. doi:10.18632/oncotarget.11751
61. Julien S, Picco G, Sewell R, Vercoutter-Edouart AS, Tarp M, Miles D, Clausen H, Taylor-Papadimitriou J, Burchell JM (2009) Sialyl-Tn vaccine induces antibody-mediated tumour protection in a relevant murine model. Br J Cancer 100(11):1746–1754. doi:10.1038/sj.bjc.6605083
62. Zhang S, Walberg LA, Ogata S, Itzkowitz SH, Koganty RR, Reddish M, Gandhi SS, Longenecker BM, Lloyd KO, Livingston PO (1995) Immune sera and monoclonal antibodies define two configurations for the sialyl Tn tumor antigen. Cancer Res 55(15):3364–3368
63. Gilewski TA, Ragupathi G, Dickler M, Powell S, Bhuta S, Panageas K, Koganty RR, Chin-Eng J, Hudis C, Norton L, Houghton AN, Livingston PO (2007) Immunization of high-risk breast cancer patients with clustered sTn-KLH conjugate plus the immunologic adjuvant QS-21. Clin Cancer Res 13(10):2977–2985. doi:10.1158/1078-0432.CCR-06-2189
64. Finn OJ, Jerome KR, Henderson RA, Pecher G, Domenech N, Magarian-Blander J, Barratt-Boyes SM (1995) MUC-1 epithelial tumor mucin-based immunity and cancer vaccines. Immunol Rev 145:61–89
65. Graham RA, Burchell JM, Taylor-Papadimitriou J (1996) The polymorphic epithelial mucin: potential as an immunogen for a cancer vaccine. Cancer Immunol Immunother 42(2):71–80
66. Blixt O, Bueti D, Burford B, Allen D, Julien S, Hollingsworth M, Gammerman A, Fentiman I, Taylor-Papadimitriou J, Burchell JM (2011) Autoantibodies to aberrantly glycosylated MUC1 in early stage breast cancer are associated with a better prognosis. Breast Cancer Res 13(2):R25. doi:10.1186/bcr2841
67. Hiltbold EM, Ciborowski P, Finn OJ (1998) Naturally processed class II epitope from the tumor antigen MUC1 primes human CD4+ T cells. Cancer Res 58(22):5066–5070
68. Hiltbold EM, Alter MD, Ciborowski P, Finn OJ (1999) Presentation of MUC1 tumor antigen by class I MHC and CTL function correlate with the glycosylation state of the protein taken up by dendritic cells. Cell Immunol 194(2):143–149. doi:10.1006/cimm.1999.1512
69. Siroy A, Abdul-Karim FW, Miedler J, Fong N, Fu P, Gilmore H, Baar J (2013) MUC1 is expressed at high frequency in early-stage basal-like triple-negative breast cancer. Hum Pathol 44(10):2159–2166. doi:10.1016/j.humpath.2013.04.010
70. Apostolopoulos V, Pietersz GA, Tsibanis A, Tsikkinis A, Drakaki H, Loveland BE, Piddlesden SJ, Plebanski M, Pouniotis DS, Alexis MN, McKenzie IF, Vassilaros S (2006) Pilot phase III immunotherapy study in early-stage breast cancer patients using oxidized mannan-MUC1 [ISRCTN71711835]. Breast Cancer Res 8(3):R27. doi:10.1186/bcr1505
71. Banchereau J, Steinman RM (1998) Dendritic cells and the control of immunity. Nature 392(6673):245–252. doi:10.1038/32588
72. Greenberg NM, Anderson JW, Hsueh AJ, Nishimori K, Reeves JJ, deAvila DM, Ward DN, Rosen JM (1991) Expression of biologically active heterodimeric bovine follicle-stimulating hormone in milk of transgenic mice. Proc Natl Acad Sci U S A 88(19):8327–8331
73. Rohrbach F, Weth R, Kursar M, Sloots A, Mittrucker HW, Wels WS (2005) Targeted delivery of the ErbB2/HER2 tumor antigen to professional APCs results in effective antitumor immunity. J Immunol 174(9):5481–5489
74. Nabekura T, Nagasawa T, Nakauchi H, Onodera M (2008) An immunotherapy approach with dendritic cells genetically modified to express the tumor-associated antigen, HER2. Cancer Immunol Immunother 57(5):611–622. doi:10.1007/s00262-007-0399-8
75. Gabrilovich DI, Nadaf S, Corak J, Berzofsky JA, Carbone DP (1996) Dendritic cells in antitumor immune responses. II. Dendritic cells grown from bone marrow precursors, but not mature DC from tumor-bearing mice, are effective antigen carriers in the therapy of established tumors. Cell Immunol 170(1):111–119. doi:10.1006/cimm.1996.0140

76. Fong L, Brockstedt D, Benike C, Wu L, Engleman EG (2001) Dendritic cells injected via different routes induce immunity in cancer patients. J Immunol 166(6):4254–4259
77. Eggert AA, Schreurs MW, Boerman OC, Oyen WJ, de Boer AJ, Punt CJ, Figdor CG, Adema GJ (1999) Biodistribution and vaccine efficiency of murine dendritic cells are dependent on the route of administration. Cancer Res 59(14):3340–3345
78. Palucka K, Banchereau J (2012) Cancer immunotherapy via dendritic cells. Nat Rev Cancer 12(4):265–277. doi:10.1038/nrc3258
79. Brossart P, Wirths S, Stuhler G, Reichardt VL, Kanz L, Brugger W (2000) Induction of cytotoxic T-lymphocyte responses in vivo after vaccinations with peptide-pulsed dendritic cells. Blood 96(9):3102–3108
80. Czerniecki BJ, Koski GK, Koldovsky U, Xu S, Cohen PA, Mick R, Nisenbaum H, Pasha T, Xu M, Fox KR, Weinstein S, Orel SG, Vonderheide R, Coukos G, DeMichele A, Araujo L, Spitz FR, Rosen M, Levine BL, June C, Zhang PJ (2007) Targeting HER-2/neu in early breast cancer development using dendritic cells with staged interleukin-12 burst secretion. Cancer Res 67(4):1842–1852. doi:10.1158/0008-5472.CAN-06-4038
81. Lowenfeld L, Mick R, Datta J, Xu S, Fitzpatrick E, Fisher CS, Fox KR, DeMichele A, Zhang PJ, Weinstein SP, Roses RE, Czerniecki BJ (2016) Dendritic cell vaccination enhances immune responses and induces regression of HER2pos DCIS independent of route: results of randomized selection design trial. Clin Cancer Res. doi:10.1158/1078-0432.CCR-16-1924
82. Sharma A, Koldovsky U, Xu S, Mick R, Roses R, Fitzpatrick E, Weinstein S, Nisenbaum H, Levine BL, Fox K, Zhang P, Koski G, Czerniecki BJ (2012) HER-2 pulsed dendritic cell vaccine can eliminate HER-2 expression and impact ductal carcinoma in situ. Cancer 118(17):4354–4362. doi:10.1002/cncr.26734
83. Dranoff G, Jaffee E, Lazenby A, Golumbek P, Levitsky H, Brose K, Jackson V, Hamada H, Pardoll D, Mulligan RC (1993) Vaccination with irradiated tumor cells engineered to secrete murine granulocyte-macrophage colony-stimulating factor stimulates potent, specific, and long-lasting anti-tumor immunity. Proc Natl Acad Sci U S A 90(8):3539–3543
84. Dranoff G (2002) GM-CSF-based cancer vaccines. Immunol Rev 188:147–154
85. Antonia SJ, Seigne J, Diaz J, Muro-Cacho C, Extermann M, Farmelo MJ, Friberg M, Alsarraj M, Mahany JJ, Pow-Sang J, Cantor A, Janssen W (2002) Phase I trial of a B7-1 (CD80) gene modified autologous tumor cell vaccine in combination with systemic interleukin-2 in patients with metastatic renal cell carcinoma. J Urol 167(5):1995–2000
86. Jaffee EM, Hruban RH, Biedrzycki B, Laheru D, Schepers K, Sauter PR, Goemann M, Coleman J, Grochow L, Donehower RC, Lillemoe KD, O'Reilly S, Abrams RA, Pardoll DM, Cameron JL, Yeo CJ (2001) Novel allogeneic granulocyte-macrophage colony-stimulating factor-secreting tumor vaccine for pancreatic cancer: a phase I trial of safety and immune activation. J Clin Oncol 19(1):145–156. doi:10.1200/JCO.2001.19.1.145
87. Soiffer R, Hodi FS, Haluska F, Jung K, Gillessen S, Singer S, Tanabe K, Duda R, Mentzer S, Jaklitsch M, Bueno R, Clift S, Hardy S, Neuberg D, Mulligan R, Webb I, Mihm M, Dranoff G (2003) Vaccination with irradiated, autologous melanoma cells engineered to secrete granulocyte-macrophage colony-stimulating factor by adenoviral-mediated gene transfer augments antitumor immunity in patients with metastatic melanoma. J Clin Oncol 21(17):3343–3350. doi:10.1200/JCO.2003.07.005
88. Salgia R, Lynch T, Skarin A, Lucca J, Lynch C, Jung K, Hodi FS, Jaklitsch M, Mentzer S, Swanson S, Lukanich J, Bueno R, Wain J, Mathisen D, Wright C, Fidias P, Donahue D, Clift S, Hardy S, Neuberg D, Mulligan R, Webb I, Sugarbaker D, Mihm M, Dranoff G (2003) Vaccination with irradiated autologous tumor cells engineered to secrete granulocyte-macrophage colony-stimulating factor augments antitumor immunity in some patients with metastatic non-small-cell lung carcinoma. J Clin Oncol 21(4):624–630. doi:10.1200/JCO.2003.03.091
89. Emens LA, Asquith JM, Leatherman JM, Kobrin BJ, Petrik S, Laiko M, Levi J, Daphtary MM, Biedrzycki B, Wolff AC, Stearns V, Disis ML, Ye X, Piantadosi S, Fetting JH, Davidson NE, Jaffee EM (2009) Timed sequential treatment with cyclophosphamide, doxorubicin, and an allogeneic granulocyte-macrophage colony-stimulating factor-secreting breast tumor vaccine: a chemotherapy dose-ranging factorial study of safety and immune activation. J Clin Oncol 27(35):5911–5918. doi:10.1200/JCO.2009.23.3494
90. Rice J, Ottensmeier CH, Stevenson FK (2008) DNA vaccines: precision tools for activating effective immunity against cancer. Nat Rev Cancer 8(2):108–120. doi:10.1038/nrc2326
91. Campos-Perez J, Rice J, Escors D, Collins M, Paterson A, Savelyeva N, Stevenson FK (2013) DNA fusion vaccine designs to induce tumor-lytic CD8+ T-cell attack via the immunodominant cysteine-containing epitope of NY-ESO 1. Int J Cancer 133(6):1400–1407. doi:10.1002/ijc.28156
92. Rice J, Buchan S, Stevenson FK (2002) Critical components of a DNA fusion vaccine able to induce protective cytotoxic T cells against a single epitope of a tumor antigen. J Immunol 169(7):3908–3913
93. Rice J, Elliott T, Buchan S, Stevenson FK (2001) DNA fusion vaccine designed to induce cytotoxic T cell responses against defined peptide motifs: implications for cancer vaccines. J Immunol 167(3):1558–1565
94. Zhu D, Williams JN, Rice J, Stevenson FK, Heckels JE, Christodoulides M (2008) A DNA fusion vaccine

induces bactericidal antibodies to a peptide epitope from the PorA porin of Neisseria meningitidis. Infect Immun 76(1):334–338. doi:10.1128/IAI.00943-07
95. Watson MA, Fleming TP (1994) Isolation of differentially expressed sequence tags from human breast cancer. Cancer Res 54(17):4598–4602
96. Tiriveedhi V, Tucker N, Herndon J, Li L, Sturmoski M, Ellis M, Ma C, Naughton M, Lockhart AC, Gao F, Fleming T, Goedegebuure P, Mohanakumar T, Gillanders WE (2014) Safety and preliminary evidence of biologic efficacy of a mammaglobin-a DNA vaccine in patients with stable metastatic breast cancer. Clin Cancer Res 20(23):5964–5975. doi:10.1158/1078-0432.CCR-14-0059
97. Galluzzi L, Buque A, Kepp O, Zitvogel L, Kroemer G (2015) Immunological effects of conventional chemotherapy and targeted anticancer agents. Cancer Cell 28(6):690–714. doi:10.1016/j.ccell.2015.10.012
98. Bezu L, Gomes-de-Silva LC, Dewitte H, Breckpot K, Fucikova J, Spisek R, Galluzzi L, Kepp O, Kroemer G (2015) Combinatorial strategies for the induction of immunogenic cell death. Front Immunol 6:187. doi:10.3389/fimmu.2015.00187

Tumor Associated Macrophages as Therapeutic Targets for Breast Cancer

16

Liyan Lao, Siting Fan, and Erwei Song

Abstract

Tumor-associated macrophages (TAMs) are the most abundant inflammatory infiltrates in the tumor stroma. TAMs promote tumor growth by suppressing immunocompetent cells, including neovascularization and supporting cancer stem cells. In the chapter, we discuss recent efforts in reprogramming or inhibiting tumor-protecting properties of TAMs, and developing potential strategies to increase the efficacy of breast cancer treatment.

Keywords

Tumor-associated Macrophages • Breast Cancer • Polarization • Immunosuppression • Metastasis • Resistance • Therapeutic target

16.1 Introduction

Macrophages, originally identified by Metchnikoff for their phagocytic capacity, are pivotal and plastic components of the innate immune system, which play essential roles in pathogen elimination, homeostasis maintenance and tissue repair [1]. In adult mammals, resident macrophages in a variety of tissues display marked transcriptional and functional diversity [2]. In spite of their antimicrobial effects in acute infectious and inflammatory diseases, macrophages adopt a suppressive phenotype at the late stages of inflammation that limits the inflammatory activities while facilitating wound healing and tissue growth, which counteracts the tissue

L. Lao
Sun Yat-Sen Memorial Hospital, Sun Yat-Sen University, 107 Yanjiang West Road, Guangzhou 510120, People's Republic of China

S. Fan
Breast Tumor Center, Sun Yat-Sen Memorial Hospital, Sun Yat-Sen University, Guangzhou 510120, China

Guangdong Provincial Key Laboratory of Malignant Tumor Epigenetics and Gene Regulation, Sun Yat-Sen Memorial Hospital, Sun Yat-Sen University, Guangzhou 510120, China

E. Song (✉)
Sun Yat-Sen Memorial Hospital, Sun Yat-Sen University, Guangzhou, Guangdong 510120, China
e-mail: songew@mail.sysu.edu.cn

E. Song, H. Hu (eds.), *Translational Research in Breast Cancer*, Advances in Experimental Medicine and Biology 1026, https://doi.org/10.1007/978-981-10-6020-5_16

impairing potential of immune response [3–5]. Accordingly, macrophages are the mainstays for maintaining the delicate balance between tissue destruction and restoration and exhibit striking heterogeneity and plasticity in response to environmental challenges.

Early seminal work by Rudolf Virchow in 1863 has elegantly linked chronic inflammation to tumorigenesis [6]. It has subsequently been confirmed and extended by numerous studies that pro-inflammatory cytokines and growth factors produced by macrophages create a mutagenic microenvironment and triggered tumor initiation [7, 8]. A study in the 1970s demonstrated that after tumor establishment, intratumoral macrophages exerted cytotoxic effects on cancer cells [9], which was soon overturned [10]. Considered as "wounds that never heal", tumors are abundantly populated by reparative macrophages which suppress antitumor immunity and promote tumor growth [11]. The seed-and-soil hypothesis of Paget proposed the central role of interactions between tumor cells and surrounding microenvironment in tumor survival and metastatic potential [12]. Various studies have remarkably advanced our understanding of tumor associated macrophages (TAMs), which are polarized into a protumoral phenotype [5] and facilitate tumor cell survival, immune evasion, vascular generation, systemic dissemination and therapeutic refractoriness [13–16]. In fact, the enrichment of TAMs correlates with poor prognosis [15].

Here we describe and discuss the dynamic interplay between tumor cells and diverse macrophage subpopulations that display tumorigenic potential in malignant progression, immune suppression and metastasis in breast cancer. Attention is paid to the significant alterations that TAMs undergo in response to various anticancer agents and their profound effects on the therapeutic efficacy. We also discuss anticipated or clinical therapeutic strategies deleting or reprogramming TAMs within tumor microenvironment(TME) as monotherapies or complemental approaches to improve patient prognosis.

16.2 Macrophage Accumulation in Breast Cancer

16.2.1 The Origins of TAMs

The historical assumption that tumor infiltrating macrophages originated exclusively from bone marrow derived monocytes was called into question. Indeed, despite some exceptions, resident macrophages in various organs like the brain, liver and lung, are yolk-sac or fetal liver-derived, which are seeded before birth and maintained by self-renewal [17, 18]. Inversely, transient monocytic input from the blood stream occurs in inflammatory settings [3, 19]. Herein, mammary gland comprises embryonic and recruited macrophages from circulating monocytes to replenish the pool [20], both of which undergo functional and phenotypic switch during carcinogenesis in response to stimulating signals in TME [21]. (see Fig. 16.1). Studies in mice unveiled that the ontogenetic source of TAM precursors had little influence on macrophage activities [22]. Nevertheless, additional investigation is needed to determine the molecules and downstream signaling that mediate the changes in these resident macrophages and their effects on tumor progression.

In fact, the recruitment of circulating precursors and their differentiation into tumor-boosting phenotype have a pivotal role [20]. These precursors include myeloid-derived suppressor cells (MDSCs), which arise from bone marrow-derived immature myeloid cells that are attracted by factors produced in the TME [23]. Monocytic (M)-MDSCs are able to convert into macrophages within tumors, where MHCII and F4/80 are progressively upregulated while Ly6C and Gr1 are downregulated [24]. Low level of *Stat3* activity was found in MDSCs in breast cancer, which represented a key process in mediating their transition into TAMs [25]. Notably, the major TAM population originates from Ly6C^{hi}CCR2^{+} inflammatory monocytes, whose maturation is dependent on recombination signal binding protein for the immunoglobulin kappa J (Rbpj), a transcriptional regulator of Notch signaling [20]. On the contrary, Ly6C^{-}/CX3CR1^{+} nonclassical monocytes patrol the lung micro-

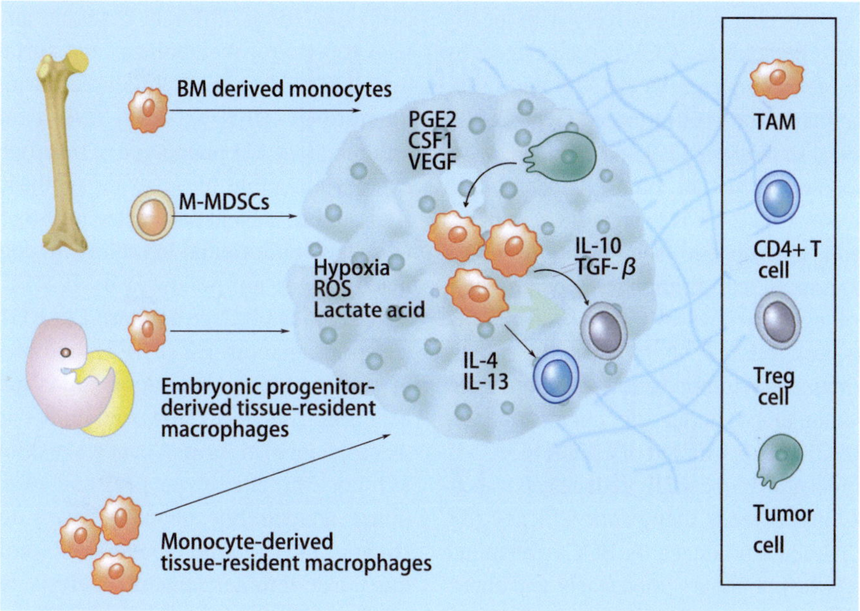

Fig. 16.1 The origin of tumor-associated macrophages and their programming in the tumor microenvironment. In response to cytokines, chemokines, and growth factors, such as CSF-1, VEGF, CCL2, CXCL12, and Sema3A, inflammatory monocytes and monocytic myeloid-derived suppressor cells (M-MDSCs) are attracted into the neoplastic region. The recruited precursors of hematopoietic stem cell origin and tissue-resident macrophages derived from either embryonic progenitors or monocytes undergo functional and phenotypic transition in the tumor microenvironment and constitute tumor-associated macrophages (TAMs) in breast cancer. Herein, IL-4 and IL-13 from $CD4^{+}$T cells, as well as IL-10 and TGF-β from regulatory T cells, induce the angiogenic and immune-evasive capacities in macrophages. Neoplasm-derived stimuli such as PGE2, CSF-1, and VEGF synergize with Th2-type immune cells to reinforce the protumoral properties of TAMs. Exposure to hypoxia and accumulation of lactic acid and superoxide have a profound impact on macrophage differentiation. On the stimulation of extrinsic signals, transcriptional regulation through JAK-STAT6 pathway, IRF4 and 5, PPARγ, KLF4, mTORC1, and PI3Kγ signaling, as well as epigenetic and RNA changes, is responsible for macrophage phenotypic conversion

vasculature and hamper metastatic tumor cell seeding [26]. In addition to circulating monocytes, M-MDSCs, embryonic precursors and self-renewal of TAMs via colony-stimulating factor 1 (CSF1) [27] also contribute to intratumoral macrophage accumulation in mouse mammary tumor [2]. Accordingly, the TAM pool depends less on peripheral monocytic input as compared with resident macrophages [20].

16.2.2 Macrophage Recruitment to Tumor

A variety of cytokines and chemokines are involved in the recruitment of circulating monocytes and M-MDSCs into neoplastic lesions. The major macrophage lineage regulator CSF1 has long been recognized as one of the paramount chemoattractants for monocytes in the bloodstream [28]. Destruction of CSF-1/CSF-1R activity serves as a primary approach of TAM depletion in diverse tumor models, which could restrain tumor progression to malignancies and their metastatic capacities [29]. Indeed, upregulation of CSF1 or CSF-1R in human breast carcinomas correlates with poor prognosis [28, 29]. In PyMT-induced mouse mammary tumors, genetic gain of function of VEGFA rescues the delayed tumor growth and angiogenesis induced by CSF1 ablation, through its effects on the formation of high-density vascular network and massive influx of macrophages [30]. Actually, these growth factors probably reinforce the recruitment and retention of TAMs by working collaboratively with locally synthesized chemokines. CCL2 synthesized by metastatic breast cancer cells (BCCs) and tissue stroma facilitates the infiltration of $Gr1^+$ $CCR2^+$ inflammatory monocytes [20] and correlates with poor outcome in human breast cancers [31]. Inhibition of CCL2-CCR2 signaling impedes intratumoral macrophage accumulation, tumor cell extravasation, and pulmonary seeding, leading to prolonged survival of tumor-bearing mice [32]. Besides, $TIE2^{Hi}CXCR4^{Hi}$ TAMs rapidly infiltrate mammary tumor following chemotherapy or radiotherapy under the chemoattractive gradient of CXCL12, a ligand of CXCR4 [33, 34]. CXCR4 blockade dramatically reduces neovascular density and extends therapeutic efficacy [33, 34]. CXCL1 and 2 produced by BCCs precipitate tumor growth, pulmonary metastasis and therapeutic unresponsiveness through the recruitment of $CD11b^+Gr1^+$ myeloid cells [35]. Attraction and migration of macrophages into the hypoxic tumor regions involve Semaphorin 3A (Sema3A), a ligand of Neuropilin-1 (Nrp1), whereas loss of *Nrp1* on TAMs curbs the growth and dissemination of orthotopic tumors [36]. Dissecting mechanisms of the expansion and retention of tumor-educated macrophages paves the way for therapies aiming to limit the protumoral activities of TAMs by inhibiting macrophage infiltration in tumor stroma.

16.3 Macrophage Plasticity in Tumor Microenvironment

A key feature of macrophages is their functional and phenotypic diversity and plasticity in response to environmental cues. The alternatively activated macrophages, M2 macrophages, induced by interleukin-4 (IL-4) were first discovered in the 1990s [37]. Mills et al. proposed a dichotomy between classically activated macrophages (M1) that facilitated T helper 1 (Th1) response and alternatively activated macrophages (M2) that enhanced Th2 activities [38]. Mantovani et al. subgrouped the M2 phenotype into M2a, M2b, and M2c on the basis of the activating stimuli [39]. The M1/M2 classification acquires a comprehensive extension, where the two populations differ in cytokine and chemokine repertoire, metabolism, and surface receptors [40]. Herein, lipopolysaccharide (LPS), interferon-γ (IFN-γ), and TNF-α induce an inflammatory M1 phenotype of macrophages that express inducible nitric oxide synthase (iNOS), IL-1, and IL-12 and participate in antigen presentation and tumoricidal immunity. In contrast, M2 macrophages are polarized by IL-4, IL-13, and IL-10, which produce arginase 1 (ARG1), VEGF, and prostaglandin E2 (PGE2), and are involved in inflammation resolution and tumorigenic activities [38, 41]. However, tumor-infiltrating macrophages in breast cancers exhibit neither M1 nor M2 phenotype [20] and display functional adaptability upon exposure to distinct stimuli [42]. Gene expression profiles and the transcriptome network analysis reveal that M1 and M2 phenotypes represent two extremes of a continuum of macrophage activation states [43, 44]. In fact, the heterogeneity and plasticity of TAMs indicate remarkable limitations in their assignment to invariant specific phenotypes according to homogeneous cell cultures in vitro [41]. Considering their diverse effects on tumors, Qian and Pollard classified TAMs into six functional subpopulations: activated, angiogenic, immunosuppressive, invasive, perivascular, and metastasis-associated macrophages (MAM) [29]. Currently, a multidimensional insight into macrophage activation suggests a more precise system to link oncology, microenvironment signals, and insult-induced stress signals to macrophage phenotypes [17]. Exploring TAM diversity taking advantage of high-resolution, single-cell, and deep phenotyping technologies presents a hopeful and challenging approach to advance our knowledge on and provide foundations for therapeutic targeting of TAMs [17, 45].

In early-stage autochthonous mammary tumors, an intermediate profile is detected in macrophages [20]. However, after the establishment of malignancies, intratumoral macrophages, resident or continuously recruited, are polarized by integrated multiple signals from the microenvironment away from the pro-inflammatory phenotype toward protumoral M2-like population [14]. These cues include regulatory factors from immune cells, cytokines produced by neoplastic cells, and signals from homeostatic imbalance [41, 46]. In the MMTV-PyMT transgenic mice, IL-4 and IL-13 produced by Th2-polarized $CD4^+T$ lymphocytes expedite tumor dissemination and reemergence after irradiation through the reinforcement of tumor-favorable properties of TAMs [47, 48]. In addition to IL-4 and IL-13 from $CD4^+T$ cells, immunoregulatory factors like IL-10 and TGF-β from regulatory T cells (Tregs) exert significant modulation on tumorigenic activities of TAMs [41, 49]. Intriguingly, these cytokines are also synthesized by cancer cells [46]. Likewise, neoplasm-derived stimuli such as PGE2, CSF-1, and VEGF synergize with Th2-type immune cells to induce angiogenic and immune-evasive capacities of macrophages [40, 50]. CSF-1R signaling blockade in mammary tumor-bearing mice, in addition to eliciting an evident reduction in TAM number, shifts macrophage phenotypic balance in favor of tumor-promoting $MHC\text{-}II^{hi}$ subtype [51]. PGE2 renders BCCs the immunomodulatory effects on bone marrow-derived mononuclear cells and fuels tumor-promoting inflammation [50]. Besides, disruption of homeostatic balance in TME leads to hypoxia, accumulation of lactic acid and superoxide. Exposure to poorly vascularized tumor regions modulates TAM enrichment and functions through the upregulation of hypoxia-inducible factors (HIF)-1α and HIF-2α [52]. Reactive oxygen species (ROS) production is crucial for macrophage differentiation into anti-inflammatory population through late-phase ERK activation [53]. Loss of HIF-1α or elimination of ROS restrains ARG1 expression and the immunosuppressive activities of TAMs in MMTV-PyMT mice [53, 54].

These extrinsic signals dictate the signaling cascades, transcriptional responses, and epigenetic changes that shape the activation and properties of the macrophages. Transcriptional regulation through JAK-STAT6 pathway, IRF4, PPARγ [55], KLF4 [40], and MerTK signaling [56] in TAMs evokes a bias away from inflammatory cytokine synthesis toward wound healing cytokines and thus favors tumor survival. The transcription factor IRF5 is a critical factor in macrophage polarization which equips macrophages with an $IL\text{-}12^{hi}IL\text{-}23^{hi}IL\text{-}10^{lo}$ cytokine profile and promotes human Th1-Th17 responses [57]. Indeed, nitric oxide (NO) derived from inducible nitric oxide synthase (iNOS) presents a hurdle to classical macrophage activation by facilitating nitration of IRF5 and leads to the impairment of IRF5-targeted activation of M1-subtype signature gene [58]. Lysosomal adaptor protein Lamtor1 is part of the amino acid sensing complex that serves as a scaffold for the activation of mechanistic target of rapamycin complex 1 (mTORC1) in response to IL-4. Lamtor1 and mTORC1 have an essential effect on macrophage phenotypic conversion by stimulating their downstream transcription factor liver X receptor (LXR) [59]. Moreover, PI3Kγ ablation reinforces a CD8+ T-cell response and hampers mammary tumor growth by provoking a critical switch of TAMs into a pro-inflammatory phenotype. Herein, PI3Kγ signaling attenuates NF-κB phosphorylation but stimulates C/EBPβ activation in a mTOR-dependent manner, thereby inducing a conversion of macrophage transcriptional program [60]. Emerging evidence suggests a consequential role for epigenetic mechanisms and small and long noncoding RNAs in modulating signaling pathways and gene expression during macrophage programming and redirection [61–63]. Dynamic reorganization of the chromatin landscape is the mainstay for macrophage maturation and functional transition, where chemical modification of lysine 4 in histone H3 (H3K4) and H3K27 regulates the open or poised state of massive enhances and transcriptional activities [64]. In addition, small RNAs like miR-146a, miR-222 [65], and miR-19a-3p [66], and long noncoding RNAs like *THRIL*, are involved in TAM polarization [62].

TAMs educated by tumor cells are characterized by increased production of IL-10, CCL18, EGF, VEGF, and TGF-β and serve as tumor promoters through their support for neovascularization, immunosuppression, invasion, as well as therapeutic resistance of breast cancer [14, 40]. The resemblance of TAMs to tissue-repairing macrophages in homeostasis suggests that the innate wound healing mechanisms are utilized by tumors to their own advantage [3]. Intratumoral macrophages are misdirected in the TME to be accomplices for breast malignancies, exploiting their tissue-remodeling and anti-inflammatory capacities [15]. Collectively, our current understanding of the molecular mechanisms of TAM polarization, the predominant signals, and the role of epigenetic regulation and RNAs in macrophage phenotypic regulation is still limited and there are plenty of exciting areas for future investigation. Furthermore, a large proportion of our knowledge comes from studies performed in mouse models. Species-specific differences between human and mouse macrophage responses might contribute to distinct conclusions. Inspiringly, the functional diversity and phenotypic plasticity of TAMs highlight promising therapeutic strategies to reeducate TAMs into tumoricidal populations and abrogate their bolster for tumor progression.

16.4 Macrophage-Mediated Immune Suppression

Recognition of tumor-specific antigens triggers immunologic regression and eradication of the incipient tumors, making it conceivable that engagement of immune system may set a critical barrier for malignant progression [67]. Antitumor efficacy of effector T lymphocytes is an active process under the modulation of cell surface inhibitory receptors, soluble factors, metabolic reprograming, and immunoregulatory cell types [68].

Escape of neoplastic cells from T-cell-dependent tumoricidal activities is typically linked to soluble anti-inflammatory cytokines derived from TAMs. Programmed by the local milieu, macrophages are possessed of impaired immunostimulatory capacities characterized by reduced production of IL-12, IL-1, TNF-α, and IFNs, as well as potentiated immunosuppressive potential with increased secretion of IL-10, TGF-β, and PGE2 [50, 69–72]. In particular, tumor-educated macrophages are the predominant provenience of IL-10 in mammary carcinomas, which accounts for the restraint of cytotoxic functions of effector T cells caused by TAMs. Herein, the suppressive effects of IL-10 are indirectly related to the hindrance of IL-12 production by dendritic cells, which results in CD8+ T-cell inactivation [73]. In addition, high expression of cyclooxygenase2 (COX2) is found in TAMs isolated from mammary tumor tissue, which indicates its correlation with poor survival outcome in breast cancer patients [74, 75]. Sustained increase of PGE2 expression provokes a tumor-promoting inflammatory profile and compromises the tumoricidal activities of T lymphocytes, dampening immunological control of tumor growth [50, 76]. In line with these data, ablation of TAMs through interference of CSF1-signaling pathway refuels antitumor immunity, resulting in delayed malignant progression and pulmonary metastasis following chemotherapy [77].

Inhibitory receptors on functional effector T cells, including programmed cell death protein 1 (PD-1) and cytotoxic T-lymphocyte antigen 4 (CTLA-4), induce self-tolerance and protect the host against autoimmunity under physiological conditions. Nevertheless, in malignancies, one of the prevalent mechanisms of immune evasion is through the expression of ligands of these negative regulators [78]. The CTLA-4 ligands B7-1 and B7-2 are ubiquitous on antigen-presenting cells like macrophages. Binding of CTLA-4 to B7-1 or B7-2 precludes the activation of T-cell costimulatory CD28 and delivers suppressive signals to effector T cells [68]. Likewise, the interaction between PD-1 and its ligands, PD-L1 and PD-L2, restrains TCR and BCR signaling, impedes lymphocyte proliferation and functions, and leads to T-cell exhaustion or apoptosis [79]. In TAMs and MDSCs, activation and direct binding of HIF-1α to hypoxia-response element (HRE) in the PD-L1 proximal promoter result in selective upregulation of PD-L1 in the low-oxygen-concentration microenvironment [80]. The engagement of Siglec-9, a sialic acid-binding protein on macrophages, by the aberrantly glycosylated mucin MUC1-ST on BCCs exerts a profound influence over macrophage differentiation and TAM formation through MEK-ERK signaling. MUC1-ST-educated macrophages show higher expression of surface receptors like PD-L1, CD206, and CD163 and secrete factors favorable to tumor progression.

The increase of PD-L1 ranges from 1.5-fold to over 7-fold, which plays a crucial role in the immune tolerance triggered by TAMs [81]. Furthermore, the protumoral macrophages are responsible for COX2 induction in cancer cells [74], which augments sialyltransferase production and MUC1-ST formation, generating a positive feedback loop in the subversion of immune system and acceleration of tumor development. Besides, nuclear translocation of c-MYC, the pleiotropic transcription factor, was observed in TAMs and involved in the induction of a series of molecules including VEGF, MMP9, and HIF-1α [82]. MYC is a direct regulator of CD47 and PD-L1 by binding to their gene promoters in leukemia and lymphomas [83]. However, it remains unclear whether other cancers share similar mechanisms. In contrast, the epigenetic "reader" protein BRD4 modulates PD-L1 expression in an MYC-independent manner in various cancers, including breast cancers. Mechanisms of checkpoint regulation in TAMs and mammary tumor cells are yet to be determined, and further investigation is required. Notably, blockade of PD-L1 and CTLA-4 damages tumor progression and dissemination by restoring proliferative capacities and tumoricidal functions of $CD8^+$T cells in breast cancer [80]. These studies provide arresting rationales for administration of checkpoint blockade to combat the immune-impeding mechanisms and facilitate robust antitumor responses in patients with breast cancer. At present, immune checkpoint inhibitors have shown promising results in some cancers, although their application to breast cancer patients is still in clinical trials.

A myriad of studies has highlighted the contributions of metabolic regulation to tumor escape of immune surveillance, through nutrient deprivation and accumulation of immunosuppressive metabolites [84]. In early-stage mammary tumor patient samples, elevated level of ARG1 is detectable in tumor-infiltrating $CD14^+$ myeloid cells [85]. Polarized TAMs augment the production of a variety of protumor molecules, among which is ARG1 [54, 69, 86], the critical enzyme that catabolizes L-arginine into urea and L-ornithine. Overexpression of ARG1 disarms the tumoricidal activity of T cells by decreasing arginine supply in the surrounding microenvironment. Extracellular L-arginine depletion causes damage to the host defense system by abating the expression of CD3 ζ chain of TCR and inducing cell cycle arrest in tumor infiltrating T cells [87]. Paradoxically, arginine deficiency triggers mitochondrial fragmentation and autophagy-dependent death of BCCs devoid of argininosuccinate synthetase 1, the enzyme in low abundance in a large proportion of breast cancer bio-samples [88]. Although it might partially be a consequence of different experimental conditions, the complex role of arginine auxotrophy in tumor immune unresponsiveness and cell death brings confusion in therapeutic utility of manipulating arginine abundance for tumor control. Intriguingly, L-arginine is also a substrate of iNOS, which is co-expressed with ARG1 on TAMs albeit it is considered as one of the hallmarks of M1 phenotype. In the TME, ARG1-mediated L-arginine paucity facilitates the functional switch of iNOS from NO synthesis to production of superoxide, including ROS and reactive nitrogen oxide species (RNOS) [76, 89]. Indeed, TAMs and MDSCs are the major sources of free radical peroxynitrite in breast cancer samples, which induces modification of MHC class I molecules on tumor cells and brings huge hindrance to the presentation and recognition of tumor-specific antigens, driving the escape of neoplastic cells from antitumor immunity [90]. In human tissues of triple-negative breast cancer, high level of endogenous iNOS is associated with worse survival, whose inhibition significantly reduces tumor development and pulmonary colonization [91]. Similarly, indoleamine 2,3-dioxygenase (IDO) is crucial to the rate-limiting step in tryptophan metabolism by converting tryptophan into kynurenine. Mammary tumor cells elevate the expression of IDO by macrophages through MUC1-ST-Siglec-9 interaction [81], limiting the availability of tryptophan to immune cells. Tryptophan starvation triggers the activation of GCN2 kinase and growth retardation of T lymphocytes, while kynurenine accumulation dampens the cytotoxic functions of effector T cells and bolsters Treg cell differentiation [92, 93]. Accordingly, IDO ablation by small-molecule inhibitors fuels immune response against neoplastic cells and augments chemotherapeutic efficacy [94]. As the TORC1 pathway is essential for both amino acid sensing and regulation of T-cell activation [95], it is tempting to determine the effects of TORC1 in metabolic reprogramming for T-cell-dependent tumor eradication. Moreover, hypoxia recapitulates the microenvironmental modulating functions of metabolic changes in regard to immune resistance of breast cancers. Inhibition of hypoxic response by HIF-1α blockade profoundly reduces the production of immunosuppressive enzymes like ARG1 and iNOS and restores T-cell proliferative capacity and tumoricidal activities [54].

Rather than directly evoking immunological unresponsiveness, TAMs recruit and interact with other immunoregulating cell types, which presents efficacious strategies for cancer treatment. Although TAMs recruit circulating Tregs by chemokines in ovarian cancer and macrophages may cooperate with BCCs to drive CCL22 production and Treg infiltration [96], the main source of intratumoral Treg cells has been recently identified in breast cancer. Herein, Tregs within mammary tumors display TCR repertoire resembling naive CD4+ T cells instead of periphery Tregs. The abundance of intratumoral naive CD4+ T cells is associated with the expansion of tumor-infiltrating Tregs as well as poor patient prognosis, suggesting that Tregs in human breast cancer primarily arise from naive CD4+ T cells that convert into Tregs in situ [97, 98]. In agreement, CCL18 derived from TAMs is responsible for the chemotaxis of naive CD4+ T cells toward the neoplastic region through its receptor PITPNM3, whose knockdown significantly hinders Treg infiltration and tumor progression in human mammary tumor xenografts in humanized mice [97]. Differentiation of naive T cells is dependent on exposure to autologous dendritic cells and tumor conditioned medium [97], where TGF-β, IL-2, and retinoic acid may play a part [98]. However, the mechanism is yet to be completely understood. Tregs subvert the antitumor immune responses not only by IL-2 deprivation, CTLA-4 expression, and IL-10 production but also by modulating the effector activities of NK cells, dendritic cells, and macrophages [98, 99]. Besides, a variety of chemokines derived from TAMs, such as PGE2, VEGF, IL-6, and CSF-1, facilitate the proliferation and recruitment of distinct leukocytes including MDSCs into the neoplastic region [23, 39]. In an interesting twist, MDSCs not only have the capacity to differentiate into TAMs but also synergize with TAMs with respect to immune-exhausting mechanisms, such as IL-10 secretion [100]. The immune-subverting activity of MDSCs depends on PD-L1, ARG1, and iNOS expression and IL-10, ROS, and RNOS production [24], which is similar to that of TAMs. Intriguingly, MDSCs are involved in Treg induction [23]. The complex interaction among TAMs, MDSCs, and Tregs may build up positive feedback loops in the immune refractoriness of tumor cells, providing a rationale for therapeutically breaking the vicious circle.

An increased understanding of the dominant mechanisms of immune regulation by TAMs and deconstruction of the modulating signaling and underlying network will undoubtedly pose additional opportunities for therapeutic immunological interventions. However, the distinction between murine and human immune system and the immunogenicity of transplanted tumor models may bring extra limitations in the studies. For the important role of host immune defense in primary tumor elimination, metastasis prevention, and therapeutic efficacy like abscopal effect, immunotherapy targeting the aforementioned mechanisms will refuel the tumoricidal activities of the immune system and improve patient prognosis.

16.5 Tumor Angiogenesis

Neovascular generation is essential for oxygen and nutrient supply and metabolic waste disposal to address the growing need of cancer cells for sustenance, without which tumor may succumb to dormancy. Actually, the "angiogenic switch" is necessary for malignant conversion and considered as one of the most important biological capabilities acquired by tumor cells during the multistep development [101]. The contributions of TAMs to neovasculature in breast cancer have been confirmed and extended by a plethora of studies [102]. TAM abundance is associated with high microvascular density and poor prognosis in human breast cancer [103].

TAMs exert great influence on the process of angiogenesis and lymphangiogenesis by producing a variety of angiogenic growth factors and proteinases. These factors have crucial effects on distinct aspects of neovascularization and include EGF, FGF2, TNF-α, COX2, PDGF-β, PIGF, matrix metalloproteinases (MMP-2, MMP-9, and MMP-14), and cysteine cathepsin proteases [104–108]. In the MMTV-PyMT murine model of mammary carcinoma, TAMs accelerate malignant progression by modulating angiogenic switch and VEGF secretion [103]. Herein, WNT family ligand WNT7B produced by TAMs in the microenvironment acts on vascular endothelial cells and enhances their production of VEGFA through Wnt/β-catenin signaling, leading to neovascularization and tumor progression. In fact, substantial upregulation of WNT7B is detectable in human mammary carcinomas and TAMs isolated from human breast cancer samples, indicating the therapeutic significance of WNT7B signaling [109]. Anti-VEGF treatment combining CCL2 inhibition protects tumor-bearing mice from substantial blood vessel formation and metastatic tumor cell proliferation [110]. A subset of MMPs, especially MMP-9 and MMP-3, cleave matrix-bound isoforms of VEGFA and regulate its bioavailability, promoting capillary dilation [108, 111]. Evidence suggests that TAMs produce and release lipocalin 2 (LCN2) in response to sphingosine-1-phosphate (S1P) secreted by apoptotic tumor cells. Macrophage-derived LCN2 induces the production of VEGFC in lymphatic endothelial cells (LEC) through PI3K signaling, generating an autocrine loop and activating VEGFR3 on the endothelium, which results in lymphatic vessel formation and tumor metastasis [112].

Cell-to-cell interaction with endothelial cell is instrumental in the proangiogenic functions of TAMs. A unique subset of macrophages, which express angiopoietin 2 (ANG2) receptor Tie2 and align along the blood vessel through expression of ANG2 on endothelial cells [113], are endowed with proangiogenic properties to induce tumor blood vessel formation [114]. In the MMTV-PyMT mouse model, targeting ANG2/Tie2 axis by anti-ANG2 monoclonal antibody impairs the proangiogenic activity of Tie2-expressing monocytes and their cross talk with endothelial cells, leading to neovasculature regression and inhibition of tumor progression and dissemination [115]. Moreover, Tie2+-expressing monocytes/macrophages (TEMs) are the predominant population of TAMs that aggressively infiltrate the metastatic lymph nodes in the biopsies of human breast cancer, but not the reactive lymph nodes [116]. Evidence shows that the TEMs identified in untreated human breast cancer express lymphatic markers like LYVE-1, VEGFR-3, and podoplanin and insert into lymphatic vessels. A combination of Tie2 and VEGFR kinase inhibitors abrogates the in vitro lymphangiogenic activity of TEMs sorted from dissociated human breast tumor [117].

During development and interaction with the environment, tumor and myeloid cells have evolved to harness low oxygen tension to their own advantage through HIF transcription factors that orchestrate metabolic and vascular accommodation. Hypoxia is a key regulator of neovasculature formation that modulates expression of various proangiogenic factors like VEGF, PIGF, and ANG2 [104]. The interaction of hypoxia-induced Sema3A with Nrp1 on macrophages leads to activation of VEGFR1 signaling and migration of TAMs to low-oxygenated region, where they boost vessel formation through the secretion of VEGF and MMP9 [36]. In addition, increasing evidence suggests that monocyte/macrophage may possess the ability to directly differentiate into blood or lymphatic endothelial cells [118–120]. The mechanism of the transdifferentiation of TAMs and their structural contributions to vasculature, as well as the communication between the macrophages and endothelium and the evolution of the mixed network, remain to be elucidated in breast cancer.

16.6 Tumor Invasion and Intravasation

16.6.1 Effects of Soluble Factors on TAM-Mediated Invasion

A combination of researches in mouse model and breast cancer cell xenografts has confirmed the crucial role of macrophages in breast cancer cell invasion and migration [121] (see Fig. 16.2). In this process, TAM-derived EGF stimulates EGFR on tumor cell surface and enhances neoplastic mobility and their production of CSF1. In turn, CSF1 expedites macrophage infiltration and their secretion of EGF, establishing a dangerous positive feedback loop between TAMs and tumor cells [121, 122]. Attracted by EGF-producing perivascular macrophages, breast tumor cells migrate along the collagen fibers in a high velocity to the blood vessels in a lockstep manner [123]. A 3D individual cell-based model directs co-migration of tumor cells and macrophages,

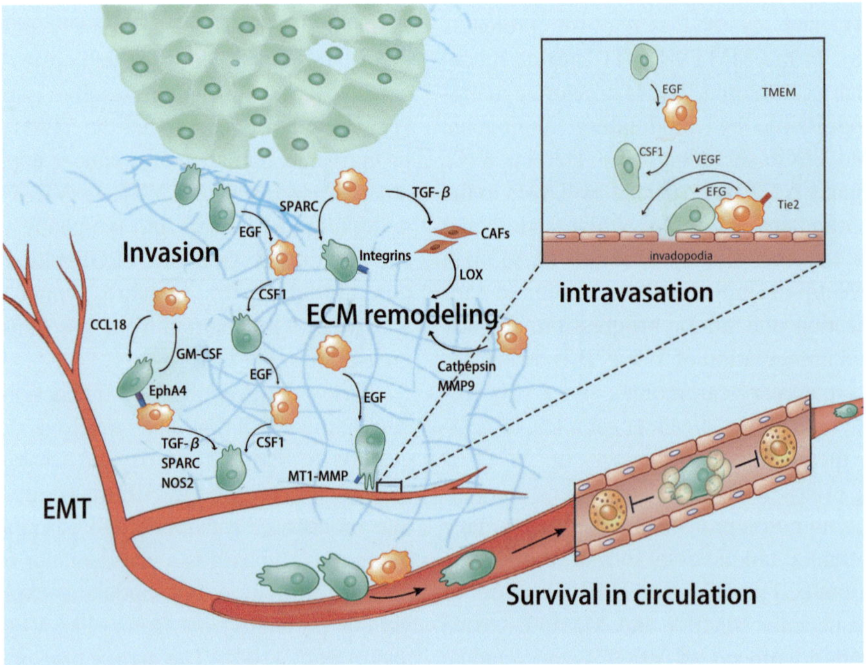

Fig. 16.2 Macrophage-dependent tumor invasion and intravasation in breast cancer. In response to tumor-derived CSF1, macrophage production of EGF enables cancer cells to release CSF1 in turn and migrate along the collagen fibers to the blood vessels in a lockstep manner. Tissue remodeling and stiffening of extracellular matrix accompany tumor invasion. TGF-β from TAMs drives the activation of cancer-associated fibroblasts (CAFs) and their production of LOX, the collagen cross-linking enzyme. SPARC from macrophages amplifies deposition of fibronectins and aggregation of integrin, while cathepsins and MMP9 pave the path through the microenvironment for tumor cell migration. The EGF/CSF-1 paracrine signaling loop has an essential effect on invadosome formation and cancer expression of MT1-MMP, enhancing tumor penetration through the adjacent stroma and basement membrane. Epithelial-to-mesenchymal transition (EMT) is a critical step that endows cancer cells with intensive migratory capacity and stem-like properties. CCL18 from TAMs is fundamental in mesenchymal conversion of tumor cells, whose secretion of GM-CSF programs macrophages to a TAM-like population. Binding to EphA4 of cancer cells and the synthesis of TGF-β, NOS2 and SPARC are significant strategies for macrophage-dependent EMT. TGF-β-induced mesenchymal transition involves stepwise activation of the several double-negative feedback loops. At the intravasating site, the cooperation among macrophages, cancer cells, and endothelial cells forms tumor microenvironment for metastasis (TMEM). Herein, VEGFA signaling in Tie2^{Hi} macrophages facilitates transient vascular permeability. Non-TMEM tumor cells undergo transendothelial migration with the help of the nonmigratory TMEM-tumor cell and the immobile TAM. In the circulation, cancer cells coated with platelet and fibrin escape from NK cell-mediated immune elimination

which sheds light on the role of this paracrine interaction [124]. Mechanistically, Wiskott-Aldrich syndrome protein (WASP) activation in TAMs in response to CSF-1R stimulation enhances macrophage migration toward cancer cells under the chemoattractive gradient and promotes release of EGF in a metalloprotease-dependent shedding manner [125]. In fact, macrophages are the dominant source of EGF in primary breast cancer [126]. EGFR-activated steroid receptor coactivator-1 (src-1) in cancer cells plays an intrinsic role in pulmonary metastasis of mammary tumor by mediating the Ets-2-dependent elevation of HER2 and increasing production of CSF-1 for macrophage recruitment [127]. At single-cell level, a real-time 3D migration test has observed morphological and cytoskeletal changes in EGF-treated mammary tumor cells, which display enhanced migratory activity and a more invasive phenotype [128]. Meanwhile, TAMs induce phosphorylation of Stat3 and upregulation of Sox-2 in neoplastic cells through EGF/EGFR axis, converting tumor cells to a cancer stem cell phenotype with increased viability and metastatic property [129]. Ablation of either EGFR or CSF-1R attenuates tumor invasiveness and aggressiveness induced by HRGβ1 or CXCL12, suggesting that the EGF/CSF-1 paracrine invasion loop exerts influence on other pro-invasive factors [130]. Accordingly, EGFR overexpression serves as a poor prognostic factor in $HER2^+$ breast cancer [131].

The CCL18/PITPNM3 axis has a profound impact on multiaspects of breast cancer angiogenesis and metastasis. By stimulating its receptor PITPNM3 on tumor cells, CCL18 provokes integrin aggregation and their adherence to extracellular fibronectins, thereby sustaining directional migration of invasive tumor cells [132]. Herein, Pyk2 forms a stable complex with CCL18-binding PITPNM3 and activates Src kinase, resulting in tumor cell α5/β1 clustering [133]. Moreover, activated PITPNM3 bolsters actin filament polymerization within tumor cells in response to CCL18 and subsequently raises their migratory capacities through LIMK/cofilin phosphorylation [134].

16.6.2 Tissue Remodeling in Tumor Migration

The extracellular matrix (ECM) orchestrates cellular structure and functions through biophysical and biochemical interactions between cells and the microenvironment [135]. Accumulating evidence suggests that matrix stiffness and tissue remodeling play a crucial role in breast cancer invasion and intravasation [136]. Contrary to their influence on cells in two dimensions, higher collagen cross-linking and ECM stiffness potently increase the migration and spreading speed of mammary tumor cells in 3D collagen gels [137]. Analysis of both murine tumors and human breast cancer samples reveals that tumor development is accompanied by an incremental elevation in collagen deposition and thickening of interstitial matrix [138, 139]. In turn, stroma remodeling and stiffening are associated with an increase in macrophage infiltration and TGF-β signaling [139]. Enrichment of myeloid cells and their secretion of TGF-β drive the activation of fibroblasts and the production of lysyl oxidase (LOX) [140]. The collagen cross-linking enzyme LOX facilitates ECM stiffening and focal adhesions, accelerating mammary tumor progression in vivo. Tissue fibrosis extends integrin clustering and activates PI3K signaling to promote cellular migration and invasion [138]. Pharmacologic deletion of LOX tempers matrix thickening and causes a significant reduction in circulating tumor cells and lung metastases in murine breast cancers [140]. Besides, macrophage-derived osteonectin, an important glycoprotein in cell-ECM interaction, augments deposition of fibronectin fiber and aggression of mammary tumor cells along the fibers through αvβ5 integrin, which contributes to cancer metastasis [141]. IL-4-induced elevation of cathepsin proteases in tumor-educated macrophages has pivotal effects on breast cancer progression and dissemination [142]. Cathepsins are involved in cleavage of cell-cell junction and ECM degradation in favor of tumor invasion and intravasation [143]. Macrophages show an increase in the production of the proteolytic enzyme MMP9 in tumor context, which remodels the ECM and paves a path through the microenvironment for tumor cell migration and spreading [107]. In addition, fibrotic environment retains chemokines and growth factors more easily, prolonging their acting duration and serving as an extra force for tumor aggression and invasion [144].

Cancer cells migrate through the neighboring stroma and basement membrane in a manner dependent on cytoskeleton rearrangement and invadosome formation, the actin-rich membrane protrusions which promote cell-matrix adhesion, matrix degradation, and cell invasion [145, 146]. With invadosomes anchoring the forefront of the cell, breast cancer cells migrate along the collagen fibers toward blood vessels as single cells or collectively as ensembles [147, 148]. The aforementioned EGF/CSF-1 paracrine signaling loops are involved in the formation of invadopodia in cancer cells and podosomes in macrophages [123] through the N-WASP-Arp2/3 pathway [149]. N-WASP is an important component of invadosomes that not only enhances the actin-nucleating activity of Arp2/3complex but also traffics MT1-MMP from endosomes to invadopodia and stabilizes it by tethering the tail to F-actin [150, 151]. EFGR stimulation also induces PLCγ-mediated release of cofilin in breast cancer cells. By binding to the cofilin-severed actin, $Mena^{INV}$ expedites actin filament elongation and invadopodia formation [152]. Interestingly, Rho GTPase signaling, activated either by soluble factors or direct contact, plays a role in the process of invadosome formation. Upon physical contact, TAMs activate RhoA pathway in breast cancer cells and modulate invadopodia formation, supporting tumor aggression through matrix barriers [153]. Cytokines and growth factor produced by TAMs like IL-6, IL-8 and EGF are also stimuli for Rho GTPase signaling, which enhance migratory capacity of breast cancer cells [154].

16.6.3 TAMs and Epithelial-to-Mesenchymal Transition

During invasion and intravasation, tumor cells undergo a reversible phenotypic change, losing their epithelial characteristics and intercellular adherence and acquiring mesenchymal traits and migratory properties. Epithelial-to-mesenchymal transition (EMT) not only endows cells with mobility and aggressiveness but also induces stem cell-like properties, including resistance to senescence and apoptosis, immune tolerance, and insensitivity to chemotherapy [155, 156]. It's elucidated that, in coculture research and humanized mice, TAMs induce EMT of mammary tumor cells through CCL18 production and activation of PITPNM3, which is a receptor for CCL18 on cancer cells. On the other hand, mesenchymal-like tumor cells program and educate the surrounding macrophages to a TAM-like phenotype by secreting granulocyte-macrophage colony-stimulating factor (GM-CSF). The abundance of lactate in the microenvironment attenuates the pro-inflammatory potential of GM-CSF in inducing M1 phenotype of macrophages. This forms a positive feedback loop between TAMs and mesenchymal-like cancer cells [157, 158]. Activated by TAM-derived CCL18, PITPNM3 induces mesenchymal properties of breast cancer cells through stabilization of Snail via Pyk2/Src/PI3K/Akt signaling pathway [133] or PI3K/Akt/GSK3β pathway [134]. Besides, CCL18 downregulates miR98 and miR27b in tumor cells at posttranscriptional level through N-Ras/ERK/PI3K/NF-κB/Lin28b signaling, resulting in enhanced tumor invasion and lung metastasis [159].

TGF-β is identified as the major inducer of mesenchymal markers, β-catenin signal, and tumor invasiveness in breast cancer [160–162]. TGF-β can either be directly secreted by TAMs [161] or produced by cancer cells in response to macrophage-derived cytokines like TNF-α, IL-1β, and IL-6 [163]. Transcription factors Snail, ZEB, and bHLH families are involved in TGF-β-induced EMT in a Smad-dependent mechanism [164–166]. Snail activity deprives E-cadherin expression and enhances vimentin production and cell mobility through upregulation of ZEB and downregulation of miR-34 [162]. Furthermore, phenotypic change in response through non-Smad signaling has a pivotal role in cancer metastasis. TGF-β1 activates the Ras effector Blimp-1 via c-Raf/Erk/AP-1 pathway. Subsequently, Blimp-1 elicits the repression of BMP-5 and upregulation of Snail, enhancing EMT signaling and tumor migration [167]. Forkhead transcription factor (Foxq1) activation induced by TGF-β1 interacts with E-box in its promoter region and endows mammary tumor cells with mesenchymal properties and invasiveness [168].

Direct contact between mesenchymal-like cancer cells and TAMs is observable in xenograft tumors and patient sections of breast tumor. Enrichment of CD90 and EphA4 protein on the stem-like cancer cells serves as an anchor for macrophage binding and delivers significant signals. TAMs trigger nuclear translocation of NF-κB by stimulating EphA4, which leads to the production of robust cytokines such as IL-6, IL-8, and GM-CSF and thus maintains the stem cell phenotype of tumor cells [169].

16.6.4 Tumor Transendothelial Migration

In addition to invading along collagens and through basement membranes, translocation of cancer cells from subluminal side of endothelium into the circulation is the next rate-limiting step of metastasis [170]. Circulating tumor cells are identified in non-metastatic breast cancer or early-state mammary tumor, suggesting that intravasation may occur early in tumor progression [171, 172]. Multiphoton intravital imaging studies have demonstrated that interaction between tumor cells and macrophages promotes breast cancer metastasis [122]. Wyckoff and colleagues directly visualized that TAMs were distributed in tumor margin and perivascular region subluminal to the endothelial cells. $Tie2^+$ perivascular macrophages assisted the transendothelial migration of mammary tumor cells in an EGF-CSF1-dependent mechanism [113]. At the intravasating site, tripartite interaction between cancer cells, macrophages, and endothelial cells is essential for tumor dissemination and forms tumor microenvironment for metastasis (TMEM) [173]. Herein, VEGFA signaling in $Tie2^{Hi}$ macrophages mediates localized interruption of vascular junctions and transient vascular permeability in TMEM [174]. Non-TMEM tumor cells undergo transendothelial migration with the help of the nonmigratory TMEM-tumor cell and the immobile TAM [173]. Moreover, direct contact between macrophages and tumor cells turns on Notch1-dependent $Mena^{INV}$ expression in breast cancer cells and provokes invadopodia-mediated transendothelial migration [175]. Genetic ablation of macrophages or macrophage-specific deletion of *vegfa* impairs blood vessel permeability and intravasation of mammary tumor cells in TMEM [173]. Currently, it is widely accepted that the TMEM score in primary breast carcinomas, which means the total number of TMEMs in ten high-power fields, is predictive of the risk of distant metastasis [176, 177].

16.7 Tumor Extravasation and Metastatic Outgrowth

16.7.1 Tumor Survival in the Circulation

After intravasation, breast cancer cells in the circulation are exposed to the immune system, oxidative stress, and mechanical shear forces, where they need to not only survive but also disseminate. In fact, only 0.01% of tumor cells that enter the bloodstream establish metastasis at distant sites. Tissue factors (TF) derived from circulating tumor cells (CTC) trigger the aggregation of platelet with tumor cells and deposition of fibrin, which gives rise to CTC/platelet/fibrin clots and protects CTC from shear stresses and immune tolerance through inhibiting recognition by NK cells in the blood vessels.

16.7.2 Seeding Distant Metastatic Sites

The process of CTC seeding at distant target organ is strongly repressed when $CD11b^+Gr1^-$ macrophages are depleted in genetic or pharmacologic manner, highlighting the important roles that TAMs play in tumor cell extravasation [32, 178]. CTC/platelet/fibrin clots arrest at the capillary of target organs by attaching the vascular endothelium through adherence and signal transduction among tumor cells, endothelial cells, and platelet [179]. In fact, after lodging in the capillary, intravascular tumor cells proliferate and form metastatic foci without exiting the vessels until they outgrow and breach the vascular walls [180] (see Fig. 16.3). CCR2-expressing inflammatory monocytes in bloodstream are recruited to the micrometastatic niche under the CCL2 chemoattractive gradient produced by breast cancer cells or other stromal cells, to take part in the seeding and growth of tumor cells [32]. In a TF-dependent mechanism, the formation and attachment of cancer cell clots also activate endothelial cells and induce their expression of vascular cell adhesion molecule-1 (VCAM-1) and vascular adhesion protein-1 (VAP-1) that are significant for macrophage recruitment and breast cancer metastasis [181]. CCL2 stimulation sets off a chemokine cascade in macrophages and induces their secretion of CCL3, which acts on CCR1+ macrophages in an autocrine manner and promotes the accumulation and prolonged retention of macrophages in the mouse lung [182]. In the context of metastatic cancer cells, recruited monocytes or macrophages differentiate into MAMs with CCR2, VEGFR1, Ly6C, and F4/80 expression. Through direct contact or in a paracrine mechanism, these macrophages facilitate breast cancer transendothelial migration and subsequent survival in the lung. Moreover, 3D reconstructed confocal images have demonstrated that physical interaction between breast cancer cells and macrophages is essential for the pulmonary seeding of extravasating metastatic cells and their persistent growth [178]. Activated by α_4 integrins on TAMs, VCAM-1 in BCC clusters upon the cell surface and downregulates the expression of pro-apoptotic cytokine TRAIL via Ezrin/PI3K/Akt pathway, thereby supporting extravasation and seeding of mammary cancer cells [183]. Macrophage-specific deletion of *vegfa* gene abrogates the permeability of vascular walls and the seeding potential of breast cancer, suggesting that CCL2-recruited CD11b$^+$Gr1$^-$ macrophages promote tumor cell extravasation partly by increasing the endothelial permeability via VEGFA [32]. As mentioned above, TAM-derived EGF plays a role in the augmentation of invadosome formation in breast tumor cells. Intravital imaging experiments illuminate that during extravasation, tumor cells project invadopodia to lung interstitial through endothelial junctions [184], which underlies another strategy for TAMs to support tumor extravasation.

16.7.3 Colonization and Overt Metastasis Formation

Even if cancer cells succeed in seeding the target organ, most of them succumb [185] or stay in quiescence in the perivascular niches because of lack of proliferative signals (cellular dormancy), inability to produce sufficient vasculature (angiogenic dormancy), and tumor clearance by immune surveillance (immunological dormancy) [186, 187].

Recruited by a variety of chemokines secreted by breast cancer cells, macrophages infiltrating the micrometastases undergo differentiation and play a significant role in the process of cancer recurrence and overt metastasis formation. The expression of transcription factor Sox2 in breast cancer promotes metastatic growth in a macrophage-dependent manner, which recruits TAMs into tumor microenvironment through upregulation of several cytokines such as CCL2, CCL3, and ICAM-1 via NF-κB and Stat3 signaling pathways [188]. Tumor cell-derived CCL5 extends the infiltration of TAMs and their secreting activity of cytokines like EGF and TGF-β indirectly, by augmenting Gfi1 expression in CD4$^+$ T cells and their polarization toward a Th2 phenotype, which modulates the metastasis of MMTV-PyMT transgenic tumors [189]. The enhanced production of the enzyme N-acetyl-galactosaminyltransferases (GALNTs) in breast cancer cells provokes TAM influx to the metastatic sites. GALNTs promote macrophage-stimulated tumor growth by transforming FGFR1 on cancer cells, which results in

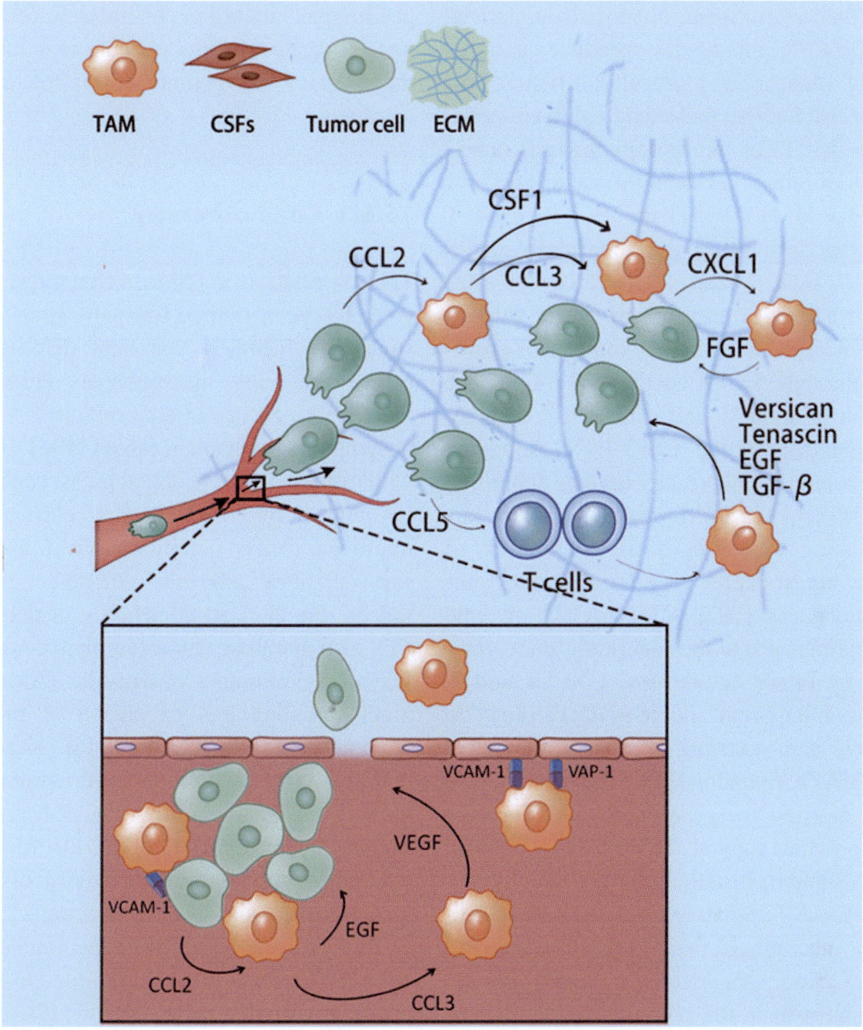

Fig. 16.3 Seeding of breast cancer cells at distant metastatic sites and their formation of overt metastases. Intravascular tumor cells lodge the capillary at distant metastatic site and attract CCR2$^+$ inflammatory monocytes through CCL2 synthesis. CCL2 stimulation induces macrophage release of CCL3, which amplifies the accumulation of macrophages in an autocrine manner. On tumor attachment, endothelial cells express VCAM-1 and VAP-1 in support of macrophage enrichment. Recruited monocytes or macrophages differentiate into MAMs, whose binding to VCAM-1 of BCCs precipitates their extravasation. MAMs mediate vascular permeability via VEGFA and tumor invadosome formation via EGF. After extravasation, GALNT14 in BCCs provokes the influx of macrophages through CXCL1 synthesis and potentiates the proliferative effect of MAM-derived FGF. Macrophage-derived CSF1 creates an autocrine pathway and modulates their expression of miR-21 and miR-29a. MAMs induce mesenchymal to epithelial transition (MET) of neoplastic cells via versican synthesis and initiate tumor outgrowth through tenascin C production. In the pulmonary microenvironment, CCL5 from cancer cells mediates the Th2 polarization of CD4$^+$ T cells, which extends macrophage infiltration and their secretion of EGF and TGF-β

augmentation of the proliferative effect of TAM-derived FGF [190]. In a metastasis dormancy model in the bone, tumor expression of VCAM-1 through NF-κB pathway interacts with integrin α4β1-expressing monocytic progenitors and enhances osteoclast activity, contributing to the formation of overt skeletal metastases of mammary tumor [191].

After their recruitment, bone marrow-derived $CD11b^{+}Ly6C^{high}$ monocytes produce a large amount of versican, an extracellular matrix proteoglycan that induces mesenchymal to epithelial transition (MET) of tumor cells through downregulation of Smad2 levels at the metastatic sites [192]. In the MMTV-PyMT model, TH2 $CD4^{+}$ T lymphocytes infiltrating the metastatic niches produce IL-4 and IL-13, polarize $CD11b^{+}Gr1^{-}F4/80^{+}$ macrophages to a tumorigenic phenotype, and enhance tumor cell proliferation through EGFR activation [47]. The expression of miR-21 and miR-29a is upregulated in MAMs by the CSF1-ETS2 signaling pathway but not in the primary tumors. Inhibiting these microRNAs in TAMs with Dicer inhibitors strongly abrogates angiogenesis and viability of metastatic breast cancer cells by activating anti-angiogenic genes [193]. $CD11b^{+}Gr1^{+}$ myeloid cells are attracted into the metastatic microenvironment by tumor cell-derived CXCL1 and 2, serving as an important source of the calprotectin S100A8/9 and exerting pro-survival effect through MAPK signaling [35]. Once permitted in the bone marrow, the gene-expression signature of c-Src provides support for survival of indolent tumor cells and metastatic outgrowth by potentiating PI3K-AKT signaling in breast cancer activated by macrophage-derived CXCL12 [194]. Tenascin C from cancer cells or stroma initiates tumor outgrowth in the pulmonary parenchyma through Notch and WNT pathway [195]. More importantly, TAMs have a profound effect on metastatic tumor colonization and transition from indolent micrometastases to overt metastases by promoting proliferative signaling, phenotypic change, angiogenesis, and immune evasion, even after long-term dormancy.

16.8 Macrophage Regulation of Therapeutic Efficacy

In addition to their involvement in tumor growth and dissemination, TAMs interact with anticancer therapies, such as chemotherapy, radiotherapy, immunotherapy, and targeted therapy. These therapeutic approaches induce functional and phenotypic alteration in intratumoral macrophages, while the latter play a role in modulating the efficacy of various forms of anticancer therapies.

16.8.1 Chemotherapy

The contribution of TAMs in chemotherapy resistance has been studied for many years. An early study demonstrated that host defense mechanisms, including macrophages, enhanced the therapeutic efficacy of doxorubicin in a leukemia or lymphoma transplant model [196]. In patients with primary invasive ductal breast cancer, tumor infiltration by macrophages correlates with improved cancer-specific survival and overall survival after adjuvant chemotherapy [197]. Indeed, the therapeutic efficacy of docetaxel, in 4T1-Neu mammary tumor implants, involves the selective elimination of M2-like TAMs through Stat3 signaling and expansion of tumoricidal macrophages, with induction of IL-12 and reduction of IL-10 [198]. Similarly, doxorubicin treatment in 4T1 mammary cancer-bearing mice triggers apoptosis of $Gr\text{-}1^{+}CD11b^{+}$MDSCs and potentiates the antitumor activity of perforin, granzyme B, and IFNγ producing cytotoxic T cells (CTL) and NK cells. Concomitantly, residual MDSCs exhibit curtailed immune-inhibiting potential with decreased ARG-1, IDO, and ROS level, which suggests myeloid cell reprogramming [199]. The infiltration of IFN-γ-producing effector T cells in tumor and the alteration of TAMs from an immunosuppressive phenotype to pro-inflammatory macrophages are also detectable following treatment with doxorubicin and lapatinib, the HER2 inhibitor to MMTV-neu mice [200]. Collectively, the cytotoxic macrophages within the tumor induced by chemotherapy play a part in immunogenic cell death (ICD), as they exhibit antigen-presenting capacity and reinforce the tumoricidal effect of T cells to acquire durable success in tumor control [201].

The opposite effects of TAMs on chemotherapeutic efficacy have been evidenced in a growing body of the literature. Paulus and colleagues have elucidated that TAM depletion by anti-CSF1 antibodies in immunodeficient mice implanted by human breast cancer xenografts reverses tumor resistance to CMF chemotherapy (cyclophosphamide, methotrexate, and 5-fluorouracil) [202]. Mechanistically, macrophage expansion following cytotoxic agents drives tumor cells to escape from the therapy-induced cell death by delivering survival signals. EGF produced by TAMs endows $EGFR^+$ cancer cells with stem-like phenotypes characterized not only by increased tumor viability and metastatic potential but also by enhanced drug-efflux capacity and post-chemotherapeutic tumorigenicity in vivo [129]. In addition to damaging cancer cells, chemotherapy agents induce TNF-α production of endothelial cells, which boosts the metastasis-promoting CXCL1/2-S100A8/9 axis, as discussed above. Herein, synthesis of S100A8/9 by recruited $CXCR2^+$ macrophages bolsters tumor regrowth and chemotherapeutic tolerance [35]. In the MMTV-PyMT model, Taxol treatment elicits an influx of $Iba1^+$TAMs, which protects tumor cells from Taxol-induced cell death by producing the lysosomal enzymes cathepsins B and S. Consistent intervention of Taxol combined with cathepsin inhibition in vivo potently improves late-stage survival [203].

Another important mechanism of chemoresistance is subverting immune clearance of tumor cells. Macrophage enrichment in the MMTV-PyMT model, in response to CSF1 production by neoplastic cells under cytotoxic stress of paclitaxel, blunts tumor chemosensitivity via impairing CTL infiltration and antitumor response [77]. In line with these findings, IL-10 derived from chemotherapy-recruited macrophages is essential in T-cell suppression by curbing IL-12 production of dendritic cells. Indeed, elevation of IL-12A in patients with breast cancer correlates with an improved pathological response to chemotherapy [73]. Vascular alterations after cytotoxic agents within tumor attenuate chemotherapeutic efficacy. Myeloid cells, including macrophages, infiltrate the transgenic mammary tumor through a stromal CCL2-CCR2 axis. MMP9 produced by the myeloid cells decreases vascular permeability, inhibits intratumoral distribution of doxorubicin, and contributes to tumor reemergence [204]. Intriguingly, decreased vascular leakage has been associated with a better prognosis in other tumor models. $TIE2^{Hi}CXCR4^{Hi}$ macrophages accumulate after administration of paclitaxel and doxorubicin in perivascular region in 4T1 and MMTV-PyMT tumor and mediate revascularization and tumor recurrence, in part, through the production of VEGFA [33].

These important findings show different effects of TAMs on chemotherapy through distinct mechanisms, in part depending on tumor subtypes and therapeutic agents. Actually, intratumoral macrophages may induce a mixture of different signaling and alterations to modulate the efficacy of the same therapeutic agent. The complex interaction between infiltrating macrophages and chemotherapeutic agents provides a compelling rationale for targeting tumorigenic population of TAMs while sparing tumoricidal ones in breast cancer combined with cytotoxic therapeutic agents.

16.8.2 Radiotherapy

As one of the mainstays of treatment for breast cancer, radiotherapy has a profound effect on tumor stroma beyond its antitumor activity through DNA damaging. Radiation-induced impairment of cancer and vascular cells triggers damage-associated molecular pattern (DAMP) signaling, which can stimulate pattern recognition receptor (PRR) on macrophages. Activated macrophages exhibit phagocytic and antigen-presenting properties critical in effective ICD of tumor cells even in distant organs (abscopal effect), which is synergistic with radiotherapy [205–207].

In contrast, increasing data suggest that TAM infiltration after irradiation contributes to tumor reoccurrence and metastasis [208]. Radiotherapy-triggered vascular destruction exacerbates hypoxia, creating a protumor microenvironment [209]. In MCa8 mouse mammary carcinomas, hypoxia-induced CXCL12 recruits a mass of myeloid cells, primarily macrophages, in the irradiated tumor, which precipitate tumor regrowth through their paracrine response on vasculature. CXCR4 blockade or bone marrow depletion by whole-body irradiation shows inhibitory effect on tumor relapse after local radiotherapy [34]. In MMTV-PyMT tumors, radiation-damaged mammary epithelial cells and subsequent influx of protumorigenic macrophages elevate the production of CSF-1 and IL-34, which results in reduction of CD8+ T cells and tumor reemergence [48]. Immunologic or pharmacologic deletion of TAMs by neutralizing the CSF-1/CSF-1R pathway delays breast cancer revascularization and recurrence [48, 206]. Intratumoral macrophages provide a plethora of molecules and signaling to establish a tumor-protective microenvironment, especially via vascular recovery [210]. In transplantable MT1A2 mammary tumors, radiotherapy induces an influx of BM-derived $CD11b^+$ myelomonocytic cells, whose production of MMP9 has pivotal roles in neovascularization by enhancing colonization of BM-derived circulating endothelial cells. Indeed, the transplanted tumor fails to grow in MMP-9 KO mice, while transplantation of wild-type BM abrogates this effect [210, 211]. Data shows that in the 4T1 model, MMP14 blockade utilizing DX-2400 synergizes with radiotherapy in attenuating tumor growth and progression. Indeed, DX-2400 intervention triggers macrophage expansion and phenotypic conversion to tumoricidal subtypes with downregulation of TGF-β and SMAD2/3 signaling [212].

16.8.3 Vascular-Targeted Therapies

Anti-angiogenic tumor therapies show limited efficacy in dampening tumor growth and metastasis in patients with breast cancer. Inasmuch as their strong support for tumor neovascularization, TAM enrichment counteracts the therapeutic interception of angiogenesis through their interaction with endothelial cells and induction of compensatory proangiogenic factors [213].

Sorafenib, which targets VEGFR2, PDGFR, and Raf kinases, exerts dinky effect on 4T1 tumors. Macrophage influx is visualized by near-infrared fluorescence (NIRF) imaging in the sorafenib-resistant tumor, whereas TAM deletion restrains tumor growth and lung metastasis [214]. Selective destruction of the vessel network and breast cancer necrosis triggered by combretastatin A4 phosphate (CA4P), the vascular-targeting agent, creates a hypoxic microenvironment and consequent upregulation of CXCL12. $CXCR4^+$ TEMs rapidly infiltrate subcutaneous N202 (Neu+) mammary carcinomas and abrogate the vascular damage and necrosis induction of this archetypal VDA. The efficacy of CA4P treatment is dramatically increased by either CXCR4 antagonist or genetic TEM depletion [215]. Accumulating evidence suggests that anti-angiogenesis strategies that normalize vasculature and alleviate tumor hypoxia provide more benefit and longer survival for the patients [216].

Ang2 exhibits proangiogenic activity by activating Tie2 and limits the antitumor efficacy of VEGF blockade. Indeed, Ang2 blockers potentiate the antivascular effect of aflibercept, a VEGF inhibitor, and lead to effective reduction in tumor vascularity and perfusion [115, 217]. Mechanistically, Ang2 endows endothelial cells with a pro-inflammatory phenotype characterized by upregulation of chemoattractant CCL2 and subsequent induction of $CCR2^+$ TAM infiltration. In the presence of Ang2, endothelium shows potentiated response to the myeloid cell-derived angiogenic cytokine Bv8. Combining Ang2 inhibitor and low-dose metronomic chemotherapy strongly delays the metastatic growth and enhances overall survival in the 4T1 orthotopic breast cancer [218]. Recent data shows that ABTAA, an Ang2-binding and Tie2-activating antibody, suppresses mammary tumor growth and metastasis not only by decreasing the infiltration of TAMs and reprogramming macrophages toward antitumor phenotype but also through its effect on vascular normalization. Restoration of structural integrity of vasculature characterized by improved pericytes and basement membrane leads to decreased leakage and hypoxia, enhanced perfusion, and anticancer drug delivery [219].

16.8.4 Monoclonal Antibodies and Immunotherapy

A growing body of literature has shown that antitumor activity of monoclonal antibody (mAb) therapy depends on the intratumoral macrophages. Interaction of the Fc fragment of mAbs with Fcγ receptors (FcγRs) on macrophages triggers the engagement of the FcγRs and leads to the activation of antibody-dependent cellular cytotoxicity/phagocytosis (ADCC/ADCP) [220, 221]. FcγR polymorphism in macrophage is associated with distinct affinity to mAbs and thus predicts the antitumor efficacy of targeted therapies in breast cancer and lymphoma [222, 223]. Early studies showed that Fc receptor-dependent macrophage cytotoxicity contributed substantially to the efficacy of trastuzumab against breast cancer [224]. Park and colleagues also demonstrated that in MMTV-neu tumors, besides interrupting oncogenic HER2 signals, trastuzumab released a significant amount of HMGB-1, which was an endogenous danger signal enhancing the FcR-mediated phagocytosis and promoting tumor-specific CD8+ T-cell responses [225]. Mammary tumor conditioning endows macrophage with not only expression of M2a markers but also M1-associated markers and activating FcγRs. The transition of TAMs from the invasion promoter to an antitumor phenotype successfully eradicates Ab-bound tumor cells in the presence of CD142, a mAb directed against tissue factor [42]. In MDA-MB-231 xenograft model, macrophage depletion significantly reduces the efficacy of anti-CD142 to restrain primary tumor growth and lung metastasis [42]. Consequently, engineering Fc domains of mAb to reinforce FcγR binding and subsequent recruitment of macrophages to mediate ADCC, such as the anti-HER2 "grababody" [226], show promise in strengthening the efficacy of targeted therapy for cancer [220].

Since the antitumor activities of chemotherapy, radiotherapy, and targeted therapy all depend on effective immune clearance and surveillance to obtain long-term efficacy, it's key to overcome the immunosuppressive tumor microenvironment and enhance the immune response [227]. MAbs targeting checkpoint blockade CTLA-4, PD1, or PD-L1, T-cell immunoglobulin and mucin domain-containing protein-3 (TIM-3), and lymphocyte activation gene (LAG-3) strengthen antitumor immunity [227, 228]. The antitumor efficacy of anti-CTLA-4 antibodies depends on ADCC mediated by tumor-infiltrating $CD11b^+$ macrophages. These antibodies bind to the high-density CTLA-4 on Treg cells and activate FcγRIV on macrophages, resulting in elimination of Treg cells and subsequent potentiated T-cell response [221]. Furthermore, blocking PD-L1 that is expressed on TAMs and serves as an important immunosuppressive mechanism will no doubt reduce the protumor properties of TAMs and expand T effector cell response.

Cancer cell destruction induced by various modalities of therapeutic agents triggers tumor-specific immune response, during which macrophages display antigen-presenting activity and potent T-cell stimulating capacity. However, a plethora of tumorigenic factors in local microenvironment provide confusing information for these recruited monocytes/macrophages and program them into an anti-inflammatory phenotype, which facilitates tumor regrowth and therapeutic resistance. There may be an equilibrium and competition between the protumor and antitumor strengths in the sophisticated cancer microenvironment following therapeutic intervention, which may account for the paradoxical roles that TAMs play in cancer treatment. Additional investigation is in demand to evaluate the potential mechanism and biomarker of the therapy-associated macrophage modulation. These findings provide a rationale for combining TAM-targeting approaches to potentiate the tumoricidal activity and reduce refractoriness of the conventional therapies (see Table 16.1).

Table 16.1 Therapeutic targeting of macrophages in breast cancer

Compound	Target	Mechanism	Clinical phase	Combination compound(s)	References
Inhibition of TAM recruitment					
Carlumab (CNTO 888)	Anti-CCL2 antibody	Targeting CCL2-CCR2 axis	Phase I	Systemic therapy	[230, 231]
Pf-04136309	CCR2 small-molecule antagonist	Targeting CCL2-CCR2 axis	Phase I	Standard chemotherapy	[232]
Plerixafor (AMD3100)	CXCR4 antagonist	Targeting CXCL12-CXCR4 axis	Phase III	G-CSF	[235]
Pexidartinib (PLX3397)	CSF1R small-molecule inhibitor	Targeting CSF1-CSF1R axis	Phase I	Paclitaxel	[238]
AMG 820	Anti-CSF-1R mAb	Targeting CSF1-CSF1R axis	Phase I	Standard treatment	[240]
Depletion of TAMs or their progenitors					
Bisphosphonate	TAMs or their progenitors	Decrease in TAM number and function	Phase IV	As adjuvant treatment	[253]
Zoledronic acid	See above	See above	Phase IV	Endocrine therapy	[254, 255]
Clodronate	See above	See above	Phase IV	As adjuvant treatment	[256]
Reprogramming macrophages toward a tumoricidal phenotype					
Imiquimod	TLR7 agonist	Activation of an antitumor phenotype of macrophages	Phase I	As adjuvant treatment	[259]
Imiquimod	See above	See above	Phase II	Albumin-bound paclitaxel	[261]
MGN1703	TLR9 agonist	Inducing pro- inflammatory polarity	Phase I	As adjuvant treatment	[262]
Aspirin	COX inhibitor	Reducing PGE2 secretion	Cohort study	Prediagnostic use	[263].
Impeding the protumor functions of TAMs					
MPDL3280A	Anti-PD-L1 mAb	Alleviating immune suppression	Phase I	Systemic regimens	[282]
Pembrolizumab	Anti-PD-1 antibody	See above	Phase Ib	Single agent	[271, 272]
Tremelimumab	Anti-CTLA-4 antibody	Alleviating immune suppression	Phase I	Exemestane	[273]
IPI549	PI3Kγ small-molecule blocker	Switching macrophage phenotype	Phase I	Alone and combined with nivolumab	
Indoximod	Broad IDO pathway blocker	Blocking IDO-mediated immune suppression	Phase I	Docetaxel	[275]
Indoximod	See above	See above	Phase II	Chemotherapy	[276]

(continued)

Table 16.1 (continued)

Compound	Target	Mechanism	Clinical phase	Combination compound(s)	References
Nesvacumab	Anti-Ang2 mAb	Angiogenesis blockade	Phase Ib	Ziv-aflibercept	[281]
MNRP1685A	Anti-NRP1 antibody	Angiogenesis blockade	Phase I	As adjuvant treatment	[280]

Abbreviation: *Ang-2* angiopoietin-2, *COX* cyclooxygenase, *CSF* colony-stimulating factor, *CTLA-4* cytotoxic T-lymphocyte-associated protein 4, *IDO* indoleamine 2,3-dioxygenase, *IL* interleukin, *MDSC* myeloid-derived suppressor cell, *NRP1* Neuropilin-1, *PD-1* programmed cell death protein 1, *PD-L1* programmed cell death 1 ligand 1, *PGE2* prostaglandin E2, *TAM* tumor-associated macrophage, *TLR* Toll-like receptor

16.9 Therapeutic Targeting of Macrophages

16.9.1 Inhibition of TAM Recruitment by CCL2-CCR2 Axis

In light of the increasing evidence for TAMs supporting tumor growth and metastasis, attempts have been made to curb their protumor activities by suppressing TAM infiltration into breast cancer. Diverse factors including chemokines, cytokines, and complement components are involved in macrophage recruitment in mammary tumor, among which CCL2-CCR2 axis plays a predominant part [32]. Preclinical studies have corroborated that interference of CCL2-CCR2 signaling through genetic manipulation or pharmacologic inactivation decreases TAM influx, lowers metastatic burden, and prolongs the survival of mice in different breast cancer models [32, 182, 204, 218, 229]. Administration of CCL2 or CCR2 blockade in combination with conventional therapeutic regimens significantly improves the efficacy of treatment [32, 204, 218, 229].

An increasing number of experimental medicines targeting CCL2-CCR2 axis have entered clinical trials. In spite of being well tolerated, the anti-CCL2 antibody carlumab (CNTO 888) shows, in a phase I trial, limited antitumor efficacy in patients with solid tumors [230]. Combinations of carlumab and standard-of-care chemotherapy agents elicit rapid but transient reduction in serum-free CCL2 in patients with advanced solid tumors, which is followed by continued increase [231]. Meanwhile, interruption of antibody-mediated CCL2 blockade precipitates an overshoot of pulmonary metastases and accelerates death of tumor-bearing mice by yielding an unexpected influx of monocytes from the bone marrow into the metastatic niches. Excessive monocytes have pivotal roles in generation of neoplastic vasculature, proliferation of metastatic cells, and the lethal rebound through the upregulation of IL-6 and VEGFA [110]. It reveals that compensatory production of CCL2 by tumor or stromal cells and worsened monocyte infiltration are main challenges for the antitumor effect of carlumab. Notably, targeting the receptor may be an attractive device to circumvent the limitations resulted from the reactive CCL2 excess. Recently, the preliminary therapeutic effect and safety of PF-04136309, an oral CCR2 small-molecule antagonist, in combination with standard chemotherapy, have been demonstrated in a phase Ib clinical trial conducted on pancreatic patients. Herein, the addition of PF-04136309 elicited a higher objective response rate than expected and a reduction in the infiltration of TAMs [232]. However, the antitumor efficacy, safety, and tolerability of anti-CCR2 drugs in patients with breast cancer remain to be elucidated in clinical trials.

The feedback mechanisms and the unexpected tissue remodeling in microenvironment can abrogate the therapeutic efficacy of drugs targeting CCL2-CCR2 signaling and evoke a deteriorated prognosis for cancer patients, especially in the case of sudden cessation of monotherapy. Besides CCL2, a great diversity of chemokines, cytokines, and growth factors are involved in macrophage recruitment, such as CSF1 and a series of CXC chemokines [233, 234], which presents another challenge for translating the strategies targeting CCL2-CCR2 into clinical benefit in patients with neoplastic diseases. The confined mobilization of monocytes from the bone marrow, which is triggered by interference with CCL2-CCR2 axis, may be partially responsible for some drug-related adverse events [110]. In addition, other chemokines emerge as hopeful targets for abating intratumoral macrophage accumulation. For example, genetic deletion of CCL3 or its receptor CCR1 confines pulmonary retention of MAMs and metastatic seeding of breast cancer cells [182]. CXCR4 has a pivotal role in the trafficking of monocytes or macrophages to CXCL12-rich tumor sites, leading to tumor reoccurrence and therapeutic bluntness [33, 34]. Antitumor efficacy of plerixafor, a CXCR4 antagonist, in combination with G-CSF, has been proved in patients with non-Hodgkin's lymphoma [235].

16.9.2 Inhibition of TAM Recruitment by CSF1-CSF1R Axis

As the major orchestrator for monocyte-macrophage lineage, CSF-1 has profound effects on macrophage growth, differentiation, and recruitment to the tumor region [29, 236]. In patients with breast cancer, CSF-1 is abundantly expressed and associated with poor prognosis [28]. Interruption of CSF1-CSF1R signaling in preclinical studies breaks the EGF-CSF1-positive feedback loop, curbs tumor metastasis, and increases the therapeutic sensitivity of cytotoxic agents and irradiation in mammary tumors [121, 123, 202].

Studies have shown tissue-specific reduction of TAMs and enrichment of intratumoral CD8+ T cells in the MMTV-PyMT model following the administration of PLX3397 [77] or BLZ945 [237], which are both selective small-molecule inhibitors of CSF1R. In a phase Ib study, combination of PLX3397 and paclitaxel is generally well tolerated, which raises plasma CSF-1 level and reduces circulating $CD16^{+}CD14^{+}$monocytes in breast and other solid tumors [238]. Unlike PLX3397 and BLZ945, emactuzumab (RG7155) is a monoclonal antibody that represses CSF1-dependent or CSF1-independent activation of CSF-1R by obstructing receptor dimerization [239]. In a phase I clinical trial, comparison of pretreatment and on-treatment biopsy samples from patients with mammary and other solid malignancies shows that administration of RG7155 attenuates CSF-$1R^{+}CD163^{+}$ TAM infiltration and reconstructs intratumoral T-cell composition by increasing CD8/CD4 ratio [239]. Besides, the monoclonal anti-CSF-1R antibody AMG 820 boosts secretion of serum CSF-1 and reduces skin macrophages in patients with various advanced solid tumors [240]. A recent study elucidates that Pexidartinib (PLX3397) mediates prolonged regression in tumor volume in patients with tenosynovial giant cell tumors, a rare type of sarcoma characterized by overexpression of CSF1R [241]. Albeit the validity of CSF1R ablation has been confirmed in several tumors, clinical trials investigating the efficacy of these drugs on tumor growth and metastasis in breast cancer are ongoing.

Unexpectedly, in 4T1.2 and EMT6.5 breast carcinomas, blockade of CSF-1R/CSF-1 elicits neutrophil expansion and accelerates spontaneous metastasis to the lung and spine, which is reversed by neutralizing anti-G-CSFR antibody treatment [242]. Although the detrimental outcome occurs in specific tumor models, the compensatory upregulation of other signals like G-CSF and infiltration of tumor-associated neutrophils may rescue anti-CSF1R treatment-mediated tumor repression. Furthermore, continuous CSF-1R inhibition results in enhanced IGF-1 secretion by TAMs into the extracellular microenvironment in response to IL-4/IL-4R pathway. Activation of IGF-1R on tumor cells sustains tumor growth and induces acquired resistance to anti-CSF-1R treatment in gliomas [243]. The mechanisms of CSF-1R blockade tolerance in breast cancer are yet to be determined, as well as whether a rise of IGF-1 production occurs during the process. Unlike CCR2 inhibition working upon CCR2+ monocytes in the bloodstream or bone marrow [232], the efficacy of anti-CSF1R drugs is dependent on and limited by their ability to access the neoplastic region. In addition, macrophage depletion in the non-tumor tissue and organs is observed in nonhuman primates following CSF1R ablation, suggesting that the off-target activity can be a significant concern for this kind of agents moving forward. Thanks to its tumor-depleting effects [48, 77, 206, 244], ablation of CSF1/CSF1R may be a promising therapeutic strategy to complement the standard-of-care therapeutic regimens by compromising the recruitment of tumor-promoting TAMs.

16.9.3 Depletion of TAMs or Their Progenitors

Depletion of TAMs or their progenitors is an attractive and promising therapeutic option to lessen their support for tumor growth, distant dissemination, and therapeutic tolerance. For instance, chemotherapeutic agents such as docetaxel [198] and doxorubicin [199, 245] can confine tumor progression partially through TAM eradication in breast cancer.

Bisphosphonates (BPs) are antiresorptive drugs for osteoporosis and skeletal complications related to metastatic cancers, which can be engulfed by and lead to apoptosis of bone macrophages [246]. Plentiful evidence, both in vitro and in vivo, shows that BPs exert great influence over TAMs by inducing apoptotic cell death, dampening their proliferation, impairing their protumoral functions, and even reprogramming TAMs to tumoricidal phenotype [247], among which zoledronic acid is the most potent one [248]. Early seminal work by Diel and coworkers elucidated the extraskeletal effect of clodronate which caused a significant reduction in both bony and visceral metastases in patients with primary breast cancer and detectable cancer cells in the bone marrow [249]. In accordance, tumor-infiltrating $CD11b^{+}F4/80^{+}$macrophages and MDSCs were significantly reduced in amino-bisphosphonate-treated BALB-neuT mice, along with dropped serum pro-MMP-9 and VEGF and impaired tumor growth [250]. Furthermore, zoledronic acid hampers the initiation and growth of ErbB-2-driven mammary carcinomas by decreasing intratumoral macrophages and VEGF release. Residual macrophages show recovery in the pro-inflammatory properties with reduced IL-10 but augmented IFN-γ secretion [251]. A study utilizing real-time intravital microscopy provides unequivocal evidence that BPs access extraskeletal breast cancer via highly leaky tumor vasculature, where they initially bind to granular micro-calcifications. These BP-coated micro-calcifications undergo rapid and efficient internalization in vivo by TAMs, which leads to their functional subversion and depletion [252].

Accumulating evidence indicates that BPs improve the survival of breast cancer patients, independent of their antiresorptive effects on the skeleton. Collaborative meta-analyses found significant reductions not only in bone recurrence but also in extraskeletal distant relapse and breast cancer mortality in a subgroup of postmenopausal women from 18,766 patients with early-stage breast cancer randomized in trials of adjuvant BPs. However, the treatment had no apparent effect on distant reemergence or cancer mortality among premenopausal women [253]. In another clinical trial with a 94.4-month median follow-up, the addition of zoledronic acid to endocrine therapy improved disease-free survival in premenopausal patients with endocrine-responsive early-stage breast cancer, who received ovarian suppressive therapy simultaneously [254, 255]. Similar results were observed in a 1:1 randomized trial in 3323 women with breast cancer receiving daily oral clodronate as adjuvant therapy, whereas the subpopulation of postmenopausal women showed postponement in bone and non-bone metastasis and tumor reemergence [256]. These findings highlight the influence of reproductive hormones on the antitumor efficacy of BPs. Even though they offer no overall benefit in a mixed group of breast cancer patients [257], BPs delay extraskeletal recurrence and improve survival outcome in patients with a naturally or medically induced menopause. Inasmuch, it is tempting to consider BPs as adjuvant agents in the administration of early breast cancer in a broader range of postmenopausal and ovarian-suppressed premenopausal women to reduce mortality and extend survival.

16.9.4 Reprograming Macrophages Toward a Tumoricidal Phenotype

As mentioned above, invasion-promoting TAMs regain Fc-dependent phagocytic and tumoricidal capacities in the presence of mAb [42]. The plasticity of TAMs provides a compelling rationale to reeducate intratumoral macrophages and exploit their antitumor properties rather than TAM depletion or destruction, for developing therapeutic approaches.

Functional skewing of macrophages from anti-inflammatory M2-like population toward a tumoricidal phenotype has been achieved via activation of Toll-like receptors (TLRs) on TAMs. In preclinical studies, topical administration of imiquimod, a TLR7 agonist, synergizes with local irradiation and low-dose cyclophosphamide in tumor regression and metastasis abrogation in both the local region and distant field (abscopal effect) [258]. A phase II trial unveiled that imiquimod treatment achieved a partial response rate of 20% in women with refractory breast cancer cutaneous metastases. The posttreatment increased tumor-infiltrating T lymphocytes and induced a Th1 cytokine profile, suggesting an immune-mediated response [259]. A case report described that the addition of topical imiquimod to systemic therapy successfully led to a partial regression of cutaneous metastases in the patient with refractory breast cancer [260]. Application of imiquimod in combination with albumin-bound paclitaxel elicits an overall objective response rate of 72% in treatment-resistant breast cancer with chest wall metastases, though the duration of response is limited [261]. Meanwhile, pretreatment elevation of PD-1$^+$ T lymphocytes and M-MDSCs predicts suboptimal or no response, raising arguments for the efficacy of combining immunotherapy with inhibition of MDSC recruitment to TLR7 agonist [261]. Similarly, the TLR9 agonist MGN1703 is effective in the retardation of disease progression in patients with various metastatic solid tumors [262]. These results support TLR stimulation as a viable therapeutic strategy to recover the antineoplastic activities of TAMs and render immune-dependent tumor growth control.

PGE2 is an instrumental prostanoid lipid in breast cancer that is associated with induction and maintenance of M2-like macrophage polarity and supports tumor evasion of immune surveillance [50]. Among TAMs isolated from breast cancer tissue, COX-2 is abundantly expressed and correlated to poor prognosis [74]. The use of COX inhibitors, particularly aspirin, is associated with protection against tumor progression and metastasis in patients with cancer. In a cohort study in women with stage I–III breast cancer, prediagnostic application of aspirin elicited dose-dependent protection against lymph node metastasis and a significant reduction in 5-year breast cancer-specific mortality among lymph node-negative subgroup of patients [263]. Moreover, aspirin administration in mammary tumors with mutant PIK3CA, which encodes catalytic subunit of PI3K, attenuates tumor viability and growth independent of the effect on COX-2 and NF-κB but via activation of AMPK pathway and restraint of mTORC1. The addition of a PI3K inhibitor synergizes with aspirin and further damages tumor progression, suggesting a combination therapeutic regimen for patients with breast cancer [264].

With the aim of circumventing unwarranted robust macrophage activation and systemic inflammation, a novel and attractive strategy to subvert the protumoral microenvironment is developed by taking advantage of the tumor-homing ability of TAMs. Genetically engineered TEMs are capable of mediating preferential transgenic expression of inflammatory stimuli within the tumor. In distinct breast cancer models, tumor-specific IFN-α delivery by transgenic TEMs blunts the angiogenic properties of macrophages and advances the recruitment of effector T cells into neoplastic sites, leading to repression of tumor development and abrogation of metastasis with limited systemic toxicity [265, 266].

Redirection of intratumoral macrophages has been achieved through manipulation of a variety of molecules or signaling. Targeted loss of MerTK from tumor-educated macrophages, which is a significant member of the TYRO3/AXL/MerTK receptor tyrosine kinase family, favors the pro-inflammatory properties of macrophage and CD8+ T-cell infiltration [56]. The host-produced histidinerich glycoprotein (HRG) in tumor stroma combats tumor progression by redirecting TAMs away from the tumor protectors, wherein deletion of TAMs counteracts the antitumor effects of HRG [267]. By hampering ROS generation, BHA triggers reduction in M2-like TAM infiltration and hinders tumorigenesis in MMTV-PyMT models [53]. Tumor local delivery of IL-21 fuels the cytotoxic properties of macrophages and subsequently reverses the extrinsic resistance of anti-Her2/Neu treatment, highlighting its therapeutic significance [268]. Conventional tumoricidal agents like DTX and DOX damage the angiogenic and pro-metastatic activities of TAMs while restoring their production of pro-inflammatory cytokines, which may partially account for the therapeutic efficacy of these regimens [198, 199, 269].

The efforts to change TAM polarization are still in their infancy, and further investigations are demanded to alleviate off-target effects. Nevertheless, the above findings imply that reeducation of TAMs and exploitation of their tumoricidal potentials present promising therapeutic approaches against both primary cancer and metastatic malignancies. Attempts are therefore prompted to further evaluate the efficacy and safety of the TAM-resetting therapies.

16.9.5 Impeding the Protumor Functions of TAMs

It's a promising therapeutic approach to selectively target the precise tumor-promoting mechanisms of macrophages while sparing the antitumor and homeostatic functions of cells within the tumor or throughout the body.

Interference of interaction between immune checkpoints and their ligands on macrophages sets up a significant and potent obstacle to the immune-blunting activities of TAMs. It's clinically proved that PD-1 blockade by MPDL3280A is efficient in 19% of patients with PD-L1-positive TNBC [270]. A phase Ib study has described the anti-PD-1 antibody pembrolizumab as an efficacious and tolerated regimen with an overall response rate of 18.5% and long-lasting response in women with refractory advanced TNBC [271, 272]. However, the anti-CTLA-4 antibody tremelimumab is less effective in patients attacked by metastatic ER-positive breast cancer in combination with exemestane [273]. Selective targeting of PI3Kγ isoform in preclinical data impedes the progression of checkpoint blockade-resistant mammary tumor by switching the suppressive phenotype of myeloid cells toward pro-inflammatory population [274]. The safety and antitumor activity of IPI549, a selective small-molecule PI3Kγ blocker, alone and in combination with PD-1 inhibitor, are currently being tested in a clinical trial (clinicaltrials.gov identifier NCT02637531).

The ubiquitous expression of IDO has a critical role in the immune evasion of cancer cells with the assistance of TAMs. Preclinical studies demonstrated that IDO ablation combined with cytotoxic agents was synergistic in murine breast cancer. Indoximod, a broad IDO pathway blocker tested in a phase I clinical trial for metastatic solid tumors, showed tolerability and therapeutic activity in combination with docetaxel [275]. A phase II randomized trial of indoximod combining chemotherapy demonstrates safety data compatible with that of docetaxel and paclitaxel without immune-specific serious adverse events. The trial research is now ongoing to examine the anticancer efficacy of indoximod in metastatic breast cancer [276].

ADCC and ADCP are essential in tumor eradication. Nevertheless, the transmembrane glycoprotein CD47, which is highly expressed on BCCs [277], delivers a "don't eat me" signal through its receptor signal-regulatory protein α (SIRPα) on macrophages. Blockade of CD47-SIRPα interaction with anti-CD47 mAbs potentiates macrophage phagocytosis and tumor suppression in an orthotopic mouse mammary tumor model as well as a xenotransplantation model established with BCCs from patient samples [277]. In concert with it, the CD47-binding recombinant fusion protein TTI-621(SIRPαFc) boosts macrophage cytotoxicity in breast and other cancers, with limited interaction with human erythrocytes [278].

The surface molecule Nrp1 on TAMs is critical for the transactivation of VEGFR1 in a

PlexinA1/4-dependent mechanism [36]. Genetic or pharmacologic ablation of Nrp1 on macrophages retains their entrapment in oxygen-rich areas and restrains their proangiogenic and immune-evading potential, leading to suppression of tumor development and dissemination [36, 279]. The human mAb MNRP1685A, which effectively blocks the VEGF-binding domain of Nrp1, is currently under clinical evaluation and found to be well tolerated in patients with advanced solid tumors in a phase I study [280].

In the murine transgenic model, targeting the Ang2/Tie2 axis compromises the angiogenic property of TAMs and normalizes the neoplastic vasculature, bringing great hindrance to tumor development and spreading [115, 219]. Preliminary anticancer efficacy and safety profile of combining nesvacumab, a selective Ang-2 mAb, to anti-VEGF therapy has also been described in a clinical study in patients with advanced solid malignancies [281].

Although many novel TAM-targeting agents display preliminary efficacy in clinical trials, it remains a significant challenge to identify consistent biomarkers and select patients who will benefit from the appropriate use of drugs modulating macrophage function. Development of nanoparticulate formulations might offer an opportunity to interfere the expression of protumoral gene for they can cross the membrane barrier while mAbs and small-molecule inhibitors cannot. With growing knowledge of neoplasm and TAMs, new regulatory node will continuously be discovered, which will give rise to the emergence of novel targeted therapies.

16.10 Conclusions and Perspectives

With ongoing attempts to decipher TAMs, their roles in tumor development and therapeutic response begin to emerge. Cells of the monocyte-macrophage lineage are key players in inflammation regulation. Intratumoral macrophages have multiple origins, while recruitment of circulating monocytes in response to a variety of chemoattracting molecules represents the predominant source. In the context of breast cancer, TAMs undergo reprogramming and adopt tumorigenic phenotype, which indicates their pivotal effects on tumor angiogenesis, immune evasion, distant metastasis, and therapeutic tolerance.

The findings discussed above provide compelling rationales for therapeutic targeting of TAMs and serve as preclinical experimental data for new drug exploitation. Indeed, some TAM-targeting agents have already been proved effective in tumor control, especially in melanoma. A plethora of clinical trials are underway testing drugs designed to inhibit TAM infiltration, delete tumor-educated macrophages, or block their protumoral activities. However, further studies are still in need to address the definition of macrophage subpopulations and the key factors regulating their various protumoral functions in different subtypes of breast cancer. The advent of single-cell analysis techniques will facilitate studies on cell heterogeneity and cellular-specific response patterns to environmental cues and therapeutic agents.

It remains a significant challenge to selectively target the protumor population of TAMs or their tumorigenic mechanisms without serious impact on tissue-resident macrophages and systemic inflammation. In addition to TAMs, tumor-associated neutrophils and mesenchymal stem cells play critical roles in neovascularization, immunosuppression, and metastasis. These cells may limit the antitumor efficacy of therapeutic agents targeting TAMs, probably through a compensatory influx. Studies are warranted to determine the therapeutic response of TAM-targeting agents in breast cancer in combination with various standard-of-care treatment or other novel interventions reprogramming the microenvironment.

References

1. Lavin Y, Mortha A, Rahman A, Merad M (2015) Regulation of macrophage development and function in peripheral tissues. Nat Rev Immunol 15(12):731–744. doi:10.1038/nri3920
2. Perdiguero EG, Geissmann F (2016) The development and maintenance of resident macrophages. Nat Immunol 17(1):2–8. doi:10.1038/ni.3341

3. Wynn TA, Chawla A, Pollard JW (2013) Macrophage biology in development, homeostasis and disease. Nature 496(7446):445–455. doi:10.1038/nature12034
4. Hagerling C, Casbon AJ, Werb Z (2015) Balancing the innate immune system in tumor development. Trends Cell Biol 25(4):214–220. doi:10.1016/j.tcb.2014.11.001
5. Glass CK, Natoli G (2016) Molecular control of activation and priming in macrophages. Nat Immunol 17(1):26–33. doi:10.1038/ni.3306
6. Balkwill F, Mantovani A (2001) Inflammation and cancer: back to Virchow? Lancet 357(9255):539–545. doi:10.1016/s0140-6736(00)04046-0
7. Grivennikov SI, Greten FR, Karin M (2010) Immunity, inflammation, and cancer. Cell 140(6):883–899. doi:10.1016/j.cell.2010.01.025
8. Balkwill FR, Mantovani A (2012) Cancer-related inflammation: common themes and therapeutic opportunities. Semin Cancer Biol 22(1):33–40. doi:10.1016/j.semcancer.2011.12.005
9. Evans R, Alexander P (1970) Cooperation of immune lymphoid cells with macrophages in tumour immunity. Nature 228(5272):620–622
10. Mantovani A (1978) Effects on in vitro tumor growth of murine macrophages isolated from sarcoma lines differing in immunogenicity and metastasizing capacity. Int J Cancer 22(6):741–746
11. Dvorak HF (1986) Tumors: wounds that do not heal. Similarities between tumor stroma generation and wound healing. N Engl J Med 315(26):1650–1659. doi:10.1056/nejm198612253152606
12. Paget S (1989) The distribution of secondary growths in cancer of the breast. 1889. Cancer Metastasis Rev 8(2):98–101
13. Joyce JA, Pollard JW (2009) Microenvironmental regulation of metastasis. Nat Rev Cancer 9(4):239–252. doi:10.1038/nrc2618
14. Mantovani A, Marchesi F, Malesci A, Laghi L, Allavena P (2017) Tumour-associated macrophages as treatment targets in oncology. Nat Rev Clin Oncol. doi:10.1038/nrclinonc.2016.217
15. Williams CB, Yeh ES, Soloff AC (2016) Tumor-associated macrophages: unwitting accomplices in breast cancer malignancy NPJ Breast Cancer:2. doi:10.1038/npjbcancer.2015.25
16. Lewis CE, Harney AS, Pollard JW (2016) The multifaceted role of perivascular macrophages in tumors. Cancer Cell 30(1):18–25. doi:10.1016/j.ccell.2016.05.017
17. Ginhoux F, Schultze JL, Murray PJ, Ochando J, Biswas SK (2016) New insights into the multidimensional concept of macrophage ontogeny, activation and function. Nat Immunol 17(1):34–40. doi:10.1038/ni.3324
18. Hashimoto D, Chow A, Noizat C, Teo P, Beasley MB, Leboeuf M, Becker CD, See P, Price J, Lucas D, Greter M, Mortha A, Boyer SW, Forsberg EC, Tanaka M, van Rooijen N, Garcia-Sastre A, Stanley ER, Ginhoux F, Frenette PS, Merad M (2013) Tissue-resident macrophages self-maintain locally throughout adult life with minimal contribution from circulating monocytes. Immunity 38(4):792–804. doi:10.1016/j.immuni.2013.04.004
19. Ginhoux F, Jung S (2014) Monocytes and macrophages: developmental pathways and tissue homeostasis. Nat Rev Immunol 14(6):392–404. doi:10.1038/nri3671
20. Franklin RA, Liao W, Sarkar A, Kim MV, Bivona MR, Liu K, Pamer EG, Li MO (2014) The cellular and molecular origin of tumor-associated macrophages. Science 344(6186):921–925. doi:10.1126/science.1252510
21. Franklin RA, Li MO (2016) Ontogeny of tumor-associated macrophages and its implication in cancer regulation. Trends in Cancer 2(1):20–34. doi:10.1016/j.trecan.2015.11.004
22. van de Laar L, Saelens W, De Prijck S, Martens L, Scott CL, Van Isterdael G, Hoffmann E, Beyaert R, Saeys Y, Lambrecht BN, Guilliams M (2016) Yolk sac macrophages, fetal liver, and adult monocytes can colonize an empty niche and develop into functional tissue-resident macrophages. Immunity 44(4):755–768. doi:10.1016/j.immuni.2016.02.017
23. Gabrilovich DI, Nagaraj S (2009) Myeloid-derived suppressor cells as regulators of the immune system. Nat Rev Immunol 9(3):162–174. doi:10.1038/nri2506
24. Bronte V, Brandau S, Chen SH, Colombo MP, Frey AB, Greten TF, Mandruzzato S, Murray PJ, Ochoa A, Ostrand-Rosenberg S, Rodriguez PC, Sica A, Umansky V, Vonderheide RH, Gabrilovich DI (2016) Recommendations for myeloid-derived suppressor cell nomenclature and characterization standards. Nat Commun 7:12150. doi:10.1038/ncomms12150
25. Kumar V, Cheng P, Condamine T, Mony S, Languino LR, McCaffrey JC, Hockstein N, Guarino M, Masters G, Penman E, Denstman F, Xu X, Altieri DC, Du H, Yan C, Gabrilovich DI (2016) CD45 phosphatase inhibits STAT3 transcription factor activity in myeloid cells and promotes tumor-associated macrophage differentiation. Immunity 44(2):303–315. doi:10.1016/j.immuni.2016.01.014
26. Hanna RN, Cekic C, Sag D, Tacke R, Thomas GD, Nowyhed H, Herrley E, Rasquinha N, McArdle S, Wu R, Peluso E, Metzger D, Ichinose H, Shaked I, Chodaczek G, Biswas SK, Hedrick CC (2015) Patrolling monocytes control tumor metastasis to the lung. Science 350(6263):985–990. doi:10.1126/science.aac9407
27. Tymoszuk P, Evens H, Marzola V, Wachowicz K, Wasmer MH, Datta S, Muller-Holzner E, Fiegl H, Bock G, van Rooijen N, Theurl I, Doppler W (2014) In situ proliferation contributes to accumulation of tumor-associated macrophages in spontaneous mammary tumors. Eur J Immunol 44(8):2247–2262. doi:10.1002/eji.201344304

28. Lin EY, Nguyen AV, Russell RG, Pollard JW (2001) Colony-stimulating factor 1 promotes progression of mammary tumors to malignancy. J Exp Med 193(6):727–740
29. Qian BZ, Pollard JW (2010) Macrophage diversity enhances tumor progression and metastasis. Cell 141(1):39–51. doi:10.1016/j.cell.2010.03.014
30. Lin EY, Jf L, Bricard G, Wang W, Deng Y, Sellers R, Porcelli SA, Pollard JW (2007) Vascular endothelial growth factor restores delayed tumor progression in tumors depleted of macrophages. Mol Oncol 1(3):288–302. doi:10.1016/j.molonc.2007.10.003
31. Ueno T, Toi M, Saji H, Muta M, Bando H, Kuroi K, Koike M, Inadera H, Matsushima K (2000) Significance of macrophage chemoattractant protein-1 in macrophage recruitment, angiogenesis, and survival in human breast cancer. Clinical cancer research : an official journal of the American Association for Cancer Research 6(8):3282–3289
32. Qian BZ, Li J, Zhang H, Kitamura T, Zhang J, Campion LR, Kaiser EA, Snyder LA, Pollard JW (2011) CCL2 recruits inflammatory monocytes to facilitate breast-tumour metastasis. Nature 475(7355):222–225. doi:10.1038/nature10138
33. Hughes R, Qian BZ, Rowan C, Muthana M, Keklikoglou I, Olson OC, Tazzyman S, Danson S, Addison C, Clemons M, Gonzalez-Angulo AM, Joyce JA, De Palma M, Pollard JW, Lewis CE (2015) Perivascular M2 macrophages stimulate tumor relapse after chemotherapy. Cancer Res 75(17):3479–3491. doi:10.1158/0008-5472.CAN-14-3587
34. Kozin SV, Kamoun WS, Huang Y, Dawson MR, Jain RK, Duda DG (2010) Recruitment of myeloid but not endothelial precursor cells facilitates tumor regrowth after local irradiation. Cancer Res 70(14):5679–5685. doi:10.1158/0008-5472.can-09-4446
35. Acharyya S, Oskarsson T, Vanharanta S, Malladi S, Kim J, Morris PG, Manova-Todorova K, Leversha M, Hogg N, Seshan VE, Norton L, Brogi E, Massagué J (2012a) A CXCL1 paracrine network links cancer chemoresistance and metastasis. Cell 150(1):165–178. doi:10.1016/j.cell.2012.04.042
36. Casazza A, Laoui D, Wenes M, Rizzolio S, Bassani N, Mambretti M, Deschoemaeker S, VanGinderachter J, Tamagnone L, Mazzone M (2013) Impeding macrophage entry into hypoxic tumor areas by Sema3A/Nrp1 signaling blockade inhibits angiogenesis and restores antitumor immunity. Cancer Cell 24(6):695–709. doi:10.1016/j.ccr.2013.11.007
37. Stein M, Keshav S, Harris N, Gordon S (1992) Interleukin 4 potently enhances murine macrophage mannose receptor activity: a marker of alternative immunologic macrophage activation. J Exp Med 176(1):287–292
38. Mills CD (2012) M1 and M2 macrophages: oracles of health and disease. Crit Rev Immunol 32(6):463–488
39. Mantovani A, Sica A, Sozzani S, Allavena P, Vecchi A, Locati M (2004) The chemokine system in diverse forms of macrophage activation and polarization. Trends Immunol 25(12):677–686. doi:10.1016/j.it.2004.09.015
40. Mantovani A, Allavena P (2015) The interaction of anticancer therapies with tumor-associated macrophages. J Exp Med 212(4):435–445. doi:10.1084/jem.20150295
41. Ostuni R, Kratochvill F, Murray PJ, Natoli G (2015) Macrophages and cancer: from mechanisms to therapeutic implications. Trends Immunol 36(4):229–239. doi:10.1016/j.it.2015.02.004
42. Grugan KD, McCabe FL, Kinder M, Greenplate AR, Harman BC, Ekert JE, Van Rooijen N, Anderson GM, Nemeth JA, Strohl WR, Jordan RE, Brezski RJ (2012) Tumor-associated macrophages promote invasion while retaining fc-dependent antitumor function. J Immunol 189(11):5457–5466. doi:10.4049/jimmunol.1201889
43. Gautier EL, Shay T, Miller J, Greter M, Jakubzick C, Ivanov S, Helft J, Chow A, Elpek KG, Gordonov S, Mazloom AR, Ma'ayan A, Chua WJ, Hansen TH, Turley SJ, Merad M, Randolph GJ (2012) Gene-expression profiles and transcriptional regulatory pathways that underlie the identity and diversity of mouse tissue macrophages. Nat Immunol 13(11):1118–1128. doi:10.1038/ni.2419
44. Xue J, Schmidt SV, Sander J, Draffehn A, Krebs W, Quester I, De Nardo D, Gohel TD, Emde M, Schmidleithner L, Ganesan H, Nino-Castro A, Mallmann MR, Labzin L, Theis H, Kraut M, Beyer M, Latz E, Freeman TC, Ulas T, Schultze JL (2014) Transcriptome-based network analysis reveals a spectrum model of human macrophage activation. Immunity 40(2):274–288. doi:10.1016/j.immuni.2014.01.006
45. Shalek AK, Satija R, Shuga J, Trombetta JJ, Gennert D, Lu D, Chen P, Gertner RS, Gaublomme JT, Yosef N, Schwartz S, Fowler B, Weaver S, Wang J, Wang X, Ding R, Raychowdhury R, Friedman N, Hacohen N, Park H, May AP, Regev A (2014) Single-cell RNA-seq reveals dynamic paracrine control of cellular variation. Nature 510(7505):363–369. doi:10.1038/nature13437
46. Ruffell B, Affara NI, Coussens LM (2012) Differential macrophage programming in the tumor microenvironment. Trends Immunol 33(3):119–126. doi:10.1016/j.it.2011.12.001
47. DeNardo DG, Barreto JB, Andreu P, Vasquez L, Tawfik D, Kolhatkar N, Coussens LM (2009) CD4(+) T cells regulate pulmonary metastasis of mammary carcinomas by enhancing protumor properties of macrophages. Cancer Cell 16(2):91–102. doi:10.1016/j.ccr.2009.06.018
48. Shiao SL, Ruffell B, De Nardo DG, Faddegon BA, Park CC, Coussens LM (2015a) Th2-polarized CD4+ T cells and macrophages limit efficacy of

radiation therapy. Cancer Immunol Res 3(5):518–525. doi:10.1158/2326-6066.CIR-14-0232
49. Pedroza-Gonzalez A, Xu K, Wu TC, Aspord C, Tindle S, Marches F, Gallegos M, Burton EC, Savino D, Hori T, Tanaka Y, Zurawski S, Zurawski G, Bover L, Liu YJ, Banchereau J, Palucka AK (2011) Thymic stromal lymphopoietin fosters human breast tumor growth by promoting type 2 inflammation. J Exp Med 208(3):479–490. doi:10.1084/jem.20102131
50. Zelenay S, van der Veen AG, Bottcher JP, Snelgrove KJ, Rogers N, Acton SE, Chakravarty P, Girotti MR, Marais R, Quezada SA, Sahai E, Reis e Sousa C (2015) Cyclooxygenase-dependent tumor growth through evasion of immunity. Cell 162(6):1257–1270. doi:10.1016/j.cell.2015.08.015
51. Van Overmeire E, Stijlemans B, Heymann F, Keirsse J, Morias Y, Elkrim Y, Brys L, Abels C, Lahmar Q, Ergen C, Vereecke L, Tacke F, De Baetselier P, Van Ginderachter JA, Laoui D (2016) M-CSF and GM-CSF receptor signaling differentially regulate monocyte maturation and macrophage polarization in the tumor microenvironment. Cancer Res 76(1):35–42. doi:10.1158/0008-5472.CAN-15-0869
52. Semenza GL (2016) The hypoxic tumor microenvironment: a driving force for breast cancer progression. BBA-Mol Cell Res 1863(3):382–391. doi:10.1016/j.bbamcr.2015.05.036
53. Zhang Y, Choksi S, Chen K, Pobezinskaya Y, Linnoila I, Liu ZG (2013a) ROS play a critical role in the differentiation of alternatively activated macrophages and the occurrence of tumor-associated macrophages. Cell Res 23(7):898–914. doi:10.1038/cr.2013.75
54. Doedens AL, Stockmann C, Rubinstein MP, Liao D, Zhang N, DeNardo DG, Coussens LM, Karin M, Goldrath AW, Johnson RS (2010a) Macrophage expression of hypoxia-inducible factor-1 alpha suppresses T-cell function and promotes tumor progression. Cancer Res 70(19):7465–7475. doi:10.1158/0008-5472.can-10-1439
55. Murray PJ, Allen JE, Biswas SK, Fisher EA, Gilroy DW, Goerdt S, Gordon S, Hamilton JA, Ivashkiv LB, Lawrence T, Locati M, Mantovani A, Martinez FO, Mege JL, Mosser DM, Natoli G, Saeij JP, Schultze JL, Shirey KA, Sica A, Suttles J, Udalova I, van Ginderachter JA, Vogel SN, Wynn TA (2014) Macrophage activation and polarization: nomenclature and experimental guidelines. Immunity 41(1):14–20. doi:10.1016/j.immuni.2014.06.008
56. Cook RS, Jacobsen KM, Wofford AM, DeRyckere D, Stanford J, Prieto AL, Redente E, Sandahl M, Hunter DM, Strunk KE, Graham DK, Earp HS 3rd (2013) MerTK inhibition in tumor leukocytes decreases tumor growth and metastasis. J Clin Invest 123(8):3231–3242. doi:10.1172/JCI67655
57. Krausgruber T, Blazek K, Smallie T, Alzabin S, Lockstone H, Sahgal N, Hussell T, Feldmann M, Udalova IA (2011) IRF5 promotes inflammatory macrophage polarization and TH1-TH17 responses. Nat Immunol 12(3):231–238. doi:10.1038/ni.1990
58. Lu G, Zhang R, Geng S, Peng L, Jayaraman P, Chen C, Xu F, Yang J, Li Q, Zheng H, Shen K, Wang J, Liu X, Wang W, Zheng Z, Qi CF, Si C, He JC, Liu K, Lira SA, Sikora AG, Li L, Xiong H (2015) Myeloid cell-derived inducible nitric oxide synthase suppresses M1 macrophage polarization. Nat Commun 6:6676. doi:10.1038/ncomms7676
59. Kimura T, Nada S, Takegahara N, Okuno T, Nojima S, Kang S, Ito D, Morimoto K, Hosokawa T, Hayama Y, Mitsui Y, Sakurai N, Sarashina-Kida H, Nishide M, Maeda Y, Takamatsu H, Okuzaki D, Yamada M, Okada M, Kumanogoh A (2016) Polarization of M2 macrophages requires Lamtor1 that integrates cytokine and amino-acid signals. Nat Commun 7:13130. doi:10.1038/ncomms13130
60. Kaneda MM, Messer KS, Ralainirina N, Li H, Leem CJ, Gorjestani S, Woo G, Nguyen AV, Figueiredo CC, Foubert P, Schmid MC, Pink M, Winkler DG, Rausch M, Palombella VJ, Kutok J, McGovern K, Frazer KA, Wu X, Karin M, Sasik R, Cohen EE, Varner JA (2016) PI3Kgamma is a molecular switch that controls immune suppression. Nature 539(7629):437–442. doi:10.1038/nature19834
61. Smale ST, Tarakhovsky A, Natoli G (2014) Chromatin contributions to the regulation of innate immunity. Annu Rev Immunol 32:489–511. doi:10.1146/annurev-immunol-031210-101303
62. Li Z, Chao TC, Chang KY, Lin N, Patil VS, Shimizu C, Head SR, Burns JC, Rana TM (2014a) The long noncoding RNA THRIL regulates TNFalpha expression through its interaction with hnRNPL. Proc Natl Acad Sci U S A 111(3):1002–1007. doi:10.1073/pnas.1313768111
63. Saeed S, Quintin J, Kerstens HH, Rao NA, Aghajanirefah A, Matarese F, Cheng SC, Ratter J, Berentsen K, van der Ent MA, Sharifi N, Janssen-Megens EM, Ter Huurne M, Mandoli A, van Schaik T, Ng A, Burden F, Downes K, Frontini M, Kumar V, Giamarellos-Bourboulis EJ, Ouwehand WH, van der Meer JW, Joosten LA, Wijmenga C, Martens JH, Xavier RJ, Logie C, Netea MG, Stunnenberg HG (2014) Epigenetic programming of monocyte-to-macrophage differentiation and trained innate immunity. Science 345(6204):1251086. doi:10.1126/science.1251086
64. Amit I, Winter DR, Jung S (2016) The role of the local environment and epigenetics in shaping macrophage identity and their effect on tissue homeostasis. Nat Immunol 17(1):18–25. doi:10.1038/ni.3325
65. Li Y, Zhao L, Shi B, Ma S, Xu Z, Ge Y, Liu Y, Zheng D, Shi J (2015a) Functions of miR-146a and miR-222 in tumor-associated macrophages in breast cancer. Sci Rep 5:18648. doi:10.1038/srep18648
66. Yang J, Zhang Z, Chen C, Liu Y, Si Q, Chuang TH, Li N, Gomez-Cabrero A, Reisfeld RA, Xiang R, Luo Y (2014) MicroRNA-19a-3p inhibits breast cancer

progression and metastasis by inducing macrophage polarization through downregulated expression of Fra-1 proto-oncogene. Oncogene 33(23):3014–3023. doi:10.1038/onc.2013.258
67. Noy R, Pollard JW (2014) Tumor-associated macrophages: from mechanisms to therapy. Immunity 41(1):49–61. doi:10.1016/j.immuni.2014.06.010
68. Joyce JA, Fearon DT (2015) T cell exclusion, immune privilege, and the tumor microenvironment. Science 348(6230):74–80. doi:10.1126/science.aaa6204
69. Ojalvo LS, King W, Cox D, Pollard JW (2009) High-density gene expression analysis of tumor-associated macrophages from mouse mammary tumors. Am J Pathol 174(3):1048–1064. doi:10.2353/ajpath.2009.080676
70. Pollard JW (2004) Tumour-educated macrophages promote tumour progression and metastasis. Nat Rev Cancer 4(1):71–78
71. Chen W, Ten Dijke P (2016) Immunoregulation by members of the TGFbeta superfamily. Nat Rev Immunol 16(12):723–740. doi:10.1038/nri.2016.112
72. Ng TH, Britton GJ, Hill EV, Verhagen J, Burton BR, Wraith DC (2013) Regulation of adaptive immunity; the role of interleukin-10. Front Immunol 4:129. doi:10.3389/fimmu.2013.00129
73. Ruffell B, Chang-Strachan D, Chan V, Rosenbusch A, Ho CM, Pryer N, Daniel D, Hwang ES, Rugo HS, Coussens LM (2014) Macrophage IL-10 blocks CD8+ T cell-dependent responses to chemotherapy by suppressing IL-12 expression in intratumoral dendritic cells. Cancer Cell 26(5):623–637. doi:10.1016/j.ccell.2014.09.006
74. Li HZ, Yang B, Huang J, Lin Y, Xiang TX, Wan JY, Li HY, Chouaib S, Ren GS (2015b) Cyclooxygenase-2 in tumor-associated macrophages promotes breast cancer cell survival by triggering a positive-feedback loop between macrophages and cancer cells. Oncotarget 6(30):29637–29650
75. Tian J, Hachim MY, Hachim IY, Dai M, Lo C, Raffa FA, Ali S, Lebrun JJ (2017) Cyclooxygenase-2 regulates TGFbeta-induced cancer stemness in triple-negative breast cancer. Sci Rep 7:40258. doi:10.1038/srep40258
76. Adams JL, Smothers J, Srinivasan R, Hoos A (2015) Big opportunities for small molecules in immuno-oncology. Nat Rev Drug Discov 14(9):603–622. doi:10.1038/nrd4596
77. DeNardo DG, Brennan DJ, Rexhepaj E, Ruffell B, Shiao SL, Madden SF, Gallagher WM, Wadhwani N, Keil SD, Junaid SA, Rugo HS, Shelley Hwang E, Jirström K, West BL, Coussens LM (2011) Leukocyte complexity predicts breast cancer survival and functionally regulates response to chemotherapy. Cancer Discov 1(1):54–67. doi:10.1158/2159-8274.CD-10-0028
78. Lau J, Cheung J, Navarro A, Lianoglou S, Haley B, Totpal K, Sanders L, Koeppen H, Caplazi P, McBride J, Chiu H, Hong R, Grogan J, Javinal V, Yauch R, Irving B, Belvin M, Mellman I, Kim JM, Schmidt M (2017) Tumour and host cell PD-L1 is required to mediate suppression of anti-tumour immunity in mice. Nat Commun 8:14572. doi:10.1038/ncomms14572
79. Wherry EJ (2011) T cell exhaustion. Nat Immunol 131(6):492–499. doi:10.1038/ni.2035
80. Noman MZ, Desantis G, Janji B, Hasmim M, Karray S, Dessen P, Bronte V, Chouaib S (2014) PD-L1 is a novel direct target of HIF-1alpha, and its blockade under hypoxia enhanced MDSC-mediated T cell activation. J Exp Med 211(5):781–790. doi:10.1084/jem.20131916
81. Beatson R, Tajadura-Ortega V, Achkova D, Picco G, Tsourouktsoglou TD, Klausing S, Hillier M, Maher J, Noll T, Crocker PR, Taylor-Papadimitriou J, Burchell JM (2016) The mucin MUC1 modulates the tumor immunological microenvironment through engagement of the lectin Siglec-9. Nat Immunol 17(11):1273–1281. doi:10.1038/ni.3552
82. Pello OM, De Pizzol M, Mirolo M, Soucek L, Zammataro L, Amabile A, Doni A, Nebuloni M, Swigart LB, Evan GI, Mantovani A, Locati M (2012) Role of c-MYC in alternative activation of human macrophages and tumor-associated macrophage biology. Blood 119(2):411–421. doi:10.1182/blood-2011-02-339911
83. Casey SC, Tong L, Li Y, Do R, Walz S, Fitzgerald KN, Gouw AM, Baylot V, Gutgemann I, Eilers M, Felsher DW (2016) MYC regulates the antitumor immune response through CD47 and PD-L1. Science 352(6282):227–231. doi:10.1126/science.aac9935
84. Speiser DE, Ho PC, Verdeil G (2016) Regulatory circuits of T cell function in cancer. Nat Rev Immunol 16(10):599–611. doi:10.1038/nri.2016.80
85. de Boniface J, Mao Y, Schmidt-Mende J, Kiessling R, Poschke I (2012) Expression patterns of the immunomodulatory enzyme arginase 1 in blood, lymph nodes and tumor tissue of early-stage breast cancer patients. OncoImmunology 1(8):1305–1312. doi:10.4161/onci.21678
86. Movahedi K, Laoui D, Gysemans C, Baeten M, Stange G, Van den Bossche J, Mack M, Pipeleers D, In't Veld P, De Baetselier P, Van Ginderachter JA (2010) Different tumor microenvironments contain functionally distinct subsets of macrophages derived from Ly6C(high) monocytes. Cancer Res 70(14):5728–5739. doi:10.1158/0008-5472.can-09-4672
87. Murray PJ (2016) Amino acid auxotrophy as a system of immunological control nodes. Nat Immunol 17(2):132–139. doi:10.1038/ni.3323
88. Qiu F, Chen YR, Liu X, Chu CY, Shen LJ, Xu J, Gaur S, Forman HJ, Zhang H, Zheng S, Yen Y, Huang J, Kung HJ, Ann DK (2014) Arginine starvation impairs mitochondrial respiratory function in ASS1-deficient breast cancer cells. Sci Signal 7(319):ra31. doi:10.1126/scisignal.2004761

89. Rodriguez PC, Ochoa AC, Al-Khami AA (2017) Arginine metabolism in myeloid cells shapes innate and adaptive immunity. Front Immunol 8:93. doi:10.3389/fimmu.2017.00093
90. Lu T, Ramakrishnan R, Altiok S, Youn JI, Cheng P, Celis E, Pisarev V, Sherman S, Sporn MB, Gabrilovich D (2011) Tumor-infiltrating myeloid cells induce tumor cell resistance to cytotoxic T cells in mice. J Clin Invest 121(10):4015–4029. doi:10.1172/jci45862
91. Granados-Principal S, Liu Y, Guevara ML, Blanco E, Choi DS, Qian W, Patel T, Rodriguez AA, Cusimano J, Weiss HL, Zhao H, Landis MD, Dave B, Gross SS, Chang JC (2015) Inhibition of iNOS as a novel effective targeted therapy against triple-negative breast cancer. Breast cancer research : BCR 17:25. doi:10.1186/s13058-015-0527-x
92. Savas P, Salgado R, Denkert C, Sotiriou C, Darcy PK, Smyth MJ, Loi S (2016) Clinical relevance of host immunity in breast cancer: from TILs to the clinic. Nat Rev Clin Oncol 13(4):228–241. doi:10.1038/nrclinonc.2015.215
93. Uyttenhove C, Pilotte L, Theate I, Stroobant V, Colau D, Parmentier N, Boon T, Van den Eynde BJ (2003) Evidence for a tumoral immune resistance mechanism based on tryptophan degradation by indoleamine 2,3-dioxygenase. Nat Med 9(10):1269–1274. doi:10.1038/nm934
94. Muller AJ, DuHadaway JB, Donover PS, Sutanto-Ward E, Prendergast GC (2005) Inhibition of indoleamine 2,3-dioxygenase, an immunoregulatory target of the cancer suppression gene Bin1, potentiates cancer chemotherapy. Nat Med 11(3):312–319. doi:10.1038/nm1196
95. Weichhart T, Hengstschlager M, Linke M (2015) Regulation of innate immune cell function by mTOR. Nat Rev Immunol 15(10):599–614. doi:10.1038/nri3901
96. Faget J, Biota C, Bachelot T, Gobert M, Treilleux I, Goutagny N, Durand I, Leon-Goddard S, Blay JY, Caux C, Menetrier-Caux C (2011) Early detection of tumor cells by innate immune cells leads to T(reg) recruitment through CCL22 production by tumor cells. Cancer Res 71(19):6143–6152. doi:10.1158/0008-5472.CAN-11-0573
97. Su S, Liao J, Liu J, Huang D, He C, Chen F, Yang L, Wu W, Chen J, Lin L, Zeng Y, Ouyang N, Cui X, Yao H, Su F, Huang JD, Lieberman J, Liu Q, Song E (2017) Blocking the recruitment of naive CD4+ T cells reverses immunosuppression in breast cancer. Cell Res. doi:10.1038/cr.2017.34
98. Sakaguchi S, Yamaguchi T, Nomura T, Ono M (2008) Regulatory T cells and immune tolerance. Cell 133(5):775–787. doi:10.1016/j.cell.2008.05.009
99. Tanaka A, Sakaguchi S (2017) Regulatory T cells in cancer immunotherapy. Cell Res 27(1):109–118. doi:10.1038/cr.2016.151
100. Parker KH, Sinha P, Horn LA, Clements VK, Yang H, Li J, Tracey KJ, Ostrand-Rosenberg S (2014) HMGB1 enhances immune suppression by facilitating the differentiation and suppressive activity of myeloid-derived suppressor cells. Cancer Res 74(20):5723–5733. doi:10.1158/0008-5472.can-13-2347
101. Hanahan D, Weinberg RA (2011) Hallmarks of cancer: the next generation. Cell 144(5):646–674. doi:10.1016/j.cell.2011.02.013
102. Biswas SK, Allavena P, Mantovani A (2013) Tumor-associated macrophages: functional diversity, clinical significance, and open questions. Semin Immunopathol 35(5):585–600. doi:10.1007/s00281-013-0367-7
103. Lin EY, Pollard JW (2007) Tumor-associated macrophages press the angiogenic switch in breast cancer. Cancer Res 67(11):5064–5066. doi:10.1158/0008-5472.CAN-07-0912
104. Rivera LB, Bergers G (2015) Intertwined regulation of angiogenesis and immunity by myeloid cells. Trends Immunol 36(4):240–249. doi:10.1016/j.it.2015.02.005
105. Murdoch C, Muthana M, Coffelt SB, Lewis CE (2008) The role of myeloid cells in the promotion of tumour angiogenesis. Nat Rev Cancer 8(8):618–631. doi:10.1038/nrc2444
106. Kim S, Choi JH, Lim HI, Lee SK, Kim WW, Cho S, Kim JS, Kim JH, Choe JH, Nam SJ, Lee JE, Yang JH (2009) EGF-induced MMP-9 expression is mediated by the JAK3/ERK pathway, but not by the JAK3/STAT-3 pathway in a SKBR3 breast cancer cell line. Cell Signal 21(6):892–898
107. Kessenbrock K, Plaks V, Werb Z (2010) Matrix metalloproteinases: regulators of the tumor microenvironment. Cell 141(1):52–67. doi:10.1016/j.cell.2010.03.015
108. Deryugina EI, Quigley JP (2015) Tumor angiogenesis: MMP-mediated induction of intravasation- and metastasis-sustaining neovasculature. Matrix Biol 44-46:94–112. doi:10.1016/j.matbio.2015.04.004
109. Yeo EJ, Cassetta L, Qian BZ, Lewkowich I, Li JF, Stefater Iii JA, Smith AN, Wiechmann LS, Wang Y, Pollard JW, Lang RA (2014) Myeloid wnt7b mediates the angiogenic switch and metastasis in breast cancer. Cancer Res 74(11):2962–2973. doi:10.1158/0008-5472.CAN-13-2421
110. Bonapace L, Coissieux MM, Wyckoff J, Mertz KD, Varga Z, Junt T, Bentires-Alj M (2014) Cessation of CCL2 inhibition accelerates breast cancer metastasis by promoting angiogenesis. Nature 515(7525):130–133. doi:10.1038/nature13862
111. Lee S, Jilani SM, Nikolova GV, Carpizo D, Iruela-Arispe ML (2005) Processing of VEGF-A by matrix metalloproteinases regulates bioavailability and vascular patterning in tumors. J Cell Biol 169(4):681–691. doi:10.1083/jcb.200409115
112. Jung M, Ören B, Mora J, Mertens C, Dziumbla S, Popp R, Weigert A, Grossmann N, Fleming I, Brüne B (2016) Lipocalin 2 from macrophages stimulated by tumor cell-derived sphingosine-1-phosphate

promotes lymphangiogenesis and tumor metastasis. Sci Signal 9(434). doi:10.1126/scisignal.aaf3241
113. Wyckoff JB, Wang Y, Lin EY, Li JF, Goswami S, Stanley ER, Segall JE, Pollard JW, Condeelis J (2007) Direct visualization of macrophage-assisted tumor cell intravasation in mammary tumors. Cancer Res 67(6):2649–2656. doi:10.1158/0008-5472.CAN-06-1823
114. De Palma M, Venneri MA, Galli R, Sergi LS, Politi LS, Sampaolesi M, Naldini L (2005) Tie2 identifies a hematopoietic lineage of proangiogenic monocytes required for tumor vessel formation and a mesenchymal population of pericyte progenitors. Cancer Cell 8(3):211–226. doi:10.1016/j.ccr.2005.08.002
115. Mazzieri R, Pucci F, Moi D, Zonari E, Ranghetti A, Berti A, Politi Letterio S, Gentner B, Brown Jeffrey L, Naldini L, De Palma M (2011) Targeting the ANG2/TIE2 Axis inhibits tumor growth and metastasis by impairing angiogenesis and disabling rebounds of Proangiogenic myeloid cells. Cancer Cell 19 (4):512–526. doi:http://dx.doi.org/10.1016/j.ccr.2011.02.005
116. Kim OH, Kang GH, Noh H, Cha JY, Lee HJ, Yoon JH, Mamura M, Nam JS, Lee DH, Kim YA, Park YJ, Kim H, Oh BC (2013) Proangiogenic TIE2(+)/CD31 (+) macrophages are the predominant population of tumor-associated macrophages infiltrating metastatic lymph nodes. Mol Cells 36(5):432–438. doi:10.1007/s10059-013-0194-7
117. Bron S, Henry L, Faes-Van't Hull E, Turrini R, Vanhecke D, Guex N, Ifticene-Treboux A, Marina Iancu E, Semilietof A, Rufer N, Lehr HA, Xenarios I, Coukos G, Delaloye JF, Doucey MA (2016) TIE-2-expressing monocytes are lymphangiogenic and associate specifically with lymphatics of human breast cancer. OncoImmunology 5(2):e1073882. doi:10.1080/2162402x.2015.1073882
118. Rehman J, Li J, Orschell CM, March KL (2003) Peripheral blood "endothelial progenitor cells" are derived from monocyte/macrophages and secrete angiogenic growth factors. Circulation 107(8):1164–1169
119. Bailey AS, Willenbring H, Jiang S, Anderson DA, Schroeder DA, Wong MH, Grompe M, Fleming WH (2006) Myeloid lineage progenitors give rise to vascular endothelium. Proc Natl Acad Sci U S A 103(35):13156–13161. doi:10.1073/pnas.0604203103
120. Kumar AH, Martin K, Turner EC, Buneker CK, Dorgham K, Deterre P, Caplice NM (2013) Role of CX3CR1 receptor in monocyte/macrophage driven neovascularization. PLoS One 8(2):e57230. doi:10.1371/journal.pone.0057230
121. Goswami S, Sahai E, Wyckoff JB, Cammer M, Cox D, Pixley FJ, Stanley ER, Segall JE, Condeelis JS (2005) Macrophages promote the invasion of breast carcinoma cells via a colony-stimulating factor-1/epidermal growth factor paracrine loop. Cancer Res 65(12):5278–5283. doi:10.1158/0008-5472.can-04-1853
122. Wyckoff J, Wang WG, Lin EY, Wang YR, Pixley F, Stanley ER, Graf T, Pollard JW, Segall J, Condeelis J (2004) A paracrine loop between tumor cells and macrophages is required for tumor cell migration in mammary tumors. Cancer Res 64(19):7022–7029. doi:10.1158/0008-5472.can-04-1449
123. Condeelis J, Pollard JW (2006) Macrophages: obligate partners for tumor cell migration, invasion, and metastasis. Cell 124(2):263–266. doi:10.1016/j.cell.2006.01.007
124. Knutsdottir H, Condeelis JS, Palsson E (2016) 3-D individual cell based computational modeling of tumor cell-macrophage paracrine signaling mediated by EGF and CSF-1 gradients. Integrative biology : quantitative biosciences from nano to macro 8(1):104–119. doi:10.1039/c5ib00201j
125. Ishihara D, Dovas A, Hernandez L, Pozzuto M, Wyckoff J, Segall J, Condeelis J, Bresnick A, Cox D (2013) Wiskott-Aldrich syndrome protein regulates leukocyte-dependent breast cancer metastasis. Cell Rep 4(3):429–436. doi:10.1016/j.celrep.2013.07.007
126. O'Sullivan C, Lewis CE, Harris AL, McGee JO (1993) Secretion of epidermal growth factor by macrophages associated with breast carcinoma. Lancet 342(8864):148–149
127. Wang S, Yuan Y, Liao L, Kuang SQ, Tien JC, O'Malley BW, Xu J (2009) Disruption of the SRC-1 gene in mice suppresses breast cancer metastasis without affecting primary tumor formation. Proc Natl Acad Sci U S A 106(1):151–156. doi:10.1073/pnas.0808703105
128. Truong D, Puleo J, Llave A, Mouneimne G, Kamm RD, Nikkhah M (2016) Breast cancer cell invasion into a three dimensional tumor-Stroma microenvironment. Sci Rep 6:34094. doi:10.1038/srep34094
129. Yang J, Liao D, Chen C, Liu Y, Chuang TH, Xiang R, Markowitz D, Reisfeld RA, Luo Y (2013) Tumor-associated macrophages regulate murine breast cancer stem cells through a novel paracrine EGFR/Stat3/sox-2 signaling pathway. Stem Cells 31(2):248–258. doi:10.1002/stem.1281
130. Hernandez L, Smirnova T, Kedrin D, Wyckoff J, Zhu L, Stanley ER, Cox D, Muller WJ, Pollard JW, Van Rooijen N, Segall JE (2009) The EGF/CSF-1 paracrine invasion loop can be triggered by heregulin beta1 and CXCL12. Cancer Res 69(7):3221–3227. doi:10.1158/0008-5472.can-08-2871
131. Lee HJ, Seo AN, Kim EJ, Jang MH, Kim YJ, Kim JH, Kim SW, Ryu HS, Park IA, Im SA, Gong G, Jung KH, Kim HJ, Park SY (2015) Prognostic and predictive values of EGFR overexpression and EGFR copy number alteration in HER2-positive breast cancer. Br J Cancer 112(1):103–111. doi:10.1038/bjc.2014.556
132. Chen J, Yao Y, Gong C, Yu F, Su S, Chen J, Liu B, Deng H, Wang F, Lin L, Yao H, Su F, Anderson KS, Liu Q, Ewen ME, Yao X, Song E (2011a) CCL18 from tumor-associated macrophages promotes breast cancer metastasis via PITPNM3. Cancer Cell 19(4):541–555. doi:10.1016/j.ccr.2011.02.006

133. Li HY, Cui XY, Wu W, Yu FY, Yao HR, Liu Q, Song EW, Chen JQ (2014b) Pyk2 and Src mediate signaling to CCL18-induced breast cancer metastasis. J Cell Biochem 115(3):596–603. doi:10.1002/jcb.24697
134. Zhang B, Yin C, Li H, Shi L, Liu N, Sun Y, Lu S, Liu Y, Sun L, Li X, Chen W, Qi Y (2013b) Nir1 promotes invasion of breast cancer cells by binding to chemokine (C-C motif) ligand 18 through the PI3K/Akt/GSK3beta/snail signalling pathway. European journal of cancer (Oxford, England : 1990) 49(18):3900–3913. doi:10.1016/j.ejca.2013.07.146
135. Mouw JK, Ou G, Weaver VM (2014) Extracellular matrix assembly: a multiscale deconstruction. Nat Rev Mol Cell Biol 15(12):771–785. doi:10.1038/nrm3902
136. Kai F, Laklai H, Weaver VM (2016) Force matters: biomechanical regulation of cell invasion and migration in disease. Trends Cell Biol 26(7):486–497. doi:10.1016/j.tcb.2016.03.007
137. Lang NR, Skodzek K, Hurst S, Mainka A, Steinwachs J, Schneider J, Aifantis KE, Fabry B (2015) Biphasic response of cell invasion to matrix stiffness in three-dimensional biopolymer networks. Acta Biomater 13:61–67. doi:10.1016/j.actbio.2014.11.003
138. Levental KR, Yu H, Kass L, Lakins JN, Egeblad M, Erler JT, Fong SF, Csiszar K, Giaccia A, Weninger W, Yamauchi M, Gasser DL, Weaver VM (2009) Matrix crosslinking forces tumor progression by enhancing integrin signaling. Cell 139(5):891–906. doi:10.1016/j.cell.2009.10.027
139. Acerbi I, Cassereau L, Dean I, Shi Q, Au A, Park C, Chen YY, Liphardt J, Hwang ES, Weaver VM (2015) Human breast cancer invasion and aggression correlates with ECM stiffening and immune cell infiltration. Integrative biology : quantitative biosciences from nano to macro 7(10):1120–1134. doi:10.1039/c5ib00040h
140. Pickup MW, Laklai H, Acerbi I, Owens P, Gorska AE, Chytil A, Aakre M, Weaver VM, Moses HL (2013) Stromally derived lysyl oxidase promotes metastasis of transforming growth factor-beta-deficient mouse mammary carcinomas. Cancer Res 73(17):5336–5346. doi:10.1158/0008-5472.can-13-0012
141. Sangaletti S, Di Carlo E, Gariboldi S, Miotti S, Cappetti B, Parenza M, Rumio C, Brekken RA, Chiodoni C, Colombo MP (2008) Macrophage-derived SPARC bridges tumor cell-extracellular matrix interactions toward metastasis. Cancer Res 68(21):9050–9059. doi:10.1158/0008-5472.CAN-08-1327
142. Gocheva V, Wang HW, Gadea BB, Shree T, Hunter KE, Garfall AL, Berman T, Joyce JA (2010) IL-4 induces cathepsin protease activity in tumor-associated macrophages to promote cancer growth and invasion. Genes Dev 24(3):241–255. doi:10.1101/gad.1874010
143. Olson OC, Joyce JA (2015) Cysteine cathepsin proteases: regulators of cancer progression and therapeutic response. Nat Rev Cancer 15(12):712–729. doi:10.1038/nrc4027
144. Hynes RO (2009) The extracellular matrix: not just pretty fibrils. Science 326(5957):1216–1219. doi:10.1126/science.1176009
145. Linder S, Wiesner C, Himmel M (2011) Degrading devices: invadosomes in proteolytic cell invasion. Annu Rev Cell Dev Biol 27:185–211. doi:10.1146/annurev-cellbio-092910-154216
146. Hall A (2009) The cytoskeleton and cancer. Cancer Metastasis Rev 28(1–2):5–14. doi:10.1007/s10555-008-9166-3
147. Giampieri S, Manning C, Hooper S, Jones L, Hill CS, Sahai E (2009) Localized and reversible TGFbeta signalling switches breast cancer cells from cohesive to single cell motility. Nat Cell Biol 11(11):1287–1296. doi:10.1038/ncb1973
148. Friedl P, Gilmour D (2009) Collective cell migration in morphogenesis, regeneration and cancer. Nat Rev Mol Cell Biol 10(7):445–457. doi:10.1038/nrm2720
149. Yamaguchi H, Lorenz M, Kempiak S, Sarmiento C, Coniglio S, Symons M, Segall J, Eddy R, Miki H, Takenawa T, Condeelis J (2005) Molecular mechanisms of invadopodium formation: the role of the N-WASP-Arp2/3 complex pathway and cofilin. J Cell Biol 168(3):441–452. doi:10.1083/jcb.200407076
150. Yu X, Zech T, McDonald L, Gonzalez EG, Li A, Macpherson I, Schwarz JP, Spence H, Futo K, Timpson P, Nixon C, Ma Y, Anton IM, Visegrady B, Insall RH, Oien K, Blyth K, Norman JC, Machesky LM (2012) N-WASP coordinates the delivery and F-actin-mediated capture of MT1-MMP at invasive pseudopods. J Cell Biol 199(3):527–544. doi:10.1083/jcb.201203025
151. Gligorijevic B, Wyckoff J, Yamaguchi H, Wang Y, Roussos ET, Condeelis J (2012) N-WASP-mediated invadopodium formation is involved in intravasation and lung metastasis of mammary tumors. J Cell Sci 125(Pt 3):724–734. doi:10.1242/jcs.092726
152. Philippar U, Roussos ET, Oser M, Yamaguchi H, Kim HD, Giampieri S, Wang Y, Goswami S, Wyckoff JB, Lauffenburger DA, Sahai E, Condeelis JS, Gertler FB (2008) A Mena invasion isoform potentiates EGF-induced carcinoma cell invasion and metastasis. Dev Cell 15(6):813–828. doi:10.1016/j.devcel.2008.09.003
153. Roh-Johnson M, Bravo-Cordero JJ, Patsialou A, Sharma VP, Guo P, Liu H, Hodgson L, Condeelis J (2014) Macrophage contact induces RhoA GTPase signaling to trigger tumor cell intravasation. Oncogene 33(33):4203–4212. doi:10.1038/onc.2013.377
154. Allen SG, Chen YC, Madden JM, Fournier CL, Altemus MA, Hiziroglu AB, Cheng YH, Wu ZF, Bao L, Yates JA, Yoon E, Merajver SD (2016) Macrophages enhance migration in inflammatory

breast cancer cells via RhoC GTPase signaling. Sci Rep 6:39190. doi:10.1038/srep39190
155. Thiery JP, Acloque H, Huang RY, Nieto MA (2009) Epithelial-mesenchymal transitions in development and disease. Cell 139(5):871–890. doi:10.1016/j.cell.2009.11.007
156. Mani SA, Guo W, Liao MJ, Eaton EN, Ayyanan A, Zhou AY, Brooks M, Reinhard F, Zhang CC, Shipitsin M, Campbell LL, Polyak K, Brisken C, Yang J, Weinberg RA (2008) The epithelial-mesenchymal transition generates cells with properties of stem cells. Cell 133(4):704–715. doi:10.1016/j.cell.2008.03.027
157. Su S, Liu Q, Chen J, Chen J, Chen F, He C, Huang D, Wu W, Lin L, Huang W, Zhang J, Cui X, Zheng F, Li H, Yao H, Su F, Song E (2014a) A positive feedback loop between mesenchymal-like cancer cells and macrophages is essential to breast cancer metastasis. Cancer Cell 25(5):605–620. doi:10.1016/j.ccr.2014.03.021
158. Su S, Wu W, He C, Liu Q, Song E (2014b) Breaking the vicious cycle between breast cancer cells and tumor-associated macrophages. OncoImmunology 3(8). doi:10.4161/21624011.2014.953418
159. Lin X, Chen L, Yao Y, Zhao R, Cui X, Chen J, Hou K, Zhang M, Su F, Chen J, Song E (2015) CCL18-mediated down-regulation of miR98 and miR27b promotes breast cancer metastasis. Oncotarget 6(24):20485–20499
160. Mak KK, Wu AT, Lee WH, Chang TC, Chiou JF, Wang LS, Wu CH, Huang CY, Shieh YS, Chao TY, Ho CT, Yen GC, Yeh CT (2013) Pterostilbene, a bioactive component of blueberries, suppresses the generation of breast cancer stem cells within tumor microenvironment and metastasis via modulating NF-kappaB/microRNA 448 circuit. Mol Nutr Food Res 57(7):1123–1134. doi:10.1002/mnfr.201200549
161. Bonde AK, Tischler V, Kumar S, Soltermann A, Schwendener RA (2012) Intratumoral macrophages contribute to epithelial-mesenchymal transition in solid tumors. BMC Cancer 12. doi:10.1186/1471-2407-12-35
162. Zhang J (2015) TGF-b–induced epithelial-to-mesenchymal transition proceeds through stepwise activation of multiple feedback loops. CELL BIOLOGY
163. Singh R, Shankar BS, Sainis KB (2014) TGF-beta1-ROS-ATM-CREB signaling axis in macrophage mediated migration of human breast cancer MCF7 cells. Cell Signal 26(7):1604–1615. doi:10.1016/j.cellsig.2014.03.028
164. Xu J, Lamouille S, Derynck R (2009) TGF-beta-induced epithelial to mesenchymal transition. Cell Res 19(2):156–172. doi:10.1038/cr.2009.5
165. Deckers M (2006) The tumor suppressor Smad4 is required for transforming growth factor -induced epithelial to mesenchymal transition and bone metastasis of breast cancer cells. Cancer Res 66(4):2202–2209. doi:10.1158/0008-5472.can-05-3560
166. De Craene B, Berx G (2013) Regulatory networks defining EMT during cancer initiation and progression. Nat Rev Cancer 13(2):97–110. doi:10.1038/nrc3447
167. Romagnoli M, Belguise K, Yu Z, Wang X, Landesman-Bollag E, Seldin DC, Chalbos D, Barille-Nion S, Jezequel P, Seldin ML, Sonenshein GE (2012) Epithelial-to-mesenchymal transition induced by TGF- 1 is mediated by blimp-1-dependent repression of BMP-5. Cancer Res 72(23):6268–6278. doi:10.1158/0008-5472.can-12-2270
168. Zhang H, Meng F, Liu G, Zhang B, Zhu J, Wu F, Ethier SP, Miller F, Wu G (2011) Forkhead transcription factor Foxq1 promotes epithelial-mesenchymal transition and breast cancer metastasis. Cancer Res 71(4):1292–1301. doi:10.1158/0008-5472.can-10-2825
169. Lu H, Clauser KR, Tam WL, Fröse J, Ye X, Eaton EN, Reinhardt F, Donnenberg VS, Bhargava R, Carr SA, Weinberg RA (2014) A breast cancer stem cell niche supported by juxtacrine signalling from monocytes and macrophages. Nat Cell Biol 16(11):1105–1117. doi:10.1038/ncb3041. http://www.nature.com/ncb/journal/v16/n11/abs/ncb3041.html#supplementary-information
170. Nguyen DX, Bos PD, Massague J (2009) Metastasis: from dissemination to organ-specific colonization. Nat Rev Cancer 9(4):274–284. doi:10.1038/nrc2622
171. Lucci A, Hall CS, Lodhi AK, Bhattacharyya A, Anderson AE, Xiao L, Bedrosian I, Kuerer HM, Krishnamurthy S (2012) Circulating tumour cells in non-metastatic breast cancer: a prospective study. Lancet Oncol 13(7):688–695. doi:10.1016/s1470-2045(12)70209-7
172. Krishnamurthy S, Cristofanilli M, Singh B, Reuben J, Gao H, Cohen EN, Andreopoulou E, Hall CS, Lodhi A, Jackson S, Lucci A (2010) Detection of minimal residual disease in blood and bone marrow in early stage breast cancer. Cancer 116(14):3330–3337. doi:10.1002/cncr.25145
173. Kadioglu E, De Palma M (2015) Cancer metastasis: perivascular macrophages under watch. Cancer Discov 5(9):906–908. doi:10.1158/2159-8290.cd-15-0819
174. Harney AS, Arwert EN, Entenberg D, Wang Y, Guo P, Qian BZ, Oktay MH, Pollard JW, Jones JG, Condeelis JS (2015) Real-time imaging reveals local, transient vascular permeability, and tumor cell Intravasation stimulated by TIE2hi macrophage-derived VEGFA. Cancer Discov 5(9):932–943. doi:10.1158/2159-8290.cd-15-0012
175. Pignatelli J, Bravo-Cordero JJ, Roh-Johnson M, Gandhi SJ, Wang Y, Chen X, Eddy RJ, Xue A, Singer RH, Hodgson L, Oktay MH, Condeelis JS (2016) Macrophage-dependent tumor cell transendothelial migration is mediated by Notch1/MenaINV-initiated invadopodium formation. Sci Rep 6:37874. doi:10.1038/srep37874

176. Rohan TE, Xue X, Lin HM, D'Alfonso TM, Ginter PS, Oktay MH, Robinson BD, Ginsberg M, Gertler FB, Glass AG, Sparano JA, Condeelis JS, Jones JG (2014) Tumor microenvironment of metastasis and risk of distant metastasis of breast cancer. J Natl Cancer Inst 106(8)
177. Robinson BD, Sica GL, Liu YF, Rohan TE, Gertler FB, Condeelis JS, Jones JG (2009) Tumor microenvironment of metastasis in human breast carcinoma: a potential prognostic marker linked to hematogenous dissemination. Clin Cancer Res 15(7):2433–2441. doi:10.1158/1078-0432.ccr-08-2179
178. Qian B, Deng Y, Im JH, Muschel RJ, Zou Y, Li J, Lang RA, Pollard JW (2009) A distinct macrophage population mediates metastatic breast cancer cell extravasation, establishment and growth. PLoS One 4(8). doi:10.1371/journal.pone.0006562
179. Reymond N, d'Água BB, Ridley AJ (2013) Crossing the endothelial barrier during metastasis. Nat Rev Cancer 13(12):858–870. doi:10.1038/nrc3628
180. AL-MEHDI AB (2000) Intravascular origin of metastasis from the proliferation of endothelium-attached tumor cells: a new model for metastasis. Nat Med
181. Ferjančič S, Gil-Bernabé AM, Hill SA, Allen PD, Richardson P, Sparey T, Savory E, McGuffog J, Muschel RJ (2013) VCAM-1 and VAP-1 recruit myeloid cells that promote pulmonary metastasis in mice. Blood 121(16):3289–3297. doi:10.1182/blood-2012-08-449819
182. Kitamura T, Qian BZ, Soong D, Cassetta L, Noy R, Sugano G, Kato Y, Li J, Pollard JW (2015) CCL2-induced chemokine cascade promotes breast cancer metastasis by enhancing retention of metastasis-associated macrophages. J Exp Med 212(7):1043–1059. doi:10.1084/jem.20141836
183. Chen Q, Zhang XF, Massagué J (2011b) Macrophage binding to receptor VCAM-1 transmits survival signals in breast cancer cells that invade the lungs. Cancer Cell 20(4):538–549. doi:10.1016/j.ccr.2011.08.025
184. Leong HS, Robertson AE, Stoletov K, Leith SJ, Chin CA, Chien AE, Hague MN, Ablack A, Carmine-Simmen K, McPherson VA, Postenka CO, Turley EA, Courtneidge SA, Chambers AF, Lewis JD (2014) Invadopodia are required for cancer cell extravasation and are a therapeutic target for metastasis. Cell Rep 8(5):1558–1570. doi:10.1016/j.celrep.2014.07.050
185. Chambers AF, Groom AC, MacDonald IC (2002) Dissemination and growth of cancer cells in metastatic sites. Nat Rev Cancer 2(8):563–572. doi:10.1038/nrc865
186. Hensel JA, Flaig TW, Theodorescu D (2013) Clinical opportunities and challenges in targeting tumour dormancy. Nat Rev Clin Oncol 10(1):41–51. doi:10.1038/nrclinonc.2012.207
187. Wheeler SE, Clark AM, Taylor DP, Young CL, Pillai VC, Stolz DB, Venkataramanan R, Lauffenburger D, Griffith L, Wells A (2014) Spontaneous dormancy of metastatic breast cancer cells in an all human liver microphysiologic system. Br J Cancer 111(12):2342–2350. doi:10.1038/bjc.2014.533
188. Mou W, Xu Y, Ye Y, Chen S, Li X, Gong K, Liu Y, Chen Y, Li X, Tian Y, Xiang R, Li N (2015) Expression of Sox2 in breast cancer cells promotes the recruitment of M2 macrophages to tumor microenvironment. Cancer Lett 358(2):115–123. doi:10.1016/j.canlet.2014.11.004
189. Zhang Q, Qin J, Zhong L, Gong L, Zhang B, Zhang Y, Gao WQ (2015) CCL5-mediated Th2 immune polarization promotes metastasis in luminal breast cancer. Cancer Res 75(20):4312–4321. doi:10.1158/0008-5472.CAN-14-3590
190. Song KH, Park MS, Nandu TS, Gadad S, Kim SC, Kim MY (2016) GALNT14 promotes lung-specific breast cancer metastasis by modulating self-renewal and interaction with the lung microenvironment. Nat Commun 7:13796. doi:10.1038/ncomms13796
191. Palus S, Schur R, Akashi YJ, Bockmeyer B, Datta R, Halem H, Dong J, Culler MD, Adams V, Anker SD, Springer J (2011) Ghrelin and its analogues, BIM-28131 and BIM-28125, improve body weight and regulate the expression of MuRF-1 and MAFbx in a rat heart failure model. PLoS One 6(11):e26865. doi:10.1371/journal.pone.0026865
192. Gao D, Joshi N, Choi H, Ryu S, Hahn M, Catena R, Sadik H, Argani P, Wagner P, Vahdat LT, Port JL, Stiles B, Sukumar S, Altorki NK, Rafii S, Mittal V (2012) Myeloid progenitor cells in the premetastatic lung promote metastases by inducing mesenchymal to epithelial transition. Cancer Res 72(6):1384–1394. doi:10.1158/0008-5472.CAN-11-2905
193. Mathsyaraja H, Thies K, Taffany DA, Deighan C, Liu T, Yu L, Fernandez SA, Shapiro C, Otero J, Timmers C, Lustberg MB, Chalmers J, Leone G, Ostrowski MC (2015) CSF1-ETS2-induced microRNA in myeloid cells promote metastatic tumor growth. Oncogene 34(28):3651–3661. doi:10.1038/onc.2014.294
194. Zhang XH, Wang Q, Gerald W, Hudis CA, Norton L, Smid M, Foekens JA, Massague J (2009) Latent bone metastasis in breast cancer tied to Src-dependent survival signals. Cancer Cell 16(1):67–78. doi:10.1016/j.ccr.2009.05.017
195. Oskarsson T, Acharyya S, Zhang XHF, Vanharanta S, Tavazoie SF, Morris PG, Downey RJ, Manova-Todorova K, Brogi E, Massague J (2011) Breast cancer cells produce tenascin C as a metastatic niche component to colonize the lungs. Nat Med 17(7):867–874. doi:http://www.nature.com/nm/journal/v17/n7/abs/nm.2379.html#supplementary-information
196. Mantovani A, Polentarutti N, Luini W, Peri G, Spreafico F (1979) Role of host defense merchanisms in the antitumor activity of adriamycin and daunomycin in mice. J Natl Cancer Inst 63(1):61–66
197. Mohammed ZM, Going JJ, Edwards J, Elsberger B, McMillan DC (2013) The relationship between

lymphocyte subsets and clinico-pathological determinants of survival in patients with primary operable invasive ductal breast cancer. Br J Cancer 109(6):1676–1684. doi:10.1038/bjc.2013.493
198. Kodumudi KN, Woan K, Gilvary DL, Sahakian E, Wei S, Djeu JY (2010) A novel Chemoimmunomodulating property of Docetaxel: suppression of myeloid-derived suppressor cells in tumor bearers. Clin Cancer Res 16(18):4583–4594. doi:10.1158/1078-0432.ccr-10-0733
199. Alizadeh D, Trad M, Hanke NT, Larmonier CB, Janikashvili N, Bonnotte B, Katsanis E, Larmonier N (2014) Doxorubicin eliminates myeloid-derived suppressor cells and enhances the efficacy of adoptive T-cell transfer in breast cancer. Cancer Res 74(1):104–118. doi:10.1158/0008-5472.CAN-13-1545
200. Hannesdottir L, Tymoszuk P, Parajuli N, Wasmer MH, Philipp S, Daschil N, Datta S, Koller JB, Tripp CH, Stoitzner P, Muller-Holzner E, Wiegers GJ, Sexl V, Villunger A, Doppler W (2013) Lapatinib and doxorubicin enhance the Stat1-dependent antitumor immune response. Eur J Immunol 43(10):2718–2729. doi:10.1002/eji.201242505
201. Kroemer G, Galluzzi L, Kepp O, Zitvogel L (2013a) Immunogenic cell death in cancer therapy. Annu Rev Immunol 31. doi:10.1146/annurev-immunol-032712-100008
202. Paulus P, Stanley ER, Schafer R, Abraham D, Aharinejad S (2006) Colony-stimulating factor-1 antibody reverses chemoresistance in human MCF-7 breast cancer xenografts. Cancer Res 66(8):4349–4356. doi:10.1158/0008-5472.can-05-3523
203. Shree T, Olson OC, Elie BT, Kester JC, Garfall AL, Simpson K, Bell-McGuinn KM, Zabor EC, Brogi E, Joyce JA (2011) Macrophages and cathepsin proteases blunt chemotherapeutic response in breast cancer. Genes Dev 25(23):2465–2479. doi:10.1101/gad.180331.111
204. Nakasone ES, Askautrud HA, Kees T, Park JH, Plaks V, Ewald AJ, Fein M, Rasch MG, Tan YX, Qiu J, Park J, Sinha P, Bissell MJ, Frengen E, Werb Z, Egeblad M (2012) Imaging tumor-stroma interactions during chemotherapy reveals contributions of the microenvironment to resistance. Cancer Cell 21(4):488–503. doi:10.1016/j.ccr.2012.02.017
205. Barker HE, Paget JT, Khan AA, Harrington KJ (2015) The tumour microenvironment after radiotherapy: mechanisms of resistance and recurrence. Nat Rev Cancer 15(7):409–425. doi:10.1038/nrc3958
206. Shiao SL, Coussens LM (2010) The tumor-immune microenvironment and response to radiation therapy. J Mammary Gland Biol Neoplasia 15(4):411–421. doi:10.1007/s10911-010-9194-9
207. Kroemer G, Galluzzi L, Kepp O, Zitvogel L (2013b) Immunogenic cell death in cancer therapy. Annu Rev Immunol 31:51–72. doi:10.1146/annurev-immunol-032712-100008
208. Sun X, Gao D, Gao L, Zhang C, Yu X, Jia B, Wang F, Liu Z (2015) Molecular imaging of tumor-infiltrating macrophages in a preclinical mouse model of breast cancer. Theranostics 5(6):597–608. doi:10.7150/thno.11546
209. Meijer TW, Kaanders JH, Span PN, Bussink J (2012) Targeting hypoxia, HIF-1, and tumor glucose metabolism to improve radiotherapy efficacy. Clinical cancer research : an official journal of the American Association for Cancer Research 18(20):5585–5594. doi:10.1158/1078-0432.ccr-12-0858
210. Russell JS, Brown JM (2013) The irradiated tumor microenvironment: role of tumor-associated macrophages in vascular recovery. Front Physiol 4:157. doi:10.3389/fphys.2013.00157
211. Ahn GO, Brown JM (2008) Matrix metalloproteinase-9 is required for tumor vasculogenesis but not for angiogenesis: role of bone marrow-derived myelomonocytic cells. Cancer Cell 13(3):193–205. doi:10.1016/j.ccr.2007.11.032
212. Ager EI, Kozin SV, Kirkpatrick ND, Seano G, Kodack DP, Askoxylakis V, Huang Y, Goel S, Snuderl M, Muzikansky A, Finkelstein DM, Dransfield DT, Devy L, Boucher Y, Fukumura D, Jain RK (2015) Blockade of MMP14 activity in murine breast carcinomas: implications for macrophages, vessels, and radiotherapy. J Natl Cancer Inst 107(4). doi:10.1093/jnci/djv017
213. Bergers G, Hanahan D (2008) Modes of resistance to anti-angiogenic therapy. Nat Rev Cancer 8(8):592–603. doi:10.1038/nrc2442
214. Zhang C, Gao L, Cai Y, Liu H, Gao D, Lai J, Jia B, Wang F, Liu Z (2016a) Inhibition of tumor growth and metastasis by photoimmunotherapy targeting tumor-associated macrophage in a sorafenib-resistant tumor model. Biomaterials 84:1–12. doi:10.1016/j.biomaterials.2016.01.027
215. Welford AF, Biziato D, Coffelt SB, Nucera S, Fisher M, Pucci F, Di Serio C, Naldini L, De Palma M, Tozer GM, Lewis CE (2011) TIE2-expressing macrophages limit the therapeutic efficacy of the vascular-disrupting agent combretastatin A4 phosphate in mice. J Clin Invest 121(5):1969–1973. doi:10.1172/jci44562
216. Jain Rakesh K (2014) Antiangiogenesis strategies revisited: from starving tumors to alleviating hypoxia. Cancer Cell 26(5):605–622. doi:10.1016/j.ccell.2014.10.006
217. Daly C, Eichten A, Castanaro C, Pasnikowski E, Adler A, Lalani AS, Papadopoulos N, Kyle AH, Minchinton AI, Yancopoulos GD, Thurston G (2013) Angiopoietin-2 functions as a Tie2 agonist in tumor models, where it limits the effects of VEGF inhibition. Cancer Res 73(1):108–118. doi:10.1158/0008-5472.CAN-12-2064
218. Srivastava K, Hu J, Korn C, Savant S, Teichert M, Kapel SS, Jugold M, Besemfelder E, Thomas M, Pasparakis M, Augustin HG (2014) Postsurgical adjuvant tumor therapy by combining anti-

angiopoietin-2 and metronomic chemotherapy limits metastatic growth. Cancer Cell 26(6):880–895. doi:10.1016/j.ccell.2014.11.005

219. Park J-S, Kim I-K, Han S, Park I, Kim C, Bae J, Oh SJ, Lee S, Kim JH, Woo D-C, He Y, Augustin HG, Kim I, Lee D, Koh GY (2016) Normalization of tumor vessels by Tie2 activation and Ang2 inhibition enhances drug delivery and produces a favorable tumor microenvironment. Cancer Cell 30(6):953–967. doi:10.1016/j.ccell.2016.10.018
220. Sliwkowski MX, Mellman I (2013) Antibody therapeutics in cancer. Science 341(6151):1192–1198. doi:10.1126/science.1241145
221. Furness AJS, Vargas FA, Peggs KS, Quezada SA (2014) Impact of tumour microenvironment and fc receptors on the activity of immunomodulatory antibodies. Trends Immunol 35(7):290–298. doi:10.1016/j.it.2014.05.002
222. Mellor JD, Brown MP, Irving HR, Zalcberg JR, Dobrovic A (2013) A critical review of the role of Fc gamma receptor polymorphisms in the response to monoclonal antibodies in cancer. J Hematol Oncol 6(1). doi:10.1186/1756-8722-6-1
223. Tamura K, Shimizu C, Hojo T, Akashi-Tanaka S, Kinoshita T, Yonemori K, Kouno T, Katsumata N, Ando M, Aogi K, Koizumi F, Nishio K, Fujiwara Y (2011) FcgammaR2A and 3A polymorphisms predict clinical outcome of trastuzumab in both neoadjuvant and metastatic settings in patients with HER2-positive breast cancer. Ann Oncol 22(6):1302–1307. doi:10.1093/annonc/mdq585
224. Clynes RA, Towers TL, Presta LG, Ravetch JV (2000) Inhibitory fc receptors modulate in vivo cytoxicity against tumor targets. Nat Med 6(4):443–446. doi:10.1038/74704
225. Park S, Jiang Z, Mortenson ED, Deng L, Radkevich-Brown O, Yang X, Sattar H, Wang Y, Brown NK, Greene M, Liu Y, Tang J, Wang S, Fu YX (2010) The therapeutic effect of anti-HER2/neu antibody depends on both innate and adaptive immunity. Cancer Cell 18(2):160–170. doi:10.1016/j.ccr.2010.06.014
226. Cai Z, Fu T, Nagai Y, Lam L, Yee M, Zhu Z, Zhang H (2013) scFv-based "grababody" as a general strategy to improve recruitment of immune effector cells to antibody-targeted tumors. Cancer Res 73(8):2619–2627. doi:10.1158/0008-5472.CAN-12-3920
227. Pardoll DM (2012) The blockade of immune checkpoints in cancer immunotherapy. Nat Rev Cancer 12(4):252–264. doi:10.1038/nrc3239
228. Makkouk A, Weiner GJ (2015) Cancer immunotherapy and breaking immune tolerance: new approaches to an old challenge. Cancer Res 75(1):5–10. doi:10.1158/0008-5472.CAN-14-2538
229. Lu X, Kang Y (2009) Chemokine (C-C motif) ligand 2 engages CCR2+ stromal cells of monocytic origin to promote breast cancer metastasis to lung and bone. J Biol Chem 284(42):29087–29096. doi:10.1074/jbc.M109.035899
230. Sandhu SK, Papadopoulos K, Fong PC, Patnaik A, Messiou C, Olmos D, Wang G, Tromp BJ, Puchalski TA, Balkwill F, Berns B, Seetharam S, De Bono JS, Tolcher AW (2013) A first-in-human, first-in-class, phase i study of carlumab (CNTO 888), a human monoclonal antibody against CC-chemokine ligand 2 in patients with solid tumors. Cancer Chemother Pharmacol 71(4):1041–1050. doi:10.1007/s00280-013-2099-8
231. Brana I, Calles A, LoRusso PM, Yee LK, Puchalski TA, Seetharam S, Zhong B, de Boer CJ, Tabernero J, Calvo E (2015) Carlumab, an anti-C-C chemokine ligand 2 monoclonal antibody, in combination with four chemotherapy regimens for the treatment of patients with solid tumors: an open-label, multicenter phase 1b study. Target Oncol 10(1):111–123. doi:10.1007/s11523-014-0320-2
232. Nywening TM, Wang-Gillam A, Sanford DE, Belt BA, Panni RZ, Cusworth BM, Toriola AT, Nieman RK, Worley LA, Yano M, Fowler KJ, Lockhart AC, Suresh R, Tan BR, Lim KH, Fields RC, Strasberg SM, Hawkins WG, DeNardo DG, Goedegebuure SP, Linehan DC (2016) Targeting tumour-associated macrophages with CCR2 inhibition in combination with FOLFIRINOX in patients with borderline resectable and locally advanced pancreatic cancer: a single-centre, open-label, dose-finding, non-randomised, phase 1b trial. Lancet Oncol 17(5):651–662. doi:10.1016/s1470-2045(16)00078-4
233. Gouwy M, Struyf S, Noppen S, Schutyser E, Springael JY, Parmentier M, Proost P, Van Damme J (2008) Synergy between coproduced CC and CXC chemokines in monocyte chemotaxis through receptor-mediated events. Mol Pharmacol 74(2):485–495. doi:10.1124/mol.108.045146
234. Gouwy M, Struyf S, Berghmans N, Vanormelingen C, Schols D, Van Damme J (2011) CXCR4 and CCR5 ligands cooperate in monocyte and lymphocyte migration and in inhibition of dual-tropic (R5/X4) HIV-1 infection. Eur J Immunol 41(4):963–973. doi:10.1002/eji.201041178
235. DiPersio JF, Micallef IN, Stiff PJ, Bolwell BJ, Maziarz RT, Jacobsen E, Nademanee A, McCarty J, Bridger G, Calandra G (2009) Phase III prospective randomized double-blind placebo-controlled trial of Plerixafor plus granulocyte Colony-stimulating factor compared with placebo plus granulocyte Colony-stimulating factor for autologous stem-cell mobilization and transplantation for patients with non-Hodgkin's lymphoma. J Clin Oncol 27(28):4767–4773. doi:10.1200/JCO.2008.20.7209
236. Pollard JW (2009) Trophic macrophages in development and disease. Nat Rev Immunol 9(4):259–270
237. Strachan DC, Ruffell B, Oei Y, Bissell MJ, Coussens LM, Pryer N, Daniel D (2013) CSF1R inhibition delays cervical and mammary tumor growth in murine models by attenuating the turnover of tumor-associated macrophages and enhancing infiltration

by CD8+ T cells. OncoImmunology 2(12):e26968. doi:10.4161/onci.26968
238. Sharma N, Wesolowski R, Reebel L, Rodal MB, Peck A, West B, Karlin DA, Dowlati A, Le MH, Coussens LM, Rugo HS (2014) A phase 1b study to assess the safety of PLX3397, a CSF-1 receptor inhibitor, and paclitaxel in patients with advanced solid tumors. J Clin Oncol 32(15_suppl):TPS3127–TPS3127. doi:10.1200/jco.2014.32.15_suppl.tps3127
239. Ries CH, Cannarile MA, Hoves S, Benz J, Wartha K, Runza V, Rey-Giraud F, Pradel LP, Feuerhake F, Klaman I, Jones T, Jucknischke U, Scheiblich S, Kaluza K, Gorr IH, Walz A, Abiraj K, Cassier PA, Sica A, Gomez-Roca C, de Visser KE, Italiano A, Le Tourneau C, Delord JP, Levitsky H, Blay JY, Ruttinger D (2014) Targeting tumor-associated macrophages with anti-CSF-1R antibody reveals a strategy for cancer therapy. Cancer Cell 25(6):846–859. doi:10.1016/j.ccr.2014.05.016
240. Papadopoulos K, Gluck L, Martin LP, Olszanski AJ, Ngarmchamnanrith G, Rasmussen E, Amore B, Nagorsen D, Hill JS, Stephenson J (2016) Abstract CT137: first-in-human study of AMG 820, a monoclonal anti-CSF-1R (c-fms) antibody, in patients (pts) with advanced solid tumors. Cancer Res 76(14 Supplement):CT137
241. Tap WD, Wainberg ZA, Anthony SP, Ibrahim PN, Zhang C, Healey JH, Chmielowski B, Staddon AP, Cohn AL, Shapiro GI, Keedy VL, Singh AS, Puzanov I, Kwak EL, Wagner AJ, Von Hoff DD, Weiss GJ, Ramanathan RK, Zhang J, Habets G, Zhang Y, Burton EA, Visor G, Sanftner L, Severson P, Nguyen H, Kim MJ, Marimuthu A, Tsang G, Shellooe R, Gee C, West BL, Hirth P, Nolop K, van de Rijn M, Hsu HH, Peterfy C, Lin PS, Tong-Starksen S, Bollag G (2015) Structure-guided blockade of CSF1R kinase in Tenosynovial Giant-cell tumor. N Engl J Med 373(5):428–437. doi:10.1056/NEJMoa1411366
242. Swierczak A, Cook AD, Lenzo JC, Restall CM, Doherty JP, Anderson RL, Hamilton JA (2014) The promotion of breast cancer metastasis caused by inhibition of CSF-1R/CSF-1 signaling is blocked by targeting the G-CSF receptor. Cancer Immunol Res 2(8):765–776. doi:10.1158/2326-6066.cir-13-0190
243. Quail DF, Bowman RL, Akkari L, Quick ML, Schuhmacher AJ, Huse JT, Holland EC, Sutton JC, Joyce JA (2016) The tumor microenvironment underlies acquired resistance to CSF-1R inhibition in gliomas. Science 352(6288):aad3018. doi:10.1126/science.aad3018
244. Zhang M, Zhang H, Tang F, Wang Y, Mo Z, Lei X, Tang S (2016b) Doxorubicin resistance mediated by cytoplasmic macrophage colony-stimulating factor is associated with switch from apoptosis to autophagic cell death in MCF-7 breast cancer cells. Experimental biology and medicine. Maywood, NJ. doi:10.1177/1535370216660399
245. Ma Y, Galluzzi L, Zitvogel L, Kroemer G (2013) Autophagy and cellular immune responses. Immunity 39(2):211–227. doi:10.1016/j.immuni.2013.07.017
246. Mundy GR (2002) Metastasis to bone: causes, consequences and therapeutic opportunities. Nat Rev Cancer 2(8):584–593. doi:10.1038/nrc867
247. Van Acker HH, Anguille S, Willemen Y, Smits EL, Van Tendeloo VF (2016) Bisphosphonates for cancer treatment: mechanisms of action and lessons from clinical trials. Pharmacol Ther 158:24–40. doi:10.1016/j.pharmthera.2015.11.008
248. Moreau MF, Guillet C, Massin P, Chevalier S, Gascan H, Basle MF, Chappard D (2007) Comparative effects of five bisphosphonates on apoptosis of macrophage cells in vitro. Biochem Pharmacol 73(5):718–723. doi:10.1016/j.bcp.2006.09.031
249. Diel IJ, Solomayer EF, Costa SD, Gollan C, Goerner R, Wallwiener D, Kaufmann M, Bastert G (1998) Reduction in new metastases in breast cancer with adjuvant clodronate treatment. N Engl J Med 339(6):357–363. doi:10.1056/nejm199808063390601
250. Melani C, Sangaletti S, Barazzetta FM, Werb Z, Colombo MP (2007) Amino-biphosphonate-mediated MMP-9 inhibition breaks the tumor-bone marrow axis responsible for myeloid-derived suppressor cell expansion and macrophage infiltration in tumor stroma. Cancer Res 67(23):11438–11446. doi:10.1158/0008-5472.can-07-1882
251. Coscia M, Quaglino E, Iezzi M, Curcio C, Pantaleoni F, Riganti C, Holen I, Monkkonen H, Boccadoro M, Forni G, Musiani P, Bosia A, Cavallo F, Massaia M (2010) Zoledronic acid repolarizes tumour-associated macrophages and inhibits mammary carcinogenesis by targeting the mevalonate pathway. J Cell Mol Med 14(12):2803–2815. doi:10.1111/j.1582-4934.2009.00926.x
252. Junankar S, Shay G, Jurczyluk J, Ali N, Down J, Pocock N, Parker A, Nguyen A, Sun S, Kashemirov B, McKenna CE, Croucher PI, Swarbrick A, Weilbaecher K, Phan TG, Rogers MJ (2015) Real-time intravital imaging establishes tumor-associated macrophages as the extraskeletal target of bisphosphonate action in cancer. Cancer Discov 5(1):35–42. doi:10.1158/2159-8290.cd-14-0621
253. Coleman R, Powles T, Paterson A, Gnant M, Anderson S, Diel I, Gralow J, von Minckwitz G, Moebus V, Bergh J, Pritchard KI, Bliss J, Cameron D, Evans V, Pan H, Peto R, Bradley R, Gray R (2015) Adjuvant bisphosphonate treatment in early breast cancer: meta-analyses of individual patient data from randomised trials. Lancet 386(10001):1353–1361. doi:10.1016/s0140-6736(15)60908-4
254. Gnant M, Mlineritsch B, Schippinger W, Luschin-Ebengreuth G, Postlberger S, Menzel C, Jakesz R, Seifert M, Hubalek M, Bjelic-Radisic V, Samonigg H, Tausch C, Eidtmann H, Steger G, Kwasny W, Dubsky P, Fridrik M, Fitzal F, Stierer M, Rucklinger E, Greil R, Marth C (2009) Endocrine therapy

plus zoledronic acid in premenopausal breast cancer. N Engl J Med 360(7):679–691. doi:10.1056/NEJMoa0806285
255. Gnant M, Mlineritsch B, Stoeger H, Luschin-Ebengreuth G, Knauer M, Moik M, Jakesz R, Seifert M, Taucher S, Bjelic-Radisic V, Balic M, Eidtmann H, Eiermann W, Steger G, Kwasny W, Dubsky P, Selim U, Fitzal F, Hochreiner G, Wette V, Sevelda P, Ploner F, Bartsch R, Fesl C, Greil R (2015) Zoledronic acid combined with adjuvant endocrine therapy of tamoxifen versus anastrozol plus ovarian function suppression in premenopausal early breast cancer: final analysis of the Austrian breast and colorectal cancer study group trial 12. Ann Oncol 26(2):313–320. doi:10.1093/annonc/mdu544
256. Paterson AH, Anderson SJ, Lembersky BC, Fehrenbacher L, Falkson CI, King KM, Weir LM, Brufsky AM, Dakhil S, Lad T, Baez-Diaz L, Gralow JR, Robidoux A, Perez EA, Zheng P, Geyer CE Jr, Swain SM, Costantino JP, Mamounas EP, Wolmark N (2012) Oral clodronate for adjuvant treatment of operable breast cancer (National Surgical Adjuvant Breast and bowel project protocol B-34): a multicentre, placebo-controlled, randomised trial. Lancet Oncol 13(7):734–742. doi:10.1016/s1470-2045(12)70226-7
257. Coleman RE, Marshall H, Cameron D, Dodwell D, Burkinshaw R, Keane M, Gil M, Houston SJ, Grieve RJ, Barrett-Lee PJ, Ritchie D, Pugh J, Gaunt C, Rea U, Peterson J, Davies C, Hiley V, Gregory W, Bell R (2011) Breast-cancer adjuvant therapy with zoledronic acid. N Engl J Med 365(15):1396–1405. doi:10.1056/NEJMoa1105195
258. Dewan MZ, Vanpouille-Box C, Kawashima N, DiNapoli S, Babb JS, Formenti SC, Adams S, Demaria S (2012) Synergy of topical toll-like receptor 7 agonist with radiation and low-dose cyclophosphamide in a mouse model of cutaneous breast cancer. Clinical cancer research : an official journal of the American Association for Cancer Research 18(24):6668–6678. doi:10.1158/1078-0432.ccr-12-0984
259. Adams S, Kozhaya L, Martiniuk F, Meng TC, Chiriboga L, Liebes L, Hochman T, Shuman N, Axelrod D, Speyer J, Novik Y, Tiersten A, Goldberg JD, Formenti SC, Bhardwaj N, Unutmaz D, Demaria S (2012) Topical TLR7 agonist imiquimod can induce immune-mediated rejection of skin metastases in patients with breast cancer. Clinical cancer research : an official journal of the American Association for Cancer Research 18(24):6748–6757. doi:10.1158/1078-0432.ccr-12-1149
260. Henriques L, Palumbo M, Guay M-P, Bahoric B, Basik M, Kavan P, Batist G (2014) Imiquimod in the treatment of breast cancer skin metastasis. J Clin Oncol 32(8):e22–e25. doi:10.1200/JCO.2012.46.4883
261. Salazar LG, Lu H, Reichow JL, Childs JS, Coveler AL, Higgins DM, Waisman J, Allison KH, Dang Y, Disis ML (2017) Topical Imiquimod plus nab-paclitaxel for breast cancer cutaneous metastases: a phase 2 clinical trial. JAMA Oncol. doi:10.1001/jamaoncol.2016.6007
262. Weihrauch MR, Richly H, von Bergwelt-Baildon MS, Becker HJ, Schmidt M, Hacker UT, Shimabukuro-Vornhagen A, Holtick U, Nokay B, Schroff M, Wittig B, Scheulen ME (2015) Phase I clinical study of the toll-like receptor 9 agonist MGN1703 in patients with metastatic solid tumours. European journal of cancer (Oxford, England : 1990) 51(2):146–156. doi:10.1016/j.ejca.2014.11.002
263. Barron TI, Flahavan EM, Sharp L, Bennett K, Visvanathan K (2014) Recent Prediagnostic aspirin use, lymph node involvement, and 5-year mortality in women with stage I–III breast cancer: a Nationwide population-based cohort study. Cancer Res 74(15):4065
264. Henry WS, Laszewski T, Tsang T, Beca F, Beck AH, McAllister SS, Toker A (2016) Aspirin suppresses growth in PI3K-mutant breast cancer by activating AMPK and inhibiting mTORC1 signaling. Cancer Res
265. De Palma M, Mazzieri R, Politi LS, Pucci F, Zonari E, Sitia G, Mazzoleni S, Moi D, Venneri MA, Indraccolo S, Falini A, Guidotti LG, Galli R, Naldini L (2008) Tumor-targeted interferon-alpha delivery by Tie2-expressing monocytes inhibits tumor growth and metastasis. Cancer Cell 14(4):299–311. doi:10.1016/j.ccr.2008.09.004
266. Escobar G, Moi D, Ranghetti A, Ozkal-Baydin P, Squadrito ML, Kajaste-Rudnitski A, Bondanza A, Gentner B, De Palma M, Mazzieri R, Naldini L (2014) Genetic engineering of hematopoiesis for targeted IFN-α delivery inhibits breast cancer progression. Sci Transl Med 6(217). doi:10.1126/scitranslmed.3006353
267. Rolny C, Mazzone M, Tugues S, Laoui D, Johansson I, Coulon C, Squadrito ML, Segura I, Li X, Knevels E, Costa S, Vinckier S, Dresselaer T, Åkerud P, De Mol M, Salomäki H, Phillipson M, Wyns S, Larsson E, Buysschaert I, Botling J, Himmelreich U, Van Ginderachter JA, De Palma M, Dewerchin M, Claesson-Welsh L, Carmeliet P (2011) HRG inhibits tumor growth and metastasis by inducing macrophage polarization and vessel normalization through downregulation of PlGF. Cancer Cell 19(1):31–44. doi:10.1016/j.ccr.2010.11.009
268. Xu M, Liu M, Du X, Li S, Li H, Li X, Li Y, Wang Y, Qin Z, Fu YX, Wang S (2015) Intratumoral delivery of IL-21 overcomes anti-Her2/Neu resistance through shifting tumor-associated Macrophages from M2 to M1 phenotype. J Immunol 194(10):4997–5006. doi:10.4049/jimmunol.1402603
269. Zhou J, Donatelli SS, Gilvary DL, Tejera MM, Eksioglu EA, Chen X, Coppola D, Wei S, Djeu JY (2016) Therapeutic targeting of myeloid-derived suppressor cells involves a novel mechanism mediated by clusterin. Sci Rep 6:29521. doi:10.1038/srep29521

270. Emens LA, Braiteh FS, Cassier P, Delord J-P, Eder JP, Fasso M, Xiao Y, Wang Y, Molinero L, Chen DS (2015a) Inhibition of PD-L1 by MPDL3280A leads to clinical activity in patients with metastatic triple-negative breast cancer (TNBC). AACR,
271. Nanda R, Chow LQM, Dees EC, Berger R, Gupta S, Geva R, Pusztai L, Pathiraja K, Aktan G, Cheng JD, Karantza V, Buisseret L (2016) Pembrolizumab in patients with advanced triple-negative breast cancer: phase Ib KEYNOTE-012 study. J Clin Oncol 34(21):2460–2467. doi:10.1200/JCO.2015.64.8931
272. Nanda R, Specht J, Dees C, Berger R, Gupta S, Geva R, Pusztai L, Pathiraja K, Ray A, Karantza V, Buisseret L (2017) Abstract P6-10-03: KEYNOTE-012: long-lasting responses in a phase Ib study of pembrolizumab for metastatic triple-negative breast cancer (mTNBC). Cancer Res 77(Suppl 4):P6-10-03
273. Vonderheide RH, LoRusso PM, Khalil M, Gartner EM, Khaira D, Soulieres D, Dorazio P, Trosko JA, Rüter J, Mariani GL (2010) Tremelimumab in combination with exemestane in patients with advanced breast cancer and treatment-associated modulation of inducible costimulator expression on patient T cells. Clin Cancer Res 16(13):3485–3494
274. De Henau O, Rausch M, Winkler D, Campesato LF, Liu C, Cymerman DH, Budhu S, Ghosh A, Pink M, Tchaicha J, Douglas M, Tibbitts T, Sharma S, Proctor J, Kosmider N, White K, Stern H, Soglia J, Adams J, Palombella VJ, McGovern K, Kutok JL, Wolchok JD, Merghoub T (2016) Overcoming resistance to checkpoint blockade therapy by targeting PI3Kgamma in myeloid cells. Nature 539(7629):443–447. doi:10.1038/nature20554
275. Soliman HH, Jackson E, Neuger T, Dees EC, Harvey RD, Han H, Ismailkhan R, Minton S, Vahanian NN, Link C (2014) A first in man phase I trial of the oral immunomodulator, indoximod, combined with docetaxel in patients with metastatic solid tumors. Oncotarget 5(18):8136–8146
276. Tang SC, Montero A, Munn D, Link C, Vahanian N, Kennedy E, Soliman H (2016) Abstract P2-11-09: a phase 2 randomized trial of the IDO pathway inhibitor indoximod in combination with taxane based chemotherapy for metastatic breast cancer: preliminary data. Cancer Res 76(Suppl 4):P2-11-09
277. Willingham SB (2012) The CD47-signal regulatory protein alpha (SIRPa) interaction is a therapeutic target for human solid tumors. PNAS
278. Petrova PS, Viller NN, Wong M, Pang X, Lin GHY, Dodge K, Chai V, Chen H, Lee V, House V, Vigo NT, Jin D, Mutukura T, Charbonneau M, Truong T, Viau S, Johnson LD, Linderoth E, Sievers EL, Maleki Vareki S, Figueredo R, Pampillo M, Koropatnick J, Trudel S, Mbong N, Jin L, Wang JCY, Uger RA (2017) TTI-621 (SIRPαFc): a CD47-blocking innate immune checkpoint inhibitor with broad antitumor activity and minimal erythrocyte binding. Clin Cancer Res 23(4):1068–1079. doi:10.1158/1078-0432.ccr-16-1700
279. Arpel A, Gamper C, Spenle C, Fernandez A, Jacob L, Baumlin N, Laquerriere P, Orend G, Cremel G, Bagnard D (2016) Inhibition of primary breast tumor growth and metastasis using a neuropilin-1 transmembrane domain interfering peptide. Oncotarget 7(34):54723–54732. doi:10.18632/oncotarget.10101
280. Weekes CD, Beeram M, Tolcher AW, Papadopoulos KP, Gore L, Hegde P, Xin Y, Yu R, Shih LM, Xiang H, Brachmann RK, Patnaik A (2014) A phase I study of the human monoclonal anti-NRP1 antibody MNRP1685A in patients with advanced solid tumors. Investig New Drugs 32(4):653–660. doi:10.1007/s10637-014-0071-z
281. Papadopoulos KP, Graham DM, Tolcher AW, Razak ARA, Patnaik A, Bedard PL, Rasco DW, Amaya A, Moore KN, Konner JA, Matei D, Martin LP, Adriaens L, Brownstein CM, Lowy I, Gao B, Kostic A, DiCioccio AT, Trail P, Siu LL (2014) A phase 1b study of combined angiogenesis blockade with nesvacumab, a selective monoclonal antibody (MAb) to angiopoietin-2 (Ang2) and ziv-aflibercept in patients with advanced solid malignancies. J Clin Oncol 32(15_Suppl):2522–2522. doi:10.1200/jco.2014.32.15_suppl.2522
282. Emens LA, Braiteh FS, Cassier P, Delord J-P, Eder JP, Fasso M, Xiao Y, Wang Y, Molinero L, Chen DS, Krop I (2015b) Abstract 2859: inhibition of PD-L1 by MPDL3280A leads to clinical activity in patients with metastatic triple-negative breast cancer (TNBC). Cancer Res 75(15 Supplement):2859

New Approaches in CAR-T Cell Immunotherapy for Breast Cancer

17

Jinghua Wang and Penghui Zhou

Abstract

Despite significant advances in surgery, chemotherapy, radiotherapy, endocrine therapy, and molecular-targeted therapy, breast cancer remains the leading cause of death from malignant tumors among women. Immunotherapy has recently become a critical component of breast cancer treatment with encouraging activity and mild safety profiles. CAR-T therapy using genetically modifying T cells with chimeric antigen receptors (CAR) is the most commonly used approach to generate tumor-specific T cells. It has shown good curative effect for a variety of malignant diseases, especially for hematological malignancies. In this review, we briefly introduce the history and the present state of CAR research. Then we discuss the barriers of solid tumors for CARs application and possible strategies to improve therapeutic response with a focus on breast cancer. At last, we outlook the future directions of CAR-T therapy including managing toxicities and developing universal CAR-T cells.

Keywords

Breast cancer • Immunotherapy • T cells • Chimeric antigen receptor

17.1 Introduction

Breast cancer is the most diagnosed cancer in women. Despite significant advances in surgery, chemotherapy, radiotherapy, endocrine therapy, and now molecular-targeted therapy, breast cancer remains the leading cause of death from malignant tumors among women [1, 2]. After decades of researches and trials, it seems that manipulation and utilization of antitumor properties of the immune system have begun to show

J. Wang
Department of Hematology, Guangdong General Hospital, Guangdong Academy of Medical Sciences, Guangzhou 510080, China

P. Zhou (✉)
Sun Yat-sen University Cancer Center, State Key Laboratory of Oncology in South China, Collaborative Innovation Center for Cancer Medicine, Guangzhou 510060, China
e-mail: zhouph@sysucc.org.cn

E. Song, H. Hu (eds.), *Translational Research in Breast Cancer*, Advances in Experimental Medicine and Biology 1026, https://doi.org/10.1007/978-981-10-6020-5_17

promise for a variety of tumors [3, 4]. Through the years, many advances have been made in the immunotherapy of breast cancer. Immunotherapy has become an important part of breast cancer treatment, along with encouraging activity and mild safety.

Immunotherapy for breast cancer involves a wide range of therapies including monoclonal antibodies (mAbs), vaccinations, immune checkpoint inhibition, and adoptive T-cell transfer immunotherapy. HER-2/neu monoclonal antibody has been successfully used in treatment for breast cancer patients. However, overexpression of HER-2/neu accounts for only 25–30% of breast cancer patients. Vaccinations induce specific antitumor immunity, but objective tumor regression is rarely observed in clinic [5]. Cytotoxic T cells play a key role in immune-mediated control of cancer [6–12]. Plenty of studies have proved that the extent of cytotoxic T cell-infiltrating tumors is a key factor in determining the natural progression of a variety of cancers [6–9, 13–15]. Over the past two decades, T-cell-based immune therapy has gained general acceptance with its curative potency for several types of malignant diseases [16]. Current T-cell-based immune therapies are generally based on two methods. The first involves the isolation of antitumor T lymphocytes from the primary tumor tissues of the patients, which is called tumor-infiltrating lymphocytes (TILs). However, due to the difficulties of TIL isolation and culture, TIL therapy is limited to a few types of tumors with high number of TIL [17]. Another way is to generate T cells with a predetermined antitumor specificity via gene therapy-based approaches. There are two gene modification strategies, including TCR gene transfer and chimeric antigen receptor (CAR) gene transfer, which are used to endow polyclonal T cells with an antigen specificity of choice. We highlight the CAR-T cell therapy in this review.

17.2 Present State of CAR-T Therapy

Genetically modifying T cells with CARs is the most common method of producing tumor-specific T cells. CARs usually consist of an extracellular ligand-binding domain of a single-chain antibody (scFv), a hinge, a transmembrane domain, a cytoplasmic signaling chain, and/or costimulatory molecules. CAR-engineered T cells combine the specificity of mAbs with the homing and killing capacity of T cells. Specifically, CAR-T cell therapy is considered to have several advantages when compared with other cellular immunotherapies. Firstly, CAR-T cells are generated using nonspecifically activated polyclonal T cells. Therefore, they overcome the difficulty of isolation and amplification of natural tumor-specific CD4+ and CD8+ T cells [18, 19]. Secondly, CAR-T cells recognize the target antigens in a MHC-independent manner. This property enables CAR-T cells to recognize target cells with reduced HLA expression or antigen processing, which are considered as an important factor in tumor immunological escape [20–22]. Thirdly, CAR-T cells can home to tumor sites actively and specifically and possess the capacity to expand and persist over a long term after tumor recognition in vivo. Therefore, CAR-T cells targeted to tumor-associated antigens (TAAs) may be more effective than mAbs in producing long-lasting tumor responses [23]. Another particular advantage of CAR-T cells is the capacity to cross the blood–brain barrier [24]. This characteristic is highly useful for treating malignant tumors that involve in or have been transferred to the central nervous system, though adverse reactions relevant to central nervous system must be considered as well.

The concept of the CAR was put forward by Gross and colleagues in 1989, who fused the antibody-binding domain Fab with the TCR signaling domain CD3ζ and named it as T body. Since then,

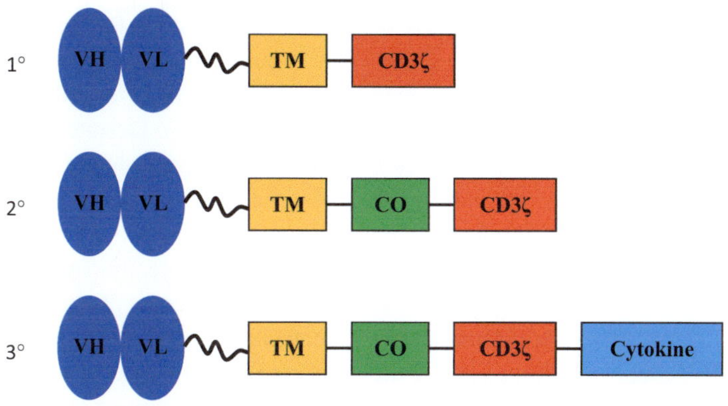

Fig. 17.1 The evolution of chimeric antigen receptors (CARs). CARs are classified into first-generation (one), second-generation (two), or third-generation (three) CARs. Abbreviation: *VH* heavy chain variable region, *VL* light chain variable region, *TM* transmembrane domain, *CO* costimulatory signaling domain

different generations of CAR-T cells have been claimed with confusing definition. In our opinion, based on the three signals required for T-cell activation, which are TCR, costimulation, and cytokine, CAR-T cells could be divided into three generations (Fig. 17.1). The first generation of CARs contained scFv and only a single signaling domain derived from CD3ζ [25]. However, the effect of the first-generation CAR trials was disappointing. Both complete T-cell activation and prevention of apoptosis required a costimulatory signal [26]. The second-generation CARs were subsequently developed, which contained two or three costimulatory signal domains of CD28 and/or 4-1BB, or other costimulatory molecules, to complete the activation signal of the CAR-T cells [27, 28]. The third-generation CARs were embedded into a cytokine cassette which endowed the CAR-T cells with a better function or survival environment. Other features such as migration, homeostatic proliferation, suppression resistance, etc. were subsequently embedded into CAR-T cells, which were described as TRUCK CAR-T cells [29, 30]. For example, the transgenic cytokine IL-12 produced by TRUCK T cells not only improves T-cell activation and modulates the immunological environment but also recruits other immune cells for the fight against those antigen-negative cancer cells that are not recognized by CAR-T cells. Other cytokines like IL-23, IL-27, and IL-15 are alternative payload for TRUCK T cells. In treatment for solid cancer, such TRUCK T cells might have an advantage to modulate the tumor environment, thus enhancing the T-cell antitumor response [31, 32].

T cells engrafted with CAR recognize a wide variety of TAAs expressed on a broad range of tumors, representing both solid and hematologic malignancies. One of the most impressive clinical results ever achieved by CAR-T cells is that polyclonal T cells express CD19-specific CARs with CD28-CD3ζ or 41BB-CD3ζ as signaling domains [24, 33–37]. Complete responses were achieved after infusion of 2nd generation CAR-T cells in patients with CD19+ hematological malignancies including NHL, acute lymphoblastic leukemia (ALL) and chronic lymphocytic leukemia (CLL). There are also clinical studies with 2nd generation CAR-T cells specific for the κ-light chain of human immunoglobulin or for CD30. Clinical responses including CRs have been observed [38, 39]. In contrast to B-cell malignancies, clinical experiences of CARs in treatment of T-cell or myeloid-derived malignancies are limited.

17.3 CAR-T Therapy for Breast Cancer: Problems and Solutions

CAR-based therapy for solid tumors involves the use of CARs targeting colorectal cancer [40, 41], ovarian cancer [42], prostate cancer [43], metastatic renal cell carcinoma, and so on [44].

There are also studies targeting HER-2, Lewis Y, mesothelin, folate receptor alpha (FR-α), and Muc1 for breast cancer in vitro and in animal models [28, 40, 45–55]. HER-2 expression is known to impact breast cancer recurrence and ultimately survival [56]. The use of anti-HER2 mAbs has significantly improved breast cancer prognosis. HER-2-targeted therapies are now a main component of HER-2 overexpressing breast cancer treatment [57, 58]. There are several clinical trials of CAR-T cells targeting HER-2 in progress, such as a phase I/II study of HER-2-targeted CAR-T cells in chemotherapy or HER-2 antibody inhibitor therapy for refractory HER-2-advanced breast cancer (NCT01935843) and a phase II study of anti-CD3 x anti-HER2/Neu-armed activated T cells after second-line chemotherapy in women with HER2/Neu (0, 1+ or 2+) metastatic breast cancer (NCT01022138). Moreover, clinical trials of CAR-T-cell therapy targeting other antigens for patients with breast cancer are ongoing, including a phase I study of CAR-T cells targeting cMet, which is aberrant activation in cancer and correlates with poor prognosis, in metastatic breast cancer refractory to at least one standard treatment or newly diagnosed patients with operable triple-negative breast cancer (TNBC) (NCT03060356), and a phase I study of CAR-T cells targeting mesothelin, a tumor antigen associated with TNBC, in metastatic HER2-negative breast cancer (NCT02580747). Despite the successes in treating hematological malignancies, CAR-T cells have encountered significant challenges for treatment of solid tumors [44, 59–62]. Some of the key problems are the rarity of target antigens, limited persistence of the CAR-T cells, inefficient homing of T cells to tumor sites, and less cytotoxicity in the local tumor immunosuppressive microenvironment [63]. The preclinical and clinical studies on treatment for breast cancer with CAR-T therapy are summarized in Table 17.1.

17.3.1 Target Antigen

Antigens currently targeted in clinical studies include HER2, mesothelin, CEA, carbonic anhydrase IX (CAIX), FR-α, CD171, GD2, EGFRvIII, fibroblast activation protein (FAP), and vascular endothelial growth factor receptor 2 (VEGF-R2) [64]. Like other forms of cancer immunotherapy, CARs should ideally target antigens that are expressed only on cancer cells but not on normal tissues. Besides, unlike the

Table 17.1 Preclinical and clinical studies on treatment for breast cancer with CAR-T therapy

Antigen	Gene transfer	Signaling domain	Clinical trial identifier	Phase	References
ERBB2	γ-retrovirus	CD28, 4-1BB, CD3ζ	–	–	[31]
ErbB	Retrovirus	CD28, CD3ζ	–	–	[36]
ErbB2	Retrovirus	CD28, 4-1BB, CD3ζ	–	–	[37]
ErbB2	Retrovirus	CD28, CD3ζ	–	–	[38]
Mesothelin	Lentivirus	4-1BB, CD3ζ	–	–	[51]
Lewis-Y	Retrovirus	CD28, CD3ζ	–	–	[43]
MUCI	Retrovirus	CD28, OX40, CD3ζ	–	–	[55]
FRα	Lentivirus	CD27, CD3ζ	–	–	[46]
Her-2		4-1BB, CD3ζ	NCT01935843	I/II	–
CD3 × HER2			NCT01022138	II	–
cMet	RNA electroporated	4-1BB, CD3ζ	NCT03060356	I	–
Mesothelin	Retrovirus	4-1BB, CD3ζ	NCT02580747	I	–

native TCR, the CARs containing scFv only recognize target antigens expressed on the cell surface, rather than internal antigens which are processed and rendered by the cells' MHC. Consequently, only few solid tumor antigens are available, though numerous antigens are being actively explored for CAR-T cell therapy. An alternative approach is to target antigen-MHC complex, which could make intracellular antigens available, though the generation of this kind of antibody is quite difficult. Conventional T cells only recognize single antigens, but CAR-T cells could be genetically modified to recognize multiple antigens, which should allow the recognition of unique antigen expression patterns on tumor cells. One example is the "split signal CARs," which limit full T-cell activation to tumors that express multiple antigens [43, 65, 66]. Other strategies for recognizing multiple antigens include tandem CARs, ectodomains of which are 2 scFvs [67], and so-called universal ectodomain CARs that incorporate avidin or a fluorescein isothiocyanate-specific scFv to identify tumor cells incubated with labeled monoclonal antibodies [43, 65, 68, 69]. Another possible concern is immune escape. Antigenic shift may cause tumor cells to produce new tumor antigens that may not be identified by the original CAR-T cells. Such escape variants are not rare because most of the cancer cells are genetically unstable [70]. Immune escape, previously described as a drug resistance mechanism in chemotherapy, may become a dilemma in cell-based therapies. The risk of immune escape can be reduced by targeting multiple antigens. Another solution is to target antigens that are expressed on the tumor stroma. The tumor stromal compartment supports tumor growth directly by secreting cytokines and growth factors, providing nutrients, and contributing to tumor-induced immunosuppression [71]. Moreover, tumor stroma is demonstrated to be genetically more stable by studies targeting FAP expressed on cancer-associated fibroblasts or VEGFR-2 expressed on the endothelial cells of the tumor vasculature [72–75].

17.3.2 Persistence

It is important to achieve high levels of CAR-T cells persisting in the peripheral circulation of patients, in order to ensure sufficient cells are available to penetrate into tumor sites. Early trials using the first generation of CAR-T cells targeting ovarian [76] and renal cell antigens [44] indicated that the lack of persistence might be induced by lack of patient preconditioning or anti-CAR immune responses. CARs were then added with costimulatory signals to improve persistence in vivo, particularly when administered to lymphodepleted hosts [36, 77, 78]. Another effort to improve the persistence of CAR-T cells focuses on the range of cytokines that are used to culture the T cells. IL-2 has been selected as an essential cytokine to drive the expansion of T cells in vitro. There are other cytokines including IL-15, IL-7, and IL-21 that can result in cultured T cells preferential to IL-2-expanded T cells. Studies show that IL-15 can promote the proliferation of T lymphocytes, prevent apoptosis and exhaustion [79, 80], reverse anergy [79], stimulate long-lasting antigen-experienced memory cells [81], and overcome Treg-mediated inhibition [82–85]. IL-7 plays an important role in maintaining the homeostasis of mature T cells and the maintenance of memory T cells [86]. Meanwhile, CAR-T cells can be genetically modified to produce cytokines to improve the expansion and persistence in vivo while avoiding systemic toxicity [30, 82, 84, 87]. The function of CAR-T cells may be enhanced not only by adding stimulatory signals (costimulation, cytokines/cytokine receptors) but also by blocking down regulatory signals. Antibodies that block the programmed death-1 (PD-1) receptor or the PD-L1 ligand or the cytotoxic T-lymphocyte-associated antigen 4 (CTLA-4) have produced encouraging clinical results as single agents [3, 88]. Convincing evidence also demonstrates their benefit for triple-negative breast cancer [89]. The combination of these antibodies with CAR-T cells prolongs the effector function of CAR-T cells at tumor sites, which is a logical evolution of current clinical strategies.

Besides, the source and phenotype of T cells used to generate CAR-T cells will affect the latter's persistence. Selecting T cells that express naive markers such as CD62L before the genetic modification may produce CAR-T cells that possess better persistence ability than effector or more differentiated T cells [90]. Alternatively, virus-specific cytotoxic T lymphocytes (CTLs) have the potential for life-long persistence, and the CTLs contain both CD4+ and CD8+ subsets, with the latter being a critical compartment for the former's long-term persistence [91, 92]. Virus-specific CTLs also feature expression of homing/chemokines receptors commensurate with their capacity for trafficking to and residing in the designated lymphoid or non-lymphoid tissues [93]. Memory T-stem cell could differentiate into memory T cells, leading to a continuous supply of CAR-T cells. On the other hand, hematopoietic stem cells could be engineered with CARs to produce CAR-T cells in a sustained way [94].

17.3.3 Homing

Since the direct binding of tumor antigen is the primary condition of CAR to display its function, the efficient migration of CAR-T cells into tumor sites is essential to the success of the CAR-based therapeutic approach. The success of CAR-T cell therapy for B-cell malignancies is probably caused by the fact that the target B cells are readily accessible to CAR-T cells and express a variety of costimulatory receptor ligands that can promote CAR-T cell function [95]. Chemokines play an important role in the migration of lymphocytes [96], as typified by recent studies [97–99]. However, the chemokine system is complex. Therefore, it is important to develop a strategy to make use of the important homing chemokines and avoid the potential regulatory effect of other tumor-expressed chemokines, in order to achieve efficient targeting of CAR-T cells [17].

17.3.4 Tumor Microenvironment

The tumor microenvironment possesses a variety of pro-tumorigenic and immunosuppressive qualities that are consistent with supporting tumor growth and proliferation and with preventing the antitumor effects of the immune system. The tumor microenvironment comprises several factors such as immunosuppressive cytokines, regulatory modulators, and coinhibitory receptors [100]. The immunosuppressive cell populations include regulatory T cells, immature myeloid cell populations, and tumor-associated macrophages [9, 101–103]. As highly complex interactions among different components in the tumor microenvironment contribute to clinical outcomes, CAR-T cells must be armed and thrive in the environment. Genetically engineering of the CAR vector to include dominant negative TGFβ receptors to overcome the adverse effects of tumor-derived TGFβ [104], and to adopt knockdown strategies to avoid apoptosis mediated by Fas/Fas ligand [105] or the expression of survival genes such as BCL-XL [106], may protect the CAR-T cells against the tumor immunosuppressive microenvironment. Besides, transgenic expression of cytokines such as IL-15 or IL-12 can reverse the immunosuppressive tumor environment. In an alternate strategy, silencing of genes that inhibit the function of T cells in the tumor microenvironment or the transgenic expression of constitutively active signaling molecules may improve CAR-T cell function [105, 107]. Lastly, a combined treatment of agents that propagate cell-based immunotherapies and agents that circumvent antitumor mechanisms may be beneficial for CAR-T cells to overcome the tumor microenvironment.

17.4 Toxicities and Management

As the potency of CARs was enhanced, toxicity induced by this immunotherapeutic approach was unfortunately observed. The continued expansion of CAR-T cells implies that the associated toxicities may show corresponding persistence and deterioration with time. "On target, off tumor" toxicity is currently a major concern, which results from the activation of CAR-T cells by targeting antigen within healthy tissues. This is a well-recognized phenomenon and has led to several different side effects. Prevention of on-target toxicity requires accurate selection of antigens that are more restricted in their expression. Another approach is to infuse CAR-T cells with transient expression of the CARs only. Thus, the expression level decreases with the cell division, and the transcription becomes diluted gradually [108–110]. Another well-documented clinical side effect is systemic inflammatory response syndrome (SIRS) or cytokine storm, which is driven by a variety of cytokines, including IFN-γ, TNF-α, IL-2 [33, 77], and the most important IL-6 [24]. To reduce the onset or severity of SIRS, researchers are modifying the dose escalation of T cells and have introduced the prompt use of antibodies that block the effects of IL-6. In addition, there are genetically modified T cells expressing a suicide or safety switch along with the CAR. These cells would retain their long-term expansion and expression capacity, but could be eliminated by activating the suicide genes once toxicity occurs [111–113]. Although the expression of multiple CARs in T cells is likely to increase safety [43, 65, 66], it remains to be proved whether the benefits can be summarized within heterogeneous human malignancies, as the patterns and levels of antigen expression may vary between different malignancies.

17.5 Universal CAR-T Cells

The current standard CAR-T cell therapy requires autologous adoptive cell transfer, which is expensive and time-consuming. For newborns and elder patients, it is often difficult to obtain enough T cells with good quality to generate patient-specific CAR-T cells. To make CAR-T therapy more accessible, it is highly desirable to develop an allogeneic adoptive transfer strategy, in which universal CAR-T cells derived from healthy donors can be applied to treat multiple patients circumventing the inherent variability of individualized patient. For this strategy to work, human leukocyte antigens class I (HLA-Is) on CAR-T cells need to be removed to minimize their immunogenicity, and the T-cell receptor (TCR) on allogeneic CAR-T cells needs to be eliminated to avoid graft-versus-host disease (GVHD) [114]. There have been studies to efficiently generate CAR-T cells with TCRα subunit constant (TRAC) and beta-2 microglobulin (β_2M) genes disrupted. However, these TRAC/β_2M-negative CAR-T cells need to be further tested for their efficacy and safety in clinical studies [114–116].

17.6 Combinatorial CAR-T Cell Therapy

It may be better to fight a war with a well-orchestrated army than a "single bullet," so combining CAR-T cells with other therapies offers the potential to improve antitumor effects. For example, combining blocking antibodies (CTLA-4, PD-1, and PD-L1) to the coinhibitory receptors, epigenetic modifiers that upregulate the expression of TAA [117], or targeted therapies that inhibit tumor cell growth without impairing T cells may be beneficial [118]. In the future, experimental treatment will be needed to determine how the CAR-T cell approach will be combined with other therapies for solid tumors, such as breast cancer.

17.7 Conclusions

The general concept of CAR-T cell was invented about 20 years ago. CAR-T cells are changing from being simply "promising" to being "effective" regimens for treating hematological malignancies. As we continue to improve the function of CAR-T cells in tumor microenvironment, broader application can be expected beyond hematological tumors and into solid tumors. Clinical trials comparing different genetic modification strategies will be important in the future for optimizing CAR-T cell therapy, which would be a potentially effective method to cure breast cancer disease.

References

1. Siegel RL, Miller KD, Jemal A (2017) Cancer statistics, 2017. CA Cancer J Clin 67(1):7–30
2. Chen W et al (2016) Cancer statistics in China, 2015. CA Cancer J Clin 66(2):115–132
3. Brahmer JR et al (2012) Safety and activity of anti-PD-L1 antibody in patients with advanced cancer. N Engl J Med 366(26):2455–2465
4. Topalian SL et al (2012) Safety, activity, and immune correlates of anti-PD-1 antibody in cancer. N Engl J Med 366(26):2443–2454
5. Zhou J, Zhong Y (2004) Breast cancer immunotherapy. Cell Mol Immunol 1(4):247–255
6. Galon J et al (2006) Type, density, and location of immune cells within human colorectal tumors predict clinical outcome. Science 313(5795):1960–1964
7. Hamanishi J et al (2007) Programmed cell death 1 ligand 1 and tumor-infiltrating CD8+ T lymphocytes are prognostic factors of human ovarian cancer. Proc Natl Acad Sci U S A 104(9):3360–3365
8. Mahmoud SM et al (2011) Tumor-infiltrating CD8+ lymphocytes predict clinical outcome in breast cancer. J Clin Oncol 29(15):1949–1955
9. Bindea G et al (2013) Spatiotemporal dynamics of intratumoral immune cells reveal the immune landscape in human cancer. Immunity 39(4):782–795
10. Matsushita H et al (2012) Cancer exome analysis reveals a T-cell-dependent mechanism of cancer immunoediting. Nature 482(7385):400–404
11. Oble DA et al (2009) Focus on TILs: prognostic significance of tumor infiltrating lymphocytes in human melanoma. Cancer Immun 9:3
12. DuPage M et al (2012) Expression of tumour-specific antigens underlies cancer immunoediting. Nature 482(7385):405–409
13. Pages F et al (2009) In situ cytotoxic and memory T cells predict outcome in patients with early-stage colorectal cancer. J Clin Oncol 27(35):5944–5951
14. Rusakiewicz S et al (2013) Immune infiltrates are prognostic factors in localized gastrointestinal stromal tumors. Cancer Res 73(12):3499–3510
15. Stumpf M et al (2009) Intraepithelial CD8-positive T lymphocytes predict survival for patients with serous stage III ovarian carcinomas: relevance of clonal selection of T lymphocytes. Br J Cancer 101(9):1513–1521
16. Dotti G et al (2001) Adenovector-induced expression of human-CD40-ligand (hCD40L) by multiple myeloma cells. A model for immunotherapy. Exp Hematol 29(8):952–961
17. Cheadle EJ et al (2014) CAR T cells: driving the road from the laboratory to the clinic. Immunol Rev 257(1):91–106
18. Pittet MJ et al (2001) Expansion and functional maturation of human tumor antigen-specific CD8+ T cells after vaccination with antigenic peptide. Clin Cancer Res 7(3 Suppl): 796s–803s
19. Valmori D et al (2000) Naturally occurring human lymphocyte antigen-A2 restricted CD8+ T-cell response to the cancer testis antigen NY-ESO-1 in melanoma patients. Cancer Res 60(16):4499–4506
20. Jakobsen MK et al (1995) Defective major histocompatibility complex class I expression in a sarcomatoid renal cell carcinoma cell line. J Immunother Emphasis Tumor Immunol 17(4):222–228
21. Lou Y et al (2008) Combining the antigen processing components TAP and Tapasin elicits enhanced tumor-free survival. Clin Cancer Res 14(5):1494–1501
22. Singh R, Paterson Y (2007) Immunoediting sculpts tumor epitopes during immunotherapy. Cancer Res 67(5):1887–1892
23. Sun M et al (2014) Construction and evaluation of a novel humanized HER2-specific chimeric receptor. Breast Cancer Res 16(3):R61
24. Grupp SA et al (2013) Chimeric antigen receptor-modified T cells for acute lymphoid leukemia. N Engl J Med 368(16):1509–1518
25. Eshhar Z et al (2001) Functional expression of chimeric receptor genes in human T cells. J Immunol Methods 248(1–2):67–76
26. Lenschow DJ, Walunas TL, Bluestone JA (1996) CD28/B7 system of T cell costimulation. Annu Rev Immunol 14:233–258
27. Carpenito C et al (2009) Control of large, established tumor xenografts with genetically retargeted human T cells containing CD28 and CD137 domains. Proc Natl Acad Sci U S A 106(9):3360–3365
28. Song DG et al (2012) CD27 costimulation augments the survival and antitumor activity of redirected human T cells in vivo. Blood 119(3):696–706
29. Chmielewski M et al (2011) IL-12 release by engineered T cells expressing chimeric antigen receptors can effectively muster an antigen-independent macrophage response on tumor cells that have shut down tumor antigen expression. Cancer Res 71(17):5697–5706
30. Zhang L et al (2011) Improving adoptive T cell therapy by targeting and controlling IL-12 expression to the tumor environment. Mol Ther 19(4):751–759
31. Hunter CA (2005) New IL-12-family members: IL-23 and IL-27, cytokines with divergent functions. Nat Rev Immunol 5(7):521–531
32. Tamzalit F et al (2014) IL-15.IL-15Ralpha complex shedding following trans-presentation is essential for the survival of IL-15 responding NK and T cells. Proc Natl Acad Sci U S A 111(23):8565–8570
33. Kochenderfer JN et al (2012) B-cell depletion and remissions of malignancy along with cytokine-

associated toxicity in a clinical trial of anti-CD19 chimeric-antigen-receptor-transduced T cells. Blood 119(12):2709–2720
34. Brentjens RJ et al (2011) Safety and persistence of adoptively transferred autologous CD19-targeted T cells in patients with relapsed or chemotherapy refractory B-cell leukemias. Blood 118(18):4817–4828
35. Porter DL et al (2011) Chimeric antigen receptor-modified T cells in chronic lymphoid leukemia. N Engl J Med 365(8):725–733
36. Kalos M et al (2011) T cells with chimeric antigen receptors have potent antitumor effects and can establish memory in patients with advanced leukemia. Sci Transl Med 3(95): 95ra73
37. Brentjens RJ et al (2013) CD19-targeted T cells rapidly induce molecular remissions in adults with chemotherapy-refractory acute lymphoblastic leukemia. Sci Transl Med 5(177):177ra38
38. Savoldo B et al (2007) Epstein Barr virus specific cytotoxic T lymphocytes expressing the anti-CD30zeta artificial chimeric T-cell receptor for immunotherapy of Hodgkin disease. Blood 110(7):2620–2630
39. Ramos CA et al (2016) Clinical responses with T lymphocytes targeting malignancy-associated kappa light chains. J Clin Invest 126(7):2588–2596
40. Morgan RA et al (2010) Case report of a serious adverse event following the administration of T cells transduced with a chimeric antigen receptor recognizing ERBB2. Mol Ther 18(4):843–851
41. Schlimper C et al (2012) Improved activation toward primary colorectal cancer cells by antigen-specific targeting autologous cytokine-induced killer cells. Clin Dev Immunol 2012:238924
42. Kandalaft LE, Powell DJ, Coukos G (2012) A phase I clinical trial of adoptive transfer of folate receptor-alpha redirected autologous T cells for recurrent ovarian cancer. J Transl Med 10:157
43. Kloss CC et al (2013) Combinatorial antigen recognition with balanced signaling promotes selective tumor eradication by engineered T cells. Nat Biotechnol 31(1):71–75
44. Lamers CH et al (2013) Treatment of metastatic renal cell carcinoma with CAIX CAR-engineered T cells: clinical evaluation and management of on-target toxicity. Mol Ther 21(4):904–912
45. Davies DM et al (2012) Flexible targeting of ErbB dimers that drive tumorigenesis by using genetically engineered T cells. Mol Med 18:565–576
46. Zhao Y et al (2009) A herceptin-based chimeric antigen receptor with modified signaling domains leads to enhanced survival of transduced T lymphocytes and antitumor activity. J Immunol 183(9):5563–5574
47. Teng MW et al (2004) Immunotherapy of cancer using systemically delivered gene-modified human T lymphocytes. Hum Gene Ther 15(7):699–708
48. Stancovski I et al (1993) Targeting of T lymphocytes to Neu/HER2-expressing cells using chimeric single chain Fv receptors. J Immunol 151(11):6577–6582
49. Moritz D et al (1994) Cytotoxic T lymphocytes with a grafted recognition specificity for ERBB2-expressing tumor cells. Proc Natl Acad Sci U S A 91(10):4318–4322
50. Altenschmidt U et al (1996) Cytolysis of tumor cells expressing the Neu/erbB-2, erbB-3, and erbB-4 receptors by genetically targeted naive T lymphocytes. Clin Cancer Res 2(6):1001–1008
51. Lanitis E et al (2012) Redirected antitumor activity of primary human lymphocytes transduced with a fully human anti-mesothelin chimeric receptor. Mol Ther 20(3):633–643
52. Westwood JA et al (2005) Adoptive transfer of T cells modified with a humanized chimeric receptor gene inhibits growth of Lewis-Y-expressing tumors in mice. Proc Natl Acad Sci U S A 102(52):19051–19056
53. Mezzanzanica D et al (1998) Transfer of chimeric receptor gene made of variable regions of tumor-specific antibody confers anticarbohydrate specificity on T cells. Cancer Gene Ther 5(6):401–407
54. Moon EK et al (2011) Expression of a functional CCR2 receptor enhances tumor localization and tumor eradication by retargeted human T cells expressing a mesothelin-specific chimeric antibody receptor. Clin Cancer Res 17(14):4719–4730
55. Wilkie S et al (2008) Retargeting of human T cells to tumor-associated MUC1: the evolution of a chimeric antigen receptor. J Immunol 180(7):4901–4909
56. O'Shaughnessy JA (2006) Molecular signatures predict outcomes of breast cancer. N Engl J Med 355(6):615–617
57. Slamon DJ et al (2001) Use of chemotherapy plus a monoclonal antibody against HER2 for metastatic breast cancer that overexpresses HER2. N Engl J Med 344(11):783–792
58. Baselga J et al (2012) Pertuzumab plus trastuzumab plus docetaxel for metastatic breast cancer. N Engl J Med 366(2):109–119
59. Lipowska-Bhalla G et al (2012) Targeted immunotherapy of cancer with CAR T cells: achievements and challenges. Cancer Immunol Immunother 61(7):953–962
60. Gilham DE et al (2012) CAR-T cells and solid tumors: tuning T cells to challenge an inveterate foe. Trends Mol Med 18(7):377–384
61. Lamers CH et al (2006) Treatment of metastatic renal cell carcinoma with autologous T-lymphocytes genetically retargeted against carbonic anhydrase IX: first clinical experience. J Clin Oncol 24(13):e20–e22
62. Park JR et al (2007) Adoptive transfer of chimeric antigen receptor re-directed cytolytic T lymphocyte clones in patients with neuroblastoma. Mol Ther 15(4):825–833

63. Han EQ et al (2013) Chimeric antigen receptor-engineered T cells for cancer immunotherapy: progress and challenges. J Hematol Oncol 6:47
64. Kakarla S, Gottschalk S (2014) CAR T cells for solid tumors: armed and ready to go? Cancer J 20(2):151–155
65. Wilkie S et al (2012) Dual targeting of ErbB2 and MUC1 in breast cancer using chimeric antigen receptors engineered to provide complementary signaling. J Clin Immunol 32(5):1059–1070
66. Lanitis E et al (2013) Chimeric antigen receptor T cells with dissociated signaling domains exhibit focused antitumor activity with reduced potential for toxicity in vivo. Cancer Immunol Res 1(1):43–53
67. Grada Z et al (2013) TanCAR: a novel Bispecific chimeric antigen receptor for cancer immunotherapy. Mol Ther Nucleic Acids 2:e105
68. Urbanska K et al (2012) A universal strategy for adoptive immunotherapy of cancer through use of a novel T-cell antigen receptor. Cancer Res 72(7):1844–1852
69. Tamada K et al (2012) Redirecting gene-modified T cells toward various cancer types using tagged antibodies. Clin Cancer Res 18(23):6436–6445
70. Janssen A, Medema RH (2013) Genetic instability: tipping the balance. Oncogene 32(38):4459–4470
71. Rabinovich GA, Gabrilovich D, Sotomayor EM (2007) Immunosuppressive strategies that are mediated by tumor cells. Annu Rev Immunol 25:267–296
72. Kakarla S et al (2013) Antitumor effects of chimeric receptor engineered human T cells directed to tumor stroma. Mol Ther 21(8):1611–1620
73. Roberts EW et al (2013) Depletion of stromal cells expressing fibroblast activation protein-alpha from skeletal muscle and bone marrow results in cachexia and anemia. J Exp Med 210(6):1137–1151
74. Niederman TM et al (2002) Antitumor activity of cytotoxic T lymphocytes engineered to target vascular endothelial growth factor receptors. Proc Natl Acad Sci U S A 99(10):7009–7014
75. Tran E et al (2013) Immune targeting of fibroblast activation protein triggers recognition of multipotent bone marrow stromal cells and cachexia. J Exp Med 210(6):1125–1135
76. Kershaw MH et al (2006) A phase I study on adoptive immunotherapy using gene-modified T cells for ovarian cancer. Clin Cancer Res 12(20 Pt 1):6106–6115
77. Brentjens R et al (2010) Treatment of chronic lymphocytic leukemia with genetically targeted autologous T cells: case report of an unforeseen adverse event in a phase I clinical trial. Mol Ther 18(4):666–668
78. Savoldo B et al (2011) CD28 costimulation improves expansion and persistence of chimeric antigen receptor-modified T cells in lymphoma patients. J Clin Invest 121(5):1822–1826
79. Li XC et al (2001) IL-15 and IL-2: a matter of life and death for T cells in vivo. Nat Med 7(1):114–118
80. Mueller K, Schweier O, Pircher H (2008) Efficacy of IL-2- versus IL-15-stimulated CD8 T cells in adoptive immunotherapy. Eur J Immunol 38(10):2874–2885
81. Ochoa MC et al (2013) Interleukin-15 in gene therapy of cancer. Curr Gene Ther 13(1):15–30
82. Perna SK et al (2013) Interleukin 15 provides relief to CTLs from regulatory T cell-mediated inhibition: implications for adoptive T cell-based therapies for lymphoma. Clin Cancer Res 19(1):106–117
83. Quintarelli C et al (2007) Co-expression of cytokine and suicide genes to enhance the activity and safety of tumor-specific cytotoxic T lymphocytes. Blood 110(8):2793–2802
84. Hoyos V et al (2010) Engineering CD19-specific T lymphocytes with interleukin-15 and a suicide gene to enhance their anti-lymphoma/leukemia effects and safety. Leukemia 24(6):1160–1170
85. Hsu C et al (2005) Primary human T lymphocytes engineered with a codon-optimized IL-15 gene resist cytokine withdrawal-induced apoptosis and persist long-term in the absence of exogenous cytokine. J Immunol 175(11):7226–7234
86. Carrette F, Surh CD (2012) IL-7 signaling and CD127 receptor regulation in the control of T cell homeostasis. Semin Immunol 24(3):209–217
87. Pegram HJ et al (2012) Tumor-targeted T cells modified to secrete IL-12 eradicate systemic tumors without need for prior conditioning. Blood 119(18):4133–4141
88. Hodi FS et al (2010) Improved survival with ipilimumab in patients with metastatic melanoma. N Engl J Med 363(8):711–723
89. Flies DB et al (2014) Coinhibitory receptor PD-1H preferentially suppresses CD4(+) T cell-mediated immunity. J Clin Invest 124(5):1966–1975
90. Wang X et al (2012) Phenotypic and functional attributes of lentivirus-modified CD19-specific human CD8+ central memory T cells manufactured at clinical scale. J Immunother 35(9):689–701
91. Rooney CM et al (1998) Infusion of cytotoxic T cells for the prevention and treatment of Epstein-Barr virus-induced lymphoma in allogeneic transplant recipients. Blood 92(5):1549–1555
92. Heslop HE et al (2010) Long-term outcome of EBV-specific T-cell infusions to prevent or treat EBV-related lymphoproliferative disease in transplant recipients. Blood 115(5):925–935
93. Hislop AD et al (2007) Cellular responses to viral infection in humans: lessons from Epstein-Barr virus. Annu Rev Immunol 25:587–617
94. Gschweng E, De Oliveira S, Kohn DB (2014) Hematopoietic stem cells for cancer immunotherapy. Immunol Rev 257(1):237–249
95. Cheadle EJ et al (2012) Ligation of the CD2 co-stimulatory receptor enhances IL-2 production from first-generation chimeric antigen receptor T cells. Gene Ther 19(11):1114–1120

96. Bromley SK, Mempel TR, Luster AD (2008) Orchestrating the orchestrators: chemokines in control of T cell traffic. Nat Immunol 9(9):970–980
97. Kershaw MH et al (2002) Redirecting migration of T cells to chemokine secreted from tumors by genetic modification with CXCR2. Hum Gene Ther 13(16):1971–1980
98. Di Stasi A et al (2009) T lymphocytes coexpressing CCR4 and a chimeric antigen receptor targeting CD30 have improved homing and antitumor activity in a Hodgkin tumor model. Blood 113(25):6392–6402
99. Craddock JA et al (2010) Enhanced tumor trafficking of GD2 chimeric antigen receptor T cells by expression of the chemokine receptor CCR2b. J Immunother 33(8):780–788
100. Gajewski TF et al (2006) Immune resistance orchestrated by the tumor microenvironment. Immunol Rev 213:131–145
101. Gabrilovich DI, Nagaraj S (2009) Myeloid-derived suppressor cells as regulators of the immune system. Nat Rev Immunol 9(3):162–174
102. Shiao SL et al (2011) Immune microenvironments in solid tumors: new targets for therapy. Genes Dev 25(24):2559–2572
103. Tanchot C et al (2013) Tumor-infiltrating regulatory T cells: phenotype, role, mechanism of expansion in situ and clinical significance. Cancer Microenviron 6(2):147–157
104. Foster AE et al (2008) Antitumor activity of EBV-specific T lymphocytes transduced with a dominant negative TGF-beta receptor. J Immunother 31(5):500–505
105. Dotti G et al (2005) Human cytotoxic T lymphocytes with reduced sensitivity to Fas-induced apoptosis. Blood 105(12):4677–4684
106. Eaton D et al (2002) Retroviral transduction of human peripheral blood lymphocytes with Bcl-X(L) promotes in vitro lymphocyte survival in pro-apoptotic conditions. Gene Ther 9(8):527–535
107. Sun J et al (2010) T cells expressing constitutively active Akt resist multiple tumor-associated inhibitory mechanisms. Mol Ther 18(11):2006–2017
108. Zhao Y et al (2010) Multiple injections of electroporated autologous T cells expressing a chimeric antigen receptor mediate regression of human disseminated tumor. Cancer Res 70(22):9053–9061
109. Barrett DM et al (2011) Treatment of advanced leukemia in mice with mRNA engineered T cells. Hum Gene Ther 22(12):1575–1586
110. Almasbak H et al (2011) Transiently redirected T cells for adoptive transfer. Cytotherapy 13(5):629–640
111. Straathof KC et al (2005) An inducible caspase 9 safety switch for T-cell therapy. Blood 105(11):4247–4254
112. Di Stasi A et al (2011) Inducible apoptosis as a safety switch for adoptive cell therapy. N Engl J Med 365(18):1673–1683
113. Arber C et al (2013) The immunogenicity of virus-derived 2A sequences in immunocompetent individuals. Gene Ther 20(9):958–962
114. Liu X et al (2017) CRISPR-Cas9-mediated multiplex gene editing in CAR-T cells. Cell Res 27(1):154–157
115. Torikai H et al (2012) A foundation for universal T-cell based immunotherapy: T cells engineered to express a CD19-specific chimeric-antigen-receptor and eliminate expression of endogenous TCR. Blood 119(24):5697–5705
116. Riolobos L et al (2013) HLA engineering of human pluripotent stem cells. Mol Ther 21(6):1232–1241
117. Chou J et al (2012) Epigenetic modulation to enable antigen-specific T-cell therapy of colorectal cancer. J Immunother 35(2):131–141
118. Liu C et al (2013) BRAF inhibition increases tumor infiltration by T cells and enhances the antitumor activity of adoptive immunotherapy in mice. Clin Cancer Res 19(2):393–403

Immune Checkpoint Blockade in Breast Cancer Therapy

18

Xia Bu, Yihui Yao, and Xiaoyu Li

Abstract

Cancer immunotherapy is emerging as the most promising novel strategy for cancer treatment. Cancer immunotherapy is broadly categorized into three forms: immune checkpoint modulation, adoptive cell transfer, and cancer vaccine. Immune checkpoint blockade is demonstrated as the most clinically effective treatment with low immune-related adverse events (irAE). Blockade of PD-1/PD-L1 and CTLA-4 has achieved remarkable success in treating various types of tumors, which sparks great interests in this therapeutic strategy and expands the role of immune checkpoint blockade in treating tumors including breast cancer. Based on the notable results obtained from clinical trials, the United States' Food and Drug Administration (FDA) has approved multiple CTLA-4 monoclonal antibodies as well as the PD-1/PD-L1 monoclonal antibodies for treatment of different types of tumors. The theories of immunoediting, T-cell exhaustions, and co-stimulatory/co-inhibitory pathways are immunological foundations for immune checkpoint blockade therapy. Breast cancers such as triple negative breast cancer and HER-2 negative breast cancer respond to immune checkpoint blockade therapy due to their high immunogenicity. PD-1/PD-L1 blockade has just received FDA approval as a standard cancer therapy for solid tumors such as breast cancer. Development of immune checkpoint blockade focuses on two directions: one is to identify proper biomarkers of immune checkpoint blockade in breast cancer, and the other is to combine therapies with PD-1/PD-L1 blockade antibodies to achieve optimal clinical outcomes.

X. Bu (✉) • Y. Yao
Department of Medical Oncology, The First Affiliated Hospital, Henan University Cancer Center, School of Medicine, Henan University, Kaifeng, People's Republic of China
e-mail: xiabumail@gmail.com

X. Li
Department of Hematology, The First Affiliated Hospital, Henan University Cancer Center, School of Medicine, Henan University, Kaifeng, People's Republic of China

E. Song, H. Hu (eds.), *Translational Research in Breast Cancer*, Advances in Experimental Medicine and Biology 1026, https://doi.org/10.1007/978-981-10-6020-5_18

Keywords
Immune checkpoint blockade • Breast cancer • PD-1 • CTLA-4 • Cancer immunotherapy

18.1 Introduction

Breast cancer is one of the leading causes of death around the world. Localized breast cancer can become metastatic in about 30% of patients [1]. Even though the improvements in radiotherapy, surgery, chemotherapy, hormone therapy, and HER-2-targeted therapy have increased the efficacy of treatment for breast cancer, no optimal option exists for treating metastatic breast cancer or triple negative breast cancer. Therefore, new approaches to treatment of these breast tumors are urgently needed. Over the past few years, immune checkpoint blockade has been attractively recognized as a promising novel strategy for cancer immunotherapy, as immune checkpoint inhibitors, such as PD-1-blocking antibodies, are capable to restore potent antitumor T-cell immunity, resulting in tumor inhibition or eradication [2].

In the current era of cancer immunotherapy, three strategies have been conducted to treat patients with a wide variety of cancer types including breast cancer. The first one is using immune checkpoint blockade to unleash the antitumor activity of the patients' immune system such as cytotoxic T-lymphocyte-associated antigen 4(CTLA-4) antibody, programmed death-1 (PD-1) antibody, and programmed death-ligand 1 (PD-L1) antibody. The second strategy is using adoptive immune cell transfer to facilitate the killing of tumor cells by active immune cells. The third one is using therapeutic vaccines to elicit immune responses against tumor-associated antigens (TAA) or tumor-specific antigens (TSA).

18.1.1 Immune Checkpoint Blockade

Breast cancer was originally thought to be poorly immunogenic and thus gained little attention from studies on cancer immunotherapy. Recently, immunotherapy with immune checkpoint blockade targeting PD-1/PD-L1 and CTLA-4 has achieved remarkable success in various cancer treatments, arising great interests in utilizing this promising therapeutic strategy to compensate current therapies for different types of tumors including breast cancer.

The US Food and Drug Administration (FDA) has approved drugs of humanized monoclonal blocking antibodies against CTLA-4 [3], PD-1 [4–6], and PD-L1 [7, 8] for cancer treatment. According to the data obtained from clinical trials with specific, potent, and long-lasting immune responses in the tumors, a total of six immune checkpoint inhibitors have received FDA approval:

1. Ipilimumab (Yervoy, Bristol-Myers Squibb), a human anti-CTLA-4 blocking antibody, was approved in 2011 for treatment of metastatic or unresectable melanoma patients [3, 9, 10].
2. Nivolumab (Opdivo, Bristol-Myers Squibb) is a human anti-PD-1 blocking antibody approved in 2014 to treat different types of metastatic tumors [4, 5, 11].
3. Pembrolizumab (Keytruda, Merck), a human anti-PD-1 blocking antibody, was initially approved for treatment of melanoma in 2014 and then approved in 2016 as the first-line treatment for non-small cell lung cancer (NSCLC) according to the results from clinical trials KEYNOTE-024 (ClinicalTrials.gov Identifier: NCT02142738) [6].
4. Atezolizumab (Tecentriq, Genentech) is a human anti-PD-L1 blocking antibody that received approval in 2016 for treatment of bladder cancer as well as NSCLC based on the results from OAK and POPLAR of two clinical trials.
5. Avelumab (Bavencio, EMD Serono, Inc), a human anti-PD-L1 antibody, was approved in 2017 for treatment of bladder cancer based on

the results from a phase II clinical trial JAVELIN Merkel 200 (ClinicalTrials.gov Identifier: NCT02155647).

6. Durvalumab (IMFINZI, AstraZeneca), a human anti-PD-L1 blocking antibody, was approved in 2017 for treating bladder cancer based on the results from a clinical trial CD-ON-MEDI4736-1108 (ClinicalTrials.gov Identifier: NCT01693562).

All these blocking antibodies have also been evaluated for their efficacy and safety in multiple clinical trials for patients with breast cancers such as HER2/neu-positive tumor and triple negative breast cancer, which will be discussed in the section of clinical trials below.

The blockade of PD-1/PD-L1 pathway is found to be the most effective treatment for various tumors with favorable clinical outcomes and low immune-related adverse events in clinical trials. Thus far, FDA has approved the application of PD-1/PD-L1 blockade immunotherapy for a total of seven types of cancers. They are melanoma, non-small cell lung cancer, renal cell carcinoma, urothelial carcinoma, classical Hodgkin's lymphoma, head and neck squamous carcinoma, and Merkel cell carcinoma. Moreover, more than 1000 clinical trials with single agent treatment or combination therapy are currently under investigation in patients with various types of cancer.

18.1.2 Adoptive Cell Transfer

The goal of the adoptive cell transfer (ACT) is to enhance antitumor immunity medicated by antigen-specific immune cells [12, 13]. Chimeric antigen receptor-engineered T-cell (CAR-T) therapy is emerging as a novel strategy for adoptive cell transfer in cancer [14–16]. CAR-T cell transfer involves modification of patients' T cells to destroy their tumors. Host T cells isolated from patients are genetically engineered to generate chimeric antigen receptors (CAR) specific for the tumor-associated/tumor-specific antigens ex vivo after expansion in culture. The CAR-T cells are infused back to the patients, allowing the modified T cells to specifically recognize and eliminate cancer cells in vivo. These T cells possess the capability to sustain as memory cells in vivo.

A CAR structure consists of an extracellular single-chain variable fragment (scFv) of TAA-/TSA-specific antibody, transmembrane domain, and intracellular T-cell-derived activation sequences. ScFv sequence is designed to specifically bind to a target of tumor-associated/tumor-specific antigen expressed on tumor cell surface. The first-generation CAR contains only CD3-zeta (CD3ζ) chain within the intracellular part without a co-stimulatory receptor domain, which mediates inefficient antitumor immunity [17, 18]. The second-generation CAR adds one co-stimulatory molecule sequence derived from CD28, OX40, or 4-1BB in the intracellular domain based on the structure of the first-generation CAR, resulting in augmented T-cell cytotoxicity, extended T-cell persistence in vivo, and elevated production of cytokines [19–22]. The third- or fourth-generation CAR adds two or more different co-stimulatory receptor domains [23, 24]. The clinical impact of these constructs is currently unclear. CD19 is currently the best studied antigen in CAR-T cell therapy in B-cell malignancies [25]. CD19-targeted CAR-T cell therapy has made great progress in B-cell lymphoblastic leukemia in clinical studies [26–29], prompting researchers to investigate therapeutic effect of CAR-T cell transfer in other types of tumor including breast cancer. An ongoing clinical trial is being conducted to evaluate the safety and antitumor activity of CAR-T therapy for TNBC (ClinicalTrials.gov Identifier: NCT02706392). Another one for HER+ breast cancer is also under investigation (ClinicalTrials.gov Identifier: NCT00228358).

18.1.3 Cancer Vaccines

Breast cancer is an immunogenic neoplasia. Cancer vaccines aim to eliminate tumors by stimulating the patient's immune system to fight cancer on its own. Therapeutic cancer vaccines are used to elicit cellular immune responses and

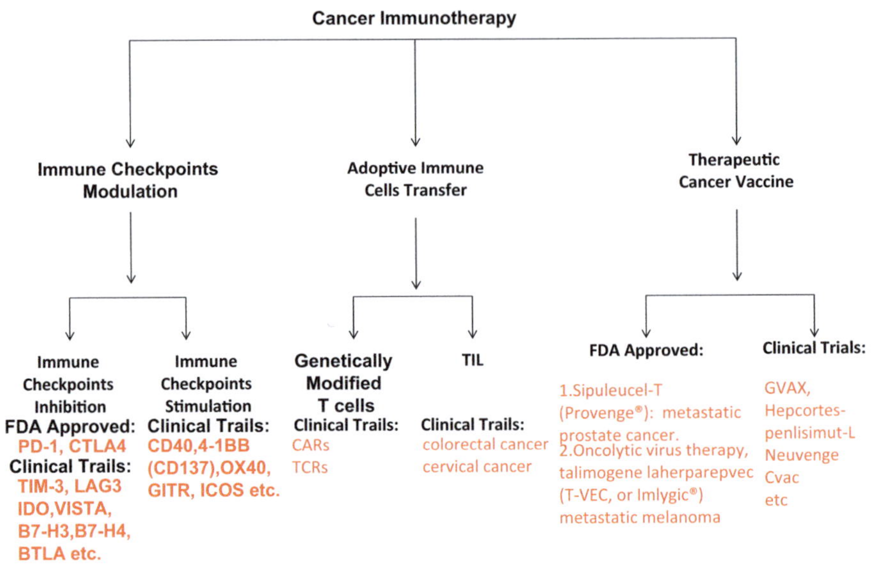

Fig. 18.1 Cancer immunotherapy at a glance

humoral immune responses, leading to recognition and destruction of tumor cells [30]. Sipuleucel-T for prostate therapeutic cancer vaccine received approval from the US Food and Drug Administration (FDA) in 2011. It is the only cancer therapeutic vaccine approved by FDA (see Fig. 18.1).

18.2 Immunologic Basis of Cancer Immunotherapy

18.2.1 The Concept of Immunoediting

The host immune system can be considered as a double-edged sword in cancer. On the one hand, immune cells mediate antitumor immunity, protecting the body from cancer. On the other hand, the immune system can promote tumor progression, shaping malignant disease. Tumors derived from RAG2-deficient mice (NKT, B and T cells are absent in the mouse strain) have higher immunogenicity than those generated in immunocompetent mice [31]. The "three Es" phases during the interactions between the immune microenvironment and tumor cells have been proposed by Dr. Schreiber with the concept of cancer immunosurveillance (also known as cancer immunoediting): elimination, equilibrium, and escape [32–34]. This concept suggests that cancer immunosurveillance by both innate and adaptive arms can eradicate highly immunogenic cancer cells but cannot reject poorly immunogenic cancer variants, resulting in the immune escape of cancer cells.

In the process of elimination phase, the innate immune effector cells, such as macrophages, natural killer cells (NK), dendritic cells (DC), gamma-delta T cells, and natural killer T cells (NKT), provide immediate responses to the tumor cells by producing pro-inflammatory cytokines such as IFN-gamma, which in turn induces a variety of cells to release chemokines including CXCL9, CXCL10, and CXCL11 in the tumor microenvironment, attracting macrophages, dendritic cells (DC), and natural killer cells (NK), etc to the tumor spot. Then, the immune system's innate immunity is followed by the adaptive immunity. Mature dendritic cells (also known as professional antigen-presenting cells, APC) uptake dead tumor cells, process the tumor-specific/tumor-associated antigens, and present these antigens to T cells by MHC class I or MHC class II molecules upon migrating to tumor-draining lymph node, educating the naïve T cells to become tumor-specific CD8+ cytotoxic T cells or CD4+ T helper cells, respectively.

During the process of equilibrium phase, cancer cell variants escape from the first phase of elimination. A mouse model of chemical carcinogenesis was employed in a study. The results from this study have showed that tumor cells are highly immunogenic (unedited) in the equilibrium phase. However, when the tumor cells are progressively growing, they become less immunogenic [35]. Another study has demonstrated that formation of carcinogen-driven sarcomas could be prevented by the synergy of IFN-gamma and T cells and that the tumor cells have greater ability to sustain in an immunocompetent host [36]. These findings prove the hypothesis of cancer immunoediting for tumor generation in the body with intact immune system. Furthermore, multiple investigations indicate that chemical carcinogen methylcholanthrene (MCA)-driven sarcomas induced in severe combined immunodeficiency (SCID) mice have higher immunogenicity than those arising in immunocompetent mice [33, 36–38], suggesting an immunoselection mediated by T cells in the immunocompetent host where the tumor is derived. In other words, tumor cells with great immunogenicity are destructed by the immune system, while tumor variants with low immunogenicity can develop multiple strategies for evading immune-mediated tumor rejection. No correlation between tumor antigenicity and lack of MHC class I expression was observed in these studies.

In the third phase of escape process, tumor cell variants that have been edited by immune cells in the second phase of equilibrium proliferate in an ungoverned manner in the immunocompetent host. Genetic and epigenetic changes in tumor cells may dampen host immune defenses, which enable tumor cells to be resistant to the immune attack. In order to achieve tumor progression and outgrowth, tumor cells may utilize diverse mechanisms to evade immune-mediated antitumor activity [33]. Various strategies of cancer immune escape are exploited by tumors including breast cancer [31, 39–42], such as (1) downregulation of major histocompatibility complex (MHC) class I molecules expressed on the surface of tumor cells; (2) upregulation of PD-L1 protein expressed on the surface of tumor cells, inducing T cell tolerance; (3) lack or mutation of tumor-specific/tumor-associated antigens, leading to the generation of tumor cell variants with low antigenicity that bypass detection by host immune system; (4) upregulation of co-inhibitory receptors that negatively regulate T-cell activation such as PD-1, CTLA-4, TIM3, and LAG3 on the exhausted tumor-specific T cells; (5) anti-inflammatory tumor microenvironment such as elevated levels of interleukin 10 (IL-10), transforming growth factor-b (TGF-b), and indoleamine 2,3 dioxygenase (IDO); and (6) immunosuppressive network, including myeloid-derived suppressor cells (MDSC) and regulatory FoxP3+ T cells (Treg cells).

18.2.2 T-Cell Exhaustion

The first evidence for T cell exhaustion resulted from studies of mice infected with lymphocytic choriomeningitis virus (LCMV), where exhausted virus-specific CD8+ T cells were identified using tetrameric MHC class I peptide complexes [43, 44]. Exhausted virus-specific CD4+ T cells were also observed during persistent viral infection [45–48].

Various studies on features of T-cell exhaustion have revealed that (a) exhausted CD4+ and/or CD8+ T cells lose effector functions [45, 49] and fail to produce cytokines such as IL-2 [50, 51], TNF-alpha, and IFN-gamma [51–56]; (b) exhausted nonfunctional T cells lose highly proliferative capability; (c) elevated expression of co-inhibitory receptors on the cell surface of exhausted T cells such as PD-1, CTLA-4, TIM3, LAG3, etc., seems to be a key player involved in the T-cell dysfunction during chronic viral infection [57–59]; and (d) the extent of T-cell exhaustion correlates positively with duration of chronic viral infection, viral burden, low CD4+ T-cell level, and the expression levels of co-inhibitory molecules [51, 60]. Subsequently, T-cell exhaustion was detected in human cancers. In the tumor microenvironment, the effector CD4+ helper T and CD8+ cytotoxic T cells are modulated by an anti-inflammatory network. The T cells progressively lose effector function and proliferative capacity, express high levels of co-inhibitory molecules such as PD-1, TIM3, LAG3, and CTLA4, secret low levels of effector cytokines, and differentiate into exhausted T cells [61–67]. Targeting immune checkpoint inhibitory receptors such as PD-1 and LAG3 can reverse the exhausted state and restore antitumor T cell immunity [62, 63, 68, 69].

The molecular mechanisms by which immune checkpoint molecules induce T-cell dysfunction remain to be clarified. Numerous studies in mice and humans have revealed that multiple inhibitory immune checkpoint receptors exhibit increased co-expression on exhausted tumor-specific CD8+ T cells and that the expression levels of these molecules are associated with the severity of T-cell exhaustion [70–73]. Results of a clinical study [70] in patients with advanced melanoma demonstrated that the level of T-cell exhaustion on Tim-3 + PD-1+ tumor antigen-specific CD8+ T cells was higher than that on Tim-3 - PD-1- or Tim-3-PD-1+ CD8+ T cells in peripheral blood and tumor-infiltrating lymphocytes. The Tim-3/PD-1 double positive dysfunctional CD8+ T cells secret less IL-2, TNF-alpha, and IFN-gamma when compared to the single positive (either Tim-3+ or PD-1+) CD8+ T cells. Furthermore, blockade of Tim-3 in combination with blockade of PD-1 signaling can synergistically reverse tumor-mediated T-cell exhaustion by augmenting proliferation and cytokine production of tumor antigen-specific CD8+ T cells.

A mouse model of HER2/neu-positive breast cancer was used to evaluate the role of CD4+ helper T cells in modulating the effector function of neu-specific CD8+ memory T cells and exhaustion of the CD8+ T cells during primary tumor challenge. The results showed a higher frequency of the exhausted helpless CD8+ T cells with upregulated expression of PD-1 when compared to helped CD8+ T cells. CD8+ T cell exhaustion can be suppressed by CD4+ helper T cells during priming phase of the antitumor immune responses [74].

18.2.3 T-Cell Co-stimulation and Co-inhibition

To fully activate T cells triggered by antigen, two signals are required. One signal is the interaction between antigens (such as peptides) bound to major histocompatibility complex (MHC) molecules on antigen-presenting cells (APC) and T-cell receptor (TCR) on T cells. The TCR antigen recognition informs specificity in the immune response which functions in a MHC-dependent manner. A second signal, called co-stimulatory signal or co-stimulation, is provided by the molecules of B7-CD28 ligand-receptor family on APC cells (e.g., CD80, CD86) that bind to co-stimulatory receptors on (e.g., CD28, OX-40, 4-1BB) T cells, inducing expansion of the primary signal and promoting effector response. Co-stimulation pathway is antigen-independent. It plays an essential role in determining the outcome of antigen-driven T-cell activation [75, 76]. TCR-medicated T-cell activation (the first signal) is regulated not only by co-stimulatory pathways but also by co-inhibitory pathways (also known as co-inhibition). Both pathways are second signals. Co-inhibition plays a critical role in keeping up peripheral tolerance and thus inhibits autoimmunity [77–79].

Multiple co-stimulatory and co-inhibitory molecules have been discovered in the immune system. The best characterized co-stimulatory and co-inhibitory pathways involve CD28 receptor and CTLA-4 receptor, respectively, which bind to their shared co-stimulatory ligands B7-1(CD80) and B7-2 (CD86) [80, 81]. Based on the structural characteristics, the molecules of these pathways are classified into two families, which are the TNF/TNF receptor superfamily and the immunoglobulin (Ig) superfamily such as B7-CD28, TIM family, LAG3, and CD226-TIGIT-CD96) [82, 83]. Co-stimulatory pathway and co-inhibitory pathway are antigen-independent. They are key players in controlling antigen-driven T-cell activation, T-cell exhaustion, and immune tolerance. Modulating these pathways provides basis for cancer immunotherapy [81].

18.2.3.1 CTLA-4/CD28 Pathway

Cytotoxic T-lymphocyte-associated antigen-4 (CTLA-4, CD152), the second member of the B7/CD28 family, is a negative regulator of T-cell activation (co-inhibitory receptor). It is homologous to the CD28 co-stimulatory receptor and competes with CD28 for binding to co-stimulatory ligands B7-1 (CD80) and B7-2 (CD86). CTLA-4 binds to B7-1 and B7-2 with a greater affinity and avidity over CD28, leading to the inactivation of TCR signaling [77, 79, 84–88]. CTLA-4 is constitutively expressed on the surface of naïve Treg cells [89–91]. However, it is upregulated on activated CD4+ and CD8+ T cells upon TCR signaling-mediated T-cell activation [92–94].

CTLA-4 is also expressed on B cells, monocytes, dendritic cells (DCs), granulocytes, and more [95–98]. In addition, CTLA-4 is capable of removing protein kinase C-theta and CARMA1 from the immune synapse [80] and limiting the time for T cells to stay in the synapse. Moreover, CTLA-4 triggers the transendocytosis of B7 [82] and promotes the function of Treg. Antibodies directed against CTLA-4 can block the interaction between CTLA-4 and CD80/CD86 and unleash the function of the immune system. Thus far, more than 240 clinical trials have been or are being conducted using CTLA-4 monoclonal antibody or combination therapy in various cancer types.

18.2.3.2 PD-1/PD-L1/PD-L2 Pathway

Programmed cell death 1 (PD-1, CD279) and its ligands PD-L1 (B7-H1, CD274) and PD-L2 (B7-DC, CD273) are components of co-inhibitory pathways in the immune system. PD-1 is expressed on T cells, B cells, NKs, NKTs, macrophages, and some DC subsets during immune activation and chronic inflammation through induction of cytokines (such as IL-2, IL-7, IL-15, IL-21, etc.) [57, 99–101]. Transcriptional regulation of PD-1 expression in T cells is controlled by transcription factors NFATc1, FoxO1, Notch, IRF9, and T-bet [102–105]. PD-L1 and PD-L2 are expressed by different cells. PD-1-deficient mice develop spontaneous autoimmune diseases such as dilated cardiomyopathy [106] and lupus-like autoimmune diseases [107]. These phenotypes suggest that PD-1, by serving as a negative immune modulator, plays a critical role in the maintenance of peripheral self-tolerance. PD-L1 is expressed on T cells, B cells, dendritic cells (DC), macrophages, mesenchymal stem cells, and bone marrow-derived mast cells [108]. PD-L2 is expressed on DCs, macrophages, bone marrow-derived mast cells, and peritoneal B1 cells [109]. Once bound by its ligand PD-L1 or PD-L2, PD-1 prevents T-cell activation and triggers immune tolerance of tumor cells in peripheral tissues. In addition, the interaction downregulates expression levels of certain transcription factors in effector cells such as T-bet, GATA-3, and Eomes [110]. The cytoplasmic tail of the PD-1 molecule is followed by two structural motifs: an immunoreceptor tyrosine-based inhibition motif (ITIM) and an immunoreceptor tyrosine-based switch motif (ITSM). The two tyrosine residues within ITIM or ITSM motif are phosphorylated by Src family protein tyrosine kinases (PTKs) upon TCR stimulation, which in turn activate and recruit cytoplasmic tyrosine phosphatases Src homology 2 (SH2) domain-containing PTPs, SHP-1 and SHP-2. The activation of phosphatases following binding to the two phosphorylated tyrosine kinases results in the phosphorylation and recruitment of signaling molecules involved in the downstream signaling pathways.

The ITSM motif plays a crucial role in the inhibitory function mediated by PD-1. Cancer cells expressing high levels of PD-L1 take advantage of this PD-1/PD-L1 signaling pathway for immune escape by binding with PD-1-expressing T cells in the tumor microenvironment [111]. Thus, blockade of PD-1 signaling pathway could be an effective antitumor therapeutic strategy, which releases a brake in the immune system, restoring T-cell-mediated tumor attack in the tumor milieu [4, 5, 112, 113].

A phase II clinical trial was conducted by Salazar et al. [114] in 15 patients with metastatic breast cancer to investigate the objective response rates of toll-like receptor (TLR)-7 alone and in combination with chemotherapy. The findings showed that non-responder patients had increased levels of preexisting circulating PD-1 + CD4+ and PD-1 + CD8+ T cells compared to the responders, suggesting that T-cell exhaustion could be used as a biomarker for predicting the clinical response.

18.2.3.3 MSI-High and MMR-Deficient Statuses as the Biomarkers of PD-1/PD-L1 Blockade Immunotherapy in Breast Cancer

In May 2017, pembrolizumab received FDA approval for treatment of metastatic microsatellite instability-high (MSI-high) or mismatch repair-deficient (MMR-deficient) solid tumors that progressed following chemotherapy. This is the first time that FDA has approved a cancer therapy based on a tumor's genetic feature rather than the type of original tissue/location. The favorable clinical outcome of patients with MSI-high and MMR-deficient tumors is correlated with the high frequency of tumor-infiltrating lymphocytes in the tumor microenvironment with high immunogenicity, where the tumors are responsive to PD-1/PD-L1 immune blockade. The most common tumor type with MSI-high and MMR-deficient statuses is colorectal cancer with Lynch syndrome (hereditary nonpolyposis colorectal carcinoma, HNPCC). Other MSI-high and MMR-deficient tumors include endometrial and gastrointestinal tumors, breast cancer, prostate cancer, bladder cancer, and thyroid tumor, etc [115].

The risk of breast cancer is associated with genomic instability such as double-stranded DNA break (DSBR)-targeted damage. In addition to DSBR, mismatch repair (MMR) proteins are a genome surveillance complex associated with BRCA1 to detect and repair replication-associated DNA damage that has escaped the DNA polymerase proofreading mechanism [116]. These damages include point mutations caused by single-base mismatches and then the incorrect incorporation of a nucleotide and frame-shift mutations induced by errors in the number of bases incorporated at repetitive sequences, which result in insertions/deletions loops (IDLs). Such mutations tend to occur in regions containing microsatellites, simple repeat sequences scattered throughout the genome. Therefore, the deficiency of MMR can be identified through detection of alterations in the number of such repeats [117].

Another very important genomic aberration is MSI (microsatellite instability), which is an early event of carcinogenesis promoted by mutation phenotype through the event of genomic instability and further gains more essential mutations to enhance tumor progression [118]. Microsatellites are defined as repeat DNA sequences consisting of two to five base pairs, which usually occur 10–60 times and are scattered throughout coding and noncoding regions of the genome. MSI refers to the replicative error phenotype caused from MMR and is subdivided into three groups: MSI-High (MSI-H), MSI-Low (MSI-L), or Microsatellite Stable (MSS) [119] Defects in the DNA mismatch repair (MMR) pathway underlie the development of MSI in colorectal cancer (CRC). Currently, neoantigen-based vaccination is being studied in a clinical trial for Lynch syndrome and in a trial for sporadic MSI CRC of advanced stage [120]. MSI testing could have an expanded role as a tool in the armamentarium of precision medicine [121]. Studies have already shown increased levels of MSI within tumors, including breast cancer, compared to normal tissues derived from the same individual, implicating defective MMR with many tumor types including breast cancer [122, 123]. Significant activity associated with mismatch repair (MMR) deficiency has been observed in hypermutated, microsatellite unstable (MSI) metastatic colorectal cancer (CRC). The evidence supports the development of immune checkpoint inhibitors in this specific subgroup of CRC patients [124, 125].

The approval of PD-1/PD-L1 blockade for treatment of MSI-high and MMR-deficient tumors makes PD-1/PD-L1 blockade a potential standard therapy for MSI-high and MMR-deficient breast cancers. It is urgently important to develop novel predictive and prognostic biomarkers to select appropriate patients for PD-1/PD-L1 blockade therapy.

18.3 Clinical Trials of Immune Checkpoint Blockade in Breast Cancer

18.3.1 Antibody Blockade of CTLA-4

In the preclinical animal models, the ability of CTLA-4 blocking antibody to eliminate tumor with low immunogenicity has been demonstrated to promote T-cell stimulation, inducing potent antitumor immune responses that can recognize and eliminate tumors [126, 127].

Based on this preclinical rationale, two CTLA-4-specific blocking antibodies including ipilimumab and tremelimumab have been extensively investigated in clinical trials in solid tumors including breast cancer [3, 128–133]. Ipilimumab and tremelimumab are fully humanized monoclonal antibodies directed against CTLA-4. The isotypes of the two antibodies are IgG1 and IgG2, respectively. Both agents are shown to stimulate immune system by blocking the binding of CTLA-4 to B7-1 and B7-2 and induce T-cell-mediated tumor rejection.

In a phase I clinical trial [133] conducted in 26 patients with advanced estrogen positive breast cancer, antitumor activity of tremelimumab was evaluated in combination with exemestane, a potent steroidal aromatase inhibitor that impeded synthesis of estrogen. Tremelimumab (3–10 mg/kg) was administered every 90 days plus exemestane 25 mg orally daily. The study showed that 11 patients (42%) had stable disease following treatment. Combination therapy correlated with elevated peripheral CD4+ and CD8+ T cells expressing inducible co-stimulator (ICOS) and a marked increase in the ratio of ICOS+ T cells to FoxP3+ regulatory T cells [133], indicating elicited antitumor T-cell immunity. Minor immune-related adverse events were observed, including fatigue, pruritus, diarrhea, and constipation, suggesting that tremelimumab plus exemestane was safe and tolerable in patients with estrogen-sensitive advanced breast cancer.

In another clinical study, combination therapy of ipilimumab and cryoablation has been investigated in 19 patients with early-stage breast cancer [132]. Cryoablation, a process that freezes tumor and induces tumor cell lysis, is considered to physically disrupt tumor via augmenting antigen presentation. ICOS-overexpressing CD4+ and CD8+ T cell subpopulations with enhanced proliferation were observed in the peripheral blood among the cohort of combination therapy. Moreover, the results showed that the increased ratio of CD8+ T cells over Tregs in tumor-infiltrating lymphocytes and elevated Th1 cytokines were correlated with the combination treatment. Currently, there are five ongoing clinical trials of anti-CTLA-4 antibody for breast cancer. They are listed in Table 18.1.

Table 18.1 Ongoing clinical trials of anti-CTLA-4 antibody in breast cancer

Phase	CTLA-4 antibody	In combination with	BC subtype
Trial ID			
II	Tremelimumab	Durvalumab	HER2-BC
NCT02536794			
I	Tremelimumab	Durvalumab	HER2-BC
NCT03132467			
I	Ipilimumab	Nivolumab/entinostat	HER2-BC
NCT02453620			
Pilot	Ipilimumab	Nivolumab/ cryoablation	Early BC
NCT02833233			
I/II	Ipilimumab	BBI608	Malignancies
NCT02467361			

BC breast cancer

18.3.2 Antibody Blockade of PD-1/PD-L1 in Breast Cancer

Tumors derived from 20% TNBC patients overexpress PD-L1 [134], indicating PD-1/PD-L1 signaling as a promising therapeutic target in breast cancer.

Nivolumab is a human IgG4 monoclonal antibody that inhibits interaction between PD-1and PD-L1, thereby restoring antitumor immunity. This agent is currently under investigation in four ongoing clinical trials for patients with breast cancer (see Table 18.2). For example, one of the clinical trials is being conducted to evaluate the safety of combination cryoablation and nivolumab in patients with early-stage breast cancer (ClinicalTrials.gov Identifier: NCT02833233). In a phase II trial with triple negative breast cancer patients, nivolumab is being tested in combination with chemotherapy/radiation therapy. Triple negative breast cancer is highly heterogeneous and has a high relapse rate. Currently, no effective therapies targeting TNBC are available. Several lines of evidence indicate that chemotherapy stimulates the activation of the immune system against cancer [64, 135–141]. Thus, it is proposed that a synergistic efficacy could be obtained with combination of nivolumab and chemotherapy in women with TNBC (ClinicalTrials.gov Identifier: NCT02499367).

Pembrolizumab is another anti-PD-1 antibody that has received the FDA approval for treatment of melanoma and non-small cell lung carcinoma (NSCLC). It is a monoclonal IgG4 designed to block the interaction between PD-1 and PD-L1/PD-L2. In a phase Ib clinical trial KEYNOTE-012 study, safety and antitumor activity of pembrolizumab has been investigated in the patients with advanced triple negative breast cancer (ClinicalTrials.gov identifier: NCT01848834). The agent was i.v. administered (10 mg/kg) every 2 weeks to patients with advanced PD-L1-positive solid tumors including TNBC. The group of TNBC recruited 32 patients. Minor immune-related advert events were observed, such as nausea, myalgia, and fatigue. This clinical trial showed evidence of a tolerable safety profile and antitumor activity of pembrolizumab patients with advanced TNBC [142].

Numerous clinical trials with anti-PD-1 antibodies and PD-L1 antibodies for patients with breast cancer are currently under investigation, which are listed in Table 18.2 and Table 18.3, respectively. At the 2017 ASCO Annual Meeting, the phase II I-SPY 2 trial results were reported for the study of pembrolizumab (PD-1 monoclonal antibody) in combination with standard chemotherapy (paclitaxel followed by doxorubicin and cyclophosphamide) as a neoadjuvant treatment for patients with locally advanced triple negative breast cancer or hormone receptor-positive/HER2-negative breast cancer (ASCO 2017 Abstract 506) [143]. The results showed that the combination of pembrolizumab with standard chemotherapy increased the estimated pathologic complete response rate from 20% to 60% in patients with triple negative breast cancer and from 13% to 34% in patients with hormone receptor-positive/HER2-negative breast cancer compared to standard chemotherapy only. Overall, based on Bayesian predictive probability of success in a confirmatory phase III trial, pembrolizumab has yielded favorable outcomes in the I-SPY 2 trial for all signatures in which it has been tested (triple negative breast cancer, all HER2-negative, and hormone receptor-positive/HER2-negative).

18.4 Conclusions and Future Perspectives

Immune checkpoint blockade therapy is currently the major type in cancer immunotherapy and will potentially be a standard therapy in solid tumors including breast cancer. Breast cancer subtypes of triple negative breast cancer and HER2-negative breast cancer are highly immunogenic and thus suitable for immune checkpoint blockade therapy. The results obtained from clinical trials evaluating PD-1/PD-L1 blockade single agent and in combination with standard chemotherapy have shown very promising outcome in the TNBC and HER2-negative breast cancer.

Table 18.2 Ongoing clinical trials of anti-PD-1 antibody in breast cancer

Phase	PD-1 antibody	In combination with	BC subtype
Trial ID			
I	Humanized anti-PD-1 antibody		Solid tumors
NCT02838823			
I /II	PDR001		TNBC
NCT02404441			
I	JS001	Radiation therapy	TNBC
NCT03151447			
I	Pembrolizumab	Radiation therapy	BC
NCT02303366			
I/II	Anti-PD-1 antibody	Chemotherapy	Malignancies
NCT02961101			
I/II	Anti-PD-1 antibody	D-CIK immunotherapy	Solid tumors
NCT02886897			
I	Pembrolizumab		HER2+ BC
NCT02129556			
I	PDR001	FAZ053	TNBC
NCT02936102			
I/II	Intratumorally dosed INT230-6 (IT-01)		Solid tumors
NCT03058289			
I	PDR001	LCL161, everolimus or panobinostat	TNBC
NCT02890069			
II	Pembrolizumab	Chemotherapy	BC
NCT03139851			
I	PDR001	NIS793	Solid tumors
NCT02947165			
II	Pembrolizumab	Chemotherapy	HER2-BC
NCT02752685			
II	Pembrolizumab	Chemotherapy	TNBC
NCT02648477			
I	Pembrolizumab	JAK2 inhibition	TNBC
NCT03012230			
II	Pembrolizumab		TNBC
NCT02644369			
II	Pembrolizumab	Radiation therapy	TNBC
NCT03004183			
I	Nivolumab	Chemotherapy	BC
NCT02309177			
I	Nivolumab	Ipilimumab/entinostat	HER2-BC
NCT02453620			
Pilot	Nivolumab	Ipilimumab/cryoablation	Early BC
NCT02833233			
II	Pembrolizumab		TNBC
NCT02447003			
II	Nivolumab	Radiation therapy/chemotherapy	TNBC
NCT02499367			
I	Pembrolizumab		TNBC
NCT01848834			

TNBC triple negative breast cancer, *BC* breast cancer

Table 18.3 Ongoing clinical trials of anti-PD-L1 antibody in breast cancer

Phase	PD-L1 antibody	In combination with	BC subtype
Trial ID			
I	Durvalumab	Tremelimumab (CTLA-4 antibody)	HER2-BC
NCT03132467			
III	Avelumab		TNBC
NCT02926196			
I-III	Durvalumab	Chemotherapies	TNBC
NCT02489448			
I	Durvalumab	Taxane-anthracycline	TNBC
NCT02685059			
III	Atezolizumab	Nab-paclitaxel/ placebo	TNBC
NCT02425891			
I	Atezolizumab	Pertuzumab/trastuzumab	HER2+ and HER2-BC
NCT02605915			
III	Atezolizumab	Chemotherapy	TNBC
NCT03125902			
I	FAZ053	PDR001	TNBC
NCT02936102			
I/II	Durvalumab	Chemotherapy	TNBC
NCT02628132			
III	Durvalumab	Tremelimumab + poly ICLC	Solid tumors
NCT02643303			
I/II	Durvalumab	Olaparib	Solid tumors
NCT02734004			
III	Durvalumab	Bevacizumab	HER2-BC
NCT02802098			
II	Durvalumab	Tremelimumab	HER2-BC
NCT02536794			
II	Atezolizumab	Veliparib	TNBC
NCT02849496			
II	Durvalumab	Olaparib	TNBC
NCT02484404			
I	Atezolizumab	Carboplatin-cyclophosphamide	Solid tumors
NCT02914470			
IiIII	Durvalumab	Adjuvant PVX-410 vaccine	TNBC
NCT02826434			
II	Atezolizumab	Paclitaxel, pertuzumab/trastuzumab	HER2 + BC
NCT03125928			
I	Avelumab		Solid tumors
NCT01772004			
I/II	Atezolizumab	Entinostat/placebo	TNBC
NCT02708680			
II	Avelumab	Utomilumab/PF-04518600/PD 0360324	Solid tumors
NCT02554812			
II	Durvalumab	Vigil	BC subtype

(continued)

Table 18.3 (continued)

Phase	PD-L1 antibody	In combination with	BC subtype
NCT02725489			
II	Atezolizumab	Chemotherapy	TNBC
NCT03164993			
II	Durvalumab		Solid tumors
NCT02669914			
I	Atezolizumab	CPI-444	TNBC
NCT02655822			

TNBC triple negative breast cancer, *BC* breast cancer

The PD-1 blocking antibody pembrolizumab has been recently approved by the FDA for treatment of MSI-high and MMR-deficient solid tumors, rendering it as the standard therapy for this subtype of breast cancer. The future directions are geared toward identifying reliable biomarkers of immune checkpoint blockade in breast cancer and combining immune checkpoint modulators with more traditional therapies such as standard chemotherapy and radiation therapy to achieve optimal clinical outcomes.

References

1. Newman LA (2009) Epidemiology of locally advanced breast cancer. Semin Radiat Oncol 19(4):195–203. doi:10.1016/j.semradonc.2009.05.003
2. Ribas A, Puzanov I, Dummer R, Schadendorf D, Hamid O, Robert C, Hodi FS, Schachter J, Pavlick AC, Lewis KD, Cranmer LD, Blank CU, O'Day SJ, Ascierto PA, Salama AK, Margolin KA, Loquai C, Eigentler TK, Gangadhar TC, Carlino MS, Agarwala SS, Moschos SJ, Sosman JA, Goldinger SM, Shapira-Frommer R, Gonzalez R, Kirkwood JM, Wolchok JD, Eggermont A, Li XN, Zhou W, Zernhelt AM, Lis J, Ebbinghaus S, Kang SP, Daud A (2015) Pembrolizumab versus investigator-choice chemotherapy for ipilimumab-refractory melanoma (KEYNOTE-002): a randomised, controlled, phase 2 trial. Lancet Oncol 16(8):908–918. doi:10.1016/S1470-2045(15)00083-2
3. Hodi FS, O'Day SJ, McDermott DF, Weber RW, Sosman JA, Haanen JB, Gonzalez R, Robert C, Schadendorf D, Hassel JC, Akerley W, van den Eertwegh AJ, Lutzky J, Lorigan P, Vaubel JM, Linette GP, Hogg D, Ottensmeier CH, Lebbe C, Peschel C, Quirt I, Clark JI, Wolchok JD, Weber JS, Tian J, Yellin MJ, Nichol GM, Hoos A, Urba WJ (2010) Improved survival with ipilimumab in patients with metastatic melanoma. N Engl J Med 363(8):711–723. doi:10.1056/NEJMoa1003466
4. Topalian SL, Hodi FS, Brahmer JR, Gettinger SN, Smith DC, McDermott DF, Powderly JD, Carvajal RD, Sosman JA, Atkins MB, Leming PD, Spigel DR, Antonia SJ, Horn L, Drake CG, Pardoll DM, Chen L, Sharfman WH, Anders RA, Taube JM, McMiller TL, Xu H, Korman AJ, Jure-Kunkel M, Agrawal S, McDonald D, Kollia GD, Gupta A, Wigginton JM, Sznol M (2012) Safety, activity, and immune correlates of anti-PD-1 antibody in cancer. N Engl J Med 366(26):2443–2454. doi:10.1056/NEJMoa1200690
5. Topalian SL, Drake CG, Pardoll DM (2012) Targeting the PD-1/B7-H1(PD-L1) pathway to activate anti-tumor immunity. Curr Opin Immunol 24(2):207–212. doi:10.1016/j.coi.2011.12.009
6. Reck M, Rodriguez-Abreu D, Robinson AG, Hui R, Csoszi T, Fulop A, Gottfried M, Peled N, Tafreshi A, Cuffe S, O'Brien M, Rao S, Hotta K, Leiby MA, Lubiniecki GM, Shentu Y, Rangwala R, Brahmer JR, Investigators K (2016) Pembrolizumab versus chemotherapy for PD-L1-positive non-small-cell lung cancer. N Engl J Med 375(19):1823–1833. doi:10.1056/NEJMoa1606774
7. Krishnamurthy A, Jimeno A (2017) Atezolizumab: a novel PD-L1 inhibitor in cancer therapy with a focus in bladder and non-small cell lung cancers. Drugs Today (Barc) 53(4):217–237. doi:10.1358/dot.2017.53.4.2589163
8. Farina MS, Lundgren KT, Bellmunt J (2017) Immunotherapy in urothelial cancer: recent results and future perspectives. Drugs. doi:10.1007/s40265-017-0748-7
9. Sharma P, Allison JP (2015) Immune checkpoint targeting in cancer therapy: toward combination strategies with curative potential. Cell 161(2):205–214. doi:10.1016/j.cell.2015.03.030
10. Robert C, Thomas L, Bondarenko I, O'Day S, Weber J, Garbe C, Lebbe C, Baurain JF, Testori A, Grob JJ, Davidson N, Richards J, Maio M, Hauschild A, Miller WH Jr, Gascon P, Lotem M, Harmankaya K, Ibrahim R, Francis S, Chen TT, Humphrey R, Hoos A, Wolchok JD (2011) Ipilimumab plus dacarbazine for previously untreated metastatic melanoma. N Engl J Med 364(26):2517–2526. doi:10.1056/NEJMoa1104621
11. Robert C, Long GV, Brady B, Dutriaux C, Maio M, Mortier L, Hassel JC, Rutkowski P, McNeil C,

Kalinka-Warzocha E, Savage KJ, Hernberg MM, Lebbe C, Charles J, Mihalcioiu C, Chiarion-Sileni V, Mauch C, Cognetti F, Arance A, Schmidt H, Schadendorf D, Gogas H, Lundgren-Eriksson L, Horak C, Sharkey B, Waxman IM, Atkinson V, Ascierto PA (2015) Nivolumab in previously untreated melanoma without BRAF mutation. N Engl J Med 372(4):320–330. doi:10.1056/NEJMoa1412082
12. June CH (2007) Adoptive T cell therapy for cancer in the clinic. J Clin Invest 117(6):1466–1476. doi:10.1172/JCI32446
13. June CH (2007) Principles of adoptive T cell cancer therapy. J Clin Invest 117(5):1204–1212. doi:10.1172/JCI31446
14. Kalos M, Levine BL, Porter DL, Katz S, Grupp SA, Bagg A, June CH (2011) T cells with chimeric antigen receptors have potent antitumor effects and can establish memory in patients with advanced leukemia. Sci Transl med 3 (95):95ra73. doi:10.1126/scitranslmed.3002842
15. Kershaw MH, Westwood JA, Darcy PK (2013) Gene-engineered T cells for cancer therapy. Nat Rev Cancer 13(8):525–541. doi:10.1038/nrc3565
16. Hay KA, Turtle CJ (2017) Chimeric antigen receptor (CAR) T cells: lessons learned from targeting of CD19 in B-cell malignancies. Drugs 77(3):237–245. doi:10.1007/s40265-017-0690-8
17. Till BG, Jensen MC, Wang J, Chen EY, Wood BL, Greisman HA, Qian X, James SE, Raubitschek A, Forman SJ, Gopal AK, Pagel JM, Lindgren CG, Greenberg PD, Riddell SR, Press OW (2008) Adoptive immunotherapy for indolent non-Hodgkin lymphoma and mantle cell lymphoma using genetically modified autologous CD20-specific T cells. Blood 112(6):2261–2271. doi:10.1182/blood-2007-12-128843
18. Savoldo B, Ramos CA, Liu E, Mims MP, Keating MJ, Carrum G, Kamble RT, Bollard CM, Gee AP, Mei Z, Liu H, Grilley B, Rooney CM, Heslop HE, Brenner MK, Dotti G (2011) CD28 costimulation improves expansion and persistence of chimeric antigen receptor-modified T cells in lymphoma patients. J Clin Invest 121(5):1822–1826. doi:10.1172/JCI46110
19. Grupp SA, Kalos M, Barrett D, Aplenc R, Porter DL, Rheingold SR, Teachey DT, Chew A, Hauck B, Wright JF, Milone MC, Levine BL, June CH (2013) Chimeric antigen receptor-modified T cells for acute lymphoid leukemia. N Engl J Med 368(16):1509–1518. doi:10.1056/NEJMoa1215134
20. Porter DL, Hwang WT, Frey NV, Lacey SF, Shaw PA, Loren AW, Bagg A, Marcucci KT, Shen A, Gonzalez V, Ambrose D, Grupp SA, Chew A, Zheng Z, Milone MC, Levine BL, Melenhorst JJ, June CH (2015) Chimeric antigen receptor T cells persist and induce sustained remissions in relapsed refractory chronic lymphocytic leukemia. Sci Transl med 7 (303):303ra139. doi:10.1126/scitranslmed.aac5415
21. Turtle CJ, Hanafi LA, Berger C, Hudecek M, Pender B, Robinson E, Hawkins R, Chaney C, Cherian S, Chen X, Soma L, Wood B, Li D, Heimfeld S, Riddell SR, Maloney DG (2016) Immunotherapy of non-Hodgkin's lymphoma with a defined ratio of CD8+ and CD4+ CD19-specific chimeric antigen receptor-modified T cells. Sci Transl med 8 (355):355ra116. doi:10.1126/scitranslmed.aaf8621
22. Maude SL, Frey N, Shaw PA, Aplenc R, Barrett DM, Bunin NJ, Chew A, Gonzalez VE, Zheng Z, Lacey SF, Mahnke YD, Melenhorst JJ, Rheingold SR, Shen A, Teachey DT, Levine BL, June CH, Porter DL, Grupp SA (2014) Chimeric antigen receptor T cells for sustained remissions in leukemia. N Engl J Med 371(16):1507–1517. doi:10.1056/NEJMoa1407222
23. Till BG, Jensen MC, Wang J, Qian X, Gopal AK, Maloney DG, Lindgren CG, Lin Y, Pagel JM, Budde LE, Raubitschek A, Forman SJ, Greenberg PD, Riddell SR, Press OW (2012) CD20-specific adoptive immunotherapy for lymphoma using a chimeric antigen receptor with both CD28 and 4-1BB domains: pilot clinical trial results. Blood 119(17):3940–3950. doi:10.1182/blood-2011-10-387969
24. Chmielewski M, Abken H (2015) TRUCKs: the fourth generation of CARs. Expert Opin Biol Ther 15(8):1145–1154. doi:10.1517/14712598.2015.1046430
25. Kalos M, June CH (2013) Adoptive T cell transfer for cancer immunotherapy in the era of synthetic biology. Immunity 39(1):49–60. doi:10.1016/j.immuni.2013.07.002
26. Ruella M, June CH (2016) Chimeric antigen receptor T cells for B cell neoplasms: choose the right CAR for you. Curr Hematol Malig Rep 11(5):368–384. doi:10.1007/s11899-016-0336-z
27. Ruella M, Barrett DM, Kenderian SS, Shestova O, Hofmann TJ, Perazzelli J, Klichinsky M, Aikawa V, Nazimuddin F, Kozlowski M, Scholler J, Lacey SF, Melenhorst JJ, Morrissette JJ, Christian DA, Hunter CA, Kalos M, Porter DL, June CH, Grupp SA, Gill S (2016) Dual CD19 and CD123 targeting prevents antigen-loss relapses after CD19-directed immunotherapies. J Clin Invest 126(10):3814–3826. doi:10.1172/JCI87366
28. Ruella M, Kenderian SS, Shestova O, Fraietta JA, Qayyum S, Zhang Q, Maus MV, Liu X, Nunez-Cruz S, Klichinsky M, Kawalekar OU, Milone M, Lacey SF, Mato A, Schuster SJ, Kalos M, June CH, Gill S, Wasik MA (2016) The addition of the BTK inhibitor Ibrutinib to anti-CD19 chimeric antigen receptor T cells (CART19) improves responses against mantle cell lymphoma. Clin Cancer Res 22(11):2684–2696. doi:10.1158/1078-0432.CCR-15-1527
29. Lee DW, Kochenderfer JN, Stetler-Stevenson M, Cui YK, Delbrook C, Feldman SA, Fry TJ, Orentas R, Sabatino M, Shah NN, Steinberg SM, Stroncek D, Tschernia N, Yuan C, Zhang H, Zhang L, Rosenberg SA, Wayne AS, Mackall CL (2015) T cells expressing CD19 chimeric antigen receptors for acute lym-

phoblastic leukaemia in children and young adults: a phase 1 dose-escalation trial. Lancet 385(9967):517–528. doi:10.1016/S0140-6736(14)61403-3
30. Greten TF, Jaffee EM (1999) Cancer vaccines. J Clin Oncol 17(3):1047–1060. doi:10.1200/JCO.1999.17.3.1047
31. Schreiber RD, Old LJ, Smyth MJ (2011) Cancer immunoediting: integrating immunity's roles in cancer suppression and promotion. Science 331(6024):1565–1570. doi:10.1126/science.1203486
32. Dunn GP, Bruce AT, Ikeda H, Old LJ, Schreiber RD (2002) Cancer immunoediting: from immunosurveillance to tumor escape. Nat Immunol 3(11):991–998. doi:10.1038/ni1102-991
33. Dunn GP, Old LJ, Schreiber RD (2004) The three Es of cancer immunoediting. Annu Rev Immunol 22:329–360. doi:10.1146/annurev.immunol.22.012703.104803
34. Dunn GP, Old LJ, Schreiber RD (2004) The immunobiology of cancer immunosurveillance and immunoediting. Immunity 21(2):137–148. doi:10.1016/j.immuni.2004.07.017
35. Koebel CM, Vermi W, Swann JB, Zerafa N, Rodig SJ, Old LJ, Smyth MJ, Schreiber RD (2007) Adaptive immunity maintains occult cancer in an equilibrium state. Nature 450(7171):903–907. doi:10.1038/nature06309
36. Shankaran V, Ikeda H, Bruce AT, White JM, Swanson PE, Old LJ, Schreiber RD (2001) IFNgamma and lymphocytes prevent primary tumour development and shape tumour immunogenicity. Nature 410(6832):1107–1111. doi:10.1038/35074122
37. Engel AM, Svane IM, Rygaard J, Werdelin O (1997) MCA sarcomas induced in scid mice are more immunogenic than MCA sarcomas induced in congenic, immunocompetent mice. Scand J Immunol 45(5):463–470
38. Svane IM, Engel AM, Nielsen MB, Ljunggren HG, Rygaard J, Werdelin O (1996) Chemically induced sarcomas from nude mice are more immunogenic than similar sarcomas from congenic normal mice. Eur J Immunol 26(8):1844–1850. doi:10.1002/eji.1830260827
39. Vesely MD, Kershaw MH, Schreiber RD, Smyth MJ (2011) Natural innate and adaptive immunity to cancer. Annu Rev Immunol 29:235–271. doi:10.1146/annurev-immunol-031210-101324
40. Smyth MJ, Dunn GP, Schreiber RD (2006) Cancer immunosurveillance and immunoediting: the roles of immunity in suppressing tumor development and shaping tumor immunogenicity. Adv Immunol 90:1–50. doi:10.1016/S0065-2776(06)90001-7
41. Zitvogel L, Tesniere A, Kroemer G (2006) Cancer despite immunosurveillance: immunoselection and immunosubversion. Nat Rev Immunol 6(10):715–727. doi:10.1038/nri1936
42. Khong HT, Restifo NP (2002) Natural selection of tumor variants in the generation of "tumor escape" phenotypes. Nat Immunol 3(11):999–1005. doi:10.1038/ni1102-999
43. Zajac AJ, Blattman JN, Murali-Krishna K, Sourdive DJ, Suresh M, Altman JD, Ahmed R (1998) Viral immune evasion due to persistence of activated T cells without effector function. J Exp Med 188(12):2205–2213
44. Gallimore A, Glithero A, Godkin A, Tissot AC, Pluckthun A, Elliott T, Hengartner H, Zinkernagel R (1998) Induction and exhaustion of lymphocytic choriomeningitis virus-specific cytotoxic T lymphocytes visualized using soluble tetrameric major histocompatibility complex class I-peptide complexes. J Exp Med 187(9):1383–1393
45. Brooks DG, Teyton L, Oldstone MB, McGavern DB (2005) Intrinsic functional dysregulation of CD4 T cells occurs rapidly following persistent viral infection. J Virol 79(16):10514–10527. doi:10.1128/JVI.79.16.10514-10527.2005
46. Oxenius A, Zinkernagel RM, Hengartner H (1998) Comparison of activation versus induction of unresponsiveness of virus-specific CD4+ and CD8+ T cells upon acute versus persistent viral infection. Immunity 9(4):449–457
47. Kaufmann DE, Kavanagh DG, Pereyra F, Zaunders JJ, Mackey EW, Miura T, Palmer S, Brockman M, Rathod A, Piechocka-Trocha A, Baker B, Zhu B, Le Gall S, Waring MT, Ahern R, Moss K, Kelleher AD, Coffin JM, Freeman GJ, Rosenberg ES, Walker BD (2007) Upregulation of CTLA-4 by HIV-specific CD4+ T cells correlates with disease progression and defines a reversible immune dysfunction. Nat Immunol 8(11):1246–1254. doi:10.1038/ni1515
48. Urbani S, Amadei B, Fisicaro P, Tola D, Orlandini A, Sacchelli L, Mori C, Missale G, Ferrari C (2006) Outcome of acute hepatitis C is related to virus-specific CD4 function and maturation of antiviral memory CD8 responses. Hepatology 44(1):126–139. doi:10.1002/hep.21242
49. Kahan SM, Wherry EJ, Zajac AJ (2015) T cell exhaustion during persistent viral infections. Virology 479-480:180–193. doi:10.1016/j.virol.2014.12.033
50. Fuller MJ, Khanolkar A, Tebo AE, Zajac AJ (2004) Maintenance, loss, and resurgence of T cell responses during acute, protracted, and chronic viral infections. J Immunol 172(7):4204–4214
51. Wherry EJ, Blattman JN, Murali-Krishna K, van der Most R, Ahmed R (2003) Viral persistence alters CD8 T-cell immunodominance and tissue distribution and results in distinct stages of functional impairment. J Virol 77(8):4911–4927
52. Virgin HW, Wherry EJ, Ahmed R (2009) Redefining chronic viral infection. Cell 138(1):30–50. doi:10.1016/j.cell.2009.06.036
53. Wherry EJ, Ahmed R (2004) Memory CD8 T-cell differentiation during viral infection. J Virol 78(11):5535–5545. doi:10.1128/JVI.78.11.5535-5545.2004

54. Fuller MJ, Zajac AJ (2003) Ablation of CD8 and CD4 T cell responses by high viral loads. J Immunol 170(1):477–486
55. Agnellini P, Wolint P, Rehr M, Cahenzli J, Karrer U, Oxenius A (2007) Impaired NFAT nuclear translocation results in split exhaustion of virus-specific CD8+ T cell functions during chronic viral infection. Proc Natl Acad Sci U S A 104(11):4565–4570. doi:10.1073/pnas.0610335104
56. Mackerness KJ, Cox MA, Lilly LM, Weaver CT, Harrington LE, Zajac AJ (2010) Pronounced virus-dependent activation drives exhaustion but sustains IFN-gamma transcript levels. J Immunol 185(6):3643–3651. doi:10.4049/jimmunol.1000841
57. Petrovas C, Casazza JP, Brenchley JM, Price DA, Gostick E, Adams WC, Precopio ML, Schacker T, Roederer M, Douek DC, Koup RA (2006) PD-1 is a regulator of virus-specific CD8+ T cell survival in HIV infection. J Exp Med 203(10):2281–2292. doi:10.1084/jem.20061496
58. Trautmann L, Janbazian L, Chomont N, Said EA, Gimmig S, Bessette B, Boulassel MR, Delwart E, Sepulveda H, Balderas RS, Routy JP, Haddad EK, Sekaly RP (2006) Upregulation of PD-1 expression on HIV-specific CD8+ T cells leads to reversible immune dysfunction. Nat Med 12(10):1198–1202. doi:10.1038/nm1482
59. Day CL, Kaufmann DE, Kiepiela P, Brown JA, Moodley ES, Reddy S, Mackey EW, Miller JD, Leslie AJ, DePierres C, Mncube Z, Duraiswamy J, Zhu B, Eichbaum Q, Altfeld M, Wherry EJ, Coovadia HM, Goulder PJ, Klenerman P, Ahmed R, Freeman GJ, Walker BD (2006) PD-1 expression on HIV-specific T cells is associated with T-cell exhaustion and disease progression. Nature 443(7109):350–354. doi:10.1038/nature05115
60. Wherry EJ (2011) T cell exhaustion. Nat Immunol 12(6):492–499
61. Baitsch L, Fuertes-Marraco SA, Legat A, Meyer C, Speiser DE (2012) The three main stumbling blocks for anticancer T cells. Trends Immunol 33(7):364–372. doi:10.1016/j.it.2012.02.006
62. Schietinger A, Greenberg PD (2014) Tolerance and exhaustion: defining mechanisms of T cell dysfunction. Trends Immunol 35(2):51–60. doi:10.1016/j.it.2013.10.001
63. Pauken KE, Wherry EJ (2015) Overcoming T cell exhaustion in infection and cancer. Trends Immunol 36(4):265–276. doi:10.1016/j.it.2015.02.008
64. Mellman I, Coukos G, Dranoff G (2011) Cancer immunotherapy comes of age. Nature 480(7378):480–489. doi:10.1038/nature10673
65. Kim PS, Ahmed R (2010) Features of responding T cells in cancer and chronic infection. Curr Opin Immunol 22(2):223–230. doi:10.1016/j.coi.2010.02.005
66. Crespo J, Sun H, Welling TH, Tian Z, Zou W (2013) T cell anergy, exhaustion, senescence, and stemness in the tumor microenvironment. Curr Opin Immunol 25(2):214–221. doi:10.1016/j.coi.2012.12.003
67. Fourcade J, Kudela P, Sun Z, Shen H, Land SR, Lenzner D, Guillaume P, Luescher IF, Sander C, Ferrone S, Kirkwood JM, Zarour HM (2009) PD-1 is a regulator of NY-ESO-1-specific CD8+ T cell expansion in melanoma patients. J Immunol 182(9):5240–5249. doi:10.4049/jimmunol.0803245
68. Pardoll DM (2012) The blockade of immune checkpoints in cancer immunotherapy. Nat Rev Cancer 12(4):252–264. doi:10.1038/nrc3239
69. Nguyen LT, Ohashi PS (2015) Clinical blockade of PD1 and LAG3--potential mechanisms of action. Nat Rev Immunol 15(1):45–56. doi:10.1038/nri3790
70. Fourcade J, Sun Z, Benallaoua M, Guillaume P, Luescher IF, Sander C, Kirkwood JM, Kuchroo V, Zarour HM Upregulation of Tim-3 and PD-1 expression is associated with tumor antigen-specific CD8+ T cell dysfunction in melanoma patients. J Exp Med 207(10):2175–2186
71. Fourcade J, Sun Z, Pagliano O, Guillaume P, Luescher IF, Sander C, Kirkwood JM, Olive D, Kuchroo V, Zarour HM (2012) CD8(+) T cells specific for tumor antigens can be rendered dysfunctional by the tumor microenvironment through upregulation of the inhibitory receptors BTLA and PD-1. Cancer Res 72(4):887–896. doi:10.1158/0008-5472.CAN-11-2637
72. Sakuishi K, Apetoh L, Sullivan JM, Blazar BR, Kuchroo VK, Anderson AC Targeting Tim-3 and PD-1 pathways to reverse T cell exhaustion and restore anti-tumor immunity. J Exp Med 207(10):2187–2194
73. Jin HT, Anderson AC, Tan WG, West EE, Ha SJ, Araki K, Freeman GJ, Kuchroo VK, Ahmed R (2010) Cooperation of Tim-3 and PD-1 in CD8 T-cell exhaustion during chronic viral infection. Proc Natl Acad Sci U S A 107(33):14733–14738. doi:10.1073/pnas.1009731107
74. Kmieciak M, Worschech A, Nikizad H, Gowda M, Habibi M, Depcrynski A, Wang E, Godder K, Holt SE, Marincola FM, Manjili MH (2011) CD4+ T cells inhibit the neu-specific CD8+ T-cell exhaustion during the priming phase of immune responses against breast cancer. Breast Cancer Res Treat 126(2):385–394. doi:10.1007/s10549-010-0942-8
75. Sharpe AH, Abbas AK (2006) T-cell costimulation--biology, therapeutic potential, and challenges. N Engl J Med 355(10):973–975. doi:10.1056/NEJMp068087
76. Appleman LJ, Boussiotis VA (2003) T cell anergy and costimulation. Immunol Rev 192:161–180
77. Grosso JF, Jure-Kunkel MN (2013) CTLA-4 blockade in tumor models: an overview of preclinical and translational research. Cancer Immun 13:5
78. Egen JG, Kuhns MS, Allison JP (2002) CTLA-4: new insights into its biological function and use in tumor immunotherapy. Nat Immunol 3(7):611–618. doi:10.1038/ni0702-611
79. Collins AV, Brodie DW, Gilbert RJ, Iaboni A, Manso-Sancho R, Walse B, Stuart DI, van der Merwe PA, Davis SJ (2002) The interaction proper-

ties of costimulatory molecules revisited. Immunity 17(2):201–210

80. Yokosuka T, Kobayashi W, Takamatsu M, Sakata-Sogawa K, Zeng H, Hashimoto-Tane A, Yagita H, Tokunaga M, Saito T (2010) Spatiotemporal basis of CTLA-4 costimulatory molecule-mediated negative regulation of T cell activation. Immunity 33(3):326–339. doi:10.1016/j.immuni.2010.09.006
81. Schneider H, Valk E, da Rocha DS, Wei B, Rudd CE (2005) CTLA-4 up-regulation of lymphocyte function-associated antigen 1 adhesion and clustering as an alternate basis for coreceptor function. Proc Natl Acad Sci U S A 102(36):12861–12866. doi:10.1073/pnas.0505802102
82. Qureshi OS, Zheng Y, Nakamura K, Attridge K, Manzotti C, Schmidt EM, Baker J, Jeffery LE, Kaur S, Briggs Z, Hou TZ, Futter CE, Anderson G, Walker LS, Sansom DM (2011) Trans-endocytosis of CD80 and CD86: a molecular basis for the cell-extrinsic function of CTLA-4. Science 332(6029):600–603. doi:10.1126/science.1202947
83. Wing K, Onishi Y, Prieto-Martin P, Yamaguchi T, Miyara M, Fehervari Z, Nomura T, Sakaguchi S (2008) CTLA-4 control over Foxp3+ regulatory T cell function. Science 322(5899):271–275. doi:10.1126/science.1160062
84. Chuang E, Lee KM, Robbins MD, Duerr JM, Alegre ML, Hambor JE, Neveu MJ, Bluestone JA, Thompson CB (1999) Regulation of cytotoxic T lymphocyte-associated molecule-4 by Src kinases. J Immunol 162(3):1270–1277
85. Carreno BM, Bennett F, Chau TA, Ling V, Luxenberg D, Jussif J, Baroja ML, Madrenas J (2000) CTLA-4 (CD152) can inhibit T cell activation by two different mechanisms depending on its level of cell surface expression. J Immunol 165(3):1352–1356
86. Cinek T, Sadra A, Imboden JB (2000) Cutting edge: tyrosine-independent transmission of inhibitory signals by CTLA-4. J Immunol 164(1):5–8
87. van der Merwe PA, Bodian DL, Daenke S, Linsley P, Davis SJ (1997) CD80 (B7-1) binds both CD28 and CTLA-4 with a low affinity and very fast kinetics. J Exp Med 185(3):393–403
88. Acuto O, Michel F (2003) CD28-mediated co-stimulation: a quantitative support for TCR signalling. Nat Rev Immunol 3(12):939–951. doi:10.1038/nri1248
89. Alegre ML, Noel PJ, Eisfelder BJ, Chuang E, Clark MR, Reiner SL, Thompson CB (1996) Regulation of surface and intracellular expression of CTLA4 on mouse T cells. J Immunol 157(11):4762–4770
90. Harper K, Balzano C, Rouvier E, Mattei MG, Luciani MF, Golstein P (1991) CTLA-4 and CD28 activated lymphocyte molecules are closely related in both mouse and human as to sequence, message expression, gene structure, and chromosomal location. J Immunol 147(3):1037–1044
91. Lindsten T, Lee KP, Harris ES, Petryniak B, Craighead N, Reynolds PJ, Lombard DB, Freeman GJ, Nadler LM, Gray GS et al (1993) Characterization of CTLA-4 structure and expression on human T cells. J Immunol 151(7):3489–3499
92. Freeman GJ, Lombard DB, Gimmi CD, Brod SA, Lee K, Laning JC, Hafler DA, Dorf ME, Gray GS, Reiser H et al (1992) CTLA-4 and CD28 mRNA are coexpressed in most T cells after activation. Expression of CTLA-4 and CD28 mRNA does not correlate with the pattern of lymphokine production. J Immunol 149(12):3795–3801
93. Linsley PS, Greene JL, Tan P, Bradshaw J, Ledbetter JA, Anasetti C, Damle NK (1992) Coexpression and functional cooperation of CTLA-4 and CD28 on activated T lymphocytes. J Exp Med 176(6):1595–1604
94. Walunas TL, Bakker CY, Bluestone JA (1996) CTLA-4 ligation blocks CD28-dependent T cell activation. J Exp Med 183(6):2541–2550
95. Kaufman KA, Bowen JA, Tsai AF, Bluestone JA, Hunt JS, Ober C (1999) The CTLA-4 gene is expressed in placental fibroblasts. Mol Hum Reprod 5(1):84–87
96. Ling V, Munroe RC, Murphy EA, Gray GS (1998) Embryonic stem cells and embryoid bodies express lymphocyte costimulatory molecules. Exp Cell Res 241(1):55–65. doi:10.1006/excr.1998.4055
97. Pioli C, Gatta L, Ubaldi V, Doria G (2000) Inhibition of IgG1 and IgE production by stimulation of the B cell CTLA-4 receptor. J Immunol 165(10):5530–5536
98. Pistillo MP, Tazzari PL, Palmisano GL, Pierri I, Bolognesi A, Ferlito F, Capanni P, Polito L, Ratta M, Pileri S, Piccioli M, Basso G, Rissotto L, Conte R, Gobbi M, Stirpe F, Ferrara GB (2003) CTLA-4 is not restricted to the lymphoid cell lineage and can function as a target molecule for apoptosis induction of leukemic cells. Blood 101(1):202–209. doi:10.1182/blood-2002-06-1668
99. Nakamoto N, Kaplan DE, Coleclough J, Li Y, Valiga ME, Kaminski M, Shaked A, Olthoff K, Gostick E, Price DA, Freeman GJ, Wherry EJ, Chang KM (2008) Functional restoration of HCV-specific CD8 T cells by PD-1 blockade is defined by PD-1 expression and compartmentalization. Gastroenterology 134(7):1927–1937, 1937 e1921–e1922. doi:10.1053/j.gastro.2008.02.033
100. Liu Y, Yu Y, Yang S, Zeng B, Zhang Z, Jiao G, Zhang Y, Cai L, Yang R (2009) Regulation of arginase I activity and expression by both PD-1 and CTLA-4 on the myeloid-derived suppressor cells. Cancer Immunol Immunother 58(5):687–697. doi:10.1007/s00262-008-0591-5
101. Nishimura H, Agata Y, Kawasaki A, Sato M, Imamura S, Minato N, Yagita H, Nakano T, Honjo T (1996) Developmentally regulated expression of the PD-1 protein on the surface of double-negative (CD4-CD8-) thymocytes. Int Immunol 8(5):773–780
102. Oestreich KJ, Yoon H, Ahmed R, Boss JM (2008) NFATc1 regulates PD-1 expression upon T cell activation. J Immunol 181(7):4832–4839

103. Staron MM, Gray SM, Marshall HD, Parish IA, Chen JH, Perry CJ, Cui G, Li MO, Kaech SM (2014) The transcription factor FoxO1 sustains expression of the inhibitory receptor PD-1 and survival of antiviral CD8(+) T cells during chronic infection. Immunity 41(5):802–814. doi:10.1016/j.immuni.2014.10.013
104. Mathieu M, Cotta-Grand N, Daudelin JF, Thebault P, Labrecque N (2013) Notch signaling regulates PD-1 expression during CD8(+) T-cell activation. Immunol Cell Biol 91(1):82–88. doi:10.1038/icb.2012.53
105. Kao C, Oestreich KJ, Paley MA, Crawford A, Angelosanto JM, Ali MA, Intlekofer AM, Boss JM, Reiner SL, Weinmann AS, Wherry EJ (2011) Transcription factor T-bet represses expression of the inhibitory receptor PD-1 and sustains virus-specific CD8+ T cell responses during chronic infection. Nat Immunol 12(7):663–671. doi:10.1038/ni.2046
106. Nishimura H, Okazaki T, Tanaka Y, Nakatani K, Hara M, Matsumori A, Sasayama S, Mizoguchi A, Hiai H, Minato N, Honjo T (2001) Autoimmune dilated cardiomyopathy in PD-1 receptor-deficient mice. Science 291(5502):319–322. doi:10.1126/science.291.5502.319
107. Nishimura H, Nose M, Hiai H, Minato N, Honjo T (1999) Development of lupus-like autoimmune diseases by disruption of the PD-1 gene encoding an ITIM motif-carrying immunoreceptor. Immunity 11(2):141–151
108. Yamazaki T, Akiba H, Iwai H, Matsuda H, Aoki M, Tanno Y, Shin T, Tsuchiya H, Pardoll DM, Okumura K, Azuma M, Yagita H (2002) Expression of programmed death 1 ligands by murine T cells and APC. J Immunol 169(10):5538–5545
109. Zhong X, Tumang JR, Gao W, Bai C, Rothstein TL (2007) PD-L2 expression extends beyond dendritic cells/macrophages to B1 cells enriched for V(H)11/V(H)12 and phosphatidylcholine binding. Eur J Immunol 37(9):2405–2410. doi:10.1002/eji.200737461
110. Nurieva R, Thomas S, Nguyen T, Martin-Orozco N, Wang Y, Kaja MK, Yu XZ, Dong C (2006) T-cell tolerance or function is determined by combinatorial costimulatory signals. EMBO J 25(11):2623–2633. doi:10.1038/sj.emboj.7601146
111. Iwai Y, Ishida M, Tanaka Y, Okazaki T, Honjo T, Minato N (2002) Involvement of PD-L1 on tumor cells in the escape from host immune system and tumor immunotherapy by PD-L1 blockade. Proc Natl Acad Sci U S A 99(19):12293–12297. doi:10.1073/pnas.192461099
112. Kantarjian HM, DeAngelo DJ, Stelljes M, Martinelli G, Liedtke M, Stock W, Gokbuget N, O'Brien S, Wang K, Wang T, Paccagnella ML, Sleight B, Vandendries E, Advani AS (2016) Inotuzumab Ozogamicin versus standard therapy for acute lymphoblastic leukemia. N Engl J Med 375(8):740–753. doi:10.1056/NEJMoa1509277
113. Brahmer JR, Tykodi SS, Chow LQ, Hwu WJ, Topalian SL, Hwu P, Drake CG, Camacho LH, Kauh J, Odunsi K, Pitot HC, Hamid O, Bhatia S, Martins R, Eaton K, Chen S, Salay TM, Alaparthy S, Grosso JF, Korman AJ, Parker SM, Agrawal S, Goldberg SM, Pardoll DM, Gupta A, Wigginton JM (2012) Safety and activity of anti-PD-L1 antibody in patients with advanced cancer. N Engl J Med 366(26):2455–2465. doi:10.1056/NEJMoa1200694
114. Salazar LG, Lu H, Reichow JL, Childs JS, Coveler AL, Higgins DM, Waisman J, Allison KH, Dang Y, Disis ML (2017) Topical Imiquimod plus nab-paclitaxel for breast cancer cutaneous metastases: a phase 2 clinical trial. JAMA Oncol. doi:10.1001/jamaoncol.2016.6007
115. Le DT, Uram JN, Wang H, Bartlett BR, Kemberling H, Eyring AD, Skora AD, Luber BS, Azad NS, Laheru D, Biedrzycki B, Donehower RC, Zaheer A, Fisher GA, Crocenzi TS, Lee JJ, Duffy SM, Goldberg RM, de la Chapelle A, Koshiji M, Bhaijee F, Huebner T, Hruban RH, Wood LD, Cuka N, Pardoll DM, Papadopoulos N, Kinzler KW, Zhou S, Cornish TC, Taube JM, Anders RA, Eshleman JR, Vogelstein B, Diaz LA Jr (2015) PD-1 blockade in tumors with mismatch-repair deficiency. N Engl J Med 372(26):2509–2520. doi:10.1056/NEJMoa1500596
116. Wang Y, Cortez D, Yazdi P, Neff N, Elledge SJ, Qin J (2000) BASC, a super complex of BRCA1-associated proteins involved in the recognition and repair of aberrant DNA structures. Genes Dev 14(8):927–939
117. Parsons R, Li GM, Longley MJ, Fang WH, Papadopoulos N, Jen J, de la Chapelle A, Kinzler KW, Vogelstein B, Modrich P (1993) Hypermutability and mismatch repair deficiency in RER+ tumor cells. Cell 75(6):1227–1236
118. Loeb LA (1994) Microsatellite instability: marker of a mutator phenotype in cancer. Cancer Res 54(19):5059–5063
119. Devaud N, Gallinger S (2013) Chemotherapy of MMR-deficient colorectal cancer. Familial Cancer 12(2):301–306. doi:10.1007/s10689-013-9633-z
120. Westdorp H, Fennemann FL, Weren RD, Bisseling TM, Ligtenberg MJ, Figdor CG, Schreibelt G, Hoogerbrugge N, Wimmers F, de Vries IJ (2016) Opportunities for immunotherapy in microsatellite instable colorectal cancer. Cancer Immunol Immunother 65(10):1249–1259. doi:10.1007/s00262-016-1832-7
121. Dudley JC, Lin MT, Le DT, Eshleman JR (2016) Microsatellite instability as a biomarker for PD-1 blockade. Clin Cancer Res 22(4):813–820. doi:10.1158/1078-0432.CCR-15-1678
122. Yee CJ, Roodi N, Verrier CS, Parl FF (1994) Microsatellite instability and loss of heterozygosity in breast cancer. Cancer Res 54(7):1641–1644
123. Shaw JA, Walsh T, Chappell SA, Carey N, Johnson K, Walker RA (1996) Microsatellite instability in early sporadic breast cancer. Br J Cancer 73(11):1393–1397
124. Bourdais R, Rousseau B, Pujals A, Boussion H, Joly C, Guillemin A, Baumgaertner I, Neuzillet C, Tournigand C (2017) Polymerase proofreading

domain mutations: new opportunities for immunotherapy in hypermutated colorectal cancer beyond MMR deficiency. Crit Rev Oncol Hematol 113:242–248. doi:10.1016/j.critrevonc.2017.03.027
125. Bupathi M, Wu C (2016) Biomarkers for immune therapy in colorectal cancer: mismatch-repair deficiency and others. J Gastrointest Oncol 7(5):713–720. doi:10.21037/jgo.2016.07.03
126. Leach DR, Krummel MF, Allison JP (1996) Enhancement of antitumor immunity by CTLA-4 blockade. Science 271(5256):1734–1736
127. van Elsas A, Hurwitz AA, Allison JP (1999) Combination immunotherapy of B16 melanoma using anti-cytotoxic T lymphocyte-associated antigen 4 (CTLA-4) and granulocyte/macrophage colony-stimulating factor (GM-CSF)-producing vaccines induces rejection of subcutaneous and metastatic tumors accompanied by autoimmune depigmentation. J Exp Med 190(3):355–366
128. Hodi FS, Mihm MC, Soiffer RJ, Haluska FG, Butler M, Seiden MV, Davis T, Henry-Spires R, MacRae S, Willman A, Padera R, Jaklitsch MT, Shankar S, Chen TC, Korman A, Allison JP, Dranoff G (2003) Biologic activity of cytotoxic T lymphocyte-associated antigen 4 antibody blockade in previously vaccinated metastatic melanoma and ovarian carcinoma patients. Proc Natl Acad Sci U S A 100(8):4712–4717. doi:10.1073/pnas.0830997100
129. Hodi FS, Butler M, Oble DA, Seiden MV, Haluska FG, Kruse A, Macrae S, Nelson M, Canning C, Lowy I, Korman A, Lautz D, Russell S, Jaklitsch MT, Ramaiya N, Chen TC, Neuberg D, Allison JP, Mihm MC, Dranoff G (2008) Immunologic and clinical effects of antibody blockade of cytotoxic T lymphocyte-associated antigen 4 in previously vaccinated cancer patients. Proc Natl Acad Sci U S A 105(8):3005–3010. doi:10.1073/pnas.0712237105
130. Phan GQ, Yang JC, Sherry RM, Hwu P, Topalian SL, Schwartzentruber DJ, Restifo NP, Haworth LR, Seipp CA, Freezer LJ, Morton KE, Mavroukakis SA, Duray PH, Steinberg SM, Allison JP, Davis TA, Rosenberg SA (2003) Cancer regression and autoimmunity induced by cytotoxic T lymphocyte-associated antigen 4 blockade in patients with metastatic melanoma. Proc Natl Acad Sci U S A 100(14):8372–8377. doi:10.1073/pnas.1533209100
131. Ribas A (2008) Overcoming immunologic tolerance to melanoma: targeting CTLA-4 with tremelimumab (CP-675,206). Oncologist 13(Suppl 4):10–15. doi:10.1634/theoncologist.13-S4-10
132. McArthur HL, Diab A, Page DB, Yuan J, Solomon SB, Sacchini V, Comstock C, Durack JC, Maybody M, Sung J, Ginsberg A, Wong P, Barlas A, Dong Z, Zhao C, Blum B, Patil S, Neville D, Comen EA, Morris EA, Kotin A, Brogi E, Wen YH, Morrow M, Lacouture ME, Sharma P, Allison JP, Hudis CA, Wolchok JD, Norton L (2016) A pilot study of preoperative single-dose Ipilimumab and/or Cryoablation in women with early-stage breast cancer with comprehensive immune profiling. Clin Cancer Res 22(23):5729–5737. doi:10.1158/1078-0432.CCR-16-0190
133. Vonderheide RH, LoRusso PM, Khalil M, Gartner EM, Khaira D, Soulieres D, Dorazio P, Trosko JA, Ruter J, Mariani GL, Usari T, Domchek SM (2010) Tremelimumab in combination with exemestane in patients with advanced breast cancer and treatment-associated modulation of inducible costimulator expression on patient T cells. Clin Cancer Res 16(13):3485–3494. doi:10.1158/1078-0432.CCR-10-0505
134. Mittendorf EA, Philips AV, Meric-Bernstam F, Qiao N, Wu Y, Harrington S, Su X, Wang Y, Gonzalez-Angulo AM, Akcakanat A, Chawla A, Curran M, Hwu P, Sharma P, Litton JK, Molldrem JJ, Alatrash G (2014) PD-L1 expression in triple-negative breast cancer. Cancer Immunol Res 2(4):361–370. doi:10.1158/2326-6066.CIR-13-0127
135. Obeid M, Tesniere A, Ghiringhelli F, Fimia GM, Apetoh L, Perfettini JL, Castedo M, Mignot G, Panaretakis T, Casares N, Metivier D, Larochette N, van Endert P, Ciccosanti F, Piacentini M, Zitvogel L, Kroemer G (2007) Calreticulin exposure dictates the immunogenicity of cancer cell death. Nat Med 13(1):54–61. doi:10.1038/nm1523
136. Kroemer G, Galluzzi L, Kepp O, Zitvogel L (2013) Immunogenic cell death in cancer therapy. Annu Rev Immunol 31:51–72. doi:10.1146/annurev-immunol-032712-100008
137. Bracci L, Schiavoni G, Sistigu A, Belardelli F (2014) Immune-based mechanisms of cytotoxic chemotherapy: implications for the design of novel and rationale-based combined treatments against cancer. Cell Death Differ 21(1):15–25. doi:10.1038/cdd.2013.67
138. Apetoh L, Ghiringhelli F, Tesniere A, Obeid M, Ortiz C, Criollo A, Mignot G, Maiuri MC, Ullrich E, Saulnier P, Yang H, Amigorena S, Ryffel B, Barrat FJ, Saftig P, Levi F, Lidereau R, Nogues C, Mira JP, Chompret A, Joulin V, Clavel-Chapelon F, Bourhis J, Andre F, Delaloge S, Tursz T, Kroemer G, Zitvogel L (2007) Toll-like receptor 4-dependent contribution of the immune system to anticancer chemotherapy and radiotherapy. Nat Med 13(9):1050–1059. doi:10.1038/nm1622
139. Apetoh L, Obeid M, Tesniere A, Ghiringhelli F, Fimia GM, Piacentini M, Kroemer G, Zitvogel L (2007) Immunogenic chemotherapy: discovery of a critical protein through proteomic analyses of tumor cells. Cancer Genomics Proteomics 4(2):65–70
140. Obeid M, Panaretakis T, Tesniere A, Joza N, Tufi R, Apetoh L, Ghiringhelli F, Zitvogel L, Kroemer G (2007) Leveraging the immune system during chemotherapy: moving calreticulin to the cell surface converts apoptotic death from "silent" to immunogenic. Cancer Res 67(17):7941–7944. doi:10.1158/0008-5472.CAN-07-1622
141. Tesniere A, Panaretakis T, Kepp O, Apetoh L, Ghiringhelli F, Zitvogel L, Kroemer G (2008) Molecular characteristics of immunogenic

cancer cell death. Cell Death Differ 15(1):3–12. doi:10.1038/sj.cdd.4402269
142. Nanda R, Chow LQ, Dees EC, Berger R, Gupta S, Geva R, Pusztai L, Pathiraja K, Aktan G, Cheng JD, Karantza V, Buisseret L (2016) Pembrolizumab in patients with advanced triple-negative breast cancer: phase Ib KEYNOTE-012 study. J Clin Oncol 34(21):2460–2467. doi:10.1200/JCO.2015.64.8931
143. Rita Nanda MCL, Yau C, Asare S, Hylton N, Van't Veer L, Jane Perlmutter, Wallace AM, Chien AJ, Forero-Torres A, Ellis E, Han H, Clark AS, Albain KS, Boughey JC, Elias AD, Berry DA, Yee D, DeMichele A, Esserman L; I-SPY Network, The University of Chicago, Chicago, IL; The University of Texas MD Anderson Cancer Center, Houston, TX; Mayo Clinic, Rochester, MN; Masonic Cancer Center, University of Minnesota, Minneapolis, MN; Abramson Cancer Center, Philadelphia, PA; Buck Institute for Age Research, Novato, CA; Quantum Leap Health Care Collaborative, San Francisco, CA; UC San Francisco, San Francisco, CA; University of California, San Francisco, San Francisco, CA; Gemini Group, Ann Arbor, MI; University of California San Diego Moores Cancer Center, La Jolla, CA; University of Alabama at Birmingham, Birmingham, AL; Swedish Cancer Inst, Seattle, WA; H. Lee Moffitt Cancer Center & Research Institute, Tampa, FL; Hospital of the University of Pennsylvania, Philadelphia, PA; Loyola University Chicago Stritch School of Medicine, Cardinal Bernardin Cancer Center, Maywood, IL; University of Colorado Comprehensive Cancer Center, Aurora, CO (2017) Pembrolizumab plus standard neoadjuvant therapy for high-risk breast cancer (BC): results from I-SPY 2. Paper presented at the 2017 ASCO annual meeting, Chicago, June 5

Strategies and Progress of Endocrine Therapy for Patients with Metastatic Breast Cancer

19

Hope S. Rugo, Huiping Li, and Xinyu Gui

Abstract

Breast cancer is one of the most prevalent cancers and the leading causes of cancer mortality in women worldwide and in China. For hormone receptor-positive (HR+) breast cancer, accounting for approximately 60–80% of breast cancer, endocrine therapy (ET) is the primary treatment strategy. For patients with HR+ metastatic breast cancer (MBC), there are many endocrine-based treatment options that can improve long-term outcomes and optimize quality of life. With the emergence and availability of new and effective agents, the options for ET have expanded in the last two decades. Although hormone therapy has been a standard of care for many decades, treatment must be individualized based on tumor biology and extent of disease. For example, the patients with impending organ failure may be treated with induction chemotherapy to improve organ function, followed by ET. For the patients who develop metastatic disease while on adjuvant ET, particularly when associated with organ failure, or for those with low expression of hormone receptors or expression of HER2, chemotherapy again may be a preferred initial treatment. ET blocks estrogen-driven tumor growth through different mechanisms; however, HR+ MBC can be intrinsically resistant or may acquire resistance to the treatment. Several targeted agents have been approved to use in combination with ET to improve response and delay development of resistance.

H.S. Rugo (✉)
University of California San Francisco Helen Diller Family Comprehensive Cancer Center, San Francisco, CA, USA
e-mail: Hope.Rugo@ucsf.edu

H. Li • X. Gui
Key laboratory of Carcinogenesis and Translational Research (Ministry of Education), Department of Breast Oncology, Peking University, Cancer Hospital & Institute, Beijing 100142, China

E. Song, H. Hu (eds.), *Translational Research in Breast Cancer*, Advances in Experimental Medicine and Biology 1026, https://doi.org/10.1007/978-981-10-6020-5_19

Keywords
HR+ Breast cancer • Endocrine therapy • ER-expressing breast cancer

19.1 Common Agents for ET of Breast Cancer

The endocrine-based treatment options for patients with HR+ MBC include selective estrogen receptor modulators, aromatase inhibitors, and selective estrogen receptor degraders.

19.1.1 Selective Estrogen Receptor Modulators (SERMs)

SERMs bind competitively to the estrogen receptor (ER) and alter the biologic actions of the receptor complex through changing conformation. Tamoxifen was the first SERM in clinical practice and has been used extensively for treatment of patients with both early- and late-stage breast cancer for decades. A review of 86 clinical trials involving over 5000 patients with MBC treated with tamoxifen concluded an overall response rate (ORR) of 34% with an additional 19% of patients achieving stable disease for at least 6 months [1].

19.1.2 Aromatase Inhibitors (AIs)

AIs reduce estrogen levels by blocking the conversion of androgens to estrogens by the aromatase enzyme in tissues other than the ovaries. Third-generation AIs include letrozole and anastrozole (nonsteroidal AIs which bind reversibly to the enzyme) and exemestane (a steroidal AI which binds irreversibly to aromatase) [2, 3]. Switching between nonsteroidal and steroidal AIs produces modest additional clinical benefits (primarily stable disease), suggesting that the two types of AIs are not fully cross-resistant [4]. A large meta-analysis including 8504 patients compared the survival benefits with standard hormonal treatment and several generations of AIs in patients with MBC. Statistically significant survival benefits with third-generation AIs have been found in published data [5].

19.1.3 Selective Estrogen Receptor Degraders (SERDs)

Fulvestrant is the only US Food and Drug Administration (FDA)-approved SERD, which blocks ER dimerization and DNA binding, inhibits nuclear uptake, increases the turnover and degradation of ER, and decreases estrogen-independent signaling. Fulvestrant (250 mg monthly) given by intramuscular injection (IM) was at least as effective as anastrozole (1 mg daily) in the second-line treatment of postmenopausal women with ABC [6]. However, the time to steady-state levels of fulvestrant with 250 mg monthly dosing was at least 3 months. Subsequent studies employed a loading dose, using 500 mg every 2 weeks for three doses, followed by 500 mg a month. The CONFIRM study compared fulvestrant 500 mg to 250 mg (with a loading dose in both arms) in postmenopausal women with ER+ ABC and demonstrated a significant improvement in both PFS and OS with the 500 mg dose [7]. This study led to FDA approval of fulvestrant at 500 mg monthly with a loading dose. More recent data suggests that higher dose of fulvestrant improves disease control and has a survival advantage compared with anastrozole in a specific subpopulation of patients with advanced disease. The phase II FIRST trial compared anastrozole to 500 mg fulvestrant and demonstrated similar results for clinical benefit rate, which was the primary endpoint [8]. With longer follow-up, time to progression (TTP) and overall survival (OS) were improved with fulvestrant compared to anastrozole. Median OS was 54.1 months for fulvestrant versus 48.4 months for anastrozole (hazard ratio 0.70; 95% CI: 0.50–0.98; $p = 0.04$). The phase III FALCON study compared fulvestrant with anastrozole as the first ET treatment in patients with HR+ locally advanced or MBC who had not received previous hormonal therapy. Progression-free survival (PFS) was significantly longer in fulvestrant group than in anastrozole group (hazard ratio 0.797; 95% CI: 0.637–0.999;

$p = 0.0486$; median PFS, 16.6 versus 13.8 months) [9]. Subset analysis demonstrated that improved PFS with fulvestrant was seen only in patients without visceral disease (PFS 22.3 versus 13.8 months, HR 0.59, 95% CI 0.42, 0.84), compared to no difference in those with visceral disease (PFS 13.8 versus 15.9 months, HR 0.99, 95% CI 0.74, 1.33). To date, there is no difference in OS. These findings raise the question about differential efficacy with hormone therapies based on disease setting and exposure to prior hormone therapy, although 34% of patients in FALCON were exposed to prior chemotherapy. Fulvestrant monotherapy may be a preferred option for patients without prior ET therapy or visceral disease. The currently approved standard dose of fulvestrant is 500 mg, and it should be administered twice in the first month and then once a month after that [10].

19.2 Improving Response to ET with Targeted Therapies

Essentially all patients with MBC receiving ET will eventually experience disease progression. Although sequential lines of ET are employed with success in many cases, resistance may develop at any time [11]. Research focusing on explaining the mechanism behind resistance to ET has led to a rapidly expanding understanding of the genomic and biochemical pathways, including the phosphatidylinositol 3-kinase (PI3K)/Akt/mammalian target of rapamycin (mTOR) pathway [12] as well as others.

19.2.1 Everolimus

Everolimus is an orally selective inhibitor of mTOR complex one which interferes with activation of the PI3K/Akt/mTOR pathway and has been demonstrated to improve antitumor efficacy when combined with ET in patients with advanced breast cancer. BOLERO-2, a randomized phase III study, compared exemestane and everolimus to exemestane and placebo in 724 postmenopausal patients with HR+ ABC recurring or progressing while receiving a nonsteroidal AI for ABC, or within 12 months of completing adjuvant nonsteroidal AI treatment. The addition of everolimus improved median PFS from 3.2 months to 7.8 months (HR 0.45, 95% CI 0.38, 0.54, $p < 0.0001$) [13]. Based on this data, everolimus plus exemestane was approved by the FDA for treatment of patients with HR+/HER2- (human epidermal growth factor receptor 2-negative) ABC recurring or progressing on prior nonsteroidal AIs. Everolimus has shown benefits with additional combinations. The phase II TAMRAD study evaluated the hypothesis that combining an mTOR inhibitor and antiestrogen could reestablish endocrine sensitivity. Tamoxifen plus everolimus increased clinical benefit rate (CBR), TTP, and OS compared with tamoxifen alone in postmenopausal women with AI-resistant MBC [14]. PrECOG 0102 is a randomized phase II trial that evaluated fulvestrant plus everolimus or placebo in 129 postmenopausal women with HR+ ABC progressing on AI therapy [15]. Investigator-assessed PFS was improved with the addition of everolimus from 5.1 to 10.4 months (HR 0.6, 95% CI 0.40, 0.92, $p < 0.02$). In all trials, the addition of everolimus was associated with more toxicity than that seen with hormone therapy alone, including a higher proportion of treatment discontinuation and grade >3 adverse effects [13]. Class-specific toxicities include stomatitis, pneumonitis, and hyperglycemia. The incidence and severity of stomatitis are markedly reduced with the use of a dexamethasone mouthwash during the first 8 weeks of therapy [16].

19.2.2 Cyclin-Dependent Kinase (CDK)4/CDK6 Inhibitors

CDK4 and CDK6 are serine/threonine kinases that bind to partner cyclins and regulate G1-S phase transition when phosphorylated. The activation of CDK4/CDK6 by cyclin D induces the phosphorylation of retinoblastoma protein (Rb) and the progression of the cell cycle into S phase [17]. Blockade of Rb hyperphosphorylation by CDK4/CDK6 inhibitors leads to G1 arrest in luminal breast cancer cells [18]. Preclinical studies in well-defined cell lines found that luminal-type cell lines were the most sensitive to CDK4/CDK6 inhibition, whereas basal-like subtypes were resistant [18]. In addition, synergistic activity was demonstrated when a CDK4/CDK6 inhibitor was combined with tamoxifen [19]. These data set the scene for combination studies with CDK4/CDK6 inhibitors and endocrine therapy to improve response and delay resistance in hormone receptor-positive breast cancer.

Palbociclib is an oral, highly selective small-molecule inhibitor of CDK4/CDK6, which has been approved for the treatment of postmenopausal ER+/HER2- ABC in combination with a nonsteroidal AI or fulvestrant [20, 21]. The PALOMA-1 trial was a randomized phase II study that randomized 165 postmenopausal women with advanced HR+/HER2- tumors with no previous systemic treatment for ABC to receive a combination of palbociclib (125 mg by mouth daily for 21 days, followed by 1 week off, then repeated) and letrozole versus letrozole alone [22]. Combination therapy improved PFS from 10.2 to 20.2 months (hazard ratio 0.488; 95% CI: 0.319–0.748; $p = 0.0004$). The primary toxicity was neutropenia (all grade 73%, grade 3/4 54%) without an increase in febrile neutropenia. Other toxicities were primarily grade 1 and 2 and included alopecia 22%, fatigue 24%, and thrombocytopenia 16%. Overall survival was similar between the two arms, at 37.5 and 34.5 months for the combination versus single-agent arms, respectively, although the trial was not powered for this secondary endpoint [23].

Based on the encouraging PFS data from PALOMA-1 combined with a modest toxicity profile, the combination of palbociclib plus letrozole received accelerated approval from the US FDA as treatment for HR+ MBC in 2015 [24]. Two-phase III trials were subsequently conducted in the first- and later-line ABC settings. PALOMA-3 randomized 521 patients with HR+ MBC progressing on prior endocrine therapy in a 2:1 ratio to receive fulvestrant with palbociclib or placebo [21]. Eligibility included progression on or within 12 months of adjuvant therapy, on therapy for metastatic disease, and up to one prior chemotherapy regimen was allowed for advanced disease. Of the enrolled population, 21% were pre- or perimenopausal, about 60% had visceral disease, and between 31 and 36% had received one line of chemotherapy for advanced disease. The addition of palbociclib improved PFS (9.5 vs 4.6 months; HR 0·46, 95% CI 0·36–0·59, $p < 0{\cdot}0001$); the most common toxicity was neutropenia without an increase in febrile neutropenia. Neither the presence of PI3K mutations nor HR expression level impacted response, and the addition of palbociclib was effective regardless of visceral involvement [25].

PALOMA-2 randomized 666 postmenopausal women without prior exposure to endocrine therapy for HR+ MBC in a 2:1 ratio to receive letrozole with either palbociclib or placebo [26]. Again, the addition of palbociclib significantly improved PFS (24.8 vs 14.5 months; HR 0.58, 95% CI 0.46–0.72); $P < 0.000001$), without new toxicities compared to prior studies. Dose reductions for palbociclib occurred in 36% of patients without apparent impact on efficacy. Palbociclib is now approved in both the first- and later-line settings in combination with an AI or fulvestrant.

Ribociclib is the second oral CDK4/CDK6 inhibitor to be approved as first-line therapy for HR+ MBC in combination with an AI in postmenopausal women, based on data from the phase III MONALEESA-2 trial [27]. Updated data is now available for the 668 patients randomized in a 1:1 ratio to letrozole with either ribociclib (300 mg by mouth for 21 days followed by a 7-day break, then repeated) or placebo [28]. The addition of ribociclib to letrozole increased PFS (35.3 vs 16 months; HR 0.568, 95% CI 0.457–0.704, $p = 0.0000001$). Efficacy was seen in all subgroups, regardless of the presence of visceral metastases or prior adjuvant endocrine or chemotherapy. The most common toxicity was neutropenia (grade 3/4 in 59.6%) without an increase in the risk of febrile neutropenia, 51% of patients had at least one dose reduction of ribociclib, and 8% of patients discontinued therapy largely due to increase in liver enzymes or emesis. Unique toxicities of ribociclib include self-limited increase in transaminases (grade 3/4 in 9–10%) and QTcF interval prolongation >480 msec in 3.3%, with one on-treatment death attributed to hypokalemia with grade 2 QT prolongation. Interestingly, both palbociclib and ribociclib cause grade 1–2 alopecia in about a third of patients.

Abemaciclib is the third highly specific oral small-molecule CDK4/CDK6 inhibitor and was approved by the US FDA in September of 2017 in combination with fulvestrant as second or later-line therapy for hormone receptor positive advanced breast cancer based on the results of the MONARCH-2 trial, and as a single agent after progression on hormone and chemotherapy based on the results of MONARCH-1. Based on encouraging single-agent activity in the phase I setting, the phase II MONARCH 1 trial evaluated abemaciclib as monotherapy in patients with HR+ MBC who had received at least two prior lines of chemotherapy with at least one line occurring in the advanced setting [29]. Abemaciclib was given at a dose of 200 mg twice a day continuously to 132 patients, 90% of whom had visceral metastases, with a single-agent confirmed response rate of 19.7%, a median duration of response 8.6 months, a median PFS of 6.0 months, and a median OS of 22.3 months [29]. Grade 3/4 neutropenia occurred in only 26.9%, but diarrhea occurred in 90%, and grade 3/4 diarrhea was reported in 19.7%.

The phase III MONARCH-2 trial randomized 669 postmenopausal women in a 2:1 ratio to fulvestrant combined with either abemaciclib or placebo [30]. Abemaciclib was dose reduced from 200 mg twice daily to 150 mg twice daily after 179 patients were enrolled due to unacceptably high rates of diarrhea. Eligibility was different than that used for PALOMA-3; patients could not have received chemotherapy and could only have received one line of prior hormone therapy for MBC; 59–60% of patients had received endocrine therapy only in the early stage setting. The addition of abemaciclib to fulvestrant improved PFS (16.4 vs 9.3 months; HR 0.553, 95% CI 0.449–0.681, $p < 0.0000001$). Similar to the phase III studies with palbociclib and ribociclib, benefit was seen in all subgroups. Combining both doses of abemaciclib, grade 3 diarrhea was reported in 13.4% and grade 2 in 31.7%, whereas grade 3/4 neutropenia was seen in only 26.5%. Following the dose reduction, the incidence of diarrhea decreased, with grade 3 seen in 11.3%.

MONARCH-3 is a phase III trial that randomized 493 post-menopausal women with untreated hormone receptor positive metastatic breast cancer at least 12 months from last adjuvant endocrine therapy to receive abemaciclib (150 mg twice a day) with letrozole or placebo in a 2:1 ratio. At a median follow-up of 17.8 months, results from the interim analysis were presented at the European Society of Medical Oncology in September of 2017 [31]. The primary endpoint of PFS was met at the interim analysis with a median PFS for the placebo arm of 14.7 months, and a median PFS not reached in the abemaciclib arm (HR 0.543; 95% CI 0.409, 0.723, $p = 0.000021$). The ORR was also significantly increased. Exploratory analyses of PFS suggested that benefit from the addition of abemaciclib was limited to patients with a treatment free interval of less than 36 months, and that benefit was greater in patients without bone-only disease than in those with bone only disease. These analysis are limited by the short follow-up in good risk subsets, as the median PFS was not yet reached.

Toxicity was similar to MONARCH 2, with a lower rate of grade 3 diarrhea at 9.5%. Follow-up is ongoing.

Ongoing Trials

A number of CDK4/CDK6 inhibitor trials are ongoing in both metastatic and early-stage diseases. These trials are evaluating ribociclib with fulvestrant in the second- or greater-line setting (MONALEESA-3; data to be presented at San Antonio in December of 2017), ribociclib in premenopausal women with chemical ovarian suppression (MONALEESA-7), palbociclib and exemestane vs capecitabine (PEARL), and abemaciclib in HR+. All three agents are clinical trials combined with hormone therapy in the adjuvant setting, and there are a number of studies that have been presented or are soon to be reported in the neoadjuvant setting (for NeoMonarch, see section on immunotherapy). The abovementioned CDK4/CDK6 inhibitor trials are summarized in Table 19.1 with the exception of MONARCH 3, which was presented only when this chapter was in proof, with results summarized above.

Table 19.1 CDK 4/6 inhibitor trials

	Design	Patients	Results	Safety
PALOMA-1	*A randomized phase II study*	Randomized 165 postmenopausal women with advanced HR+/HER2- tumors with no previous systemic treatment for ABC to receive a combination of palbociclib (125 mg by mouth daily for 21 days, followed by one week off, then repeated) and letrozole versus letrozole alone.	Combination therapy improved PFS from 10.2 to 20.2 months (hazard ratio 0.488; 95% CI: 0.319–0.748; p=0.0004).	The primary toxicity was neutropenia (all grade 73%, grade 3/4 54%) without an increase in febrile neutropenia Other toxicities were primarily grade 1 and 2, and included alopecia 22%, fatigue 24%, thrombocytopenia 16%.
			Overall survival was similar between the two arms, at 37.5 and 34.5 months for the combination versus single agent arms, respectively, although the trial was not powered for this secondary endpoint.	
PALOMA-3	*A randomized phase III study*	Randomized 521 patients with HR+ MBC progressing on prior endocrine therapy in a 2:1 ratio to receive fulvestrant with palbociclib or placebo.	The addition of palbociclib improved PFS (9.5 vs 4.6 months; HR 0·46, 95% CI 0·36–0·59, p<0·0001).	The most common toxicity was neutropenia without an increase in febrile neutropenia.
	Eligibility included progression on or within 12 months of adjuvant therapy, on therapy for metastatic disease, and up to one prior chemotherapy regimen was allowed for advanced disease.	Of the enrolled population, 21% were pre- or perimenopausal, about 60% had visceral disease, and between 31 and 36% had received one line of chemotherapy for advanced disease.	Neither the presence of PI3K mutations, or HR expression level impacted response, and the addition of palbociclib was effective regardless of visceral involvement.	
PALOMA-2	*A randomized phase III study*	Randomized 666 postmenopausal women without prior exposure to endocrine therapy for HR+ MBC in a 2:1 ratio to receive letrozole with either palbociclib or placebo.	The addition of palbociclib significantly improved PFS (24.8 vs. 14.5 months; HR 0.58, 95% CI 0.46–0.72); P<0.000001).	No new toxicities compared to prior studies.
			Dose reductions for palbociclib occurred in 36% of patients without apparent impact on efficacy.	

MonaLEESA-2	*A randomized phase III study*	Updated data is now available for the 668 patients randomized 1:1 to letrozole with either ribociclib (300 mg by mouth for 21 days followed by a 7 day break, then repeated) or placebo.	The addition of ribociclib to letrozole increased PFS (35.3 vs 16 months; HR 0.568, 95% CI 0.457–0.704, p=0.0000001). Efficacy was seen in all subgroups, regardless of presence of visceral metastases or prior adjuvant endocrine or chemotherapy.	The most common toxicity was neutropenia (grade 3/4 in 59.6%) without an increase in the risk of febrile neutropenia, 51% of patients had at least one dose reduction of ribociclib, and 8% of patients discontinued therapy largely due to increase in liver enzymes or emesis. Unique toxicities of ribociclib include self-limited increase in transaminases (grade 3/4 in 9-10%) and QTcF interval prolongation > 480 msec in 3.3%, with one on-treatment death attributed to hypokalemia with grade 2 QT prolongation. Interestingly, both palbociclib and ribociclib cause grade 1–2 alopecia in about a third of patients.
MONARCH-1	*A randomized phase II study* Evaluated abemaciclib as monotherapy in patients with HR+ MBC who had received at least two prior lines of chemotherapy with at least one line occurring in the advanced setting.	Abemaciclib was given at a dose of 200 mg twice a day continuously to 132 patients, 90% of whom had visceral metastases.	With a single agent confirmed response rate of 19.7%, a median duration of response 8.6 months, median PFS 6.0 months and median OS 22.3 months.	Grade ¾ neutropenia occurred in only 26.9%, but diarrhea occurred in 90%, and grade 3/4 diarrhea was reported in 19.7%.
MONARCH-2	*A randomized phase III study* Eligibility was different than that used for PALOMA-3; patients could not have received chemotherapy and could only have received one line of prior hormone therapy for MBC; 59–60% of patients had received endocrine therapy only in the early stage setting.	Randomized 669 postmenopausal women in a 2:1 ratio to fulvestrant combined with either abemaciclib or placebo. Abemaciclib was dose reduced from 200 mg twice daily to 150 mg twice daily after 179 patients were enrolled due to unacceptably high rates of diarrhea.	The addition of abemaciclib to fulvestrant improved PFS (16.4 vs. 9.3 months; HR 0.553, 95% CI 0.449–0.681, p<.0000001). Similar to the phase III studies with palbociclib and ribociclib, benefit was seen in all subgroups.	Combining both doses of abemaciclib, grade 3 diarrhea was reported in 13.4%, and grade 2 in 31.7%, whereas grade 3/4 neutropenia was seen in only 26.5%. Following the dose reduction, the incidence of diarrhea decreased, with grade 3 seen in 11.3%.

(continued)

Table 19.1 (continued)

	Design	Patients	Results	Safety
MONARCH 3	*An ongoing phase III trial*	See text for summary of results presented at ESMO in September of 2017.		
	Compared letrozole plus abemaciclib versus letrozole plus placebo in postmenopausal women with untreated HR+ MBC.			
MonaLEESA-3	*An ongoing phase III trial*	Investigates the combination of ribociclib with fulvestrant as first- or second-line treatment.	Primary outcome measures:	
		Enrollment: 725	Progression free survival (PFS)	
			Secondary outcome measures:	
			Overall survival (OS)	
			Progression free survival (PFS) per blinded independent review committee (BICR)	
			Overall response rate (ORR)	
			Time to definitive deterioration of ECOG performance status in one category of the score	
			Safety	
			Time to definitive 10% deterioration in the global health status/quality of life (QOL) scale score of the EORTC QLQ-C30	
			Change from baseline in the global health status/QoL scale score of the EORTC QLQ-C30	
			Clinical benefit ratio (CBR)	
			Time to response (TTR)	
			Duration of response (DOR)	

MonaLEESA-7	*An ongoing phase III trial*	First-line patients are being randomized to receive ribociclib or placebo either in combination with tamoxifen plus goserelin or an aromatase-inhibitor plus goserelin.	Primary outcome measures:
	Study of efficacy and safety in premenopausal women with HR+/HER2- ABC.	Enrollment: 672	Progression free survival (PFS)
			Secondary outcome measures:
	The only trial evaluating a CDK 4/6 inhibitor exclusively in premenopausal women.		Overall survival (OS)
			Clinical benefit rate (CBR)
			Safety and tolerability of LEE011
			Time to response (TTR)
			Duration of response (DOR)
			Time to definitive deterioration of the ECOG PS from baseline
			Time to 10% deterioration in the global health status/QOL scale score of the EORTC QLQ-C30
			Change from baseline in the global health status/QOL scale score of the EORTC QLQ-C30
			Overall Response Rate (ORR)
PEARL study	*An ongoing phase III trial*	Patients will be randomized 1:1 to endocrine therapy (cohort 1: exemestane 25 mg daily, cohort 2: fulvestrant 500 mg days 1 and 15 cycle 1 and then day 1 every 4 weeks) plus palbociclib (125 mg daily x 3 weeks every 4 weeks) vs. capecitabine (1,250 mg/m2 twice daily x 2 weeks every 3 weeks).	Primary outcome measures:
	Study of palbociclib in combination with endocrine therapy (exemestane or fulvestrant) versus chemotherapy (capecitabine) in HR+/HER2-MBC patients with resistance to nonsteroidal aromatase inhibitors.		Progression-free survival (PFS)
			Secondary outcome measures:
			Objective response
			Clinical benefit
			Response duration
			Overall survival (OS)
			Safety

19.2.3 Vascular Endothelial Growth Factor (VEGF) Inhibitor

The VEGF inhibitor bevacizumab has been evaluated in combination with endocrine agents in several clinical trials, based on data that high VEGF levels in breast cancer are associated with a decreased response to ET and worse outcome [32]. A recent systematic review of 14 phase III trials evaluating bevacizumab including over 4400 MBC patients showed reduced relapse rate (RR) and improvement in PFS, without OS benefit [33]. The phase III randomized LEA trial evaluated ET (letrozole or fulvestrant) with bevacizumab vs AI alone as first-line therapy in 374 patients with HR+ MBC; the addition of bevacizumab failed to produce a statistically significant increase in PFS, although PFS was numerically longer in the bevacizumab arm (19.3 vs 14.4 months; HR 0.83; 95% CI, 0.65–1.06; $P = 0.126$) [34]. However, CALGB 40503 trial employed a similar design in 343 women and found a significant improvement in PFS with the addition of bevacizumab (20.2 vs 15.6 months; HR 0.75; 95% CI, 0.59 to 0.96; $P = 0.016$) [35]. Neither trial showed an improvement in OS with the addition of bevacizumab, and toxicity including hypertension and proteinuria was increased in the combination arms. Based on these data, bevacizumab is not being pursued as treatment for HR+ breast cancer, although alternative anti-angiogenic therapies are under development.

19.2.4 HER2-Targeted Therapy

HER2 positivity has been shown to be associated with relative resistance to endocrine therapy [36], sparking interest in the combination of HER2-targeted therapy and ET. Two-phase III trials have evaluated this approach, using either the antibody trastuzumab or the oral tyrosine kinase inhibitor lapatinib, in combination with an AI as first-line therapy for HR+/HER2+ MBC. The phase III TAnDEM trial randomized 207 women to receive anastrozole with or without trastuzumab and demonstrated improved PFS with the addition of trastuzumab (4.8 vs 2.4 months; HR 0.63, 95% CI, 0.47 to 0.84, $p = 0.0016$) [37]. A second phase III compared letrozole plus lapatinib to letrozole plus placebo in 219 women with HR+/HER2+ MBC and found a significant improvement in PFS with the addition of lapatinib (8.2 vs 3.0 months; HR 0.71; 95% CI 0.53 to 0.96; $p = 0.019$) [36], although toxicity including diarrhea and rash was modestly increased with lapatinib. Neither trial showed an improvement in OS, and the combination of taxanes, trastuzumab, and pertuzumab has demonstrated the most significant improvement in OS ever demonstrated in a phase III trial in breast cancer [38]. In addition, the PERTAIN trial demonstrated improved PFS when pertuzumab was added to the combination of trastuzumab and an AI as first-line therapy for HR+/HER2+ MBC [15]. Therefore, the primary utilization of hormone therapy combined with HER2-targeted therapy for MBC is generally in the maintenance setting, although patients with limited metastatic disease or who are not candidates for chemotherapy can clearly benefit.

19.2.5 Emerging Therapies

PI3K Inhibitors

Two-phase III studies have evaluated the addition of the pan-PI3K inhibitor buparlisib to fulvestrant as second- or greater-line therapy for HR+ MBC. BELLE-2 randomized 1147 patients with HR+ MBC progressing on an AI and with up to one line of prior chemotherapy for advanced disease to receive fulvestrant with either buparlisib or placebo [39]. In the overall trial population, there was no difference in PFS, but in the 372 patients with known activation of the PI3K pathway, PFS was modestly improved with buparlisib (6.8 vs 4.0 months, HR 0.76, 95% CI 0·60–0·97, one sided $p = 0{\cdot}014$). Treatment with buparlisib was associated with an increase in hepatic enzymes, hyperglycemia, and rash. The BELLE-3 trial utilized a similar study design with a 2:1 randomization in 432 women, but all patients had to have progressed on an mTOR inhibitor [15]. PFS was again only modestly improved with buparlisib (3.9 vs 1.8 months; HR 0.67 (0.53–0.84),

<0.001). PI3K mutation status was assessed in a subset of patients in both primary tumor tissue and blood with 34 and 39%, respectively, having mutations. Interestingly, the presence of a mutation was associated with a significant improvement in PFS, but there was no difference in PFS in those with wild-type PI3K. Again, buparlisib was associated with an increase in hepatic enzymes, with one case of Hy's law, and an increase in depression, with three cases of suicide attempts. A phase II randomized trial with another pan-PI3K inhibitor (the FERGI trial, pictilisib) showed increased toxicity without an improvement in PFS [40].

Due to toxicity and limited efficacy, further development of the pan-PI3K inhibitors has been discontinued. However, encouraging data with less toxicity has been seen with the addition of the more alpha-specific PI3K inhibitors to endocrine therapy, including alpelisib and taselisib. Phase III studies have completed accrual with PI3K mutation status assessed in all patients; data is expected to be presented in late 2017 or 2018, and data from neoadjuvant trials will be presented in late 2017.

Histone Deacetylase (HDAC) Inhibitors

Based on preclinical data suggesting that the oral HDAC inhibitor entinostat could inhibit ER+ tumor growth and restore hormone sensitivity [41], a phase II trial compared the exemestane plus entinostat to exemestane plus placebo in 130 patients with HR+ MBC progressing on a nonsteroidal AI [42]. Entinostat improved PFS (4.3 vs 2.3 months; HR 0.73; 95% CI, 0.50–1.07, one sided $P = 0.055$; two sided $P = 0.11$), as well as the exploratory endpoint of OS (28.1 vs 19.8 months; HR 0.59, 95% CI, 0.36–0.97, $P = 0.036$). Toxicity included grade 3/4 fatigue and neutropenia. A randomized phase III trial is ongoing with a similar study design (ECOG 2112).

Immunotherapy

Hormone receptor-positive breast cancer has traditionally been thought of as less immunogenic, but there may be significant differences in subsets based on proliferation. The KEYNOTE-028 basket trial treated 25 patients with single-agent pembrolizumab (a PD-1 inhibitor) and reported an overall response rate (ORR) of just 12% [43]. The NeoMonarch trial evaluated the CDK4/CDK6 inhibitor abemaciclib alone or in combination with anastrozole as neoadjuvant therapy for HR+ breast cancer and demonstrated a significant increase in the number of patients with complete cell cycle suppression defined as a Ki-67 of ≤2.7% with abemaciclib compared to anastrozole alone [44]. Interestingly, this trial also demonstrated a marked increase in tumor bed infiltration with CD8 T cells in patients treated with abemaciclib. These data have stimulated interest in combination therapies that can enhance the host antitumor immune response. One small phase II study is evaluating the combination of abemaciclib and pembrolizumab.

19.3 The Strategies of ET for ABC

It is widely accepted that HR+ breast cancer may in some cases be treated as a chronic disease with reasonably long survival, although identifying relatively better prognosis disease may be difficult. ET has been associated with significant clinical benefits in the majority of patients with HR+ disease, and international guidelines recommend sequential ET as the primary treatment strategy for this disease [45]. Patients with immediately life-threatening disease or with pending organ failure should be treated with chemotherapy first, and hormone therapy can be used as maintenance after the disease control is achieved. The combination of chemotherapy with endocrine therapy is not recommended, as there is no data to suggest that this is a beneficial approach, and it could result in both increased toxicity and worse outcome. The main treatment objective for the patients with ABC is to palliate symptoms and prolong survival while minimizing the adverse effects of the therapy.

19.3.1 Menopausal Status and ET

Menopausal status should be considered when choosing ET, although menopause has been defined differently in various breast cancer clinical trials. One set of definitions is provided by the NCCN guidelines: (1) bilateral ovariectomy; (2) ≥60 years old; (3) <60 years old, last menstrual period occurring >12 months ago without prior chemotherapy, tamoxifen, toremifene, or ovarian suppression therapy, and FSH and estradiol levels in the postmenopausal range; (4) <60 years old, taking tamoxifen or toremifene, FSH and estradiol levels in the postmenopausal range; (5) receiving a LHRH agonist or antagonist to induce chemical menopause; (6) for premenopausal women receiving adjuvant chemotherapy, cessation of menstruation cannot be defined as menopause; and (7) although patients may stop ovulation and menstruation after chemotherapy, estrogen production may continue, and ovarian function may still recover. If AIs are considered as endocrine therapy for patients with chemotherapy-induced menopause, oophorectomy or continuous monitoring of estradiol levels (and FSH as indicted) is needed to ensure that the patients remain in menopause [46]. The chance of ovarian function recovery from chemotherapy-induced menopause is dependent on the chemotherapy regimen, patient age, and ovarian reserve [47, 48]. Therefore, the use of AIs in young patients should be undertaken with caution and with careful monitoring of ovarian function, with consideration given for concomitant use of chemical ovarian suppression with LHRH agonists.

19.3.2 Selection of ET with or Without Targeted Agents

ET is the preferred first-line therapy for HR+ metastatic breast cancer without immediately life-threatening disease. Depending on response and extent of disease, sequential hormone therapy should be employed. The choice of specific treatment depends on prior treatment and response as well as exposure to endocrine therapy in the early-stage setting [47].

In general, first-line therapy for HR+ MBC should be an AI, combined with ovarian suppression in premenopausal women. Tamoxifen can be considered in premenopausal women with limited extent of disease, where ovarian suppression after diagnosis of MBC is not feasible. Based on the FALCON data, patients with HR+ MBC without prior exposure to ET and without visceral disease could be considered for treatment with fulvestrant, with AI used in the second-line setting.

The use of targeted therapies has to involve consideration of prior treatment, treatment goals, financial burden, and extent of disease. Interestingly, the randomized phase III trials that have evaluated targeted agents combined with ET have failed to identify a subgroup of patients with either more or less benefit; all patient subgroups have had improved PFS from the addition of the targeted agent to ET. Indeed, it is difficult to determine which patients will have prolonged PFS without targeted therapy. In the first-line setting, two CDK4/CDK6 inhibitors have demonstrated improved PFS with modest toxicity which is relatively easily managed by dose delays and dose reductions. However, the cost of therapy is substantial, and given the availability of these agents in countries contributing to the phase III trials, widespread crossover on progression makes the detection of survival benefits unlikely.

The majority of patients with HR+ MBC should be considered for treatment with combined ET and targeted therapy, at least at some point during their treatment course. In the first-line setting, the choice is an AI with a CDK4/CDK6 inhibitor versus fulvestrant alone in the subset of patients without prior exposure to ET and no evidence of visceral disease. There is not yet data with the combination of fulvestrant and a CDK4/CDK6 inhibitor in the first-line setting, but this approach could be utilized in patients intolerant of AIs. Following progression on an AI, fulvestrant (with a CDK4/CDK6 inhibitor if not previously used) or exemestane and everolimus are the next treatment options, and these options can be used in sequence. Everolimus can also be given in combination with fulvestrant based on the data from the recent prECOG study. Everolimus should be given in combination with a dexamethasone-based mouthwash to prevent stomatitis.

For HER2+ disease, generally chemotherapy plus trastuzumab and pertuzumab is the preferred first-line therapy, with endocrine therapy used in combination with antibodies as maintenance. Patients with limited soft tissue or bone disease, or those intolerant of chemotherapy, could be considered for first-line therapy with an AI plus HER2-targeted therapy.

Primary resistance to ET is defined as recurrence within the first 2 years of adjuvant therapy or progressive disease within 6 months of starting ET in MBC, while secondary resistance is defined as a recurrence after the first 2 years of adjuvant ET or disease progression more than 6 months after initiation of ET in ABC [49]. These definitions are practical and useful for the therapy process of patients. Although the presence or absence of resistance to ET did not predict differential benefit to any targeted therapy, this could be used in a resource-limited setting to determine timing of use of targeted agents.

The diagram below summarizes suggested sequencing alternatives in ET for patients with HR+/HER2-negative MBC.

1ST LINE
Nonsteroidal AI with or without a CDK 4/6 inhibitor
•Fulvestrant (in patients without prior exposure to ET and without visceral disease as a single agent, or in those progressing on adjuvant AI with or without a CDK 4/6 inhibitor)
•Exemestane or fulvestrant and everolimus (in patients progressing on AI and aduvant CDK4/6 inhibitor)

2ND LINE
•Fulvestrant (with a CDK 4/6 inhibitor if not previously given, or with everolimus)
•Exemestane with or without everolimus
•AI with a CDK 4/6 inhibitor if treated in the first-line setting with fulvestrant alone

3RD and later LINES
•Exemestane and everolimus (if not previously given, preferred)
•Tamoxifen or toremifene
•Megestrol acetateDefine according to the previous two lines and response

References

1. Litherland S, Jackson IM (1988) Antioestrogens in the management of hormone-dependent cancer. Cancer Treat Rev 15:183–194
2. Buzdar AU, Robertson JF, Eiermann W, Nabholtz JM (2002) An overview of the pharmacology and pharmacokinetics of the newer generation aromatase inhibitors anastrozole, letrozole, and exemestane. Cancer 95:2006–2016
3. Li HP, Ji JF, Hou KY, Jia TZ, Zhao HM, Xiao Y, Wang MP, Wang YF (2007) Clinical study of aromatase inhibitors in advanced breast cancer. J Peking Univ Health Sci 39:193–196
4. Miller WR, Bartlett J, Brodie AMH, Brueggemeier RW, di Salle E, Lonning PE, Llombart A, Maass N, Maudelonde T, Sasano H, Goss PE (2008) Aromatase inhibitors: are there differences between steroidal and nonsteroidal aromatase inhibitors and do they matter? Oncologist 13:829–837
5. Mauri D, Pavlidis N, Polyzos NP, Ioannidis JP (2006) Survival with aromatase inhibitors and inactiva-

tors versus standard hormonal therapy in advanced breast cancer: meta-analysis. J Natl Cancer Inst 98:1285–1291
6. Robertson JF, Osborne CK, Howell A, Jones SE, Mauriac L, Ellis M, Kleeberg UR, Come SE, Vergote I, Gertler S, Buzdar A, Webster A, Morris C (2003) Fulvestrant versus anastrozole for the treatment of advanced breast carcinoma in postmenopausal women: a prospective combined analysis of two multicenter trials. Cancer 98:229–238
7. Di Leo A, Jerusalem G, Petruzelka L, Torres R, Bondarenko IN, Khasanov R, Verhoeven D, Pedrini JL, Smirnova I, Lichinitser MR, Pendergrass K, Garnett S, Lindemann JP, Sapunar F, Martin M (2010) Results of the CONFIRM phase III trial comparing fulvestrant 250 mg with fulvestrant 500 mg in postmenopausal women with estrogen receptor-positive advanced breast cancer. J Clin Oncol 28:4594–4600
8. Ellis MJ, Llombart-Cussac A, Feltl D, Dewar JA, Jasiowka M, Hewson N, Rukazenkov Y, Robertson JF (2015) Fulvestrant 500 mg versus anastrozole 1 mg for the first-line treatment of advanced breast cancer: overall survival analysis from the phase II FIRST study. J Clin Oncol 33:3781–3787
9. Robertson J, Bondarenko IM, Trishkina E, Dvorkin M, Panasci L, Manikhas A, Shparyk Y, Cardona-Huerta S, Cheung KL, Philco-Salas MJ, Ruiz-Borrego M, Shao Z, Noguchi S, Rowbottom J, Stuart M, Grinsted LM, Fazal M, Ellis MJ (2016) Fulvestrant 500 mg versus anastrozole 1 mg for hormone receptor-positive advanced breast cancer (FALCON): an international, randomised, double-blind, phase 3 trial. Lancet 388:2997–3005
10. Boer K (2017) Fulvestrant in advanced breast cancer: evidence to date and place in therapy. Ther Adv Med Oncol 9:465–479
11. Gong C, Zhao Y, Wang B, Hu X, Wang Z, Zhang J, Zhang S (2017) Efficacy and safety of everolimus in Chinese metastatic HR positive, HER2 negative breast cancer patients: a real-world retrospective study. Oncotarget
12. Seidman AD (2016) When to combine endocrine therapy with a new agent for hormone receptor-positive metastatic breast cancer in postmenopausal women. Oncology (Williston Park) 30:224–228
13. Yardley DA, Noguchi S, Pritchard KI, Burris HR, Baselga J, Gnant M, Hortobagyi GN, Campone M, Pistilli B, Piccart M, Melichar B, Petrakova K, Arena FP, Erdkamp F, Harb WA, Feng W, Cahana A, Taran T, Lebwohl D, Rugo HS (2013) Everolimus plus exemestane in postmenopausal patients with HR(+) breast cancer: BOLERO-2 final progression-free survival analysis. Adv Ther 30:870–884
14. Bachelot T, Bourgier C, Cropet C, Ray-Coquard I, Ferrero JM, Freyer G, Abadie-Lacourtoisie S, Eymard JC, Debled M, Spaeth D, Legouffe E, Allouache D, El KC, Pujade-Lauraine E (2012) Randomized phase II trial of everolimus in combination with tamoxifen in patients with hormone receptor-positive, human epidermal growth factor receptor 2-negative metastatic breast cancer with prior exposure to aromatase inhibitors: a GINECO study. J Clin Oncol 30:2718–2724
15. Gampenrieder SP, Rinnerthaler G, Greil R (2017) SABCS 2016: systemic therapy for metastatic breast cancer. Memo 10:86–89
16. Rugo HS, Seneviratne L, Beck JT, Glaspy JA, Peguero JA, Pluard TJ, Dhillon N, Hwang LC, Nangia C, Mayer IA, Meiller TF, Chambers MS, Sweetman RW, Sabo JR, Litton JK (2017) Prevention of everolimus-related stomatitis in women with hormone receptor-positive, HER2-negative metastatic breast cancer using dexamethasone mouthwash (SWISH): a single-arm, phase 2 trial. Lancet Oncol 18:654–662
17. Shah AN, Cristofanilli M (2017) The growing role of CDK4/6 inhibitors in treating hormone receptor-positive advanced breast cancer. Curr Treat Options in Oncol 18:6
18. Finn RS, Dering J, Conklin D, Kalous O, Cohen DJ, Desai AJ, Ginther C, Atefi M, Chen I, Fowst C, Los G, Slamon DJ (2009) PD 0332991, a selective cyclin D kinase 4/6 inhibitor, preferentially inhibits proliferation of luminal estrogen receptor-positive human breast cancer cell lines in vitro. Breast Cancer Res 11:R77
19. Musgrove EA, Caldon CE, Barraclough J, Stone A, Sutherland RL (2011) Cyclin D as a therapeutic target in cancer. Nat Rev Cancer 11:558–572
20. Walker AJ, Wedam S, Amiri-Kordestani L, Bloomquist E, Tang S, Sridhara R, Chen W, Palmby TR, Fourie ZJ, Fu W, Liu Q, Tilley A, Kim G, Kluetz PG, McKee AE, Pazdur R (2016) FDA approval of Palbociclib in combination with fulvestrant for the treatment of hormone receptor-positive, HER2-negative metastatic breast cancer. Clin Cancer Res 22:4968–4972
21. Turner NC, Ro J, Andre F, Loi S, Verma S, Iwata H, Harbeck N, Loibl S, Huang BC, Zhang K, Giorgetti C, Randolph S, Koehler M, Cristofanilli M (2015) Palbociclib in hormone-receptor-positive advanced breast cancer. N Engl J Med 373:209–219
22. Finn RS, Crown JP, Lang I, Boer K, Bondarenko IM, Kulyk SO, Ettl J, Patel R, Pinter T, Schmidt M, Shparyk Y, Thummala AR, Voytko NL, Fowst C, Huang X, Kim ST, Randolph S, Slamon DJ (2015) The cyclin-dependent kinase 4/6 inhibitor palbociclib in combination with letrozole versus letrozole alone as first-line treatment of oestrogen receptor-positive, HER2-negative, advanced breast cancer (PALOMA-1/TRIO-18): a randomised phase 2 study. Lancet Oncol 16:25–35
23. Finn RS, Crown JP, Ettl J, Schmidt M, Bondarenko IM, Lang I, Pinter T, Boer K, Patel R, Randolph S, Kim ST, Huang X, Schnell P, Nadanaciva S, Bartlett CH, Slamon DJ (2016) Efficacy and safety of palbociclib in combination with letrozole as first-line treatment of ER-positive, HER2-negative, advanced breast cancer: expanded analyses of subgroups from the randomized pivotal trial PALOMA-1/TRIO-18. Breast Cancer Res 18:67

24. Lu J (2015) Palbociclib: a first-in-class CDK4/CDK6 inhibitor for the treatment of hormone-receptor positive advanced breast cancer. J Hematol Oncol 8:98
25. Cristofanilli M, Turner NC, Bondarenko I, Ro J, Im SA, Masuda N, Colleoni M, DeMichele A, Loi S, Verma S, Iwata H, Harbeck N, Zhang K, Theall KP, Jiang Y, Bartlett CH, Koehler M, Slamon D (2016) Fulvestrant plus palbociclib versus fulvestrant plus placebo for treatment of hormone-receptor-positive, HER2-negative metastatic breast cancer that progressed on previous endocrine therapy (PALOMA-3): final analysis of the multicentre, double-blind, phase 3 randomised controlled trial. Lancet Oncol 17:425–439
26. Finn RS, Martin M, Rugo HS, Jones S, Im SA, Gelmon K, Harbeck N, Lipatov ON, Walshe JM, Moulder S, Gauthier E, DR L, Randolph S, Dieras V, Slamon DJ (2016) Palbociclib and letrozole in advanced breast cancer. N Engl J Med 375:1925–1936
27. Hortobagyi GN, Stemmer SM, Burris HA, Yap YS, Sonke GS, Paluch-Shimon S, Campone M, Blackwell KL, Andre F, Winer EP, Janni W, Verma S, Conte P, Arteaga CL, Cameron DA, Petrakova K, Hart LL, Villanueva C, Chan A, Jakobsen E, Nusch A, Burdaeva O, Grischke EM, Alba E, Wist E, Marschner N, Favret AM, Yardley D, Bachelot T, Tseng LM, Blau S, Xuan F, Souami F, Miller M, Germa C, Hirawat S, O'Shaughnessy J (2016) Ribociclib as first-line therapy for HR-positive, advanced breast cancer. N Engl J Med 375:1738–1748
28. Hortobagyi GN, Stemmer SM, Burris HA, Yap YS, Sonke GS, Paluch-Shimon S, Campone M, Petrakova K, Blackwell KL, Winer EP, Janni W, Verma S, Conte PF, Arteaga CL, Cameron DA, Xuan F, Miller MK, Germa C, Hirawat S, O'Shaughnessy J (2017) Updated results from MONALEESA-2, a phase 3 trial of first-line ribociclib + letrozole in hormone receptor-positive (HR+), HER2-negative (HER2–), advanced breast cancer (ABC). J Clin Oncol:1038
29. Dickler MN, Tolaney SM, Rugo HS, Cortes J, Dieras V, Patt D, Wildiers H, Hudis CA, O'Shaughnessy J, Zamora E, Yardley DA, Frenzel M, Koustenis A, Baselga J (2017) MONARCH 1, A phase II study of abemaciclib, a CDK4 and CDK6 inhibitor, as a single agent, in patients with refractory HR+/HER2- metastatic breast cancer. Clin Cancer Res 23(17):5218–5224
30. Sledge GJ, Toi M, Neven P, Sohn J, Inoue K, Pivot X, Burdaeva O, Okera M, Masuda N, Kaufman PA, Koh H, Grischke EM, Frenzel M, Lin Y, Barriga S, Smith IC, Bourayou N, Llombart-Cussac A (2017) MONARCH 2: abemaciclib in combination with fulvestrant in women with HR+/HER2- advanced breast cancer who had progressed while receiving endocrine therapy. J Clin Oncol:O2017737585
31. Di Leo A, Toi M, Campone M et al. MONARCH 3: abemaciclib as initial therapy for patients with HR+, HER2- advanced breast cancer. In: Proceedings of the ESMO 2017
32. Murphy CE, Lansdown MR, Speirs V, Carder PJ (2003) Correspondence re: K. Heer et al., serum vascular endothelial growth factor in breast cancer: its relation with cancer type and estrogen receptor status. Clin Cancer Res 7:3491–3494. 2001. Clin Cancer Res 9:3514, 3515
33. Kumler I, Christiansen OG, Nielsen DL (2014) A systematic review of bevacizumab efficacy in breast cancer. Cancer Treat Rev 40:960–973
34. Martin M, Loibl S, von Minckwitz G, Morales S, Martinez N, Guerrero A, Anton A, Aktas B, Schoenegg W, Munoz M, Garcia-Saenz JA, Gil M, Ramos M, Margeli M, Carrasco E, Liedtke C, Wachsmann G, Mehta K, De la Haba-Rodriguez JR (2015) Phase III trial evaluating the addition of bevacizumab to endocrine therapy as first-line treatment for advanced breast cancer: the letrozole/fulvestrant and avastin (LEA) study. J Clin Oncol 33:1045–1052
35. Dickler MN, Barry WT, Cirrincione CT, Ellis MJ, Moynahan ME, Innocenti F, Hurria A, Rugo HS, Lake DE, Hahn O, Schneider BP, Tripathy D, Carey LA, Winer EP, Hudis CA (2016) Phase III trial evaluating letrozole as first-line endocrine therapy with or without bevacizumab for the treatment of postmenopausal women with hormone receptor-positive advanced-stage breast cancer: CALGB 40503 (Alliance). J Clin Oncol 34:2602–2609
36. Johnston S, Pippen JJ, Pivot X, Lichinitser M, Sadeghi S, Dieras V, Gomez HL, Romieu G, Manikhas A, Kennedy MJ, Press MF, Maltzman J, Florance A, O'Rourke L, Oliva C, Stein S, Pegram M (2009) Lapatinib combined with letrozole versus letrozole and placebo as first-line therapy for postmenopausal hormone receptor-positive metastatic breast cancer. J Clin Oncol 27:5538–5546
37. Kaufman B, Mackey JR, Clemens MR, Bapsy PP, Vaid A, Wardley A, Tjulandin S, Jahn M, Lehle M, Feyereislova A, Revil C, Jones A (2009) Trastuzumab plus anastrozole versus anastrozole alone for the treatment of postmenopausal women with human epidermal growth factor receptor 2-positive, hormone receptor-positive metastatic breast cancer: results from the randomized phase III TAnDEM study. J Clin Oncol 27:5529–5537
38. Swain SM, Kim SB, Cortes J, Ro J, Semiglazov V, Campone M, Ciruelos E, Ferrero JM, Schneeweiss A, Knott A, Clark E, Ross G, Benyunes MC, Baselga J (2013) Pertuzumab, trastuzumab, and docetaxel for HER2-positive metastatic breast cancer (CLEOPATRA study): overall survival results from a randomised, double-blind, placebo-controlled, phase 3 study. Lancet Oncol 14:461–471
39. Baselga J, Im SA, Iwata H, Cortes J, De Laurentiis M, Jiang Z, Arteaga CL, Jonat W, Clemons M, Ito Y, Awada A, Chia S, Jagiello-Gruszfeld A, Pistilli B, Tseng LM, Hurvitz S, Masuda N, Takahashi M, Vuylsteke P, Hachemi S, Dharan B, Di Tomaso E, Urban P, Massacesi C, Campone M (2017) Buparlisib plus fulvestrant versus placebo plus fulvestrant in

postmenopausal, hormone receptor-positive, HER2-negative, advanced breast cancer (BELLE-2): a randomised, double-blind, placebo-controlled, phase 3 trial. Lancet Oncol 18:904–916

40. Krop IE, Mayer IA, Ganju V, Dickler M, Johnston S, Morales S, Yardley DA, Melichar B, Forero-Torres A, Lee SC, de Boer R, Petrakova K, Vallentin S, Perez EA, Piccart M, Ellis M, Winer E, Gendreau S, Derynck M, Lackner M, Levy G, Qiu J, He J, Schmid P (2016) Pictilisib for oestrogen receptor-positive, aromatase inhibitor-resistant, advanced or metastatic breast cancer (FERGI): a randomised, double-blind, placebo-controlled, phase 2 trial. Lancet Oncol 17:811–821
41. Sabnis GJ, Goloubeva O, Chumsri S, Nguyen N, Sukumar S, Brodie AM (2011) Functional activation of the estrogen receptor-alpha and aromatase by the HDAC inhibitor entinostat sensitizes ER-negative tumors to letrozole. Cancer Res 71:1893–1903
42. Yardley DA, Ismail-Khan RR, Melichar B, Lichinitser M, Munster PN, Klein PM, Cruickshank S, Miller KD, Lee MJ, Trepel JB (2013) Randomized phase II, double-blind, placebo-controlled study of exemestane with or without entinostat in postmenopausal women with locally recurrent or metastatic estrogen receptor-positive breast cancer progressing on treatment with a nonsteroidal aromatase inhibitor. J Clin Oncol 31:2128–2135
43. Rugo HS, Delord JP, Im SA, Ott PA, PihaPaul SA, Bedard PL, Sachdev J, Tourneau CL, Brummelen EV, Varga A, Saraf S, Pietrangelo D, Karantza V, Tan A (2016) Preliminary efficacy and safety of pembrolizumab (MK-3475) in patients with PD-L1–positive, estrogen receptor-positive (ER+)/HER2-negative advanced breast cancer enrolled in KEYNOTE-028. Cancer Res 76:S5–S7
44. Hurvitz S, Martin M, Fernández Abad M, Chan D, Rostorfer R, Petru E, Barriga S, Costigan TM, Caldwell CW, Nguyen T, Press M, Slamon D (2017) Biological effects of abemaciclib in a phase 2 neoadjuvant study for postmenopausal patients with HR+, HER2- breast cancer. Cancer Res 4(suppl):S4–S6
45. Rugo HS, Rumble RB, Macrae E, Barton DL, Connolly HK, Dickler MN, Fallowfield L, Fowble B, Ingle JN, Jahanzeb M, Johnston SR, Korde LA, Khatcheressian JL, Mehta RS, Muss HB, Burstein HJ (2016) Endocrine therapy for hormone receptor-positive metastatic breast cancer: American society of clinical oncology guideline. J Clin Oncol 34:3069–3103
46. Gradishar WJ, Anderson BO, Balassanian R, Blair SL, Burstein HJ, Cyr A, Elias AD, Farrar WB, Forero A, Giordano SH, Goetz M, Goldstein LJ, Hudis CA, Isakoff SJ, Marcom PK, Mayer IA, McCormick B, Moran M, Patel SA, Pierce LJ, Reed EC, Salerno KE, Schwartzberg LS, Smith KL, Smith ML, Soliman H, Somlo G, Telli M, Ward JH, Shead DA, Kumar R (2015) Breast cancer version 2.2015. J Natl Compr Cancer Netw 13:448–475
47. Higgins MJ, Wolff AC (2008) Therapeutic options in the management of metastatic breast cancer. Oncology (Williston Park) 22:614–623, 623, 627–629
48. Li HP, Ma LW, Zhang SL, Jia TZ, Deng HJ, Zhang ZH, Liang L, Wang MP, Xiao Y, Cao BS, Chen S, Wang YF (2006) Observation and clinical significance of adjuvant chemotherapy-induced amenorrhea in premenopausal breast cancer patients. China Oncol 28:848–851
49. Reinert T, Barrios CH (2015) Optimal management of hormone receptor positive metastatic breast cancer in 2016. Ther Adv Med Oncol 7:304–320

MIX
Papier aus verantwortungsvollen Quellen
Paper from responsible sources
FSC® C105338

If you have any concerns about our products,
you can contact us on
ProductSafety@springernature.com

In case Publisher is established outside the EU,
the EU authorized representative is:
Springer Nature Customer Service Center GmbH
Europaplatz 3, 69115 Heidelberg, Germany

Printed by Libri Plureos GmbH
in Hamburg, Germany